Alzheimer's Disease

Alzheimer's disease continues to present one of the foremost scientific and clinical challenges in contemporary neuroscience. Despite significant advances in molecular biology and neuroimaging, its complex etiology and progressive course remain only partially understood, and truly effective disease-modifying therapies remain elusive.

This scholarly volume offers a comprehensive and integrative examination of Alzheimer's disease, combining cutting-edge research with clinical perspectives. *Alzheimer's Disease: Diagnosis, Molecular Targets, and Therapeutic Modalities* assembles contributions from leading experts, providing in-depth analysis across a wide range of critical topics:

- Innovative diagnostic biomarkers for early detection, prognosis, and therapeutic monitoring
- Exploration of molecular targets, including secretases, SIRT2, RAGE, calpain, caspases, NMDA receptors, and others implicated in AD pathophysiology
- Emerging therapeutic interventions encompassing enzyme inhibitors, immunotherapeutic strategies, metal chelation, neurotrophic modulation, and aggregation inhibitors
- Advances in translational research featuring nanotechnology-based delivery systems and bioactive natural compounds
- Evidence-based lifestyle interventions integrating nutrition and physical activity in disease management

Spanning 21 rigorously curated chapters, this volume offers a multidisciplinary framework that bridges fundamental research, translational applications, and clinical relevance. It is an essential reference for neuroscientists, clinical researchers, pharmacologists, graduate students, and healthcare professionals engaged in the investigation and management of neurodegenerative disorders.

By synthesizing current scientific knowledge with emerging therapeutic strategies, this volume aims to advance understanding and stimulate further research toward effective interventions for Alzheimer's disease.

Alzheimer's Disease
Diagnosis, Molecular Targets, and Therapeutic Modalities

Edited by
Devendra Kumar, Shraddha Manish Gupta,
Nagendra Singh Chauhan, and Kamal Shah

CRC Press
Taylor & Francis Group
Boca Raton London New York

CRC Press is an imprint of the
Taylor & Francis Group, an informa business

Designed cover image: iStock by Getty Images (1450703720)

First edition published 2026
by CRC Press
2385 NW Executive Center Drive, Suite 320, Boca Raton FL 33431

and by CRC Press
4 Park Square, Milton Park, Abingdon, Oxon, OX14 4RN

CRC Press is an imprint of Taylor & Francis Group, LLC

ISBN: 9781032991290 (hbk)
ISBN: 9781032991306 (pbk)
ISBN: 9781003602477 (ebk)

DOI: 10.1201/9781003602477

Typeset in Times
by KnowledgeWorks Global Ltd.

In loving memory of a dear friend Dr. Manish Gupta,
whose presence brought light, laughter, and strength to our lives.
Though you are no longer with us, your spirit lives on in every page,
guiding us with the kindness, courage, and warmth you shared so freely.
This book is for you—always remember.

Contents

Foreword

Alzheimer's is a progressive neurodegenerative disease that lacks legitimate pathophysiology apprehension and effective treatment. Dementia is one of the world wide problems. It is assumed to raise double in the next 20 years. This book *Alzheimer's Disease: Diagnosis, Molecular Targets and Therapeutic Modalities* has compiled various therapeutic modalities, molecular targets, and recent diagnosis methods. This book analyzes the fundamentals and more advanced aspects of Alzheimer's disease focusing on Tau, Tangle, Plaque hypotheses. Blood-brain barrier: its structure, function, carriers, transport, fate, and strategy to cross BBB, vaccine, active and passive immunotherapy, γ-secretase inhibitors and modulators, β-secretase inhibitors, α-secretase enhancers, NSAID-derived carboxylic acid analogs, non-NSAID imidazole compounds, natural products as GSMs, SIRT2 inhibitors, caspase inhibitors, glycogen synthase kinase-3 (GSK-3) inhibitors, calpains, acetylcholinesterase inhibitors, noncompetitive N-methyl-D-aspartate antagonist, receptor for advanced glycation end-products (RAGE) inhibitors, phosphodiesterase (PDE) inhibitors, apolipoprotein, biomarkers, non-peptidic anti-aggregate compounds, etc.

This book would create a better understanding of the pathophysiology of the disease. It will provide deep insight into various molecular targets and therapeutic modalities available to combat this disease. This book would be assets for both master students and researchers working in the area of nondegenerative disorders.

I congratulate the editors, Dr. Devendra, Dr. Shraddha, Dr. Chauhan, and Dr. Kamal for the dedication and hard work they have put it to give final shape to this venture. The book has excellent blend of contributors across the globe. All chapters are unique in their content but take on single objective of the treatment of Alzheimer's disease. This book is an attempt to bring together the knowledge, expertise, and experiences of researchers working in the area of neurodegenerative disorders. The authors successfully compiled the chapter contents with updated literature. I believe 'Alzheimer's Disease: Diagnosis, Molecular Targets, and Therapeutic Modalities' will remain a valuable resource for years to come.

Prof. Sushil Kumar Singh
Professor and former Head of the Department,
Department of Pharmaceutical Engineering & Technology,
Indian Institute of Technology (Banaras Hindu University),
Varanasi-221005.

Preface

Alzheimer's disease (AD) continues to pose a formidable challenge to the global healthcare community. Despite significant advances in neuroscience and molecular biology, the pathogenesis and progression of AD remain incompletely understood, and effective disease-modifying treatments are still elusive. The multifaceted nature of Alzheimer's demands a comprehensive exploration of its molecular mechanisms, diagnostic challenges, and evolving therapeutic strategies.

This book, *Alzheimer's Disease: Diagnosis, Molecular Targets, and Therapeutic Modalities*, is a collaborative effort aimed at providing an in-depth overview of the current landscape in AD research and treatment. It brings together insights from a wide spectrum of biomedical research, covering emerging diagnostic biomarkers, molecular targets, such as secretases, SIRT2, and RAGE, and a variety of therapeutic approaches, including enzyme inhibitors, immunotherapy, metal chelators, and lifestyle interventions like nutrition and exercise.

The chapters have been thoughtfully curated and authored by experts in the field, each contributing to a better understanding of the complex pathophysiology of AD and the promise of novel intervention strategies. From fundamental concepts and diagnostic challenges to advanced therapeutic approaches such as nanotechnology and neurotrophic agents, this book is intended to serve as a valuable resource for researchers, clinicians, postgraduate students, and all professionals involved in neuroscience and drug discovery.

This book consists of 21 chapters. Chapter 1 provides a foundational understanding of AD, including its pathophysiology and diagnostic aspects. Chapters 2, 4, 5, 6, 7, 8, 9, 10, 11, 12, 13, 14, 15, and 17 delve into the exploration of molecular and pharmacological targets such as the blood-brain barrier, secretase enzymes, SIRT2, caspases, calpain, AChE, NMDA receptors, PDEs, autophagy, RAGE, neurotrophic agents, metal chelators, APOE, and anti-aggregate compounds for therapeutic intervention. Chapters 3 and 16 address advancements in diagnostics and immunotherapy, focusing on active and passive vaccination strategies and the role of biomarkers for diagnosis, prognosis, and therapeutic monitoring of AD. Chapters 18 and 19 discuss translational and mechanistic aspects, including nanotechnology-based interventions and natural molecules with potent therapeutic targets. Chapters 20 and 21 highlight integrative and lifestyle-based approaches, such as therapeutic nutrition strategies and the importance of exercise in AD management.

This comprehensive collection offers multidisciplinary insights into diagnostic innovations, emerging molecular targets, and current and future therapeutic modalities. It is intended to serve as a valuable reference for researchers, clinicians, and students engaged in neuroscience and neurodegenerative disease research.

We hope that this comprehensive volume not only enhances knowledge but also inspires continued innovation and collaboration in the pursuit of effective solutions for AD.

A special note of love and appreciation to Aayansh, Agam, Harshita, Ishita, Khushi, Satyansh, Sparsh, and Soumya—the bright young stars who brought joy, warmth, and light, whose innocent smiles and boundless curiosity have been a constant source of inspiration throughout this journey. We extend our deepest gratitude to the contributors and reviewers for their time, expertise, and dedication.

Editors
Devendra Kumar,
Shraddha Manish Gupta,
Nagendra Singh Chauhan,
and Kamal Shah

About the Editors

Devendra Kumar is currently working as Assistant Professor at School of Pharmacy & Technology Management, NMIMS University, India. He was Post-Doctoral Fellow at Department of Cellular and Integrative Physiology, University of Texas, Health Science Center, San Antonio, USA. And he obtained his Ph.D. from Indian Institute of Technology (Banaras Hindu University), Varanasi, India. He received Extramural Grant from Indian Council of Medical Research (ICMR, file #IIRP-2023-5582/F1, Amount of project: 39,44,085) and has written more than 45 articles published in national and international journals, nine patents (three granted, six filed), two book chapters. He has citations of more than 1123 with h-index 19 and i10 index of 27.

Shraddha Manish Gupta has more than 15 years of research and teaching experience and currently is Assistant Professor at School of Health Sciences and Technology, UPES, Dehradun, India. She completed B.Pharm and M.Pharm from NDMVP'S College of Pharmacy, Nasik (M.H.) in 2003 and 2006, respectively, and Ph.D. from Oriental University Indore, India. She secured 96 percentile in GATE-2004 conducted by IIT, Delhi. She worked as Research Assistant in School of Pharmacy, Monash University, Malaysia. She was Mentor in CSIR-SRTP (2020) program under Interdisciplinary Approach for Sustainable Use of Resources Project-13. She has written more than ten articles published in Q1 international journals, two patents (one granted, one published), and eight book chapters.

Nagendra Singh Chauhan obtained his M.Pharm and Ph.D. from the Department of Pharmaceutical Sciences, Dr. H.S. Gour University, Sagar in 2006 and 2011. He is currently working as Senior Scientific Officer Grade-II and Government Analyst at the Drugs Testing Laboratory Avam Anusandhan Kendra, Raipur, Chhattisgarh, India. He has professional expertise in natural product isolation and phytopharmacology. He has written more than 75 articles published in national and international journals, 32 book chapters, and edited five books. He has citations of more than 2700 with h-index 42 and i10 index of 38 (Google Scholar) and Scopus h-index of 15 and 669 citations.

Kamal Shah has more than 18 years of research and teaching experience and currently working as Professor at the Institute of Pharmaceutical Research, GLA University, Mathura, India. He completed B.Pharm from Rajiv Gandhi Proudyogiki Vishwavidyalaya, Bhopal (M.P.) in 2003. He was a gold medalist in B.Pharm. He completed his M.Pharm from the Department of Pharmaceutical Sciences, Dr. Hari Singh Gour University, Sagar, India, and earned his Ph.D. from Dr. A. P. J. Abdul Kalam Technical University, Lucknow, India. He has written more than 90 articles published in national and international journals, 20 book chapters, and five edited books. His Scopus h-index is 17, with 905 citations.

Contributors

Akib Ali
Biological Science and Technology Division, CSIR-North East Institute of Science and Technology
Jorhat, Assam, India

Shubham Anand
Department of Pharmacology, School of Pharmacy
ITM University
Gwalior, Madhya Pradesh, India

Mandeep Kumar Arora
Faculty of Pharmacy
School of Pharmaceutical and Population Health Informatics
DIT University
Dehradun, India

Alina Arulsamy
Jeffrey Cheah School of Medicine and Health Sciences
Monash University, Malaysia, Subang Jaya, Malaysia

Rajendra Awasthi
Department of Pharmaceutical Sciences
School of Health Sciences and Technology
UPES
Dehradun, India

Sayani Banerjee
Department of Pharmacology and Toxicology
National Institute of Pharmaceutical Education and Research
Kolkata, India

Sugato Banerjee
Department of Pharmacology and Toxicology
National Institute of Pharmaceutical Education and Research
Kolkata, India

Amritansh Bhanot
Department of Pharmaceutical Sciences
School of Health Sciences and Technology
UPES
Dehradun, Uttarakhand, India

Pankaj Bharali
Biological Science and Technology Division
CSIR-North East Institute of Science and Technology
Jorhat, Assam, India
Academy of Scientific and Innovative Research (AcSIR)
Ghaziabad, India

Hiranmoy Bhattacharya
Advanced Pharmacognosy Research Laboratory
Department of Pharmaceutical Technology
Jadavpur University
Kolkata, West Bengal, India

Sreejita Chatterjee
Department of Dietetics and Applied Nutrition
Amity University Kolkata
West Bengal, India

Sushil Kumar Chaudhary
BRIC-Institute of Bioresources and Sustainable Development
Upper Shillong, Meghalaya, India

Havagiray R. Chitme
Amity Institute of Pharmacy
Amity University
Noida, Uttar Pradesh, India

Saikat Dewanjee
Advanced Pharmacognosy Research Laboratory
Department of Pharmaceutical Technology
Jadavpur University
Kolkata, West Bengal, India

Sylvia F. Fawzi
Department of Pharmacology and Toxicology
Faculty of Pharmacy
Badr University in Cairo (BUC)
Cairo, Egypt

Mina Y. George
Department of Pharmacology and Toxicology
Faculty of Pharmacy
Ain Shams University
Cairo, Egypt

Satish Kumar Gupta
Faculty of Pharmacy
School of Pharmaceutical and Population Health Informatics
DIT University
Dehradun, India

Shraddha Manish Gupta
Department of Pharmaceutical Sciences
School of Health Science and Technology (SOHST)
UPES
Dehradun, Uttarakhand, India

Prakash Haloi
Amity Institute of Pharmacy
Amity University
Noida, Uttar Pradesh, India

Hemant R. Jadhav
Department of Pharmacy
Birla Institute of Technology and Science Pilani
Pilani, Rajasthan, India

Abhijeet Joshi
Department of Pharmacy
Birla Institute of Technology and Sciences Pilani,
Hyderabad Campus, Telangana State, India

Ripunjay Kalita
Department of Medicinal Chemistry
National Institute of Pharmaceutical Education and Research
Guwahati, Assam, India

Aaditi Karnik
Department of Pharmacy
Birla Institute of Technology and Sciences Pilani
Hyderabad Campus, Telangana State, India

Iram Khan
Department of Pharmaceutical Sciences
School of Health Science and Technology (SOHST)
UPES
Dehradun, Uttarakhand, India

Tamanna K. Khan
Department of Chemistry
GN Khalsa College of Art Science and Commerce (Autonomous)
Matunga (East), Mumbai, Maharashtra, India

Giriraj T Kulkarni
Faculty of Pharmacy
School of Pharmaceutical and Population Health Informatics
DIT University
Dehradun, Uttarakhand, India

Amit Kumar
Pharmacology and Toxicology Laboratory
Dietetics and Nutrition Technology Division
CSIR-Institute of Himalayan Bioresource Technology
Palampur, Himachal Pradesh, India
Academy of Scientific and Innovative Research (AcSIR)
Ghaziabad, India

Devendra Kumar
Department of Pharmaceutical Chemistry
School of Pharmacy & Technology Management
SVKM's NMIMS University
Mukesh Patel Technology Park, Shirpur, Dhule, Maharashtra, India

Babita Kumari
Biological Science and Technology Division
CSIR-North East Institute of Science and Technology
Jorhat, Assam, India

Vijeta Kumari
Department of Pharmacy
Birla Institute of Technology and Science Pilani, Pilani Campus
Vidya Vihar, Pilani, Rajasthan, India

Daniel E. Mansour
Medical School
University of Minnesota
Minneapolis, Minnesota

Pierre Mansour
Ohio State University College of Medicine
Columbus, Ohio, USA

Chanchal Mittal
Department of Pharmacology and Toxicology
National Institute of Pharmaceutical Education and Research
Kolkata, West Bengal, India

Ehsan Ullah Mughal
Department of Chemistry
University of Gujrat
Gujrat, Pakistan

Riya Mukherjee
Graduate Institute of Biomedical Sciences
Chang Gung University
Taoyuan, Taiwan
Department of Medical Biotechnology and Laboratory Science
Chang Gung University
Taoyuan, Taiwan

Nafeesa Naeem
Department of Chemistry
University of Gujrat
Gujrat, Pakistan

Dilek Özbeyli
Department of Medical Services and Techniques
Vocational School of Health-Related Services
Marmara University
Istanbul, Türkiye

Ramendra Pati Pandey
Department of Biotechnology
SRM University
Delhi-NCR, Sonipat, Haryana, India

Kalpana Pandya
Department of Pharmaceutical Chemistry
School of Pharmacy & Technology Management
SVKM's NMIMS University
Shirpur, Dhule, Maharashtra, India
Shobhaben Pratapbhai Patel School of Pharmacy & Technology Management
SVKM's Narsee Monjee Institute of Management Studies (NMIMS) Deemed to be University
Mumbai, Maharashtra, India

Debjyoti Paul
Department of Dietetics and Applied Nutrition
Amity University Kolkata
West Bengal, India

Sonima Prasad
University Institute of Pharma Sciences (UIPS)
Chandigarh University
Mohali, Punjab, India

Shubham Nilkanth Rahmatkar
Pharmacology and Toxicology Laboratory
Dietetics and Nutrition Technology Division
CSIR-Institute of Himalayan Bioresource Technology
Palampur, Himachal Pradesh, India
Academy of Scientific and Innovative Research (AcSIR)
Ghaziabad, Uttar Pradesh, India

Ashi Ramavat
Department of Biotechnology
SRM University
Delhi-NCR, Sonipat, Haryana, India

Ankit Rathee
University Institute of Pharma Sciences (UIPS)
Chandigarh University
Mohali, Punjab, India

Faisal Rehman
School of Biochemistry and Bioscience
University College Cork
University College Cork, Cork, Ireland

Amina Sadiq
Department of Chemistry
Govt. College Women University
Sialkot, Pakistan

Esra Tansu Sarıyer
Department of Nutrition and Dietetics
Faculty of Health Science
University of Health Sciences
Istanbul, Türkiye

Mohmad Farooq Shaikh
Jeffrey Cheah School of Medicine and Health Sciences
Monash University Malaysia
School of Dentistry and Medical Sciences
Monash University, Malaysia, Subang Jaya, Malaysia

Mushtaque A. S. Shaikh
Department of Pharmaceutical Chemistry
Vivekanand Education Society's College of Pharmacy (Autonomous)
Chembur (East), Mumbai, Maharashtra, India

Amit Sharma
Department of Pharmaceutical Chemistry
Manipal College of Pharmaceutical Sciences
Manipal Academy of Higher Education (MAHE)
Manipal, Karnataka, India
Department of Pharmacy
Birla Institute of Technology and Science Pilani
Pilani, Rajasthan, India

Bhupesh Sharma
Department of Pharmaceutical Sciences
Faculty of Life Sciences
Gurugram University
Gurugram, Haryana, India

Arti Singh
Department of Pharmaceutical Sciences
School of Health Science and Technology (SOHST)
UPES
Dehradun, Uttarakhand, India

Charan Singh
Department of Pharmaceutical Sciences
HNB Garhwal University (A Central University)
Chauras Campus, Tehri Garhwal, Uttarakhand, India

Damanpreet Singh
Pharmacology and Toxicology Laboratory
Dietetics and Nutrition Technology Division
CSIR-Institute of Himalayan Bioresource Technology
Palampur, Himachal Pradesh, India
Academy of Scientific and Innovative Research (AcSIR)
Ghaziabad, Uttar Pradesh, India

Geetanjali Singh
Department of Pharmaceutical Sciences
School of Health Science and Technology (SOHST)
UPES
Dehradun, Uttarakhand, India

Avanish Tripathi
Department of Pharmacology
School of Pharmacy
ITM University
Gwalior, Madhya Pradesh, India

Garima Varshney
Faculty of Pharmacy
School of Pharmaceutical and Population Health Informatics
DIT University
Dehradun, Uttarakhand, India

Meral Yüksel
Department of Medical Services and Techniques
Vocational School of Health-Related Services
Marmara University
Istanbul, Türkiye

Hadri Hadi Md Yusof
Jeffrey Cheah School of Medicine and Health Sciences
Monash University
Monash University, Malaysia, Subang Jaya, Malaysia

1 Understanding Alzheimer's Disease

An Introduction, Pathophysiology, and Diagnosis

Avanish Tripathi and Shubham Anand

1.1 INTRODUCTION

Alzheimer's disease (AD) is a progressive neurodegenerative disorder and the leading cause of dementia worldwide, accounting for 60–80% of dementia cases among older adults. It is characterised by a gradual decline in memory, thinking, reasoning, and behavioural abilities, eventually interfering with daily life and independence (La Barbera et al., 2022). While memory loss is often the earliest and most prominent symptom, Alzheimer's also affects language, visual-spatial skills, judgement, and personality as the disease advances (Aisen et al., 2017).

Named after Dr. Alois Alzheimer, who first described its hallmark brain changes in 1906, the disorder is defined pathologically by the accumulation of amyloid-beta (Aβ) plaques and neurofibrillary tangles (NFTs) of tau protein in the brain (Golmohammadi et al., 2024). These changes disrupt communication between neurons, ultimately leading to cell death and brain shrinkage.

Alzheimer's is not a normal part of ageing, though advancing age remains the greatest known risk factor (Anand et al., 2017). Most cases develop after age 65 (late-onset), but a small proportion occur earlier (early-onset) (Wang et al., 2024). The disease currently affects over 35 million people globally, with incidence expected to rise sharply as populations age (Hoogmartens et al., 2021). Women are disproportionately affected, accounting for nearly two-thirds of cases (Singh et al., 2023).

Despite decades of research, the precise causes of Alzheimer's remain incompletely understood, involving complex interactions between genetic, environmental, and lifestyle factors (Haghighijoo et al., 2022). There is currently no cure, but advances in diagnosis and treatment are ongoing, aiming to slow progression and improve quality of life for patients and caregivers (Faiyaz et al., 2021). The cholinergic hypothesis suggests that loss of cholinergic neurons reduces acetylcholine, impairing memory and cognition, which is managed with acetylcholinesterase inhibitors such as donepezil. The amyloid hypothesis highlights amyloid-β accumulation as the trigger for neurotoxicity and cognitive decline, with anti-amyloid therapies like aducanumab under use. The tau hypothesis points to hyperphosphorylated tau forming neurofibrillary tangles, treated with anti-tau antibodies and kinase inhibitors. The oxidative stress hypothesis emphasises reactive oxygen species and mitochondrial dysfunction causing neuronal damage, targeted by antioxidants and mitochondrial stabilisers. The excitotoxicity hypothesis describes excessive glutamate overstimulating NMDA receptors, leading to neuronal death, for which memantine is prescribed. Finally, the type 3 diabetes hypothesis links brain insulin resistance and reduced glucose metabolism to AD-like pathology, with experimental therapies including intranasal insulin and antidiabetic drugs are shows in Tables 1.1–1.3.

TABLE 1.1
Major Hypotheses in Alzheimer's Disease

Hypothesis	Main Idea	Key Pathological Feature	Therapeutic Focus
Cholinergic hypothesis	Loss of cholinergic neurons reduces ACh, impairing memory and cognition	↓ ACh, ↓ ChAT, and AChE	Acetylcholinesterase inhibitors (e.g., donepezil)
Amyloid hypothesis	Aβ accumulation initiates neurotoxicity and cognitive decline	Aβ plaques, Aβ oligomers	Anti-amyloid therapies (e.g., aducanumab)
Tau hypothesis	Hyperphosphorylated tau disrupts microtubules and causes neurodegeneration	Neurofibrillary tangles	Anti-tau antibodies, kinase inhibitors
Oxidative stress hypothesis	ROS generation damages neurons, worsens Aβ and tau pathology	ROS, mitochondrial dysfunction	Antioxidants, mitochondrial stabilisers
Excitotoxicity hypothesis	Excessive glutamate leads to NMDA receptor overactivation and neuronal death	↑ Ca^{2+}, oxidative stress	Memantine (NMDA receptor antagonist)
Type 3 diabetes hypothesis	Insulin resistance in the brain causes AD-like pathology	↓ insulin signalling, ↓ glucose metabolism	Intranasal insulin, antidiabetics (experimental)

DOI: 10.1201/9781003602477-1

TABLE 1.2
Global Alzheimer's Disease Statistics

Region	Prevalence Rate per 100,000	Projection by 2050 (Millions)
East Asia	781.4	~30–35
North Africa and Middle East	777.6	~15
South Asia	428.4	~20
Western Sub-Saharan Africa	469.8	~10
United States	682.5	~13.8
Global Total	682.5 (Avg)	~139

TABLE 1.3
Genetic Factors in Alzheimer's Disease

Gene	Type of AD	Function	Effect
APP	Familial (early-onset)	Amyloid precursor protein	↑ Aβ42 production
PSEN1/PSEN2	Familial (early-onset)	Part of γ-secretase complex	Altered APP cleavage → ↑ Aβ42
APOE ε4	Sporadic (late-onset)	Lipid transport and Aβ clearance	↓ Aβ clearance, ↑ risk 3–15x
MAPT	Related to tauopathy	Codes tau protein	↑ tau aggregation (NFT formation)

1.2 EPIDEMIOLOGY OF ALZHEIMER'S DISEASE

AD is a major and escalating public health concern globally, with its epidemiology marked by rapidly increasing prevalence, significant mortality, and profound socioeconomic impact. By 2020, over 55 million people worldwide had been living with dementia, with Alzheimer's accounting for the majority of these cases. This figure is projected to almost double every 20 years, reaching 78 million by 2030 and 139 million by 2050, primarily due to the ageing global population (Abadio et al., 2011). In 2019 alone, there were an estimated 51.6 million prevalent cases of AD and other dementias worldwide, with an age-standardised prevalence rate of 682.5 per 100,000 individuals, representing a 5.7% increase since 1990 (Chi et al., 2014). The highest regional prevalence rates are observed in East Asia (781.4 per 100,000) and North Africa and the Middle East (777.6 per 100,000), while the lowest are found in South Asia (428.4 per 100,000) and Western Sub-Saharan Africa (469.8 per 100,000) (Eratne et al., 2018).

In the United States, the burden is particularly significant. In 2024, an estimated 6.9 million Americans aged 65 and older were living with AD, with about one in nine people (10.9%) in this age group affected. Nearly three-quarters of these individuals are aged 75 or older, and almost two-thirds are women, reflecting both the longer life expectancy of women and possible biological susceptibilities (Khubetova et al., 2023). The prevalence is higher among certain ethnic groups, like older Black Americans are about twice as likely, and older Hispanics about one and a half times as likely, to have Alzheimer's or other dementias compared to older Whites. The number of Americans living with Alzheimer's is expected to rise sharply, reaching a projected 12.7 million by 2050 and potentially 13.8 million by 2060, barring major advances in prevention or cure (Mielke et al., 2014). Mortality associated with Alzheimer's is also rising. In 2019, there were 1.6 million deaths globally due to AD and other dementias. The age-standardised death rate varies by region, with the highest rates in high-income Asia Pacific and Central Sub-Saharan Africa (around 27 per 100,000), and the lowest in South Asia and the Caribbean (around 19–21 per 100,000) (Chan et al., 2013). In the United States, Alzheimer's is the sixth leading cause of death, and one in three seniors dies with Alzheimer's or another dementia, more than breast cancer and prostate cancer combined. Deaths due to Alzheimer's have more than doubled between 2000 and 2021, even as deaths from other major diseases like heart disease have declined (Dartigues & Letenneur, 2000). After diagnosis at age 65, individuals typically survive an average of four to eight years, although some may live as long as 20 years. The disease also imposes a heavy burden of disability. In 2019, Alzheimer's and other dementias accounted for 25.3 million disability-adjusted life years (DALYs) globally, with the highest rates in Tropical Latin America and North Africa and the Middle East, and the lowest in South Asia and the Caribbean (Sosa-Ortiz et al., 2012). These trends are compounded by the immense caregiving demands in the United States alone, more than 11 million caregivers provided an estimated 18 billion hours of unpaid care in 2022, valued at $339.5 billion.

1.2.1 Pathophysiological Mechanisms of Alzheimer's Disease (AD)

AD is primarily driven by three interrelated pathophysiological mechanisms. The accumulation of Aβ peptides leads to the formation of extracellular amyloid plaques, tau protein NFTs intracellularly, and reduced acetylcholine synthesis. These factors contribute to neurodegeneration and cognitive decline.

1.3 ALZHEIMER'S DISEASE'S MOLECULAR MECHANISMS

AD is characterised by complex molecular mechanisms involving Aβ accumulation, tau pathology, mitochondrial dysfunction, and neuroinflammation. Below is a detailed synthesis of these processes are shown in Figure 1.1.

1.3.1 Cholinergic Hypothesis

The cholinergic hypothesis, a cornerstone of AD research since the 1970s, attributes cognitive decline to the selective degeneration of cholinergic neurons in the basal forebrain, particularly the nucleus basalis of Meynert (NbM), which supplies acetylcholine (ACh) to memory-related regions like the hippocampus and neocortex (Francis et al., 1999). This hypothesis emerged from postmortem studies revealing marked reductions in choline acetyltransferase (ChAT), the enzyme synthesising ACh, and acetylcholinesterase (AChE), which degrades ACh, in AD brains deficits correlating strongly with the severity of memory impairment. The basal forebrain cholinergic neurons (BFCNs) are especially vulnerable to early degeneration, with their loss disrupting synaptic plasticity and long-term potentiation (LTP) critical for learning (Craig et al., 2011). ACh deficiency impairs attentional processes, information encoding, and cortical arousal, while also exacerbating tau pathology through dysregulated kinase activity (e.g., GSK3β) and reduced neurotrophic support, such as nerve growth factor (NGF), which normally sustains BFCN survival. Aβ oligomers further target cholinergic synapses, inducing oxidative stress (OS) and calcium dysregulation, which accelerate neuronal death. Clinically, the hypothesis covers the development of acetylcholinesterase inhibitors (AChEIs) like donepezil, which temporarily boost synaptic ACh levels, providing modest symptomatic relief (Hampel et al., 2019). However, the cholinergic hypothesis alone fails to explain non-cholinergic neuron loss or the primacy of Aβ and tau pathology, leading to its reinterpretation as a downstream consequence of broader neurodegenerative processes. Despite this, it remains vital for understanding early synaptic dysfunction and guiding therapies that complement emerging anti-amyloid and anti-tau strategies (Babic et al., 1999).

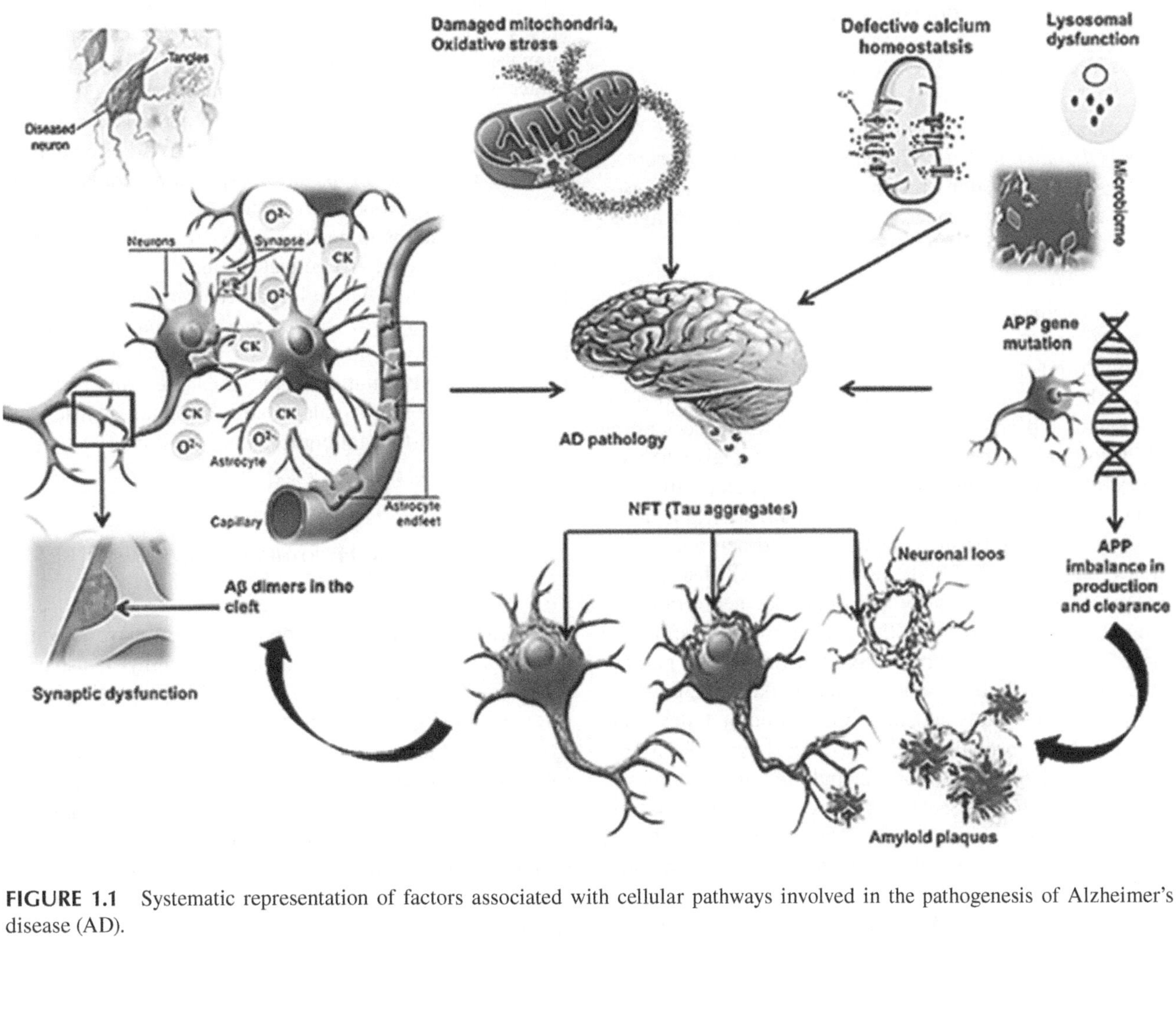

FIGURE 1.1 Systematic representation of factors associated with cellular pathways involved in the pathogenesis of Alzheimer's disease (AD).

1.3.2 Amyloid-Beta (Aβ) Hypothesis

The amyloid hypothesis posits that the accumulation of Aβ peptides in the brain is the primary driver of AD, initiating a cascade of pathological events leading to neurodegeneration. Aβ peptides, particularly $A\beta_{42}$, are generated through sequential cleavage of amyloid precursor protein (APP) by β-secretase (BACE1) and γ-secretase (Hillen, 2019; Selkoe & Hardy, 2016). In familial AD (FAD), mutations in APP, PSEN1, or PSEN2 genes increase Aβ42 production or its aggregation propensity, while in sporadic cases, impaired clearance mechanisms (e.g., ApoE4-linked deficits) promote Aβ buildup. Soluble Aβ oligomers, rather than insoluble plaques, are now recognised as the most neurotoxic species, disrupting synaptic function by binding to neuronal membranes, inducing calcium dysregulation, and impairing LTP, which underlies memory deficits (Levin & Vasenina, 2017). Aβ peptides are generated via sequential cleavage of APP by β-secretase (BACE1) and γ-secretase. The amyloidogenic pathway produces Aβ isoforms (e.g., $A\beta_{42}$), which are aggregation-prone due to hydrophobic residues (Hampel et al., 2021).

- *Genetic factors*: Mutations in APP, PSEN1, and PSEN2 increase Ae_{42} production or alter cleavage specificity, promoting aggregation (Šerý et al., 2013).
- *Aggregation stages*: Soluble Aβ monomers → oligomers (neurotoxic) → protofibrils → insoluble fibrils → plaques (Murphy & Levine, 2010).
- *Oligomer toxicity*: Aβ oligomers (AβOs) disrupt synaptic plasticity by binding GM1 ganglioside on neuronal membranes, inducing calcium influx and OS (Sehar et al., 2022).

These oligomers also trigger mitochondrial dysfunction, generating reactive oxygen species (ROS) (Roy & Rauk, 2005) that exacerbate OS and neuronal apoptosis. Aβ pathology further synergises with tau hyperphosphorylation, as Aβ activates kinases like GSK3β, accelerating NFT formation and neurodegeneration. Aβ also triggers neuroinflammation by activating microglia and astrocytes, which release pro-inflammatory cytokines (e.g., IL-1β, TNF-α) that exacerbate synaptic loss and neurodegeneration (Hampel et al., 2020). Additionally, Aβ disrupts the balance between excitatory and inhibitory neurotransmission, causing hyperexcitability and network dysfunction observed in early AD stages. Synergistically, Aβ potentiates tau pathology by activating kinases like GSK3β, which hyperphosphorylate tau, destabilise microtubules, and promote NFT formation. This cascade of synaptic failure, OS, and inflammatory damage underlies the progressive cognitive decline characteristic of AD, with synaptic loss correlating most strongly with clinical symptoms. While recent anti-Aβ immunotherapies (e.g., lecanemab, aducanumab) reduce amyloid plaques, their modest clinical benefits suggest that Aβ removal may need to occur early in the disease and reach low levels (approx 20 centiloids) to halt downstream tau-mediated damage. Despite challenges, the hypothesis remains central to AD research, supported by genetic evidence (e.g., Down syndrome's Aβ overproduction) and neuropathological data, though it is increasingly viewed as part of a multifactorial cascade involving tau, neuroinflammation, and vascular dysfunction (Karran & De Strooper, 2022; Paroni et al., 2019).

1.3.3 Tau Hypothesis of Alzheimer's Disease

The tau hypothesis postulates that hyperphosphorylation and aggregation of tau protein are central to AD pathogenesis, driving neurodegeneration through microtubule destabilisation and synaptic dysfunction. Tau, a microtubule-associated protein, stabilises axonal microtubules under normal conditions, but abnormal phosphorylation, triggered by kinases like GSK3β and CDK5, causes it to dissociate and form paired helical filaments (PHFs), which aggregate into NFTs (Kametani & Hasegawa, 2018). These NFTs disrupt axonal transport, impair synaptic plasticity, and induce neuronal death, correlating strongly with cognitive decline severity and disease progression. Critically, tau pathology often emerges before Aβ plaques in sporadic AD, particularly in the transentorhinal cortex during Braak stages I–II, spreading to limbic and neocortical regions as symptoms worsen (Gulisano et al., 2018). Unlike Aβ, which shows weak correlation with dementia severity, tau burden aligns closely with neurodegeneration patterns and clinical symptoms, as evidenced by PET imaging. Additionally, tau exhibits prion-like propagation, spreading trans-synaptically between neurons, with misfolded tau seeds templating further aggregation in healthy cells. Cross-talk between tau and Aβ exists, e.g., Aβ activates kinases that hyperphosphorylate tau, and APOE4 exacerbates both pathologies, but tau is increasingly viewed as the primary executor of neuronal toxicity. The hypothesis is bolstered by genetic links (e.g., MAPT mutations in tauopathies) and therapeutic failures targeting Aβ alone, suggesting tau-centric strategies (e.g., anti-tau antibodies, kinase inhibitors) may be critical for halting AD progression (Fišar, 2022).

1.3.4 Excitotoxic Hypothesis in Alzheimer's Disease

The excitotoxic hypothesis proposes that excessive activation of glutamate receptors, particularly N-methyl-D-aspartate receptors (NMDARs), drives neuronal degeneration in AD through prolonged calcium influx, OS, and mitochondrial dysfunction. Glutamate, the CNS's primary excitatory neurotransmitter (Cacabelos et al., 2000) normally mediates synaptic plasticity and learning. However, in AD, Aβ oligomers disrupt glutamate homeostasis, leading to chronic NMDAR overactivation and calcium overload, which activate proteases (e.g., calpains), generate ROS, and trigger apoptosis(Olney et al., 1997).

1.3.5 Oxidative Stress

The OS hypothesis posits that imbalanced redox homeostasis, driven by excessive ROS and impaired antioxidant defences, is a central mechanism in AD pathogenesis, contributing to synaptic dysfunction, protein misfolding, and neuronal death (Nunomura, 2013). OS arises from mitochondrial dysfunction, metal ion accumulation (e.g., iron, copper), and Aβ-induced ROS generation, leading to lipid peroxidation, protein oxidation, and DNA damage in the AD brain. Key evidence includes elevated levels of oxidative markers like 4-hydroxynonenal (a lipid peroxidation product) and 8-hydroxyguanosine (oxidised DNA) in postmortem AD brains, particularly in regions vulnerable to Aβ and tau pathology. Notably, oxidative damage is most pronounced in early-stage AD and even in mild cognitive impairment (MCI), suggesting OS precedes Aβ plaque and NFT formation. Aβ oligomers exacerbate OS by disrupting mitochondrial electron transport chains, generating ROS, and binding transition metals (e.g., Fe^{2+}, Cu^{2+}) that catalyse Fenton reactions, amplifying free radical production. Conversely, OS promotes Aβ accumulation by upregulating β-secretase (BACE1) activity and APP processing, creating a vicious cycle (Du et al., 2018). Synaptic dysfunction, a hallmark of early AD, is directly linked to ROS-mediated impairment of NMDA receptor signalling and calcium overload, which disrupt LTP and trigger excitotoxicity (Teixeira et al., 2019). Furthermore, OS activates kinases like GSK3β, promoting tau hyperphosphorylation and NFT formation. Genetic and environmental risk factors (e.g., APOE4, ageing, diabetes) converge on OS pathways, while antioxidant deficits (e.g., reduced glutathione) fail to mitigate damage. Therapeutic strategies targeting OS include antioxidants (e.g., vitamin E) and metal chelators, though clinical efficacy remains limited, underscoring the need for early intervention and multimodal approaches addressing Aβ, tau, and inflammation alongside oxidative pathways (Praticò, 2008).

1.3.6 Apolipoprotein E (ApoE)

The ApoE hypothesis underscores the pivotal role of the APOE ε4 allele as the strongest genetic risk factor for late-onset AD (LOAD), mediating neuropathology through both Aβ-dependent and Aβ-independent pathways. ApoE, a cholesterol transporter critical for lipid homeostasis, exists in three isoforms (E2, E3, E4), with ApoE4 conferring a 3–15-fold increased AD risk per allele. ApoE4 exacerbates Aβ pathology by impairing clearance mechanisms: it binds Aβ less efficiently than ApoE3, reducing its degradation via low-density lipoprotein receptors (LDLR/LRP1) and promoting fibrillar Aβ aggregation into plaques (Kim et al., 2009). Concurrently, ApoE4's poor lipidation (due to deficient ABCA1-mediated cholesterol efflux) disrupts synaptic lipid delivery, compromising membrane integrity and repair, while elevating neuronal cholesterol levels that upregulate β-secretase (BACE1) and γ-secretase activity, amplifying Aβ production. Beyond amyloid, ApoE4 drives neuroinflammation by polarising microglia towards a pro-inflammatory state (elevating IL-1β, TNF-α) and reducing phagocytic Aβ clearance, while impairing astrocytic support for neurons (He et al., 2024). It also accelerates tau pathology by activating kinases like GSK3β, which hyperphosphorylate tau, and by promoting mitochondrial dysfunction and OS. Cerebrovascular contributions of ApoE4 are significant. It disrupts blood-brain barrier (BBB) integrity, increasing permeability to neurotoxins and impairing perivascular Aβ drainage, while elevating cerebral amyloid angiopathy (CAA) risk through vascular Aβ deposition (Soininen & Riekkinen, 1996). Intriguingly, populations like Nigerians exhibit an ApoE4 paradox, showing high ε4 prevalence but low AD incidence, potentially due to protective factors like low cholesterol or distinct genetic modifiers. Therapeutically, strategies targeting ApoE4 include structure correctors to improve lipidation, ABCA1 agonists to enhance lipid transport, and gene editing to convert ε4 to ε3 (El Haj et al., 2016). ApoE4's Aβ-independent role is that it disrupts glucose metabolism, synaptic plasticity, and mitochondrial function, highlighting its multifaceted impact, necessitating combinatorial therapies addressing lipid dysregulation, neuroinflammation, and tau pathology alongside amyloid. This hypothesis positions ApoE4 as a central node integrating genetic, metabolic, and inflammatory cascades in AD pathogenesis (Koutsodendris et al., 2021).

1.3.7 GSK-3 Hypothesis

The GSK-3 hypothesis proposes that dysregulated glycogen synthase kinase-3 (GSK-3) activity, particularly the GSK-3β isoform, drives AD pathogenesis by exacerbating tau hyperphosphorylation, Aβ production, and synaptic dysfunction. GSK-3β, a constitutively active serine-threonine kinase, becomes hyperactive in AD due to factors like Aβ-induced OS, APOE4-related lipid dysregulation, and neuroinflammation. Aβ oligomers directly or indirectly enhance GSK-3β activity by promoting tyrosine phosphorylation (pY216), a marker of kinase activation, while reducing inhibitory serine phosphorylation (pS9). This hyperactive GSK-3β phosphorylates tau at multiple sites (e.g., Ser396, Thr231), destabilising microtubules and promoting NFT formation, a hallmark of AD neuropathology. GSK-3β also amplifies Aβ toxicity by upregulating BACE1 and γ-secretase activity, increasing Aβ production, while impairing Aβ clearance pathways. Further, amyloid and tau, GSK-3β disrupts synaptic plasticity by inhibiting LTP and promoting neuroinflammation via microglial activation and pro-inflammatory cytokine release (e.g., IL-1β, TNF-α). In transgenic models, GSK-3β overexpression accelerates tauopathy and neuronal death, while its inhibition reduces tau phosphorylation, Aβ toxicity, and cognitive deficits. Notably, GSK-3β's role extends to synaptic homeostasis, where its dysregulation impairs neurogenesis and memory consolidation, linking it to early cognitive decline. Therapeutic strategies targeting

GSK-3β, such as small-molecule inhibitors or gene modulation, aim to restore kinase balance, mitigate tau/Aβ pathology, and improve synaptic function, positioning GSK-3β as a central node in AD's multifactorial cascade (Takashima, 2006).

1.3.8 CREB Signalling Pathways

The cAMP response element-binding protein (CREB) signalling pathway is a central regulator of synaptic plasticity, memory formation, and neuronal survival, with its dysfunction implicated in AD pathogenesis. CREB, activated via phosphorylation at Ser133, is regulated by multiple kinases, including protein kinase A (PKA), calcium/calmodulin-dependent kinases (CaMKs), and mitogen-activated protein kinases (MAPKs), in response to synaptic activity, neurotrophins (e.g., BDNF), and calcium influx. In AD, Aβ oligomers disrupt CREB activation by impairing calcium homeostasis and downregulating PKA/CaMK pathways, leading to reduced phosphorylation of CREB and diminished transcription of target genes critical for synaptic plasticity (e.g., BDNF, c-Fos, Arc). This suppression of CREB-mediated gene expression correlates with synaptic loss, impaired LTP, and cognitive decline. Notably, FAD mutations in presenilin (PS1/PS2) cause constitutive CREB activation via exaggerated inositol trisphosphate receptor (InsP3R)-mediated calcium release, paradoxically promoting apoptosis and sensitising neurons to Aβ toxicity (Saura & Valero, 2011). CREB dysfunction also exacerbates tau pathology by failing to regulate pro-survival genes and synaptic maintenance factors, while neuroinflammation further suppresses CREB activity through cytokine-driven inhibition of PKA. Conversely, BDNF-CREB interactions are crucial for neuronal resilience, as BDNF activates CREB via TrkB receptors and the MAPK/ERK pathway, but Aβ disrupts BDNF signalling, creating a feedback loop of synaptic failure. Peripheral CREB dysregulation in blood mononuclear cells (PBMCs) has been proposed as an early biomarker of AD, reflecting systemic signalling defects. Therapeutic strategies aim to restore CREB activity through PKA/CaMK activators, BDNF mimetics, or CREB gene therapy, targeting synaptic repair and cognitive resilience in AD (Sharma & Singh, 2021).

1.3.9 Alzheimer's Disease as Type 3 Diabetes

The concept of Type 3 diabetes posits that AD represents a brain-specific metabolic disorder characterised by insulin resistance, impaired insulin signalling, and glucose dysregulation in the brain, distinct from systemic diabetes (Type 1 or 2). This hypothesis stems from evidence that AD brains exhibit reduced insulin sensitivity and diminished insulin-like growth factor (IGF) activity, leading to impaired glucose metabolism, synaptic dysfunction, and neurodegeneration. Insulin resistance in the brain disrupts neuronal survival pathways, exacerbates OS, and promotes tau hyperphosphorylation and Aβ accumulation. APOE4, the strongest genetic risk factor for LOAD, impairs insulin signalling by impairing lipid metabolism and Aβ clearance, linking metabolic dysfunction to AD progression. Systemic diabetes includes chronic inflammation, vascular damage, and mitochondrial dysfunction, though AD's metabolic disturbances are localised to the brain (Michailidis et al., 2022). Research suggests that insulin resistance may precede cognitive decline, with clinical trials testing intranasal insulin to restore brain insulin signalling and slow AD progression. However, the hypothesis remains controversial, as AD's aetiology involves multifactorial pathways beyond insulin dysfunction, including genetic (e.g., APOE4) and environmental factors (Kciuk et al., 2024).

1.4 RISK FACTORS FOR ALZHEIMER'S DISEASE

AD's risk factors are broadly categorised into non-modifiable (age, genetics, sex) and modifiable (lifestyle, health conditions, environmental exposures). Age is the strongest risk factor, with incidence doubling every five years after 65 due to cumulative cellular damage, synaptic loss, and reduced Aβ clearance (Fenclová et al., 2020). Genetic factors include FAD mutations (APP, PSEN1, PSEN2), which account for less than 1% of cases but guarantee early-onset AD, and risk genes like APOE4 present in ~65% of AD patients, which increases lifetime risk 3–15-fold by impairing lipid metabolism and Aβ clearance. Sex disparities show women over 80 have higher AD prevalence than men, partly due to longevity and postmenopausal hormonal changes (Xu et al., 2015). Modifiable risk factors include cardiovascular conditions (hypertension, diabetes, obesity), which promote cerebrovascular damage and Aβ accumulation via chronic hypoxia and inflammation (Xia et al., 2018). Lifestyle factors such as physical inactivity, smoking, excessive alcohol, and unhealthy diets exacerbate OS and neuroinflammation. Environmental exposures like air pollution and traumatic brain injuries (TBIs) accelerate Aβ deposition and tau pathology (Luchsinger & Mayeux, 2004). Hearing loss, depression, and low education further elevate risk by reducing cognitive reserve and social engagement (Speck et al., 1995). Emerging evidence highlights infections (e.g., herpes simplex virus) and sleep disorders as contributors to neuroinflammation and Aβ dysregulation. Preventive strategies targeting modifiable factors managing blood pressure, increasing physical activity, could reduce AD risk (Lindsay et al., 2002).

1.5 TREATMENT OF ALZHEIMER'S DISEASE

AD's treatment focuses on symptomatic management and disease-modifying therapies, with current options targeting cognitive decline, behavioural symptoms, and underlying pathology (Briggs et al., 2016).

1.5.1 Symptomatic Pharmacotherapy

Cholinesterase inhibitors (donepezil, rivastigmine, galantamine) boost acetylcholine levels by inhibiting its breakdown, temporarily improving cognition and daily function in mild-to-moderate AD. Donepezil is FDA-approved for all stages (mild to severe), while rivastigmine is available as a transdermal patch to reduce gastrointestinal side effects. Memantine, an NMDA receptor antagonist, regulates glutamate activity to slow cognitive decline in moderate-to-severe AD, often combined with cholinesterase inhibitors for additive benefits (Lleó et al., 2006).

1.5.2 Disease-Modifying Immunotherapies

Lecanemab and donanemab, FDA-approved monoclonal antibodies, target Aβ plaques in early AD, slowing cognitive decline by 27–35% over 18 months. These require pre-treatment biomarker confirmation (PET/CSF amyloid testing) and MRI monitoring for amyloid-related imaging abnormalities (ARIA), including brain swelling/bleeding. Aducanumab, previously approved, is no longer marketed due to limited efficacy and safety concerns (Ruthirakuhan et al., 2016).

1.6 CONCLUSION

AD is a complex interplay of genetic, molecular, and environmental factors, with no single hypothesis fully capturing its pathogenesis. While the cholinergic, amyloid, and tau hypotheses remain foundational, newer perspectives involving OS, metabolic dysfunction, and neuroinflammation underscore the multifactorial nature of AD. Current therapeutic strategies, including acetylcholinesterase inhibitors and anti-amyloid agents, provide only modest benefits, highlighting the need for early intervention and combination approaches targeting multiple pathological pathways. Ongoing research into biomarkers, personalised medicine, and novel targets such as GSK-3 and CREB signalling holds promise for altering the disease course. A deeper understanding of the mechanisms at play is essential for developing effective interventions that address symptom management and disease modification.

REFERENCES

Abadio, A. K. R., Kioshima, E. S., Martins, N. F., Maigret, B., Felipe, M. S. S., Freitas, S. M., Maigret, B., Felipe, M. S. S., Abagyan, R. A., Kufareva, I., Totrov, M., Abbasi, M., Sadeghi-Aliabadi, H., Hassanzadeh, F., Amanlou, M., Abdelmonsef, A. H., Dulapalli, R., Dasari, T., Padmarao, L. S., … Olson, A. J. A. J. (2011). A pharmacophore based virtual screening approach for the discovery of *Trypanosoma cruzi* GAPDH inhibitors. *Journal of Chemical Information and Modeling*, *5*, 2019). https://doi.org/10.4155/fmc.13.166

Aisen, P. S., Cummings, J., Jack, C. R., Morris, J. C., Sperling, R., Frölich, L., Jones, R. W., Dowsett, S. A., Matthews, B. R., Raskin, J., Scheltens, P., & Dubois, B. (2017). On the path to 2025: Understanding the Alzheimer's disease continuum. *Alzheimer's Research and Therapy*. https://doi.org/10.1186/s13195-017-0283-5

Anand, A., Patience, A. A., Sharma, N., & Khurana, N. (2017). The present and future of pharmacotherapy of Alzheimer's disease: A comprehensive review. *European Journal of Pharmacology*. https://doi.org/10.1016/j.ejphar.2017.09.043

Babic, T., Francis, P. T., Palmer, A. M., Snape, M., & Wilcock, G. K. (1999). The cholinergic hypothesis of Alzheimer's disease: A review of progress (multiple letters) [11]. *Journal of Neurology Neurosurgery and Psychiatry*. https://doi.org/10.1136/jnnp.67.4.558

Briggs, R., Kennelly, S. P., & O'Neill, D. (2016). Drug treatments in Alzheimer's disease. *Clinical Medicine, Journal of the Royal College of Physicians of London*. https://doi.org/10.7861/clinmedicine.16-3-247

Cacabelos, R., Alvarez, A., Lombardi, V., Fernandez-Novoa, L., Corzo, L., Perez, P., Laredo, M., Pichel, V., Hernandez, A., Varela, M., Figueroa, J., Prous, J. J., Windisch, M., & Vigo, C. (2000). Pharmacological treatment of Alzheimer disease: From psychotropic drugs and cholinesterase inhibitors to pharmacogenomics. *Drugs of Today*. *36*(7), 415–499.

Chan, K. Y., Wu, J. J., Liu, L., Theodoratou, E., Car, J., Middleton, L., Russ, T. C., Deary, I. J., Campbell, H., Wang, W., & Rudan, I. (2013). Epidemiology of Alzheimer's disease and other forms of dementia in China, 1990–2010 381 p-2016-2023: A systematic review and analysis. *The Lancet*. https://doi.org/10.1016/S0140-6736(13)60221-4

Chi, S., Yu, J. T., Tan, M. S., & Tan, L. (2014). Depression in Alzheimer's disease: Epidemiology, mechanisms, and management. *Journal of Alzheimer's Disease*. https://doi.org/10.3233/JAD-140324

Craig, L. A., Hong, N. S., & McDonald, R. J. (2011). Revisiting the cholinergic hypothesis in the development of Alzheimer's disease. *Neuroscience and Biobehavioral Reviews*. https://doi.org/10.1016/j.neubiorev.2011.03.001

Dartigues, J. F., & Letenneur, L. (2000). Genetic epidemiology of Alzheimer's disease. *Current Opinion in Neurology*. https://doi.org/10.1097/00019052-200008000-00004

Du, X., Wang, X., & Geng, M. (2018). Alzheimer's disease hypothesis and related therapies. *Translational Neurodegeneration*. https://doi.org/10.1186/s40035-018-0107-y

El Haj, M., Antoine, P., Amouyel, P., Lambert, J. C., Pasquier, F., & Kapogiannis, D. (2016). Apolipoprotein E (APOE) ε4 and episodic memory decline in Alzheimer's disease: A review. *Ageing Research Reviews*. https://doi.org/10.1016/j.arr.2016.02.002

Eratne, D., Loi, S. M., Farrand, S., Kelso, W., Velakoulis, D., & Looi, J. C. L. (2018). Alzheimer's disease: Clinical update on epidemiology, pathophysiology and diagnosis. *Australasian Psychiatry*. https://doi.org/10.1177/1039856218762308

Faiyaz, M., Ganayee, M. A., Akhtar, S., Krishnan, S., Flora, B., Dogra, D., Jha, N. K., Chellappan, D. K., Negi, P., Dua, K., Kesari, K. K., & Gupta, P. K. (2021). Nanomaterials in Alzheimer's disease treatment: A comprehensive review. *Frontiers in Bioscience – Landmark*. https://doi.org/10.52586/4992

Fenclová, E., Albrecht, J., Harsa, P., & Jirák, R. (2020). Risk factors for Alzheimer's disease. *Ceska a Slovenska Psychiatrie*. https://doi.org/10.15354/si.20.re036

Fišar, Z. (2022). Linking the amyloid, tau, and mitochondrial hypotheses of Alzheimer's disease and identifying promising drug targets. *Biomolecules*. https://doi.org/10.3390/biom12111676

Francis, P. T., Palmer, A. M., Snape, M., & Wilcock, G. K. (1999). The cholinergic hypothesis of Alzheimer's disease: A review of progress. *Journal of Neurology Neurosurgery and Psychiatry*. https://doi.org/10.1136/jnnp.66.2.137

Golmohammadi, M., Mahmoudian, M., Hasan, E. K., Alshahrani, S. H., Romero-Parra, R. M., Malviya, J., Hjazi, A., Najm, M. A. A., Almulla, A. F., Zamanian, M. Y., Kadkhodaei, M., & Mousavi, N. (2024). Neuroprotective effects of riluzole in Alzheimer's disease: A comprehensive review. *Fundamental and Clinical Pharmacology*. https://doi.org/10.1111/fcp.12955

Gulisano, W., Maugeri, D., Baltrons, M. A., Fà, M., Amato, A., Palmeri, A., D'Adamio, L., Grassi, C., Devanand, D. P., Honig, L. S., Puzzo, D., & Arancio, O. (2018). Role of amyloid-β and tau proteins in Alzheimer's disease: Confuting the amyloid cascade. *Journal of Alzheimer's Disease*. https://doi.org/10.3233/JAD-179935

Haghighijoo, Z., Zamani, L., Moosavi, F., & Emami, S. (2022). Therapeutic potential of quinazoline derivatives for Alzheimer's disease: A comprehensive review. *European Journal of Medicinal Chemistry*. https://doi.org/10.1016/j.ejmech.2021.113949

Hampel, H., Hardy, J., Blennow, K., Chen, C., Perry, G., Kim, S. H., Villemagne, V. L., Aisen, P., Vendruscolo, M., Iwatsubo, T., Masters, C. L., Cho, M., Lannfelt, L., Cummings, J. L., & Vergallo, A. (2021). The amyloid-β pathway in Alzheimer's disease. *Molecular Psychiatry*. https://doi.org/10.1038/s41380-021-01249-0

Hampel, H., Lista, S., Vanmechelen, E., Zetterberg, H., Giorgi, F. S., Galgani, A., Blennow, K., Caraci, F., Das, B., Yan, R., Afshar, M., Aguilar, L. F., Akman-Anderson, L., Arenas, J., Ávila, J., Babiloni, C., Baldacci, F., Batrla, R., Benda, N., … Zugaza, J. L. (2020). β-Secretase1 biological markers for Alzheimer's disease: State-of-art of validation and qualification. *Alzheimer's Research and Therapy*. https://doi.org/10.1186/s13195-020-00686-3

Hampel, H., Mesulam, M. M., Cuello, A. C., Khachaturian, A. S., Vergallo, A., Farlow, M. R., Snyder, P. J., Giacobini, E., & Khachaturian, Z. S. (2019). Revisiting the cholinergic hypothesis in Alzheimer's disease: Emerging evidence from translational and clinical research. *The Journal of Prevention of Alzheimer's Disease*. https://doi.org/10.14283/jpad.2018.43

He, M., Lian, T., Guo, P., Zhang, W., Zhang, Y., Huang, Y., Liu, G., Guan, H., Li, J., Luo, D., Zhang, W., Zhang, W., Qi, J., Yue, H., Wang, X., & Zhang, W. (2024). The roles of apolipoprotein E ε4 on neuropathology and neuroinflammation in patients with Alzheimer's disease. *CNS Neuroscience and Therapeutics*. https://doi.org/10.1111/cns.14440

Hillen, H. (2019). The beta amyloid dysfunction (BAD) hypothesis for Alzheimer's disease. *Frontiers in Neuroscience*. https://doi.org/10.3389/fnins.2019.01154

Hoogmartens, J., Cacace, R., & Van Broeckhoven, C. (2021). Insight into the genetic etiology of alzheimer's disease: A comprehensive review of the role of rare variants. *Alzheimer's and Dementia: Diagnosis, Assessment and Disease Monitoring*. https://doi.org/10.1002/dad2.12155

Kametani, F., & Hasegawa, M. (2018). Reconsideration of amyloid hypothesis and tau hypothesis in Alzheimer's disease. *Frontiers in Neuroscience*. https://doi.org/10.3389/fnins.2018.00025

Karran, E., & De Strooper, B. (2022). The amyloid hypothesis in Alzheimer disease: New insights from new therapeutics. *Nature Reviews Drug Discovery*. https://doi.org/10.1038/s41573-022-00391-w

Kciuk, M., Kruczkowska, W., Gałęziewska, J., Wanke, K., Kałuzińska-Kołat, Ż, Aleksandrowicz, M., & Kontek, R. (2024). Alzheimer's disease as type 3 diabetes: Understanding the link and implications. *International Journal of Molecular Sciences*, *25*(22), 11955. https://doi.org/10.3390/ijms252211955

Khubetova, I. V., Voloshchuk, D. A., Kosenko, KA, & Vorokhta, Y. M. (2023). Epidemiology of Alzheimer's disease in the Odesa region. *Zaporozhye Medical Journal*. https://doi.org/10.14739/2310-1210.2023.2.270277

Kim, J., Basak, J. M., & Holtzman, D. M. (2009). The role of apolipoprotein E in Alzheimer's disease. *Neuron*. https://doi.org/10.1016/j.neuron.2009.06.026

Koutsodendris, N., Nelson, M. R., Rao, A., & Huang, Y. (2021). Apolipoprotein E and Alzheimer's disease: Findings, hypotheses, and potential mechanisms. *Annual Review of Pathology: Mechanisms of Disease*. https://doi.org/10.1146/annurev-pathmechdis-030421-112756

La Barbera, L., Mauri, E., D'Amelio, M., & Gori, M. (2022). Functionalization strategies of polymeric nanoparticles for drug delivery in Alzheimer's disease: Current trends and future perspectives. *Frontiers in Neuroscience*. https://doi.org/10.3389/fnins.2022.939855

Levin, O. S., & Vasenina, E. E. (2017). 25 years of the amyloid hypothesis of the origin of Alzheimer's disease: Advances, failures, and new perspectives. *Neuroscience and Behavioral Physiology*. https://doi.org/10.1007/s11055-017-0513-0

Lindsay, J., Laurin, D., Verreault, R., Hébert, R., Helliwell, B., Hill, G. B., & McDowell, I. (2002). Risk factors for Alzheimer's disease: A prospective analysis from the Canadian study of health and aging. *American Journal of Epidemiology*. https://doi.org/10.1093/aje/kwf074

Lleó, A., Greenberg, S. M., & Growdon, J. H. (2006). Current pharmacotherapy for Alzheimer's disease. *Annual Review of Medicine*. https://doi.org/10.1146/annurev.med.57.121304.131442

Luchsinger, J. A., & Mayeux, R. (2004). Cardiovascular risk factors and Alzheimer's disease. *Current Atherosclerosis Reports*. https://doi.org/10.1007/s11883-004-0056-z

Michailidis, M., Moraitou, D., Tata, D. A., Kalinderi, K., Papamitsou, T., & Papaliagkas, V. (2022). Alzheimer's disease as type 3 diabetes: Common pathophysiological mechanisms between Alzheimer's disease and type 2 diabetes. *International Journal of Molecular Sciences*. https://doi.org/10.3390/ijms23052687

Mielke, M. M., Vemuri, P., & Rocca, W. A. (2014). Clinical epidemiology of Alzheimer's disease: Assessing sex and gender differences. *Clinical Epidemiology*. https://doi.org/10.2147/CLEP.S37929

Murphy, M. P., & Levine, H. (2010). Alzheimer's disease and the amyloid-β peptide. *Journal of Alzheimer's Disease*. https://doi.org/10.3233/JAD-2010-1221

Nunomura, A. (2013). Oxidative stress hypothesis for Alzheimer's disease and its potential therapeutic implications. *Clinical Neurology*. https://doi.org/10.5692/clinicalneurol.53.1043

Olney, J. W., Wozniak, D. F., & Farber, N. B. (1997). Excitotoxic neurodegeneration in Alzheimer disease: New hypothesis and new therapeutic strategies. *Archives of Neurology*. https://doi.org/10.1001/archneur.1997.00550220042012

Paroni, G., Bisceglia, P., & Seripa, D. (2019). Understanding the amyloid hypothesis in Alzheimer's disease. *Journal of Alzheimer's Disease*. https://doi.org/10.3233/JAD-180802

Praticò, D. (2008). Oxidative stress hypothesis in Alzheimer's disease: A reappraisal. *Trends in Pharmacological Sciences*. https://doi.org/10.1016/j.tips.2008.09.001

Roy, S., & Rauk, A. (2005). Alzheimer's disease and the "ABSENT" hypothesis: Mechanism for amyloid β endothelial and neuronal toxicity. *Medical Hypotheses*. https://doi.org/10.1016/j.mehy.2004.08.031

Ruthirakuhan, M., Herrmann, N., Suridjan, I., Abraham, E. H., Farber, I., & Lanctôt, K. L. (2016). Beyond immunotherapy: New approaches for disease modifying treatments for early Alzheimer's disease. *Expert Opinion on Pharmacotherapy*. https://doi.org/10.1080/14656566.2016.1258060

Saura, C. A., & Valero, J. (2011). The role of CREB signaling in Alzheimer's disease and other cognitive disorders. *Reviews in the Neurosciences*. https://doi.org/10.1515/RNS.2011.018

Sehar, U., Rawat, P., Reddy, A. P., Kopel, J., & Reddy, P. H. (2022). Amyloid beta in aging and Alzheimer's disease. *International Journal of Molecular Sciences*. https://doi.org/10.3390/ijms232112924

Selkoe, D. J., & Hardy, J. (2016). The amyloid hypothesis of Alzheimer's disease at 25 years. *EMBO Molecular Medicine*. https://doi.org/10.15252/emmm.201606210

Šerý, O., Povová, J., Míšek, I., Pešák, L., & Janout, V. (2013). Molecular mechanisms of neuropathological changes in Alzheimer's disease: A review. *Folia Neuropathologica*. https://doi.org/10.5114/fn.2013.34190

Sharma, V. K., & Singh, T. G. (2021). CREB: A multifaceted target for Alzheimer's disease. *Current Alzheimer Research*. https://doi.org/10.2174/1567205018666210218152253

Singh, S., Mahajan, M., Kumar, D., Singh, K., Chowdhary, M., & Amit (2023). An inclusive study of recent advancements in Alzheimer's disease: A comprehensive review. *Neuropeptides*. https://doi.org/10.1016/j.npep.2023.102369

Soininen, H. S., & Riekkinen, P. J. (1996). Apolipoprotein E, memory and Alzheimer's disease. *Trends in Neurosciences*. https://doi.org/10.1016/0166-2236(96)10027-8

Sosa-Ortiz, A. L., Acosta-Castillo, I., & Prince, M. J. (2012). Epidemiology of dementias and Alzheimer's disease. *Archives of Medical Research*. https://doi.org/10.1016/j.arcmed.2012.11.003

Speck, C. E., Kukull, W. A., Brenner, D. E., Bowen, J. D., McCormick, W. C., Teri, L., Pfanschmidt, M. L., Thompson, J. D., & Larson, E. B. (1995). History of depression as a risk factor for Alzheimer's disease. *Epidemiology*. https://doi.org/10.1097/00001648-199507000-00006

Takashima, A. (2006). GSK-3 is essential in the pathogenesis of Alzheimer's disease. *Journal of Alzheimer's Disease*. https://doi.org/10.3233/jad-2006-9s335

Teixeira, J. P., de Castro, A. A., Soares, F. V., da Cunha, E. F. F., & Ramalho, T. C. (2019). Future therapeutic perspectives into the Alzheimer's disease targeting the oxidative stress hypothesis. *Molecules*. https://doi.org/10.3390/molecules24234410

Wang, X., Li, H., Sheng, Y., He, B., Liu, Z., Li, W., Yu, S., Wang, J., Zhang, Y., Chen, J., Qin, L., & Meng, X. (2024). The function of sphingolipids in different pathogenesis of Alzheimer's disease: A comprehensive review. *Biomedicine and Pharmacotherapy*. https://doi.org/10.1016/j.biopha.2023.116071

Xia, X., Jiang, Q., McDermott, J., & Han, J. D. J. (2018). Aging and Alzheimer's disease: Comparison and associations from molecular to system level. *Aging Cell*. https://doi.org/10.1111/acel.12802

Xu, W., Tan, L., Wang, H. F., Jiang, T., Tan, M. S., Tan, L., Zhao, Q. F., Li, J. Q., Wang, J., & Yu, J. T. (2015). Meta-analysis of modifiable risk factors for Alzheimer's disease. *Journal of Neurology, Neurosurgery and Psychiatry*. https://doi.org/10.1136/jnnp-2015-310548

2 Alzheimer's Disease
The Role of Blood-Brain Barrier and Strategy to Cross BBB

Hadri Hadi Md Yusof, Alina Arulsamy, and Mohmad Farooq Shaikh

2.1 INTRODUCTION

Alzheimer's disease (AD) predominantly impacts the elderly population by gradual degradation of memory and cognitive abilities. The disease accounts for the commonest form of dementia (60–80%), characterized by the formation of amyloid-beta (Aβ) plaques and tau protein neurofibrillary tangles (NFTs) that lead to synaptic dysfunction and neural loss (Knopman et al., 2021). The aggregation of Aβ peptides produced by abnormal processing and cleavage of the amyloid precursor protein (APP) provokes disruption in synaptic communication and initiates inflammation leading to neuronal damages (Hardy & Higgins, 1992). Additionally, NFTs of tau proteins in the intracellular space caused by hyperphosphorylation of tau proteins are also strongly linked to the pathology of AD (Querfurth & LaFerla, 2010). Another hypothesis that is being researched intensively is from a study done by McGeer and McGeer in 1995 that suggested autotoxic loop caused by the aggregation of Aβ plaques and NFTs could trigger inflammatory response past a threshold where the positive feedback from the toxicity of cell death could exacerbate the pathological decline to a point where the disease could be observed clinically (McGeer & McGeer, 1995). Emerging evidence also underscores vascular contributions, particularly blood-brain barrier (BBB) dysfunction, in AD pathology (Iturria-Medina et al., 2016). Compromised BBB integrity increases permeability, where neurotoxic substances could infiltrate the brain, thereby accelerating neurodegeneration. We aim to explore the role and structure of BBB and how it is affected in AD and ways to cross the BBB for neurodegenerative diseases, specifically AD.

2.2 ALZHEIMER'S DISEASE AND THE BLOOD-BRAIN BARRIER

2.2.1 Pathophysiology and History of Alzheimer's Disease

The pathogenesis of AD has been widely researched with multiple competing theories attempting to state the mechanisms contributing to the neurodegeneration. Among the studies, Braak and Braak in 1991 have introduced the six-stage model for AD progression based on the distribution of NFTs and neuropil threads (NTs) (Braak & Braak, 1991). This staging begins in the transentorhinal region and proceeds to the limbic and isocortical areas, making NFTs a potential and predictable biomarker of disease progression. Opposite to Aβ with variable distribution, NFTs exhibit a consistent pattern that is correlated to the disease severity. The same study also identified a "pre-clinical phase" of AD where neuropathological abnormalities occur before any clinical symptoms, marking an interesting avenue for early diagnosis and intervention research (Beach et al., 2012; Braak & Braak, 1991).

The "Amyloid Cascade Hypothesis", first introduced in 1992 by Hardy and Higgins, marked a significant chapter in AD understanding and research, and the suggestion provided a basal framework that has significantly influenced the current understanding and direction of AD research (Hardy & Higgins, 1992). The authors suggested that the accumulation of Aβ protein produced via proteolytic cleavage of the APP acts as the trigger of the event in downstream AD pathology. This pathology had challenged previous theories proposing that Aβ is the main culprit for subsequent neurodegenerative processes such as the formation of NFTs from hyperphosphorylated tau proteins, neuronal cell death and eventually the development of AD. According to the model, the misprocessing of APP leads to the aggregation of Aβ which exerts toxic effects on neurons by disrupting the calcium homeostasis and activating tau hyperphosphorylation. These cellular events lead to synaptic dysfunction and neuronal loss, both hallmarks of AD progression.

This revolutionary perspective has highlighted potential genetic and molecular mechanisms of Aβ deposition. Firstly, the mutation of APP gene, especially ones linked to familial AD, is associated with increased Aβ production and reduction in Aβ clearance, further supporting the hypothesis of amyloid pathology and its influence in the disease. The authors further speculated that cerebrovascular damage, chronic inflammation and other pathological features commonly associated with AD arise as a secondary effect of Aβ accumulation, making Aβ the apex etiological contributor to the disease.

The "Amyloid Cascade Hypothesis" not only provided a theoretical foundation for understanding AD but also reoriented research efforts towards targeting Aβ as a therapeutic strategy. Its introduction marked a paradigm shift in the field, inspiring decades of investigation into amyloid

DOI: 10.1201/9781003602477-2

biology, APP processing pathways and the development of anti-amyloid therapies. Although the hypothesis has faced challenges in light of findings that suggest a more complex interplay of factors in AD progression, it remains a cornerstone of modern AD research and a testament to the enduring impact of Hardy and Higgins' seminal contribution.

2.2.2 Structure and Function of the Blood-Brain Barrier

Blood vessels create an extensive network that links the heart to every organ and tissue in the body, delivering oxygen and nutrients while aiding in the removal of carbon dioxide and metabolic waste. This vascular system, often referred to as the vascular tree, consists of arteries and arterioles that carry blood to tissues, capillary beds where exchange occurs, and venules and veins that return blood to the heart. Each segment of this network is uniquely adapted to its specific location and the organ it serves (Daneman & Prat, 2015).

Capillaries are separated into three main types. The first, continuous non-fenestrated capillaries, the least permeable of all capillary types, can be found in skin and lungs, characterized by the presence of tightly joined endothelial cells, a complete basement membrane and the lack of any fenestrations (Aird, 2007). Next, continuous fenestrated capillaries can be found in the intestinal villi and the endocrinal glands and both share very similar structure but include a diaphragm cover fenestration in the plasma membrane. Finally, the discontinuous capillaries found in liver feature large gaps between endothelial cell and incomplete basement membrane and are the most permeable (Aird, 2007).

The BBB is a collection of continuous, non-fenestrated capillaries that serves as a selective membrane for the movement of molecules, ions and cells between the blood circulation system and the central nervous system (CNS). The BBB also serves to maintain homeostasis, supporting neuronal functions and protecting the brain from toxins, inflammation, pathogens and injuries. Although vital, the restrictive nature of the BBB presents a significant challenge to deliver and design a therapeutic drug to treat CNS disorders and this has been a deep area of research in finding ways to bypass or modulate the BBB (Daneman & Prat, 2015; Larsen et al., 2014; Zenaro et al., 2017).

The first conceptual idea of BBB was first suggested by Paul Ehrlich in 1885 where the bloodstream was injected with dyes and all organs were stained except the brain. This finding was attributed to the brain's low affinity for the dye (Ehrlich, 1885). In 1913, the assumption was challenged by his student, Edwin E. Goldman, who directly injected the dye into the cerebrospinal fluid (CSF) of rabbits and dogs and observed a clear separation between the CNS and the blood (Dyrna et al., 2013).

Structurally, the BBB's make-up is divided into several cellular components: firstly, the endothelial cells and the cellular junctions contribute to the physical barrier of BBB. This includes the tight junction and adherens junction proteins, inhibition of non-selective fenestrae, pinocytosis and bulk flow transcytosis on top of the suppression of leukocyte adhesion molecules (Obermeier et al., 2013). With this, any possible movements across the BBB are regulated by specific transporters, allowing nutrient uptake and toxin secretions (Obermeier et al., 2013).

The restricted permeability and high transendothelial electrical resistance of BBB are maintained by tight junctions formed by integral membrane proteins like occludin and claudins (claudin-1, -3, -5 and -12) with adherens junctions supported by cadherins and junctional adhesion molecules (JAMs) that provide strength and support to the endothelial layer (Zhao et al., 2015).

To add to that, key supporters of the BBB endothelium, such as pericytes, encapsulate the abluminal surface of the cerebral vessels such as capillaries, pre-capillary arterioles and post-capillary venules help in maintaining the barrier integrity, assist in the building of new vessels, ensure microvascular stability, phagocytose toxic metabolites and modulate capillary diameter (Zhao et al., 2015).

On the other hand, the astrocytes serve their supporting role by providing nutrients to neurons, being involved in the regulation of extracellular potassium balance, clearing and recycling neurotransmitters and controlling immune reactions as part of the regulation of the BBB (Obermeier et al., 2013). The perivascular astrocyte end feet are highly specialized and polarized structures. They encircle the abluminal side of the cerebral vessel and contain the most abundant water channel (aquaporin-4, AQP4) and ATP-sensitive inward rectifier potassium channel Kir4.1 (Hosokawa et al., 2011). This tripartite structure (Figure 2.1)—composed of endothelial cells, pericytes and astrocytes—ensures selective permeability, allowing essential nutrients to reach the brain while restricting harmful substances (Sweeney et al., 2018).

It is in the nature of the BBB to restrict movement of bulk fluid, attributed to its low rates of transcytosis and pinocytosis. Nevertheless, small molecules such as lipids and gases such as oxygen and carbon dioxide can pass through the BBB due to their size and solubility (Pardridge, 2007). Solute carrier-mediated transporters (CMTs) are responsible for the transport of macromolecules or other polar substances such as glucose, amino acids and fatty acids. Receptor-mediated transcytosis (RMTs) allows larger proteins and peptides such as insulin and transferrin to cross the BBB (Pardridge, 2015).

2.2.3 Mechanisms of Blood-Brain Barrier Dysfunction in Alzheimer's Disease

In AD, BBB disruption is highly regarded as an important factor in the disease's pathogenesis, as shown in Figure 2.2. In this section, we'll dive deeper into the various potential mechanisms that could underlie BBB dysfunction in AD, specifically focusing on the molecular and cellular irregularities.

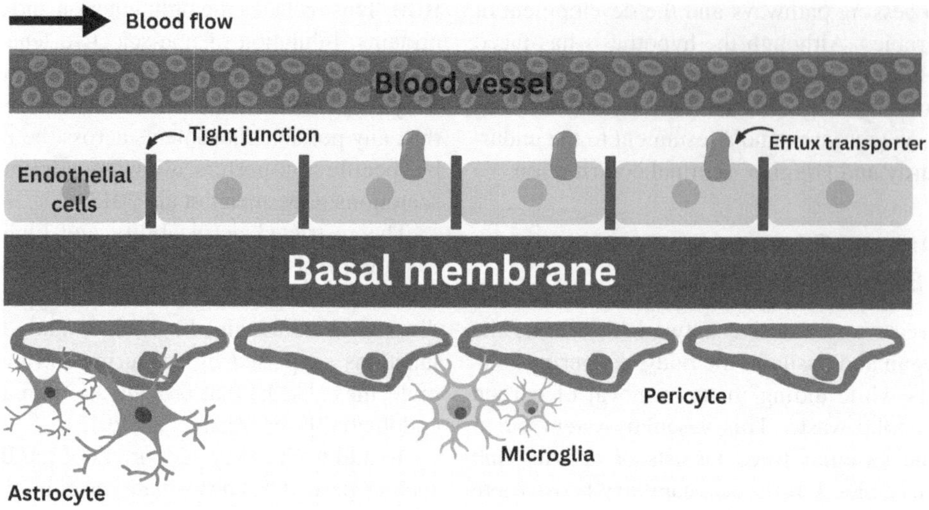

FIGURE 2.1 Basic structure of the blood-brain barrier.

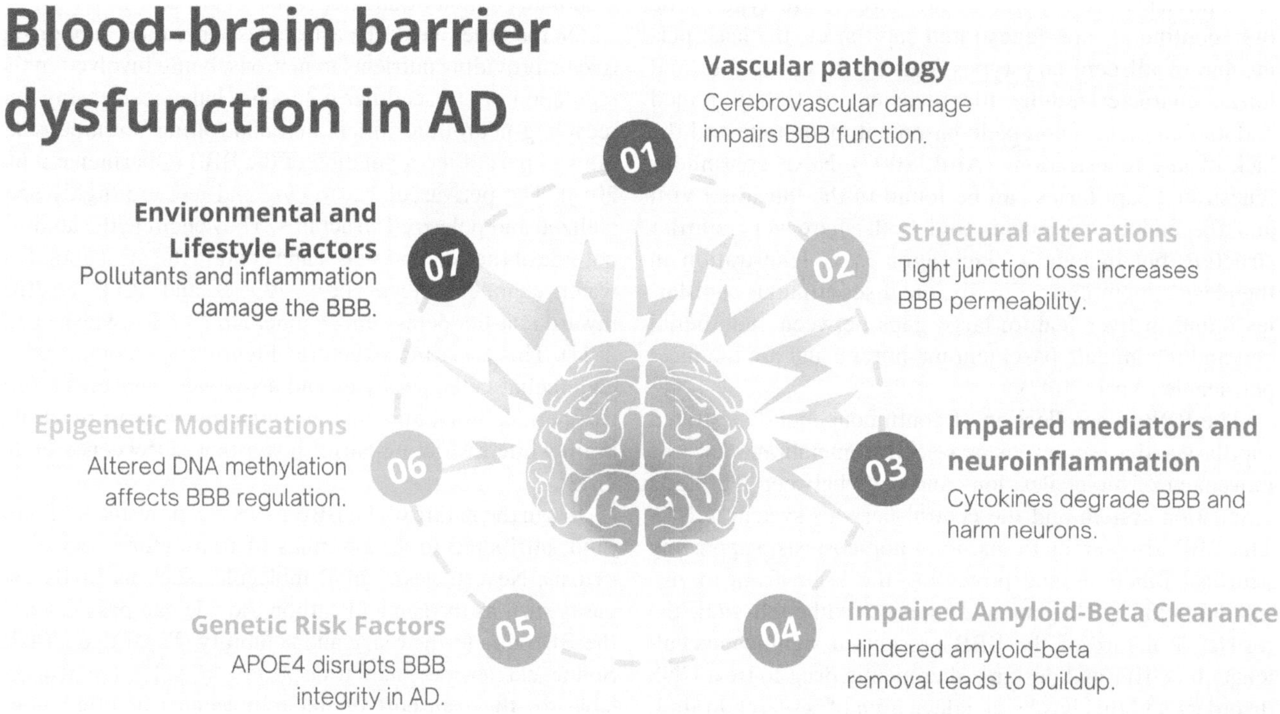

FIGURE 2.2 List of factors that cause BBB alterations in AD.

2.2.3.1 Vascular Pathology

In a study by Toledo and colleagues in 2013, in patients with an autopsy-based diagnosis under the US National Alzheimer's Disease Coordinating Centre Database (5,715 patients), 80% out of 4,629 patients who were diagnosed with AD with no evidence of mixed-dementia had vascular pathology (Toledo et al., 2013). This number of cases encompasses cerebrovascular disease, lacunae, multiple microinfarcts indicative of small vessel disease, haemorrhage, atherosclerosis, arteriolosclerosis and cerebral amyloid angiopathy (CAA) (Toledo et al., 2013). CAA is known as a significant contributor to the disruption of the BBB and is acknowledged as one of the principal pathological features of AD (Hardy & Higgins, 1992). In a preclinical, pre-symptomatic detection of AD, alterations in vascular biomarkers can appear even before the emergence of cognitive deficits and AD-related biomarkers such as Aβ accumulation, reduced CSF levels of $A\beta_{42}$ and an increase in tau and hyperphosphorylated tau in CSF (Iturria-Medina et al., 2016).

Based on the two-hit vascular hypothesis suggested by Nelson and colleagues in 2016, cerebrovascular damage is attributed as the first insult that is not only enough to initiate the neuronal injury and neurodegeneration but can also accentuate the accumulation of Aβ toxins in the brain (second hit) (Nelson et al., 2016). The first hit damages the blood vessels, resulting in the disruption of the BBB and reduced cerebral perfusion, which in turn allow neurotoxic substances and inflammatory cells to enter the brain, exacerbating neuronal injury and inflammation. The second hit which is the accumulation of Aβ toxins as a result of the first, on top of the impaired clearance of Aβ toxins, contributes to the further build up in the brain followed by increased production of antibodies against the toxin and inflammation ensues leading to the hallmark pathologies seen in AD (Nelson et al., 2016; Zhao et al., 2015).

2.2.3.2 Structural Alterations in the BBB

In neurodegenerative diseases, the BBB is structurally changed to a point where its function is impaired. These changes include degeneration of endothelial layers, which form the inner lining of the blood vessels, and a decreased expression of tight and adherens junctions, which in normal circumstances would assist in maintaining the integrity of BBB (Sweeney et al., 2018). The damage to these structures and junctions leads to over permeability of the barrier, permitting unwanted substances and toxins to leak into the brain. On top of that, the increase in bulk flow transcytosis, where big volumes of molecules could then move through the endothelial layer, bypassing the selective transport mechanisms. The breakdown of pericytes, which normally provide structural support to the BBB, further weakens the barrier which ultimately impair the efficiency of solute transport and decreases the effectiveness of therapeutic agents, slowing down their potential to reach neuronal targets (Sagare et al., 2013; Zhao et al., 2015).

2.2.3.3 Impaired Mediators and Neuroinflammation

Neuroinflammation is one of the main hallmarks of AD and is crucial contributor in the event of BBB breakdown. Activated microglia and astrocytes would generate pro-inflammatory cytokines, like tumour necrosis factor-alpha (TNF-α), interleukin-1 beta (IL-1β) and interleukin-6 (IL-6), that could affect the expression of tight junction proteins, further disrupting BBB's permeability. Additionally, these inflammatory cytokines activate the recruitment of peripheral immune response to the CNS, further exacerbating the damages. Similarly, the inflammation process could lead to the increased generation of reactive oxygen species (ROS) which plays a similar role in inducing oxidative stress followed by damage to the BBB (Nelson et al., 2016; Sweeney et al., 2018; Zhao et al., 2015).

In conditions like AD, the inflammatory response by peripheral macrophages and neutrophils would signal an activated innate immune response. Additionally, diseases like multiple sclerosis showed that BBB breakdown leads to T and B lymphocytes to enter the brain initiating adaptive immune response. Further damages to the barrier could allow pathogens into the brain causing direct and indirect immune response worsening the plaque formation (Nelson et al., 2016; Sweeney et al., 2018).

2.2.3.4 Impaired Clearance Mechanism for Amyloid Beta

Mawuenyega and colleagues in 2010 stated that the increased concentration of Aβ in patients with sporadic AD can be attributed to the impaired clearance mechanism of Aβ in the CNS (Mawuenyega et al., 2010). *In vivo* studies showed that Aβ is mostly cleared from the CNS via the transvascular clearance mechanism in the BBB which contributes to about 70–85% of its overall clearance as opposed to the outflow via the interstitial fluids (Bading et al., 2002; Deane et al., 2004; Tarasoff-Conway et al., 2015). In recent years, a deeper understanding of the molecular aspects of Aβ clearance has surfaced. Specifically, the interaction between Aβ and the lipoprotein receptor-related protein 1 (LRP1) located on the abluminal surface of the BBB triggers the internalization of Aβ by endothelial cells followed by the outflow via the circulatory system (Bell et al., 2007). The internalization of LRP1-Aβ complex into the endothelial cells was facilitated by the phosphatidylinositol-binding clathrin assembly protein (PICALM) that is also involved in the intracellular trafficking of Aβ-containing vesicles through endocytosis, subsequently guiding them through Rab5-positive early endosomes and Rab11-positive sorting endosomes. This process assists in the exocytosis of Aβ at the luminal surface of the BBB, hence completing the transcytosis cycle (Deane et al., 2004).

The soluble form of LRP1 (sLRP1) in the bloodstream produced by the proteolytic cleavage by β-secretase binds and sequesters free Aβ40 and Aβ42 and facilitates the transport of Aβ through excretory organs (Sagare et al., 2013). In the brain's ISF, $A\beta_{42}$ forms apoJ-Aβ complex via association with clusterin (CLU or apoJ) which would then be transported across the BBB into the blood circulation via LRP2 (Bell et al., 2007). Additionally, the clearance of Aβ could also be done via enzymatic degradation processes, including neprilysin, insulin-degrading enzyme, matrix metalloproteinases, plasmin and tissue plasminogen activator (Querfurth & LaFerla, 2010).

2.2.3.5 Genetic Factors and Molecular Mechanisms

Several gene mutations have since been associated with an increased risk of sporadic and familial AD and often associated with cerebrovascular irregularities. For example, apoE4 is widely regarded as the most significant genetic factor for late-onset AD (LOAD) and is known for its detrimental effects on cerebrovascular integrity (Zlokovic, 2013). All three alleles of human apoE gene can be found on a single gene locus on human chromosome 19q13, encoded for three apoE isoforms, apoE2, apoE3 and apoE4 (Zlokovic, 2013) where apoE4 poses a major genetic risk factor for AD where it has an arginine at residue 112 as opposed to cysteine in apoE2 and apoE3. This difference in the amino acid residue is unstable and could have multiple pathogenic effects. The apoE4 variant showed a significant impact on the BBB, greater incidence of CAA and an increase in accumulation of fibrinogen and iron in the brain of individuals with AD (Halliday et al., 2013; Zipser et al., 2007; Zonneveld et al., 2014). In vivo transgenic mouse model associated with mutated apoE4 gene showcased vascular abnormalities prior to neuronal damage and changes in behaviour (Bell et al., 2012). Different apolipoprotein E (apoE) variants interact differently with Aβ. While apoE2 and apoE3 complexed with Aβ are efficiently transported across the BBB through LRP1, apoE4-Aβ complexes are transported slower due to inefficient transcytosis via low-density lipoprotein receptor (VLDLR) (Bell et al., 2007; Deane et al., 2008).

Next, PICALM is also shown to play a significant role in Aβ clearance in the brain (Parikh et al., 2014). PICALM plays a role in endocytosis, receptor internalization and protein trafficking within cells (Treusch et al., 2011). PICALM is expressed in high number in the endothelial cells of the BBB and study has demonstrated that a specific variant of the rs3851179 gene correlates with reduced expression of the gene in endothelial cells resulting in impaired Aβ clearance in in vitro models of BBB suggesting that specific PICALM variations could influence AD progression via cerebrovascular pathways (Zhao et al., 2015).

Furthermore, changes in clusterin (CLU), also known as apoJ, have been associated with sporadic AD where it interacts with and inhibits the aggregation of Aβ and assists in the clearance across the BBB (Bell et al., 2007). Alterations in presenilin (PSEN) genes are linked to autosomal dominant variants of AD leading to the increase in Aβ synthesis (Querfurth & LaFerla, 2010); specifically, mutations in PSEN1 are noted to alter cerebrovascular integrity, demonstrated by the impairment of pericytes and mural cells, disruption of the BBB and deposition of Aβ within the cerebral microvessels. Similar pathology has been demonstrated by Gama Sosa and colleagues in 2010 in transgenic mouse model of PSEN1 mutation (Gama Sosa et al., 2010). However, the specific link to the mutations and the accumulation of Aβ during AD progression is still unclear.

2.2.3.6 Epigenetic Modifications

Sanchez and Graff published a review in 2015 where they explored a wide range of epigenetic modifications commonly linked to AD, specifically looking at significant changes in DNA methylation and histone modifications (Sanchez-Mut & Graff, 2015). One key finding revealed that AD is associated with extensive modification in DNA methylation, evidenced by in vitro AD cell line. For example, the H4 glioblastoma cell line, which contains the Swedish mutation of the APP (K670M/N671L), a common mutation in Swedish family, showed increased Aβ production (Citron et al., 1992; Haass et al., 1995; Mullan et al., 1992). DNA microarray data exhibited trends of hypomethylation following a bisulfite treatment on the same cell line (Sung et al., 2011). Exposure of SH-SY5Y neuronal-like cell line to a conditioned media derived from cells containing the Indiana mutation (V717F) (Murrell et al., 1991; Suzuki et al., 1994) in APP linked to increased Aβ levels also led to notable DNA hypomethylation (Hodgson et al., 2013). Furthermore, exposing synthetic Aβ on brain microvascular endothelial cells showed decreased DNA methylation assessed by HPLC (Chen et al., 2009) although exposing IMR-32 cell lines with synthetic Aβ did not show any notable alterations in DNA methylation evaluated through DNA microarrays (Taher et al., 2014). These results demonstrated the intricacy of DNA methylation responses to Aβ exposure on different cell types.

On the flip side, the realm of histone modification in AD is not as comprehensive as the connection between modifications and the pathologies is mostly circumstantial. Several investigations have indicated that histone deacetylase (HDAC) inhibitors may provide neuroprotective effects on AD specifically on the enhancement of dendritic spine density and the improvements in learning and memory as seen in animal models of AD (Fischer et al., 2007; Francis et al., 2009; Ricobaraza et al., 2009). Nevertheless, the mechanism at which HDAC inhibitors are directly involved with improving AD pathologies remains unclear. It has been established that HDAC2 levels increase in old age in human and murine subjects (Chouliaras et al., 2013; Singh & Thakur, 2014) and elevation in HDAC2 level is seen in several AD mouse models such as APP/PS1 (Gonzalez-Zuniga et al., 2014), p25/Cdk5 (Cruz et al., 2006) and 5xFAD (Oakley et al., 2006), all of which showcased increased Aβ formation and neurodegeneration. Similarly, post-mortem samples from patients with AD showed increased levels of HDAC2 in hippocampus and entorhinal cortex (Graff et al., 2012). Mechanistically, HDAC2 interacts with and modulates the expression of genes involved in learning and neuroplasticity. Pharmacological and viral-mediated inhibition of HDAC2 has shown positive effect in restoring cognitive functions in p25/Cdk5 mice (Graff et al., 2012; Wagner et al., 2015), providing strong link to the role of HDAC2 in cognitive decline seen in AD. However, one study by Mastroeni and colleagues in 2011 showed depletion of HDAC2 levels in AD patients (Mastroeni et al., 2011).

Alternatively, although the role of HDAC2 in AD is relatively definitive, conflicting evidences have since arisen regarding changes in basal histone acetylation level in AD. Zhang and colleagues in 2012 showed lowered levels of acetylation, Rao and colleagues in 2012 showed unchanged levels and Narayan and colleagues in 2015 showed increased levels of histone acetylation in post-mortem AD samples and no significant changes were seen in Tg2576 and 3xTg-AD mouse models (Cadena-del-Castillo et al., 2014; Francis et al., 2009). However, there are evidences of increased H3 and H4 acetylation levels in primary cultures of 3xTg-AD mice (Walker et al., 2013). These discrepancies could be due to the fact that instead of global histone acetylation changes, AD could possibly have diminished ability to modify epigenetic patterns during learning and memory formation. In these cases, HDAC inhibitors that elevate acetylation could in turn "prime" acetylation levels, thus increasing gene expression (Graff et al., 2012). To support this argument, basal levels of H4K12ac remained stable in ageing mice and increased in young mice in the process of learning and memory formation (Peleg et al., 2010). In Tg2576 mice that carry the Swedish APP mutation together with M146V PSEN1 mutation, global h4 acetylation showed no changes, whereas wild-type mice showed elevated acetylation following learning and memory tasks (Chishti et al., 2001). To conclude this section, there is a possibility that histone acetylation changes are locus-specific and highly sensitive to HDAC inhibitors without reflecting much global changes across chromatin. Genome-wide studies on histone modifications could be done to garner deeper insights on the mechanisms.

2.2.3.7 Lifestyle and Environment

Brain health can severely be impacted by environmental toxicants (solvents, cigarette smoke, industrial waste etc.) (Santibanez et al., 2007). Multiple studies have indicated the exposure effects of toxicant on the brain could increase the risk of developing neurodegenerative diseases such as AD and Parkinson's disease (PD) (Kukull et al., 1995) by affecting the cerebral microvasculature and penetrating the BBB, leading to increased inflammation, inducing cytotoxic effects and neuronal disruption. Increased risk of hospitalizations for AD and PD has been linked to exposure of patients to fine particulate matter (PM2.5) (Shi et al., 2020) that is further evidenced by longitudinal studies that indicated higher exposure to PM2.5 led to greater potential of developing these conditions. Furthermore, cognitive decline correlated to disruption of cerebral microvasculature is vulnerable to the damaging effects of environmental pollutants.

Another potential effect of toxic pollutant exposure is inducing mitochondrial dysfunction which leads to decrease in cellular energy production and metabolism, promotes free radical production and ultimately causes apoptosis (Moulton & Yang, 2012). Additionally, activation of microglia and neuroinflammation by environmental toxicants leads to cellular senescence and protein aggregation, ultimately leading to damage and cell death, all of which would accelerate the onset and progression of neurodegenerative diseases.

Alterations in the microbiota-gut-brain axis (MGBA) are now highly recognized as significant cause of the degeneration processes, especially in AD (Pistollato et al., 2016). The MGBA communication is complex and influences the brain function and immune responses which could be severely affected by pollutant exposure and poor diet choices. As discussed earlier, the triggering of neuroinflammation could lead to the compromise of the BBB, thus accelerating the neurotoxic agent uptake into the CNS, and this cycle is repeated in the degeneration process. Dysbiosis within the gut microbiota significantly plays a role in the pathogenesis of AD and BBB disruption (Kowalski & Mulak, 2019).

In summary, addressing the effects of environmental factors and improving gut microbiota health is vital in maintaining a healthy brain.

2.3 STRATEGIES FOR CROSSING THE BLOOD-BRAIN BARRIER

2.3.1 Nanoparticle-Based Drug Delivery Systems

One promising avenue in drug delivery systems in crossing the BBB is the nanoparticle-based approach. This is done by taking advantage of the physiochemical properties of nanoparticles; their small size, surface charge and flexibility in modifications could enhance the chance of transporting a drug across the BBB. One strategy is via encapsulation of active pharmacological ingredients in nanoparticles or by adding targeted ligand functionalities, which could greatly improve the stability, permeability and site-specific delivery to the brain. This strategy has since showed promising potential in treating CNS disorders such as AD, PD and certain brain tumours where delivering CNS-targeted drug delivery effectively has remained a challenge.

A study by Georgieva and colleagues in 2023 showed improved efficacy by using chitosan nanoparticles, specifically in their cationic nature, which could open the tight junction of BBB and facilitate drug transport across the brain. The galantamine-loaded chitosan nanoparticles had a mean particle size of about 240 nm (Georgieva et al., 2023), whereas the same drug loaded in chitosan/alginate nanoparticles had a mean size of about 286 nm. These sizes are beneficial in improving the BBB penetrance. These nanoparticles were synthesized through ionic gelation methods with a high drug loading efficiency of 67% for chitosan-based nanoparticles and 70% for chitosan/alginate complex. Another potential benefit of nanoparticle complexation is increased stability and controlled release of the drug where it was demonstrated in the same study that the galantamine release could be modulated which is crucial in maintaining therapeutic levels in the brain.

Another study done by Wadhwa and colleagues in 2023 showed that polymer-lipid hybrid nanoparticles (PLHNPs) showcased significantly improved delivery of repaglinide (REP) to the brain, addressing the short half-life and poor BBB permeability issue of the drug. The nanoparticles demonstrated improved permeability to the brain as much as 1.6-fold compared to the lone drug and this could be attributed to the PEGylated core-shell structure, which improved the interactions between the brain endothelium, increasing circulation time by evading reuptake by the reticuloendothelial system (RES) (Wadhwa et al., 2023).

Additionally, by including components like soy phosphatidylcholine (SPC) and choline enables facilitated diffusion across the BBB further improving the penetrance. In vivo pharmacokinetic studies have showed improved maximum concentration and area under the curve, indicating better overall exposure of REP in the body. The hybrid nature of PLHNPs also improved gastrointestinal stability and controlled release of the drug, maintaining therapeutic levels over time. Ultimately, these data suggested PLHNPs as a promising avenue to effectively target the brain and cross the BBB (Wadhwa et al., 2023).

2.3.2 Drug Repurposing and Small Molecule Development

While discovering new pharmacological knowledge and new therapy targets is important, exploring existing pharmacological knowledge and identifying compounds with favourable CNS penetration properties are gaining more attention. The major benefit of working with drug repurposing is the existing pharmacokinetics (PK), pharmacodynamics (PD) and drug safety profiles which could greatly accelerate the preclinical and clinical phases of drug development.

One example of drugs that have been successfully repurposed for new indication are aspirin (Sandler et al., 2003), raloxifene, thalidomide, regorafenib and metformin. Sandler and colleagues in 2003 found favourable evidence of aspirin on colorectal adenomas. Administering 325 mg of aspirin daily could significantly reduce the occurrence of colorectal adenomas where 17% of the patients had one detected adenoma as opposed to 27% in the placebo cohort. Furthermore, the average count of adenomas identified decreased (0.30 ± 0.87) compared to the placebo cohort (0.49 ± 0.99, P = 0.003). Moreover, the relative risk for recurrent adenomas was 0.65 with 95% confidence interval (CI) ranging from 0.46 to 0.91, highlighting the protective effect of aspirin. The interval to the initial identified adenoma was extended with a hazard ratio of 0.64 (95% CI, 0.43–0.94; P = 0.022), which suggests postponement of adenoma progression. All in all, the study suggested that daily aspirin use significantly decreases the incidence of colorectal adenoma in patients with existing history of colorectal cancer.

The challenge of delivering drugs across BBB still poses a major challenge. One way to overcome this is through chemical modification; altering the chemical structure of drugs could improve the lipophilicity of the drug, thus enhancing the BBB crossing ability. For example, the esterification of pyrroloquinoline quinone (PQQ) to create PQQ-trimethylester (PQQ-TME) could increase the lipophilicity and enhance the BBB permeability, demonstrating a two-fold increase compared to PQQ which has a similar profile to sodium fluorescein, a well-known tracer used in BBB studies (Tsukakoshi et al., 2018). PQQ-TME is also able to improve inhibitory activity against fibrillation of amyloidogenic proteins, including α-synuclein and Aβ, proving itself as a promising candidate for neurodegenerative disease treatment. The effectiveness of a drug could be improved by chemical modifications with the intent of crossing the BBB, thus addressing the current obstacle of targeting CNS disorders.

2.3.3 Natural Compounds and Their Therapeutic Potential

One field of drug discovery study that has recently gained significant attention is in natural compounds and their potential in crossing the BBB from plants, microorganisms and other natural sources, which often exhibit various potential benefits such as neuroprotective, anti-inflammatory and antioxidant properties. Unfortunately, many of these compounds have limited BBB permeability and advances in formulation technology such as encapsulation in nanoparticles or conjugation with BBB-targeting moieties have shown promising results in enhancing the delivery to the brain. Exploring these compounds' potential is an interesting space in the development of novel treatment strategies for neurodegenerative diseases like AD and PD, leveraging their multifaceted biological therapeutic effects.

The ability of natural compounds to cross the BBB is relative to its molecular weight. In a study done by Tang and colleagues in 2022, several neuroprotective natural compounds were assessed for their ability to cross the BBB when encapsulated in exosomes. It was found that compounds with sizes bigger than 1109 Da such as ginsenoside Rb1 were unable to efficiently cross the BBB as opposed to compounds with low molecular weight such as baicalin (MW: 446.36 Da) and neferine (MW: 624.77 Da). These data have shed light on the indispensable role of selecting the right natural compounds to target CNS disorders. The same study also critically suggested the benefits of encapsulation, enhancing delivery of compounds across the BBB to reduce accumulation of Aβ and α-synuclein (Tang et al., 2022).

Ganesan and colleagues in 2015 published a review on the issues of low bioavailability of phytobioactive compounds such as curcumin and resveratrol in PD treatments (Ganesan et al., 2015) and suggested that nanosized compounds in the range of 1–100 nm could improve the permeability across the BBB, thus improving their therapeutic potential. The same study also shed light on the necessity of co-delivery of the compounds with nanodelivery systems targeting CNS disorders such as solid lipid nanoparticles, nanostructured lipid carriers and nanoliposomes.

2.4 CONCLUSION

2.4.1 Summary of Key Findings

We have discussed extensively the current views of the role of BBB in AD with special focus on its dual role as protective structure and contributor to neurodegeneration. The degeneration of BBB in AD, strongly driven by cerebrovascular disorders, neuroinflammation and altered accumulation of Aβ and its clearance, accelerates the injury and progression. We have also discussed the innovative ways to cross the BBB, including nanoparticle-based delivery systems, repurposing of drugs and small molecules, and natural compounds as a worthy therapeutic candidate with each posing potential to overcome the restrictive nature of BBB while minimizing off-target effects. Advances in molecular technologies emphasized the importance of targeting BBB disruption as a main focus in AD therapy, suggesting the need for a multi-approach strategy that combines restorative and treatment focus.

2.4.2 Future Research Directions

Even though the current research on AD and BBB pathology and treatment is extensive, several main challenges still remain. These are several future research recommendations:

1. *Understanding the Molecular Mechanisms*: An extensive investigation into the cellular and molecular mechanisms of BBB dysfunction in AD and their role in tight junction degradation, loss in pericyte and neuroinflammation cascade are important in identifying specific therapeutic targets.
2. *Innovating the Field of Nanotherapeutics*: Further expanding the development of multifunctional nanoparticles focused on exploiting RMTs or tight junction modulation for improved BBB penetration and site-specific drug delivery by incorporating biomimetic elements could further enhance the therapeutic efficacy and biocompatibility.
3. *Engineering Natural Compounds*: By engineering structural modifications and optimization of natural compounds together with enhanced encapsulation strategies could potentially overcome the bioavailability issues in crossing the BBB while taking advantage of their multifactorial biological effects.
4. *Systems-Level Integration*: Understanding the dynamic interplay between BBB disruption, systemic inflammation and the gut-brain axis is crucial in future development of more holistic therapeutic strategies. Multimodal strategies that target the microbiota-gut-brain-BBB axis could be transformative.
5. *Personalized and Translational Approach:* Understanding the link between preclinical studies and their translatable potential is important in mimicking the human BBB and AD pathophysiology. On top of that, incorporating the genetic and epigenetic components into therapeutic development will make treatment more personalized to individual differences in BBB and AD pathologies and progressions.
6. *Validating Clinical Dataset*: Clinical trials should address long-term efficacy, safety and pharmacokinetic effects of BBB-linked interventions while also addressing the scaling challenges for nanoparticle synthesis and achieving the necessary regulatory approval.

Addressing these key challenges could greatly improve the modulation of BBB-related diseases, paving the way for novel interventions that could slow down AD progression and improve patient outcomes.

REFERENCES

Aird, W. C. (2007). Phenotypic heterogeneity of the endothelium: II. Representative vascular beds. *Circulation Research*, *100*(2), 174–190. https://doi.org/10.1161/01.RES.0000255690.03436.ae

Bading, J. R., Yamada, S., Mackic, J. B., Kirkman, L., Miller, C., Calero, M., Ghiso, J., Frangione, B., & Zlokovic, B. V. (2002). Brain clearance of Alzheimer's amyloid-beta40 in the squirrel monkey: A SPECT study in a primate model of cerebral amyloid angiopathy. *Journal of Drug Targeting*, *10*(4), 359–368. https://doi.org/10.1080/10611860290031831

Beach, T. G., Monsell, S. E., Phillips, L. E., & Kukull, W. (2012). Accuracy of the clinical diagnosis of Alzheimer disease at National Institute on Aging Alzheimer Disease Centers, 2005–2010. *Journal of Neuropathology & Experimental Neurology*, *71*(4), 266–273. https://doi.org/10.1097/NEN.0b013e31824b211b

Bell, R. D., Sagare, A. P., Friedman, A. E., Bedi, G. S., Holtzman, D. M., Deane, R., & Zlokovic, B. V. (2007). Transport pathways for clearance of human Alzheimer's amyloid beta-peptide and apolipoproteins E and J in the mouse central nervous system. *Journal of Cerebral Blood Flow & Metabolism*, *27*(5), 909–918. https://doi.org/10.1038/sj.jcbfm.9600419

Bell, R. D., Winkler, E. A., Singh, I., Sagare, A. P., Deane, R., Wu, Z., Holtzman, D. M., Betsholtz, C., Armulik, A., Sallstrom, J., Berk, B. C., & Zlokovic, B. V. (2012). Apolipoprotein E controls cerebrovascular integrity via cyclophilin A. *Nature*, *485*(7399), 512–516. https://doi.org/10.1038/nature11087

Braak, H., & Braak, E. (1991). Neuropathological stageing of Alzheimer-related changes. *Acta Neuropathologica*, *82*(4), 239–259. https://doi.org/10.1007/BF00308809

Cadena-del-Castillo, C., Valdes-Quezada, C., Carmona-Aldana, F., Arias, C., Bermudez-Rattoni, F., & Recillas-Targa, F. (2014). Age-dependent increment of hydroxymethylation in the brain cortex in the triple-transgenic mouse model of Alzheimer's disease. *Journal of Alzheimer's Disease*, *41*(3), 845–854. https://doi.org/10.3233/JAD-132285

Chen, K. L., Wang, S. S., Yang, Y. Y., Yuan, R. Y., Chen, R. M., & Hu, C. J. (2009). The epigenetic effects of amyloid-beta(1–40) on global DNA and neprilysin genes in murine cerebral endothelial cells. *Biochemical and Biophysical Research Communications*, *378*(1), 57–61. https://doi.org/10.1016/j.bbrc.2008.10.173

Chishti, M. A., Yang, D. S., Janus, C., Phinney, A. L., Horne, P., Pearson, J., Strome, R., Zuker, N., Loukides, J., French, J., Turner, S., Lozza, G., Grilli, M., Kunicki, S., Morissette, C., Paquette, J., Gervais, F., Bergeron, C., Fraser, P. E., … Westaway, D. (2001). Early-onset amyloid deposition and cognitive deficits in transgenic mice expressing a double mutant form of amyloid precursor protein 695. *Journal of Biological Chemistry*, *276*(24), 21562–21570. https://doi.org/10.1074/jbc.M100710200

Chouliaras, L., van den Hove, D. L., Kenis, G., Draanen, M., Hof, P. R., van Os, J., Steinbusch, H. W., Schmitz, C., & Rutten, B. P. (2013). Histone deacetylase 2 in the mouse hippocampus: Attenuation of age-related increase by caloric restriction. *Current Alzheimer Research*, *10*(8), 868–876. https://doi.org/10.2174/1567205011310080009

Citron, M., Oltersdorf, T., Haass, C., McConlogue, L., Hung, A. Y., Seubert, P., Vigo-Pelfrey, C., Lieberburg, I., & Selkoe, D. J. (1992). Mutation of the beta-amyloid precursor protein in familial Alzheimer's disease increases beta-protein production. *Nature*, *360*(6405), 672–674. https://doi.org/10.1038/360672a0

Cruz, J. C., Kim, D., Moy, L. Y., Dobbin, M. M., Sun, X., Bronson, R. T., & Tsai, L. H. (2006). p25/cyclin-dependent kinase 5 induces production and intraneuronal accumulation of amyloid beta in vivo. *Journal of Neuroscience*, *26*(41), 10536–10541. https://doi.org/10.1523/JNEUROSCI.3133-06.2006

Daneman, R., & Prat, A. (2015). The blood-brain barrier. *Cold Spring Harbor Perspectives in Biology*, *7*(1), a020412. https://doi.org/10.1101/cshperspect.a020412

Deane, R., Sagare, A., Hamm, K., Parisi, M., Lane, S., Finn, M. B., Holtzman, D. M., & Zlokovic, B. V. (2008). apoE isoform-specific disruption of amyloid beta peptide clearance from mouse brain. *Journal of Clinical Investigation*, *118*(12), 4002–4013. https://doi.org/10.1172/JCI36663

Deane, R., Wu, Z., Sagare, A., Davis, J., Du Yan, S., Hamm, K., Xu, F., Parisi, M., LaRue, B., Hu, H. W., Spijkers, P., Guo, H., Song, X., Lenting, P. J., Van Nostrand, W. E., & Zlokovic, B. V. (2004). LRP/amyloid beta-peptide interaction mediates differential brain efflux of Abeta isoforms. *Neuron*, *43*(3), 333–344. https://doi.org/10.1016/j.neuron.2004.07.017

Dyrna, F., Hanske, S., Krueger, M., & Bechmann, I. (2013). The blood-brain barrier. *Journal of Neuroimmune Pharmacology*, *8*(4), 763–773. https://doi.org/10.1007/s11481-013-9473-5

Ehrlich, P. (1885). *Das Sauerstoff-Bedürfniss des Organismus: eine farbenanalytische Studie/von P.* Ehrlich. https://wellcomecollection.org/works/bk8py7q7

Fischer, A., Sananbenesi, F., Wang, X., Dobbin, M., & Tsai, L. H. (2007). Recovery of learning and memory is associated with chromatin remodelling. *Nature*, *447*(7141), 178–182. https://doi.org/10.1038/nature05772

Francis, Y. I., Fa, M., Ashraf, H., Zhang, H., Staniszewski, A., Latchman, D. S., & Arancio, O. (2009). Dysregulation of histone acetylation in the APP/PS1 mouse model of Alzheimer's disease. *Journal of Alzheimer's Disease*, *18*(1), 131–139. https://doi.org/10.3233/JAD-2009-1134

Gama Sosa, M. A., Gasperi, R. D., Rocher, A. B., Wang, A. C., Janssen, W. G., Flores, T., Perez, G. M., Schmeidler, J., Dickstein, D. L., Hof, P. R., & Elder, G. A. (2010). Age-related vascular pathology in transgenic mice expressing presenilin 1-associated familial Alzheimer's disease mutations. *American Journal of Pathology*, *176*(1), 353–368. https://doi.org/10.2353/ajpath.2010.090482

Ganesan, P., Ko, H. M., Kim, I. S., & Choi, D. K. (2015). Recent trends in the development of nanophytobioactive compounds and delivery systems for their possible role in reducing oxidative stress in Parkinson's disease models. *International Journal of Nanomedicine*, *10*, 6757–6772. https://doi.org/10.2147/IJN.S93918

Georgieva, D., Nikolova, D., Vassileva, E., & Kostova, B. (2023). Chitosan-based nanoparticles for targeted nasal galantamine delivery as a promising tool in Alzheimer's disease therapy. *Pharmaceutics*, *15*(3). https://doi.org/10.3390/pharmaceutics15030829

Gonzalez-Zuniga, M., Contreras, P. S., Estrada, L. D., Chamorro, D., Villagra, A., Zanlungo, S., Seto, E., & Alvarez, A. R. (2014). c-Abl stabilizes HDAC2 levels by tyrosine phosphorylation repressing neuronal gene expression in Alzheimer's disease. *Molecular Cell*, *56*(1), 163–173. https://doi.org/10.1016/j.molcel.2014.08.013

Graff, J., Rei, D., Guan, J. S., Wang, W. Y., Seo, J., Hennig, K. M., Nieland, T. J., Fass, D. M., Kao, P. F., Kahn, M., Su, S. C., Samiei, A., Joseph, N., Haggarty, S. J., Delalle, I., & Tsai, L. H. (2012). An epigenetic blockade of cognitive functions in the neurodegenerating brain. *Nature*, *483*(7388), 222–226. https://doi.org/10.1038/nature10849

Haass, C., Lemere, C. A., Capell, A., Citron, M., Seubert, P., Schenk, D., Lannfelt, L., & Selkoe, D. J. (1995). The Swedish mutation causes early-onset Alzheimer's disease by beta-secretase cleavage within the secretory pathway. *Nature Medicine*, *1*(12), 1291–1296. https://doi.org/10.1038/nm1295-1291

Halliday, M. R., Pomara, N., Sagare, A. P., Mack, W. J., Frangione, B., & Zlokovic, B. V. (2013). Relationship between cyclophilin a levels and matrix metalloproteinase 9 activity in cerebrospinal fluid of cognitively normal apolipoprotein E4 carriers and blood-brain barrier breakdown. *JAMA Neurology*, *70*(9), 1198–1200. https://doi.org/10.1001/jamaneurol.2013.3841

Hardy, J. A., & Higgins, G. A. (1992). Alzheimer's disease: The amyloid cascade hypothesis. *Science*, *256*(5054), 184–185. https://doi.org/10.1126/science.1566067

Hodgson, N., Trivedi, M., Muratore, C., Li, S., & Deth, R. (2013). Soluble oligomers of amyloid-beta cause changes in redox state, DNA methylation, and gene transcription by inhibiting EAAT3 mediated cysteine uptake. *Journal of Alzheimer's Disease*, *36*(1), 197–209. https://doi.org/10.3233/JAD-130101

Hosokawa, T., Nakajima, H., Doi, Y., Sugino, M., Kimura, F., Hanafusa, T., & Takahashi, T. (2011). Increased serum matrix metalloproteinase-9 in neuromyelitis optica: Implication of disruption of blood-brain barrier. *Journal of Neuroimmunology*, *236*(1–2), 81–86. https://doi.org/10.1016/j.jneuroim.2011.04.009

Iturria-Medina, Y., Sotero, R. C., Toussaint, P. J., Mateos-Perez, J. M., Evans, A. C., & Alzheimer's Disease Neuroimaging, I. (2016). Early role of vascular dysregulation on late-onset Alzheimer's disease based on multifactorial data-driven analysis. *Nature Communications*, *7*, 11934. https://doi.org/10.1038/ncomms11934

Knopman, D. S., Amieva, H., Petersen, R. C., Chetelat, G., Holtzman, D. M., Hyman, B. T., Nixon, R. A., & Jones, D. T. (2021). Alzheimer disease. *Nature Reviews Disease Primers*, *7*(1), 33. https://doi.org/10.1038/s41572-021-00269-y

Kowalski, K., & Mulak, A. (2019). Brain-gut-microbiota axis in Alzheimer's disease. *Journal of Neurogastroenterology and Motility*, *25*(1), 48–60. https://doi.org/10.5056/jnm18087

Kukull, W. A., Larson, E. B., Bowen, J. D., McCormick, W. C., Teri, L., Pfanschmidt, M. L., Thompson, J. D., O'Meara, E. S., Brenner, D. E., & van Belle, G. (1995). Solvent exposure as a risk factor for Alzheimer's disease: A case-control study. *American Journal of Epidemiology*, *141*(11), 1059–1071; discussion 1072–1059. https://doi.org/10.1093/oxfordjournals.aje.a117370

Larsen, J. M., Martin, D. R., & Byrne, M. E. (2014). Recent advances in delivery through the blood-brain barrier. *Current Topics in Medicinal Chemistry*, *14*(9), 1148–1160. https://doi.org/10.2174/1568026614666140329230311

Mastroeni, D., Grover, A., Delvaux, E., Whiteside, C., Coleman, P. D., & Rogers, J. (2011). Epigenetic mechanisms in Alzheimer's disease. *Neurobiology of Aging*, *32*(7), 1161–1180. https://doi.org/10.1016/j.neurobiolaging.2010.08.017

Mawuenyega, K. G., Sigurdson, W., Ovod, V., Munsell, L., Kasten, T., Morris, J. C., Yarasheski, K. E., & Bateman, R. J. (2010). Decreased clearance of CNS beta-amyloid in Alzheimer's disease. *Science*, *330*(6012), 1774. https://doi.org/10.1126/science.1197623

McGeer, P. L., & McGeer, E. G. (1995). The inflammatory response system of brain: Implications for therapy of Alzheimer and other neurodegenerative diseases. *Brain Research Reviews*, *21*(2), 195–218. https://doi.org/10.1016/0165-0173(95)00011-9

Moulton, P. V., & Yang, W. (2012). Air pollution, oxidative stress, and Alzheimer's disease. *Journal of Environmental and Public Health*, *2012*, 472751. https://doi.org/10.1155/2012/472751

Mullan, M., Crawford, F., Axelman, K., Houlden, H., Lilius, L., Winblad, B., & Lannfelt, L. (1992). A pathogenic mutation for probable Alzheimer's disease in the APP gene at the n-terminus of beta-amyloid. *Nature Genetics*, *1*(5), 345–347. https://doi.org/10.1038/ng0892-345

Murrell, J., Farlow, M., Ghetti, B., & Benson, M. D. (1991). A mutation in the amyloid precursor protein associated with hereditary Alzheimer's disease. *Science*, *254*(5028), 97–99. https://doi.org/10.1126/science.1925564

Nelson, A. R., Sweeney, M. D., Sagare, A. P., & Zlokovic, B. V. (2016). Neurovascular dysfunction and neurodegeneration in dementia and Alzheimer's disease. *Biochimica et Biophysica Acta*, *1862*(5), 887–900. https://doi.org/10.1016/j.bbadis.2015.12.016

Oakley, H., Cole, S. L., Logan, S., Maus, E., Shao, P., Craft, J., Guillozet-Bongaarts, A., Ohno, M., Disterhoft, J., Van Eldik, L., Berry, R., & Vassar, R. (2006). Intraneuronal beta-amyloid aggregates, neurodegeneration, and neuron loss in transgenic mice with five familial Alzheimer's disease mutations: Potential factors in amyloid plaque formation. *Journal of Neuroscience*, *26*(40), 10129–10140. https://doi.org/10.1523/JNEUROSCI.1202-06.2006

Obermeier, B., Daneman, R., & Ransohoff, R. M. (2013). Development, maintenance and disruption of the blood-brain barrier. *Nature Medicine*, *19*(12), 1584–1596. https://doi.org/10.1038/nm.3407

Pardridge, W. M. (2007). Blood-brain barrier delivery. *Drug Discovery Today*, *12*(1–2), 54–61. https://doi.org/10.1016/j.drudis.2006.10.013

Pardridge, W. M. (2015). Targeted delivery of protein and gene medicines through the blood-brain barrier. *Clinical Pharmacology & Therapeutics*, *97*(4), 347–361. https://doi.org/10.1002/cpt.18

Parikh, I., Fardo, D. W., & Estus, S. (2014). Genetics of PICALM expression and Alzheimer's disease. *PLoS ONE*, *9*(3), e91242. https://doi.org/10.1371/journal.pone.0091242

Peleg, S., Sananbenesi, F., Zovoilis, A., Burkhardt, S., Bahari-Javan, S., Agis-Balboa, R. C., Cota, P., Wittnam, J. L., Gogol-Doering, A., Opitz, L., Salinas-Riester, G., Dettenhofer, M., Kang, H., Farinelli, L., Chen, W., & Fischer, A. (2010). Altered histone acetylation is associated with age-dependent memory impairment in mice. *Science*, *328*(5979), 753–756. https://doi.org/10.1126/science.1186088

Pistollato, F., Sumalla Cano, S., Elio, I., Masias Vergara, M., Giampieri, F., & Battino, M. (2016). Role of gut microbiota and nutrients in amyloid formation and pathogenesis of Alzheimer disease. *Nutrition Reviews*, *74*(10), 624–634. https://doi.org/10.1093/nutrit/nuw023

Querfurth, H. W., & LaFerla, F. M. (2010). Alzheimer's disease. *New England Journal of Medicine*, *362*(4), 329–344. https://doi.org/10.1056/NEJMra0909142

Ricobaraza, A., Cuadrado-Tejedor, M., Perez-Mediavilla, A., Frechilla, D., Del Rio, J., & Garcia-Osta, A. (2009). Phenylbutyrate ameliorates cognitive deficit and reduces tau pathology in an Alzheimer's disease mouse model. *Neuropsychopharmacology*, *34*(7), 1721–1732. https://doi.org/10.1038/npp.2008.229

Sagare, A. P., Bell, R. D., Zhao, Z., Ma, Q., Winkler, E. A., Ramanathan, A., & Zlokovic, B. V. (2013). Pericyte loss influences Alzheimer-like neurodegeneration in mice. *Nature Communications*, *4*, 2932. https://doi.org/10.1038/ncomms3932

Sanchez-Mut, J. V., & Graff, J. (2015). Epigenetic alterations in Alzheimer's disease. *Frontiers in Behavioral Neuroscience*, *9*, 347. https://doi.org/10.3389/fnbeh.2015.00347

Sandler, R. S., Halabi, S., Baron, J. A., Budinger, S., Paskett, E., Keresztes, R., Petrelli, N., Pipas, J. M., Karp, D. D., Loprinzi, C. L., Steinbach, G., & Schilsky, R. (2003). A randomized trial of aspirin to prevent colorectal adenomas in patients with previous colorectal cancer. *New England Journal of Medicine*, *348*(10), 883–890. https://doi.org/10.1056/NEJMoa021633

Santibanez, M., Bolumar, F., & Garcia, A. M. (2007). Occupational risk factors in Alzheimer's disease: A review assessing the quality of published epidemiological studies. *Occupational and Environmental Medicine*, *64*(11), 723–732. https://doi.org/10.1136/oem.2006.028209

Shi, L., Wu, X., Danesh Yazdi, M., Braun, D., Abu Awad, Y., Wei, Y., Liu, P., Di, Q., Wang, Y., Schwartz, J., Dominici, F., Kioumourtzoglou, M. A., & Zanobetti, A. (2020). Long-term effects of PM(2.5) on neurological disorders in the American Medicare population: A longitudinal cohort study. *Lancet Planetary Health*, *4*(12), e557–e565. https://doi.org/10.1016/S2542-5196(20)30227-8

Singh, P., & Thakur, M. K. (2014). Reduced recognition memory is correlated with decrease in DNA methyltransferase1 and increase in histone deacetylase2 protein expression in old male mice. *Biogerontology*, *15*(4), 339–346. https://doi.org/10.1007/s10522-014-9504-5

Sung, H. Y., Choi, E. N., Ahn Jo, S., Oh, S., & Ahn, J. H. (2011). Amyloid protein-mediated differential DNA methylation status regulates gene expression in Alzheimer's disease model cell line. *Biochemical and Biophysical Research Communications*, *414*(4), 700–705. https://doi.org/10.1016/j.bbrc.2011.09.136

Suzuki, N., Cheung, T. T., Cai, X. D., Odaka, A., Otvos, L. Jr., Eckman, C., Golde, T. E., & Younkin, S. G. (1994). An increased percentage of long amyloid beta protein secreted by familial amyloid beta protein precursor (beta APP717) mutants. *Science*, *264*(5163), 1336–1340. https://doi.org/10.1126/science.8191290

Sweeney, M. D., Sagare, A. P., & Zlokovic, B. V. (2018). Blood-brain barrier breakdown in Alzheimer disease and other neurodegenerative disorders. *Nature Reviews Neurology*, *14*(3), 133–150. https://doi.org/10.1038/nrneurol.2017.188

Taher, N., McKenzie, C., Garrett, R., Baker, M., Fox, N., & Isaacs, G. D. (2014). Amyloid-beta alters the DNA methylation status of cell-fate genes in an Alzheimer's disease model. *Journal of Alzheimer's Disease*, *38*(4), 831–844. https://doi.org/10.3233/JAD-131061

Tang, B., Zeng, W., Song, L. L., Wang, H. M., Qu, L. Q., Lo, H. H., Yu, L., Wu, A. G., Wong, V. K. W., & Law, B. Y. K. (2022). Extracellular vesicle delivery of neferine for the attenuation of neurodegenerative disease proteins and motor deficit in an Alzheimer's disease mouse model. *Pharmaceuticals (Basel)*, *15*(1). https://doi.org/10.3390/ph15010083

Tarasoff-Conway, J. M., Carare, R. O., Osorio, R. S., Glodzik, L., Butler, T., Fieremans, E., Axel, L., Rusinek, H., Nicholson, C., Zlokovic, B. V., Frangione, B., Blennow, K., Menard, J., Zetterberg, H., Wisniewski, T., & de Leon, M. J. (2015). Clearance systems in the brain-implications for Alzheimer disease. *Nature Reviews Neurology*, *11*(8), 457–470. https://doi.org/10.1038/nrneurol.2015.119

Toledo, J. B., Arnold, S. E., Raible, K., Brettschneider, J., Xie, S. X., Grossman, M., Monsell, S. E., Kukull, W. A., & Trojanowski, J. Q. (2013). Contribution of cerebrovascular disease in autopsy confirmed neurodegenerative disease cases in the National Alzheimer's Coordinating Centre. *Brain*, *136*(Pt 9), 2697–2706. https://doi.org/10.1093/brain/awt188

Treusch, S., Hamamichi, S., Goodman, J. L., Matlack, K. E., Chung, C. Y., Baru, V., Shulman, J. M., Parrado, A., Bevis, B. J., Valastyan, J. S., Han, H., Lindhagen-Persson, M., Reiman, E. M., Evans, D. A., Bennett, D. A., Olofsson, A., DeJager, P. L., Tanzi, R. E., Caldwell, K. A., … Lindquist, S. (2011). Functional links between Abeta toxicity, endocytic trafficking, and Alzheimer's disease risk factors in yeast. *Science*, *334*(6060), 1241–1245. https://doi.org/10.1126/science.1213210

Tsukakoshi, K., Yoshida, W., Kobayashi, M., Kobayashi, N., Kim, J., Kaku, T., Iguchi, T., Nagasawa, K., Asano, R., Ikebukuro, K., & Sode, K. (2018). Esterification of PQQ enhances blood-brain barrier permeability and inhibitory activity against amyloidogenic protein fibril formation. *ACS Chemical Neuroscience*, *9*(12), 2898–2903. https://doi.org/10.1021/acschemneuro.8b00355

Wadhwa, G., Krishna, K. V., Dubey, S. K., & Taliyan, R. (2023). PEGylated polymer–lipid hybrid nanoparticles to enhance in vivo exposure and uptake of repaglinide in brain cells to treat diabetes-linked neurodegenerative disorders. *ACS Applied Nano Materials*, *6*(5), 3497–3512. https://doi.org/10.1021/acsanm.2c05272

Wagner, F. F., Zhang, Y. L., Fass, D. M., Joseph, N., Gale, J. P., Weiwer, M., McCarren, P., Fisher, S. L., Kaya, T., Zhao, W. N., Reis, S. A., Hennig, K. M., Thomas, M., Lemercier, B. C., Lewis, M. C., Guan, J. S., Moyer, M. P., Scolnick, E., Haggarty, S. J., … Holson, E. B. (2015). Kinetically selective inhibitors of histone deacetylase 2 (HDAC2) as cognition enhancers. *Chemical Science*, *6*(1), 804–815. https://doi.org/10.1039/C4SC02130D

Walker, M. P., LaFerla, F. M., Oddo, S. S., & Brewer, G. J. (2013). Reversible epigenetic histone modifications and Bdnf expression in neurons with aging and from a mouse model of Alzheimer's disease. *Age (Dordr)*, *35*(3), 519–531. https://doi.org/10.1007/s11357-011-9375-5

Zenaro, E., Piacentino, G., & Constantin, G. (2017). The blood-brain barrier in Alzheimer's disease. *Neurobiology of Disease*, *107*, 41–56. https://doi.org/10.1016/j.nbd.2016.07.007

Zhao, Z., Nelson, A. R., Betsholtz, C., & Zlokovic, B. V. (2015). Establishment and dysfunction of the blood-brain barrier. *Cell*, *163*(5), 1064–1078. https://doi.org/10.1016/j.cell.2015.10.067

Zipser, B. D., Johanson, C. E., Gonzalez, L., Berzin, T. M., Tavares, R., Hulette, C. M., Vitek, M. P., Hovanesian, V., & Stopa, E. G. (2007). Microvascular injury and blood-brain barrier leakage in Alzheimer's disease. *Neurobiology of Aging*, *28*(7), 977–986. https://doi.org/10.1016/j.neurobiolaging.2006.05.016

Zlokovic, B. V. (2013). Cerebrovascular effects of apolipoprotein E: Implications for Alzheimer disease. *JAMA Neurology*, *70*(4), 440–444. https://doi.org/10.1001/jamaneurol.2013.2152

Zonneveld, H. I., Goos, J. D., Wattjes, M. P., Prins, N. D., Scheltens, P., van der Flier, W. M., Kuijer, J. P., Muller, M., & Barkhof, F. (2014). Prevalence of cortical superficial siderosis in a memory clinic population. *Neurology*, *82*(8), 698–704. https://doi.org/10.1212/WNL.0000000000000150

3 The Application of Vaccines in Alzheimer's Disease

Active and Passive Immunotherapies

Sreejita Chatterjee, Riya Mukherjee, Ashi Ramavat, Debjyoti Paul, and Ramendra Pati Pandey

3.1 INTRODUCTION

Neurodegenerative illnesses, including Alzheimer's disease (AD), now afflict roughly one in every three people worldwide, with a significant increase in incidence recorded after COVID-19, worsening the global disease burden. AD, in particular, poses a serious healthcare issue because of its complicated pathophysiology and fast-changing epidemiological trends (Ferrari & Sorbi, 2021; Zhang et al., 2024).

AD is a progressive, multifaceted disorder marked by a diverse pathogenic landscape that includes oxidative stress, metal dyshomeostasis (for example, calcium imbalance), decreased acetylcholinesterase function, reduced autophagy, and neuronal death. These pathways affect neurovascular coupling, resulting in impaired clearance of interstitial and cerebrospinal fluid (CSF) and contributing to toxic accumulation in the brain (Kapadia et al., 2020).

AD includes loss of memory, behavioral changes, cognitive and visuospatial skill impairment (Vashisth et al., 2024). A major hallmark of central pathogenic features includes the presence of extracellular cordially deposited amyloid-beta (Aβ) plaques and further intracellular neurofibrillary Tangels (NFT) containing high proportions of phosphorylated tau. Symptoms include aphasia, spatial orientation, mood changes, forgetfulness, etc. Most of these anomalies are observed in the cortical and limbic regions (Spencer & Masliah, 2014). There are sporadic (late onset) and familial (early onset) forms of AD, as these mutations have been linked with early onset AD through mutations in amyloid precursor protein (APP), presenilin 1, and presenilin 2 upon correlation with autosomal dominant inheritance. The pathogenic Aβ peptides are made by γ secretase complex genes (Alshamrani et al., 2023; Zhou et al., 2021). Most cases of late-onset AD evolve from the Apolipoprotein E (APOE) ε4 allele along with other genes that affect neuroinflammation (e.g., TREM2, CD33, TYROBP) and synaptic function (e.g., PICALM, BIN1, CR1) (Bhattacherjee et al., 2021; Chen et al., 2021; Franzmeier et al., 2021; Guglielmotto et al., 2019; Mitsumori et al., 2020; Tamagno et al., 2021; Wolfe et al., 2018). The amyloid cascade hypothesis alludes to neural degeneration and death arising due to Aβ deposition, APP cleavage into Aβ1-40 and Aβ1-42. Small Aβ oligomers are rendered to be more toxic than mature fibrils (Ballard et al., 2011). AD has a pathological course that includes hippocampal atrophy, enlargement of ventricles and cortical thinning, particularly in the temporal and frontal bones, and widespread neuron loss. APP can be proteolytically cleaved by the β-secretase and the γ-secretase to produce a variety of Aβ isoforms (Aβ38, Aβ40, and Aβ42) (Sisodia et al., 2002), wherein Aβ42 is the most aggregation-prone and neurotoxic. α-Secretase cleavage also occurs within the Aβ domain to release non-amyloidogenic soluble APPα fragments, which possess neurotrophic and neuroprotective properties (Anderson et al., 1992; Kojro & Fahrenholz, 2005; Spencer et al., 2014).

After over 200 ineffectual clinical trials during the last decade, immunotherapy emerges as a novel attempt in the treatment of AD via improving, suppressing, or introducing immune responses, for instance, through vaccines constituting mutant oligomer-Aβ activated dendritic cell (Aljassabi et al., 2024). Currently, FDA-approved treatments, such as acetylcholinesterase inhibitors and NMDA receptor antagonists, only provide symptomatic relief. Immunotherapeutic techniques, including active and passive immunization, aim to influence disease progression by targeting core pathogenic proteins such as Aβ, tau, and α-synuclein (α-syn). The first-generation Aβ vaccine, AN1792, used a pro-inflammatory QS-21 adjuvant but was terminated due to T-cell-mediated meningoencephalitis. To reduce unfavorable immune reactions, second-generation vaccines targeting B-cell epitopes of Aβ (e.g., Aβ1-42) were developed. T-cell epitopes were left out Athar et al., 2021.

Monoclonal antibodies (mAbs) targeting Aβ and phosphorylated tau (p-tau) are becoming more popular immunotherapies. The recent approval of Aducanumab and Lecanemab—humanized mAbs that traverse the blood-brain barrier (BBB) and selectively diminish amyloid plaques—marks a watershed moment in AD treatment paradigms. Aducanumab, while approved, is still under post-marketing surveillance (Phase IV), with final study results expected by 2030. Parallel efforts have also targeted α-syn pathology, which is implicated in overlapping neurodegenerative illnesses. Immunoblot analysis employing sequential extraction techniques demonstrated that α-syn mAb treatment reduced Lewy body (LB)-like and Lewy neurite (LN)-like inclusions in neuronal cells. These treatments alter α-syn solubility profiles, decreasing TX-soluble fractions and increasing SDS-extractable, higher molecular weight, post-translationally changed α-syn species (Tran et al., 2014).

DOI: 10.1201/9781003602477-3

Despite hopeful developments, obstacles remain. High treatment costs, heterogeneity in patient responses, and the complexities of breaching the BBB continue to stymie overall therapeutic success. Nonetheless, immunotherapy has the potential to alter management, particularly when paired with early detection and multimodal therapies such as stem cell therapy, neurogenesis-promoting drugs, and anti-inflammatory strategies (Jucker & Walker, 2023).

This chapter delves into recent clinical and translational advances in Alzheimer's immunotherapy, critically assessing the existing landscape, mechanistic basis, and future directions. Through immunization medicine can initiate a transformative shift in Alzheimer's treatment based on immunologic specialization, which leads toward disease modification above symptom management.

3.2 OVERVIEW OF THE ENIGMA OF ALZHEIMER'S DISEASE PATHOGENESIS AS A NEURODEGENERATIVE DISORDER

The neurological disease AD exhibits three distinctive pathological features among which the Aβ protein generates extracellular plaques, while NFTs form from hyperphosphorylated tau protein in neural cells, and neuroinflammation impairs brain health continuously. The combination of oxidative stress causes rapid AD progression through modifications in APP processing together with alterations in secretase activities that subsequently produce greater amounts of Aβ (Heneka et al., 2015). The process increases both the levels of hyperphosphorylated tau protein and the formation of NFTs. Through oxidative stress by off-target immune response via astrogliosis and microgliosis, microglia become activated, and this causes ROS and inflammatory cytokines to release in a detrimental cycle that makes neuronal damage more severe (Terry et al., 1994). The combination of free metal ions and Aβ-metal complexes generates ROS, leading to neuronal cell death (Spencer et al., 2014).

Despite decades of research, current AD therapies only provide symptomatic relief with modest success in slowing disease progression. This emphasizes the critical need for disease-modifying medicines (Mangialasche et al., 2010). Immunotherapy, particularly vaccine-based techniques, has shown promise in addressing fundamental pathogenic components such as Aβ, tau, and α-syn (Vroom & Dodart, 2024).

Passive immunotherapy involves the direct infusion of mAbs that target pathogenic proteins, whereas active immunotherapy aims to stimulate the patient's immune system to manufacture endogenous antibodies via vaccination (Foster et al., 2009).

3.3 IMMUNOLOGICAL TARGETS IN AD AND ACTIVE IMMUNOTHERAPY STRATEGIES

Active immunotherapy employs vaccines to elicit adaptive immune responses against AD-specific antigens, primarily Aβ and tau. These vaccines aim to generate antigen-specific B-cell responses without inducing harmful T-cell-mediated inflammation. The primary goal is to reduce the burden of pathological protein aggregates and mitigate neurodegeneration and/or autoimmune toxicity (Wisniewski & Goni, 2015).

3.3.1 Targeting Amyloid-Beta (Aβ): Vaccines and Mechanisms

Aβ peptides, particularly soluble oligomers, are neurotoxic and implicated in synaptic dysfunction, memory impairment, and disease progression. The immunological rationale for Aβ-targeting vaccines is to neutralize or facilitate the clearance of Aβ, thereby reducing plaque formation and protecting neuronal integrity. Aggregated Aβ42 vaccine imparted antibodies that selectively targets β-amyloid and vascular amyloid without cross-reacting with denatured or indigenous full-length APP/its derivative among AD patients (Hock et al., 2002). Novel vaccine development strategies for AD focus on targeting the N-terminal of Aβ, avoiding T-cell epitopes in the C-terminal region like short N-terminal Aβ peptides possessing Th2 adjuvants, DNA vaccines, viral vector vaccines. Th2 immune response-inducing vaccines are considered promising, minimizing inflammatory and encephalitogenic T cell activation risks (Tabira, 2010).

- ***AN1792***: The first-generation Aβ vaccine showed significant preclinical efficacy in reducing amyloid deposits. However, clinical trials were discontinued due to T-cell-mediated meningoencephalitis in participants, highlighting the risks of full-length Aβ42 peptide vaccines that can activate autoreactive T cells and central nervous system (CNS) infiltration (Gelinas et al., 2004; Gilman et al., 2005).

To overcome these limitations, second-generation vaccines were developed:

- ***Amilomotide (CAD106)***: CAD106 launched by Novartis Pharmaceuticals, elicits Aβ-specific antibodies without activating T cells. It targets B cell epitope alone, the minor N-terminal Aβ fragment (Aβ1–6), conjugated to an adjuvant carrier procured from numerous coat protein copies of bacteriophage Qβ. Phase I/II trials demonstrated immunogenicity and improved safety (Winblad et al., 2012).
- ***UB-311***: This synthetic peptide vaccine devoid of T-lymphocyte epitope is designed to generate antibodies against Aβ with minimal inflammatory risk. Early trials suggest favorable immune response modulation and safety profiles (Wang et al., 2017).

Affiris AG has initiated a Phase I trial employing AFFITOM(R) technology for neutralizing a pyroglutamic-3-modified Aβ N-terminus, a post-translational modified Aβ conformer that generates greater susceptibility to aggregation. Hypothetically, it occurs after amassing over plaques or vascular amyloid (Frost et al., 2013; Saido et al., 1995). Mechanistic refinements in these newer vaccines include shortening Aβ peptides (to <8 amino acids) to prevent MHC binding and T-cell activation, and incorporating specific T-helper epitopes to bypass immune tolerance without inciting inflammation (Ahmad et al., 2024). Other ongoing trials targeting Aβ include sodium oligo-mannurarate (GV-971), BACE1 inhibitors like verubecestat, elenbecestat, atabecestat and CNP520 (Panza et al., 2019).

3.3.2 Tau Pathology and Tau-Targeting Vaccines

Tau, a microtubule-binding protein, if hyperphosphorylated can undergo aggregation into a β-sheet rich form of tau oligomers. Upon precipitation, they give rise to primary helical filaments (PHFs) in the form of NFTs, disrupting microtubule dynamics and neuronal function (Wisniewski & Goni, 2015). As tau pathology correlates more closely with cognitive decline than Aβ, it has become an increasingly attractive immunotherapeutic target. Key pharmacotherapeutic targets incorporate microtubule-stabilizing components, tau acetylation, aggregation and protein kinase inhibitors (Panza et al., 2016).

- ***AADvacl (Axon Neuroscience SE)***: It constitutes a tau fragment conjugated to a keyhole limpet hemocyanin (KLH) that elicits pathological tau-tau interplay further triggering its misfolding as well as oligomerization. KLH combined with adjuvant Al(OH)3 incites Th2 immune response with augmented antibody titers (Kontsekova et al., 2014). This active tau vaccine promotes antibody production against pathological tau forms. Phase I/II studies have reported immunogenicity and potential for slowing tau accumulation (Yu and Watts, 2013). Both Phase 1 study with 3 monthly AADvacl doses among subjects with mild-to-moderate AD and phase 2 studies reported better immunotolerance and safety profiles (Novak et al., 2021).
- Other candidates, including **ACI-35 (targeting phosphorylated tau)**, **ACI-24**, **AD02**, and **V-950**, Lu AF20513 are under investigation for their capacity to reduce tau pathology and improve cognitive outcomes. Examples of tau-targeting antibodies in clinical trials include gosuranemab, tilavonemab, semorinemab, zagotenemab (MC1), lecanemab, bepranemab, and donanemab (Sha et al., 2025).

These strategies aim to halt tau aggregation, promote its clearance, and prevent neurodegenerative spread via antibody-mediated mechanisms.

3.3.3 Synucleins in Alzheimer's Disease and Immunotherapy

The protein α-syn has been linked to Parkinson's disease and dementia with Lewy bodies (DLB), but it also contributes to AD pathogenesis. Elevated levels of α-syn in CSF have been seen in both asymptomatic persons at risk of AD and familial AD cases. α-syn is associated with tau hyperphosphorylation and Aβ-related toxicity, especially in APOEε4 carriers (Twohig & Nielsen, 2019). mAbs targeting α-syn have been shown to diminish insoluble protein aggregates and modify disease pathophysiology, opening up a new avenue for AD immunotherapy (Figure 3.1).

3.4 Clinical Insights and Efficacy of Active Immunotherapy

Clinical trials investigating active immunotherapies have yielded mixed outcomes, highlighting both promise and challenges:

- *Efficacy*: Many vaccine candidates have shown a capacity to reduce Aβ or tau burden in preclinical models and early-phase human trials. They exhibit brief and temporary drug inclusion generating durable antibodies being cost-effective. Cognitive benefits, however, have been modest and variable.
- *Adverse Events*: The AN1792 trial revealed a critical safety concern with T-cell activation leading to encephalitis. This has driven the design of newer vaccines with improved safety profiles (Orgogozo et al., 2003).
- *Heterogeneity in Immune Response*: Elderly populations, who are most affected by AD, often exhibit weakened or inconsistent immune responses to vaccines. Addressing this variability is essential for therapeutic success (Brody & Holtzman, 2008).

Second-generation vaccines such as CAD106, (liposome-based vaccine), ACC-001, AFFITOPE, and tau-targeting vaccines like AADvacl and ACI-35 aim to balance immunogenicity with safety, paving the way for more targeted, durable, and patient-friendly interventions (Kohyama & Matsumoto, 2015).

Despite the therapeutic potential of active immunotherapy, several challenges persist:

- Achieving optimal immune activation without triggering harmful inflammation, autoreactive T-cell and autoimmune responses or amyloid-related imaging abnormalities (ARIA). It also depends on the patient's self immunity and signalling mechanisms (Winblad et al., 2014).
- Enhancing vaccine efficacy in aging or immunocompromised populations (Lemere, 2009).
- Addressing the heterogeneity of AD progression and underlying pathophysiology.

Nevertheless, advancements in epitope design, adjuvant selection, and delivery platforms offer hope for

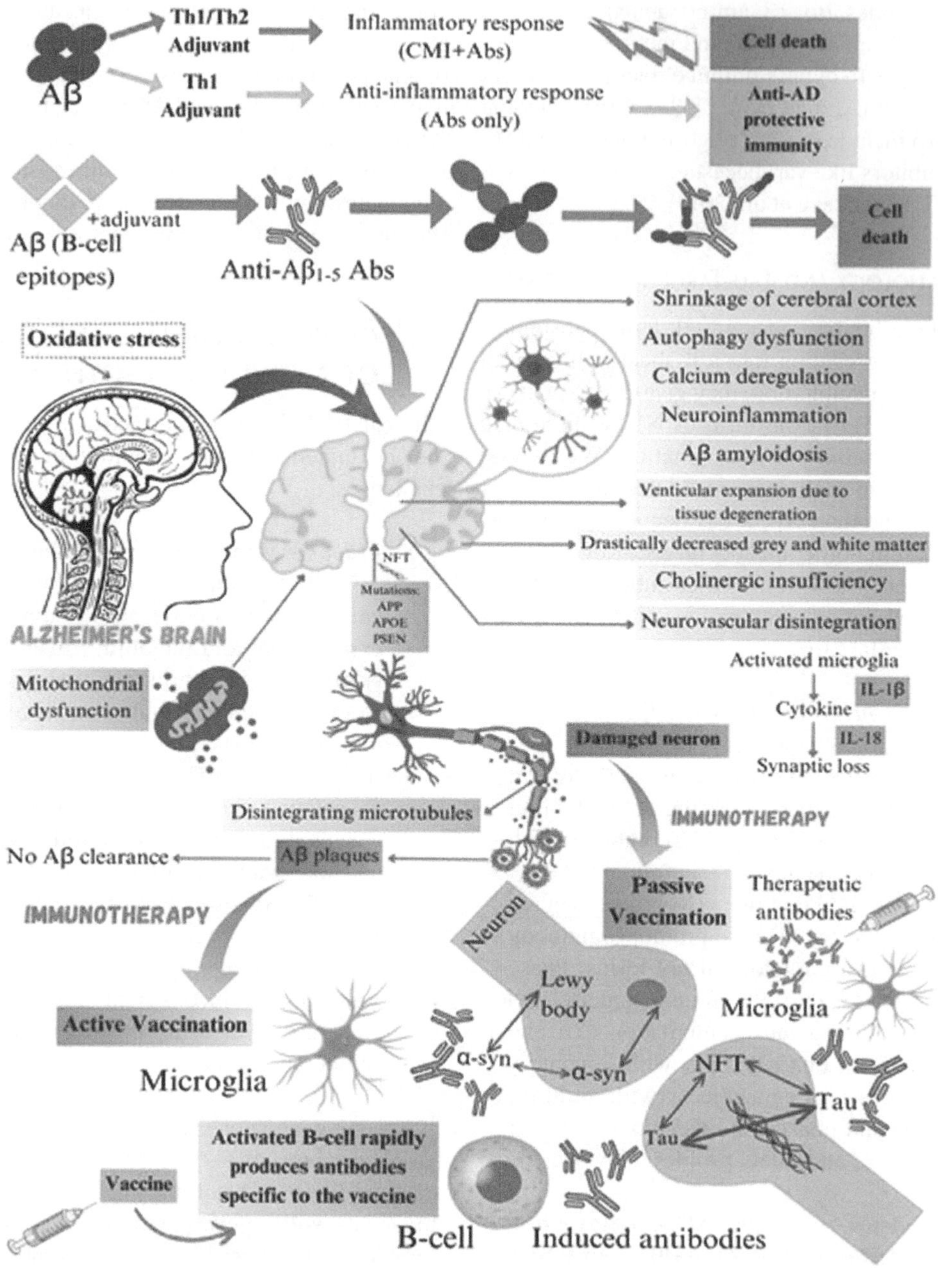

FIGURE 3.1 Schematic diagram illustrating the pathophysiology of AD and approach to immunotherapy. Abbreviations: CMI: cell-mediated immunity, Abs: antibodies, APP: amyloid precursor protein, APOE: apolipoprotein E, PSEN: presenilin, IL-18: interlukin-18, IL-1β: interlukin 1β, NFT: neurofibrillary tangles.

next-generation vaccines. Combined with early diagnosis and multimodal approaches, including neuroprotective agents, anti-inflammatory drugs, and regenerative strategies like immunotherapy, may redefine the treatment landscape for AD (Fulop et al., 2022).

3.5 PASSIVE IMMUNOTHERAPIES

Passive immunotherapy involves the direct administration of exogenously synthesized mAbs that specifically target pathological proteins such as Aβ or tau. Unlike active immunotherapy, which relies on the host's immune system to generate antibodies, passive immunotherapy delivers pre-formed antibodies, offering immediate therapeutic potential (Guo et al., 2024; Loureiro et al., 2020; Marković et al., 2023; Usman et al., 2021; Wisniewski & Goñi, 2015). This strategy is particularly advantageous in individuals with compromised or variable immune responses, such as the elderly or immunocompromised patients (Bittar et al., 2018). A distinct differentiation between active and passive immunotherapies in AD has been enlisted in Table 3.1.

TABLE 3.1
Difference between Active and Passive Immunotherapies

Parameter	Active Immunotherapy	Passive Immunotherapy
Mechanism	Stimulates the body to produce its own antibodies	Administers pre-made antibodies
Duration of Action	Can lead to long-lasting immune response	Requires repeated dosing
Safety	Risk of overactivation of the immune system (e.g., inflammation)	Risk of immune reactions to external antibodies
Patient Suitability	Less effective in individuals with compromised immune systems	Suitable for patients with weakened immune responses
Clinical Progress	Promising but with some safety concerns	Some successes, but clinical efficacy varies

3.5.1 Amyloid-Beta-Targeting Monoclonal Antibodies

3.5.1.1 Aducanumab (Aduhelm®)

Aducanumab is the first FDA-approved disease-modifying therapy for AD, granted approval in 2021 and developed by analyzing B-memory cells from healthy older adults. It is a fully human IgG1 mAb that selectively binds to aggregated forms of Aβ, including soluble oligomers and insoluble fibrils. Studies in transgenic mice showed that its analog effectively traversed the BBB, facilitating dose-dependent decline of soluble as well as insoluble Aβ. A Phase 1b trial spanned over a year in Aβ-PET positive prodromal-to-mild AD patients demonstrated a dose- and time-dependent decrease in brain fibrillar Aβ. However, APOEε4 carriers exhibited a high incidence of ARIA-E (Sevigny et al., 2016). Clinical trials have demonstrated that aducanumab effectively reduces Aβ plaque burden in a dose- and time-dependent manner. However, its impact on clinical outcomes, particularly cognitive function, has been subject to significant debate. The approval was based on biomarker reduction rather than consistent cognitive improvement, necessitating ongoing Phase IV confirmatory trials to validate its long-term efficacy and safety (Budd Haeberlein et al., 2023; Miculas et al., 2023).

3.5.1.2 Bapineuzumab

Bapineuzumab (NCT00575055) was one of the earliest mAbs developed to target Aβ plaques. While initial studies showed reductions in plaque load, later phase trials failed to demonstrate significant improvements in cognitive performance and ARIA risks. This highlights the challenge of translating biomarker changes into meaningful clinical outcomes (Janssen Alzheimer Immunotherapy Research & Development, LLC, 2012; Salloway et al., 2014).

3.5.1.3 Solanezumab

Solanezumab (NCT02008357) differs mechanistically from aducanumab and bapineuzumab by binding to soluble, monomeric forms of Aβ rather than aggregated plaques. It is a humanized IgG1 mAb targeting the mid-domain of Aβ. A phase 3 A4 investigation, administering 1600 mg intravenous solanezumab among 1169 preclinical AD patients between 65-85 years did not exhibit delayed cognitive decline over 240 weeks. The mean alteration in primary end-point PACC score was −1.43 in solanezumab vs. −1.13 in placebo (P=0.26). Post-hoc analysis of 18F-florbetapir amyloid positron-emission tomography (PET) and 18F-flortaucipir tau PET substudy revealed escalated brain amyloid (rising by mean 11.6 centiloids in solanezumab vs. 19.3 centiloids in placebo) as well as elevated neocortical and medial temporal tau in both groups respectively. Adverse event, i.e. ARIA coupled with microhemorrhage or hemosiderosis occured in both groups (Sperling et al., 2023). Clinical studies showed only modest cognitive benefits in patients with early-stage AD, which were not sufficient to secure regulatory approval. Nonetheless, it has contributed valuable insights into early intervention and Aβ clearance dynamics.

3.5.1.4 Donanemab

Donanemab targets N-terminal pyroglutamate-modified Aβ, a form associated with early plaque seeding. Trials in early AD populations showed notable plaque clearance and modest cognitive benefit, highlighting its potential in early intervention strategies (Lowe et al., 2021; Sims et al., 2023).

3.5.2 Other Aβ-Targeting mAbs

- *Gantenerumab (RO4909832, RG1450; IgG1)*: With a high affinity for agglomerated Aβ adherence, this humanized mAb promulgates Aβ clearance via Fc receptor-regulated phagocytosis (Bohrmann et al., 2012). Phase 3 trials could not delay cognitive decline (via addressing the priliminary clinical total dementia score endpoint) with 510 mg subcutaneous intravenous administration of gantenerumab every 2 weeks, aiming for 1020 mg monthly target dose spanned over 116 weeks (Ostrowitzki et al., 2017). CSF biomarkers like phosphorylated tau181, neurogranin and total tau protein were persistently escalated post treatment, exhibiting low immunogenicity and safety. Open-label multicentre studies (NCT04339413, NCT04374253) of long-term gantenerumab administration through subcutaneous injection are underway to determine its tolerability and safety profile.
- *Ponezumab (IgG2)*: Demonstrated better tolerability with limited effector function (Freeman et al., 2012).
- *Crenezumab (IgG4)*: Exhibited reduced ARIA rates. In one study, only two asymptomatic ARIA-E and ARIA-H cases were observed in the 15 mg/kg group (n = 247) (Cummings et al., 2018).
- *BAN2401 (Lecanemab)*: This mAb has selective adherence affinity to soluble Aβ protofibrils (Logovinsky et al., 2016). Recently

FDA-approved, targets Aβ protofibrils and showed robust plaque reduction. However, ARIA was again more prevalent in APOE-ε4 carriers (Swanson et al., 2021).

SAR-228810: A humanized IgG4 antibody stemming from murine antibody 13C3 (created by immunization employing incubated synthetic Aβ42) (Schupf et al., 2008)). SAR-228810 binds to soluble protofibrillar and insoluble fibrillar Aβ aggregates monomer in ELISA assays with nearly 100-fold selectivity over Aβ. (Ravetch and Fukuyama (2009)). The antibody is conformationally selective and binds amyloid plaques (Pradier et al. (2018)). Escalated plasma Aβ42 levels augment AD risks while decrease in certain forms of Aβ with disease progression indicates the tendency of compartmentalization of peptides in the brain. It possesses a low binding affinity to stimulate FcγRs present on human microglia without adherence to C1q complement (Pradier et al. (2013)). SAR-228810 consistently demonstrated endurablility and a favourable safety profile during phase I clinical trials (NCT01485302) among 44 single-dose and 48 multiple-dose patients, without any adverse events like reported ARIA-E cases, however, having only a single reported ARIA-H (single-dose subject) case.

3.5.3 Mechanistic Insights and Safety Profiles

Immunohistochemical analyses reveal that IgG1 and IgG4 antibodies targeting Aβ can engage microglia via CR3 and FcγRIIb, promoting engulfment of Aβ–antibody–complement complexes. This process can lead to synapse loss, particularly when C1q and C3 components are activated and deposited on synapses. Studies demonstrated colocalization of Aβ, C1q, C3, and synaptic markers (PSD95, synaptophysin) in microglia, confirming microglial synapse elimination pathways.

Furthermore, Aβ may upregulate C1q expression in astrocytes and microglia, amplifying the complement cascade and leading to neurodegeneration. This was shown to be exacerbated in vitro and in vivo by Aβ-targeting mAbs, especially those with effector functions (Sun et al., 2023).

3.5.4 Tau-Targeting Antibodies

Given the strong correlation between tau pathology and cognitive decline, passive immunotherapies targeting tau protein are gaining momentum. Growing bodies of evidence suggest that tau pathology emerge before amyloid plaque formation, initially cropping up in the locus coeruleus and subsequently progressing to other brain stem nuclei and entorhinal cortex (Jack et al., 2013; Wisniewski et al., 2015). These antibodies are designed to prevent tau aggregation and intercellular propagation or enhance tau clearance via immune-mediated mechanisms. Several candidates, including gosuranemab, semorinemab, and tilavonemab, are currently in early-phase clinical trials, with some showing initial signs of target engagement and safety.

- *Gosuranemab and Tilavonemab*: Target the mid-region or N-terminal domains of tau to block propagation (Florian et al., 2023; Sopko et al., 2020).
- *Semorinemab, Zagotenemab, and other mAbs*: Investigated for preventing the spread and promoting the clearance of pathogenic tau species. (Cai et al., 2025).
- *Lecanemab and Donanemab*: While primarily Aβ-targeting, emerging evidence suggests potential cross-utility in tau pathology (Frost et al., 2025).

Phosphokinases as the likes of extracellular signal-related kinase 2 (ERK2), glycogen GSK3β, cyclin-dependent kinase 5 (CDK5), have all been exploited for decreasing tau phosphorylation (Ballard et al., 2011; Takshima, 2009). Tau-targeting mAbs face several notable challenges in the context of ADimmunotherapy. One of the primary limitations lies in the intracellular localization of tau, which restricts antibody accessibility, as most mAbs act extracellularly. Although pathological tau can be released and transmitted between neurons, facilitating antibody engagement, effective inhibition of tau propagation has proven difficult. Moreover, early-stage clinical trials of tau-directed antibodies have largely failed to demonstrate consistent cognitive benefits, raising concerns about their therapeutic translatability. Compounding these issues are emerging safety concerns, including inflammatory responses and cerebral amyloid angiopathy (CAA), especially in cases where tau pathology coexists with vascular involvement.

Passive immunotherapy strategies targeting both Aβ and tau are also associated with several adverse effects that compromise their clinical utility. Among the most frequent complications are ARIA-E (vasogenic edema) and ARIA-H (microhemorrhages), which often occur when antibodies bind to vascular Aβ deposits (Sun et al., 2023). These events are believed to result from the activation of the complement cascade and microglial receptors such as CR3 and FcγR, leading to vascular inflammation and permeability changes. The pathological condition of CAA that usually affects AD patients enhances these effects while significantly contributing to developing ARIA-related complications (Brenowitz et al., 2015). When Aβ clearance mechanism fails, the overtriggered microglial cells create an inflammation-based environment through their release of IL-1β, IL-6, and TNF-α cytokines (Deardorff et al., 2016; Heneka et al., 2013). The inflammatory brain state, together with ARIA-related conditions, leads to the breakdown of neural synapses while also causing damage to vulnerable nervous tissue areas. A MultiTEP vaccine, AV-1980R/A, is evinced to target the phosphatase activating domain (PAD) of tau in non-human primates, thereby eliciting an immune response and reducing tau tangles in AD and other tauopathies (Hovakimyan et al., 2022).

To achieve clinical success, passive immunotherapy treatments require additional studies that optimize their safety profiles through risk minimization of inflammatory and vascular complications.

3.5.5 Advantages of Passive Immunotherapy

- *Controlled Dosing and Specificity*: Passive immunotherapy allows for precise control over antibody concentration and binding specificity, enabling dose modulation based on individual patient response.
- *Bypassing Host Immune Activation*: This approach is suitable for people with poor or inconsistent immune responses and becomes an option for active immunization if the latter is unsuccessful (Triplett et al., 2025).
- *Rapid Onset of Action*: As antibodies are directly administered, therapeutic effects on target proteins can be observed relatively quickly compared to vaccine-induced responses. (Morgan, 2010).

3.5.6 Key Findings and Efficacy of Passive Immunotherapy

Several clinical trials of passive immunotherapy in AD have been performed with both a positive and negative spectrum of results. Toward reducing Aβ burden, several mAbs have shown great efficacy in PET scan and CSF biomarker analysis. Across all of these four drugs, they have been able to clear Aβ plaques out of the brain and to varying degrees. For example, the human IgG1 mAb gantenerumab (RO4909832) activates microglial process engulfment following its binding to N-terminal (3–11) and mid-region (18–27) epitope regions of the Aβ peptide (Bohrmann et al., 2012). Of note, gantenerumab can cross the BBB and clear amyloid deposits from the CNS that predate the disease (Bateman et al., 2022). Despite these promising biomarker and imaging results, cognitive improvements have, in the main, been modest and apparently inconsistent, especially for patients with advanced disease stages. This is a critical gap in the field that remains between biomarker clearance and cognitive benefit.

Passive immunotherapy has the advantage of the ability to dose customize and specificity, but also clinical challenges. ARIA, ARI, and ARIB frequently occurs following Fc receptor binding in endothelium. However, these immune-related events can cause neuroinflammation, and more so in APOE 4 carriers, and are associated with potential safety issues as initiated with the treatment (Jung et al., 2022). In addition, mAbs have transient effects which require repeated administration, which is expensive and inconvenient to the patient, causing poor long-term compliance. As well, the little evidence for cognitive improvement, especially in the late stages of AD, emphasizes the fact that more sensitive subgroups or earlier disease phases need to be identified in which to initiate treatment.

3.5.7 Limitations and Challenges of Passive Immunotherapy

While much progress has been made in targeting Aβ pathology, passive immunotherapy is limited by a number of factors. However, these practical barriers to widespread clinical application are the result of high treatment costs, need for frequent dosing, and complex administration protocol. Specifically, in response to antibody-Aβ complexes, the immune system's overactivation through Fc receptor mechanisms may cause inflammation, including ARIA, which remains a major safety concern across all clinical trials. For some people the body's defense mechanisms become overfired, leading to unintended structural damage to neurologic structures (Bhadane et al., 2024). Additionally, while Aβ load decrease is a potentially measurable and reproducible response, cognitive performance improvements have generally been limited by the translation of these biomarker improvements. This is especially true in patients with moderate to severe AD, where neurodegeneration has already advanced and may be less amenable to therapeutic rescue (Marković et al., 2024).

Additionally, the heterogeneity of Aβ isoforms and their conformations complicates targeted therapy. For example, novel antibodies like TAP01, a murine mAb targeting a pseudo β-hairpin conformation of Aβ, are currently under investigation for their potential to selectively neutralize toxic Aβ conformers. However, many of these newer agents remain in early developmental stages, primarily limited to Phase I or II clinical trials, and their long-term safety and efficacy are yet to be established. Overall, while passive immunotherapy has reshaped the landscape of AD treatment by enabling disease-targeted interventions, its future success hinges on enhancing cognitive efficacy, reducing adverse effects, and improving accessibility through optimized delivery strategies (Guo et al., 2024). Most of the major events relating the preclinical and clinical trials in the active and passive immunotherapy of AD are enlisted in Table 3.2. Other predisposing factors owing to the contemporary hurdles in passive immunotherapy of AD are as follows:

- *Short Half-Life and Repeat Dosing*: Since exogenous antibodies are eventually cleared from the system, repeated infusions are necessary to maintain therapeutic efficacy, increasing treatment burden and cost.
- *Adverse Immune Reactions*: Potential side effects include vasogenic edema, superficial siderosis, cerebral macrohemorrhages and microhemorrhages, particularly at higher doses or in APOEε4 carriers (van Dyck et al., 2022). ARIA incidences were remarkably high in patients with prodromal dementia and clinical manifestation, that eventually superseded the advantages (Loureiro et al., 2020).
- *Variable Clinical Outcomes*: Despite promising biomarker changes, the translation to consistent cognitive benefits remains elusive in many trials, emphasizing the complexity of AD pathology (Yi et al., 2024).

TABLE 3.2
A Conspectus Illustrating AD Immunotherapy Targeting Aβ and Tau

Type of Immunotherapy	Drug (for Active Immunotherapy)/ Antibody (for Passive Immunotherapy)	Company	Phase	Status
Active Aβ immunotherapy	UB-311	United Neuroscience	II	Completed
	V950	Merck	I	Completed
	Vanities cridificer (ACC-001)	Pfizer/Janssen	II	Completed
	AV-1959D	Nuravax	I	Active, recruiting
	ABvac40	Araclon Biotech	II	Active, not recruiting
	AN1792	Wyeth/ELAN (Dublin, Ireland)	II	Terminated, no improvement (NCT00021723)
	Amilomotide (CAD106)	Amgen/ Novartis (Basel, Switzerland)	III	Terminated
Active tau immunotherapy	AADvac1	Axon Neuroscience	II	Completed
	ACI-35	AC Immune/Janssen	I/II	Active, not recruiting
Passive Aβ immunotherapy	Lecanemab (BAN2401)	Biogen/Eisai/BioArctic	IV	Approved
	Donanemab	Eli Lilly	III	Active, not recruiting
	Aducanumab (BIIB037)	Neurimmune/Eisai/Biogen	III	Terminated
	Crenezumab	Roche/Genentech	III	Terminated
	Solanezumab	Eli Lilly	III	Terminated
	Ganteneruma b	Hoffman-La Roche	III	Terminated
Passive tau immunotherapy	Semorinemab (RO705705)	AC Immune and Genentech	I	Completed
	Gosuranemab	Bristol-Myers Squibb	II	Completed
	RG7345	Hoffman-La Roche	I	Completed
	Bepranemab	Hoffman-La Roche	II	Active, not recruiting
	Zagotenemab (LY3303560)	Eli Lilly	II	Terminated
	Tilavonemab (ABBV-8E12)	C2N Diagnostics and AbbVie	II	Terminated

3.6 EMERGING APPROACHES IN IMMUNOTHERAPY

3.6.1 Combination Therapies

A compelling rationale for combination immunotherapy in AD lies in the multifactorial nature of the disease, wherein simultaneous targeting of both Aβ and tau pathologies may yield superior clinical outcomes compared to single-target approaches. Active vaccines designed to induce immune responses against both Aβ and hyperphosphorylated tau have demonstrated potential, although heightened T-helper cell responses and elevated antibody titers present immunological challenges that must be carefully addressed. In one preclinical study, combination vaccination did not increase the risk of microhemorrhages, as evidenced by Perl's Prussian blue staining, when compared to vector controls. There were no significant differences in hemorrhage incidence across various treatment groups, indicating the safety of such combinatorial immunotherapy. Notably, the vaccine alleviated aberrant inflammatory responses and led to the clearance of Aβ and phosphorylated tau in 3xTg mouse models. (Feng et al., 2024; Hovakimyan et al., 2022; Jucker & Walker, 2023; Novak et al., 2021; Song et al., 2022; Song et al., 2024; Valiukas et al., 2022).

In parallel, plant-based immunogen production platforms targeting pathological AD proteins—such as Aβ and β-secretase—have been explored over the past decade. Several plant-derived immunogen configurations have shown favorable outcomes in animal models. Furthermore, nanotechnology-based strategies, such as the delivery of iron chelators via nanoparticles, have shown promise in crossing the BBB while mitigating metal-induced oxidative stress, a recognized contributor to AD pathology (Liu et al., 2010). Future clinical trials may explore combination therapies that co-deliver glutathione and iron chelators to simultaneously reduce oxidative burden and protein aggregation (Haddad et al., 2021). Multi-targeted approaches, including in silico virtual screening of small molecules , AI-driven tools and immunogenic epitopes, are actively being investigated in ongoing clinical trials, aiming to integrate systems biology with immunotherapeutic design (Lim et al., 2023).

3.6.2 Personalized Immunotherapy

Personalized immunotherapy represents a transformative shift in AD treatment, wherein therapeutic strategies—whether vaccines or mAbs—are tailored to individual patients based on their genetic, biochemical, and pathological profiles. The use of advanced biomarkers, including PET imaging and CSF assays, enables patient stratification and early detection of disease-specific alterations, particularly in Aβ and tau dynamics. Genetic markers, such as APOE4 status, also serve as crucial determinants in predicting treatment responses. Personalized vaccine design holds the potential to overcome immune tolerance and variability seen in aged populations, thereby optimizing therapeutic efficacy while minimizing adverse effects (He et al., 2025; Herline et al., 2018).

3.6.3 Gene Therapy Approaches

Gene therapy presents an innovative avenue in AD immunotherapy by employing viral vector delivery systems to induce sustained, endogenous production of therapeutic antibodies. Long-term expression within the CNS has recently been examined for the delivery of genes encoding anti-Aβ or anti-tau antibodies using adeno-associated viruses (AAVs). Such approaches eliminate the requirement for repeated dosing, enhance BBB penetration, and achieve regional protein distribution to the pathogenic protein. But existing challenges to viral vector concern immunogenicity, potential off-target effects, and delivery efficiency, and they must be overcome through such an optimized vector design and regulatory control (Nilsson et al., 2010). As such, the use of CRISPR-Cas9 technology comes to the forefront to address the pragmatic hurdles besides other therapies like curcumin-loaded nanoparticles, anti-sense therapy, and/or precision medicine (Ahmad et al., 2014).

3.7 CHALLENGES AND CONSIDERATIONS IN VACCINE APPLICATION FOR ALZHEIMER'S DISEASE

3.7.1 Blood-Brain Barrier (BBB) Penetration

Immunotherapeutic agents persistently need to be delivered across the BBB, an effective delivery of which remains a major barrier to clinical translation. As large-size mAbs usually have poor BBB permeability, only 0.1% proportion of the systemically administered antibody can get to the CNS. Some of them are metabolized in the liver or excreted via the kidneys, diminishing clinical efficacy. To provide some examples, bi-functional antibodies, engineered so that an anti-Aβ IgG is fused at the Fc region to a Fab fragment directed against the transferrin receptor, showed extraordinarily high brain penetration and much lower load of Aβ than do conventional mAbs (Niewoehner et al., 2014). Likewise, the ApoB's LDLR binding domain also promotes both BBB crossing and neuronal uptake by LDLR-mediated endocytosis and endosomal sorting complex required for transport (ESCRT)-dependent degradation; again, bypassing cytotoxic Fc receptor activation. AD therapy by targeting of Aβ by single-chain variable fragment (scFv) antibodies has also been demonstrated without eliciting an immune response (Fukuchi et al., 2006; Kasturirangan and Sierks, 2010; Kasturirangan et al., 2009; Liu et al., 2004).

Efforts are in place for CNS targeted delivery via alternative delivery strategies like intranasal administration and focused ultrasound. In addition, anti-Aβ antibodies are ex vivo shown to activate microglia to phagocytose and degrade Aβ plaques through the Fc receptors (Bard et al., 2000).

3.7.2 Age and Immune System Factors in Alzheimer's Patients

Immunotherapies are dramatically less effective in AD concomitant with aging and immune senescence. Patients with elderly age, senile and advanced dementia are usually in an immunocompromised state, representing a reduced functionality of CD4+ and CD8+ T cells (Uyemura et al., 2002) or increased production of autoreactive antibodies in CD5+ B cells (Kotb et al., 2019). In addition, age-linked immunosenescence might also negatively modulate the capacity to mount a rapid and vigorous vaccine response, that plausibly fosters tolerance of the immunotherapy, thus being ineffectual to initiate immunogenicity (Crooke et al., 2019). In addition, there is genetic variability, as well as additional comorbidities, such as the presence of APOE4 alleles, which can make treatment more difficult. Upon designing personalized or population-based immunotherapeutic regimens, such factors must be carefully considered.

Other critical challenges in AD immunotherapy incorporates:

- Selective targeting of the misfolded and pathological protein amidst their more plentiful healthier counterparts.
- Designing constructs to effectively develop immunization strategies for an animal to elicit antibodies with precise conformational selectivity (Plotkin & Cashman, 2020).

3.7.3 Safety Concerns and Adverse Reactions

The development of immunotherapy continues to be a safety consideration. Some adverse immune reactions have been reported from clinical trials, including neuroinflammation, autoimmune responses, and ARIA. These complications tend to be associated with Fc receptor-activated complement pathways and activation of microglia. Pathological protein clearance prior to prion-like cellular proliferation of misfolded protein may also provoke irreparable neurodegeneration. Such clearance of protein species should occur without undesirable inflammatory responses that automatically emerge during xenobiotic clearance by the innate immune system (Plotkin & Cashman,2020). Early intervention in AD cannot be overemphasized since starting treatment during the mild cognitive impairment stages could prevent the irreversible neurodegeneration. The strategies aimed toward improvement of safety include: dose titration, patient monitoring, and design and development of vaccines having lower inflammatory potential and novel antigen presentation (Valiukas et al., 2022).

3.7.4 Ethical and Regulatory Considerations

Immunotherapy trials in AD also raise important ethical and regulatory questions. During the course of the disease, cognitive impairments often make informed consent from the participant a challenge (Kim, 2011). Additionally, such an assessment of balance among potential therapeutic benefits and risk is particularly stringent when evaluating subjects with potential susceptibility, such as children. The approval of aducanumab by the FDA is notable, not just for what was approved but for the change in paradigm, the acceptance of biomarker-based surrogates as the endpoint of efficacy (Usman et al., 2021). But ongoing debates demonstrate clearly that regulatory procedures need to be transparent, and post-marketing surveillance should be used.

3.8 CONCLUSION AND FUTURE DIRECTIONS IN ALZHEIMER'S VACCINE RESEARCH

In conjunction, AD immunotherapy is rushing toward increasingly more personalized, combination, and more multiparametric approaches. Fortunately, emerging strategies, such as vaccines aimed at the two moieties (Aβ and tau), present hope for penning across these limits. More recently, trials have also successfully presumed these novel targets such as TNF-α, TREM2, and CD38 (Butchart et al., 2015; Ewers et al., 2019). Increasingly, single-cell RNA sequencing and spatial transcriptomics are broadening the range of options for immunotherapeutic targets and guiding precision medicine efforts.

Continued innovation in delivery systems, particularly those enhancing BBB permeability, remains a priority. Furthermore, immunotherapy should be harmonized with supportive interventions, such as antioxidant therapy, music-based neurostimulation, and cognitive rehabilitation, to optimize patient outcomes. Importantly, future therapies must account for the heterogeneity of AD, encompassing its clinical, pathological, and molecular subtypes. The development of vaccines should simulate physiological immune signaling rather than override it with artificial constructs, thereby preserving immune homeostasis (Marciani, 2016).

In summary, while substantial challenges remain, the field is steadily progressing toward safer, more effective, and more personalized immunotherapies. Achieving this will require a collaborative, interdisciplinary effort that bridges immunology, neurology, molecular biology, and patient-centered care.

REFERENCES

Ahmad, F., Karan, A., Sharma, R., Sharma, N. S., Sundar, V., Jayaraj, R., Mukherjee, S., & DeCoster, M. A. (2024). Evolving therapeutic interventions for the management and treatment of Alzheimer's disease. *Ageing Research Reviews*, *95*, 102229.

Aljassabi, A., Zieneldien, T., Kim, J., Regmi, D., Cao, C., & Citron, B. (2024). Alzheimer's disease immunotherapy: Current strategies and future prospects. *Journal of Alzheimer's Disease*, *98*(3), 755–772.

Alshamrani, M. (2023). Recent trends in active and passive immunotherapies of Alzheimer's disease. *Antibodies*, *12*(2), 41.

Anderson, J. P., Chen, Y., Kim, K. S., & Robakis, N. K. (1992). An alternative secretase cleavage produces soluble Alzheimer amyloid precursor protein containing a potentially amyloidogenic sequence. Journal of Neurochemistry. 59(6), 2328–31.

Athar, T., Al Balushi, K., & Khan, S. A. (2021). Recent advances on drug development and emerging therapeutic agents for Alzheimer's disease. Molecular Biology Reports, 48, 5629–5645. FDA's Decision to Approve New Treatment for Alzheimer's Disease.

Ballard, C., Gauthier, S., Corbett, A., Brayne, C., Aarsland, D., & Jones, E. (2011). Alzheimer's disease. *Lancet*, *9770*, 1019–1031.

Bard, F., Cannon, C., Barbour, R., Burke, R. L., Games, D., Grajeda, H., Guido, T., Hu, K., Huang, J., Johnson-Wood, K., Khan, K., Kholodenko, D., Lee, M., Lieberburg, I., Motter, R., Nguyen, M., Soriano, F., Vasquez, N., Weiss, K., Welch, B., Seubert, P., Schenk, D., & Yednock, T. (2000). Peripherally administered antibodies against amyloid beta-peptide enter the central nervous system and reduce pathology in a mouse model of Alzheimer disease. *Nature Medicine*, *6*(8), 916–919.

Bateman, R. J., Cummings, J., Schobel, S., Salloway, S., Vellas, B., Boada, M., Black, S. E., Blennow, K., Fontoura, P., Klein, G., Assunção, S. S., Smith, J., & Doody, R. S. (2022). Gantenerumab: An anti-amyloid monoclonal antibody with potential disease-modifying effects in early Alzheimer's disease. *Alzheimer's Research & Therapy*, *14*(1), 178.

Bhadane, P., Roul, K., Belemkar, S., & Kumar, D. (2024). Immunotherapeutic approaches for Alzheimer's disease: Exploring active and passive vaccine progress. *Brain Research*, *1840*, 149018.

Bhattacherjee, A., Jung, J., Zia, S., Ho, M., Eskandari-Sedighi, G., St Laurent, C. D., McCord, K. A., Bains, A., Sidhu, G., Sarkar, S., Plemel, J. R., & Macauley, M. S. (2021). The CD33 short isoform is a gain-of-function variant that enhances Aβ1-42 phagocytosis in microglia. *Molecular Neurodegeneration*, *16*, 19.

Bittar, A., Sengupta, U., & Kayed, R. (2018). Prospects for strain-specific immunotherapy in Alzheimer's disease and tauopathies. *npj Vaccines*, *3*, 9.

Bohrmann, B., Baumann, K., Benz, J., Gerber, F., Huber, W., Knoflach, F., Messer, J., Oroszlan, K., Rauchenberger, R., Richter, W. F., Rothe, C., Urban, M., Bardroff, M., Winter, M., Nordstedt, C., & Loetscher, H. (2012). Gantenerumab: a novel human anti-Abeta antibody demonstrates sustained cerebral amyloid-beta binding and elicits cell-mediated removal of human amyloid-beta. Journal of Alzheimer's Disease, 28(1), 49–69.

Brenowitz, W. D., Nelson, P. T., Besser, L. M., Heller, K. B., & Kukull, W. A. (2015). Cerebral amyloid angiopathy and its co-occurrence with Alzheimer's disease and other cerebrovascular neuropathologic changes. Neurobiology of Aging, 36(10), 2702–2708.

Brody, D. L., & Holtzman, D. M. (2008). Active and passive immunotherapy for neurodegenerative disorders. *Annual Review of Neuroscience*, *31*, 175–193.

Budd Haeberlein, S., Aisen, P. S., Barkhof, F., Chalkias, S., Chen, T., Cohen, S., Dent, G., Hansson, O., Harrison, K., von Hehn, C., Iwatsubo, T., Mallinckrodt, C., Mummery, C. J., Muralidharan, K. K., Nestorov, I., Nisenbaum, L., Rajagovindan, R., Skordos, L., Tian, Y., ... Sandrock, A. (2022). Two randomized phase 3 studies of aducanumab in early Alzheimer's disease. Journal of Prevention of Alzheimer's Disease, 9(2), 197–210.

Butchart, J., Brook, L., Hopkins, V., Teeling, J., Püntener, U., Culliford, D., Sharples, R., Sharif, S., McFarlane, B., Raybould, R., Thomas, R., Passmore, P., Perry, V. H., & Holmes, C. (2015). Etanercept in Alzheimer disease: A randomized, placebo-controlled, double-blind, phase 2 trial. *Neurology*, *84*(21), 2161–2168.

Cai, W., Zhang, H., Wu, Y., Yao, Y., & Zhang, J. (2025). Comparative the efficacy and safety of Gosuranemab, Semorinemab, Tilavonemab, and Zagotenemab in patients with Alzheimer's disease: a systematic review and network meta-analysis of randomized controlled trials. Frontiers in Aging Neuroscience, 16, 1465871.

Chen, M. J., Ramesha, S., Weinstock, L. D.; Gao, T., Ping, L., Xiao, H., Dammer, E. B., Duong, D. D., Levey, A. I., Lah, J. J., Seyfried, N. T., Wood, L. B., & Rangaraju, S. (2021). Extracellular signal-regulated kinase regulates microglial immune responses in Alzheimer's disease. Journal of Neuroscience Research, 99(6), 1704–1721.

Crooke, S. N., Ovsyannikova, I. G., Poland, G. A., & Kennedy, R. B. (2019). Immunosenescence and human vaccine immune responses. *Immunity & Ageing*, *16*, 25.

Cummings, J. L., Cohen, S., van Dyck, C. H., Brody, M., Curtis, C., Cho, W., Ward, M., Friesenhahn, M., Rabe, C., Brunstein, F., Quartino, A., Honigberg, L. A., Fuji, R. N., Clayton, D., Mortensen, D., Ho, C., & Paul, R. (2018). ABBY: A phase 2 randomized trial of crenezumab in mild to moderate Alzheimer disease. Neurology, 90(21), e1889–e1897.

Deardorff, W. J., & Grossberg, G. T. (2016). Targeting neuroinflammation in Alzheimer's disease: evidence for NSAIDs and novel therapeutics. Expert Review of Neurotherapeutics, 17(1), 17–32.

Ewers, M., Franzmeier, N., Suárez-Calvet, M., Morenas-Rodriguez, E., Caballero, M. A. A., Kleinberger, G., Piccio, L., Cruchaga, C., Deming, Y., Dichgans, M., Trojanowski, J. Q., Shaw, L. M., Weiner, M. W., Haass, C., & Alzheimer's Disease Neuroimaging Initiative. (2019). Increased soluble TREM2 in cerebrospinal fluid is associated with reduced cognitive and clinical decline in Alzheimer's disease. *Science Translational Medicine*, *11*(507), eaav6221.

Feng, X., Hou, Y., Liu, J., Yan, F., Dai, M., Chen, M., Wang, J., Li, J., Liu, Z., Sun, D., Zhang, Y., Yu, X., Kong, W., & Wu, H. (2024). A multi-targeting immunotherapy ameliorates multiple facets of Alzheimer's disease in 3xTg mice. *npj Vaccines*, 9(1), 153.

Ferrari, C., & Sorbi, S. (2021). The complexity of Alzheimer's disease: An evolving puzzle. *Physiological Reviews*, *101*(3), 1047–1081.

Florian, H., Wang, D., Arnold, S. E., Boada, M., Guo, Q., Jin, Z., Zheng, H., Fisseha, N., Kalluri, H. V., Rendenbach-Mueller, B., Budur, K., & Gold, M. (2023). Tilavonemab in early Alzheimer's disease: results from a phase 2, randomized, double-blind study. Brain, 146(6), 2275–2284.

Foster, J. K., Verdile, G., Bates, K. A., & Martins, R. N. (2009). Immunization in Alzheimer's disease: Naïve hope or realistic clinical potential? *Molecular Psychiatry*, *14*(3), 239–251.

Franzmeier, N., Ossenkoppele, R., Brendel, M., Rubinski, A., Smith, R., Kumar, A., Mattsson-Carlgren, N., Strandberg, O., Duering, M., Buerger, K., Dichgans, M., Hansson, O., Ewers, M., Alzheimer's Disease Neuroimaging Initiative (ADNI)*, & the Swedish BioFINDER study. (2022). The BIN1 rs744373 Alzheimer's disease risk SNP is associated with faster Aβ-associated tau accumulation and cognitive decline. *Alzheimer's & Dementia*, *18*(1), 103–115.

Freeman, G. B., Lin, J. C., Pons, J., & Raha, N. M. (2012). 39-week toxicity and toxicokinetic study of ponezumab (PF-04360365) in cynomolgus monkeys with 12-week recovery period. Journal of Alzheimer's Disease, 28(3), 531–541.

Frost, B., Kolb, H., Algeciras-Schimnich, A., Betthauser, T. J., DeVos, S., Edelmayer, R. M., Elwood, F., Fleisher, A. S., Henley, D., Horie, K., Hyman, B., Kreisl, W. C., Kulic, L., Leuzy, A., Palma, J. A., Parmentier-Batteur, S., Patru, M. M., Rabinovici, G. D., Reyderman, L., ..., Weber, C. J. (2025). Tau biology, biomarkers, and therapeutics. Alzheimer's & Dementia, 11(4), e70165.

Frost, J. L., Le, K. X., Cynis, H., Ekpo, E., Kleinschmidt, M., Palmour, R. M., Ervin, F. R., Snigdha, S., Cotman, C. W., Saido, T. C., Vassar, R. J., St George-Hyslop, P., Ikezu, T., Schilling, S., Demuth, H. U., & Lemere, C. A. (2013). Pyroglutamate-3 amyloid-β deposition in the brains of humans, non-human primates, canines, and Alzheimer disease-like transgenic mouse models. The American Journal of Pathology, 183(2), 369–81.

Fukuchi, K., Accavitti-Loper, M. A., Kim, H. D., Tahara, K., Cao, Y., Lewis, T. L., Caughey, R. C., Kim, H., & Lalonde, R. (2006). Amelioration of amyloid load by anti-Abeta single-chain antibody in Alzheimer mouse model. Biochemical and Biophysical Research Communications, 344(1), 79–86.

Fulop, T., Larbi, A., Pawelec, G., Cohen, A. A., Provost, G., Khalil, A., Lacombe, G., Rodrigues, S., Desroches, M., Hirokawa, K., Franceschi, C., & Witkowski, J. M. (2022). Immunosenescence and altered vaccine efficiency in older subjects: A myth difficult to change. *Vaccines*, *10*(4), 607.

Gelinas, D. S., DaSilva, K., Fenili, D., St George-Hyslop, P., & McLaurin, J. (2004). Immunotherapy for Alzheimer's disease. Proceedings of the National Academy of Sciences U S A, 101 Suppl 2(Suppl 2), 14657–14662.

Gilman, S., Koller, M., Black, R. S., Jenkins, L., Griffith, S. G., Fox, N. C., Eisner, L., Kirby, L., Rovira, M. B., Forette, F., Orgogozo, J. M., & AN1792(QS-21)-201 Study Team. (2005). Clinical effects of Abeta immunization (AN1792) in patients with AD in an interrupted trial. *Neurology*, *64*(9), 1553–1562.

Guglielmotto, M., Repetto, I. E., Monteleone, D., Vasciaveo, V., Franchino, C., Rinaldi, S., Tabaton, M., & Tamagno, E. (2019). Stroke and amyloid-β downregulate TREM-2 and Uch-L1 expression that synergistically promote the inflammatory response. *Journal of Alzheimer's Disease*, *71*(3), 907–920.

Guo, X., Yan, L., Zhang, D., & Zhao, Y. (2024). Passive immunotherapy for Alzheimer's disease. *Ageing Research Reviews*, *94*, 102192.

Haddad, M., Herve V., Ben Khedher, M. R., Rabanel, J. M., & Ramassamy, C. (2021). Glutathione: An Old and Small Molecule with Great Functions and New Applications in the Brain and in Alzheimer's Disease. Antioxidants & Redox Signaling, 35(4), 270–292.

He, C., Shen, Y., Zhang, M., & Zhou, X. (2025). Integrating genetic and immune profiles for personalized immunotherapy in Alzheimer's disease. Frontiers in Medicine, 12, 1603553.

Heneka, M. T., Carson, M. J., El Khoury, J., Landreth, G. E., Brosseron, F., Feinstein, D. L., Jacobs, A. H., Wyss-Coray, T., Vitorica, J., Ransohoff, R. M., Herrup, K., Frautschy, S. A., Finsen, B., Brown, G. C., Verkhratsky, A., Yamanaka, K., Koistinaho, J., Latz, E., Halle, A., Petzold, G. C., Town, T., Morgan, D., Shinohara, M. L., Perry, V. H., Holmes, C., Bazan, N. G., Brooks, D. J., Hunot, S., Joseph, B., Deigendesch, N., Garaschuk, O., Boddeke, E., Dinarello, C. A., Breitner, J. C., Cole, G. M., Golenbock, D. T., & Kummer, M. P. (2015). Neuroinflammation in Alzheimer's disease. *The Lancet Neurology*, *14*(4), 388–405.

Heneka, M. T., Kummer, M. P., Stutz, A., Delekate, A., Schwartz, S., Vieira-Saecker, A., Griep, A., Axt, D., Remus, A., Tzeng, T. C., Gelpi, E., Halle, A., Korte, M., Latz, E., & Golenbock, D. T. (2013). NLRP3 is activated in Alzheimer's disease and contributes to pathology in APP/PS1 mice. Nature, 493(7434), 674–8.

Herline, K., Drummond, E., & Wisniewski, T. (2018). Recent advancements toward therapeutic vaccines against Alzheimer's disease. *Expert Review of Vaccines*, *17*(8), 707–721.

Hock, C., Konietzko, U., Papassotiropoulos, A., Wollmer, A., Streffer, J., von Rotz, R. C., Davey, G., Moritz, E., & Nitsch, R. M. (2002). Generation of antibodies specific for beta-amyloid by vaccination of patients with Alzheimer disease. Nature Medicine, 8(11), 1270–1275.

Hovakimyan, A., Zagorski, K., Chailyan, G., Antonyan, T., Melikyan, L., Petrushina, I., Batt, D. G., King, O., Ghazaryan, M., Donthi, A., Foose, C., Petrovsky, N., Cribbs, D. H., Agadjanyan, M. G., & Ghochikyan, A. (2022). Immunogenicity of MultiTEP platform technology-based tau vaccine in non-human primates. *npj Vaccines*, *7*(1), 117.

Jack, C. R. Jr, Knopman, D. S., Jagust, W. J., Petersen, R. C., Weiner, M. W., Aisen, P. S., Shaw, L. M., Vemuri, P., Wiste, H. J., Weigand, S. D., Lesnick T. G., Pankratz, V. S., Donohue, M. C., & Trojanowski, J. Q. (2013). Tracking pathophysiological processes in Alzheimer's disease: an updated hypothetical model of dynamic biomarkers. Lancet Neurology, 12(2), 207–16.

Janssen Alzheimer Immunotherapy Research & Development, LLC. (2012). Bapineuzumab in patients with mild to moderate Alzheimer's disease (ApoE4 carrier). Available at: http://clinicaltrials.gov/show/NCT00575055.

Jucker, M., & Walker, L. C. (2023). Alzheimer's disease: From immunotherapy to immunoprevention. *Cell*, *186*(20), 4260–4270.

Jung, H., Lee, S. Y., Lim, S., Choi, H. R., Choi, Y., Kim, M., Kim, S., Lee, Y., Han, K. H., Chung, W. S., & Kim, C. H. (2022). Anti-inflammatory clearance of amyloid-β by a chimeric Gas6 fusion protein. *Nature Medicine*, *28*(9), 1802–1812.

Kapadia, A., Mirrahimi, A., & Dmytriw, A. A. (2020). Intersection between sleep and neurovascular coupling as the driving pathophysiology of Alzheimer's disease. *Medical Hypotheses*, *144*, 110283.

Kasturirangan, S., Brune, D., & Sierks, M. (2009). Promoting alpha-secretase cleavage of beta-amyloid with engineered proteolytic antibody fragments. Biotechnology Progress, 25(4), 1054–1063.

Kasturirangan, S., & Sierks, M. (2010). Targeted hydrolysis of Beta-amyloid with engineered antibody fragment. Current Alzheimer Research, 7(3), 214–222.

Kim, S. Y. (2011). The ethics of informed consent in Alzheimer disease research. Nature Reviews Neurology, 7(7), 410–414.

Kohyama, K., & Matsumoto, Y. (2015). Alzheimer's disease and immunotherapy: What is wrong with clinical trials? *ImmunoTargets and Therapy*, *4*, 27–34.

Kontsekova, E., Zilka, N., Kovacech, B., Skrabana, R., & Novak, M. (2014). Identification of structural determinants on tau protein essential for its pathological function: novel therapeutic target for tau immunotherapy in Alzheimer's disease. Alzheimer's Research & Therapy, 6(4), 45.

Kotb, A., Ismail, S., Kimito, I., Mohamed, W., Salman, A., & Mohammed, A. A. (2019). Increased CD5+ B-cells are associated with autoimmune phenomena in lepromatous leprosy patients. *Journal of Infection and Public Health*, *12*, 656–659.

Lemere, C. A. (2009). Developing novel immunogens for a safe and effective Alzheimer's disease vaccine. *Progress in Brain Research*, *175*, 83–93.

Lim, C. M., González Díaz, A., Fuxreiter, M., Pun, F. W., Zhavoronkov, A., & Vendruscolo, M. (2023). Multiomic prediction of therapeutic targets for human diseases associated with protein phase separation. Proceedings of the National Academy of Sciences of the United States of America, 120(40), e2300215120.

Liu G., Men P., Perry G., & Smith, M. A. (2010). Nanoparticle and iron chelators as a potential novel Alzheimer therapy. Methods in Molecular Biology, 610, 123–144.

Liu, R., Mcallister, C., Lyubchenko, Y., & Sierks, M. R. (2004). Proteolytic antibody light chains alter beta-amyloid aggregation and prevent cytotoxicity. Biochemistry, 43, 9999–10007.

Logovinsky, V., Satlin, A., Lai, R., Swanson, C., Kaplow, J., Osswald, G., Basun, H., & Lannfelt, L. (2016). Safety and tolerability of BAN2401–a clinical study in Alzheimer's disease with a protofibril selective Aβ antibody. Alzheimer's Research & Therapy, 8(1), 14.

Loureiro, J. C., Pais, M. V., Stella, F., Radanovic, M., Teixeira, A. L., Forlenza, O. V., & de Souza, L. C. (2020). Passive anti-amyloid immunotherapy for Alzheimer's disease. *Current Opinion in Psychiatry*, *33*(3), 284–291.

Lowe, S. L., Duggan Evans, C., Shcherbinin, S., Cheng, Y. J., Willis, B. A., Gueorguieva, I., Lo, A. C., Fleisher, A. S., Dage, J. L., Ardayfio, P., Aguiar, G., Ishibai, M., Takaichi, G., Chua, L., Mullins, G., & Sims, J. R. (2021). Donanemab (LY3002813) Phase 1b Study in Alzheimer's Disease: Rapid and Sustained Reduction of Brain Amyloid Measured by Florbetapir F18 Imaging. Journal of Prevention of Alzheimer's Disease, 8(4), 414–424.

Mangialasche, F., Solomon, A., Winblad, B., Mecocci, P., & Kivipelto, M. (2010). Alzheimer's disease: Clinical trials and drug development. *Lancet Neurology*, *9*, 702–716.

Marciani, D. J. (2016). A retrospective analysis of the Alzheimer's disease vaccine progress – The critical need for new development strategies. *Journal of Neurochemistry*, *137*(5), 687–700.

Marković, M., Milošević, J., Wang, W., & Cao, Y. (2023). Passive immunotherapies targeting amyloid-*β* in Alzheimer's disease: A quantitative systems pharmacology perspective. *Molecular Pharmacology*, *105*(1), 1–13.

Miculas, D. C., Negru, P. A., Bungau, S. G., Behl, T., Hassan, S. S., & Tit, D. M. (2023). Pharmacotherapy evolution in Alzheimer's disease: Current framework and relevant directions. *Cells*, *12*(1), 131.

Mitsumori, R., Sakaguchi, K., Shigemizu, D., Mori, T., Akiyama, S., Ozaki, K., Niida, S., & Shimoda, N. (2020). Lower DNA methylation levels in CpG island shores of CR1, CLU, and

PICALM in the blood of Japanese Alzheimer's disease patients. *PLoS ONE*, *15*, e0239196.
Morgan, D. (2011). Immunotherapy for Alzheimer's disease. *Journal of Internal Medicine*, *269*, 54–63.
Niewoehner, J., Bohrmann, B., Collin, L., Urich, E., Sade, H., Maier, P., Rueger, P., Stracke, J. O., Lau, W., Tissot, A. C., Loetscher, H., Ghosh, A., & Freskgård, P. O. (2014). Increased brain penetration and potency of a therapeutic antibody using a monovalent molecular shuttle. *Neuron*, *81*(1), 49–60. https://doi.org/10.1016/j.neuron.2013.10.061
Nilsson, P., Iwata, N., Muramatsu, S., Tjernberg, L. O., Winblad, B., & Saido, T. C. (2010). Gene therapy in Alzheimer's disease - potential for disease modification. Journal of Cellular and Molecular Medicine, 14(4), 741–757.
Novak, P., Kovacech, B., Katina, S., Schmidt, R., Scheltens, P., Kontsekova, E., Ropele, S., Fialova, L., Kramberger, M., Paulenka-Ivanovova, N., Smisek, M., Hanes, J., Stevens, E., Kovac, A., Sutovsky, S., Parrak, V., Koson, P., Prcina, M., Galba, J., Cente, M., Hromadka, T., Filipcik, P., Piestansky, J., Samcova, M., Prenn-Gologranc, C., Sivak, R., Froelich, L., Fresser, M., Rakusa, M., Harrison, J., Hort, J., Otto, M., Tosun, D., Ondrus, M., Winblad, B., Novak, M., & Zilka, N. (2021). ADAMANT: A placebo-controlled randomized phase 2 study of AADvac1, an active immunotherapy against pathological tau in Alzheimer's disease. *Nature Aging*, *1*(6), 521–534.
Orgogozo, J. M., Gilman, S., Dartigues, J. F., Laurent, B., Puel, M., Kirby, L. C., Jouanny, P., Dubois, B., Eisner, L., Flitman, S., Michel, B. F., Boada, M., Frank, A., & Hock, C. (2003). Subacute meningoencephalitis in a subset of patients with AD after Abeta42 immunization. Neurology, 61(1), 46–54.
Ostrowitzki, S., Lasser, R. A., Dorflinger, E., Scheltens, P., Barkhof, F., Nikolcheva, T., Ashford, E., Retout, S., Hofmann, C., Delmar, P., Klein, G., Andjelkovic, M., Dubois, B., Boada, M., Blennow, K., Santarelli, L., & Fontoura, P. (2017). A phase III randomized trial of gantenerumab in prodromal Alzheimer's disease. Alzheimer's Research & Therapy, 9(1), 95.
Panza, F., Lozupone, M., Logroscino, G., & Imbimbo, B. P. (2019). A critical appraisal of amyloid-β-targeting therapies for Alzheimer disease. Nature Reviews Neurology, 15(2), 73–88.
Panza, F., Solfrizzi, V., Seripa, D., Imbimbo, B. P., Lozupone, M., Santamato, A., Tortelli, R., Galizia, I., Prete, C., Daniele, A., Pilotto, A., Greco, A., & Logroscino, G. (2016). Tau-based therapeutics for Alzheimer's disease: Active and passive immunotherapy. Immunotherapy, 8(9), 1119–1134.
Plotkin, S. S., & Cashman, N. R. (2020). Passive immunotherapies targeting Aβ and tau in Alzheimer's disease. Neurobiology of Disease, 144, 105010.
Pradier, L., Blanchard-Brégeon, V., Bohme, A., Debeir, T., Menager, J., Benoit, P., Barneoud, P., Taupin, V., Bertrand, P., Dugay, P., Cameron, B., Shi, Y., Naimi, S., Duchesne, M., Gagnaire, M., Weeden, T., Travaline, T., Reczek, D., Khiroug, L., Slaoui, M., Brunel, P., Fukuyama, H., Ravetch, J., Canton, T., & Cohen, C. (2018). SAR228810: an antibody for protofibrillar amyloid β peptide designed to reduce the risk of amyloid-related imaging abnormalities (ARIA). Alzheimer's Research &Therapy, 10(1), 117.
Pradier, L., Cohen, C., Blanchard, V., Debeir, T., Barneoud, P., Canton, T., Menager, J., Bohme, A., Rooney, T., Guillet, M. C., Cameron, B., Shi, Y., Naimi, S., Ravetch, J., Claudel, S., & Alam, J (2013). P4-285: SAR228810: An antiprotofibrillar beta-amyloid antibody designed to reduce risk of amyloid-related imaging abnormalities (ARIA). Alzheimer's & Dementia, 9, P808–P809.
Ravetch, J., & Fukuyama, H. (2009). Antibodies specific for the protofibril form of beta-amyloid protein. US Patent, WO2009-065054-A2.
Saido, T. C., Iwatsubo, T., Mann, D. M., Shimada, H., Ihara, Y., & Kawashima, S. (1995). Dominant and differential deposition of distinct beta-amyloid peptide species, A beta N3(pE), in senile plaques. Neuron, 14(2), 457–66.
Salloway, S., Sperling, R., Fox, N. C., Blennow, K., Klunk, W., Raskind, M., Sabbagh, M., Honig, L. S., Porsteinsson, A. P., Ferris, S., Reichert, M., Ketter, N., Nejadnik, B., Guenzler, V., Miloslavsky, M., Wang, D., Lu, Y., Lull, J., Tudor, I. C., ..., & Brashear, H. R. (2014). Two phase 3 trials of bapineuzumab in mild-to-moderate Alzheimer's disease. New England Journal of Medicine, 370(4), 322–333.
Schupf, N., Tang, M. X., Fukuyama, H., Manly, J., Andrews, H., Mehta, P., Ravetch, J., & Mayeux, R. (2008). Peripheral Aβ subspecies as risk biomarkers of Alzheimer's disease. Proceedings of the National Academy of Sciences of the United States of America, 105(37), 14052–14057.
Sevigny, J., Chiao P., Bussière, T., Weinreb, P. H., Williams, L., Maier, M., Dunstan, R., Salloway, S., Chen, T., Ling, Y., O'Gorman J., Qian F., Arastu, M., Li, M., Chollate, S., Brennan, M. S., Quintero-Monzon, O., Scannevin, R. H., Arnold, H. M., Engber, T., & Sandrock, A. (2016). The antibody aducanumab reduces Aβ plaques in Alzheimer's disease. Nature, 537(7618), 50–6.
Sims, J. R., Zimmer, J. A., Evans, C. D., Lu, M., Ardayfio, P., Sparks, J., Wessels, A. M., Shcherbinin, S., Wang, H., Monkul Nery, E. S., Collins, E. C., Solomon, P., Salloway, S., Apostolova, L. G., Hansson, O., Ritchie, C., Brooks, D. A., Mintun, M., Skovronsky, D. M., & TRAILBLAZER-ALZ 2 Investigators. (2023). Donanemab in early symptomatic Alzheimer disease: The TRAILBLAZER-ALZ 2 randomized clinical trial. JAMA, 330(6), 512–527.
Sisodia, S. S., & St George-Hyslop, P. H. (2002). γ-Secretase, notch, Aβ and alzheimer's disease: Where do the presenilins fit in?. Nature Reviews Neuroscience, 3(4), 281–290.
Song, C., Shi, J., Zhang, P., Zhang, Y., Xu, J., Zhao, L., Zhang, R., Wang, H., & Chen, H. (2022). Immunotherapy for Alzheimer's disease: Targeting β-amyloid and beyond. *Translational Neurodegeneration*, *11*(1), 18.
Sopko, R., Golonzhka, O., Arndt, J., Quan, C., Czerkowicz, J., Cameron, A., Smith, B., Murugesan, Y., Gibbons, G., Kim, S. J., Trojanowski, J. Q., Lee, V. M. Y., Brunden, K. R., Graham, D. L., Weinreb, P. H., & Hering, H. (2020). Characterization of tau binding by gosuranemab. Neurobiology of Disease, 146, 105120.
Song, Y., Dai, C. L., Shinohara, M., Chyn Tung, Y., Zhou, S., Huang, W. C., Seffouh, A., Luo, Y., Willadsen, M., Jiao, Y., Morishima, M., Saito, Y., Koh, S. H., Ortega, J., Gong, C. X., & Lovell, J. F. (2024). A pentavalent peptide vaccine elicits Aβ and tau antibodies with prophylactic activity in an Alzheimer's disease mouse model. *Brain, Behavior, and Immunity*, *122*, 185–201.
Spencer, B., & Masliah, E. (2014). Immunotherapy for Alzheimer's disease: Past, present and future. *Frontiers in Aging Neuroscience*, *6*, 114.
Sperling, R. A., Donohue, M. C., Raman, R., Rafii, M. S., Johnson, K., Masters, C. L., van Dyck, C. H., Iwatsubo, T., Marshall, G. A., Yaari, R., Mancini, M., Holdridge, K. C., Case, M., Sims, J. R., Aisen, P. S., & A4 Study Team. (2023). Trial of Solanezumab in Preclinical Alzheimer's Disease. The New England Journal of Medicine, 389(12), 1096–1107.

Sun, X., Yu, X., Zhu, J., Li, L., Zhang, L., Huang, Y., Liu, D., Ji, M., Sun, X., Zhang, L., Zhou, W., Zhang, D., Jiao, J., & Liu, R. (2023). Fc effector of anti-Aβ antibody induces synapse loss and cognitive deficits in Alzheimer's disease-like mouse model. *Signal Transduction and Targeted Therapy*, *8*, 30.

Swanson, C. J., Zhang, Y., Dhadda, S., Wang, J., Kaplow, J., Lai, R. Y. K., Lannfelt, L., Bradley, H., Rabe, M., Koyama, A., Reyderman, L., Berry, D. A., Berry, S., Gordon, R., Kramer, L. D., & Cummings, J. L. (2021). A randomized, double-blind, phase 2b proof-of-concept clinical trial in early Alzheimer's disease with lecanemab, an anti-Aβ protofibril antibody. Alzheimer's Research & Therapy, 13(1), 80.

Tabira, T. (2010). Immunization therapy for Alzheimer disease: A comprehensive review of active immunization strategies. *The Tohoku Journal of Experimental Medicine*, *220*(2), 95–106.

Takashima, A. (2009). Amyloid-beta, tau, and dementia. Journal of Alzheimer's Disease, 17(4), 729–36.

Tamagno, E., Guglielmotto, M., Vasciaveo, V., & Tabaton, M. (2021). Oxidative Stress and Beta Amyloid in Alzheimer's Disease. Which Comes First: The Chicken or the Egg? Antioxidants, 10(9), 1479.

Tran, H. T., Chung, C. H., Iba, M., Zhang, B., Trojanowski, J. Q., Luk, K. C., & Lee, V. M. (2014). A-synuclein immunotherapy blocks uptake and templated propagation of misfolded α-synuclein and neurodegeneration. *Cell Reports*, *7*(6), 2054–2065.

Triplett, O., Varda, N., Decourt, B., Vasconcellos, R., & Sabbagh, M. N. (2025). Active Immunization Targeting Amyloid β for the Treatment of Alzheimer's Disease. Neurodegenerative Diseases, 1–16.

Twohig, D., & Nielsen, H. M. (2019). α-synuclein in the pathophysiology of Alzheimer's disease. Molecular Neurodegeneration, 14(1), 23.

Usman, M. B., Bhardwaj, S., Roychoudhury, S., Kumar, D., Alexiou, A., Kumar, P., Ambasta, R. K., Prasher, P., Shukla, S., Upadhye, V., Khan, F. A., Awasthi, R., Shastri, M. D., Singh, S. K., Gupta, G., Chellappan, D. K., Dua, K., Jha, S. K., Ruokolainen, J., (…)N. K. (2021). Immunotherapy for Alzheimer's disease: Current scenario and future perspectives. *The Journal of Prevention of Alzheimer's Disease*, *8*(4), 534–551.

Uyemura, K., Castle, S. C., & Makinodan, T. (2002). The frail elderly: Role of dendritic cells in the susceptibility of infection. *Mechanisms of Ageing and Development*, *123*, 955–962.

Valiukas, Z., Ephraim, R., Tangalakis, K., Davidson, M., Apostolopoulos, V., & Feehan, J. (2022). Immunotherapies for Alzheimer's disease – A review. *Vaccines (Basel)*, *10*(9), 1527.

van Dyck, C. H., Swanson, C. J., Aisen, P., Bateman, R. J., Chen, C., Gee, M., Kanekiyo, M., Li, D., Reyderman, L., Cohen, S., Froelich, L., Katayama, S., Sabbagh, M., Vellas, B., Watson, D., Dhadda, S., Irizarry, M., Kramer, L. D., & Iwatsubo, T. (2023). Lecanemab in early Alzheimer's disease. The New England Journal of Medicine, 388(1), 9–21.

Vashisth, K., Sharma, S., Ghosh, S., Babu, M. A., Ghosh, S., Iqbal, D., Kamal, M., Almutary, A. G., Jha, S. K., Ojha, S., Bhaskar, R., Jha, N. K., & Sinha, J. K. (2024). Immunotherapy in Alzheimer's disease: Current status and future directions. *Journal of Alzheimer's Disease*, *101*(s1), S23–S39.

Vroom, M. M., & Dodart, J. C. (2024). Active immunotherapy for the prevention of Alzheimer's and Parkinson's disease. *Vaccines*, *12*(9), 973.

Winblad, B., Andreasen, N., Minthon, L., Floesser, A., Imbert, G., Dumortier, T., Maguire, R. P., Blennow, K., Lundmark, J., Staufenbiel, M., Orgogozo, J. M., & Graf, A. (2012). Safety, tolerability, and antibody response of active Aβ immunotherapy with CAD106 in patients with Alzheimer's disease: randomised, double-blind, placebo-controlled, first-in-human study. The Lancet Neurology, 11(7), 597–604.

Winblad, B., Graf, A., Riviere, M. E., Andreasen, N., & Ryan, J. M. (2014). Active immunotherapy options for Alzheimer's disease. *Alzheimer's Research & Therapy*, *6*(1), 7.

Wisniewski, T., & Goñi, F. (2015). Immunotherapeutic approaches for Alzheimer's disease. *Neuron*, *85*(6), 1162–1176.

Wolfe, C. M., Fitz, N. F., Nam, K. N., Lefterov, I., & Koldamova, R. (2018). The Role of APOE and TREM2 in Alzheimer's Disease-Current Understanding and Perspectives. International Journal of Molecular Sciences, 20(1), 81.

Yi, L. X., Tan, E. K., Zhou, Z. D. (2024). Passive immunotherapy for Alzheimer's disease: challenges & future directions. Journal of Translational Medicine, 22(1), 430.

Yu, Y. J., & Watts, R. J. (2013). Developing therapeutic antibodies for neurodegenerative disease. *Neurotherapeutics*, *10*, 459–472.

Zhang, J., Zhang, Y., Wang, J., Xia, Y., Zhang, J., & Chen, L. (2024). Recent advances in Alzheimer's disease: Mechanisms, clinical trials and new drug development strategies. Signal Transduction and Targeted Therapy, 9(1), 211.

Zhou, X., Fu, A. K., & Ip, N. Y. (2021). APOE signaling in neurodegenerative diseases: An integrative approach targeting APOE coding and noncoding variants for disease intervention. *Current Opinion in Neurobiology*, *69*, 58–67.

4 Secretase-Related Interventions in Alzheimer's

α, β, γ-Secretase, γ-Secretase Inhibitors (GSIs), γ-Secretase Modulators (GSMs) and Natural Products as GSMs

Aaditi Karnik and Abhijeet Joshi

4.1 INTRODUCTION

Alzheimer's disease (AD) an intense form of dementia is delineated by cognitive loss affecting the elderly population. As per a predicted estimation of AD cases, it is expected that by 2050, there would be over 150 million AD and other dementia cases (Nichols et al., 2019). The ever-increasing global burden of AD is due to the failure of available drug interventions in rescuing neuronal senescence associated with AD. The broad range of interventions tested for AD includes enzyme inhibitors and activators, monoclonal antibodies, natural compounds, psychobiotics, and strategies such as neuromodulation. Even with this wide spectrum of available treatment alternatives, there is no improvement in the AD cases. This drives the need for enhanced research in the field of AD for finding potential treatment strategies. One such strategy includes targeting the secretases. Secretases are proteases that are engaged in maintaining the balance of amyloid beta (Aβ) production in the brain. These secretases include α, β, and γ secretases that participate in the Aβ processing and are promising drug targets (Lichtenthaler, 2011). α-secretase activity is seen most commonly in the ADAM (a disintegrin and metalloprotease) family, with the ability to cleave the amyloid precursor protein within its sequence. This renders protection in AD due to the prevention of the formation of complete Aβ. α-secretase enzymes govern APP cleavage in a regulated manner in the Golgi complex and in a constitutive manner at the cell membrane. β-secretase, also known as β-site APP cleaving enzyme (BACE-1) cleaves APP to produce a soluble extracellular fragment similar to that of α-secretase and also generates a membrane-attached C-terminal fragment that is acted upon by γ-secretase to create toxic Aβ fragments (MacLeod et al., 2015). Besides, δ and η secretases have been found and have been reported to have less effect on Aβ production. Thus, targeting secretases provides a promising approach for curing AD. The major concern regarding the development of agents targeting secretases is the ability of these secretases to cross-talk with each other, and second, they act on a large number of proteins other than APP. This creates a challenge in developing interventions that can selectively inhibit the secretase-mediated Aβ production. Moreover, these secretases show a complex and intricate structure, and thus, any drugs targeting these tend to be of a size difficult to traverse the blood-brain barrier (BBB). Several findings indicate toxic reactions to the β- and γ-secretase inhibitors and agents augmenting α-secretase activity. Newly developed γ-secretase modulators tend to overcome these adverse effects and display a better safety margin. Present research focuses on developing more agents acting upon the secretase with the aid of computational drug modelling to facilitate treatment for AD (Schaduangrat et al., 2019).

γ-secretase is the key highlight of drug targeting in AD owing to its assembly of neurotoxic Aβ peptides. Complexity in the structure is the major lacuna in understanding the potential target sites within the γ-secretase complex. Over the period of time, conformation and stoichiometry of the subunits of the γ-secretase complex have been deciphered to aid the drug development process targeting γ-secretase (Hitzenberger et al., 2020). The main motive of developing drugs acting on γ secretase includes the potential reduction of toxic Aβ formation and sparing the other critical substrates of γ-secretase. Some of the γ-secretase inhibitors synthesized have been proven to be effective at the preclinical stages but have failed at the clinical trials due to severe toxicities due to their non-selectivity. To improve the selectivity and reduce adverse events, γ-secretase modulators were further evaluated and found to be better than γ-secretase inhibitors. Moreover, structural modifications were performed to further create drug candidates suitable for BBB penetration and improve efficacy. Certain natural compounds have also been evaluated as potential γ-secretase modulators that showed a reduction in Aβ production (Hur, 2022). This chapter highlights the drug trials associated with the γ-secretase and future perspectives within the area.

DOI: 10.1201/9781003602477-4

4.2 AMYLOID PRECURSOR PROTEIN IN AD PATHOLOGY

APP acts in several ways to be crucially involved in the AD. Firstly, it acts as a membrane-bound receptor and as a secretory mediator. The most critical feature of APP is its ability to undergo endoproteolysis within its transmembrane domains (Selkoe, 1998). Besides, APP also functions in synaptogenesis, neurite outgrowth, cell adhesions, and extracellular matrix connections (Small et al., 1999). APP contains other homologs such as APP-like protein 1 (*APLP1*) and *APLP2* (Wasco et al., 1992, 1993). All three proteins show conservation of sequences except that of Aβ, which is characteristic of APP (Zheng & Koo, 2011). APP synthesized in the endoplasmic reticulum (ER), generates different species of Aβ depending upon its trafficking and localization from the ER to the trans-Golgi-network (Greenfield et al., 1999).

APP processing is a critical phenomenon and leads to AD when the equilibrium is disrupted. The key enzymes that cleave APP include α-secretase, leading to non-amyloidogenic pathway, and β, γ secretases leading to the amyloidogenic pathway (Kojro & Fahrenholz, 2005; Vetrivel & Thinakaran, 2006). The α-secretase-mediated cleavage of APP leads to the formation of soluble APP fragment (sAPPα) and a C83 fragment at the cell membrane (Postina, 2008). The sAPPα fragment that is released extracellularly renders neuroprotective effects by maintaining neuronal architecture and synaptic connections (Dar & Glazner, 2020). The C83 fragment further gets cleaved by γ-secretase to produce p3/Aβ $_{(17\text{-}40/42)}$ and APP intracellular domain (AICD), where p3 accumulation has been shown to be associated with AD and Down's syndrome (Kuhn & Raskatov, 2020; Nhan et al., 2015). Second, APP is processed by β-secretase to form sAPPβ and C99 fragments, whereby sAPPβ plays a role in neuronal and axonal outgrowth (Chasseigneaux & Allinquant, 2012). On the contrary, C99 fragment results in dysfunction of endosome, lysosome, and cause neuroinflammation (Lauritzen et al., 2019). Moreover, a study conducted on 3xTgAD mice suggests that C99 accumulation in the hippocampus is actually involved in the AD pathology in young mice. Additionally, inhibition of γ-secretase by ELND006 increases this assemblage of toxic C99 (Lauritzen et al., 2012). C99 is then cleaved by γ-secretase to generate Aβ and AICD, wherein AICD takes part in apoptosis (Müller et al., 2008). The Aβ monomers that are generated can be converted into several different forms, such as Aβ oligomers via the process of oligomerization, then to Aβ fibrils (protofibrils to mature fibrils). These insoluble fibrils can further be converted to Aβ plaques (Chen et al., 2017) (Figure 4.1).

4.3 SECRETASES: KEY REGULATORS OF Aβ PATHWAY

4.3.1 Alpha Secretase

α-secretase is a metalloproteinase connected to the plasma membrane to cleave APP at the extracellular domain. It participates in the non-amyloidogenic pathway of APP breakdown, which begins at the plasma membrane to generate

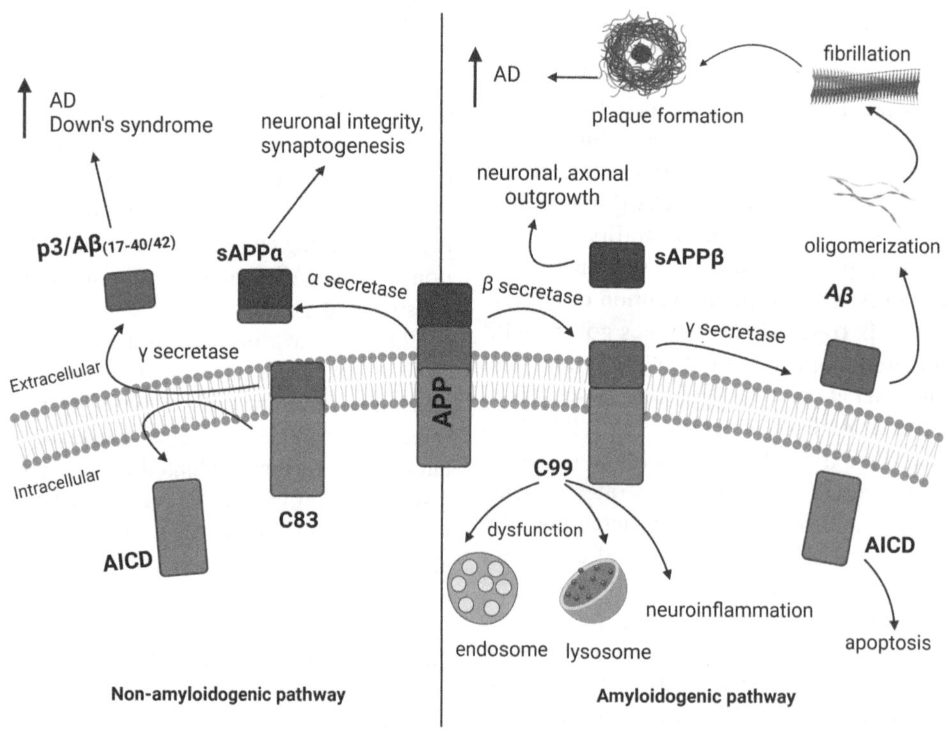

FIGURE 4.1 APP processing and role of cleaved fragments.

sAPPα by cleaving APP and the carboxy-terminal fragment of 83 amino acids (CTFα). Major members of the ADAM family, including ADAM9, ADAM10, and ADAM17, have been reported to have α-secretase activity. α-secretase cutting site between lysine 16 and leucine 17 of the Aβ domain prevents formation of toxic Aβ fragments (Yuksel & Tacal, 2019). sAPPα is a crucial factor during development due to its potential role in synaptogenesis. In neurodegenerative disorders such as AD, levels of sAPPα decline, which may contribute to the loss in cognitive coupling. Even if positive impacts of sAPPα have been reported, it is critical that the homeostasis of sAPPα levels is maintained, as disruption in its balance has been linked with Fragile X syndrome and Autism spectrum disorder (Corbett & Hooper, 2018). Moreover, sAPPα maintains neuronal integrity, improves memory, and also acts as a neurotrophic factor (Vingtdeux & Marambaud, 2012).

Of the known ADAMs acting on Aβ, ADAM10 is the primitive one in neurons, and at synaptic areas, ADAM10 is responsible for reconstructing synaptic connections (Manzine et al., 2019). Expression and activity of ADAM10 decrease in AD patients, and hence, it serves as a biosignature for AD (Marcello et al., 2017). Owing to its neuroprotective dimensions, targeting ADAM10 is often believed to have therapeutic potential against AD. Certain therapies, such as memantine, melatonin, statins, resveratrol, curcumin, quercetin, and berberine, have been proven to have neuroprotection due to their ability to positively regulate ADAM10 expression (Khezri et al., 2023). Another therapeutic strategy involves inhibition of ADAM9, which can cleave ADAM10 due to its protease activity. Thus, inhibiting ADAM9 can prevent downregulation of ADAM10 and thus restore sAPPα to mediate synaptic processing in AD (Chou et al., 2020). ADAM17 not only being an α-secretase associated with Aβ, it is also associated with the microglia-mediated neuroinflammation seen in AD patients. The major cytokines such as TNFα, IL-1, and IL-6 receptors, TGFα, and many others are targets of ADAM17 and are largely involved in the neuroinflammatory AD pathology. Thus, ADAM17 also serves as a prominent candidate for AD research (Qian et al., 2016).

4.3.2 Beta Secretase

β-secretase, also popularly referred to as BACE-1, is a member of the pepsin family with protease activity. β-secretase is abundant in neuronal cells with some expression in oligodendrocytes and astrocytes. β-secretase-mediated cleavage of APP results in the liberation of soluble N-terminus APP and a membrane-bound CTF-β. It participates in the amyloidogenic pathway and, along with the γ-secretase, produces toxic Aβ fragments (Hampel et al., 2021). BACE-1 acts as an early prognostic marker for AD, and an increase in its plasma levels may suggest conversion of mild cognitive impairment to AD (Nicsanu et al., 2022; Shen et al., 2018). This high expression of BACE-1 in the plasma correlates with the augmented levels of BACE-1 in CSF as well as brain areas in the AD patients (Grimmer et al., 2012; Zhang et al., 2020). CSF BACE-1 levels also associate with the $A\beta_{42}$ accumulation, decrease in the hippocampal volume, and tau pathology driving neurotoxicity, hippocampal atrophy, and axonal damage, respectively (Hampel et al., 2020).

The generation of noxious Aβ fragments by BACE-1 put forth the strategy of inhibiting BACE-1 to abolish the neurotoxic effect of Aβ. The development of BACE-1 inhibitors poses a major challenge of non-selectivity towards BACE-1, which results in unwanted effects. Another hurdle includes diversion of APP processing to the non-amyloidogenic pathway due to BACE-1 inhibition, causing adverse toxicities on account of increased α-secretase activity (Koelsch, 2017). Another fundamental for developing BACE-1 inhibitors includes the ability of molecules to traverse the BBB since BACE-1 is predominant in the CNS neurons. Some of the commonly known BACE-1 inhibitors being developed are Verubecestat, which marks the first initial small molecule inhibitor of BACE-1. Other BACE-1 inhibitors developed include of Lanabecestat, Atabecestat, Elenbecestat, and CNP520 (Das & Yan, 2019).

4.3.3 Gamma Secretase

Dating back to the 20th century, several studies critically evaluated the γ-secretase complex that can cleave APP within the transmembrane domain. One of the first discovered units of the multiprotein γ-secretase complex includes presenilin (PS) at the catalytic site of γ-secretase. The other three subunits of the complex comprise nicastrin, anterior pharynx-defective phenotype (Aph), and presenilin enhancer (Pen-2). All four subunits are extremely important for the γ-secretase activity (Strooper et al., 2012). The γ-secretase complex ranges from 170 to 240 kDa, depending on the non-glycosylated or glycosylated state of nicastrin, respectively, with 19 transmembranes (TMs) (Lu et al., 2014). Formation of the γ-secretase complex is a multi-step process that is initiated with the formation of APH-1 and nicastrin adduct, followed by the binding of PS to nicastrin. Further, PEN-2 gets added at the fourth TM of the PS. Lastly, the loop between TM6 and TM7 of PS-1 gets cleaved by endoproteolysis. Besides, the adduct of APH-1 and nicastrin can directly bind to PS1-PEN-2 adduct (Zhang et al., 2014). A detailed understanding of these subunits becomes essential in order to aid drug discovery and generate hits that can bind these subunits and affect the γ-secretase activity to prevent AD.

Evidences of the role of PS in γ-secretase activity of the complex comes from the studies where mutations in PS resulted into the accumulation of $A\beta_{42}$ in the brain. Besides, interaction of PS with the Notch is critical for Notch processing (Brunkan & Goate, 2005). PS1 and PS2 have their TM domains conserved, yet have divergent roles. PS1 serves important role in the embryonic development, whereas PS2 plays significant contribution to the pigmentation and mitochondrial functioning (Escamilla-Ayala

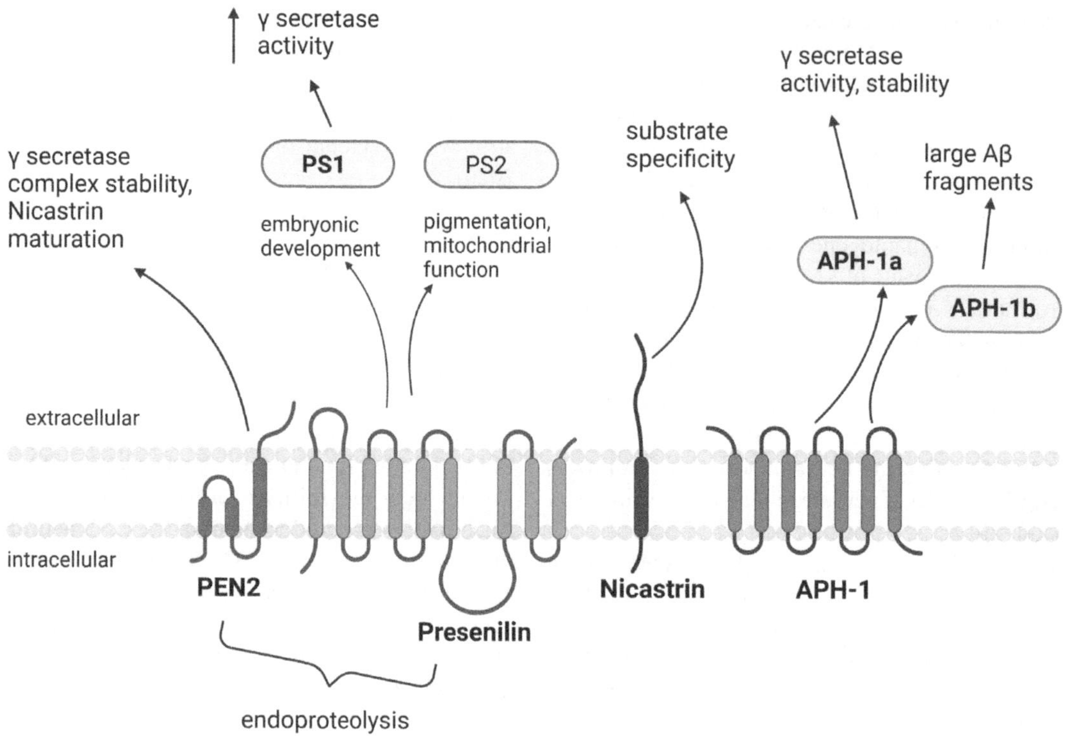

FIGURE 4.2 γ-secretase complex and function of its subunits.

et al., 2020). Owing to their ability for γ-secretase activity, PS1 has greater γ-secretase activity compared to PS2 (Lai et al., 2003). Nicastrin, the other γ-secretase subunit is often referred to as the gatekeeper of the γ-secretase activity and binds to the γ-secretase substrates directly in stoichiometry via its ectodomain (Shah et al., 2005). One of the major sites governing critical role of nicastrin ectodomain in γ-secretase activity includes Glu-333 in humans. Glu-333 is needed for the maturation of the γ-secretase complex (Chávez-Gutiérrez et al., 2008) and involves in γ-secretase activity, and together with free N-terminal it also participates in cleavage of C83 generated by α-secretase (Dries et al., 2009). Another factor affecting nicastrin-mediated γ-secretase activity includes its glycosylation. Level of nicastrin glycosylation influences its function in γ-secretase-mediated cleavage and also impacts the substrate specificity of γ-secretase (Moniruzzaman et al., 2018). Additionally, nicastrin also determines the length of Aβ by interaction between its ectodomain and APP_{C99}, and this interaction can be a potential strategic target for drug discovery (Petit et al., 2019). Furthermore, talking about *aph-1* gene, two forms, viz. *aph-1a* and *aph-1b* have been identified (Francis et al., 2002; Goutte et al., 2002). A detailed analysis of these homologs showed that APH-1a and APH-1b participate in different γ-secretase complexes with APH-1a being critical for optimum γ-secretase activity and APH-1b not being essential (Shirotani et al., 2004). Deficiency or any alterations to APH-1a affects APP processing, Notch pathway, and γ-secretase complex stability (Serneels et al., 2005). Recently it has been shown that there is involvement of APH-1b in AD pathology depicting *aph-1b* gene as a high-risk gene (Schwartzentruber et al., 2021). Along with this, it has been indicated that APH-1b mediates generation of larger Aβ fragments (Acx et al., 2014) and a full knockout of *aph-1b* prevents AD pathology (Serneels et al., 2009). PEN-2 helps in stabilizing the γ-secretase complex by interacting with PS, which allows for the endoproteolysis (Holmes et al., 2014; Prokop et al., 2004). Residues in the hydrophobic domain 1 of Pen-2, viz. glycine 22 and proline 27, lead to the γ-secretase complex formation and the stability of the complex (Bammens et al., 2011). Besides, PEN-2 also participates in nicastrin maturation and Aβ generation (Hasegawa et al., 2004), and overexpression of PEN-2 leads to increased γ-secretase activity and $Aβ_{42}$ (Seo et al., 2007) (Figure 4.2).

4.4 GAMMA SECRETASE: AS A PROMISING TARGET FOR AD

As discussed earlier, γ-secretase complex fragments Aβ into neurotoxic Aβ species leading to AD pathology. This makes γ-secretase a promising target to prevent neuronal damage associated with AD. Structural elucidation conducted till date has proven to be extremely beneficial in determining the active sites, probable target sites, and structure-activity relationships of the entire γ-secretase complex (Bai et al., 2015; Yang et al., 2019). These studies paved the way for deciding drug strategies targeting γ-secretase, which includes γ-secretase modulators and inhibitors (Lundkvist & Näslund, 2007).

4.4.1 Gamma Secretase Inhibitors

Early strategies to inhibit γ-secretase in AD began with the development of peptidomimetics that can lower toxic Aβ accumulation by inhibition of γ-secretase activity (Wolfe et al., 1998). Following this, several GSIs were discovered and grouped into transition state analogues, docking site inhibitors, and those acting at other sites (Wolfe, 2008). (Kreft et al., 2009) summarizes the structural modifications that are conducted to this class of drugs to ultimately cause effective lowering of Aβ. Several pre-clinical studies have been conducted on GSIs to evaluate their safety and potency. DAPT, a GSI tested in PDAPP mice, zebrafish resulted in a reduction of cortical Aβ, but caused severe toxicities on account of Notch impairment (Dovey et al., 2001; Geling et al., 2002). Another pre-clinical trial using LY-411,575 in TgCRND8 APP mice showed lowering of brain Aβ levels but caused thymus atrophy and abnormalities in the gastrointestinal lining on account of Notch signaling impairment (Wong et al., 2004). Begacestat when administered to Tg2576 mice decreased $A\beta_{40}$ and $A\beta_{42}$ in the brain with reversal of cognitive deficit. Begacestat also showed Notch sparing effect, making it a potential nominee for clinical trials (Mayer et al., 2008).

Translating these data from different GSIs to the clinical setup marked the beginning of clinical trials on GSIs. Some of the GSIs that were a part of clinical trials include BMS-299897, Semagacestat (LY-450139), Begacestat (GSI-953), Avagacestat (BMS-708163), Nirogacestat (PF-3084014), MK-0752, and E2012 (Kumar et al., 2018). The major concern with the GSIs related toxicities is due to the association of γ-secretase with the Notch signaling. γ-secretase-mediated cleavage of Notch is necessary for its signaling, and thus inhibiting Notch leads to potent toxicities (Barten et al., 2006). As the timeline of drug development progresses, recent research focuses on (1) the development of safer, Notch-sparing GSIs; (2) the development of GSMs.

4.4.2 Gamma Secretase Modulators

GSMs include the drugs that can modulate the generation of Aβ forms without actually altering the APP processing (Oehlrich et al., 2011). The first key insight in discovering the GSMs began with the results obtained from certain NSAIDs such as indomethacin, sulindac sulfide, and ibuprofen that could lower $A\beta_{42}$ (Weggen et al., 2001). One of the NSAIDs, namely R-flurbiprofen (tarenflurbil) showed potential for lowering $A\beta_{42}$ in AD mice with fewer side effects and was taken forward for clinical investigation (Eriksen et al., 2003; Wilcock et al., 2008). But at the clinical investigation of tarenflurbil did not refine the cognitive status of the AD patients (Green et al., 2009), which further drew attention towards developing analogues that can be potent in human trials. A further addition to the GSMs began with the generation of next-generation NSAIDs, non-NSAIDs by several pharmaceutical companies. At the forefront of non-NSAID GSMs was E2012, which reduced the proportion of neurotoxic Aβ species both pre-clinically and clinically (Nagy et al., 2010). The major mechanism of action of the first generation GSMs includes (1) conformational alteration of PS1; (2) substrate selectivity; (3) targeting the interface of enzyme-substrate. Besides, the second-generation modulators act by (1) binding PS1 and inducing conformational changes in γ-secretase complex; (2) binding Pen-2; (3) altering conversions between Aβ isoforms (Crump et al., 2013). The second-generation GSMs comprises NSAID-based carboxylic acid GSMs, heterocyclic (non-NSAID-based), which further can be divided into triazole, imidazole, etc. (Mekala et al., 2020). Investigations on non-NSAIDs showcased different docking sites for binding of these molecules on the γ-secretase complex. E2012 binds to PS1-NTF as seen in HeLa cells (Pozdnyakov et al., 2013) and also to the interface between PS1-NCT as reported by using HEK293F cells (Yang et al., 2021). Another novel GSM viz. NGP555 showed docking points at PEN-2/PS1, PS1/Nicastrin interface, and hydrocarbon pocket of PS1, thus mediating γ-secretase modulation (Kounnas et al., 2017). Table 4.1 highlights the pre-clinical and clinical investigations on some of the reported GSMs. Besides, other pharmaceutical companies working on GSMs include Roche (GSM RG6289) and AlzeCure Pharma (Nordvall et al., 2023). Besides, many natural and synthetic compounds having steroidal moieties also show potent GSM activity (Jung et al., 2013).

4.4.3 Natural Products as Forthcoming GSMs

The chemically synthesized compounds often show good efficacy as treatment but fail to render a good safety profile. This necessitates finding safer alternatives to treat the disease condition. Natural compounds offer a relatively safer alternative to the synthetic drug entities. It has long been known that natural compounds, their extracts can render neuroprotection in AD via antioxidant, anti-inflammatory, and antiapoptotic mechanisms (Ansari & Khodagholi, 2013; Essa et al., 2012). Natural compounds tend to provide a polytherapeutic approach over the modern drug molecules, which are monotherapeutic. This makes natural compounds a viable option for severe diseases such as AD (Habtemariam, 2019). Dihydroergocristine, a natural compound shows reduction in Aβ accumulation via γ-secretase inhibition (Lei et al., 2015). Considering this, several crude extracts or natural products have been tested for their potential to modulate γ-secretase activity. The major drawback in developing any molecule or finding any natural compound as a potential GSM is that, out of thousands of moieties that are screened, only a few emerge as actual hits. This makes drug discovery a tedious and slow process. Nevertheless, some of the studies on natural products have shown promising results as GSMs. Triterpene glycoside in *Actaea racemosa* extract showed high potency against $A\beta_{42}$, moderate against $A\beta_{40}$. Moreover, it increased $A\beta_{37}$ and $A\beta_{39}$, but a decline in $A\beta_{38}$ (Findeis et al., 2012; Hubbs et al., 2012). Satori had provided the initial lead from

TABLE 4.1
Pharmacokinetic, Safety, and Toxicity Analysis of Non-NSAID GSMs

Compound	Study Model	Observations	Safety and Toxicity	References
E2012	SD rats	Lowering of brain toxic Aβ species	Notch sparing; development of cataract due to higher desmosterol and lower cholesterol in the lens on account of off-target effect on DHCR24	Nagy et al. (2010), Nakano-Ito et al. (2014)
E2212	Humans	Reduction in $A\beta_{40}$, $A\beta_{42}$, reduction in plasma $A\beta_{(x\text{-}42)}$	Safer than E2012; Notch sparing	Yu et al. (2014)
FRM-36143	129S6 mice, Wistar rats	Reduction in $A\beta_{40}$, $A\beta_{42}$ in brain and CSF with increase in $A\beta_{37}$ and $A\beta_{38}$	No alteration in cholesterol metabolism i.e. no effect on DHCR24	Blain et al. (2016)
BMS-932481	SD rats, beagle dogs, cynomolgus monkeys, humans	Reduction in $A\beta_{40}$, $A\beta_{42}$ in brain and CSF with increase in $A\beta_{37}$ and $A\beta_{38}$	Notch sparing	Toyn et al. (2016), Soares et al. (2016)
NGP555	SD rats, Tg2576	Lowering CSF $A\beta_{42}$, increasing $A\beta_{38}$; prevention of cognitive decline	No goblet cell metaplasia, Notch sparing	Kounnas et al. (2017)
PF-06648671	Humans	Higher reduction in $A\beta_{42}$ than $A\beta_{40}$, more increase in $A\beta_{37}$ than $A\beta_{38}$	Good safety profile. Mild headache, nausea, and dizziness	Ahn et al. (2020)
UCSD-776890	C57BL/6J mice	Higher reduction in $A\beta_{42}$ than $A\beta_{40}$ in brain and plasma	No serious toxicities yet reported	Rynearson et al. (2021)
	PSAPP mice	Decrease in brain and plasma $A\beta_{42}$ and $A\beta_{40}$		

several natural compounds screened, and based on this, several other GSMs were developed (Fuller et al., 2012). Satori GSMs such as SPI-1865 derived from SPI-014 (triterpene glycoside) showed lowering of $A\beta_{42}$ and $A\beta_{38}$ with no effect on $A\beta_{40}$ (Loureiro et al., 2013). Several such investigations are needed in the future to elicit the potential of natural compounds as GSMs.

4.5 CONCLUSION AND FUTURE REMARKS

γ-secretase presents a potential target for drug discovery in AD. With a lot of drug candidates targeting γ-secretase being in the pipeline, it is always crucial in deciding the safety: efficacy ratio of the lead hits. The side effects associated with the utilization of GSIs lead to the development of safer, Notch-sparing GSMs that can be translated to the clinical setup. Several lead hits have proven to be effective in their γ modulation activity and are now being tested at both pre-clinical and clinical levels. Since, the synthetic compounds are majorly associated with various adverse effects, current research focuses on searching for natural compounds as safer alternatives. Some of the triterpene glycosides from *A. racemosa* have shown their efficacy in reducing Aβ burden, which further drives the need for more such studies to be conducted. What makes secretases, especially γ-secretase modulation or inhibition, a very promising approach to cure AD is their ability to prevent the generation of toxic Aβ. As it is extremely difficult to reverse the neuronal damage that has already been done in AD, it is often a better option to go for preventive measures in AD. Targeting γ-secretase is one such preventive alternative. In the future, several works need to be conducted in order to develop GSIs and GSMs that can be a promising alternative for AD.

REFERENCES

Acx, H., Chávez-Gutiérrez, L., Serneels, L., Lismont, S., Benurwar, M., Elad, N., & De Strooper, B. (2014). Signature amyloid β profiles are produced by different γ-secretase complexes. *Journal of Biological Chemistry*, *289*(7), 4346–4355. https://doi.org/10.1074/jbc.M113.530907

Ahn, J. E., Carrieri, C., Dela Cruz, F., Fullerton, T., Hajos-Korcsok, E., He, P., Kantaridis, C., Leurent, C., Liu, R., Mancuso, J., Mendes da Costa, L., & Qiu, R. (2020). Pharmacokinetic and pharmacodynamic effects of a γ-secretase modulator, PF-06648671, on CSF amyloid-β peptides in randomized phase II studies. *Clinical Pharmacology & Therapeutics*, *107*(1), 211–220. https://doi.org/10.1002/cpt.1570

Ansari, N., & Khodagholi, F. (2013). Natural products as promising drug candidates for the treatment of Alzheimer's disease: Molecular mechanism aspect. *Current Neuropharmacology*, *11*(4), 414–429.

Bai, X.-c., Yan, C., Yang, G., Lu, P., Ma, D., Sun, L., Zhou, R., Scheres, S. H. W., & Shi, Y. (2015). An atomic structure of human γ-secretase. *Nature*, *525*(7568), 212–217. https://doi.org/10.1038/nature14892

Bammens, L., Chávez-Gutiérrez, L., Tolia, A., Zwijsen, A., & De Strooper, B. (2011). Functional and topological analysis of Pen-2, the fourth subunit of the γ-secretase complex. *Journal of Biological Chemistry*, *286*(14), 12271–12282. https://doi.org/10.1074/jbc.M110.216978

Barten, D. M., Meredith, J. E., Zaczek, R., Houston, J. G., & Albright, C. F. (2006). γ-secretase inhibitors for Alzheimer's disease. *Drugs in R&D*, *7*(2), 87–97. https://doi.org/10.2165/00126839-200607020-00003

Blain, J.-F., Bursavich, M. G., Freeman, E. A., Hrdlicka, L. A., Hodgdon, H. E., Chen, T., Costa, D. E., Harrison, B. A., Kapadnis, S., Murphy, D. A., Nolan, S., Tu, Z., Tang, C., Burnett, D. A., Patzke, H., & Koenig, G. (2016). Characterization of FRM-36143 as a new γ-secretase modulator for the potential treatment of familial Alzheimer's disease. *Alzheimer's Research & Therapy*, *8*(1), 34. https://doi.org/10.1186/s13195-016-0199-5

Brunkan, A. L., & Goate, A. M. (2005). Presenilin function and γ-secretase activity. *Journal of Neurochemistry*, *93*(4), 769–792. https://doi.org/10.1111/j.1471-4159.2005.03099.x

Chasseigneaux, S., & Allinquant, B. (2012). Functions of Aβ, sAPPα and sAPPβ : Similarities and differences. *Journal of Neurochemistry*, *120 Suppl 1*(January), 99–108. https://doi.org/10.1111/j.1471-4159.2011.07584.x

Chávez-Gutiérrez, L., Tolia, A., Maes, E., Li, T., Wong, P. C., & de Strooper, B. (2008). Glu332 in the nicastrin ectodomain is essential for γ-secretase complex maturation but not for its activity. *Journal of Biological Chemistry*, *283*(29), 20096–20105. https://doi.org/10.1074/jbc.M803040200

Chen, G.-f., Xu, T.-h., Yan, Y., Zhou, Y.-r., Jiang, Y., Melcher, K., & Xu, H. E. (2017). Amyloid beta: structure, biology and structure-based therapeutic development. *Acta Pharmacologica Sinica*, *38*(9), 1205–1235. https://doi.org/10.1038/aps.2017.28

Chou, C.-W., Huang, Y.-K., Kuo, T.-T., Liu, J.-P., & Sher, Y.-P. (2020). An overview of ADAM9: Structure, activation, and regulation in human diseases. *International Journal of Molecular Sciences*, *21*(20), 7790. https://doi.org/10.3390/ijms21207790

Corbett, N. J., & Hooper, N. M. (2018). Soluble amyloid precursor protein α: Friend or foe? In K. Chattopadhyay & S. C. Basu (Eds.), *Biochemical and biophysical roles of cell surface molecules* (pp. 177–183). Springer. https://doi.org/10.1007/978-981-13-3065-0_13

Crump, C. J., Johnson, D. S., & Li, Y.-M. (2013). Development and mechanism of γ-secretase modulators for Alzheimer's disease. *Biochemistry*, *52*(19), 3197–3216. https://doi.org/10.1021/bi400377p

Dar, N. J., & Glazner, G. W. (2020). Deciphering the neuroprotective and neurogenic potential of soluble amyloid precursor protein alpha (sAPPα). *Cellular and Molecular Life Sciences*, *77*(12), 2315–2330. https://doi.org/10.1007/s00018-019-03404-x

Das, B., & Yan, R. (2019). A close look at BACE1 inhibitors for Alzheimer's disease treatment. *CNS Drugs*, *33*(3), 251–263. https://doi.org/10.1007/s40263-019-00613-7

Dovey, H. F., John, V., Anderson, J. P., Chen, L. Z., De Saint Andrieu, P., Fang, L. Y., Freedman, S. B., Folmer, B., Goldbach, E., Holsztynska, E. J., Hu, K. L., Johnson-Wood, K. L., Kennedy, S. L., Kholodenko, D., Knops, J. E., Latimer, L. H., Lee, M., Liao, Z., Lieberburg, I. M., … Audia, J. E. (2001). Functional gamma-secretase inhibitors reduce beta-amyloid peptide levels in brain. *Journal of Neurochemistry*, *76*(1), 173–181. https://doi.org/10.1046/j.1471-4159.2001.00012.x

Dries, D. R., Shah, S., Han, Y.-H., Yu, C., Yu, S., Shearman, M. S., & Yu, G. (2009). Glu-333 of nicastrin directly participates in γ-secretase activity. *Journal of Biological Chemistry*, *284*(43), 29714–29724. https://doi.org/10.1074/jbc.M109.038737

Eriksen, J. L., Sagi, S. A., Smith, T. E., Weggen, S., Das, P., McLendon, D. C., Ozols, V. V., Jessing, K. W., Zavitz, K. H., Koo, E. H., & Golde, T. E. (2003). NSAIDs and enantiomers of flurbiprofen target gamma-secretase and lower Abeta 42 in vivo. *The Journal of Clinical Investigation*, *112*(3), 440–449. https://doi.org/10.1172/JCI18162

Escamilla-Ayala, A., Wouters, R., Sannerud, R., & Annaert, W. (2020). Contribution of the presenilins in the cell biology, structure and function of γ-secretase. *Seminars in Cell & Developmental Biology, Gamma Secretase*, *105*(September), 12–26. https://doi.org/10.1016/j.semcdb.2020.02.005

Essa, M. M., Vijayan, R. K., Castellano-Gonzalez, G., Memon, M. A., Braidy, N., & Guillemin, G. J. (2012). Neuroprotective effect of natural products against Alzheimer's disease. *Neurochemical Research*, *37*(9), 1829–1842. https://doi.org/10.1007/s11064-012-0799-9

Findeis, M. A., Schroeder, F., McKee, T. D., Yager, D., Fraering, P. C., Creaser, S. P., Austin, W. F., Clardy, J., Wang, R., Selkoe, D., & Eckman, C. B. (2012). Discovery of a novel pharmacological and structural class of gamma secretase modulators derived from the extract of Actaea racemosa. *ACS Chemical Neuroscience*, *3*(11), 941–951. https://doi.org/10.1021/cn3000857

Francis, R., McGrath, G., Zhang, J., Ruddy, D. A., Sym, M., Apfeld, J., Nicoll, M., Maxwell, M., Hai, B., Ellis, M. C., Parks, A. L., Xu, W., Li, J., Gurney, M., Myers, R. L., Himes, C. S., Hiebsch, R., Ruble, C., Nye, J. S., & Curtis, D. (2002). Aph-1 and pen-2 are required for notch pathway signaling, γ-secretase cleavage of βAPP, and presenilin protein accumulation. *Developmental Cell*, *3*(1), 85–97. https://doi.org/10.1016/S1534-5807(02)00189-2

Fuller, N. O., Hubbs, J. L., Austin, W. F., Creaser, S. P., McKee, T. D., Loureiro, R. M. B., Tate, B., Xia, W., Ives, J. L., Findeis, M. A., & Bronk, B. S. (2012). Initial optimization of a new series of γ-secretase modulators derived from a triterpene glycoside. *ACS Medicinal Chemistry Letters*, *3*(11), 908–913. https://doi.org/10.1021/ml300256p

Geling, A., Steiner, H., Willem, M., Bally-Cuif, L., & Haass, C. (2002). A Γ-secretase inhibitor blocks notch signaling in vivo and causes a severe neurogenic phenotype in zebrafish. *EMBO Reports*, *3*(7), 688–694. https://doi.org/10.1093/embo-reports/kvf124

Goutte, C., Tsunozaki, M., Hale, V. A., & Priess, J. R. (2002). APH-1 is a multipass membrane protein essential for the notch signaling pathway in *Caenorhabditis elegans* embryos. *Proceedings of the National Academy of Sciences*, *99*(2), 775–779. https://doi.org/10.1073/pnas.022523499

Green, R. C., Schneider, L. S., Amato, D. A., Beelen, A. P., Wilcock, G., Swabb, E. A., & Zavitz, K. H., & for the Tarenflurbil Phase 3 Study Group. (2009). Effect of tarenflurbil on cognitive decline and activities of daily living in patients with mild Alzheimer disease: A randomized controlled trial. *JAMA*, *302*(23), 2557–2564. https://doi.org/10.1001/jama.2009.1866

Greenfield, J. P., Tsai, J., Gouras, G. K., Hai, B., Thinakaran, G., Checler, F., Sisodia, S. S., Greengard, P., & Xu, H. (1999). Endoplasmic reticulum and trans-Golgi network generate distinct populations of Alzheimer β-amyloid peptides. *Proceedings of the National Academy of Sciences*, *96*(2), 742–747. https://doi.org/10.1073/pnas.96.2.742

Grimmer, T., Alexopoulos, P., Tsolakidou, A., Guo, L.-H., Henriksen, G., Yousefi, B. H., Förstl, H., Sorg, C., Kurz, A., Drzezga, A., & Perneczky, R. (2012). Cerebrospinal fluid BACE1 activity and brain amyloid load in Alzheimer's disease. *The Scientific World Journal*, *2012*(1), 712048. https://doi.org/10.1100/2012/712048

Habtemariam, S. (2019). Natural products in Alzheimer's disease therapy: Would old therapeutic approaches fix the broken promise of modern medicines? *Molecules*, *24*(8), 1519. https://doi.org/10.3390/molecules24081519

Hampel, H., Lista, S., Vanmechelen, E., Zetterberg, H., Giorgi, F. S., Galgani, A., Blennow, K., Caraci, F., Das, B., Yan, R., & Vergallo, A. (2020). β-Secretase1 biological markers for Alzheimer's disease: State-of-art of validation and qualification. *Alzheimer's Research & Therapy*, *12*(1), 1. https://doi.org/10.1186/s13195-020-00686-3

Hampel, H., Vassar, R., De Strooper, B., Hardy, J., Willem, M., Singh, N., Zhou, J., Yan, R., Vanmechelen, E., De Vos, A., Nisticò, R., Corbo, M., Imbimbo, B. P., Streffer, J., Voytyuk, I., Timmers, M., Tahami Monfared, A. A., Irizarry, M., Albala, B., … Vergallo, A. (2021). The β-Secretase BACE1 in Alzheimer's Disease. *Biological Psychiatry*, *89*(8), 745–756. https://doi.org/10.1016/j.biopsych.2020.02.001

Hasegawa, H., Sanjo, N., Chen, F., Gu, Y.-J., Shier, C., Petit, A., Kawarai, T., Katayama, T., Schmidt, S. D., Mathews, P. M., Schmitt-Ulms, G., Fraser, P. E., & George-Hyslop, P. S. (2004). Both the sequence and length of the C terminus of PEN-2 are critical for intermolecular interactions and function of presenilin complexes. *Journal of Biological Chemistry*, *279*(45), 46455–46463. https://doi.org/10.1074/jbc.M406289200

Hitzenberger, M., Götz, A., Menig, S., Brunschweiger, B., Zacharias, M., & Scharnagl, C. (2020). The dynamics of γ-secretase and its substrates. *Seminars in Cell & Developmental Biology*, *105*(September), 86–101. https://doi.org/10.1016/j.semcdb.2020.04.008

Holmes, O., Paturi, S., Selkoe, D. J., & Wolfe, M. S. (2014). Pen-2 is essential for γ-secretase complex stability and trafficking but partially dispensable for endoproteolysis. *Biochemistry*, *53*(27), 4393–4406. https://doi.org/10.1021/bi500489j

Hubbs, J. L., Fuller, N. O., Austin, W. F., Shen, R., Creaser, S. P., McKee, T. D., Loureiro, R. M. B., Tate, B., Xia, W., Ives, J., & Bronk, B. S. (2012). Optimization of a natural product-based class of γ-secretase modulators. *Journal of Medicinal Chemistry*, *55*(21), 9270–9282. https://doi.org/10.1021/jm300976b

Hur, J.-Y. (2022). γ-secretase in Alzheimer's disease. *Experimental & Molecular Medicine*, *54*(4), 433–446. https://doi.org/10.1038/s12276-022-00754-8

Jung, J. I., Ladd, T. B., Kukar, T., Price, A. R., Moore, B. D., Koo, E. H., Golde, T. E., & Felsenstein, K. M. (2013). Steroids as γ-secretase modulators. *The FASEB Journal*, *27*(9), 3775–3785. https://doi.org/10.1096/fj.12-225649

Khezri, M. R., Mohebalizadeh, M., & Ghasemnejad-Berenji, M. (2023). Therapeutic potential of ADAM10 modulation in Alzheimer's disease: A review of the current evidence. *Cell Communication and Signaling*, *21*(1), 60. https://doi.org/10.1186/s12964-023-01072-w

Koelsch, G. (2017). BACE1 function and inhibition: Implications of intervention in the amyloid pathway of Alzheimer's disease pathology. *Molecules*, *22*(10), 1723. https://doi.org/10.3390/molecules22101723

Kojro, E., & Fahrenholz, F. (2005). The non-amyloidogenic pathway: Structure and function of α-secretases. In J. R. Harris & F. Fahrenholz (Eds.), *Alzheimer's disease: Cellular and molecular aspects of amyloid β* (pp. 105–127). Springer US. https://doi.org/10.1007/0-387-23226-5_5

Kounnas, M. Z., Lane-Donovan, C., Nowakowski, D. W., Herz, J., & Comer, W. T. (2017). NGP 555, a γ-secretase modulator, lowers the amyloid biomarker, Aβ42, in cerebrospinal fluid while preventing Alzheimer's disease cognitive decline in rodents. *Alzheimer's & Dementia: Translational Research & Clinical Interventions*, *3*(1), 65–73. https://doi.org/10.1016/j.trci.2016.09.003

Kreft, A. F., Martone, R., & Porte, A. (2009). Recent advances in the identification of gamma-secretase inhibitors to clinically test the Abeta oligomer hypothesis of Alzheimer's disease. *Journal of Medicinal Chemistry*, *52*(20), 6169–6188. https://doi.org/10.1021/jm900188z

Kuhn, A. J., & Raskatov, J. (2020). Is the P3 (Aβ17–40, Aβ17–42) peptide relevant to the pathology of Alzheimer's disease? *Journal of Alzheimer's Disease : JAD*, *74*(1), 43–53. https://doi.org/10.3233/JAD-191201

Kumar, D., Ganeshpurkar, A., Kumar, D., Modi, G., Kumar Gupta, S., & Singh, S. K. (2018). Secretase inhibitors for the treatment of Alzheimer's disease: Long road ahead. *European Journal of Medicinal Chemistry*, *148*(March), 436–452. https://doi.org/10.1016/j.ejmech.2018.02.035

Lai, M.-T., Chen, E., Crouthamel, M.-C., DiMuzio-Mower, J., Xu, M., Huang, Q., Price, E., Register, R. B., Shi, X.-P., Donoviel, D. B., Bernstein, A., Hazuda, D., Gardell, S. J., & Li, Y.-M. (2003). Presenilin-1 and presenilin-2 exhibit distinct yet overlapping γ-secretase activities *. *Journal of Biological Chemistry*, *278*(25), 22475–22481. https://doi.org/10.1074/jbc.M300974200

Lauritzen, I., Pardossi-Piquard, R., Bauer, C., Brigham, E., Abraham, J.-D., Ranaldi, S., Fraser, P., St-George-Hyslop, P., Thuc, O. L., Espin, V., Chami, L., Dunys, J., & Checler, F. (2012). The β-secretase-derived C-terminal fragment of βAPP, C99, but not Aβ, is a key contributor to early intraneuronal lesions in triple-transgenic mouse hippocampus. *Journal of Neuroscience*, *32*(46), 16243–16255. https://doi.org/10.1523/JNEUROSCI.2775-12.2012

Lauritzen, I., Pardossi-Piquard, R., Bourgeois, A., Bécot, A., & Checler, F. (2019). Does intraneuronal accumulation of carboxyl-terminal fragments of the amyloid precursor protein trigger early neurotoxicity in Alzheimer's disease? *Current Alzheimer Research*, *16*(5), 453–457. https://doi.org/10.2174/1567205016666190325092841

Lei, X., Yu, J., Niu, Q., Liu, J., Fraering, P. C., & Wu, F. (2015). The FDA-approved natural product dihydroergocristine reduces the production of the Alzheimer's disease amyloid-β peptides. *Scientific Reports*, *5*(1), 16541. https://doi.org/10.1038/srep16541

Lichtenthaler, S. F. (2011). Alpha-secretase in Alzheimer's disease: Molecular identity, regulation and therapeutic potential. *Journal of Neurochemistry*, *116*(1), 10–21. https://doi.org/10.1111/j.1471-4159.2010.07081.x

Loureiro, R. M., Dumin, J. A., McKee, T. D., Austin, W. F., Fuller, N. O., Hubbs, J. L., Shen, R., Jonker, J., Ives, J., Bronk, B. S., & Tate, B. (2013). Efficacy of SPI-1865, a novel gamma-secretase modulator, in multiple rodent models. *Alzheimer's Research & Therapy*, *5*(2), 19. https://doi.org/10.1186/alzrt173

Lu, P., Bai, X., Ma, D., Xie, T., Yan, C., Sun, L., Yang, G., Zhao, Y., Zhou, R., Scheres, S. H. W., & Shi, Y. (2014). Three-dimensional structure of human γ-secretase. *Nature*, *512*(7513), 166–170. https://doi.org/10.1038/nature13567

Lundkvist, J., & Näslund, J. (2007). γ-secretase: A complex target for Alzheimer's disease. *Current Opinion in Pharmacology*, *7*(1), 112–118. https://doi.org/10.1016/j.coph.2006.10.002

MacLeod, R., Hillert, E.-K., Cameron, R. T., & Baillie, G. S. (2015). The role and therapeutic targeting of α-, β- and γ-secretase in Alzheimer's disease. *Future Science OA*, *1*(3), FSO11. https://doi.org/10.4155/fso.15.9

Manzine, P. R., Ettcheto, M., Cano, A., Busquets, O., Marcello, E., Pelucchi, S., Di Luca, M., Endres, K., Olloquequi, J., Camins, A., & Cominetti, M. R. (2019). ADAM10 in Alzheimer's disease: Pharmacological modulation by natural compounds and its role as a peripheral marker. *Biomedicine & Pharmacotherapy*, *113*, 108661. https://doi.org/10.1016/j.biopha.2019.108661

Marcello, E., Borroni, B., Pelucchi, S., Gardoni, F., & Di Luca, M. (2017). ADAM10 as a therapeutic target for brain diseases: From developmental disorders to Alzheimer's disease. *Expert Opinion on Therapeutic Targets*, *21*(11), 1017–1026. https://doi.org/10.1080/14728222.2017.1386176

Mayer, S. C., Kreft, A. F., Harrison, B., Abou-Gharbia, M., Antane, M., Aschmies, S., Atchison, K., Chlenov, M., Cole, D. C., Comery, T., Diamantidis, G., Ellingboe, J., Fan, K., Galante, R., Gonzales, C., Ho, D. M., Hoke, M. E., Hu, Y., Huryn, D., … Jacobsen, J. S. (2008). Discovery of begacestat, a Notch-1-sparing γ-secretase inhibitor for the treatment of Alzheimer's disease. *Journal of Medicinal Chemistry*, *51*(23), 7348–7351. https://doi.org/10.1021/jm801252w

Mekala, S., Nelson, G., & Li, Y.-M. (2020). Recent developments of small molecule γ-secretase modulators for Alzheimer's disease. *RSC Medicinal Chemistry*, *11*(9), 1003–1022. https://doi.org/10.1039/D0MD00196A

Moniruzzaman, M., Ishihara, S., Nobuhara, M., Higashide, H., & Funamoto, S. (2018). Glycosylation status of nicastrin influences catalytic activity and substrate preference of γ-secretase. *Biochemical and Biophysical Research Communications*, *502*(1), 98–103. https://doi.org/10.1016/j.bbrc.2018.05.126

Müller, T., Meyer, H. E., Egensperger, R., & Marcus, K. (2008). The amyloid precursor protein intracellular domain (AICD) as modulator of gene expression, apoptosis, and cytoskeletal dynamics-relevance for Alzheimer's disease. *Progress in Neurobiology*, *85*(4), 393–406. https://doi.org/10.1016/j.pneurobio.2008.05.002

Nagy, C., Schuck, E., Ishibashi, A., Nakatani, Y., Rege, B., & Logovinsky, V. (2010). P3-415: E2012, a novel gamma-secretase modulator, decreases plasma amyloid-beta (Aβ) levels in humans. *Alzheimer's & Dementia*, *6*(4S_Part_19), S574. https://doi.org/10.1016/j.jalz.2010.05.1958

Nakano-Ito, K., Fujikawa, Y., Hihara, T., Shinjo, H., Kotani, S., Suganuma, A., Aoki, T., & Tsukidate, K. (2014). E2012-induced cataract and its predictive biomarkers. *Toxicological Sciences*, *137*(1), 249–258. https://doi.org/10.1093/toxsci/kft224

Nhan, H. S., Chiang, K., & Koo, E. H. (2015). The multifaceted nature of amyloid precursor protein and its proteolytic fragments: Friends and foes. *Acta Neuropathologica*, *129*(1), 1–19. https://doi.org/10.1007/s00401-014-1347-2

Nichols, E., Szoeke, C. E. I., Vollset, S. E., Abbasi, N., Abd-Allah, F., Abdela, J., Aichour, M. T. E., Akinyemi, R. O., Alahdab, F., Asgedom, S. W., Awasthi, A., Barker-Collo, S. L., Baune, B. T., Béjot, Y., Belachew, A. B., Bennett, D. A., Biadgo, B., Bijani, A., Sayeed, M. S. B., … Murray, C. J. L. (2019). Global, regional, and national burden of Alzheimer's disease and other dementias, 1990–2016: A systematic analysis for the global burden of disease study 2016. *The Lancet Neurology*, *18*(1), 88–106. https://doi.org/10.1016/S1474-4422(18)30403-4

Nicsanu, R., Cervellati, C., Benussi, L., Squitti, R., Zanardini, R., Rosta, V., Trentini, A., Ferrari, C., Saraceno, C., Longobardi, A., Bellini, S., Binetti, G., Zanetti, O., Zuliani, G., & Ghidoni, R. (2022). Increased serum Beta-secretase 1 activity is an early marker of Alzheimer's disease. *Journal of Alzheimer's Disease*, *87*(1), 433–441. https://doi.org/10.3233/JAD-215542

Nordvall, G., Lundkvist, J., & Sandin, J. (2023). Gamma-secretase modulators: A promising route for the treatment of Alzheimer's disease. *Frontiers in Molecular Neuroscience*, *16*(October). https://doi.org/10.3389/fnmol.2023.1279740

Oehlrich, D., Berthelot, D. J.-C., & Gijsen, H. J. M. (2011). γ-secretase modulators as potential disease modifying anti-Alzheimer's drugs. *Journal of Medicinal Chemistry*, *54*(3), 669–698. https://doi.org/10.1021/jm101168r

Petit, D., Hitzenberger, M., Lismont, S., Zoltowska, K. M., Ryan, N. S., Mercken, M., Bischoff, F., Zacharias, M., & Chávez-Gutiérrez, L. (2019). Extracellular interface between APP and nicastrin regulates Aβ length and response to Γ-secretase modulators. *The EMBO Journal*, *38*(12), e101494. https://doi.org/10.15252/embj.2019101494

Postina, R. (2008). A closer look at α-secretase. *Current Alzheimer Research*, *5*(2), 179–186. https://doi.org/10.2174/156720508783954668

Pozdnyakov, N., Murrey, H. E., Crump, C. J., Pettersson, M., Ballard, T. E., am Ende, C. W., Ahn, K., Li, Y.-M., Bales, K. R., & Johnson, D. S. (2013). γ-secretase modulator (GSM) photoaffinity probes reveal distinct allosteric binding sites on presenilin. *Journal of Biological Chemistry*, *288*(14), 9710–9720. https://doi.org/10.1074/jbc.M112.398602

Prokop, S., Shirotani, K., Edbauer, D., Haass, C., & Steiner, H. (2004). Requirement of PEN-2 for stabilization of the presenilin N-/C-terminal fragment heterodimer within the γ-secretase complex. *Journal of Biological Chemistry*, *279*(22), 23255–23261. https://doi.org/10.1074/jbc.M401789200

Qian, M., Shen, X., & Wang, H. (2016). The distinct role of ADAM17 in APP proteolysis and microglial activation related to Alzheimer's disease. *Cellular and Molecular Neurobiology*, *36*(4), 471–482. https://doi.org/10.1007/s10571-015-0232-4

Rynearson, K. D., Ponnusamy, M., Prikhodko, O., Xie, Y., Zhang, C., Nguyen, P., Hug, B., Sawa, M., Becker, A., Spencer, B., Florio, J., Mante, M., Salehi, B., Arias, C., Galasko, D., Head, B. P., Johnson, G., Lin, J. H., Duddy, S. K., … Wagner, S. L. (2021). Preclinical validation of a potent γ-secretase modulator for Alzheimer's disease prevention. *Journal of Experimental Medicine*, *218*(4), e20202560. https://doi.org/10.1084/jem.20202560

Schaduangrat, N., Prachayasittikul, V., Choomwattana, S., Wongchitrat, P., Phopin, K., Suwanjang, W., Malik, A. A., Vincent, B., & Nantasenamat, C. (2019). Multidisciplinary approaches for targeting the secretase protein family as a therapeutic route for Alzheimer's disease. *Medicinal Research Reviews*, *39*(5), 1730–1778. https://doi.org/10.1002/med.21563

Schwartzentruber, J., Cooper, S., Liu, J. Z., Barrio-Hernandez, I., Bello, E., Kumasaka, N., Young, A. M. H., Franklin, R. J. M., Johnson, T., Estrada, K., Gaffney, D. J., Beltrao, P., & Bassett, A. (2021). Genome-wide meta-analysis, fine-mapping and integrative prioritization implicate new Alzheimer's disease risk genes. *Nature Genetics*, *53*(3), 392–402. https://doi.org/10.1038/s41588-020-00776-w

Selkoe, D. J. (1998). The cell biology of β-amyloid precursor protein and presenilin in Alzheimer's disease. *Trends in Cell Biology*, *8*(11), 447–453. https://doi.org/10.1016/S0962-8924(98)01363-4

Seo, S. J., Hwang, D. Y., Cho, J. S., Chae, K. R., Kim, C. K., Shim, S. B., Jee, S. W., Lee, S. H., Sin, J. S., Choi, S. Y., Kim, J., & Kim, Y. K. (2007). PEN-2 overexpression induces γ-secretase protein and its activity with amyloid β-42 production. *Neurochemical Research*, *32*(6), 1016–1023. https://doi.org/10.1007/s11064-006-9262-0

Serneels, L., Dejaegere, T., Craessaerts, K., Horré, K., Jorissen, E., Tousseyn, T., Hébert, S., Coolen, M., Martens, G., Zwijsen, A., Annaert, W., Hartmann, D., & De Strooper, B. (2005). Differential contribution of the three Aph1 genes to γ-secretase activity in vivo. *Proceedings of the National Academy of Sciences of the United States of America*, *102*(5), 1719–1724. https://doi.org/10.1073/pnas.0408901102

Serneels, L., Van Biervliet, J., Craessaerts, K., Dejaegere, T., Horré, K., Van Houtvin, T., Esselmann, H., Paul, S., Schäfer, M. K., Berezovska, O., Hyman, B. T., Sprangers, B., Sciot, R., Moons, L., Jucker, M., Yang, Z., May, P. C., Karran, E., Wiltfang, J., … De Strooper, B. (2009). γ-secretase heterogeneity in the Aph1 subunit: Relevance for Alzheimer's disease. *Science*, *324*(5927), 639–642. https://doi.org/10.1126/science.1171176

Shah, S., Lee, S.-F., Tabuchi, K., Hao, Y.-H., Yu, C., LaPlant, Q., Ball, H., Dann, C. E., Südhof, T., & Yu, G. (2005). Nicastrin functions as a γ-secretase-substrate receptor. *Cell*, *122*(3), 435–447. https://doi.org/10.1016/j.cell.2005.05.022

Shen, Y., Wang, H., Sun, Q., Yao, H., Keegan, A. P., Mullan, M., Wilson, J., Lista, S., Leyhe, T., Laske, C., Rujescu, D., Levey, A., Wallin, A., Blennow, K., Li, R., & Hampel, H. (2018). Increased plasma beta-secretase 1 may predict conversion to Alzheimer's disease dementia in individuals with mild cognitive impairment. *Biological Psychiatry*, *83*(5), 447–455. https://doi.org/10.1016/j.biopsych.2017.02.007

Shirotani, K., Edbauer, D., Prokop, S., Haass, C., & Steiner, H. (2004). Identification of distinct γ-secretase complexes with different APH-1 variants. *Journal of Biological Chemistry*, *279*(40), 41340–41345. https://doi.org/10.1074/jbc.M405768200

Small, D. H., Clarris, H. L., Williamson, T. G., Reed, G., Key, B., Mok, S. S., Beyreuther, K., Masters, C. L., & Nurcombe, V. (1999). Neurite-outgrowth regulating functions of the amyloid protein precursor of Alzheimer's disease. *Journal of Alzheimer's Disease*, *1*(4–5), 275–285. https://doi.org/10.3233/JAD-1999-14-508

Soares, H. D., Gasior, M., Toyn, J. H., Wang, J.-S., Hong, Q., Berisha, F., Furlong, M. T., Raybon, J., Lentz, K. A., Sweeney, F., Zheng, N., Akinsanya, B., Berman, R. M., Thompson, L. A., Olson, R. E., Morrison, J., Drexler, D. M., Macor, J. E., Albright, C. F., … AbuTarif, M. (2016). The γ-secretase modulator, BMS-932481, modulates Aβ peptides in the plasma and cerebrospinal fluid of healthy volunteers. *The Journal of Pharmacology and Experimental Therapeutics*, *358*(1), 138–150. https://doi.org/10.1124/jpet.116.232256

Strooper, B. D., Iwatsubo, T., & Wolfe, M. S. (2012). Presenilins and γ-secretase: Structure, function, and role in Alzheimer disease. *Cold Spring Harbor Perspectives in Medicine*, *2*(1), a006304. https://doi.org/10.1101/cshperspect.a006304

Toyn, J. H., Boy, K. M., Raybon, J., Meredith, J. E., Robertson, A. S., Guss, V., Hoque, N., Sweeney, F., Zhuo, X., Clarke, W., Snow, K., Denton, R. R., Zuev, D., Thompson, L. A., Morrison, J., Grace, J., Berisha, F., Furlong, M., Wang, J.-S., … Ahlijanian, M. K. (2016). Robust translation of γ-secretase modulator pharmacology across preclinical species and human subjects. *The Journal of Pharmacology and Experimental Therapeutics*, *358*(1), 125–137. https://doi.org/10.1124/jpet.116.232249

Vetrivel, K. S., & Thinakaran, G. (2006). Amyloidogenic processing of β-amyloid precursor protein in intracellular compartments. *Neurology*, *66*(1_suppl_1), S69–S73. https://doi.org/10.1212/01.wnl.0000192107.17175.39

Vingtdeux, V., & Marambaud, P. (2012). Identification and biology of α-secretase. *Journal of Neurochemistry*, *120*(s1), 34–45. https://doi.org/10.1111/j.1471-4159.2011.07477.x

Wasco, W., Bupp, K., Magendantz, M., Gusella, J. F., Tanzi, R. E., & Solomon, F. (1992). Identification of a mouse brain cDNA that encodes a protein related to the Alzheimer disease-associated amyloid beta protein precursor. *Proceedings of the National Academy of Sciences*, *89*(22), 10758–10762. https://doi.org/10.1073/pnas.89.22.10758

Wasco, W., Gurubhagavatula, S., Paradis, M., Romano, D. M., Sisodia, S. S., Hyman, B. T., Neve, R. L., & Tanzi, R. E. (1993). Isolation and characterization of APLP2 encoding a homologue of the Alzheimer's associated amyloid β protein precursor. *Nature Genetics*, *5*(1), 95–100. https://doi.org/10.1038/ng0993-95

Weggen, S., Eriksen, J. L., Das, P., Sagi, S. A., Wang, R., Pietrzik, C. U., Findlay, K. A., Smith, T. E., Murphy, M. P., Bulter, T., Kang, D. E., Marquez-Sterling, N., Golde, T. E., & Koo, E. H. (2001). A subset of NSAIDs lower amyloidogenic Aβ42 independently of cyclooxygenase activity. *Nature*, *414*(6860), 212–216. https://doi.org/10.1038/35102591

Wilcock, G. K., Black, S. E., Hendrix, S. B., Zavitz, K. H., Swabb, E. A., & Laughlin, M. A. (2008). Efficacy and safety of tarenflurbil in mild to moderate Alzheimer's disease: A randomised Phase II trial. *The Lancet Neurology*, *7*(6), 483–493. https://doi.org/10.1016/S1474-4422(08)70090-5

Wolfe, M. S. (2008). γ-secretase inhibition and modulation for Alzheimer's disease. *Current Alzheimer Research*, *5*(2), 158–164.

Wolfe, M. S., Citron, M., Diehl, T. S., Xia, W., Donkor, I. O., & Selkoe, D. J. (1998). A substrate-based difluoro ketone selectively inhibits Alzheimer's γ-secretase activity. *Journal of Medicinal Chemistry*, *41*(1), 6–9. https://doi.org/10.1021/jm970621b

Wong, G. T., Manfra, D., Poulet, F. M., Zhang, Q., Josien, H., Bara, T., Engstrom, L., Pinzon-Ortiz, M., Fine, J. S., Lee, H.-J. J., Zhang, L., Higgins, G. A., & Parker, E. M. (2004). Chronic treatment with the γ-secretase inhibitor LY-411,575 inhibits β-amyloid peptide production and alters lymphopoiesis and intestinal cell differentiation. *Journal of Biological Chemistry*, *279*(13), 12876–12882. https://doi.org/10.1074/jbc.M311652200

Yang, G., Zhou, R., Guo, X., Yan, C., Lei, J., & Shi, Y. (2021). Structural basis of γ-secretase inhibition and modulation by small molecule drugs. *Cell*, *184*(2), 521–533.e14. https://doi.org/10.1016/j.cell.2020.11.049

Yang, G., Zhou, R., Zhou, Q., Guo, X., Yan, C., Ke, M., Lei, J., & Shi, Y. (2019). Structural basis of notch recognition by human γ-secretase. *Nature*, *565*(7738), 192–197. https://doi.org/10.1038/s41586-018-0813-8

Yu, Y., Logovinsky, V., Schuck, E., Kaplow, J., Chang, M.-k., Miyagawa, T., Wong, N., & Ferry, J. (2014). Safety, tolerability, pharmacokinetics, and pharmacodynamics of the novel γ-secretase modulator, E2212, in healthy human subjects. *The Journal of Clinical Pharmacology*, *54*(5), 528–536. https://doi.org/10.1002/jcph.249

Yuksel, M., & Tacal, O. (2019). Trafficking and proteolytic processing of amyloid precursor protein and secretases in Alzheimer's disease development: An up-to-date review. *European Journal of Pharmacology*, *856*(August), 172415. https://doi.org/10.1016/j.ejphar.2019.172415

Zhang, M., Zhong, X., Shi, H., Vanmechelen, E., De Vos, A., Liu, S., Chen, B., Mai, N., Peng, Q., Chen, X., Wu, Z., Hou, L., Zhou, H., Ouyang, C., Zhang, W., Liang, W., Dai, C., & Ning, Y. (2020). BACE1 and other Alzheimer's-related biomarkers in cerebrospinal fluid and plasma distinguish alzheimer's disease patients from cognitively-impaired neurosyphilis patients. *Journal of Alzheimer's Disease*, *77*(1), 313–322. https://doi.org/10.3233/JAD-200362

Zhang, X., Li, Y., Xu, H., & Zhang, Y.-W. (2014). The γ-secretase complex: From structure to function. *Frontiers in Cellular Neuroscience*, *8*(December). https://doi.org/10.3389/fncel.2014.00427

Zheng, H., & Koo, E. H. (2011). Biology and pathophysiology of the amyloid precursor protein. *Molecular Neurodegeneration*, *6*(1), 27. https://doi.org/10.1186/1750-1326-6-27

5 A Potential Modulator in Alzheimer's Disease

SirT2 Inhibitors

Mushtaque A. S. Shaikh and Tamanna K. Khan

5.1 INTRODUCTION

By the time the readers arrive at this chapter, they should be aware that Alzheimer's disease (AD) is a prevalent form of dementia and a progressive neurological disorder. Symptoms include memory loss and cognitive impairment. A well-known stream of genetic science *Epigenetics* has been employed to understand that methylation of DNA and acetylation of histone are involved in the onset of AD (Johnstone & Baylin, 2010). Histones are nuclear proteins found in the nuclei of eukaryotic cells and in most archaeal phyla. They are responsible for structural support for chromosomes by helping to compact DNA into a more manageable size. Histones play a role in controlling gene activity and form spools around which DNA winds to create nucleosomes, which are the beads that make up chromatin (Figure 5.1).

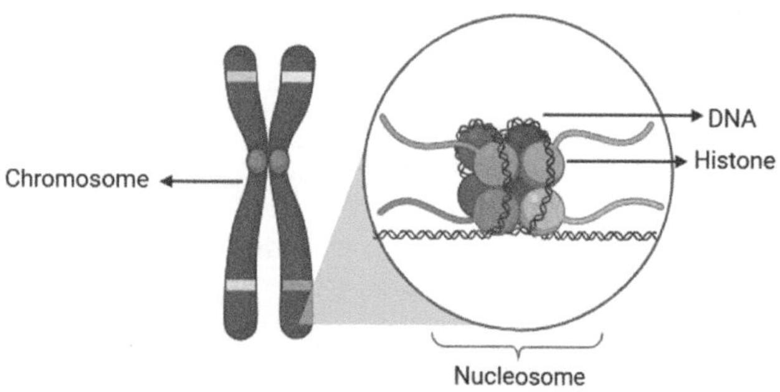

FIGURE 5.1 Placement and function of histone.

The histones are highly regulated proteins and have been known to have a role in the activation and deactivation of several genes. Structurally the histones are proteins with high basicity due to the abundance of arginine and lysine residues in their core. Positive charges on these basic residues cause significant contact with the negatively charged phosphate backbone of DNA helix, resulting in a tight connection. It has been known that alterations like acetylation and methylation can occur at these residues (specially Lys). Such modifications loosen the association between the genetic material and histone and is referred to as *relaxation* which consequently makes space for other enzymes to interact with DNA to bring about various important events like the expression of certain genes. The acetylation is a reversible process and forward and backward reaction are catalysed by two separate enzymes, *histone acetyltransferases* (HATs) and *histone deacetylases* (HDACs), respectively. As mentioned earlier there are a number of Lys residues in the protein sequence of histones and hence there exist a number of acetylation and deacetylation sites, it has been established that they may have different consequences. Thus, the degree and location of histone acetylation regulates condensation of chromatin and transcription of genes (Roth et al., 2001). It is interesting to note here that uncontrolled acetylation of histone is pivotal in the pathophysiology of AD. Thus, HDAC appears to be an important target for improving cognitive abilities and memory in AD. This idea has led to the identification and development of many non-selective HDAC inhibitors referred to as pan-HDAC inhibitors, as described in Table 5.1.

TABLE 5.1
HDAC Subtypes, Location, and Pan-HDAC Inhibitors

Subtype	Catalysis	Specific Proteins	Location	Inhibitors
Class I	Zn^{++}	HDAC1-3,8	Nucleus	Valproic acid, vorinostat, trichostatin A
Class IIa	Zn^{++}	HDAC4-7,9	Nucleus + cytoplasm	Trichostatin A, phenylbutyrate
Class IIb	Zn^{++}	HDAC6,10	Cytoplasm	Trichostatin A, tubacin
Class III	**NAD^{+}**	**SirT1-7**	**Nucleus + cytoplasm + mitochondria**	**Nicotinamide, suramin (SirT1 and SirT2)**
Class IV	Zn^{++}	HDAC11	Nucleus	–

Xu et al. (2011).
Note: The class III of HDACs is also referred to as sirtuins

DOI: 10.1201/9781003602477-5

TABLE 5.2
Sirtuin Subtypes, Location, and Catalytic Activities

Subtype	Location	Catalytic activities
SirT1	Nucleus + cytoplasm	Deacetylase
SirT2	**Nucleus + cytoplasm**	**Deacetylase**
SirT3	Mitochondria	Deacetylase
SirT4	Mitochondria	ADP-ribosyl transferase
SirT5	Mitochondria	Deacetylase, desuccinylase, and demalonylase
SirT6	Nucleus	ADP-ribosyl transferase deacetylase
SirT7	Nucleus	Deacetylase

Fernando & Wijayasinghe (2021).

The Table 5.1, gets us to the main topic of this chapter's theme, sirtuins, i.e., *silent information regulator proteins.* At this point, we would like to clarify that, though we have begun our discussion with this process involving histone and its deacetylation through sirtuins, in reality there are many other facts and processes which we will explore a little later in this chapter, elaborating connection between sirtuins, and AD.

Let us focus our attention towards the sirtuins to realize that they too are lysine deacetylases which catalyses deacetylation reaction through NAD^+ as a cofactor. However, sirtuins do not just restrict to deacetylation reactions but are also know to carry out other functions like *ADP-ribosyl transferase*, *desuccinylase*, and *demalonylase* activities as indicated in Table 5.2.

Sirtuins are actually a kind of anti-ageing factors which are found across the species and are conserved in sequence, structure, and function. As shown in Table 5.2, they can exist in a variety of subcellular locations and thus perform a wide range of functions, including the regulation of various enzymes that modulate critical proteins involved in repair of DNA, cell cycle and development, cellular ageing, neuroprotection, survival, and metabolism. Taking the lead from the introduction of this chapter, i.e., their connection to deacetylation both HDACs and sirtuins are also deacetylators to various non-histone compounds. The sirtuins have been associated with many other important biomolecules in body like P53, forkhead transcription factors (FOXO), PGC1α, and NF-κB. Sirtuins have been linked to reduce ageing and prolonged lifespan in mammals through caloric restriction (CR).

At this point let us further narrow down our focus to the role of sirtuins in AD. The extensive research on this topic has revealed that only SirT1 and SirT2 among the sirtuins, are relevant to AD. Both SirT1 and SirT2 have been established to have a role in aetiology and treatment of AD. When comparing the roles of SirT1 and SirT2 in AD, mRNA expression levels of SirT2 are higher in AD patients, while SirT1 levels are lower. SirT1 deacetylates tau protein, leading it to ubiquitination and reducing neurofibrillary tangles (NFT). In addition, blocking SirT1 has been shown to produce tau hyperphosphorylation via increased ApoE4 and OGT activity. SirT2 deacetylates α-tubulin and insulin/TGF-1, providing neuronal protection.

5.2 SirT2: ITS BIOLOGICAL FUNCTIONS AND ROLE IN ALZHEIMER'S DISEASE PATHOLOGY

SirT2 is the most common sirtuin isoform in the nervous system, with expression restricted to cytoplasmic neurites and postmitotic cell growth cones. SirT2, a conserved lysine deacetylase, has an impact on ageing, energy generation, and life extension. High SirT2 levels have been linked to neurodegeneration in AD and Parkinson's disease (PD), among other neurological illnesses. However, it has been observed that in SH-SY5Y cells (*in vitro* models of neuronal function and differentiation), increased SirT2 protects cells against rotenone- or diquat-induced cell death, whereas SirT2 inhibition accelerates cell death.

The SirT2 protein is thought to guard against oxidative stress and may aid in neurodegeneration by compensating for neuronal damage (Singh et al., 2017). The phenomenon of autophagy is diminished in neural system in case of AD. An increase in autophagy was utilized as an indication for SirT2 inhibition, which confirmed its active engagement. SirT2 controls autophagy and has been connected to AD by removing misfolded proteins and aggregating proteins such as tau and amyloid-beta (Aβ) (Silva et al., 2017). In addition to this SirT2 has been linked to Aβ, tau phosphorylation, and autophagic influx in AD (Esteves et al., 2019). Further, Silva et al. also identified that SirT2 inhibition may lead to elimination of toxic oligomers formed from Aβ (Silva et al., 2017). Some of these processes are illustrated in Figure 5.5.

SirT2 is primarily located in the cytoplasm but has the ability to penetrate into the nucleus to interact with histone subtype H4 at Lys16 residue during the process of mitosis, thereby influencing chromatin condensation. It's noteworthy, SirT2 targets a variety of biomolecules, including α-tubulin, and mitochondrial malfunction may contribute to AD pathogenesis (Vaquero et al., 2006). The correlation between the two is further supported by the discovery that SirT2 levels are increased while tubulin AD cell acetylation at Lys40 is decreased. Taylor et al. proposed that SirT2 deacetylase inhibition decreases total cholesterol in primary striatal neurons by stimulating cytoplasmic retention of sterol regulatory element binding protein-2, which subsequently causes downregulation of cholesterol production genes (Taylor et al., 2011).

The genetic investigations through SirT2 SNP models also indicated its association with human AD risk. In the same connection, APOE ε4-negative individuals in Europe are susceptible to encounter AD. Genetic polymorphism in case of SirT2 is another important concern and has recently been found to be associated with AD in a European population (Wei et al., 2014). The functional consequences of these polymorphisms are still being explored, but preliminary evidence suggests that certain variants may reduce the neuroprotective effects of SirT2 inhibition, making personalized therapeutic approaches necessary. Understanding these genetic variations can aid in the development of more targeted treatment strategies and improve patient stratification in clinical trials. According to an intriguing study on genotyping conducted by Cacabelos and his team, SirT2 may have little effect on the pathophysiology of AD, but its interactions with other genes, such as APOE and CYP2D6, may have an impact on cognitive decline and pharmacoepigenetic outcomes. This further supports that SirT2 may contribute to the treatment of AD in the future (Cacabelos et al., 2019).

Before we move on to the next section, it is crucial to note that SirT2 has been proved to be related to neurodegeneration using the technique of whole-brain radiotherapy (WBRT). Interestingly, all three disorders, Huntington's, Parkinson's, and Alzheimer's, have been shown to be acutely affected by brain radiation (Shukla et al., 2015). Sirtuin inhibition models for Parkinson's (Chen et al., 2015) and Huntington's (Chopra et al., 2012) disorders, in particular, demonstrated neuroprotective effects, so we may fairly conclude that this effect will benefit management of AD as well.

5.3 DEVELOPMENT OF SirT2 INHIBITORS

5.3.1 Drug Discovery Perspective

As discussed in previous sections, scientists have been exploring the potential of a selective SirT2 as a therapeutic target for AD, as it may benefit the condition. While SirT2 inhibitors are not yet available in the market for Alzheimer's, continued research is likely to lead to advancements in this area. However, several molecules are under preclinical and clinical investigation. As seen in Figure 5.2,

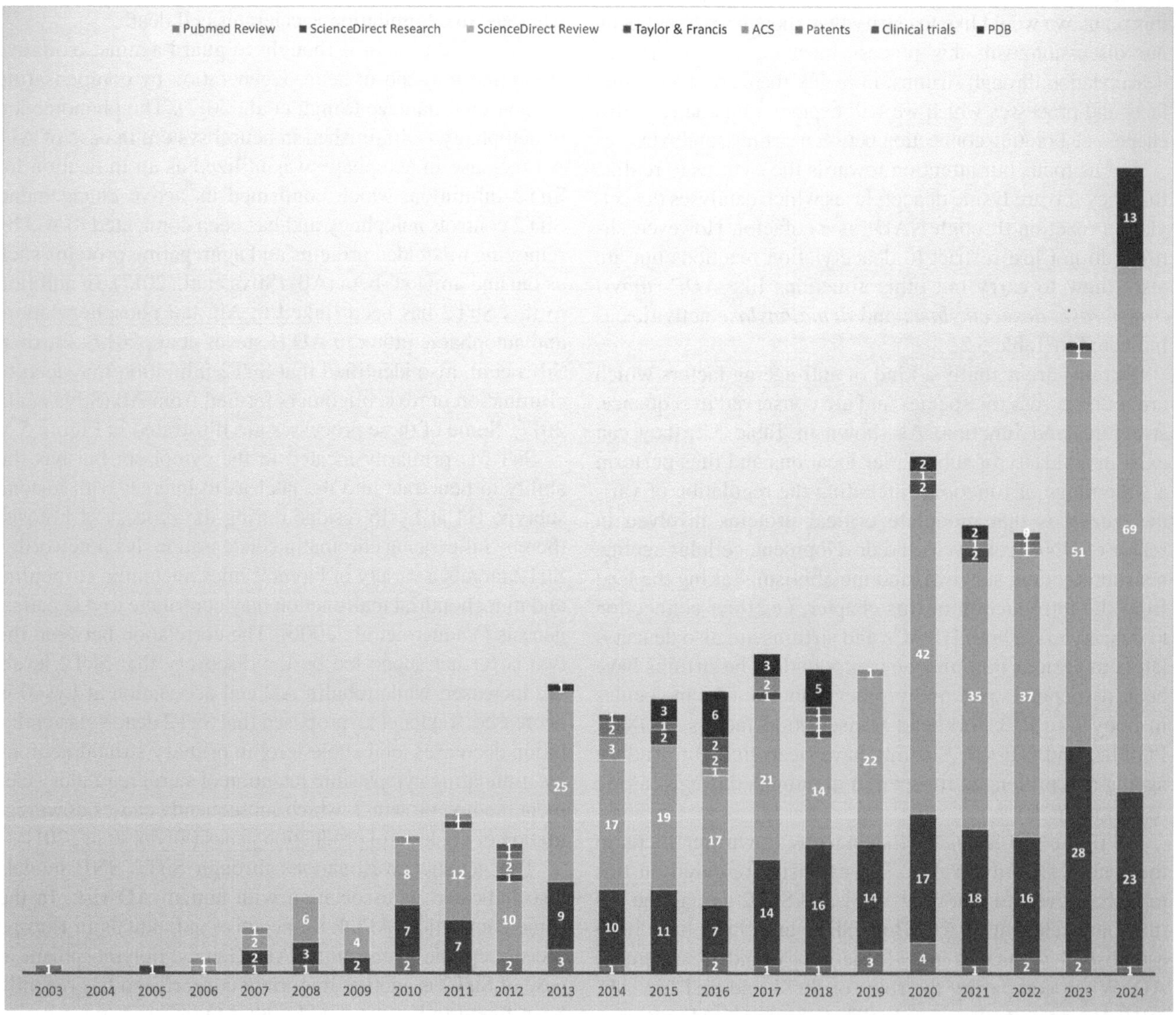

FIGURE 5.2 Status of research on SirT2 inhibitors as searched on PubMed, ScienceDirect, ACS platform, Taylor & Francis for research and review articles, Google Patents for patent filed, SirT2 structures from PDB, and https://clinicaltrials.gov/ for clinical trials in Nov 2024.

the interest of scientists has been consistently growing in this field of research as evident from the incremental number of publications, patents, clinical trials, and structural information in repositories. With this pace we can hope to see medicines from this category available in the market sooner.

The development of SirT2 inhibitors for Alzheimer's stems from their capacity to regulate critical cellular processes, including autophagy, inflammation, and neuronal protection. Early studies identified compounds like AGK2 as promising neuroprotective agents. Over the last decade, researchers have focused on optimizing these inhibitors to enhance their selectivity and efficacy. This has initiated design and testing of various chemotypes and leads in preclinical studies, demonstrating their potential to mitigate Alzheimer's-related pathology. Recent advances in computational drug discovery, such as virtual screening and structure-based design, have facilitated the identification of more selective and drug-like SirT2 inhibitors. These approaches continue to refine their therapeutic potential by improving target specificity and pharmacokinetic properties.

For developing inhibitors for any known target, it will be required to understand its structure and function. As mentioned at the beginning of this section and depicted in Figure 5.2, three-dimensional structures of various SiT2-inhibitor complexes have been deposited in protein data bank (PDB) over last few years. These PDB repositories store the 3D coordinates of many such complexes, which have been majorly studied using X-ray crystallography. The human SirT2 is enzyme comprising 389 amino acid. The SirT2 consists of six parallel β-strands (β1-3 and β7-9) creating a β-sheet, bordered by six α-helices (α1, 7, 8, and 10–12) arranged in a typical β-α-β configuration (Figure 5.3a). The catalytic core is made up of two globular domains: the Rossmann fold, or big domain, and a smaller one containing a zinc-binding module. The Rossmann-fold domain is found in a variety of NAD(H)/NADP(H) binding proteins and plays a role in their interaction. The junction of these two domains generates a large cleft between the α1-helix, which houses the enzyme's catalytic residues and serves as binding sites for the substrate and co-substrate, N-acylated lysine-containing peptide, and NAD+. SirT2 is an unusual histone deacetylase with an active site bearing a wide hydrophobic pocket to accept long-chain fatty acyl groups, and it catalyses the removal of long-chain fatty acyl groups (Teng et al., 2015). This is the same hydrophobic region where N-acylated lysine residues from acylated proteins bind, and it is also known as the Ac-K pocket. This pocket has the ability to interact with a variety of substrates, which accounts for its broad spectrum of function. This typical scenario makes SirT2 a promising target for medicinal research. Several research groups have generated peptidomimetics and tiny molecule inhibitors. A few researchers, such as Suzuki et al., used hydrophobic contacts and investigated water-bridging hydrogen-bond interactions with Phe96, Gln167, Asp95, and His187 to achieve SirT2 selective inhibition, following up on our previous discussion of SirT1 and SirT2 selectivity for AD control. All three pockets had been utilized to design the inhibitors of SirT2, i.e., substrate-binding site (i.e., the "selectivity pocket"), the NAD+-binding site, and allosteric site. Figure 5.3b shows the binding pocket for a selective molecule and residues and ligand within it.

Table 5.3 reflects some examples of the application of different principles of medicinal (Kudo et al., 2018) chemistry explored for the design of potential leads and hits for the inhibition of SirT2. Despite continual efforts a more focused approach is required to make the safe medicines reach the patient in need.

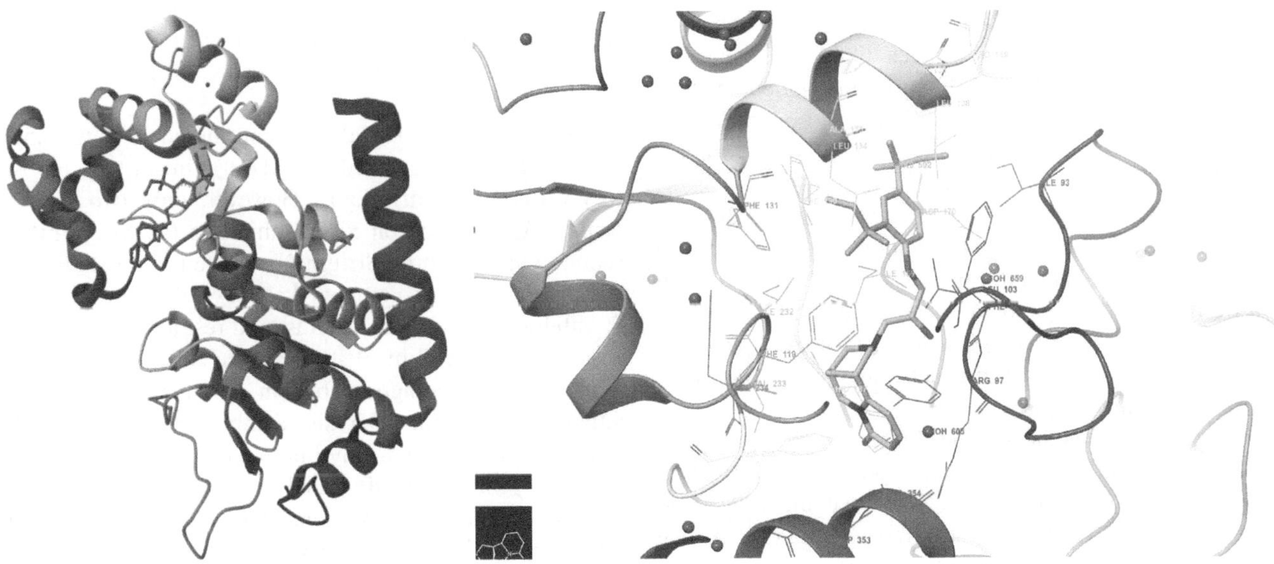

FIGURE 5.3 (a) The 3D structure (PDB code 5Y0Z) of SirT2 with a specific inhibitor in the pocket and (b) active site of the SirT2 (Kudo et al., 2018).

TABLE 5.3
Various Approaches Exploited for Discovery of Inhibitors of SirT2

Approaches	Brief
Competitive inhibition	Novel drug-like small molecule SirReal2 binding to the "selectivity pocket" which otherwise accommodates the long-chain fatty acyl groups of substrates (Rumpf et al., 2015) also exhibited local induced fit effect.
Non-competitive/allosteric inhibition	No example found for AD but as an anticancer or antiviral agent, e.g., FLS-359 and AGK2 (Shenk et al., 2024).
Mechanism-based inhibition	2-Anilinobenzamide (Mellini et al., 2017) formed π–π and H–π interactions and hydrogen bonds with water molecules.
Bio-isosteric replacement	Modification of the benzopyrone core structure resulted in other bicyclic scaffolds, such as quinolin-4(1H)-ones, benzothiazine-1,1-dioxides, benzothiadiazine-1,1-dioxides, and saccharin (Seifert et al., 2020).
Mutual drug (prodrug approach)	SirT2 and HDAC6 inhibitors were developed to inhibit tubulin deacetylation in tandem. Using biochemical in vitro assays and cell-based approaches for target engagement, molecules like Mz325 were designed (Sinatra et al., 2023).
Peptides and peptidomimetics	Design of non-cyclic peptides like Peptide 1: YGGNNVKRRTNYYC, Peptide 2: YMGEWVKRRTNYYC, and Peptide 3: YGGNGVKRRTNYYC (Alkhatabi et al., 2024).
Structure-based drug design (molecular docking)	SBVS work leveraged the abundance of X-ray structures, including SIRT2 (Scarano et al., 2023; Zhou et al., 2018).
Pharmacophore mapping	A ligand-based pharmacophore containing one hydrogen bond acceptor and three aromatic rings from a set of 31 SirReal2 analogues (Eren et al., 2019).
QSAR	Verified QSAR analysis was paired with structure-based pharmacophore modelling to identify structurally distinct SirT2 inhibitors in the natural product database (Khanfar & Alqtaishat, 2021).

5.3.2 Known SirT2 Inhibitors

As illustrated in Figure 5.2, much efforts had been put into the development of SirT2 inhibitors since the turn of the millennium. It's noteworthy that SirT2 inhibitors have already been under the focus of research studies because of their further applications, which include, among other things, anticancer, anti-infective, antiparkinsonian, and Huntington's disease (HD) properties. We will only discuss inhibitors that have been shown to be useful in AD, though. At large it can be classified as potential hits from research endeavours, leads, and useful compounds in further research (Tables 5.4 and 5.5) and those which have undergone or at preclinical studies (Table 5.6). Further potential hits can be classified as those designed, synthesized, biologically evaluated and finally those which have been deposited at protein databank (Table 5.4) as complex with SirT2 enzyme.

As far as PDB structures are concerned, some of them were specifically designed to target anticancer and antiviral activities. However, the information gathered can be fruitful for designing the drugs for centrally located SirT2 particularly for treating AD. The molecules range from peptides, peptidomimetics, natural products, and small molecule inhibitors in this category.

Nowadays, few of the designed SirT2 inhibitors are available commercially (*e.g.*, thiazoles like SirReal2 (Rumpf et al., 2015), furans like AGK-2 (Outeiro et al., 2007), and tenovin-6). However, they are still lead compounds only due to various reasons such as lack of selectivity towards its isozyme, lack of potency (revealed by limited ability to inhibit demyristoylase activity), and/or poor solubility. Classes of compounds like 1,2,4-oxadiazoles (Moniot et al., 2017), nicotinamide (Avalos et al., 2005), sirtinol (Grozinger et al., 2001), combinol (Heltweg et al., 2006), 3-pyridylbenzamides (Sakai et al., 2015), 3-(benzylthio)benzamide (Khanfar et al., 2015), tetrahydroindazole, and thienopyrimidines (Sundriyal et al., 2017) are some of the leads as inhibitors of SirT2. For 1,2,4-oxadiazole derivatives, the uncompetitive inhibition was shown to be working in the presence of both the peptide substrate and NAD+. However, it must be noted that most of the molecules mentioned here are active in micromolar range. However, molecules like nicotinamide, 2-pyridyl-benzamides, 3-(benzylthio)benzamide, tetrahydroindazole, and thienopyrimidines are very promising as their activities are below 10 μM (Sundriyal et al., 2017).

As far as clinical trials are considered to date, 3,563 studies had been reported for AD at clinical trial repository https://clinicaltrials.gov. However, only two clinical trials were reported on SirT2 inhibitors for AD. While a search with keyword "Sirtuin 2" resulted in 46 studies, of which 19 have been completed successfully, 3 terminated, 3 withdrawn, 8 unknown status, and rest are either in process. Though the clinical studies were conducted for acute kidney injury, ageing/metabolism, ammonia metabolism, amyotrophic lateral sclerosis, asthma/inflammation, atherosclerosis, circadian rhythm, cancer, heart problems, diabetes, psoriasis, COPD, and influenza, they shall pave the way for further development of SirT2 inhibitors for AD. Some of the compounds which can be mentioned here Ro-318220 for Immunosuppressants (Roche) Cambinol as oncolytic (Fred Hutchinson Cancer Research Center)

TABLE 5.4
The Inhibitors of SirT2 Deposited in Protein Data Bank Repository

PDB ID	Molecule Name	Structure	Reference
4L3O	Macrocyclic peptide S2iL5		Yamagata et al. (2014)
4RMG	SirReal2	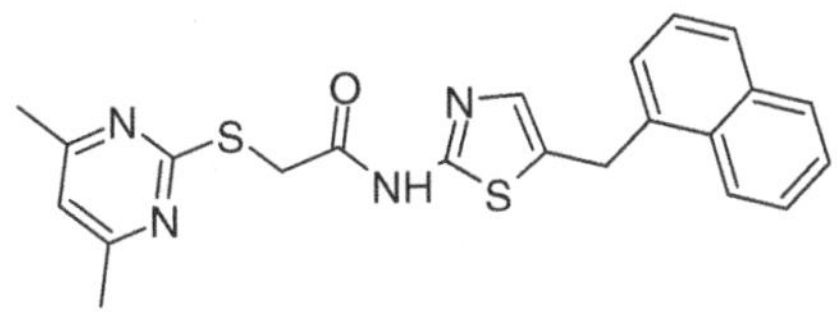	Rumpf et al. (2015)
4RMJ	Nicotinamide	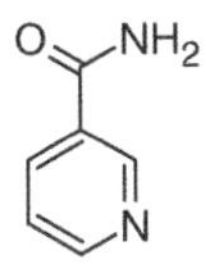	
4RMH	Human SirT2 in complex with SirReal2 and Ac-Lys-H3 peptide	–	
4RMI	Human SirT2 in complex with SirReal1 and Ac-Lys-OTC peptide	–	
4R8M	Human SIRT2 crystal structure in complex with BHJH-TM1	–	
5D7Q	CHIC35	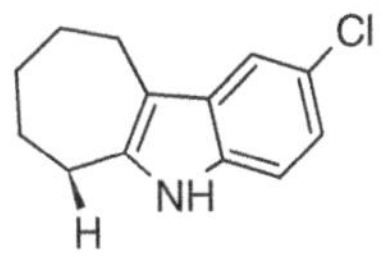	Rumpf et al. (2015)
5D7P	EX-243	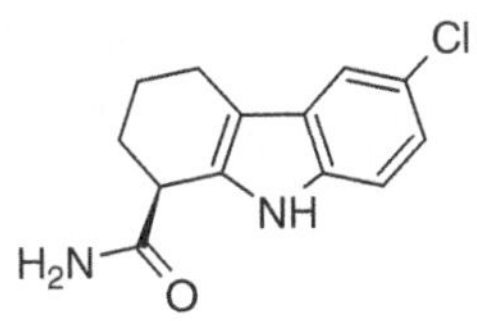	
5DY4	Brominated second-generation SirReal inhibitor	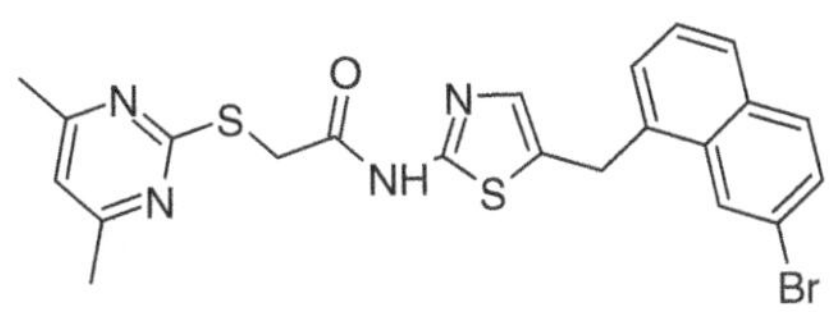	Schiedel et al. (2016)
5DY5	Crystal structure of human SirT2 in complex with a SirReal probe fragment	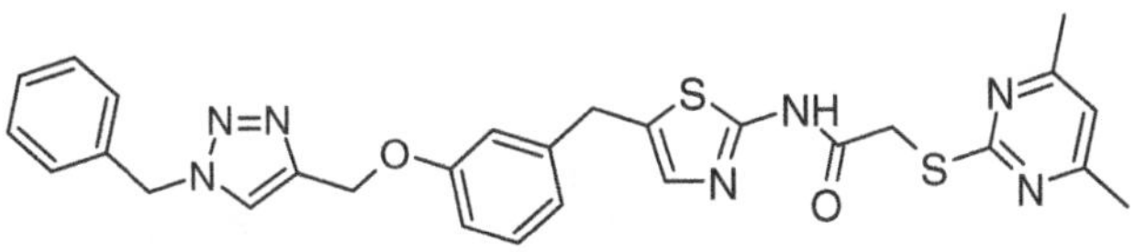	Schiedel et al. (2016)
5FYQ	SirT2 in complex with a 13-mer trifluoroacetylated Ran peptide	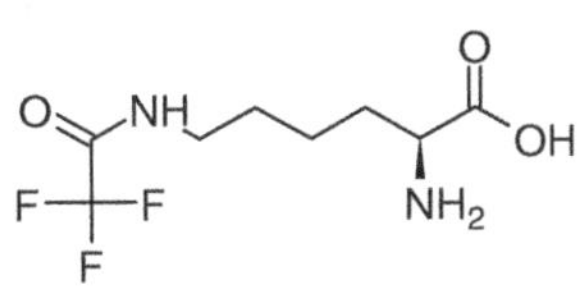	Knyphausen et al. (2016)
5MAT	Selective thienopyrimidinone-based inhibitor	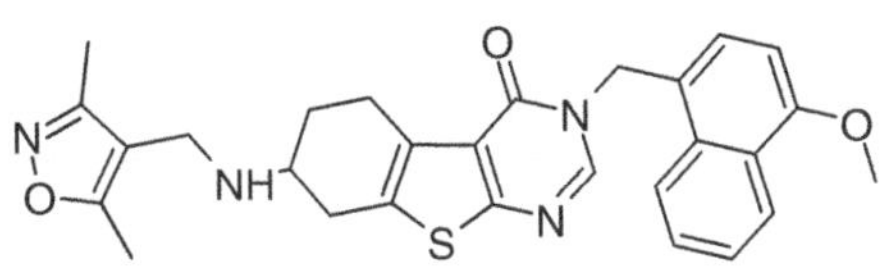	Sundriyal et al. (2017)

(Continued)

TABLE 5.4 ***(Continued)***
The Inhibitors of SirT2 Deposited in Protein Data Bank Repository

PDB ID	Molecule Name	Structure	Reference
5MAR	1,2,4-Oxadiazole derivative	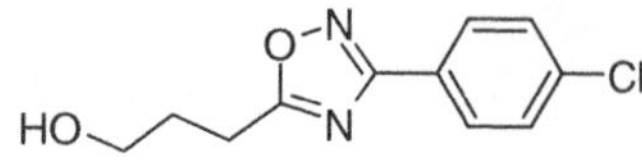	Moniot et al. (2017)
5Y5N	Selective benzamide-based inhibitor (anticancer)	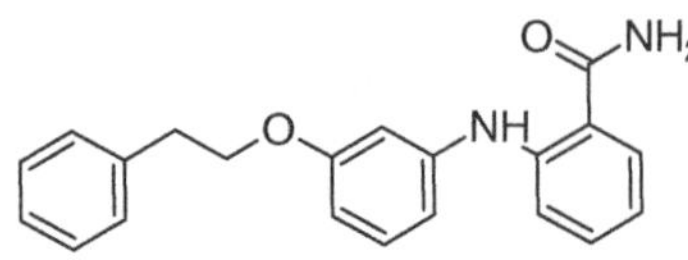	Mellini et al. (2017)
5Y0Z	NPD11033 (anticancer)	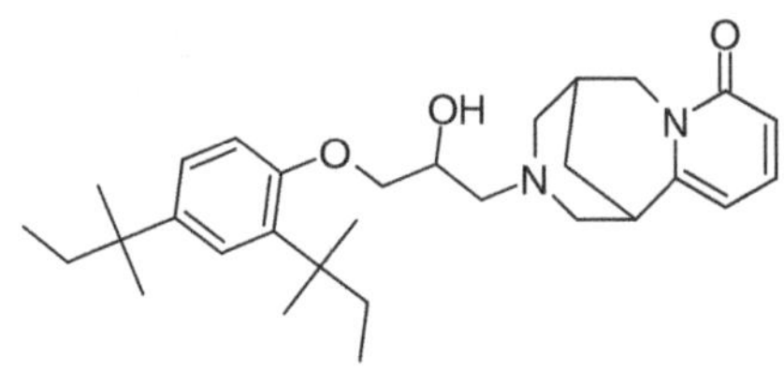	Kudo et al. (2018)
5YQN	SirT2 in complex with selective inhibitor L55 (anticancer)	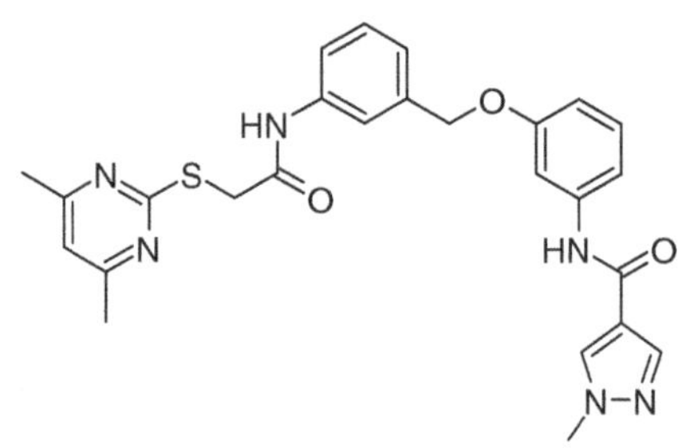	Yang et al. (2018)
5YQO	SirT2 in complex with selective inhibitor L5C (anticancer)	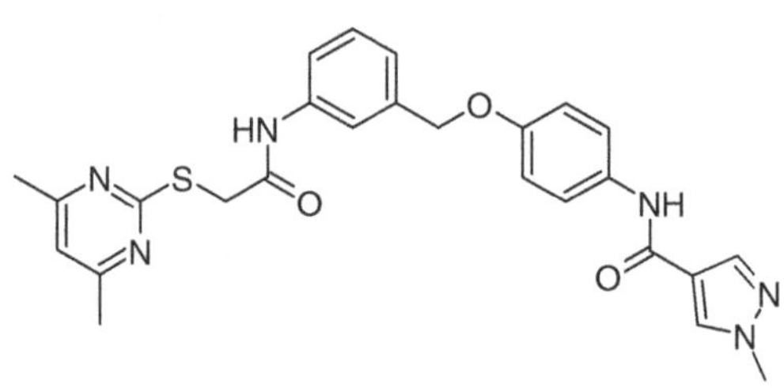	
5YQL	Selective inhibitor A2I (anticancer)	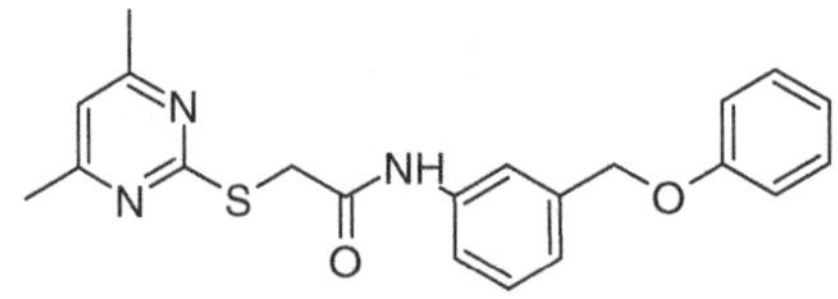	
5YQM	Selective inhibitor A29 (anticancer)	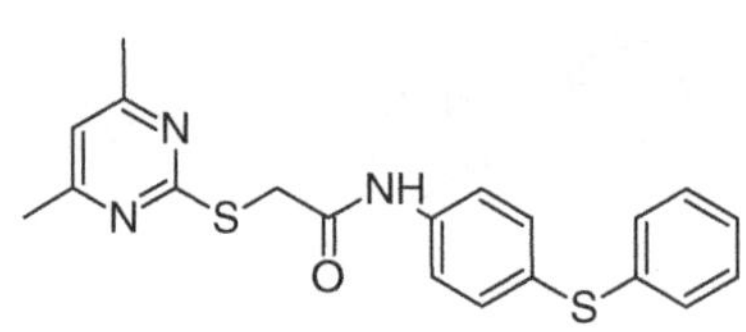	
6QCN	Quercetin	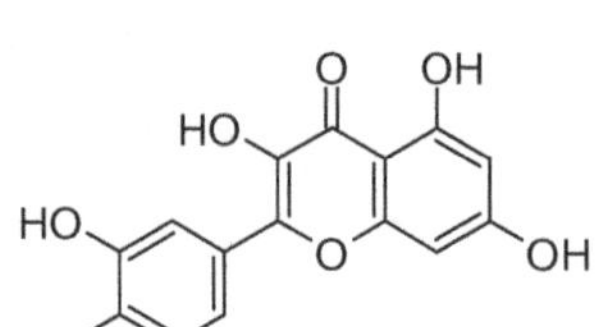	You et al. (2019)

(Continued)

TABLE 5.4 ***(Continued)***
The Inhibitors of SirT2 Deposited in Protein Data Bank Repository

PDB ID	Molecule Name	Structure	Reference
6NR0	Covalent mechanism-based inhibitor Glucose-TM-1beta and 1′-SH ADP-ribose	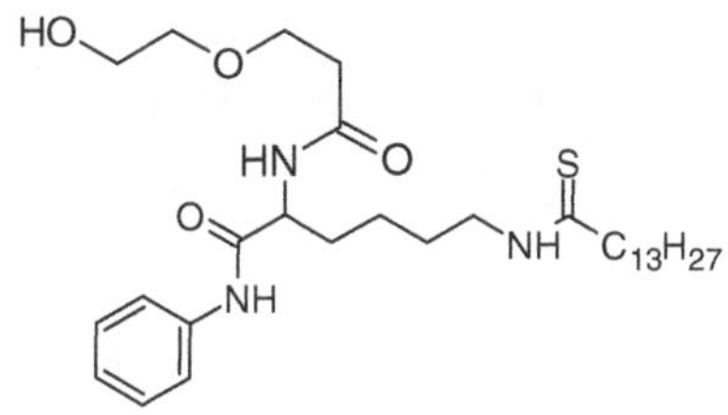	Hong et al. (2019)
6L66	Sirtuin 2 protein with H3K18 myristoylated peptide	PRO-ARG-LYS-GLN-LEU-ALA	Sinatra et al. (2023)
6L65	Sirtuin 2 protein with H3K18 myristoylated peptide	PRO-ARG-LYS-GLN-LEU	–
7BOS	Myristoyl thiourea inhibitor, No.13	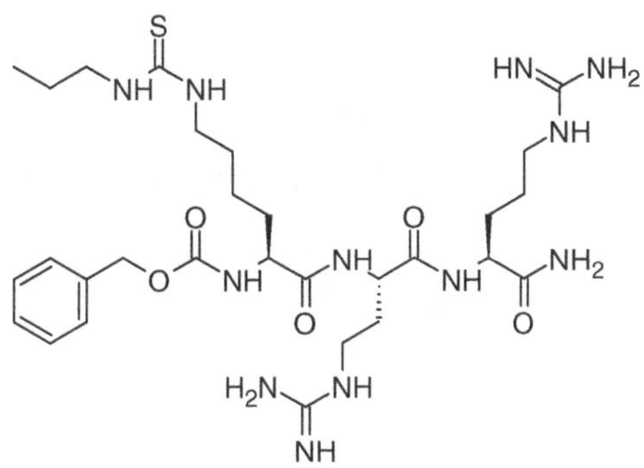	Nielsen et al. (2021)
7T1D	Molecule 359 (antiviral)	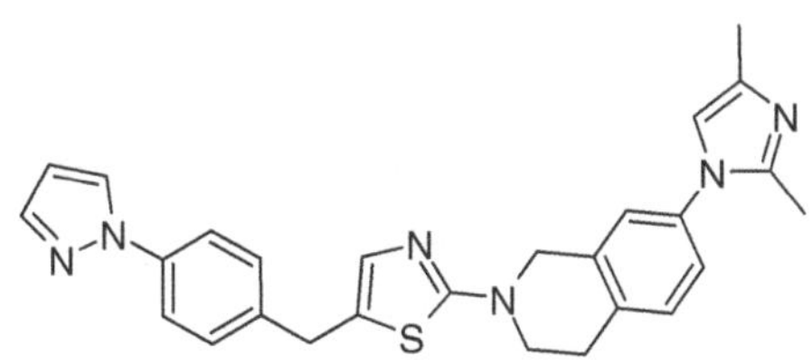	Roche et al. (2023)
8OWZ	Triazole-based SirReal	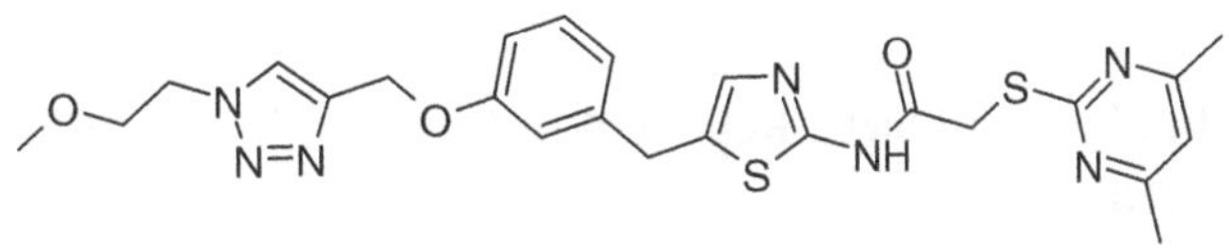	Sinatra et al. (2023)
8TGP	FAM-PEG4-H4K16(myristoyl) peptide	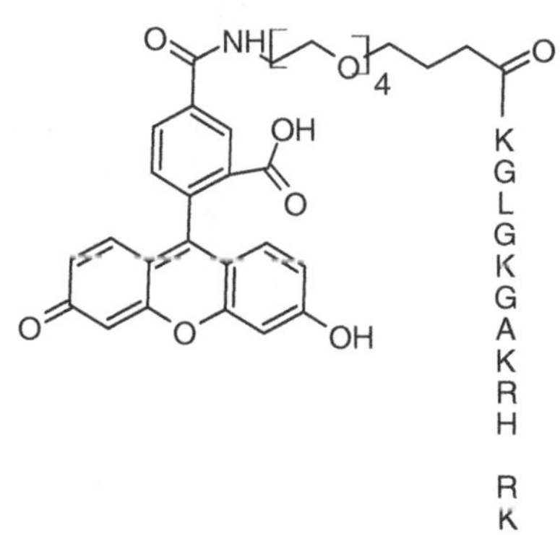	Yang et al. (2023)

TABLE 5.5
The Inhibitors of SirT2 Reported in Literature of Medicinal Chemistry and Pharmacology

Molecule Name	Structure	Reference
SirReal2	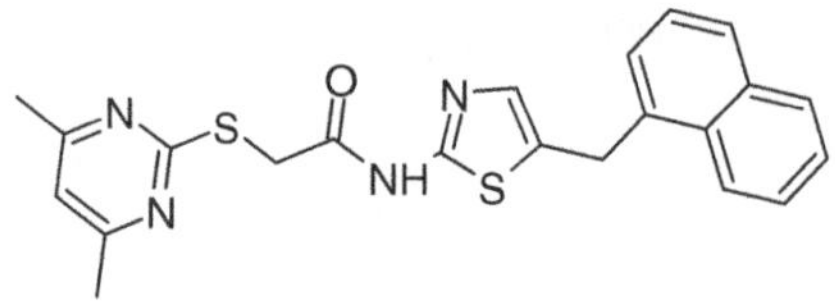	Sinatra et al. (2023)
AGK-2	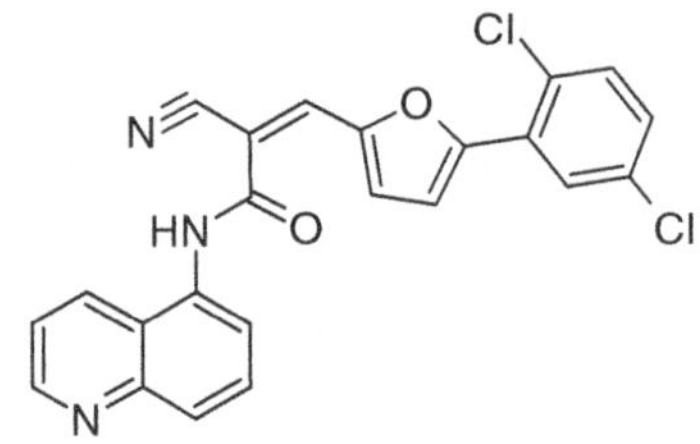	Outeiro et al. (2007)
Tenovin-6	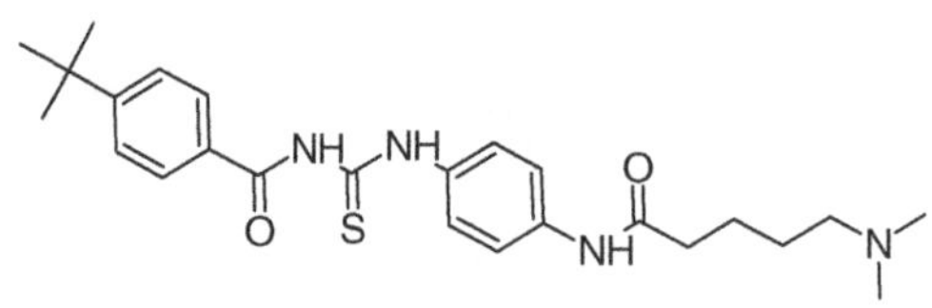	Pirrie et al. (2012)
1,2,4-oxadiazoles	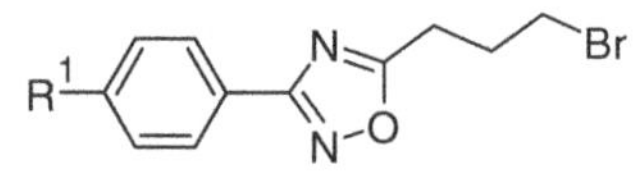	Moniot et al. (2017)
Thienopyrimidines	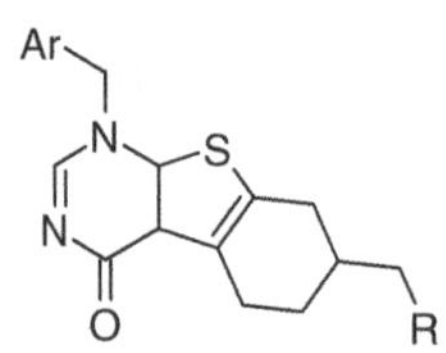	Sundriyal et al. (2017)

TABLE 5.6
Some of the Inhibitors of SirT2 Reported to Have Undergone Clinical Studies (Zhou et al. 2018)

Molecule Name	Structure	Reference
Ro-318220	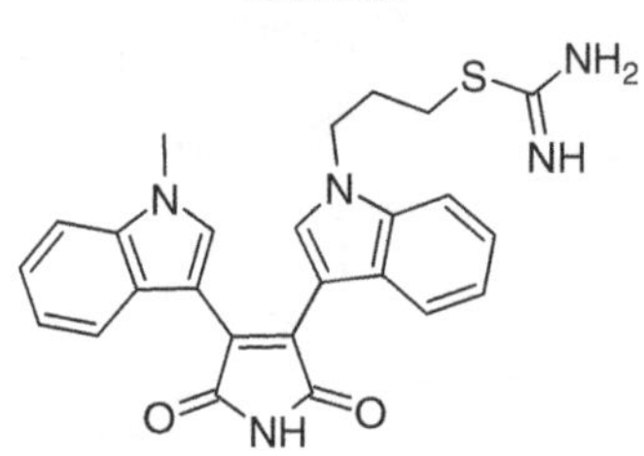	Trapp et al. (2006)
Toxoflavin	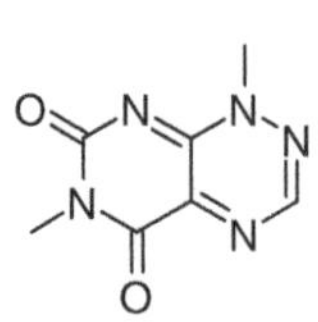	Choi et al. (2013)

(Continued)

TABLE 5.6 ***(Continued)***
Some of the Inhibitors of SirT2 Reported to Have Undergone Clinical Studies (Zhou et al. 2018)

Molecule Name	Structure	Reference
Salermide	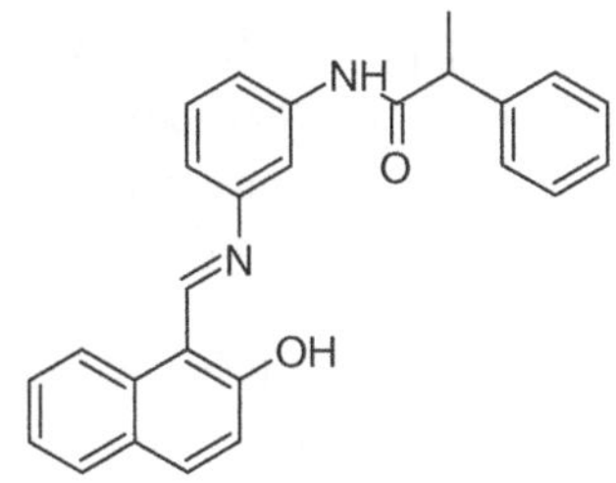	Lara et al. (2009)
AK-7	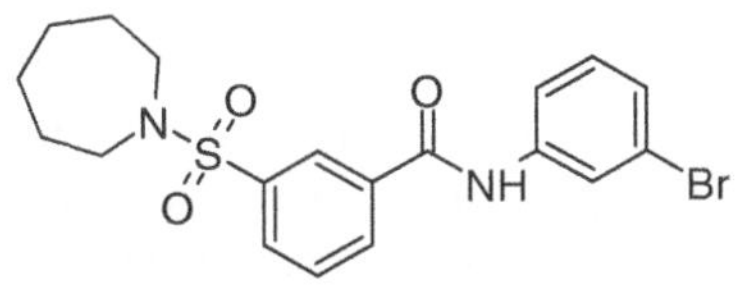	Chopra et al. (2012)
Tenovin-1	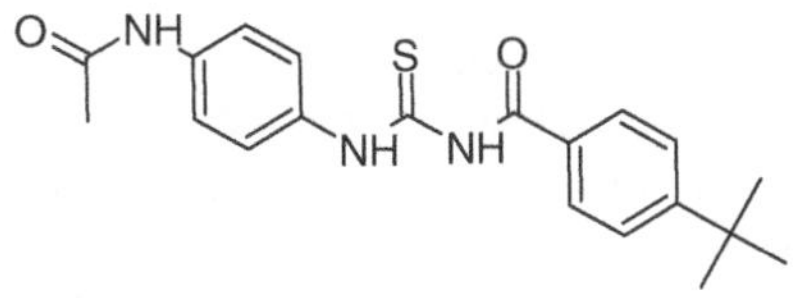	Marx et al. (2018)
781090	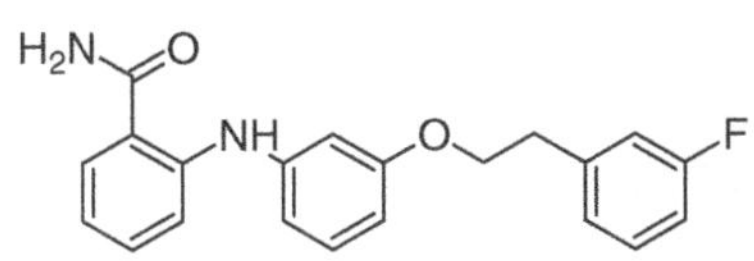	Suzuki et al. (2012)
BPDZ-711	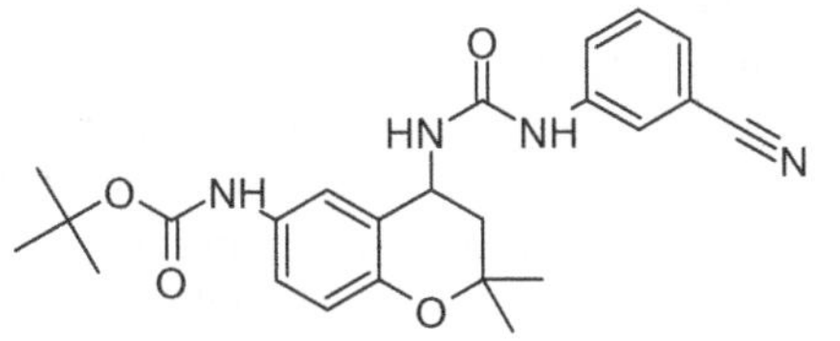	Schnekenburger et al. (2025)

AGK-2 as antiparkinsonian agent (General Hospital Corp) Toxoflavin as anticancer (Novartis), Tenovin-6 as anticancer (University of Dundee; University of St. Andrews) Salermide as anticancer (Harvard College), AK-7 for Huntington's and PD (University of London King's College) PEP-1-SIRT2 for inflammation, Tenovin-1 (Marx et al., 2018) and BPDZ-711 as anticancer, and 781090 for management of neurodegenerative diseases and antidepressants (Zhou et al., 2018). However, further clinical evidences and studies will be required for the drugs working on sites other than central nervous system (CNS), while drugs working on CNS like antiparkinsonism drugs can be directly adopted as therapy for AD in a briefer time.

Over the years, trends in drug discovery and development indicate a continuous decline in the number of small molecule chemical entities being discovered. In contrast, the proportion of drugs originating from biotechnological approaches has been steadily increasing. This shift reflects the growing emphasis on innovative therapies that leverage advances in molecular biology and genetic engineering. Given the multifaceted and often irreversible nature of diseases like Alzheimer's, which are rooted in both genetic predispositions and CNS dysfunctions, therapeutic strategies targeting genetic mechanisms hold immense promise. Gene therapies, RNA-based treatments, and biologics that modulate disease pathways at the genetic or molecular level are likely to dominate the field due to their specificity, potential to address underlying causes, and ability to achieve long-lasting effects. This paradigm shift underscores the importance of integrating genetic and biotechnological insights into drug development for CNS diseases, offering a path towards transformative treatments that go beyond symptom management to address the root causes of these disorders.

5.3.3 Mechanisms of Action, Selectivity, and ADMET Aspects

5.3.3.1 Catalytic Mechanism of SirT2

SirT2 is quite similar to SirT1 and SirT3 in terms of sequence homology. It contains a smaller Zn^{2+} binding domain, a bigger Rossmann folding domain, and a catalytic core domain with substrate and NAD^+ binding sites connected by a loop (Finnin et al., 2001). As such, all sirtuins are NAD^+-dependent class III histone deacetylases that can remove a range of acyl groups from acylated lysine residues, including myristoyl, glutaryl, and succinyl (Zhou et al., 2018). As previously stated, the active pocket is large enough to tolerate several substrates with bulkier and hydrophobic replacements. Two primary domains make up SirT2's catalytic core: a short domain that contains structural zinc ions and a larger domain that is a Rossmann fold NAD^+ binding pocket made up of six β-strands (β1-3 and β7-9) and six α-helixes (α1, α7, α8, and α10–α12). Four crossings connect these two domains, creating a wide groove. The groove is intersected by a pocket at the active site that is encircled by several hydrophobic residues, suggesting that it is a class-specific binding site that may identify particular substrates. The big groove contains sirtuin-conserved residues and the NAD+ binding site. This groove is thought to represent the catalytic core since mutations in this area have been demonstrated to cause a lack of deacetylation of catalytic activity.

Additionally, the N-terminal extension of SirT2 contains 19 residues from an amphipathic α-helix that do not have any contacts with the protein. This suggests that the N-terminal extension is not necessary for SirT2's catalytic activity, but these amphipathic helixes may play a significant role in transcriptional regulation through protein-protein interactions. The SN2 reaction-type mechanism catalyses the process.

As shown in Figure 5.4, activated anomeric carbon of ribose sugar is an active carbon owing to be in proximity of nitrogen of the NAD^+. The carbonyl amido/ester group of the acetylated protein is activated by histidine near the active site and makes it vulnerable to being attacked by anomeric carbon of the NAD^+. It leads to formation of

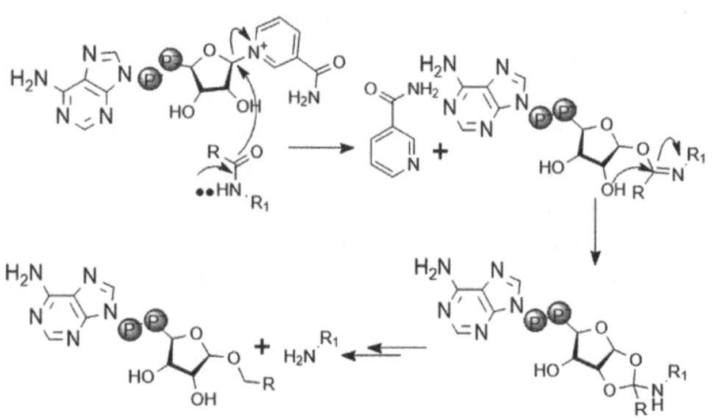

FIGURE 5.4 Mechanism of catalysis for deacetylation reaction by SirT2 enzyme.

O-alkyl amidino adduct of the target protein followed by intramolecular nucleophilic assault from 2-hydroxy group, resulting in 1-,2-cyclic intermediate which in a few steps converts to deacetylated protein (Wang et al., 2017; Zhou et al., 2018).

5.3.3.2 Mechanism of Action

SirT2 affects a wide range of physiological and pathological consequences and is essential for many biological systems. In this chapter, we concentrate on SirT2 inhibition's functions and sequence of events alleviating symptoms of AD. Curiously, conditions including PD, stroke, and brain damage also seem to be related to Alzheimer's symptoms and therapies, especially when it comes to SirT2's role.

5.3.3.2.1 General Deacetylating Actions

The role of histones in various genetic processes was addressed in the chapter's introduction. To be more specific, research has demonstrated that histone acetylation H4 plays a role in the pathogenesis of AD, and HDAC inhibitors could impact some essential gene expression via controlling histone acetylation in relation to AD.

In the context of AD, SirT2 has been implicated in several critical cellular processes that may influence disease progression. The SirT2 and its mono-ADP-ribosyl transferase activity play a significant role in regulating microtubule acetylation, a process essential for maintaining neuronal integrity. SirT2 affects the stability of microtubules by reducing the degree of α-tubulin acetylation (Green et al., 2008). The neurodegenerative alterations seen in AD may be exacerbated by dysregulation of this mechanism. Additionally, SirT2's involvement in myelination within the CNS may impact neuronal communication, further influencing disease pathology.

Though sirtuins have been found to be deeply involved in the pathology of AD, we must begin our discussion with some widely known events and their relevance to SirT2 specifically. The simultaneous presence of Aβ and tau remains a hallmark of AD pathology (Scheltens et al., 2016). β-secretase and γ-secretase sequentially cleave amyloid precursor protein (APP) to produce Aβ peptides, specifically Aβ40 and Aβ42. Among these, Aβ42 is highly aggregation-prone and forms toxic plaques, contributing to neuronal damage and inflammation. The microtubule-associated protein tau, on the other hand, is hyperphosphorylated in AD, which results in the development of NFT that impair intracellular transport and lead to neurodegeneration. Inactivation of SirT2 has been found to improve levels of stabilizing tau proteins in healthy neurons (Biella et al., 2016). Lately, Esteves et al. have also shown a reduction in phosphorylation of tau proteins (Esteves et al., 2019).

SirT2 regulates APP processing and tau phosphorylation. SirT2 inhibition has shown promising results in treating certain pathogenic characteristics. For instance, in H4-SW cells, the SirT2 inhibitor AGK2 suppressed the concentrations of Aβ40, Aβ42, and soluble AβPPβ (sAβPPβ), a β-secretase cleavage product. In tests on transgenic mice (APP23 and 3xTg versions), it was shown that SirT2 inhibitor AK-7 enhanced amounts of soluble AβPPα (sAβPPα), a non-amyloidogenic cleavage product produced by α-secretase and presumed to have neuroprotective characteristics. These findings suggest that SirT2 regulates the function or expression of α- and β-secretases, influencing the balance between amyloidogenic and non-amyloidogenic pathways and potentially mitigating AD pathology.

5.3.3.2.2 *Cell cycle Regulation and Autophagy*

Emerging data suggests that SirT2 is involved in chromatin dynamics, genomic stability, and cell cycle regulation, all of which may contribute to AD pathogenesis. As noted in Table 5.2, SirT2 is predominantly found in the cytosol, but it may migrate to the nucleus during the G2/M phase of the cell cycle. In this phase, SirT2 mediates deacetylation of histone H4 lysine 16 (H4K16ac) (Vaquero et al., 2006), facilitating H4K20 methylation and chromatin compaction, essential processes for mitotic chromosomal condensation (Inoue et al., 2007; Serrano et al., 2013). SirT2 is also associated with FOXO1 (forkhead box O1), a transcription factor that plays a crucial role in oxidative stress response, metabolism, and autophagy. Oxidative stress and starvation may cause dissociation of FOXO-1-SirT2 complex and thus cause hyper-acylation of FOXO1. In turn acetylated FOXO1 interacts with Atg7 (autophagy-related protein 7), an essential E1-like enzyme involved in autophagosome formation, required for LC3 lipidation and autophagic flux. This interaction has been speculated to cause induction of autophagy (Zhao et al., 2010). Consequently, enhancing autophagy through SirT2 inhibition may reduce tau and Aβ accumulation.

A common feature of Alzheimer's pathology is that in response to DNA damage SirT2 deacetylates histone H3 lysine 56 (H3K56), influencing DNA repair mechanisms critical for maintaining genomic stability (Vempati et al., 2010). Furthermore, SirT2 suppresses the transcriptional co-activator p300, which also have acetyltransferase activity and affects gene expression related to neuronal survival and function (Black et al., 2008).

SirT2 also interacts with mitotic components such as the centrosome, spindle, and midbody to ensure normal cell division. However, overexpression prolongs the cell cycle, which may contribute to cellular senescence or malfunction. SirT2's complex involvement in chromatin remodelling, DNA repair, and cell cycle regulation emphasize its importance in AD and the potential for therapy through targeting SirT2 at least to slow disease development.

As mentioned earlier, SirT2 affects microtubule stability by lowering α-tubulin acetylation levels (North & Verdin, 2004). SirT2-mediated deacetylation disrupts the microtubule network, impairing critical cellular processes such as autophagy. Microtubule instability caused by SirT2 activity interferes with the transport of autophagic vacuoles and their subsequent fusion with lysosomes, thereby compromising cellular homeostasis (Cappelletti et al., 2021).

Notably, inhibition of SirT2 has been shown to restore microtubule stability, reorganizing the microtubule network and facilitating axonal transport. This restoration also enhances autophagic flux by enabling the fusion of autophagic vacuoles with lysosomes. The observed increase in autophagy upon SirT2 inhibition points towards its efficacy in conditions where microtubule dynamics and autophagy are disrupted (Li et al., 2013).

Recent research reveals that SirT2 has a role in α-synuclein aggregation. Inhibiting SirT2 has been demonstrated to diminish α-synuclein toxicity and aggregation in PD models. Similar principles may apply to AD. α-synuclein, a synaptic protein involved in vesicle trafficking and neurotransmitter release, forms fibrils linked to neurodegenerative illnesses such as Parkinson's and Alzheimer's. In AD, these fibrils co-localize with tau tangles and Aβ plaques, contributing to neurotoxicity, mitochondrial dysfunction, oxidative stress, and neuroinflammation. They impede synaptic function, disrupt proteostasis, and spread prion-like particles, exacerbating disease development (de Oliveira et al., 2017).

The regulation of microtubule-related proteins, as well as the toxicity of misfolded proteins, causes apoptosis (Chen et al., 2021). Silva et al. found that suppressing SirT2 with AK-1 or gene deletion improved microtubule stability and autophagy, reducing harmful Aβ oligomers and increasing cell survival (Silva et al., 2017).

Furthermore, the involvement of the miRNA-SirT2 pathway in the process of autophagy has been investigated. The treatment with miR-212-5p, which targets SirT2, has been shown to prevent neuronal loss, which is a key aspect of Alzheimer's pathology. This method involves the increase of nuclear acetylated p53, which is consistent with data that p53 is an essential substrate for SirT2 for deacetylation. Reduced p53 levels in cytoplasm after SirT2 inhibition further augment autophagy, which is consistent with studies in PD models (Sun et al., 2018; van Leeuwen et al., 2013).

5.3.3.2.3 *Metabolic Actions*

There are findings that highlight SirT2 as a critical modulator of inflammation and oxidative stress, further underscoring its therapeutic potential in neurodegenerative diseases like Alzheimer's.

SirT2 suppresses inflammatory responses by deacetylation action at the p65 subunit site of NF κB (Gomes et al., 2015), leading to the inhibition of its activity. Given the well-established link between chronic inflammation and Alzheimer's pathology, this regulatory mechanism may influence the neuroinflammatory processes associated with the disease.

Additionally, SirT2 is involved in mitigating oxidative stress, a major factor to cause damage to neurons in AD. By deacetylating and activating glucose-6-phosphate dehydrogenase (G6PD), SirT2 triggers the pentose phosphate pathway, providing cytosolic NADPH necessary to counteract oxidative damage. This antioxidative function could play a neuroprotective role, reducing damage by oxidative species to neurons and potentially slowing AD progression (Wang et al., 2014).

5.3.3.2.4 Neurodegeneration

The outcome and consequences of SirT2 inhibition in HD and PD highlight the potential neuroprotective effects of SirT2 inhibition in AD, which often shares pathology and biochemistry with other neurodegenerative disorders. According to studies in cell and invertebrate models of PD and HD, SirT2 inhibition, unlike other sirtuin family members, may provide considerable neuroprotection (Luthi-Carter et al., 2010; Outeiro et al., 2007).

Furthermore, in vivo evidence demonstrates that SirT2 inhibition protects against neuronal loss induced by MPTP, a neurotoxin commonly used to model PD (Chen et al., 2015). These findings, coupled with the connection of SirT2 with inflammation, oxidative stress, and chromatin regulation, suggest that targeting SirT2 might also have therapeutic potential in AD by mitigating neuronal loss and enhancing neuroprotection.

Also, SirT2 deacetylates the transcription factor Foxo3a, leading to the activation of the pro-apoptotic protein Bim, which subsequently triggers cellular death. This pathway contributes to neuronal loss, a hallmark of Alzheimer's pathology. Inhibition of SirT2 could alleviate this process by preventing Foxo3a deacetylation and reducing Bim activation, thereby promoting neuronal survival. This mechanism, combined with its roles in inflammation suppression, oxidative stress regulation, and chromatin dynamics, highlights the multifaceted therapeutic utility of targeting SirT2 in AD to mitigate neurodegeneration and enhance neuronal resilience.

The relationship between SirT2 and hypoxia-inducible factor 1-alpha (HIF-1α) has noteworthy implications for AD, particularly in the context of neuronal survival under the conditions of hypoxia. HIF-1α is a critical transcription factor that orchestrates cellular responses to hypoxia by activating genes involved in angiogenesis, glucose metabolism, and oxidative stress defence (Semenza, 2000). SirT2 modulates HIF-1α activity through its deacetylase function, directly impacting its stability and transcriptional activity. This can happen under hypoxic conditions, influencing the ability of HIF-1α to mediate adaptive responses. Dysregulation of this interaction may impair hypoxic signalling, exacerbating neuronal damage and contributing to the pathogenesis of AD. Conversely, inactivation of SirT2 has been shown to augment HIF-1α transcriptional activity, leading to upregulation of glycolytic enzymes and antioxidant defences, thereby reducing oxidative stress and promoting neuronal resilience (Kaitsuka et al., 2020).

These findings underscore the dual roles of SirT2 in regulating HIF-1α and its downstream pathways, highlighting the therapeutic effectivity of SirT2 inactivation in mitigating hypoxia-induced damage and enhancing neuroprotection in AD.

The key sequence of events leading to the onset of AD, along with the potential of SirT2 inhibitors to mitigate or reverse these pathological processes, is summarized in Figure 5.5.

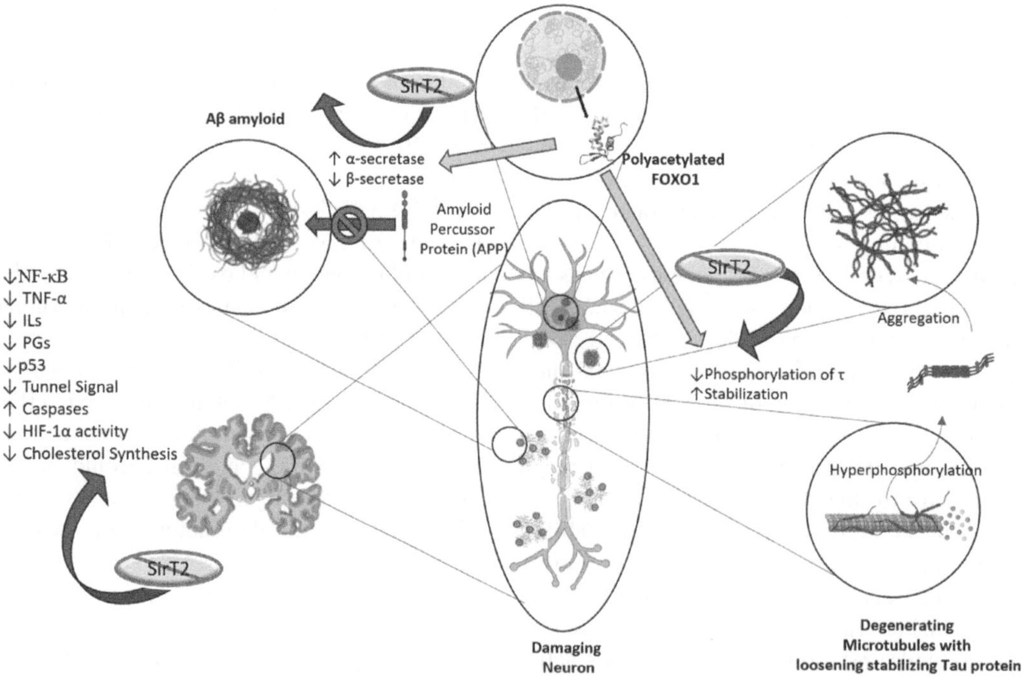

FIGURE 5.5 Major factors involved in the onset and pathology of AD and the mechanism of working of SirT2 inhibitors in anti-Alzheimer therapy.

5.3.3.3 Pharmacokinetic Aspects

The pharmacokinetic properties of SirT2 inhibitors would be critical in determining their therapeutic efficacy and safety profiles. Key considerations include the following:

a. *Absorption and Distribution*: Effective SirT2 inhibitors must be absorbed in good amounts and then cross the blood-brain barrier (BBB) to exert their effects within the CNS. Compounds like AK-7 have demonstrated the ability to penetrate the BBB, leading to improved cognitive outcomes in AD mouse models (Taylor et al., 2011).
b. *Selectivity and Off-Target Effects*: Enhanced selectivity towards SirT2 over other isoforms of sirtuin specially SirT1, SirT3, etc. is essential to minimize off-target effects. Non-selective inhibition may lead to unintended consequences, given the diverse roles of sirtuins in cellular physiology.
c. *Metabolic Stability and Clearance*: The metabolic stability of SirT2 inhibitors influences their half-life and dosing frequency. Rapid metabolism may necessitate frequent dosing, potentially increasing the risk of side effects. SirT2 inhibitors are required to reach the CNS and hence ought to be hydrophobic. As the natural tendency of the human body is to metabolize hydrophobic substances, optimization would be needed to improve the **half-life and bioavailability** of inhibitors.
d. *Peripheral vs. Central Effects*: Systemic inhibition of SirT2 can have peripheral effects, including alterations in inflammation and metabolism. For instance, while central SirT2 inhibition may be beneficial in AD, peripheral inhibition could exacerbate systemic inflammation (Sola-Sevilla et al., 2023).

Optimizing these pharmacokinetic parameters is crucial for developing SirT2 inhibitors with favourable therapeutic profiles in AD. Balancing effective CNS activity with minimal peripheral side effects will enhance the efficacy of SirT2 inhibitors as viable treatments for AD.

5.4 THERAPEUTIC POTENTIAL OF SirT2 INHIBITORS IN AD

Although a number of sirtuin isoforms seem to offer promise as therapeutic targets, only SirT1 modulators have made it into clinical trials thus far. Numerous strong SirT2 inhibitors and SirT1 activators that have successfully completed the initial clinical trials and shown safety and efficacy have resulted from early drug research efforts centred on SirT1 and SirT2. Studies have demonstrated that SirT2 inhibitors (e.g., AK-1, AGK2) enhance neuronal survival in AD models. Preclinical studies show SirT2 inhibition reverses memory deficits in AD models. Combining SirT2 inhibition with other neuroprotective strategies (e.g., HDAC inhibitors, autophagy modulators) may provide enhanced benefits. In addition several drugs with known safety profiles have been identified to possess SirT2-inhibitory activity or modulate SirT2-related pathways. SirT2 inhibitors for AD therapy offer a cost-effective and time-saving approach. Along with AD, repurposing is also another avenue for the application of SirT2 inhibitors like AK-7 (for HD) and AGK2 (for cancer).

5.5 CHALLENGES AND FUTURE DIRECTIONS

5.5.1 Limitations of SirT2 Inhibitors and Their Development

Long-term safety concerns may be involved with SirT2 inhibitors with reference to metabolism, immunity, and ageing; long-term inhibition might have unintended effects. As a matter of fact, the functioning of brain is very complex and slight changes can lead to vast effects. SirT2 is no exception to this. The effects of SirT2 inhibition have been discussed all over this chapter to benefit the treatment of AD. However, one study suggested a number of untoward effects due to decrease or non-functionality of SirT2. Microglia with reduced SirT2 can express more inflammatory markers and free radicals, which can lead to neurotoxicity. When SirT2 is not present, NF-κB may become hyper-acylated. Additionally, it has been demonstrated that changes in microglia's SirT2 levels can cause toll-like receptor (TLR) activation, which in turn causes CNS inflammation (Pais et al., 2013). It was discussed in one earlier section that a few clinical trials have been conducted on SirT2 inhibitors for the management of AD. Still most of the reported studies are preclinical in nature and require translation into human trials.

5.5.2 Translational Challenges from Bench to Bedside

Translating SirT2 inhibitors from laboratory research to clinical application in AD presents several challenges.

5.5.2.1 Complex Role of SirT2 in AD Pathogenesis

Though SirT2 inhibition has shown effectiveness in reducing Aβ production and tau hyperphosphorylation, its functions are multifaceted and need further studies. For instance, SirT2 influences autophagy, microtubule dynamics, and neuroinflammation, which are all implicated in AD progression. Balancing these effects is crucial to avoid unintended consequences (Sola-Sevilla & Puerta, 2024).

5.5.2.2 Genetic Variability and Patient Response

Genetic differences among individuals can alter the efficiency of SirT2 inhibitors. Variations in the SirT2 gene (e.g., rs10410544) have been associated with differing susceptibilities to AD and may influence treatment outcomes. Personalized approaches considering these genetic factors are essential for effective therapy (Cacabelos et al., 2019).

5.5.2.3 Potential for Adverse Effects

SirT2 is involved in a number of vital physiological processes, including metabolism and cell cycle regulation. Inhibiting SirT2 could disrupt these processes, leading to side effects. Comprehensive preclinical studies are necessary to assess the safety profile of SirT2 inhibitors.

5.5.2.4 Blood-Brain Barrier (BBB) Penetration

Effective AD treatments must cross the BBB to reach neural targets. Developing SirT2 inhibitors with adequate BBB permeability is challenging but essential for clinical efficacy (Sola-Sevilla et al., 2023).

5.5.2.5 Microglial Function Modulation

Recent research suggests that SirT2 activity in microglia, the brain's resident immune cells, can affect neuroinflammation. Targeted suppression of microglia may exacerbate AD progression, highlighting the need for cell-type-specific treatment methods.

Addressing these challenges requires a multidisciplinary approach, integrating insights from molecular biology, pharmacology, and clinical sciences to develop safe and effective SirT2-targeted therapies for AD.

5.5.3 Future Perspectives

To advance SirT2 inhibitors for AD management, future research should focus on developing highly selective and CNS-penetrant molecules. Also long-term toxicity and efficacy studies in animal models before clinical trials will be need of hour. With cutting-edge and evolving technology, exploring biomarkers for patient stratification to identify subgroups may benefit most from SirT2-targeted therapies. Biotechnological advances and a paradigm shift are expected in future where potential of gene therapy approaches will be utilized to modulate SirT2 expression in a controlled manner. Development and deployment of anti-amyloid monoclonal antibodies is an example in this direction. Also, at this point we can speculate that future research could explore biopharmaceutical formulations, such as fusion proteins or antibody-drug conjugates (ADCs), for selective targeting of SirT2-related pathways. AI and machine learning are and will continue contributing to drug discovery, and specifically for designing SirT2 inhibitor, a machine learning-based tool ***SIRT2i_Predictor*** is already available (Djokovic et al., 2023). As mentioned earlier BBB penetration is an important obstacle, and hence, employment of NDDS formulation techniques, prodrug methodologies, and target drug delivery approaches can open a new paradigm in the delivery of SirT2 inhibitors to the brain.

5.5.4 Combination Therapies

Given that AD is a complex disorder of the CNS and given its multifactorial nature, combination therapies which target multiple pathways simultaneously are a promising approach for management of this disease. This approach aims to enhance treatment efficacy, delay disease progression, and improve patient outcomes. Monotherapies targeting single pathological aspects of AD, such as acetylcholinesterase inhibitors (donepezil, rivastigmine) or anti-Aβ monoclonal antibodies (aducanumab, lecanemab), have shown limited efficacy in modifying disease progression. A combinatorial approach can address multiple aspects of AD pathogenesis, providing synergistic neuroprotective effects (Kabir et al., 2020). For AD both combinations with disease-modifying therapies and combinations with symptomatic therapies are already in practice and both can eventually be adopted with SirT2 inhibitors (Cummings et al., 2019).

Also it has been elaborated earlier that SirT2 inhibition has been shown to be effective in reducing tau protein and improving neuronal resilience. When combined with autophagy enhancers like rapamycin or metformin, SirT2 inhibitors may facilitate protein clearance and neuroprotection. Additionally, combining SirT2 inhibitors with anti-inflammatory agents like resveratrol or sirtuin-activating compounds (STACs) could further enhance neuronal survival and reduce neurodegeneration (Donmez & Outeiro, 2013). Since SirT2 inhibition has been linked to improved synaptic function, its combination with acetylcholinesterase inhibitors such as donepezil could provide cognitive benefits while slowing disease progression.

Despite the promise and various benefits of combination therapy, several hurdles have to be addressed, including drug interactions, patient stratification, BBB penetration, and medication tailoring. Synergistic effects of combination remedies must be optimized to avoid adverse interactions. Also it has been known to have altered pharmacokinetics and BBB penetration of combination drugs; hence, identifying effective combinations which cross the BBB with adequate bioavailability will be a crucial step. As the ideology of combination or synergistic approach rests on the fact that different targets in the same biochemical pathways or pathological process must be identified, it will be essential to identify common biomarkers to tailor combination therapies to patient subgroups. In addition it will be even more crucial for personalized medicine. To counter all such difficulties, scientists nowadays are working on the concept of multi-targeting with the same new chemical entities (Turgutalp & Kizil, 2024).

Combination therapies hold great promise for treating AD by targeting multiple pathological mechanisms simultaneously. With further advancements in drug discovery and biomarker identification, multi-targeted approaches may lead to more effective and personalized treatment strategies referred to as *Precision Medicine Initiative* for AD patients (Arafah et al., 2023). The next step in this direction will be to focus on large-scale clinical studies testing the safety and efficacy of combination medications while integrating precision medicine methodologies (Cummings et al., 2024). Regarding the combination and repurposing of medications in AD with immediate clinical relevance, ***DeepDrug*** is a ground-breaking AI and big data-guided mechanism for finding novel drug combinations and repurposing medications for AD and other neurodegenerative conditions (Li et al., 2025).

5.6 CONCLUSION

SirT2 inhibitors represent a promising avenue for AD therapy, targeting Aβ, tau pathology, oxidative stress, and neuroinflammation. With continued optimization in selectivity, BBB penetration, and clinical testing, they hold potential as disease-modifying treatments for AD.

REFERENCES

Alkhatabi, H. A., Naemi, F. M. A., Alsolami, R., & Alatyb, H. N. (2024). Computational design and optimization of peptide inhibitors for SIRT2. *Pharmaceuticals*, *17*(9). https://doi.org/10.3390/ph17091120

Arafah, A., Khatoon, S., Rasool, I., Khan, A., Rather, M. A., Abujabal, K. A., Faqih, Y. A. H., Rashid, H., Rashid, S. M., Bilal Ahmad, S., Alexiou, A., & Rehman, M. U. (2023). The future of precision medicine in the cure of Alzheimer's disease. *Biomedicines*, *11*(2). https://doi.org/10.3390/biomedicines11020335

Avalos, J. L., Bever, K. M., & Wolberger, C. (2005). Mechanism of sirtuin inhibition by nicotinamide: Altering the NAD+ cosubstrate specificity of a Sir2 enzyme. *Molecular Cell*, *17*(6), 855–868. https://doi.org/10.1016/j.molcel.2005.02.022

Biella, G., Fusco, F., Nardo, E., Bernocchi, O., Colombo, A., Lichtenthaler, S. F., Forloni, G., & Albani, D. (2016). Sirtuin 2 inhibition improves cognitive performance and acts on amyloid-beta protein precursor processing in two Alzheimer's disease mouse models. *Journal of Alzheimer's Disease*, *53*(3), 1193–1207. https://doi.org/10.3233/JAD-151135

Black, J. C., Mosley, A., Kitada, T., Washburn, M., & Carey, M. (2008). The SIRT2 deacetylase regulates autoacetylation of p300. *Molecular Cell*, *32*(3), 449–455. https://doi.org/10.1016/j.molcel.2008.09.018

Cacabelos, R., Carril, J. C., Cacabelos, N., Kazantsev, A. G., Vostrov, A. V., Corzo, L., Cacabelos, P., & Goldgaber, D. (2019). Sirtuins in Alzheimer's disease: SIRT2-related GenoPhenotypes and implications for PharmacoEpiGenetics. *International Journal of Molecular Sciences*, *20*(5). https://doi.org/10.3390/ijms20051249

Cappelletti, G., Calogero, A. M., & Rolando, C. (2021). Microtubule acetylation: A reading key to neural physiology and degeneration. *Neuroscience Letters*, *755*, 135900. https://doi.org/10.1016/j.neulet.2021.135900

Chen, X., Lu, W., & Wu, D. (2021). Sirtuin 2 (SIRT2): Confusing roles in the pathophysiology of neurological disorders. *Frontiers in Neuroscience*, *15*, 614107. https://doi.org/10.3389/fnins.2021.614107

Chen, X., Wales, P., Quinti, L., Zuo, F., Moniot, S., Herisson, F., Rauf, N. A., Wang, H., Silverman, R. B., Ayata, C., Maxwell, M. M., Steegborn, C., Schwarzschild, M. A., Outeiro, T. F., & Kazantsev, A. G. (2015). The sirtuin-2 inhibitor AK7 is neuroprotective in models of Parkinson's disease but not amyotrophic lateral sclerosis and cerebral ischemia. *PLoS ONE*, *10*(1), e0116919. https://doi.org/10.1371/journal.pone.0116919

Choi, G., Lee, J., Ji, J. Y., Woo, J., Kang, N. S., Cho, S. Y., Kim, H. R., Ha, J. D., & Han, S. Y. (2013). Discovery of a potent small molecule SIRT1/2 inhibitor with anticancer effects. *International Journal of Oncology*, *43*(4), 1205–1211. https://doi.org/10.3892/ijo.2013.2035

Chopra, V., Quinti, L., Kim, J., Vollor, L., Narayanan, K. L., Edgerly, C., Cipicchio, P. M., Lauver, M. A., Choi, S. H., Silverman, R. B., Ferrante, R. J., Hersch, S., & Kazantsev, A. G. (2012). The sirtuin 2 inhibitor AK-7 is neuroprotective in Huntington's disease mouse models. *Cell Reports*, *2*(6), 1492–1497. https://doi.org/10.1016/j.celrep.2012.11.001

Cummings, J. L., Osse, A. M. L., Kinney, J. W., Cammann, D., & Chen, J. (2024). Alzheimer's disease: Combination therapies and clinical trials for combination therapy development. *CNS Drugs*, *38*(8), 613–624. https://doi.org/10.1007/s40263-024-01103-1

Cummings, J. L., Tong, G., & Ballard, C. (2019). Treatment combinations for Alzheimer's disease: Current and future pharmacotherapy options. *Journal of Alzheimer's Disease*, *67*(3), 779–794. https://doi.org/10.3233/JAD-180766

de Oliveira, R. M., Vicente Miranda, H., Francelle, L., Pinho, R., Szego, E. M., Martinho, R., Munari, F., Lazaro, D. F., Moniot, S., Guerreiro, P., Fonseca-Ornelas, L., Marijanovic, Z., Antas, P., Gerhardt, E., Enguita, F. J., Fauvet, B., Penque, D., Pais, T. F., Tong, Q., Becker, S., Kugler, S., Lashuel, H. A., Steegborn, C., Zweckstetter, M., & Outeiro, T. F. (2017). The mechanism of sirtuin 2-mediated exacerbation of alpha-synuclein toxicity in models of Parkinson disease. *PLoS Biology*, *15*(3), e2000374. https://doi.org/10.1371/journal.pbio.2000374

Djokovic, N., Rahnasto-Rilla, M., Lougiakis, N., Lahtela-Kakkonen, M., & Nikolic, K. (2023). SIRT2i_predictor: A machine learning-based tool to facilitate the discovery of novel SIRT2 inhibitors. *Pharmaceuticals (Basel)*, *16*(1). https://doi.org/10.3390/ph16010127

Donmez, G., & Outeiro, T. F. (2013). SIRT1 and SIRT2: Emerging targets in neurodegeneration. *EMBO Molecular Medicine*, *5*(3), 344–352. https://doi.org/10.1002/emmm.201302451

Eren, G., Bruno, A., Guntekin-Ergun, S., Cetin-Atalay, R., Ozgencil, F., Ozkan, Y., Gozelle, M., Kaya, S. G., & Costantino, G. (2019). Pharmacophore modeling and virtual screening studies to identify novel selective SIRT2 inhibitors. *Journal of Molecular Graphics and Modelling*, *89*, 60–73.

Esteves, A. R., Palma, A. M., Gomes, R., Santos, D., Silva, D. F., & Cardoso, S. M. (2019). Acetylation as a major determinant to microtubule-dependent autophagy: Relevance to Alzheimer's and Parkinson disease pathology. *Biochimica et Biophysica Acta. Molecular Basis of Disease*, *1865*(8), 2008–2023. https://doi.org/10.1016/j.bbadis.2018.11.014

Fernando, K. K. M., & Wijayasinghe, Y. S. (2021). Sirtuins as potential therapeutic targets for mitigating neuroinflammation associated with Alzheimer's disease. *Frontiers in Cellular Neuroscience*, *15*, 746631. https://doi.org/10.3389/fncel.2021.746631

Finnin, M. S., Donigian, J. R., & Pavletich, N. P. (2001). Structure of the histone deacetylase SIRT2. *Nature Structural Biology*, *8*(7), 621–625. https://doi.org/10.1038/89668

Gomes, P., Fleming Outeiro, T., & Cavadas, C. (2015). Emerging role of sirtuin 2 in the regulation of mammalian metabolism. *Trends in Pharmacological Sciences*, *36*(11), 756–768. https://doi.org/10.1016/j.tips.2015.08.001

Green, K. N., Steffan, J. S., Martinez-Coria, H., Sun, X., Schreiber, S. S., Thompson, L. M., & LaFerla, F. M. (2008). Nicotinamide restores cognition in Alzheimer's disease transgenic mice via a mechanism involving sirtuin inhibition and selective reduction of Thr231-phosphotau. *Journal of Neuroscience*, *28*(45), 11500–11510. https://doi.org/10.1523/jneurosci.3203-08.2008

Grozinger, C. M., Chao, E. D., Blackwell, H. E., Moazed, D., & Schreiber, S. L. (2001). Identification of a class of small molecule inhibitors of the sirtuin family of NAD-dependent deacetylases by phenotypic screening. *Journal of Biological Chemistry*, *276*(42), 38837–38843. https://doi.org/10.1074/jbc.M106779200

Heltweg, B., Gatbonton, T., Schuler, A. D., Posakony, J., Li, H., Goehle, S., Kollipara, R., Depinho, R. A., Gu, Y., Simon, J. A., & Bedalov, A. (2006). Antitumor activity of a small-molecule inhibitor of human silent information regulator 2 enzymes. *Cancer Research*, *66*(8), 4368–4377. https://doi.org/10.1158/0008-5472.CAN-05-3617

Hong, J. Y., Price, I. R., Bai, J. J., & Lin, H. (2019). A glycoconjugated SIRT2 inhibitor with aqueous solubility allows structure-based design of SIRT2 inhibitors. *ACS Chemical Biology*, *14*(8), 1802–1810. https://doi.org/10.1021/acschembio.9b00384

Inoue, T., Hiratsuka, M., Osaki, M., Yamada, H., Kishimoto, I., Yamaguchi, S., Nakano, S., Katoh, M., Ito, H., & Oshimura, M. (2007). SIRT2, a tubulin deacetylase, acts to block the entry to chromosome condensation in response to mitotic stress. *Oncogene*, *26*(7), 945–957. https://doi.org/10.1038/sj.onc.1209857

Johnstone, S. E., & Baylin, S. B. (2010). Stress and the epigenetic landscape: A link to the pathobiology of human diseases? *Nature Reviews Genetics*, *11*(11), 806–812. https://doi.org/10.1038/nrg2881

Kabir, M. T., Uddin, M. S., Mamun, A. A., Jeandet, P., Aleya, L., Mansouri, R. A., Ashraf, G. M., Mathew, B., Bin-Jumah, M. N., & Abdel-Daim, M. M. (2020). Combination drug therapy for the management of Alzheimer's disease. *International Journal of Molecular Sciences*, *21*(9). https://doi.org/10.3390/ijms21093272

Kaitsuka, T., Matsushita, M., & Matsushita, N. (2020). SIRT2 inhibition activates hypoxia-inducible factor 1α signaling and mediates neuronal survival. *Biochemical and Biophysical Research Communications*, *529*(4), 957–962. https://doi.org/10.1016/j.bbrc.2020.06.159

Khanfar, M. A., & Alqtaishat, S. (2021). Discovery of potent natural-product-derived SIRT2 inhibitors using structure-based exploration of SIRT2 pharmacophoric space coupled with QSAR analyses. *Anti-Cancer Agents in Medicinal Chemistry*, *21*(16), 2278–2286.

Khanfar, M. A., Quinti, L., Wang, H., Nobles, J., Kazantsev, A. G., & Silverman, R. B. (2015). Design and evaluation of 3-(benzylthio)benzamide derivatives as potent and selective SIRT2 inhibitors. *ACS Medicinal Chemistry Letters*, *6*(5), 607–611. https://doi.org/10.1021/acsmedchemlett.5b00075

Knyphausen, P., de Boor, S., Kuhlmann, N., Scislowski, L., Extra, A., Baldus, L., Schacherl, M., Baumann, U., Neundorf, I., & Lammers, M. (2016). Insights into lysine deacetylation of natively folded substrate proteins by sirtuins. *Journal of Biological Chemistry*, *291*(28), 14677–14694. https://doi.org/10.1074/jbc.M116.726307

Kudo, N., Ito, A., Arata, M., Nakata, A., & Yoshida, M. (2018). Identification of a novel small molecule that inhibits deacetylase but not defatty-acylase reaction catalysed by SIRT2. *Philosophical Transactions of the Royal Society of London. Series B, Biological Sciences*, *373*(1748). https://doi.org/10.1098/rstb.2017.0070

Lara, E., Mai, A., Calvanese, V., Altucci, L., Lopez-Nieva, P., Martinez-Chantar, M. L., Varela-Rey, M., Rotili, D., Nebbioso, A., Ropero, S., Montoya, G., Oyarzabal, J., Velasco, S., Serrano, M., Witt, M., Villar-Garea, A., Imhof, A., Mato, J. M., Esteller, M., & Fraga, M. F. (2009). Salermide, a sirtuin inhibitor with a strong cancer-specific proapoptotic effect. *Oncogene*, *28*(6), 781–791. https://doi.org/10.1038/onc.2008.436

Li, V. O. K., Han, Y., Kaistha, T., Zhang, Q., Downey, J., Gozes, I., & Lam, J. C. K. (2025). DeepDrug as an expert guided and AI driven drug repurposing methodology for selecting the lead combination of drugs for Alzheimer's disease. *Scientific Reports*, *15*(1), 2093. https://doi.org/10.1038/s41598-025-85947-7

Li, Y., Nie, H., Wu, D., Zhang, J., Wei, X., & Ying, W. (2013). Poly(ADP-ribose) polymerase mediates both cell death and ATP decreases in SIRT2 inhibitor AGK2-treated microglial BV2 cells. *Neuroscience Letters*, *544*, 36–40. https://doi.org/10.1016/j.neulet.2013.03.032

Luthi-Carter, R., Taylor, D. M., Pallos, J., Lambert, E., Amore, A., Parker, A., Moffitt, H., Smith, D. L., Runne, H., Gokce, O., Kuhn, A., Xiang, Z., Maxwell, M. M., Reeves, S. A., Bates, G. P., Neri, C., Thompson, L. M., Marsh, J. L., & Kazantsev, A. G. (2010). SIRT2 inhibition achieves neuroprotection by decreasing sterol biosynthesis. *Proceedings of the National Academy of Sciences of the United States of America*, *107*(17), 7927–7932. https://doi.org/10.1073/pnas.1002924107

Marx, C., Marx-Blumel, L., Lindig, N., Thierbach, R., Hoelzer, D., Becker, S., Wittig, S., Lehmann, R., Slevogt, H., Heinzel, T., Wang, Z. Q., Beck, J. F., & Sonnemann, J. (2018). The sirtuin 1/2 inhibitor tenovin-1 induces a nonlinear apoptosis-inducing factor-dependent cell death in a p53 null Ewing's sarcoma cell line. *Investigational New Drugs*, *36*(3), 396–406. https://doi.org/10.1007/s10637-017-0541-1

Mellini, P., Itoh, Y., Tsumoto, H., Li, Y., Suzuki, M., Tokuda, N., Kakizawa, T., Miura, Y., Takeuchi, J., Lahtela-Kakkonen, M., & Suzuki, T. (2017). Potent mechanism-based sirtuin-2-selective inhibition by an in situ-generated occupant of the substrate-binding site, "selectivity pocket" and NAD+-binding site. *Chemical Science*, *8*(9), 6400–6408. https://doi.org/10.1039/C7SC02738A

Moniot, S., Forgione, M., Lucidi, A., Hailu, G. S., Nebbioso, A., Carafa, V., Baratta, F., Altucci, L., Giacche, N., Passeri, D., Pellicciari, R., Mai, A., Steegborn, C., & Rotili, D. (2017). Development of 1,2,4-oxadiazoles as potent and selective inhibitors of the human deacetylase sirtuin 2: Structure-activity relationship, X-ray crystal structure, and anticancer activity. *Journal of Medicinal Chemistry*, *60*(6), 2344–2360. https://doi.org/10.1021/acs.jmedchem.6b01609

NCBI. (2025). *SIRT2 sirtuin 2 [Homo sapiens (human)]*. National Library of Medicine. https://www.ncbi.nlm.nih.gov/gene?Db=gene&Cmd=ShowDetailView&TermToSearch=22933

Nielsen, A. L., Rajabi, N., Kudo, N., Lundo, K., Moreno-Yruela, C., Baek, M., Fontenas, M., Lucidi, A., Madsen, S., Yoshida, M., & Olsen, C. A. (2021). Mechanism-based inhibitors of SIRT2: Structure-activity relationship, X-ray structures, target engagement, regulation of alpha-tubulin acetylation and inhibition of breast cancer cell migration. *RSC Chemical Biology*, *2*(2), 612–626. https://doi.org/10.1039/d0cb00036a

North, B. J., & Verdin, E. (2004). Sirtuins: Sir2-related NAD-dependent protein deacetylases. *Genome Biology*, *5*(5), 224. https://doi.org/10.1186/gb-2004-5-5-224

Outeiro, T. F., Kontopoulos, E., Altmann, S. M., Kufareva, I., Strathearn, K. E., Amore, A. M., Volk, C. B., Maxwell, M. M., Rochet, J. C., McLean, P. J., Young, A. B., Abagyan, R., Feany, M. B., Hyman, B. T., & Kazantsev, A. G. (2007). Sirtuin 2 inhibitors rescue alpha-synuclein-mediated toxicity in models of Parkinson's disease. *Science*, *317*(5837), 516–519. https://doi.org/10.1126/science.1143780

Pais, T. F., Szego, E. M., Marques, O., Miller-Fleming, L., Antas, P., Guerreiro, P., de Oliveira, R. M., Kasapoglu, B., & Outeiro, T. F. (2013). The NAD-dependent deacetylase sirtuin 2 is a suppressor of microglial activation and brain inflammation. *The EMBO Journal*, *32*(19), 2603–2616. https://doi.org/10.1038/emboj.2013.200

Pirrie, L., McCarthy, A. R., Major, L. L., Morkunaite, V., Zubriene, A., Matulis, D., Lain, S., Lebl, T., & Westwood, N. J. (2012). Discovery and validation of SIRT2 inhibitors based on tenovin-6: Use of a (1)H-NMR method to assess deacetylase activity. *Molecules*, *17*(10), 12206–12224. https://doi.org/10.3390/molecules171012206

Roche, K. L., Remiszewski, S., Todd, M. J., Kulp, J. L. 3rd, Tang, L., Welsh, A. V., Barry, A. P., De, C., Reiley, W. W., Wahl, A., Garcia, J. V., Luftig, M. A., Shenk, T., Tonra, J. R., Murphy, E. A., & Chiang, L. W. (2023). An allosteric inhibitor of sirtuin 2 deacetylase activity exhibits broad-spectrum antiviral activity. *Journal of Clinical Investigation*, *133*(12). https://doi.org/10.1172/JCI158978

Roth, S. Y., Denu, J. M., & Allis, C. D. (2001). Histone acetyltransferases. *Annual Review of Biochemistry*, *70*, 81–120. https://doi.org/10.1146/annurev.biochem.70.1.81

Rumpf, T., Gerhardt, S., Einsle, O., & Jung, M. (2015). Seeding for sirtuins: Microseed matrix seeding to obtain crystals of human Sirt3 and Sirt2 suitable for soaking. *Acta Crystallographica Section F, Structural Biology Communications*, *71*(Pt 12), 1498–1510. https://doi.org/10.1107/S2053230X15019986

Rumpf, T., Schiedel, M., Karaman, B., Roessler, C., North, B. J., Lehotzky, A., Olah, J., Ladwein, K. I., Schmidtkunz, K., Gajer, M., Pannek, M., Steegborn, C., Sinclair, D. A., Gerhardt, S., Ovadi, J., Schutkowski, M., Sippl, W., Einsle, O., & Jung, M. (2015). Selective Sirt2 inhibition by ligand-induced rearrangement of the active site. *Nature Communications*, *6*, 6263. https://doi.org/10.1038/ncomms7263

Sakai, T., Matsumoto, Y., Ishikawa, M., Sugita, K., Hashimoto, Y., Wakai, N., Kitao, A., Morishita, E., Toyoshima, C., Hayashi, T., & Akiyama, T. (2015). Design, synthesis and structure-activity relationship studies of novel sirtuin 2 (SIRT2) inhibitors with a benzamide skeleton. *Bioorganic & Medicinal Chemistry*, *23*(2), 328–339. https://doi.org/10.1016/j.bmc.2014.11.027

Scarano, N., Abbotto, E., Musumeci, F., Salis, A., Brullo, C., Fossa, P., Schenone, S., Bruzzone, S., & Cichero, E. (2023). Virtual screening combined with enzymatic assays to guide the discovery of novel SIRT2 inhibitors. *International Journal of Molecular Sciences*, *24*(11), 9363.

Scheltens, P., Blennow, K., Breteler, M. M. B., de Strooper, B., Frisoni, G. B., Salloway, S., & Van der Flier, W. M. (2016). Alzheimer's disease. *The Lancet*, *388*(10043), 505–517. https://doi.org/10.1016/S0140-6736(15)01124-1

Schiedel, M., Rumpf, T., Karaman, B., Lehotzky, A., Gerhardt, S., Ovadi, J., Sippl, W., Einsle, O., & Jung, M. (2016). Structure-based development of an affinity probe for sirtuin 2. *Angewandte Chemie International Edition in English*, *55*(6), 2252–2256. https://doi.org/10.1002/anie.201509843

Schiedel, M., Rumpf, T., Karaman, B., Lehotzky, A., Olah, J., Gerhardt, S., Ovadi, J., Sippl, W., Einsle, O., & Jung, M. (2016). Aminothiazoles as potent and selective Sirt2 inhibitors: A structure-activity relationship study. *Journal of Medicinal Chemistry*, *59*(4), 1599–1612. https://doi.org/10.1021/acs.jmedchem.5b01517

Schnekenburger, M., Goffin, E., Schoumacher, M., Tumanov, N., Mouithys-Mickalad, A., de Tullio, P., Wouters, J., Lebrun, P., Diederich, M., & Pirotte, B. (2025). Preparation and pharmacological characterization of the enantiomers of the anticancer agent R/S-N-3-cyanophenyl-N′-(6-tert-butoxycarbonylamino-3,4-dihydro-2,2-dimethyl-2H-1-benzopyran-4-yl)urea: Identification of the R-enantiomer as the active eutomer. *European Journal of Medicinal Chemistry Reports*, *13*, 100244. https://doi.org/10.1016/j.ejmcr.2024.100244

Seifert, T., Malo, M., Kokkola, T., Stéen, E. J. L., Meinander, K., Wallén, E. A. A., Jarho, E. M., & Luthman, K. (2020). A scaffold replacement approach towards new sirtuin 2 inhibitors. *Bioorganic & Medicinal Chemistry*, *28*(2), 115231. https://doi.org/10.1016/j.bmc.2019.115231

Semenza, G. L. 2000. HIF-1: Mediator of physiological and pathophysiological responses to hypoxia. *Journal of Applied Physiology (1985)*, *88*(4), 1474–1480. https://doi.org/10.1152/jappl.2000.88.4.1474

Serrano, L., Martinez-Redondo, P., Marazuela-Duque, A., Vazquez, B. N., Dooley, S. J., Voigt, P., Beck, D. B., Kane-Goldsmith, N., Tong, Q., Rabanal, R. M., Fondevila, D., Munoz, P., Kruger, M., Tischfield, J. A., & Vaquero, A. (2013). The tumor suppressor SirT2 regulates cell cycle progression and genome stability by modulating the mitotic deposition of H4K20 methylation. *Genes & Development*, *27*(6), 639–653. https://doi.org/10.1101/gad.211342.112

Shenk, T., Kulp III, J. L., & Chiang, L. W. (2024). Drugs targeting sirtuin 2 exhibit broad-spectrum anti-infective activity. *Pharmaceuticals (Basel)*, *17*(10), 1298. https://doi.org/10.3390/ph17101298

Shukla, S., Shankavaram, U. T., Nguyen, P., Stanley, B. A., & Smart, D. K. (2015). Radiation-induced alteration of the brain proteome: Understanding the role of the sirtuin 2 deacetylase in a murine model. *Journal of Proteome Research*, *14*(10), 4104–4117. https://doi.org/10.1021/acs.jproteome.5b00083

Silva, D. F., Esteves, A. R., Oliveira, C. R., & Cardoso, S. M. (2017). Mitochondrial metabolism power SIRT2-dependent deficient traffic causing Alzheimer's-disease related pathology. *Molecular Neurobiology*, *54*(6), 4021–4040. https://doi.org/10.1007/s12035-016-9951-x

Sinatra, L., Vogelmann, A., Friedrich, F., Tararina, M. A., Neuwirt, E., Colcerasa, A., König, P., Toy, L., Yesiloglu, T. Z., Hilscher, S., Gaitzsch, L., Papenkordt, N., Zhai, S., Zhang, L., Romier, C., Einsle, O., Sippl, W., Schutkowski, M., Gross, O., Bendas, G., Christianson, D. W., Hansen, F. K., Jung, M., & Schiedel, M. (2023). Development of first-in-class dual Sirt2/HDAC6 inhibitors as molecular tools for dual inhibition of tubulin deacetylation. *Journal of Medicinal Chemistry*, *66*(21), 14787–14814. https://doi.org/10.1021/acs.jmedchem.3c01385

Singh, P., Hanson, P. S., & Morris, C. M. (2017). Sirtuin-2 protects neural cells from oxidative stress and is elevated in neurodegeneration. *Parkinson's Disease*, *2017*, 2643587. https://doi.org/10.1155/2017/2643587

Sola-Sevilla, N., Mesa-Lombardo, A., Aleixo, M., Exposito, S., Diaz-Perdigon, T., Azqueta, A., Zamani, F., Suzuki, T.,

Maioli, S., Eroli, F., Matton, A., Ramirez, M. J., Solas, M., Tordera, R. M., Martin, E. D., & Puerta, E. (2023). SIRT2 inhibition rescues neurodegenerative pathology but increases systemic inflammation in a transgenic mouse model of Alzheimer's disease. *Journal of Neuroimmune Pharmacology*, *18*(3), 529–550. https://doi.org/10.1007/s11481-023-10084-9

Sola-Sevilla, N., & Puerta, E. (2024). SIRT2 as a potential new therapeutic target for Alzheimer's disease. *Neural Regeneration Research*, *19*(1), 124–131. https://doi.org/10.4103/1673-5374.375315

Sun, S., Han, X., Li, X., Song, Q., Lu, M., Jia, M., Ding, J., & Hu, G. (2018). MicroRNA-212-5p prevents dopaminergic neuron death by inhibiting SIRT2 in MPTP-induced mouse model of Parkinson's disease. *Frontiers in Molecular Neuroscience*, *11*, 381. https://doi.org/10.3389/fnmol.2018.00381

Sundriyal, S., Moniot, S., Mahmud, Z., Yao, S., Di Fruscia, P., Reynolds, C. R., Dexter, D. T., Sternberg, M. J. E., Lam, E. W. F., Steegborn, C., & Fuchter, M. J. (2017). Thienopyrimidinone based sirtuin-2 (SIRT2)-selective inhibitors bind in the ligand induced selectivity pocket. *Journal of Medicinal Chemistry*, *60*(5), 1928–1945. https://doi.org/10.1021/acs.jmedchem.6b01690

Suzuki, T., Khan, M. N., Sawada, H., Imai, E., Itoh, Y., Yamatsuta, K., Tokuda, N., Takeuchi, J., Seko, T., Nakagawa, H., & Miyata, N. (2012). Design, synthesis, and biological activity of a novel series of human sirtuin-2-selective inhibitors. *Journal of Medicinal Chemistry*, *55*(12), 5760–5773. https://doi.org/10.1021/jm3002108

Taylor, D. M., Balabadra, U., Xiang, Z., Woodman, B., Meade, S., Amore, A., Maxwell, M. M., Reeves, S., Bates, G. P., Luthi-Carter, R., Lowden, P. A., & Kazantsev, A. G. (2011). A brain-permeable small molecule reduces neuronal cholesterol by inhibiting activity of sirtuin 2 deacetylase. *ACS Chemical Biology*, *6*(6), 540–546. https://doi.org/10.1021/cb100376q

Teng, Y. B., Jing, H., Aramsangtienchai, P., He, B., Khan, S., Hu, J., Lin, H., & Hao, Q. (2015). Efficient demyristoylase activity of SIRT2 revealed by kinetic and structural studies. *Scientific Reports*, *5*, 8529. https://doi.org/10.1038/srep08529

Trapp, J., Jochum, A., Meier, R., Saunders, L., Marshall, B., Kunick, C., Verdin, E., Goekjian, P., Sippl, W., & Jung, M. (2006). Adenosine mimetics as inhibitors of NAD+-dependent histone deacetylases, from kinase to sirtuin inhibition. *Journal of Medicinal Chemistry*, *49*(25), 7307–7316. https://doi.org/10.1021/jm060118b

Turgutalp, B., & Kizil, C. (2024). Multi-target drugs for Alzheimer's disease. *Trends in Pharmacological Sciences*, *45*(7), 628–638. https://doi.org/10.1016/j.tips.2024.05.005

van Leeuwen, I. M., Higgins, M., Campbell, J., McCarthy, A. R., Sachweh, M. C., Navarro, A. M., & Lain, S. (2013). Modulation of p53 C-terminal acetylation by mdm2, p14ARF, and cytoplasmic SirT2. *Molecular Cancer Therapeutics*, *12*(4), 471–480. https://doi.org/10.1158/1535-7163.MCT-12-0904

Vaquero, A., Scher, M. B., Lee, D. H., Sutton, A., Cheng, H. L., Alt, F. W., Serrano, L., Sternglanz, R., & Reinberg, D. (2006). SirT2 is a histone deacetylase with preference for histone H4 Lys 16 during mitosis. *Genes & Development*, *20*(10), 1256–1261. https://doi.org/10.1101/gad.1412706

Vempati, R. K., Jayani, R. S., Notani, D., Sengupta, A., Galande, S., & Haldar, D. (2010). p300-mediated acetylation of histone H3 lysine 56 functions in DNA damage response in mammals. *Journal of Biological Chemistry*, *285*(37), 28553–28564. https://doi.org/10.1074/jbc.M110.149393

Wang, Y., Fung, Y. M. E., Zhang, W., He, B., Chung, M. W. H., Jin, J., Hu, J., Lin, H., & Hao, Q. (2017). Deacylation mechanism by SIRT2 revealed in the 1′-SH-2′-O-myristoyl intermediate structure. *Cell Chemical Biology*, *24*(3), 339–345. https://doi.org/10.1016/j.chembiol.2017.02.007

Wang, Y. P., Zhou, L. S., Zhao, Y. Z., Wang, S. W., Chen, L. L., Liu, L. X., Ling, Z. Q., Hu, F. J., Sun, Y. P., Zhang, J. Y., Yang, C., Yang, Y., Xiong, Y., Guan, K. L., & Ye, D. (2014). Regulation of G6PD acetylation by SIRT2 and KAT9 modulates NADPH homeostasis and cell survival during oxidative stress. *The EMBO Journal*, *33*(12), 1304–1320. https://doi.org/10.1002/embj.201387224

Wei, W., Xu, X., Li, H., Zhang, Y., Han, D., Wang, Y., Yan, W., Wang, X., Zhang, J., Liu, N., & You, Y. (2014). The SIRT2 polymorphism rs10410544 and risk of Alzheimer's disease: A meta-analysis. *Neuromolecular Medicine*, *16*(2), 448–456. https://doi.org/10.1007/s12017-014-8291-0

Xu, K., Dai, X. L., Huang, C., & Jiang, Z. F. (2011). Targeting HDACs: A promising therapy for Alzheimer's disease. *Oxidative Medicine and Cellular Longevity*, *2011*, 143269. https://doi.org/10.1155/2011/143269

Yamagata, K., Goto, Y., Nishimasu, H., Morimoto, J., Ishitani, R., Dohmae, N., Takeda, N., Nagai, R., Komuro, I., Suga, H., & Nureki, O. (2014). Structural basis for potent inhibition of SIRT2 deacetylase by a macrocyclic peptide inducing dynamic structural change. *Structure*, *22*(2), 345–352. https://doi.org/10.1016/j.str.2013.12.001

Yang, J., Nicely, N. I., & Weiser, B. P. (2023). Effects of dimerization on the deacylase activities of human SIRT2. *Biochemistry*, *62*(23), 3383–3395. https://doi.org/10.1021/acs.biochem.3c00381

Yang, L. L., Wang, H. L., Zhong, L., Yuan, C., Liu, S. Y., Yu, Z. J., Liu, S., Yan, Y. H., Wu, C., Wang, Y., Wang, Z., Yu, Y., Chen, Q., & Li, G. B. (2018). X-ray crystal structure guided discovery of new selective, substrate-mimicking sirtuin 2 inhibitors that exhibit activities against non-small cell lung cancer cells. *European Journal of Medicinal Chemistry*, *155*, 806–823. https://doi.org/10.1016/j.ejmech.2018.06.041

You, W., Zheng, W., Weiss, S., Chua, K. F., & Steegborn, C. (2019). Structural basis for the activation and inhibition of Sirtuin 6 by quercetin and its derivatives. *Scientific Reports*, *9*(1), 19176. https://doi.org/10.1038/s41598-019-55654-1

Zhao, Y., Yang, J., Liao, W., Liu, X., Zhang, H., Wang, S., Wang, D., Feng, J., Yu, L., & Zhu, W. G. (2010). Cytosolic FoxO1 is essential for the induction of autophagy and tumour suppressor activity. *Nature Cell Biology*, *12*(7), 665–675. https://doi.org/10.1038/ncb2069

Zhou, Z., Ma, T., Zhu, Q., Xu, Y., & Zha, X. (2018). Recent advances in inhibitors of sirtuin1/2: An update and perspective. *Future Medicinal Chemistry*, *10*(8), 907–934. https://doi.org/10.4155/fmc-2017-0207

6 Therapeutic Potential of Caspase Inhibitors in Alzheimer's Disease

Amit Sharma, Vijeta Kumari, Ripunjay Kalita, Hemant R. Jadhav, and Amritansh Bhanot

6.1 INTRODUCTION

6.1.1 Alzheimer's Disease (AD)

Alzheimer's disease (AD) has been recognized as a predominant form of dementia. It is referred to as a severe neurodegenerative disease that leads to loss of memory and cognitive dysfunction (Nussbaum, 2003). As per the WHO Report, at present, more than 6 million Americans are living with the disease, and it is anticipated that this number will to increase to more than 13 million by 2060. There are different forms of dementia explored yet, depending on their pathologies, i.e. Lewy body dementia (LBD), frontotemporal dementia (FTD), and vascular cognitive impairment and dementia (VCID), or dementia with different fundamental pathologies (referred to as "mixed dementia"). Among them AD-related dementias (ADRD) are more colossal, and are to be investigated from their root factors to convey successful prevention and treatment (Sharma et al., 2024a, b).

Intensity of AD is diagnosed by assessing the extent of accumulation of senile plaque, and made of beta-amyloid (Aβ) and neurofibrillary tangles (NFTs) containing abnormally phosphorylated tau protein (Sharma et al., 2025). The amyloid cascade hypothesis is one of the known theories to support the pathophysiological role of beta-amyloid (Aβ), depending on which various clinical trials have been conducted aiming to modify Aβ production or its clearance (Sharma et al., 2024a, b).

Despite these recent advances, there is no specific drug regimen for a permanent cure for AD. Hence, it becomes important to identify new potential drug targets for the treatment of AD (Breijyeh & Karaman, 2020). The complexity of the disease and multiple pathological intrinsic and extrinsic pathways opens new ways to explore a combination of drug regimens targeting different neurogenerative pathways at the same time (Perneczky et al., 2010). Various studies emphasized towards the role of caspases and certain critical proteins towards neuropathology in AD. Such studies could provide critical evidence that caspases are a viable target for pharmaceutical development, and positive results may stimulate the development and testing of caspase inhibitors for their potential in treating this disease (Mortimer et al., 2005).

6.1.2 AD Associated Pathologies

Pathology of AD has been depicted via various theories. A few of them include tau theory, the amyloid cascade speculation, excitotoxicity, the cholinergic speculation, and others, such as the chart hypothesis (Jellinger, 2022). Each of these theories relates to the each other's in one or the other manner. In common they explain the increased transit of nerve cells within a body for which various components are responsible (Huber et al., 2017). As neurons are cells with exceptionally restricted multiplication, their increased transit can effortlessly cause a decrease in their number; hence, increased apoptosis of neurons, which is observed as a major consequence of AD (Muralidar et al., 2020). It has a very crucial effect on the working of the brain as an entirety. Despite the fact that apoptosis has a negative effect on the functioning of the AD brain, in common, it may be a physiological phenomenon of cell transit for the expulsion of fragile or damaged cells. Besides, apoptosis does not cause inflammation, as cell buildups are evacuated and phagocytosed by the macrophages (Kolarova et al., 2012). Furthermore, it is additionally fundamental for body improvement, immunoregulation, and cell security against DNA damage, which may otherwise result in tumour development. Still, in spite of the fact that apoptosis may have a positive effect on the working of the human body. Amid a few neurotic conditions, its sudden increase may lead to the degeneration of organs and tissues (Buccellato et al., 2021). Amid obsessive conditions, other components of cell passing can be actuated. Hence, as both insufficient and over-the-top apoptosis can have disastrous results for the life form, apoptosis should be accurately directed, which includes the activity of apoptosis inhibitors and activators. Although all the theories explain the role of the above-mentioned pathways, which are triggered in AD, increased mobilization of caspases 2, 8, 9, and 3 in AD has also been reported. It is exceptionally difficult to decide which pathway is the first one to be actuated or which is the most critical one. Expression of actuated caspases 3 and 8 is observed in the brains, confirming the role of the outward pathway within the apoptosis of neurons in AD, which can also be connected with Aβ stores (Magi et al., 2016). An increased level of enacted caspase 9 is observed within the

DOI: 10.1201/9781003602477-6

hippocampus of Alzheimer's disease patients. Additionally, there's moreover a relationship between the enacted caspases 8, 9, and 3 within the same neurons, proposing that the natural pathway is enacted by the outward pathway in the case of AD. On the other hand, the co-localization of DNA harm and caspase 9 in AD recommends that DNA damage is the driving force causing caspase 9 actuation in AD. This moreover compares to the upregulation of p53, which may be a DNA damage sensor and apoptosis activator in AD. Particularly, IRE1 may be included within the control of Aβ-induced cell passing, as a positive relationship between IRE1 movement and the movement of AD is found in AD patients, as well as expanded caspase 2 activation. This unequivocally proposes that the enactment of caspase 2 in AD is interceded by IRE1. Besides, all of these pathways lead to the same common atomic highlights, which are moreover observed in AD (Bekdash, 2021).

Enacted initiator caspases actuate killer caspases 3, 6, and 7, which are responsible for the degradation of intracellular proteins. Among them, caspase 3 is believed to be the foremost important, and its enactment is one of the markers of apoptosis; subsequently, expanded caspase 3 activity in AD neurons is accepted to be a trademark of expanded apoptosis of these cells Among the different proteins that are proteolysed by caspase 3, DFF45/ICAD, an inhibitor of DFF45/CAD, appears to be particularly critical. Debasement of DFF45/ICAD leads to the activation of DFF45/CAD, which is mindful for the fracture of DNA into 180 bp or bigger parts that are products of 180, such as 360 bp or 540 bp. This can be another vital atomic aspect of apoptosis, as in other processes of cell division, DNA is divided in an arbitrary way. Subsequently, a few strategies of evaluating apoptosis, like TUNEL (Terminal deoxynucleotidyl transferase UTP scratch conclusion labelling) strategies, are based on the location of these parts, and TUNEL evaluation also confirmed expanded apoptosis of AD neurons. Another vital marker of apoptosis is the externalization of phosphatidylserine, a molecule that ordinarily is shown as it were within the inward layer of the cell membrane (Wang et al., 2015).

Furthermore, upregulation of NLRP1 and NLRP3 is watched in numerous cell populaces in AD patients as well as in a mouse show of AD, which recommends that the upregulation of these proteins is the reason for the expanded pyroptosis in AD. By the by, there is not as it were proved that pyroptosis is overactivated in AD, but too that it plays an imperative part in AD pathophysiology. Investigate on a mouse show of AD illustrates that the knockdown of caspase 1 not as it were leads to diminished pyroptosis, as watched by lower expressions of NLRP3, caspase-1, and GSDMD, but moreover leads to a change within the cognitive capacities of creatures (Shimohama, 2000). Still, in spite of the reality that expanded neuronal death and expanded caspase actuation are broadly known features of AD, the precise atomic components that lead to this are not totally caught on. One of the reasons may be Aβ or tau activities, as these proteins are known for their pro-apoptotic activities. The reply to these questions appears to be crucial in understanding the part of caspases within the pathophysiology of AD and way better understanding this illness.

6.2 CASPASES

6.2.1 Introduction to Caspases

Caspases, a group of cysteine proteases, play a vital role in controlling apoptosis, a natural process of programmed cell death that occurs during development. This family's members break down various cellular structures by proteolysis. Beyond apoptosis, caspases have a significant role in various molecular processes too. Abnormal activation or deactivation of these important members of cellular cascades can have serious consequences and contribute to numerous diseases. Similarly, its anomalous activation is one of the majorly explored factors contributing to the neuropathology's associated with AD (Kumar et al., 2022).

There have been 18 mammalian caspases reported till date. Although placental mammals lack caspases 15, 17, and 18, whereas caspase 11 and 13 are reported in bovine and murine, which is orthologue to human caspase 4. Placental mammals lack caspases 15, 17, and 18; however, caspases 11 and 13 have been identified in bovine and murine species as orthologues of human caspase 4. Mammalian caspases, based on their functional roles and significance, are broadly categorized into two groups: apoptotic caspases and inflammatory mediators. Inflammatory mediators include caspases-1, -4, -5, -11, and -12, while apoptotic caspases comprise caspases-2, -8, -9, -10, -3, -6, and -7. Apoptotic caspases act as both initiator (-1, -4, -5, -11) and executioners (-3, -6, -7) (Lippens et al., 2000).

6.2.2 Caspases: Structure, Features and Activation

Each catalytic subunit comprises αβα sandwiches that fold into a compact barrel-like structure. This barrel includes a central core of six β-strands organized into sheets, flanked on either side by five helices oriented inversely to the plane of the β-sheets (Wójcik et al., 2024). Although some structures contain additional strands, the overall configuration remains consistent. The catalytic dimer consists of symmetrically arranged monomers, with the two active sites situated at opposite ends of the structure. The N-terminus of the small subunit and the cleaved C-terminus of the large subunit are positioned near each other. This configuration creates a structure in which the two subunits of the active caspase molecule originate from separate subunits of each zymogen through a domain-swapping mechanism (Walker et al., 1994). This model proposes that the zymogen's conformation closely mirrors that of the active form. Notably, each subunit of the active molecule is derived from a proenzyme, where cleavage of the linker between domains modifies the orientation of the subunits. Key Features and Activation: caspases are synthesized as inactive precursors, called zymogens, composed of a large subunit, a small subunit, and a prodomain (Wilson et al., 1994). According to crystallographic analyses of caspase-1 and caspase-3, the active enzyme assembles into a heterotetramer made up of one large subunit and two tiny subunits. Three major subfamilies of caspases have been identified by phylogenetic analysis: the ICE (interleukin (IL)-1ß-converting enzyme) subfamily, which includes caspase-1, caspase-4, and caspase-5; the CED-3 (*Caenorhabditis elegans* cell death protein) and CPP32 (32-kDa cysteine protease) subfamilies, which include caspase-3, caspase-6, caspase-7, caspase-8, caspase-9, and -10; and the ICH-1 subfamily, which is represented by Nedd2. The pentapeptide in the dynamic region of all caspases has the same structure, QACXG, where X might be R, Q, or G. The amino acids Cys-285 and His-237, which are involved in catalysis, as well as Arg-179, Gln-283, Arg-341, and Ser-347, which are part of the P-carboxylate authoritative take in caspase-1, are maintained inside the infinite bigger portion of subtypes of caspase. A certain subtypes of caspase are more essential than others, and not all caspases are necessary for cell passage, according to studies using caspase-1–/– and caspase-3–/– mice (Cohen, 1997). Apoptosis has frequently been caused by the overexpression of a certain caspase, which results from acceptance of this preparation. In any case, this overexpression may cause the caspase to cleave substrates that it would not normally recognize. While some caspases (caspase-1, caspase-2, caspase-4, caspase-5, caspase-8, caspase-9, and caspase-10) have extensive prodomains, others (caspase-3, caspase-6, and caspase-7) have brief prodomains. The nature of the S4 take is the main factor of specificity that distinguishes utilitarian caspase bunches from one another (Kumar, 1999). Apical caspase-8 and caspase-9 lean towards hydrophobic buildups like Leu and Ile, cytokine activators lean towards fragrant buildups (Trp or Tyr) in S4, and the official caspases, caspase-3 and caspase-7, lean towards Asp in this position (Figure 6.1). These preferences primarily allow caspases to be selective for common substrates (Figure 6.1).

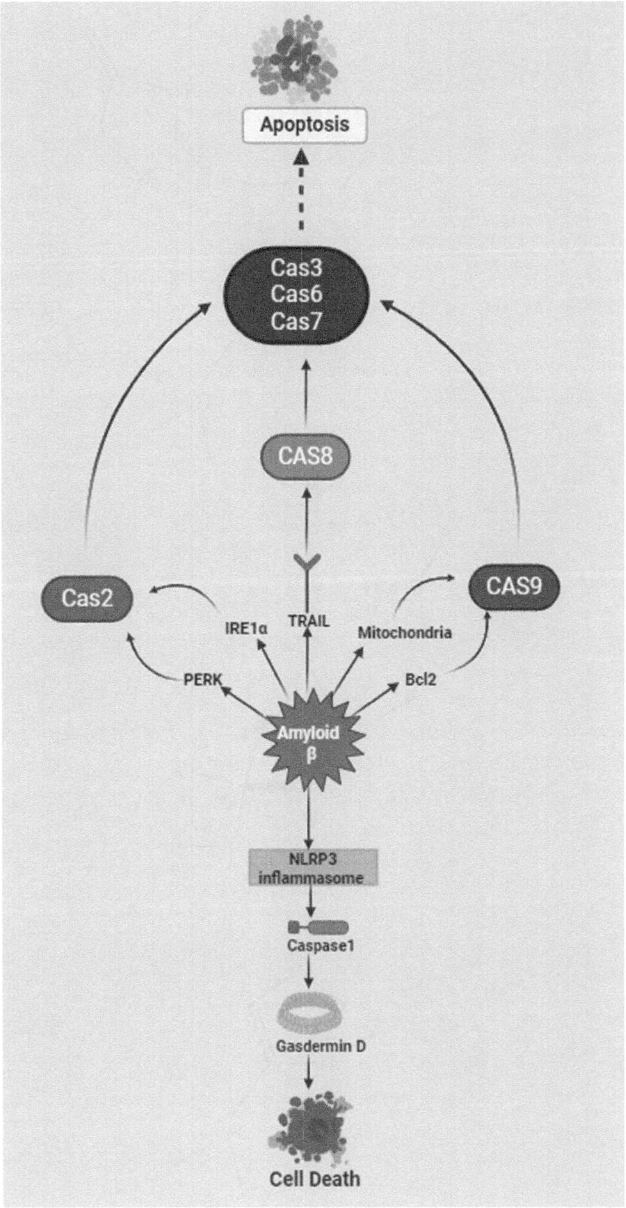

FIGURE 6.1 Activation of different caspases through Aβ. (Pop & Salvesen, 2009).

6.2.3 Caspases: Action and Catalysis

The most intense activity of caspases is shown at cytosolic ionic quality, under lowering circumstances, and at unbiased pH (6.8–7.2) (Stennicke & Salvesen, 1997). Catalytic histidine would polarize activated cysteine (Cys) at an unbiased pH. This suggests that His237 may be essential for protonating the α-amine, clearing out the bunch, and generating the nucleophilic water atom needed for diacylation, but it is not explicitly involved in the formation of the catalytic nucleophile Cys. For the aspartate accumulation in the S1 substrate cache, caspases seem to have stringent selectivity. At least three buildups (Arg179, Gln283, and Arg341) participate in the formation of the S1 stash and form hydrogen bonds with the aspartate P1 carboxylate group. Other amino acids, including glutamate, are prevented from precisely authoritatively reaching the S1 site by this tightly regulated environment (Garcia-Calvo et al., 1999). The function of the catalytic location is to restrict the organization of a tetrahedral midway, just like in other cysteine and serine proteases. Following substrate binding, the carbonyl oxygen receives a hydrogen bond from the NH spine of Gly238 and the Cys285 that shapes the oxyanionic gap, which polarizes the carbonyl gather of the shine bond. The electrophilic carbon can currently be attacked by the nucleophilic Cys285. The proton of the thiol bunch can be switched to the nearby His237 during or shortly after this. By protonating the amino bunch and removing the P1′ accumulation, this proton can currently function as a catalyst. His237 has the ability to polarize the water particle needed for complete hydrolysis during deacylation and create a moment tetrahedral middle of the road (Figure 6.2).

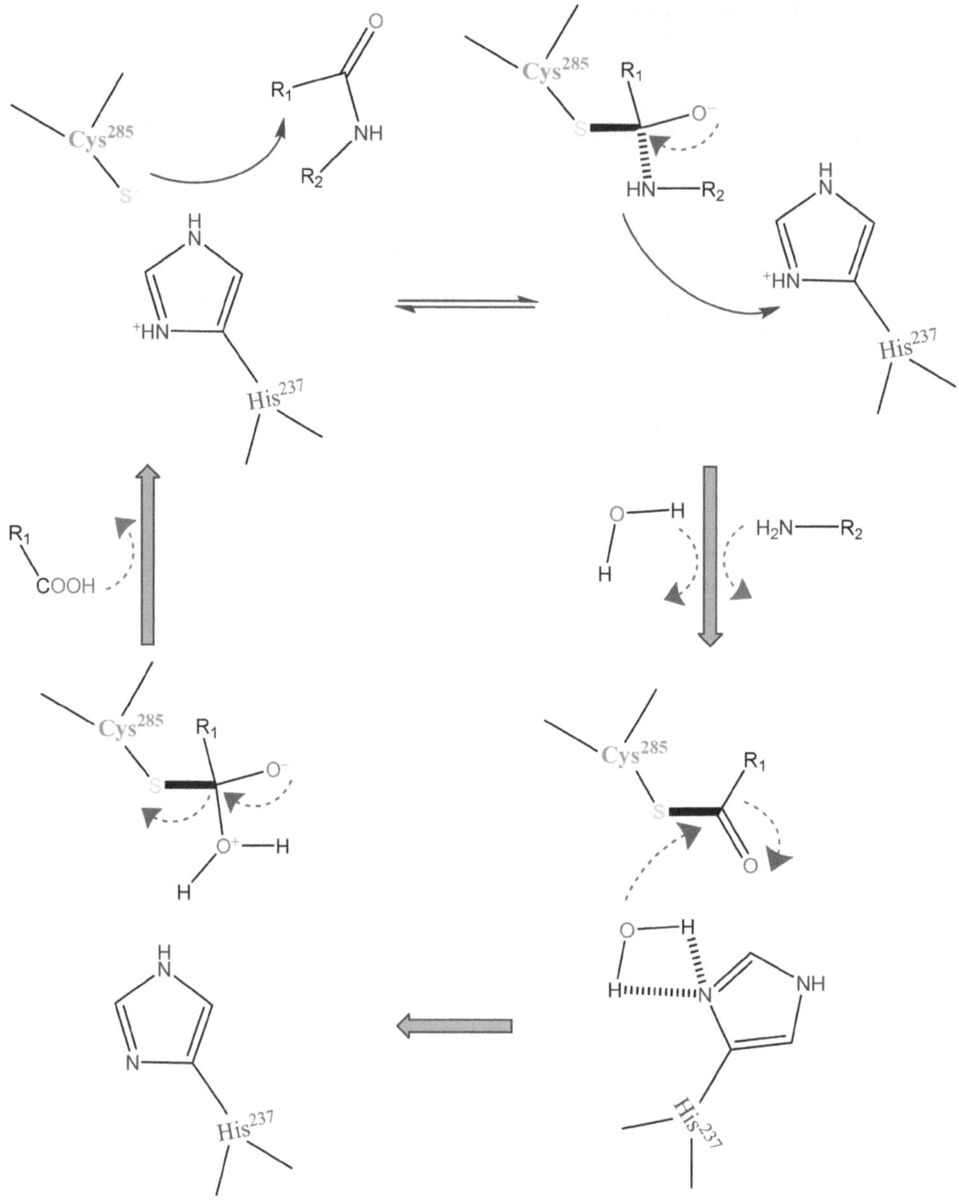

FIGURE 6.2 Catalytic activity of caspases. (Garcia-Calvo et al., 1999).

6.3 ROLE OF CASPASES IN AD OCCURRENCE AND PROGRESSION

6.3.1 Caspase Activation and Cleavage of Target Proteins in AD

The part caspases play in apoptosis is well known. Internal causes including oxidative stress and DNA damage can start this process by activating caspase-9 and the intrinsic apoptotic pathway (Riedl & Shi, 2004). The buildup of misfolded, broken, or oxidized proteins in the endoplasmic reticulum (ER) can also cause apoptosis by activating caspase-2 and the ER stress-induced pathway (Fuentes-Prior & Salvesen, 2004). Extracellular signals can also trigger apoptosis, such as the extrinsic pathway, which activates the initiator caspase-8. These pathways often occur simultaneously, with one pathway activating others in a coordinated manner. The extrinsic apoptosis pathway is triggered by the binding of death ligands to their corresponding death receptors (Yang et al., 1998). Death receptors, including tumour necrosis factor receptors (TNFRs), TNF-related apoptosis-inducing ligand receptors (TRAIL-Rs), and CD95 (also known as Fas), are key components involved in apoptosis. These receptors are distinguished by the presence of a highly conserved structural domain known as the death domain. Activation of these death receptors causes trimerization of their death domains (Gervais et al., 1999). For TNFRs, the tumour necrosis factor receptor 1-associated death domain protein (TRADD) and receptor-interacting serine/threonine-protein kinase 1 (RIPK1) bind to the death domains, forming complex I. This complex also involves cellular inhibitors of apoptosis proteins (cIAPs) and TNF (Rohn et al., 2001).

6.3.2 Action of Mammalian Caspases in Occurrence and Progression of AD

There is an increased number of evidences reporting the involvement of caspases in progression of AD. They paly crucial role in inducing neuronal death via Aβ accumulation, contribution in abnormal tau protein cleavage and NFTs formation. Several studies have shown that caspases-1, -2, -3, -6, -8, -9, and -12 contribute to the pathogenesis of AD (Lamkanfi et al., 2008). However, the involvement of caspase-12 in AD has only been observed in murine models. In humans, mutations in the caspase-12 gene lead to a premature stop in translation (Kesavardhana et al., 2020). Hence its functional significance has not been validated in humans till date. However, caspase-12 of mouse and caspase- have homology in their functionality Caspase-2 are further categorized as an effector (with caspase-3 and -7) and an initiator. Caspases-6 are found to be abundant in neurites and tau tangles as compared to caspases 3 (Pozueta et al., 2013). Although caspases is not found prominently in nuclei of the AD neuron but it has very great prominence in tangles causing lethal damage of various substrate. Damage to neuronal intermediate filaments disrupts the structural integrity of neurons, causing them to become dysfunctional without necessarily triggering cell death (Troy et al., 2000).

6.3.3 Evidence that Caspases Contribute to Pathology Associated with AD

A technique can only be considered truly effective if it is demonstrated that caspase activation is not confined to the final stages of AD.

Building on previous studies, a hypothesis was proposed suggesting that caspases may play a role earlier in the series of events associated with AD than previously recognized (Rohn et al., 2002). Caspase activation and tau cleavage may relate Aβ to NFT production. Since the publication of the notion that caspase cleavage of tau may be a proximal event in NFT production, multiple investigations have now supported this perspective (Guo et al., 2004).

Gastard et al. provide additional support for caspase activation as an early event in the evolution of AD by demonstrating caspase-3 activation within juvenile tangles in persons with moderate AD (Gamblin et al., 2003). Furthermore, Guillozet Bongaarts et al. found that caspase-mediated tau truncation may occur rather early in the illness process, before the crucial folding stages that contribute to the creation of PHF. Other studies have revealed that tau truncated at Asp421 aggregates faster and to a larger extent than full-length tau (Yin & Kuret, 2006).

This shows that caspase cleavage of tau may aid in filament production, a critical step in the genesis of NFTs associated with AD. These data suggest that caspase cleavage of tau may be an early event linking Aβ and NFTs in AD. Thus, treatments that limit tau caspase cleavage may be advantageous in not just reducing NFT development but also decreasing cognitive deterioration (Ding et al., 2006).

6.3.4 Evidence that Caspase Inhibition Is Protective in Other Neurodegenerative Disorder's

Overall, the research reviewed imply that caspases play an important role in the aetiology of AD. Can targeting this class of proteases be a viable treatment for this disease? Before answering this issue, it may be useful to review research that have investigated the pharmacological suppression of caspases to treat neurodegenerative disorders other than Alzheimer's (Li et al., 2000). Many of these investigations were conducted in the late 1990s, providing impetus for the development of caspase inhibitors as therapeutics. In a work by Li et al, the scientists employed the pan-caspase inhibitor N-benzyloxycarbonyl-Val-Ala-Asp fluoromethyl-ketone (Z-VAD-fmk) to cure transgenic mice expressing mutant human SOD1, a model of a disease called amyotrophic lateral sclerosis (ALS) (Martinou et al., 1994).

In this case, treatment of ZVAD-fmk delayed illness onset and death in ALS animals. Supporting a function for caspases in ALS was a parallel study by which transgenic ALS mice were bred with Bcl-2 OE mice generated by Martinou et al. These were the same mice utilized in our study to breed with 3xTg-AD mice, as mentioned above (Vukosavic et al., 2000). In this case, the scientists discovered that overexpressing Bcl-2 reduced neurodegeneration, delayed caspase activation, and inhibited beta-actin cleavage in ALS animals. The same authors used a similar strategy to demonstrate that intracerebroventricular treatment of a caspase inhibitor reduced disease progression and death in a Huntington's disease mice model (Ona et al., 1999).

Collectively, investigations like the aforementioned encouraged hopes that medicines might follow soon. Unfortunately, these promising early experiments with caspase inhibitors have not resulted in clinically viable anti-apoptotic therapies. The causes for this are unclear, however, they could be toxic side effects generated by the fluoromethyl-ketone residue and inadequate tissue penetration of Z-VAD-fmk (Friedlander, 2003).

These studies, however, reinforce the idea that using caspase inhibitors could be an effective approach for treating certain neurodegenerative diseases. Similar research should be carried out in animal models of AD to explore this further. The critical next step is to utilize animal models of AD to assess whether inhibitors of apoptotic machinery can prevent neuronal cell death or other aspects of the disease's pathogenesis (Schierle et al., 1999).

6.3.5 Involvement of Caspase in Cellular Process

Caspases play a crucial role in other processes beyond apoptosis. Their involvement has been seen in a variety of biological functions (Monie & Bryant, 2015). Beyond apoptosis, caspases are also vital in regulating non-apoptotic forms of cell death, such as necroptosis (which involves the deactivation of caspase-8), pyroptosis (regulated by caspases-1 and -11), and autophagy (which is inhibited by

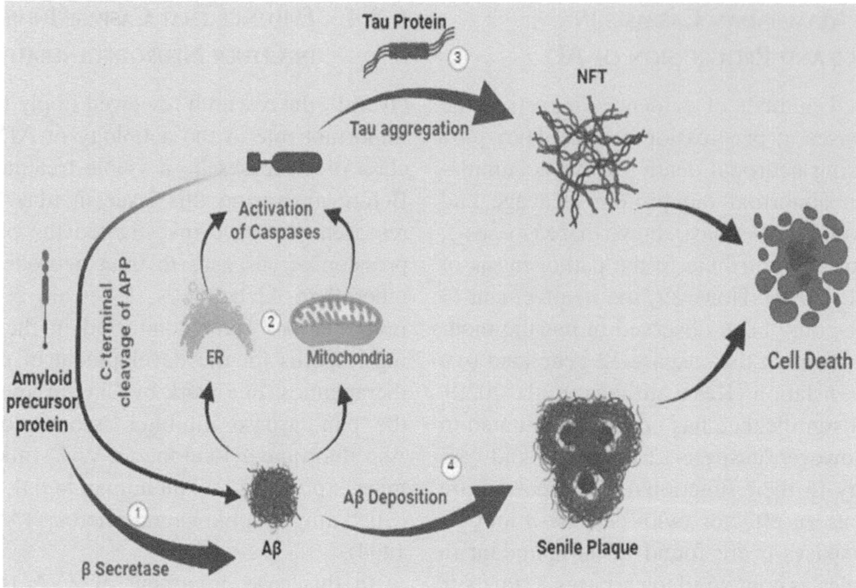

FIGURE 6.3 In first step, (1) β secretase acts on APP to form Aβ. (2) Formation of Aβ initiates apoptotic process leading to activation of caspases through ER and mitochondria, which cause degradation of APP further actuating formation of Aβ, ensuing vicious cycle of Aβ formation. (3) Activated caspases also lead to cleavage of tau protein and formation of NFT ultimately resulting in cell death. (4) Another pathway causing cell death involves deposition of Aβ and formation of senile plaques. (Farmer et al., 2020).

caspases-2, -3, -6, and -8). They influence these processes by promoting or suppressing key factors involved (Norman et al., 2010).

Caspases have a crucial role in inflammatory processes. IL-1β and IL-18, members of the IL cytokine family, play key roles in inflammation. The activation of caspase-1 is a critical step in the processing and maturation of these mediators. Inflammasomes are thought to have an important protagonist in caspase 1 activation. Caspase-8 plays a role in activating the inflammasome, maturing pro-inflammatory cytokines, and cleaving pro-IL-1β (Tenev et al., 2011). Mammalian caspases are associated with the final stages of differentiation in various cell types. Caspase-3, in particular, is crucial in this process, as it contributes to the differentiation of erythroid cells, keratinocytes, lens fibre cells, muscle progenitor cells, bone marrow stromal stem cells, osteoblasts, and monocytes (Oral et al., 2012).

Caspases that play a role in cell differentiation include caspase-2 (involved in erythroid cells and osteoblasts), caspase-7 (in odontoblasts), caspase-8 (in osteoblasts and monocytes within the bone marrow), and caspase-9 (in erythroid cells, muscle progenitor cells, and monocytes in the bone marrow). Caspase-14 is unique to terrestrial mammals, with its expression restricted to the epidermis and epithelial tissues (Denecker et al., 2008). It is found mostly in differentiating keratinocytes and not in proliferating keratinocytes. Caspase-14 is also linked to inflammatory skin conditions like psoriasis.

Mammals have limited regenerating abilities, with the exception of the liver and skin, which regenerate extensively. Studies in mice have demonstrated that caspase-3 and caspase-7 inhibit the activity of calcium-independent phospholipase A2. iPLA2 is involved in controlling the release of PGE2, a molecule that facilitates stem cell proliferation, tissue regeneration, and repair. Studies have emphasized the crucial role of mammalian caspases in managing cell proliferation and apoptosis during development and tissue regeneration (Figure 6.3). Caspases-1, -6, and -8 are particularly important for regulating B-cell proliferation and maintaining lymphocyte homeostasis (Ramos-Tovar & Muriel, 2020).

6.4 CROSS TALK BETWEEN AMYLOID β AND CASPASES IN AD

According to the amyloid cascade theory, dementia is linked to the buildup of amyloid-beta peptides and β-plaques in the central nervous system, which also plays a major role in the onset of AD. The amyloid precursor protein (APP), a transmembrane protein that is sequentially cleaved by β-secretase (BACE1) and γ-secretase, is the source of the peptide Aβ, which has 40 or 42 amino acids. A protein called τ interacts with microtubules to help them polymerize and stay stable inside the cytoskeleton of the cell (Sanabria-Castro et al., 2017). Recent research shows that amyloid plaques form in healthy brains as people age and that the accumulation of Aβ peptides and the subsequent deposition of these peptides are not always linked to cognitive decline or cell death (Kametani & Hasegawa, 2018).

The theory has changed to imply that Aβ is a contributing component to the illness rather than being trapped in plaques. The breakdown of its precursor protein, APP, produces Aβ protein, which is the main component of the cortical plaques seen in AD brains. Although APP's precise physiological functions are still unknown, it is generally accepted to be involved in proper neuronal activity and perhaps brain development. According to the amyloid hypothesis, Aβ peptides (Aβ40 and Aβ42) accumulate as a result of secretases' decreased ability to degrade APP-derived Aβ with ageing or in pathological situations. A higher Aβ42 to Aβ40 ratio encourages the development of Aβ amyloid fibrils, which results in neurotoxicity and τ disease. In the end, this leads to the death and degeneration of neurones. APP is processed in phases, with α-secretase-active enzymes coming first, then β-secretase BACE1. Alzheimer's patients have reduced amounts of APPs α, a peptide with neuroprotective qualities that is produced by α-secretase cleavage (Paroni et al., 2019).

Common variables associated with ageing and neurodegenerative disorders may influence BACE1 expression. However, caspases may also cleave APP, a mechanism associated with cognitive decline and synapse loss. It has been demonstrated that blocking caspase-3, which is thought to be involved in this cleavage, decreases APP processing (Puzzo et al., 2008). Researchers created an APP D664A KI mouse strain by homologous recombination, inserting a mutation from Asp (D) to Ala (A), in order to examine if enhanced APP cleavage by caspases is a causative or correlational component. The protein becomes insensitive to caspase cleavage when this amino acid is substituted since caspases act on Asp residues (Gervais et al., 1999). These mice's hippocampus organotypic slice cultures (OTSCs) showed resistance to Aβ-induced cytotoxicity in vitro. Aβ deposition in the brain is decreased when both caspase-8 and RIPK3 are deleted, but this effect is not achieved when RIPK3 is deleted alone. This implies that the relevance of caspases in AD extends beyond the conventional knowledge of their role in cell death (Park et al., 2020). Proteins called caspases are frequently thought to have a role in the last phases of cell death. Their significance in AD, however, appears to go beyond this, since they contribute significantly to the early stages of the illness by encouraging the development of Aβ plaques, which can lead to further difficulties (Kumar et al., 2023). Since increased activation of caspases-2, -3, -8, and -9 is a hallmark of AD, the precise caspases causing this process are yet unknown. According to in vitro research, caspases-3 and -8 are principally in charge of this process (McIlwain et al., 2013; Shalini et al., 2015). Thus, it appears that the extrinsic route, which activates caspases 3 and 8, must be activated. Since caspase-3 activity comes after caspase-8 activation, it's possible that caspase-8's impact on caspase-3 rather than caspase-8 itself is what caused the alterations in Aβ metabolism. As a result, the exact function of caspase-8 remains unknown and needs more research. But according to our research, neuroinflammation is a major factor in the aetiology of AD. Neuroinflammation and elevated levels of pro-inflammatory mediators, especially TNFα, are hallmarks of AD. Death receptors are activated by TNFα, which sets off caspases 3 and 8 and the extrinsic apoptotic pathway (Fossati et al., 2012). Aβ may have a multi-targeted effect on AD by directly binding to death receptors. One of the main concerns is that amyloid β is transported to the brain through blood-brain barrier receptors, particularly RAGEs. RAGE and amyloid β combine to cause endothelial inflammation and cell death, which has a major impact on the neurovascular alterations that occur in AD. Aβ has the ability to cause apoptosis in neurones as well as endothelial cells. A therapy resulted in caspase-8-dependent cell death in a rat neural cell line (PC12) (Zlokovic, 2004). When comparing Aβ-treated cells to non-transfected controls, knockdown experiments employing siRNA to silence caspase-8 showed a significant decrease in apoptosis as well as a decrease in the activation of caspases-9 and -3. Despite evidence of increased necroptosis in AD, caspase-8 silencing did not affect Aβ-induced necroptosis, even while it decreased apoptosis (Yan et al., 2009).

This implies that rather than shielding cells from necroptosis, the increased caspase-8 activity linked to AD encourages cell death (Qian et al., 2015). The precise process by which Aβ triggers caspase-8 is yet unclear (Richard & Mousa, 2022). According to certain research, Aβ may cause cells to produce TRAIL, which would activate death receptors. Western blot study shows that Aβ binds to TRAIL receptors 1 and 2, which co-localize with Aβ (Caccamo et al., 2017). Cells exhibit protection against Aβ-induced apoptosis when TRAIL receptors are turned down, suggesting that TRAIL signalling is necessary for caspase-8 activation in AD. Aβ may trigger the extrinsic apoptosis pathway by activating death receptors, specifically TRAIL, which in turn triggers caspases 8 and 3, finally resulting in neuronal degeneration, according to other studies. Moreover, Aβ activates caspase-9 (Paradis et al., 1996). Additionally, Aβ puts off the equilibrium between the pro-apoptotic Bax and anti-apoptotic Bcl2 proteins, which leads to caspase-9 activation and Bax activation. The Bcl2/Bax balance plays a crucial role in Aβ-mediated caspase activation, as demonstrated by the partial resistance of Bax–/– mice to Aβ-induced death (Steckley et al., 2007). It is also thought that the protein PUMA controls neuronal death. When Aβ is administered, PUMA expression rises, but when it is silenced, cells are protected from Aβ-induced apoptosis. Nevertheless, there are several studies that dispute the activation of caspase-9 after exposure to Aβ. Aβ binding to procaspase-9 is thought to impede caspase-9 activation and hinder apoptosome formation. It has also been demonstrated that tau and Aβ cause different human cell types to undergo the unfolded protein response (UPR) (Saha et al., 2021).

In addition to encouraging death, Aβ also increases IL-1 production from cells, most likely through pyroptosis and caspase-1 activation (Islam et al., 2017). The activation of NLRP3, a pattern recognition receptor that Aβ activates, is what propels this process. IL-1 release is decreased by deleting genes for caspase-1, NLRP3, or ASC, underscoring the critical roles that inflammation and caspase-1 activation play in IL-1 release after Aβ exposure (Sharoar et al., 2014). Furthermore, after Aβ injection, NLRP3 expression is reduced when caspase 8 is eliminated, indicating that caspase 8 plays a role in this process (Lee et al., 2010; Poirier et al., 2019). These findings point to a complex interaction between AD, Aβ, and caspases. In the context of AD, caspases break and deposit Aβ, which in turn sets off more caspases, albeit the precise mechanisms are still unclear. When caspase 8 is activated by Aβ, caspase 3 activation and other downstream consequences are triggered (Lorton et al., 1996; Masters & O'Neill, 2011; Yates et al., 2000). There is still disagreement on the intrinsic mechanism; some research indicate that Aβ activates caspase 9, while others do not. The intricacy is increased by the fact that caspase 9 activation is influenced by three distinct methods. Aβ first affects Bcl2 family proteins, including as Bax and Bcl2. Secondly, Aβ affects mitochondrial proteins, and third, caspase 8 cleaves the Bid protein to activate caspase 9.

Although a direct link between the UPR and Aβ-induced caspase activation has not been indicated, it is possible. Aβ can also affect pyroptosis, a complicated process involving several caspases. When caspase 8 is activated by Aβ, more death receptors are produced. Furthermore, NLRP3 and Aβ interact to activate caspase 1 and initiate pyroptosis. This causes inflammation, which in turn triggers caspase 8, which in turn triggers apoptosis and Aβ deposition. These results highlight how important caspase 8 activation is in relation to the Aβ hypothesis. Despite the widespread acceptance of the amyloid cascade theory, new evidence indicates that amyloid β is not the only factor contributing to the pathophysiology of AD (Gurung et al., 2014).

6.5 CROSS TALK BETWEEN TAU PROTEIN AND CASPASES IN AD

The τ hypothesis builds upon the amyloid cascade hypothesis. In vitro studies using neuronal cell lines, primary hippocampal and cortical neurons, and organotypic preparations of the hippocampus have revealed that amyloid β causes alterations in τ protein (Götz et al., 2001; Wang et al., 2003). The primary causes of neuronal dysfunction in AD are changes in τ protein and amyloid β oligomers. Although τ pathology is more closely linked to cognitive impairment than Aβ damage, there is no clear explanation for its significance The τ protein is extremely soluble and binds to microtubules, promoting structural alterations, axonal transport, and neuronal development (Stoothoff & Johnson, 2005). The "limbic-predominant type" of AD is characterized by a significant accumulation of τ protein in the hippocampus, whereas "hippocampal sparing" refers to a weaker accumulation of τ in the hippocampus compared to predicted pathology (Jellinger, 2012).

The τ hypothesis proposes that hyper-phosphorylated τ protein forms as a secondary pathogenic event, causing its displacement from microtubules and subsequent aggregation into paired helical filaments (PHFs) and NFTs, culminating in neurodegeneration before plaque A development (Fan et al., 2020). The Apo epsilon 4 (APOE4) allele of the apolipoprotein (APO) gene, which codes for a protein involved in cholesterol metabolism and lipid transport, is thought to play an essential role (de Castro et al., 2019). Caspases are also thought to play a function in the development of AD (Rissman et al., 2004). In vitro studies have revealed that caspases 2, 3, 7, and 8 can cleave tau into two fragments, and caspase-3-cleaved tau clumps more easily than full length tau, boosting the aggregation of full-length tau (Quintanilla et al., 2009). Furthermore, cleaved tau has been shown to be harmful to neurons. Immunostaining of AD brains verifies the importance of caspases and caspase-cleaved tau in the pathogenesis of AD, as the caspase-cleaved carboxy-terminus of tau is abundant in AD patient brains. Caspase 3, cleaved tau, and fibrillary tau are all co-localized in Alzheimer's brains, and active caspases are seen in the majority of NFT-containing cells (Quintanilla et al., 2012; Spires-Jones et al., 2008). These findings support the hypothesis that caspase-dependent tau cleavage leads to the development of cognitive impairment and AD. Animal studies have partially confirmed this, as higher caspase activity is associated with tau expression and cognitive impairment. Similar findings can be obtained for caspase 2, where the synthesis of tau that is resistant to caspases or has caspase 2 deficiency performs cytoprotectively in vitro and improves cognitive abilities in mice models (Means et al., 2016; Pozueta et al., 2013). However, the significance of tau in caspase activation is uncertain. On the one hand, tau can activate caspase 2 by triggering an unfolded protein reaction. On the other side, some researchers believe that tau can prevent caspase 9 activation. Both of these processes could be attributable to the same cause (Zhao et al., 2016). UPRs can cause p53 activation. However, new evidence reveals that, despite overexpression in AD, p53's activity is reduced due to aggregation with tau, which prevents its translocation into the nucleus. As a result, p53's beneficial effect on DNA damage repair may be inhibited. However, p53 activation can also occur during oxidative stress or owing to mitochondrial failure.

6.6 CONTRIBUTION OF CASPASES AND ROS TO THE MITOCHONDRIAL CASCADE HYPOTHESIS

As per the mitochondrial cascade hypothesis, a similar type of mechanism has been reported for brain aging and AD. It suggests that mitochondrial dysfunction is not only because of neurodegeneration. Moreover, it also results amyloidosis, tau phosphorylation, and further enters again into the cell cycle. It also briefs the role of somatic mitochondrial DNA (mtDNA) damage, and added on the information in free radical theory of aging (Swerdlow et al., 2010). It explains various mechanism, including oxidative stress and proteasome dysfunction, which further leads to mitochondrial dysfunction in neurodegenerative diseases such as AD (Chen & Yan, 2007). Generally, in the patients with neurodegeneration the production of reactive oxygen species (ROS) is increased (Pavlov et al., 2011). Mitochondrial function and amyloidosis are in connection to each other in case of AD, because mitochondrial electron transport chain (ETC) dysfunction raises the free radical production (Lin

& Beal, 2006). In addition to this APP, Aβ, and the entire γ-secretase complex are found in mitochondria or mitochondrial membranes. Due to the disturbance in effective antioxidant system the ROS production is increased within the body which results in neurodegeneration, i.e. AD. Due to the presence of free radical in the AD Patients ROS interact with biological molecules and damage cells (Peña-Bautista et al., 2019; Reddy, 2006). The abnormal accumulation of free radicals activates the NMDA receptors which further spies up the level of ROS formation Increase in the ROS level results in increased production, aggregation, τ phosphorylation, and polymerization of amyloid β. Hence the increases ROS level triggers the self-sustained mechanism of the body to initiate neurodegeneration contributing to the development of AD. It also activates the caspases induces apoptosis (Makhaeva et al., 2015). As per the above theory summarizes, the mitochondrial dysfunction and formation of ROS is crucially related to induction of apoptosis via intrinsic pathway. Mitochondria are very commonly termed as powerhouse of cell, since neurons have high energy demand, mitochondrial damage releases of pro-apoptotic factors. In addition to this, increased Aβ also interacts with mitochondrial proteins like cytochrome C oxidase, which further accelerated the mitochondrial damage and releases various factors from mitochondria, which induces apoptosis. ROS also interacts with many cellular protein which results in ER stress. Caspases act differently in AD (Grivennikova & Vinogradov, 2006). It is directly related to the apoptosis which results in neural degeneration and cognitive impairment. In AD, activation of caspases and apoptotic pathways are directly related to each other. Various data suggests that the activation of caspases 9 and intrinsic pathway occurs via activation of caspases 8, whereas many data suggest that their activation is independent of each (Zhu et al., 2006). Any how their activation plays important role to support various theory to initiate AD. The activation of intrinsic pathway act as a significant marker to induce mitochondrial damage. It also results in oxidative stress, DNA damage, or mitochondrial failure. If caspase 9 activation were only due to the downstream effect of caspase 8 activation and not due to oxidative stress, it would be suggesting that mitochondria and ROS have a negligible effect on neurons in AD, hence proving the mitochondrial theory of Alzheimer's development incorrect. Even if caspase 9 is activating complete independently, it does not conclude the significant role of mitochondria as it may also be due to the effect of Aβ on mitochondria and proteins regulating the intrinsic pathway. As per the various theories the role of the ER stress-induced pathway could not be undervalued. Even if the activation of caspase 9 is confirmed its significant role could not be precisely postulated. It has been reported that the accumulation of Aβ triggers the UPR and caspase 2. On the other hand, it is also being explains that the pathway activates due to oxidatively damaged proteins, hence mitochondrial damage and ROS release would also be activating this pathway (Figure 6.4) (Cha et al., 2012). A long with Apoptosis, AD is also accompanied by pyro ptosis and necroptosis, which

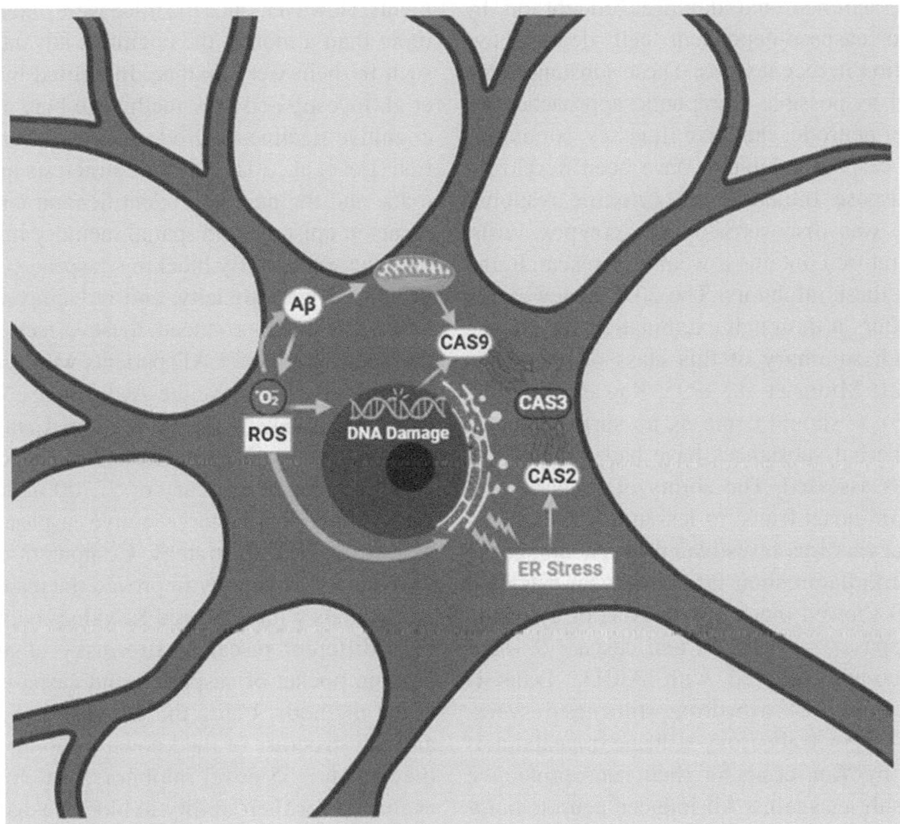

FIGURE 6.4 Role of ROS and mitochondria in activation of caspases. (Fossati et al., 2016).

are also regulated by caspases (1 and 8, respectively). Hence not only to neuronal loss but also to inflammation has also been reported. Neuroinflammation activates the, pro-inflammatory factors, i.e. TNFα. The factors are activated by the caspase-1- and 8-mediated TNFR activation. This may result into a positive feedback loop where pyroptosis and necroptosis activate themselves along with the extrinsic pathway of apoptosis, ultimately leading to neuronal loss (Hernandez-Zimbron et al., 2012). Caspases also results in converting the Aβ and tau proteins into theirs more toxic derivatives, that are more prone to deposition within the indicating the second mechanism by which caspases interact with AD pathophysiology. This also illustrates the initiation of an additional positive feedback loop in which increased Aβ and tau deposition results from caspase activation, which in turn raises caspase activity and produces more pathogenic Aβ and tau. In fact, a little trigger calculation, such pro-inflammatory cytokines or ROS, that causes mild caspase activation might ultimately cause massive neuronal death and caspase enactment. Consequently, it has been suggested that starting clinical trials of caspase inhibitors as possible medications for AD might be a restorative strategy against AD.

6.7 SIGNIFICANCE OF CASPASE INHIBITORS IN TREATMENT OF AD

6.7.1 Preclinical Studies on Caspase Inhibitors

Compounds having neuroprotective qualities, especially those that can lessen neuronal damage brought on by inflammation and caspase-dependent cell death, have drawn more attention in recent years. These substances are being investigated as possible therapeutic approaches for the management of neurodegenerative illnesses. Numerous naturally derived caspase inhibitors have been developed since the first caspase inhibitor, the cytokine response modifier (CrmA), was discovered in the cowpox virus. The scientific literature contains a wealth of research and documentation on these inhibitors. The 2021 review paper by Dhani et al. offers a thorough examination for anyone seeking a thorough summary of this class of chemicals (Dhani et al., 2021; Miura et al., 1995; Ray et al., 1992; Tesch et al., 2005). It should come as no surprise that a large number of herbal substances have had their neuroprotective qualities assessed. The ability of hesperidin, a flavonoid present in citrus fruits, to lessen AlCl3-induced apoptosis in Wistar rats was investigated by Thenmozhi et al. Rats with neuroinflammation brought on by extended exposure to AlCl3 showed much lower levels of cytosolic cytochrome c, caspase-3, caspase-8, and caspase-9 when hesperidin was co-administered with AlCl3. Trans-4-hydroxystilbene, trans-3,5,4′-trihydroxy-stilbene (resveratrol), and trans-3′,4′,3,5-tetrahydroxy-stilbene (piceatannol) were investigated by Wen et al. for their neuroprotective and antioxidant qualities against Aβ-induced neurotoxicity in rat primary cortical neurones. Their results demonstrated that these chemicals significantly decreased the activation of caspase-9 and other caspases, confirming their neuroprotective properties (Justin-Thenmozhi et al., 2018). The effect of pinoresinol diglucoside on neuroinflammation and neuronal death in mice models of AD was next investigated by Lei et al. According to Western blot research, this substance prevents neuronal death by decreasing the generation of cytochrome c and cleaved caspase-3 and increasing the Bcl-2/Bax ratio. However, these effects were caused by the indirect activation of other pathways rather than by direct contact with any one caspase subtype. Huang et al.'s 2020 study demonstrated that vitamin K2 guards against amyloid-β-induced neurotoxicity. According to the research, vitamin K2 suppresses amyloid-β-induced caspase-3-mediated apoptosis and lowers ROS. Moreno et al. recently reported on their study of new caspase-1 inhibitors obtained from marine invertebrates in Brazil that belong to two different phyla. In enzymatic tests, the researchers discovered an extract from *Chiropsalmus quadrumanus* that inhibited caspase-1. Following more purification, they observed that the chemical ingredient causing the suppression of caspase-1 (Wen et al., 2018) could be trigonelline, an alkaloid often present in plants. There has been a discernible transition from natural products to synthetic chemicals, much like other problems encountered in rational medication discovery. Consequently, a wide range of peptide inhibitors and small compounds that can target different caspase subtypes have been developed. In animal models of AD, Flores et al. found that by blocking caspase-1, the blood-brain barrier-crossing medication VX-765 can improve episodic and spatial memory impairments. However, once the mice were taken off the therapy for more than a month, the cognitive advantages that had been seen in them were undone. In animal models of AD, Zhou et al. investigated how methylene blue affected age-related cognitive decline, neurodegeneration, and neuroinflammation (Lei et al., 2021). Behavioural tests including the Barnes maze and the new item identification task shown improvements in episodic and spatial memory impairments brought on by caspase-6. By blocking caspase-6, lowering caspase-6-induced neurotoxicity, and reducing neuroinflammation, methylene blue produced these effects, which may have therapeutic value for AD patients who suffer from cognitive impairment. Tubeleviciute-Aydin et al. conducted an investigation on a related caspase subtype. Using virtual screening against the target's suspected allosteric pocket, the researchers first reduced a library of 57,700 molecules. Novel allosteric inhibitors that target active caspase-6 were discovered as a result of this strategy. Computational analysis and in vitro data significantly improved the results, revealing novel compounds with IC50 and Ki values between 2 and 13 μM. In a different research, Bresinsky et al. investigated the binding pocket of caspase-2 and caspase-3 using computational methods. Using the tau cleavage sequence YKPVD and the structure of the known inhibitor AcVDVAD-CHO, they created 35 novel inhibitors, which were subsequently evaluated for their ability to block caspase-2 and caspase-3 (Huang et al., 2021).

As anticipated, several peptides, such as AcVDVKD-CHO [casp-2 pKi: 7.63, casp-3 pKi: 6.91] and AcVDV(Dab) D-CHO [casp-2 pKi: 7.26, casp-3 pKi: 5.82], showed higher selectivity for caspase-2 than caspase-3. It's important to emphasize that using computational tools makes it easier to find leads and encourages a more focused and effective strategy. Although preclinical research has made great strides, it is still very difficult to turn these discoveries into safe and efficient therapies for Alzheimer's patients. More investigation and clinical trials are required to better understand the underlying causes of Alzheimer's and provide treatments (Moreno et al., 2022).

6.7.2 Natural and Synthetic Caspase Inhibitors

Numerous studies have demonstrated the critical role of aberrant caspase activation in a variety of disease disorders. The facts presented above clearly demonstrate its serious repercussions in AD. Thus, it is well recognized that regulating and inhibiting improperly produced caspases can significantly influence disease pathogenesis. Caspase inhibitors can be used as pharmacological assets to reduce disease pathology and partially restore normal functioning. Several caspase inhibitors have been identified or engineered and manufactured to treat caspase-mediated diseases. Caspase inhibitors are widely characterized as naturally occurring (viral and cellular inhibitors) CrmA, p35 family, inhibitor of apoptosis (IAP) proteins, and synthetic caspase inhibitors (Callus & Vaux, 2007; Deveraux & Reed, 1999Ray et al., 1992).

Synthetic agents that inhibit caspase can be classified as peptide-based or non-peptide substances. Several powerful phytoconstituents have also demonstrated caspase inhibitory action. The literature offers numerous approaches to overcoming apoptosis-related diseases in AD, however, our study will solely concentrate on caspase inhibitors in AD. Several caspase inhibitors are employed in AD-related investigations. Due to various barriers, only a few of these inhibitors (Q-VD-OPh, Z-DEVD-FMK, and VX-765) have made it to clinical trials. Drug delivery becomes a major limiting element in AD pathogenesis. Any inhibitor must be supplied constantly for several months, which is a difficult and costly task (Table 6.1) (Tesch et al., 2005).

TABLE 6.1
Caspase Inhibitors and Their Mechanisms of Action

S. No	Drug Name	Structure	MOA
		Broad Spectrum	
1.	Z-VAD-FMK	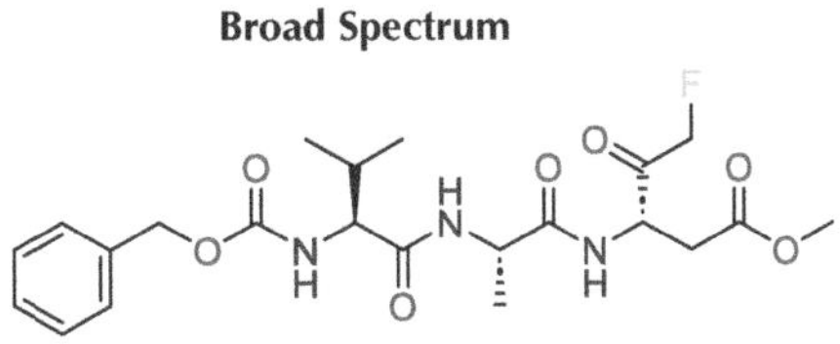	Reduction of: a. neuronal apoptosis b. both N and C-terminal tau truncation c. Aβ induced neurotoxicity
2.	Q-VD-OPh	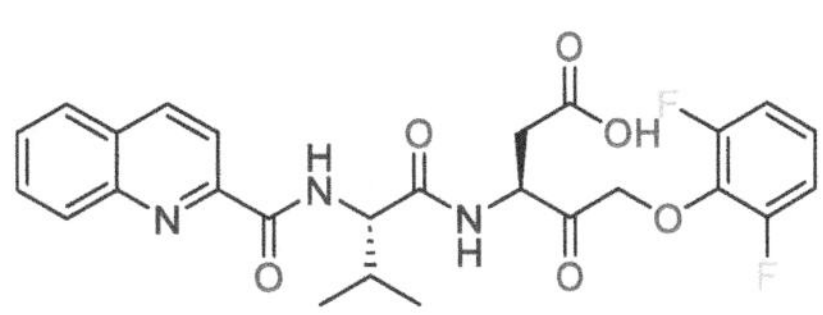	Reduction of: a. caspase cleavage. b. caspase-7 activation. c. pathological changes associated with tau
3.	Z-VAD-CH2-DCB	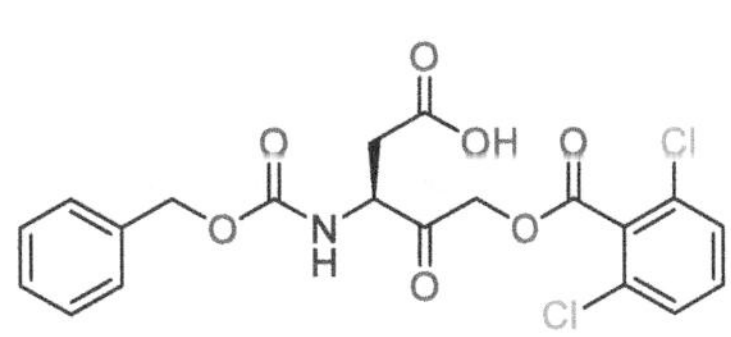	Inhibition of: a. caspase-3. b. cleavage of poly ADPribose polymerase (PARP)
		Caspase-1	
4.	Z-YVAD-FMK	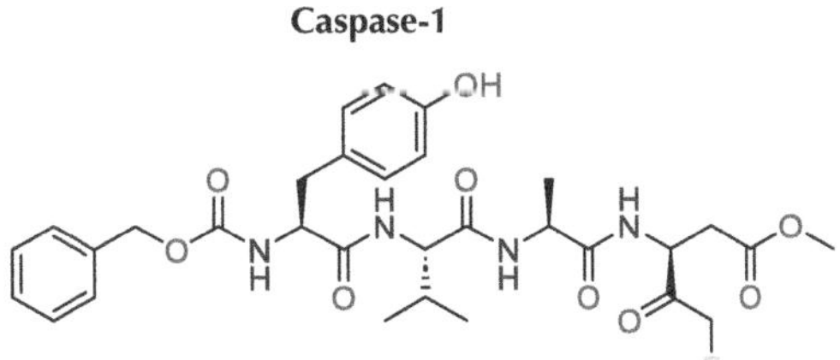	Inhibition of: a. caspase-1. b. caspase-6. c. IL-1β production in Aβ-stimulated microglia. Increased neuronal viability

(Continued)

TABLE 6.1 ***(Continued)***
Caspase Inhibitors and Their Mechanisms of Action

S. No	Drug Name	Structure	MOA
5.	VX-765	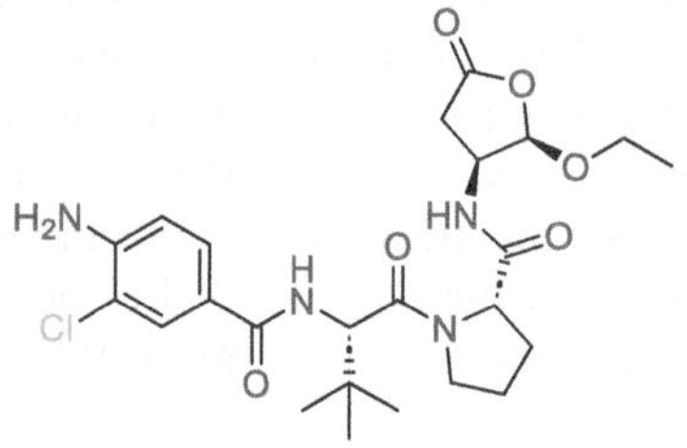	a. Inhibition of caspase 1. b. Reversal of both microglial and astroglial activation. c. Prevention of Aβ accumulation d. Reduction of IL-1β and TNF-α levels
6.	MCC950 (NLRP3 inflammasome Inhibitor)	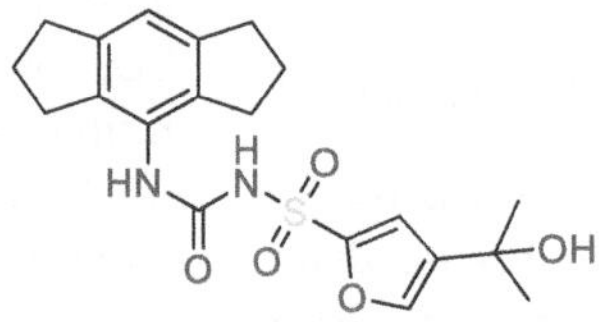	Inhibition of: a. caspase-1 activation in microglia. b. inflammasome and microglial activation c. Aβ accumulation
7.	OLT1177 (Dapansutrile) (NLRP3 inflammasome Inhibitor)	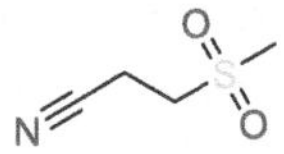	Acts of NLRP3 inflammasome preventing activation of caspase-1
8.	JC124 (NLRP3 inflammasome Inhibitor)	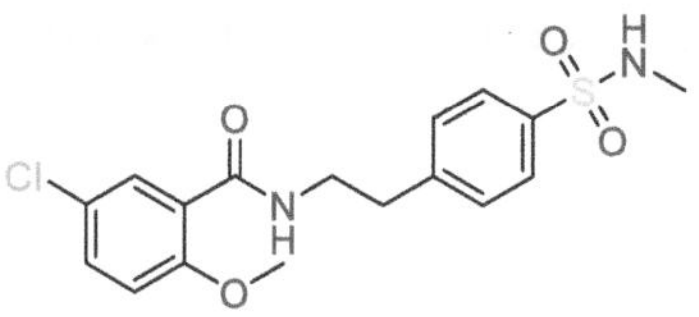	Acts on NLRP3 inflammasome thus blocking caspase-1 activation, IL-1β secretion, and adaptor apoptosis speck-like protein (ASC) aggregation
		Caspase-2	
9.	Z-VDVAD-FMK	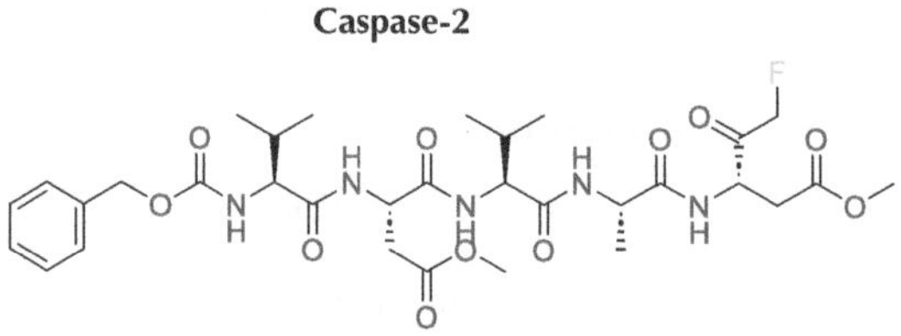	Increased cell viability
		Caspase-3	
10.	Z-DEVD-FMK	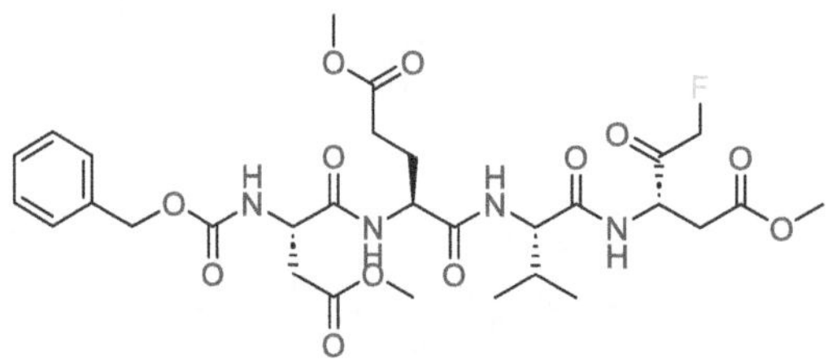	Increased protection against neuronal damage
11.	Ladostigil (Acetylcholine-butyrylcholine esterase and brain selective monoamine oxidase (MAO)-A and -B inhibitor)	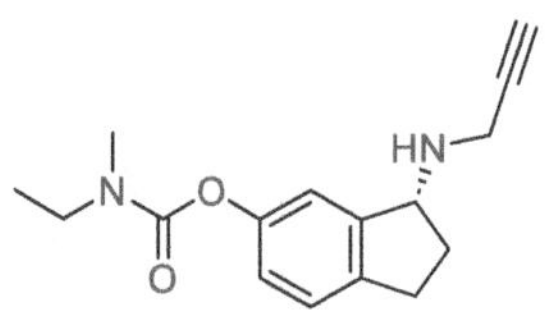	Prevents activation of caspase 3 and decreases intracellular APP levels
12.	[Gly14]-humanin (HNG)	–	Reduction of: a. caspase 3 expression. b. Aβ-induced disruption
13.	Tetrahedral DNA nanostructures (TDNs)	–	Activation and upregulation of extracellular signal-regulated protein kinase (ERK)1/2

(Continued)

TABLE 6.1 ***(Continued)***
Caspase Inhibitors and Their Mechanisms of Action

S. No	Drug Name	Structure	MOA
14.	Nanodiamond	–	a. Increased levels of phosphorylated STAT3 (pSTAT3) and Bcl-2 b. Reduced NF-ƙB and caspase-3 expression
15.	3-Hydroxybutyrate methyl ester (HBME)	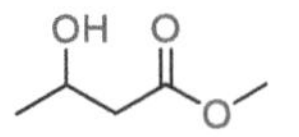	a. Reduction of ROS production b. Stabilizes the mitochondrial membrane potential
		Caspase-6	
16.	Z-VEID-FMK	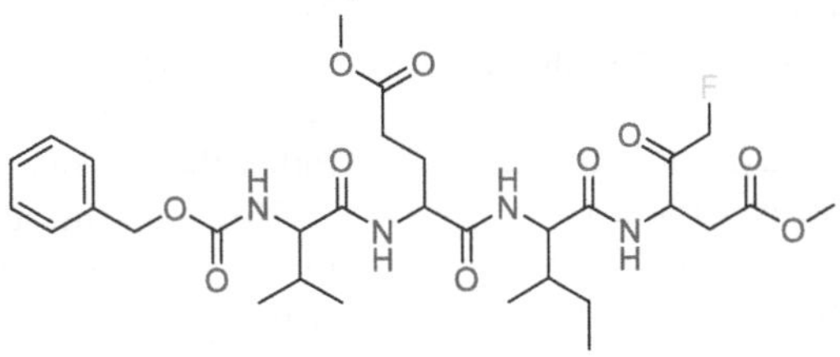	Protection of cerebellar granule cells from Aβ toxicity
		Caspase-8	
17.	IETD-CHO	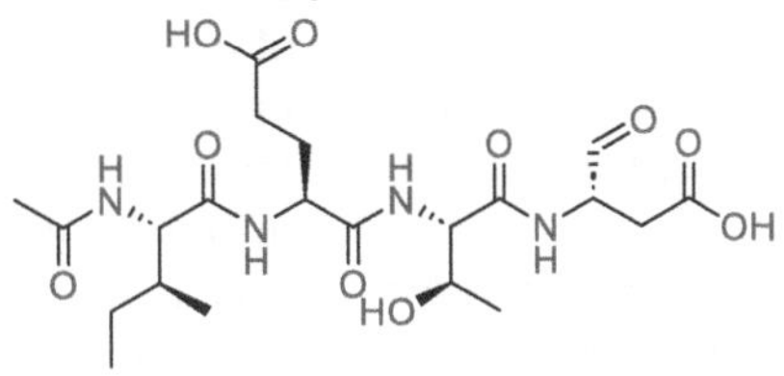	a. Inhibition of caspase-8 b. Alters the cleavage of APP and protects cells from Aβ induced cell death.
18.	Memantine	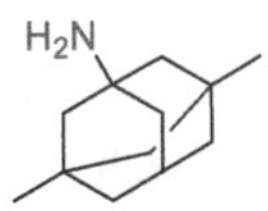	a. Inhibition of caspase 8 and Bcl-2 b. reduction of Aβ induced injury
		Caspase-9	
19.	Ac-LEHD-CHO	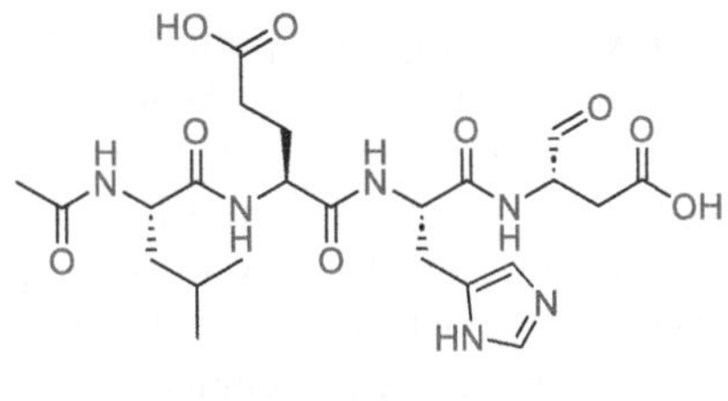	Protects from Aβ-induced cell death
		Multiple Caspase	
20.	Minocycline (Caspase-3,9,12)	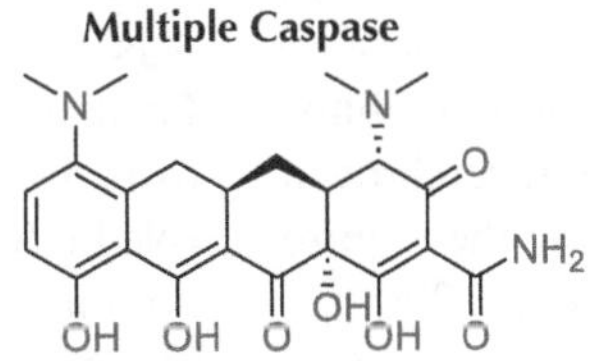	a. Inhibition of caspase-3. b. Decreased tau fragmentation c. Inhibition of caspase-9 activity
21.	DOPET (3,4-dihydroxyphenylethanol) (Caspase-3, 9)	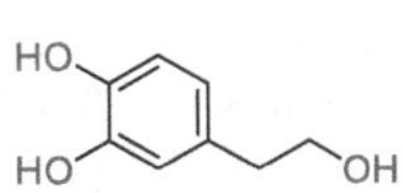	Increased ERK activity and decrease in levels of caspase 3 and 9
22.	Ac-DEVD-CHO (Caspase-3/7)	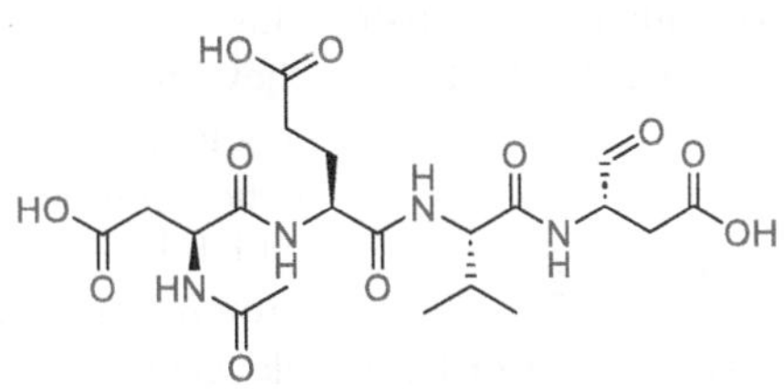	Protects from Aβ-induced: a. cell death b. DNA fragmentation and cleavage of α-Fodrin

TABLE 6.2
Caspase Classification and Functional Roles

Group	Caspase	Function
		Inflammation
1	1	Apoptosis, pyroptosis, metabolism
1	4	Apoptosis, ER stress
1	5	Not determined
1	11	Apoptosis, pyroptosis, leukocyte migration
		Non-Inflammatory group 1
1	12	ER stress
1	14	Keratinocyte differentiation
		Initiator Caspases
2	2	Autophagy, cell cycle, lipid metabolism, aging
2	8	Terminal keratinocyte differentiation, inflammasome, necroptosis, platelet formation, T and B cell proliferation
2	9	Erythropoiesis, synaptic loss, embryonic development
2	10	Interacts with caspase 8
		Executioner Caspases
3	3	Neuronal differentiation, apoptosis, stem cell differentiation, osteoblast differentiation, autophagy, plasticity
3	6	B-cell proliferation, cell cycle, synaptic loss
3	7	Bone development, dentinogenesis

Furthermore, the inhibitor should be able to pass through the blood-brain barrier, which is problematic for peptide-based compounds. The peptide-based inhibitors would require being directly injected intracerebroventricularly or by miniosmotic pumps, generating the need for surgery and rehabilitation. Non-peptide small molecules can overcome this constraint because they can be taken orally and cross the blood-brain barrier (Miura et al., 1995). However, the primary issue with these inhibitors is selectivity. Apoptosis is a common and necessary part of normal physiology. These medications are unable of differentiating between "required" and "abnormal" caspase activation inhibitors would necessitate being directly delivered intracerebroventricularly or by miniosmotic pumps, increasing the requirement for surgery and rehabilitation (Strasser et al., 2000).

Non-peptide small molecules may bypass this constraint because they can be taken orally and cross the blood-brain barrier. However, the primary issue with these inhibitors is selectivity. Apoptosis is a common and necessary part of normal physiology. These medications cannot distinguish between "required" and "abnormal" caspase activity (Table 6.2) (Xue & Horvitz, 1995).

6.8 CONCLUSION AND PERSPECTIVES

Neurodegenerative illnesses like AD have grown to be a major medical concern in the modern era due to population aging. At the brain level, death of neurons is seen, which is most likely the source of the cognitive dysfunction, even though outward symptoms include memory loss and cognitive dysfunction. The precise cause of neuronal loss is yet unknown, though. Although several theories have been proposed to explain this occurrence, the most significant and generally acknowledged ones are the mitochondrial failure, tau protein, and β-amyloid theories. Insoluble forms of amyloid β accumulate and form plaques in extracellular spaces and blood vessel walls in AD. In neurons, the tau microtubule protein also forms NFTs. The pathophysiology of AD begins with the deposition of Aβ in the cortex or NFTs, which can occur years before symptoms appear. Furthermore, because neurons require a lot of energy, they are particularly vulnerable to mitochondrial disorders. According to this idea, mitochondrial damage is a cause rather than an outcome of neurodegeneration.

However, the activation of caspases is a common theme in all of these ideas, as it eventually leads to cell death. Although caspase in-dependant methods can cause cell death, a body of evidence suggests that in AD, pyroptosis and apoptosis are the primary processes, and these types of cell death are completely caspase reliant. Still, only the mitochondrial explanation limits caspases to their role in the final step of cell death. According to this idea, the release of ROS from the mitochondria, which causes DNA damage, as well as the direct release of pro-apoptotic proteins from the mitochondria, activates the intrinsic route and caspase 9. Furthermore, by destroying proteins, ROS cause ER stress, which activates caspase 2. Caspases play dual roles in the β-amyloid and tau protein theories. Caspases, like the mitochondrial theory, are responsible for cell death because amyloid and tau are both extremely toxic. As a result, these proteins can trigger caspases in a variety of ways.

When they engage with death receptors, they trigger the extrinsic route, which activates caspase 8. Their buildup in the ER can produce ER stress, triggering caspase 2, and by interacting with mitochondrial proteins, they can activate the intrinsic route and caspase 9. However, genetic engineering experiments indicate that activation of caspase 8 is the most critical factor in causing apoptosis in AD. Furthermore, Aβ and tau can cause necroptosis and pyroptosis. Pyroptosis, which activates caspase 1, plays a crucial role in the pathogenesis of AD. Unlike apoptosis, pyroptosis not only intensifies but also significantly increases inflammation, leading to caspase activation (particularly caspase 8, as death receptors can be activated by pro-inflammatory TNFα). However, according to these hypotheses, caspases, particularly caspases 8 and 3, play a role in the early stages of AD formation because, as proteolytic enzymes, they break APP and tau protein into fragments that easily assemble. A positive feedback loop may be seen here, as the concentration of these proteins promotes caspase activity, which leads to further and stronger accumulation of these proteins and apoptosis if caspase activation is sufficiently elevated. Given the significant role caspases appear to play in AD, it is not surprise that caspase inhibitors are being investigated and evaluated as AD treatments. However, this comes with significant problems.

For starters, the blood-brain barrier prevents medications from entering the brain, and secondly, the action of caspases is required for the human body to operate; therefore, high doses or stronger drugs appear to be ineffective. Nonetheless, caspase inhibitors with significant therapeutic promise are still being investigated, and the most promising appear to interact specifically with caspases 3, 8, or 1. Inhibition of caspases 3 and 8 inhibits the extrinsic apoptotic pathway, which appears to be a critical mechanism in AD. Inhibiting caspases, which process Aβ and tau proteins, may help prevent AD at an early stage. On the other hand, caspase 1 inhibitors prevent pyroptosis, which not only reduces neuronal loss but also neuroinflammation.

REFERENCES

Bekdash, R. A. (2021). The cholinergic system, the adrenergic system and the neuropathology of Alzheimer's disease. *International Journal of Molecular Sciences*, *22*(3), 1273.

Breijyeh, Z., & Karaman, R. (2020). Comprehensive review on Alzheimer's disease: Causes and treatment. *Molecules*, *25*(24), 5789.

Buccellato, F. R., D'Anca, M., Fenoglio, C., Scarpini, E., & Galimberti, D. (2021). Role of oxidative damage in Alzheimer's disease and neurodegeneration: From pathogenic mechanisms to biomarker discovery. *Antioxidants*, *10*(9), 1353.

Caccamo, A., Branca, C., Piras, I. S., Ferreira, E., Huentelman, M. J., Liang, W. S., Readhead, B. et al (2017). Necroptosis activation in Alzheimer's disease. *Nature Neuroscience*, *20*(9), 1236–1246.

Callus, B. A., & Vaux, D. L. (2007). Caspase inhibitors: Viral, cellular and chemical. *Cell Death & Differentiation*, *14*(1), 73–78.

Cha, M.-Y., Han, S.-H., Son, S. M., Hong, H.-S., Choi, Y.-J., Byun, J., & Mook-Jung, I. (2012). Mitochondria-specific accumulation of amyloid β induces mitochondrial dysfunction leading to apoptotic cell death. *PLoS ONE*, *7*(4), e34929.

Chen, J. X., & Yan, S. D. (2007). Amyloid-β-induced mitochondrial dysfunction. *Journal of Alzheimer's Disease*, *12*(2), 177–184.

Cohen, G. M. (1997). Caspases: The executioners of apoptosis. *Biochemical Journal*, *326*(1), 1–16.

de Castro, A. A., Soares, F. V., Pereira, A. F., Polisel, D. A., Caetano, M. S., Leal, D. H. S., da Cunha, E. F. F., Nepovimova, E., Kuca, K., & Ramalho, T. C. (2019). Non-conventional compounds with potential therapeutic effects against Alzheimer's disease. *Expert Review of Neurotherapeutics*, *19*(5), 375–395.

Denecker, G., Ovaere, P., Vandenabeele, P., & Declercq, W. (2008). Caspase-14 reveals its secrets. *The Journal of Cell Biology*, *180*(3), 451–458.

Deveraux, Q. L., & Reed, J. C. (1999). IAP family proteins—Suppressors of apoptosis. *Genes & Development*, *13*(3), 239–252.

Dhani, S., Zhao, Y., & Zhivotovsky, B. (2021). A long way to go: Caspase inhibitors in clinical use. *Cell Death & Disease*, *12*(10), 949.

Ding, H., Matthews, T. A., & Johnson, G. V. W. (2006). Site-specific phosphorylation and caspase cleavage differentially impact tau-microtubule interactions and tau aggregation. *Journal of Biological Chemistry*, *281*(28), 19107–19114.

Fan, L., Mao, C., Hu, X., Zhang, S., Yang, Z., Hu, Z., & Sun, H. et al (2020). New insights into the pathogenesis of Alzheimer's disease. *Frontiers in Neurology*, *10*, 1312.

Farmer, K. M., Ghag, G., Puangmalai, N., Montalbano, M., Bhatt, N., & Kayed, R. (2020). P53 aggregation, interactions with tau, and impaired DNA damage response in Alzheimer's disease. *Acta Neuropathologica Communications*, *8*, 1–21.

Fossati, S., Ghiso, J., & Rostagno, A. (2012). TRAIL death receptors DR4 and DR5 mediate cerebral microvascular endothelial cell apoptosis induced by oligomeric Alzheimer's Aβ. *Cell Death & Disease*, *3*(6), e321.

Fossati, S., Giannoni, P., Solesio, M. E., Cocklin, S. L., Cabrera, E., Ghiso, J., & Rostagno, A. (2016). The carbonic anhydrase inhibitor methazolamide prevents amyloid beta-induced mitochondrial dysfunction and caspase activation protecting neuronal and glial cells in vitro and in the mouse brain. *Neurobiology of Disease*, *86*, 29–40.

Friedlander, R. M. (2003). Apoptosis and caspases in neurodegenerative diseases. *New England Journal of Medicine*, *348*(14), 1365–1375.

Fuentes-Prior, P., & Salvesen, G. S. (2004). The protein structures that shape caspase activity, specificity, activation and inhibition. *Biochemical Journal*, *384*(2), 201–232.

Gamblin, T. C., Chen, F., Zambrano, A., Abraha, A., Lagalwar, S., Guillozet, A. L., Lu, M., et al. (2003). Caspase cleavage of tau: Linking amyloid and neurofibrillary tangles in Alzheimer's disease. *Proceedings of the National Academy of Sciences*, *100*(17), 10032–10037.

Garcia-Calvo, M., Peterson, E. P., Rasper, D. M., Vaillancourt, J. P., Zamboni, R., Nicholson, D. W., & Thornberry, N. A. (1999). Purification and catalytic properties of human caspase family members. *Cell Death & Differentiation*, *6*(4), 362–369.

Gervais, F. G., Xu, D., Robertson, G. S., Vaillancourt, J. P., Zhu, Y., Huang, J., LeBlanc, A., et al. (1999). Involvement of caspases in proteolytic cleavage of Alzheimer's amyloid-β precursor protein and amyloidogenic Aβ peptide formation. *Cell*, *97*(3), 395–406.

Götz, J., Chen, F. V., Van Dorpe, J., & Nitsch, R. M. (2001). Formation of neurofibrillary tangles in P301L tau transgenic mice induced by Aβ42 fibrils. *Science*, *293*(5534), 1491–1495.

Grivennikova, V. G., & Vinogradov, A. D. (2006). Generation of superoxide by the mitochondrial Complex I. *Biochimica et Biophysica Acta (BBA) - Bioenergetics*, *1757*(5–6), 553–561.

Guo, H., Albrecht, S., Bourdeau, M., Petzke, T., Bergeron, C., & LeBlanc, A. C. (2004). Active caspase-6 and caspase-6-cleaved tau in neuropil threads, neuritic plaques, and neurofibrillary tangles of Alzheimer's disease. *The American Journal of Pathology*, *165*(2), 523–531.

Gurung, P., Anand, P. K., Malireddi, R. K., Vande Walle, L., Van Opdenbosch, N., Dillon, C. P., Weinlich, R., Green, D. R., Lamkanfi, M., & Kanneganti, T.-D. (2014). FADD and caspase-8 mediate priming and activation of the canonical and noncanonical Nlrp3 inflammasomes. *The Journal of Immunology*, *192*(4), 1835–1846.

Hernandez-Zimbron, L. F., Luna-Muñoz, J., Mena, R., Vazquez-Ramirez, R., Kubli-Garfias, C., Cribbs, D. H., Manoutcharian, K., & Gevorkian, G. (2012). Amyloid-β peptide binds to cytochrome C oxidase subunit 1. *PLoS ONE*, *7*(4), e42344.

Huang, S.-H., Fang, S.-T., & Chen, Y.-C. (2021). Molecular mechanism of vitamin K2 protection against amyloid-β-induced cytotoxicity. *Biomolecules*, *11*(3), 423.

Huber, C. M., Yee, C., May, T., Dhanala, A., & Mitchell, C. S. (2017). Cognitive decline in preclinical Alzheimer's disease: Amyloid-beta versus tauopathy. *Journal of Alzheimer's Disease*, *61*(1), 265–281.

Islam, M. I., Sharoar, M. G., Ryu, E.-K., & Park, I.-S. (2017). Limited activation of the intrinsic apoptotic pathway plays a main role in amyloid-β-induced apoptosis without eliciting the activation of the extrinsic apoptotic pathway. *International Journal of Molecular Medicine*, *40*(6), 1971–1982.

Jellinger, K. A. (2012). Neuropathological subtypes of Alzheimer's disease. *Acta Neuropathologica*, *123*(1), 153–154.

Jellinger, K. A. (2022). Recent update on the heterogeneity of the Alzheimer's disease spectrum. *Journal of Neural Transmission*, *129*(1), 1–24.

Justin-Thenmozhi, A., Dhivya Bharathi, M., Kiruthika, R., Manivasagam, T., Borah, A., & Essa, M. M. (2018). Attenuation of aluminum chloride-induced neuroinflammation and caspase activation through the AKT/GSK-3β pathway by hesperidin in Wistar rats. *Neurotoxicity Research 34*, 463–476.

Kametani, F., & Hasegawa, M. (2018). Reconsideration of amyloid hypothesis and tau hypothesis in Alzheimer's disease. *Frontiers in Neuroscience*, *12*, 25.

Kesavardhana, S., Malireddi, R. S., & Kanneganti, T.-D. (2020). Caspases in cell death, inflammation, and pyroptosis. *Annual Review of Immunology*, *38*, 567–595.

Kolarova, M., García-Sierra, F., Bartos, A., Ricny, J., & Ripova, D. (2012). Structure and pathology of tau protein in Alzheimer disease. *International Journal of Alzheimer's Disease*, *2012*, 731526.

Kumar, S. (1999). Mechanisms mediating caspase activation in cell death. *Cell Death & Differentiation*, *6*(11), 1060–1066.

Kumar, S., Budhathoki, S., Oliveira, C. B., Kahle, A. D., Calhan, O. Y., Lukens, J. R., & Deppmann, C. D. (2023). Role of the caspase-8/RIPK3 axis in Alzheimer's disease pathogenesis and Aβ-induced NLRP3 inflammasome activation. *JCI Insight*, *8*(3).

Kumar, S., Dorstyn, L., & Lim, Y. (2022). The role of caspases as executioners of apoptosis. *Biochemical Society Transactions*, *50*(1), 123–136.

Lamkanfi, M., Kanneganti, T.-D., Van Damme, P., Vanden Berghe, T., Vanoverberghe, I., Vandekerckhove, J., Vandenabeele, P., Gevaert, K., & Nuñez, G. (2008). Targeted peptidecentric proteomics reveals caspase-7 as a substrate of the caspase-1 inflammasomes. *Molecular & Cellular Proteomics*, *7*(12), 2350–2363.

Lee, D. Y., Lee, K.-S., Lee, H. J., Kim, D. H., Noh, Y. H., Yu, K., Jung, H.-Y., et al. (2010). Activation of PERK signaling attenuates Aβ-mediated ER stress. *PLoS ONE*, *5*(5), e10489.

Lei, S., Wu, S., Wang, G., Li, B., Liu, B., & Lei, X. (2021). Pinoresinol diglucoside attenuates neuroinflammation, apoptosis and oxidative stress in a mice model with Alzheimer's disease. *Neuroreport*, *32*(3), 259–267.

Li, M., Ona, V. O., Guegan, C., Chen, M., Jackson-Lewis, V., Andrews, L. J., Olszewski, A. J., et al. (2000). Functional role of caspase-1 and caspase-3 in an ALS transgenic mouse model. *Science*, *288*(5464), 335–339.

Lin, M. T., & Beal, M. F. (2006). Mitochondrial dysfunction and oxidative stress in neurodegenerative diseases. *Nature*, *443*(7113), 787–795.

Lippens, S., Kockx, M., Knaapen, M., Mortier, L., Polakowska, R., Verheyen, A., Garmyn, M., et al. (2000). Epidermal differentiation does not involve the pro-apoptotic executioner caspases, but is associated with caspase-14 induction and processing. *Cell Death & Differentiation*, *7*(12), 1218–1224.

Lorton, D., Kocsis, J.-M., King, L., Madden, K., & Brunden, K. R. (1996). β-Amyloid induces increased release of interleukin-1β from lipopolysaccharide-activated human monocytes. *Journal of Neuroimmunology*, *67*(1), 21–29.

Magi, S., Castaldo, P., Macrì, M. L., Maiolino, M., Matteucci, A., Bastioli, G., Gratteri, S., Amoroso, S., & Lariccia, V. (2016). Intracellular calcium dysregulation: Implications for Alzheimer's disease. *BioMed Research International*, *2016*, 6701324.

Makhaeva, G. F., Lushchekina, S. V., Boltneva, N. P., Sokolov, V. B., Grigoriev, V. V., Serebryakova, O. G., Vikhareva, E. A., et al. (2015). Conjugates of γ-carbolines and phenothiazine as new selective inhibitors of butyrylcholinesterase and blockers of NMDA receptors for Alzheimer disease. *Scientific Reports*, *5*(1), 13164.

Martinou, J.-C., Dubois-Dauphin, M., Staple, J. K., Rodriguez, I., Frankowski, H., Missotten, M., Albertini, P., et al. (1994). Overexpression of BCL-2 in transgenic mice protects neurons from naturally occurring cell death and experimental ischemia. *Neuron*, *13*(4), 1017–1030.

Masters, S. L., & O'Neill, L. A. J. (2011). Disease-associated amyloid and misfolded protein aggregates activate the inflammasome. *Trends in Molecular Medicine*, *17*(5), 276–282.

McIlwain, D. R., Berger, T., & Mak, T. W. (2013). Caspase functions in cell death and disease. *Cold Spring Harbor Perspectives in Biology*, *5*(4), a008656.

Means, J. C., Gerdes, B. C., Kaja, S., Sumien, N., Payne, A. J., Stark, D. A., Borden, P. K., Price, J. L., & Koulen, P. (2016). Caspase-3-dependent proteolytic cleavage of tau causes neurofibrillary tangles and results in cognitive impairment during normal aging. *Neurochemical Research*, *41*, 2278–2288.

Miura, M., Friedlander, R. M., & Yuan, J. (1995). Tumor necrosis factor-induced apoptosis is mediated by a CrmA-sensitive cell death pathway. *Proceedings of the National Academy of Sciences*, *92*(18), 8318–8322.

Monie, T. P., & Bryant, C. E. (2015). Caspase-8 functions as a key mediator of inflammation and pro-IL-1β processing via both canonical and non-canonical pathways. *Immunological Reviews*, *265*(1), 181–193.

Moreno, R. I., Zambelli, V. O., Picolo, G., Cury, Y., Morandini, A. C., Marques, A. C., & Sciani, J. M. (2022). Caspase-1 and cathepsin B inhibitors from marine invertebrates, aiming at a reduction in neuroinflammation. *Marine Drugs*, *20*(10), 614.

Mortimer, J. A., Borenstein, A. R., Gosche, K. M., & Snowdon, D. A. (2005). Very early detection of Alzheimer neuropathology and the role of brain reserve in modifying its clinical expression. *Journal of Geriatric Psychiatry and Neurology*, *18*(4), 218–223.

Muralidar, S., Ambi, S. V., Sekaran, S., Thirumalai, D., & Palaniappan, B. (2020). Role of tau protein in Alzheimer's disease: The prime pathological player. *International Journal of Biological Macromolecules*, *163*, 1599–1617.

Norman, J. M., Cohen, G. M., & Bampton, E. T. W. (2010). The in vitro cleavage of the hAtg proteins by cell death proteases. *Autophagy*, *6*(8), 1042–1056.

Nussbaum, R. L. (2003). Alzheimer's disease and Parkinson's disease. *The New England Journal of Medicine*, *348*(14), 1356–1364.

Ona, V. O., Li, M., Vonsattel, J. P. G., Andrews, L. J., Khan, S. Q., Chung, W. M., Frey, A. S., et al (1999). Inhibition of caspase-1 slows disease progression in a mouse model of Huntington's disease. *Nature*, *399*(6733), 263–267.

Oral, O., Oz-Arslan, D., Itah, Z., Naghavi, A., Deveci, R., Karacali, S., & Gozuacik, D. (2012). Cleavage of Atg3 protein by caspase-8 regulates autophagy during receptor-activated cell death. *Apoptosis, 17*, 810–820.

Paradis, E., Douillard, H., Koutroumanis, M., Goodyer, C., & LeBlanc, A. (1996). Amyloid β peptide of Alzheimer's disease downregulates Bcl-2 and upregulates Bax expression in human neurons. *Journal of Neuroscience, 16*(23), 7533–7539.

Park, G., Nhan, H. S., Tyan, S. H., Kawakatsu, Y., Zhang, C., Navarro, M., & Koo, E. H. (2020). Caspase activation and caspase-mediated cleavage of APP is associated with amyloid β-protein-induced synapse loss in Alzheimer's disease. *Cell Reports, 31*(13), 107839.

Paroni, G., Bisceglia, P., & Seripa, D. (2019). Understanding the amyloid hypothesis in Alzheimer's disease. *Journal of Alzheimer's Disease, 68*(2), 493–510.

Pavlov, P. F., Wiehager, B., Sakai, J., Frykman, S., Behbahani, H., Winblad, B., & Ankarcrona, M. (2011). Mitochondrial γ-secretase participates in the metabolism of mitochondria-associated amyloid precursor protein. *The FASEB Journal, 25*(1), 78–88.

Peña-Bautista, C., Baquero, M., Vento, M., & Cháfer-Pericás, C. (2019). Free radicals in Alzheimer's disease: Lipid peroxidation biomarkers. *Clinica Chimica Acta, 491*, 85–90.

Perneczky, R., Wagenpfeil, S., Lunetta, K. L., Cupples, L. A., Green, R. C., Decarli, C., Farrer, L. A., & Kurz, A. (2010). Head circumference, atrophy, and cognition: Implications for brain reserve in Alzheimer disease. *Neurology, 75*(2), 137–142.

Poirier, Y., Grimm, A., Schmitt, K., & Eckert, A. (2019). Link between the unfolded protein response and dysregulation of mitochondrial bioenergetics in Alzheimer's disease. *Cellular and Molecular Life Sciences, 76*, 1419–1431.

Pop, C., & Salvesen, G. S. (2009). Human caspases: Activation, specificity, and regulation. *Journal of Biological Chemistry, 284*(33), 21777–21781.

Pozueta, J., Lefort, R., Ribe, E. M., Troy, C. M., Arancio, O., & Shelanski, M. (2013). Caspase-2 is required for dendritic spine and behavioural alterations in J20 APP transgenic mice. *Nature Communications, 4*(1), 1939.

Puzzo, D., Privitera, L., Leznik, E., Fa, M., Staniszewski, A., Palmeri, A., & Arancio, O. (2008). Picomolar amyloid-β positively modulates synaptic plasticity and memory in hippocampus. *Journal of Neuroscience, 28*(53), 14537–14545.

Qian, M., Liu, J., Yao, J., Wang, W., Yang, J., Wei, L., Shen, Y., & Chen, W. (2015). Caspase-8 mediates amyloid-β-induced apoptosis in differentiated PC12 cells. *Journal of Molecular Neuroscience, 56*, 491–499.

Quintanilla, R. A., Dolan, P. J., Jin, Y. N., & Johnson, G. V. W. (2012). Truncated tau and Aβ cooperatively impair mitochondria in primary neurons. *Neurobiology of Aging, 33*(3), 619.e25–619.e35.

Quintanilla, R. A., Matthews-Roberson, T. A., Dolan, P. J., & Johnson, G. V. W. (2009). Caspase-cleaved tau expression induces mitochondrial dysfunction in immortalized cortical neurons: Implications for the pathogenesis of Alzheimer disease. *Journal of Biological Chemistry, 284*(28), 18754–18766.

Ramos-Tovar, E., & Muriel, P. (2020). Molecular mechanisms that link oxidative stress, inflammation, and fibrosis in the liver. *Antioxidants, 9*(12), 1279.

Ray, C. A., Black, R. A., Kronheim, S. R., Greenstreet, T. A., Sleath, P. R., Salvesen, G. S., & Pickup, D. J. (1992). Viral inhibition of inflammation: Cowpox virus encodes an inhibitor of the interleukin-1β converting enzyme. *Cell, 69*(4), 597–604.

Reddy, P. H. (2006). Amyloid precursor protein-mediated free radicals and oxidative damage: Implications for the development and progression of Alzheimer's disease. *Journal of Neurochemistry, 96*(1), 1–13.

Richard, R., & Mousa, S. (2022). Necroptosis in Alzheimer's disease: Potential therapeutic target. *Biomedicine & Pharmacotherapy, 152*, 113203.

Riedl, S. J., & Shi, Y. (2004). Molecular mechanisms of caspase regulation during apoptosis. *Nature Reviews Molecular Cell Biology, 5*(11), 897–907.

Rissman, R. A., Poon, W. W., Blurton-Jones, M., Oddo, S., Torp, R., Vitek, M. P., LaFerla, F. M., Rohn, T. T., & Cotman, C. W. (2004). Caspase-cleavage of tau is an early event in Alzheimer disease tangle pathology. *The Journal of Clinical Investigation, 114*(1), 121–130.

Rohn, T. T., Head, E., Su, J. H., Anderson, A. J., Bahr, B. A., Cotman, C. W., & Cribbs, D. H. (2001). Correlation between caspase activation and neurofibrillary tangle formation in Alzheimer's disease. *The American Journal of Pathology, 158*(1), 189–198.

Rohn, T. T., Rissman, R. A., Head, E., & Cotman, C. W. (2002). Caspase activation in the Alzheimer's disease brain: Tortuous and torturous. *Drug News & Perspectives, 15*(9), 549–557.

Saha, A., Saleem, S., Paidi, R. K., & Biswas, S. C. (2021). BH3-only proteins Puma and Beclin1 regulate autophagic death in neurons in response to amyloid-β. *Cell Death Discovery, 7*(1), 356.

Sanabria-Castro, A., Alvarado-Echeverría, I., & Monge-Bonilla, C. (2017). Molecular pathogenesis of Alzheimer's disease: An update. *Annals of Neurosciences, 24*(1), 46–54.

Schierle, G. S., Hansson, O., Leist, M., Nicotera, P., Widner, H., & Brundin, P. (1999). Caspase inhibition reduces apoptosis and increases survival of nigral transplants. *Nature Medicine, 5*(1), 97–100.

Shalini, S., Dorstyn, L., Dawar, S., & Kumar, S. (2015). Old, new and emerging functions of caspases. *Cell Death & Differentiation, 22*(4), 526–539.

Sharma, A., Rudrawar, S., Bharate, S. B., & Jadhav, H. R. (2025). Recent advancements in the therapeutic approaches for Alzheimer's disease treatment: Current and future perspective. *RSC Medicinal Chemistry, 16*(2):652–693.

Sharma, A., Rudrawar, S., Sharma, A., Bharate, S. B., & Jadhav, H. R. (2024a). Unveiling the potential of novel indol-3-yl-phenyl allylidene hydrazine carboximidamide derivatives as AChE/BACE1 dual inhibitors: A combined *in silico*, synthesis and *in vitro* study. *RSC Advances, 14*(33), 23853–23872.

Sharma, A., Rudrawar, S., Sharma, A., Bharate, S. B., & Jadhav, H. R. (2024b). Design, synthesis, *in silico*, and *in vitro* evaluation of pyrrol-2-yl-phenyl allylidene hydrazine carboximidamide derivatives as AChE/BACE1 dual inhibitors. *RSC Advances, 14*(37), 26703–26722.

Sharoar, M. G., Islam, M. I., Shahnawaz, M., Shin, S. Y., & Park, I.-S. (2014). Amyloid β binds procaspase-9 to inhibit assembly of Apaf-1 apoptosome and intrinsic apoptosis pathway. *Biochimica et Biophysica Acta (BBA) - Molecular Cell Research, 1843*(4), 685–693.

Shimohama, S. (2000). Apoptosis in Alzheimer's disease—An update. *Apoptosis, 5*(1), 9–16.

Spires-Jones, T. L., De Calignon, A., Matsui, T., Zehr, C., Pitstick, R., Wu, H.-Y., & Osetek, J. D., et al. (2008). In vivo imaging reveals dissociation between caspase activation and acute neuronal death in tangle-bearing neurons. *Journal of Neuroscience*, *28*(4), 862–867.

Steckley, D., Karajgikar, M., Dale, L. B., Fuerth, B., Swan, P., Drummond-Main, C., Poulter, M. O., Ferguson, S. S. G., Strasser, A., & Cregan, S. P. (2007). Puma is a dominant regulator of oxidative stress induced Bax activation and neuronal apoptosis. *Journal of Neuroscience*, *27*(47), 12989–12999.

Stennicke, H. R., & Salvesen, G. S. (1997). Biochemical characteristics of caspases-3, -6, -7, and -8. *Journal of Biological Chemistry*, *272*(41), 25719–25723.

Stoothoff, W. H., & Johnson, G. V. W. (2005). Tau phosphorylation: Physiological and pathological consequences. *Biochimica et Biophysica Acta (BBA) - Molecular Basis of Disease*, *1739*(2–3), 280–297.

Strasser, A., O'Connor, L., & Dixit, V. M. (2000). Apoptosis signaling. *Annual Review of Biochemistry*, *69*, 217–245.

Swerdlow, R. H., Burns, J. M., & Khan, S. M. (2010). The Alzheimer's disease mitochondrial cascade hypothesis. *Journal of Alzheimer's Disease*, *20*(s2), S265–S279.

Tenev, T., Bianchi, K., Darding, M., Broemer, M., Langlais, C., Wallberg, F., Zachariou, A., et al. (2011). The Ripoptosome, a signaling platform that assembles in response to genotoxic stress and loss of IAPs. *Molecular Cell*, *43*(3), 432–448.

Tesch, L. D., Raghavendra, M. P., Bedsted-Faarvang, T., Gettins, P. G. W., & Olson, S. T. (2005). Specificity and reactive loop length requirements for crmA inhibition of serine proteases. *Protein Science*, *14*(2), 533–542.

Troy, C. M., Rabacchi, S. A., Friedman, W. J., Frappier, T. F., Brown, K., & Shelanski, M. L. (2000). Caspase-2 mediates neuronal cell death induced by β-amyloid. *Journal of Neuroscience*, *20*(4), 1386–1392.

Vukosavic, S., Stefanis, L., Jackson-Lewis, V., Guegan, C., Romero, N., Chen, C., Dubois-Dauphin, M., & Przedborski, S. (2000). Delaying caspase activation by Bcl-2: A clue to disease retardation in a transgenic mouse model of amyotrophic lateral sclerosis. *Journal of Neuroscience*, *20*(24), 9119–9125.

Walker, N. P. C., Talanian, R. V., Brady, K. D., Dang, L. C., Bump, N. J., Ferenza, C. R., Franklin, S., et al. (1994). Crystal structure of the cysteine protease interleukin-1β-converting enzyme: A $(p20/p10)_2$ homodimer. *Cell*, *78*(2), 343–352.

Wang, H.-Y., Li, W., Benedetti, N. J., & Lee, D. H. S. (2003). α7 nicotinic acetylcholine receptors mediate β-amyloid peptide-induced tau protein phosphorylation. *Journal of Biological Chemistry*, *278*(34), 31547–31553.

Wang, W.-Y., Tan, M.-S., Yu, J.-T., & Tan, L. (2015). Role of pro-inflammatory cytokines released from microglia in Alzheimer's disease. *Annals of Translational Medicine*, *3*(10), 136.

Wen, H., Fu, Z., Wei, Y., Zhang, X., Ma, L., Gu, L., & Li, J. (2018). Antioxidant activity and neuroprotective activity of stilbenoids in rat primary cortex neurons via the PI3K/Akt signalling pathway. *Molecules*, *23*(9), 2328.

Wilson, K. P., Black, J. A. F., Thomson, J. A., Kim, E. E., Griffith, J. P., Navia, M. A., Murcko, M. A., et al. (1994). Structure and mechanism of interleukin-1β converting enzyme. *Nature*, *370*(6487), 270–275.

Wójcik, P., Jastrzębski, M. K., Zięba, A., Matosiuk, D., & Kaczor, A. A. (2024). Caspases in Alzheimer's disease: Mechanism of activation, role, and potential treatment. *Molecular Neurobiology*, *61*(7), 4834–4853.

Xue, D., & Horvitz, H. R. (1995). Inhibition of the *Caenorhabditis elegans* cell-death protease CED-3 by a CED-3 cleavage site in baculovirus p35 protein. *Nature*, *377*(6546), 248–251.

Yan, S. D., Bierhaus, A., Nawroth, P. P., & Stern, D. M. (2009). RAGE and Alzheimer's disease: A progression factor for amyloid-β-induced cellular perturbation? *Journal of Alzheimer's Disease*, *16*(4), 833–843.

Yang, F., Sun, X., Beech, W., Teter, B., Wu, S., Sigel, J., Vinters, H. V., Frautschy, S. A., & Cole, G. M. (1998). Antibody to caspase-cleaved actin detects apoptosis in differentiated neuroblastoma and plaque-associated neurons and microglia in Alzheimer's disease. *The American Journal of Pathology*, *152*(2), 379–389.

Yates, S. L., Burgess, L. H., Kocsis-Angle, J., Antal, J. M., Dority, M. D., Embury, P. B., Piotrkowski, A. M., & Brunden, K. R. (2000). Amyloid β and amylin fibrils induce increases in proinflammatory cytokine and chemokine production by THP-1 cells and murine microglia. *Journal of Neurochemistry*, *74*(3), 1017–1025.

Yin, H., & Kuret, J. (2006). C-terminal truncation modulates both nucleation and extension phases of τ fibrillization. *FEBS Letters*, *580*(1), 211–215.

Zhao, X., Kotilinek, L. A., Smith, B., Hlynialuk, C., Zahs, K., Ramsden, M., Cleary, J., & Ashe, K. H. (2016). Caspase-2 cleavage of tau reversibly impairs memory. *Nature Medicine*, *22*(11), 1268–1276.

Zhu, X., Perry, G., Moreira, P. I., Aliev, G., Cash, A. D., Hirai, K., & Smith, M. A. (2006). Mitochondrial abnormalities and oxidative imbalance in Alzheimer disease. *Journal of Alzheimer's Disease*, *9*(2), 147–153.

Zlokovic, B. V. (2004). Clearing amyloid through the blood–brain barrier. *Journal of Neurochemistry*, *89*(4), 807–811.

7 Calpain Inhibitors
A Promising Therapeutic Approach in Alzheimer's Disease

Ehsan Ullah Mughal, Nafeesa Naeem, and Amina Sadiq

7.1 INTRODUCTION

Alzheimer's disease (AD) is a progressive neurodegenerative disorder that predominantly affects older adults, causing cognitive decline, decreased awareness, and changes in behavior (Scheltens et al., 2021). The underlying pathology of AD is characterized by the accumulation of amyloid-beta (Aβ) plaques and the presence of hyperphosphorylated tau proteins, which aggregate to form neurofibrillary tangles. These abnormal protein deposits impair synaptic activity, leading to neuronal death, brain shrinkage, and the onset of clinical manifestations associated with the disease (Breijyeh & Karaman, 2020; Cummings et al., 2020; Veitch et al., 2019).

7.1.1 Overview of Alzheimer's Disease

The molecular mechanisms underlying AD are complex and involve a combination of biological, environmental, and genetic factors (Tahami Monfared et al., 2022). One of the most widely accepted theories, the amyloid cascade hypothesis, suggests that Aβ peptide accumulation triggers a cascade of events leading to neurodegeneration (Weiner et al., 2017). The tau hypothesis suggests that the abnormal aggregation and hyperphosphorylation of tau proteins disrupt microtubule stability, leading to neurodegeneration in AD. Along with amyloid and tau pathologies, chronic neuroinflammation, oxidative stress, and mitochondrial dysfunction are also key factors that drive the progression of AD (Obaid et al., 2022). Despite numerous studies, the exact etiology of AD remains elusive, posing significant challenges to the development of effective therapies (Khan et al., 2022; Knopman et al., 2021; Metwally et al., 2023; Obaid et al., 2022).

7.2 CALPAIN: STRUCTURE AND FUNCTION

Calpains are a family of calcium-dependent cysteine proteases that play significant roles in many cellular processes (Donkor, 2020; Sorimachi & Suzuki, 2001). Calpains modulate their activity by selectively cleaving target proteins. Localization serves as a key intermediary in the cellular signaling pathway (Mahaman et al., 2019). Abnormalities in calpain activity are involved in many diseases, including abnormalities of the nervous system, cancer, and cardiovascular conditions, emphasizing their importance in health and disease (Chen et al., 2018).

7.2.1 Calpain Family Overview

The calpain family consists of several isoforms, which are classified according to the structure and tissue distribution (Prajapati & Singh, 2024). Calpains are categorized into distinct subclasses, each with unique structural and functional properties. Classical calpains, such as μ-calpain and m-calpain, are well-known for their calcium-dependent protease activity. Tissue-specific calpains exhibit restricted expression in specific tissues, contributing to specialized cellular functions. Non-classical calpains, unlike classical forms, have altered calcium requirements or lack typical regulatory mechanisms. Atypical calpains possess distinct structural features, influencing their protease activity and substrate specificity. Lastly, calpain-like proteins share similarities with calpains but may not exhibit protease activity, suggesting diverse roles in cellular processes (Prajapati & Singh, 2024). Additionally, the two most studied isoforms are calpain-1 (μ-calpain) and calpain-2 (m-calpain), which are ubiquitously expressed in tissues and differ mainly in calcium requirements for stimulation (Wang, Liu, et al., 2020). Calpain-1 requires micromolar concentrations of calcium, while calpain-2 is activated at millimolar levels (Figure 7.1) (Fa et al., 2016). Both isoforms exist as heterodimers. It consists of a large catalytic subunit (~80 kDa) and a small regulatory subunit (~30 kDa). The catalytic subunit consists of four domains: an *N*-terminal anchor helix, a protease core with active-site cysteine, a C2-like domain involved in calcium binding,

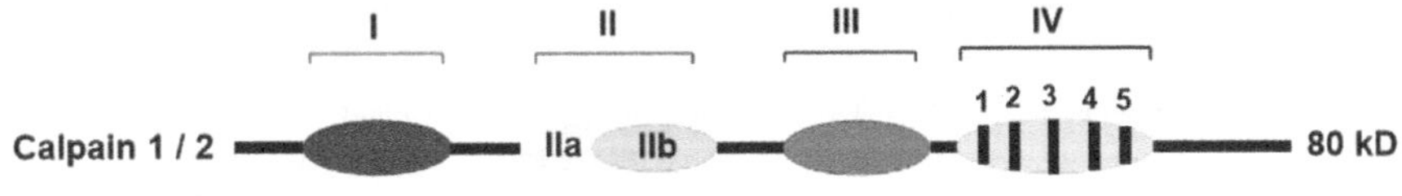

FIGURE 7.1 Structure of calpain-1 and calpain-2 isoforms.

DOI: 10.1201/9781003602477-7

and a penta-EF domain required for dimerization and calcium-dependent activation (Etwebi et al., 2018; Wang et al., 2018).

Other members of the calpain family, such as calpain-3 (p94), calpain-5, and calpain-10, have tissue-specific expression and have unique regulatory mechanisms. For example, calpain-3 is expressed in skeletal muscle and is critical for muscle function and maintenance (Zhang et al., 2024). Calpain-5 and Calpain-6 often deviate from the classical calcium-stimulated, proteolytic functions found in classical calpains. Calpain-6, for example, lacks protease activity but still plays a role in cell organization (Kotova et al., 2023). The diversity within the calpain family (Table 7.1) allows for precise spatial and temporal regulation of proteolytic activity, ensuring that calpain-mediated signaling is tightly controlled.

TABLE 7.1
Different Subclasses of Calpain Enzymes

Subclass	Major Isoforms	Tissue Expression	Function	Activation Features	Associated Diseases/ Pathologies
Classical Calpains	Calpain-1 (μ-calpain)	Ubiquitous (most tissues)	Cytoskeletal remodeling, signal transduction, apoptosis, and necrosis	Activated by micromolar concentrations of Ca^{2+} (μM); heterodimeric structure with large catalytic and small regulatory subunits	Alzheimer's disease, cancer, ischemic damage, myocardial infarction
	Calpain-2 (m-calpain)	Ubiquitous (most tissues)	Similar to Calpain-1	Requires millimolar concentrations of Ca^{2+} (mM); also heterodimeric with catalytic/regulatory subunits	Cardiac hypertrophy, cancer, neurodegenerative disorders
Tissue-Specific Calpains	Calpain-3 (p94)	Primarily in skeletal muscle	Muscle maintenance, sarcomere organization, myofibril stability	Autolyzes in the absence of calcium, involved in muscle repair; mutations lead to muscular dystrophies	Limb-girdle muscular dystrophy type 2A (LGMD2A), muscular disorders
	Calpain-8 (nCL-2)	Stomach	Gastric function, digestive tract-specific protease	Low calcium requirement for activation, functions in non-apoptotic signaling pathways	Stomach ulcers, gastric diseases
	Calpain-9 (nCL-4)	Intestines, stomach	Cell adhesion, tissue integrity in digestive system	Involved in gastrointestinal tissue-specific processes; potential tumor suppressor in gastric cancer	Gastric cancer
Non-Classical Calpains	Calpain-5 (hTRA3)	Retina	Retinal function, neuronal signaling	Lacks calcium-binding EF-hand domains, still active; plays a critical role in maintaining photoreceptor cells	Autosomal dominant neovascular inflammatory vitreoretinopathy (ADNIV), retinal degeneration
	Calpain-6	Placenta, skeletal muscle	Cytoskeletal dynamics, non-proteolytic	Non-proteolytic; lacks catalytic activity due to missing active site cysteine; involved in cytoskeletal organization	Implicated in tumor development, cancer metastasis
Atypical Calpains	Calpain-7 (PalBH)	Ubiquitous	Intracellular protein degradation, membrane trafficking	Non-conventional calpain with different domain structure, less dependent on calcium for activity	Possible role in neurodegenerative diseases
	Calpain-10	Pancreas, skeletal muscle	Mitochondrial function, glucose metabolism	Involved in insulin signaling pathways, mitochondrial processing; non-classical activation mechanism	Type 2 diabetes mellitus, metabolic disorders
Calpain-Like Proteins	Calpain-12	Hair follicles, skin	Hair follicle development	Poorly understood but involved in skin and hair growth processes	Alopecia (hair loss), skin abnormalities
	Calpain-15	Brain, kidneys	Unknown	Limited catalytic activity, similar structure to other calpains	Neurodegenerative disorders, kidney diseases
	Calpain-16	Testis, brain	Spermatogenesis, neural development	Putative calpain with limited functional characterization; may have structural roles rather than proteolytic activity	Male infertility, neurological abnormalities

7.2.2 Physiological Roles of Calpains

Calpain is involved in many physiological processes. It includes cytoskeletal changes and signal transduction, cell motility, and cell death (Murachi, 2019). One of their primary functions is the regulation of cytoskeletal dynamics, where they cleave components such as spectrin, talin, and filamin to facilitate changes in cell shape, migration, and adhesion (Hyatt & Powers, 2020). Calpain contributes to synaptic plasticity by modulating synaptic protein cleavage and regulating neurotransmitter release, processes essential for learning and memory (Briz & Baudry, 2017).

In addition to its role in regulating cell structure, calpain is also a key player in cell cycle progression and apoptosis (Shoeman & Traub, 2019). By targeting cell cycle regulators such as cyclins, calpain affects the increase in cell number. The involvement of calpain in the cleavage of proapoptotic and antiapoptotic proteins reveals its role in promoting and preventing programmed cell death. This proteolytic fine-tuning ensures that calpains maintain cellular homeostasis under normal conditions (Miyazaki et al., 2020).

However, when dysregulated, calpain activity can lead to pathological outcomes. Excessive or uncontrolled calpain stimulation is associated with nerve damage in neurodegenerative diseases such as Alzheimer's and Parkinson's diseases, where it contributes to tau hyperphosphorylation, synaptic dysfunction, and neuronal loss (Dewitt & Hallett, 2020).

7.3 CALPAIN ACTIVATION IN ALZHEIMER'S DISEASE

Calpain activation has emerged as a key factor contributing to AD due to calcium-dependent proteases (Mahaman et al., 2019). Calpain is therefore highly sensitive to intracellular calcium levels, and its abnormal activation has been implicated in the molecular pathways that drive neurodegeneration in AD (Schultz et al., 2021). Increasing evidence indicates that dysregulated calpain-activity: AD enhances central pathological processes. Including the accumulation of Aβ, tau hyperphosphorylation, and synaptic dysfunction (Schultz et al., 2021). This section discusses the current understanding of calpain dysregulation in AD and the mechanisms through which calpains contribute to disease progression.

7.3.1 Evidence of Calpain Dysregulation in Alzheimer's Disease

Studies have consistently shown that calpain activity in the brains of AD patients increases, especially in areas such as the hippocampus and cortex, which are important for memory and cognition (Wang et al., 2017).

Post-mortem analysis of AD brains revealed elevated levels of calpain-cleaved substrates, as well as the presence of activated calpain isoforms. Furthermore, animal models of AD have shown impaired calcium homeostasis, a hallmark of this neurodegenerative disease. This imbalance leads to excessive activation of calpain, which is associated with cognitive impairment and neurological changes (Magi et al., 2016; Popugaeva et al., 2017).

7.3.2 Mechanisms by Which Calpain Contributes to AD Pathogenesis

Calpains play an important role in the development of AD through various mechanisms, mainly by regulating key proteins involved in neuronal function and survival. One of the most important functions of calpain in AD is the interaction with tau protein (Zhang et al., 2021). Calpain directly cleaves tau, leading to the accumulation of neurofibrillary tangles. Furthermore, calpain indirectly influences tau phosphorylation through kinases such as cyclin-dependent kinase 5 (CDK5) by phosphorylating tau at abnormal sites, further enhancing its toxic effects (Lloret et al., 2015).

Calpain activation plays an important role in the pathogenesis of AD through several mechanisms. It affects Aβ synthesis that cleaves amyloid precursor protein (APP) and its regulatory proteins, disrupting the balance of Aβ synthesis and clearance, leading to aggregation of toxic Aβ species. This aggregation is stimulated by a feedback loop of Aβ-induced calcium regulation, calpains, increased activation, thereby increasing Neurodegeneration. In addition, calpain cleaves tau, promotes its aggregation and forms neurofibrillary tangles, impairs microtubule stability, and impairs axonal transport. Calpain also affects synaptic function by cleaving synaptic proteins such as PSD-95 and spectrin, thus decreasing synaptic plasticity and contributing to cognitive decline. Additionally, calpain increases mitochondrial dysfunction well, leading to decreased muscle mass and oxidative stress (Chen et al., 2018; Li et al., 2022; Metwally et al., 2023).

Calpain activation causes proteolytic cleavage of tau at various sites, producing tau fragments that accumulate to form tangles (Figure 7.2). This cleavage impairs tau's ability to stabilize microtubules, resulting in cytoskeletal destabilization and disrupted axonal transport, both of which contribute to neurodegeneration. Similarly, Aβ peptides arise from the proteolytic processing of APP (Wang, Zhao, et al., 2020). Dysregulation in APP processing can lead to the accumulation of Aβ plaques, which are neurotoxic and significantly contribute to AD progression. Studies indicate that calpain activation promotes Aβ production by facilitating APP cleavage at specific sites, resulting in the generation of harmful Aβ fragments. For example, in the prefrontal cortex of AD patients, the activation ratio of calpain-1 is significantly elevated, tripling compared to healthy controls, thus linking excessive calpain activity with neurodegenerative changes. In preclinical studies, calpain inhibitors decrease tau cleavage, reduce Aβ production, and protect synaptic integrity in animal models of AD. They have shown promising results but to develop calpain inhibitors, selection remains

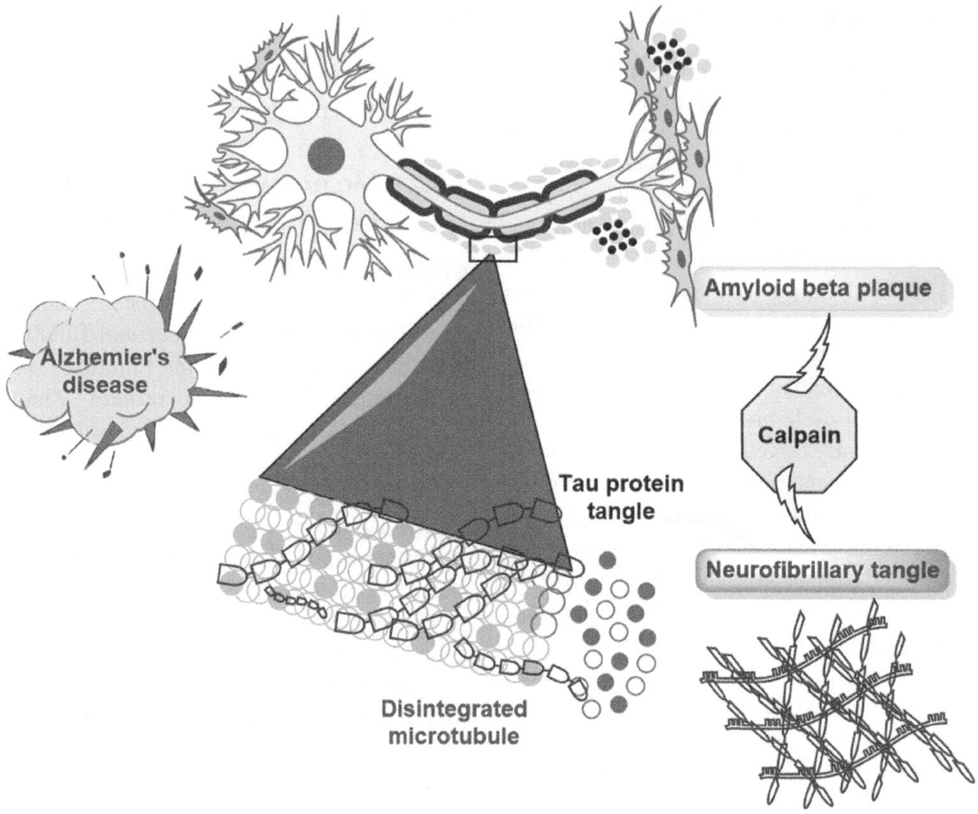

FIGURE 7.2 Mechanisms of calpain activation in the pathogenesis of AD.

challenging due to the need to preserve the essential physiological function of calpain. Despite these challenges, calpain inhibitors remain a potential therapeutic option for AD (Esteras & Abramov, 2020; Li et al., 2022; Uddin et al., 2020; Wang, Zhao, et al., 2020).

7.4 RATIONALE FOR TARGETING CALPAIN IN AD THERAPY

7.4.1 Therapeutic Potential of Calpain Inhibition

The potential of calpain inhibition as a therapeutic strategy for AD lies in its ability to target multiple pathways associated with neurodegeneration. By inhibiting calpain activity, it is possible to interfere with tau and Aβ pathology, synaptic function, and mitochondrial health were also preserved (Shapovalov et al., 2022). Preclinical studies have shown that calpain inhibitors can reduce tau breakdown and hyperphosphorylation, preventing tau aggregation and the formation of neurofibrillary tangles. These effects play an important role in the delayed progression of AD, as tau pathology is a key characteristic of the disease (Cagmat et al., 2015). In addition, calpain inhibitors have the potential to protect synaptic plasticity by preserving synaptic proteins such as PSD-95, which are important for learning and memory processes (Hassen et al., 2018; Tamtaji et al., 2019) (Table 7.2).

TABLE 7.2
Role of Calpain in AD-Related Pathways

Pathway	Effect of Calpain Dysregulation
Amyloid precursor protein (APP)	Cleavage leading to Aβ plaque formation
Tau hyperphosphorylation	Formation of neurofibrillary tangles
Synaptic function	Disruption of synaptic plasticity
Oxidative stress	Amplification of neuronal damage

7.4.2 Advantages Over Existing Therapeutic Targets

Current AD treatments, such as acetylcholinesterase inhibitors (AChEIs) and NMDA receptor antagonists, primarily aim to alleviate symptoms rather than address the disease's underlying mechanisms (Lee et al., 2019). Calpain inhibition offers several advantages over these existing approaches. It targets multiple aspects of AD pathology, including both tau and Aβ accumulation, as well as synaptic and mitochondrial dysfunction, providing a comprehensive therapeutic approach (Wu et al., 2020). In contrast, Aβ-focused treatments tend to address only one aspect of the disease, with limited success in improving clinical outcomes. Furthermore, calpain inhibitors may provide neuroprotection by preventing synaptic and neuronal damage, unlike AChEIs, which temporarily enhance neurotransmitter levels without addressing the underlying neurodegenerative processes. Calpain inhibition could directly slow disease progression and help preserve cognitive function (Li et al., 2018; Wang, Liu, et al., 2020).

Additionally, targeting calpain activity may offer a better safety profile than existing treatments. It reduces the possibility of widespread systemic side effects. Developing isoform-specific calpain inhibitors could improve therapeutic efficacy while minimizing unwanted effects, given the diverse roles of calpain isoforms across different tissues (Hassen et al., 2018).

7.5 CALPAIN INHIBITORS: CLASSIFICATION AND MECHANISM OF ACTION

Calpain inhibitors can be classified according to their chemical structure and inhibitory effect (Kalash et al., 2018).

7.5.1 Types of Calpain Inhibitors

7.5.1.1 Peptide-Based Inhibitors

These inhibitors are designed to mimic the enzyme's natural substrates, allowing competitive binding to its active site. These inhibitors typically have a peptide backbone linked to a functional group that interacts with the catalytic cysteine residue in the active site. One example is calpain inhibitor I (N-acetyl-Leu-Leu-Norleucinal), which inhibits calpain-1 and calpain-2 by forming a reversible complex with the active site. However, despite their effectiveness, these inhibitors often face challenges such as low bioavailability and stability, which restrict their clinical use (Metwally et al., 2023).

7.5.1.2 Non-Peptide Small Molecule Inhibitors

These inhibitors bind to the active site of calpain or interact with regions outside the catalytic core, changing the structure of the enzyme and retard its function. For example, MDL28170 is a reversible non-peptide inhibitor. This has shown neuroprotective effects in previous clinical models of AD. The main advantage of small molecules like MDL28170 is their ability to cross the blood-brain barrier. This makes it ideal for treating neurological conditions (Dhani et al., 2021).

7.5.1.3 Endogenous Inhibitors

In addition to synthetic inhibitors, calpain activity is naturally regulated by endogenous proteins, such as calpastatin. Calpastatin specifically binds to the large catalytic subunit of calpain, inhibiting both calpain-1 and calpain-2. While endogenous inhibitors are not used directly as therapeutic agents, they provide a blueprint for the design of calpain-selective inhibitors (Table 7.3) and serve as an important biological control of calpain activity under normal physiological conditions (Tušar et al., 2021).

7.5.2 Mechanism by Which Calpain Inhibitors Modulate Calpain Activity

Calpain inhibitors exert their effects by targeting the active site of the enzyme or modulating its conformational states, thereby preventing calcium-induced proteolysis. The mechanism of calpain inhibition can be understood through the following modes of action (Table 7.4).

7.5.2.1 Competitive Inhibition at the Active Site

Many calpain inhibitors, particularly peptide-based and small molecule inhibitors, function by competitively binding to the active site of the enzyme. Calpains are cysteine

TABLE 7.3
Calpain Inhibitors and Their Mechanisms of Action

Inhibitor Type	Mechanism	Examples
Synthetic	Competitive inhibition	Calpain inhibitor I, PD150606
Natural compounds	Allosteric modulation	Epigallocatechin gallate (EGCG)
Peptidic	Calpastatin-based inhibition	Calpastatin peptides

TABLE 7.4
Mechanisms through Which Calpain Inhibitors Regulate Calpain Activity

Mechanism	Description	Examples/Details
Competitive Inhibition	Inhibitors bind to the active site, competing with natural substrates and blocking proteolysis.	Peptide-based inhibitors like Calpain Inhibitor I and MDL28170 bind to the catalytic cysteine.
Allosteric Inhibition	Inhibitors bind outside the active site, causing conformational changes that prevent enzyme activation.	Allosteric inhibitors stabilize the inactive enzyme form, reducing activity even with high calcium levels.
Modulation of Calcium Binding	Inhibitors disrupt calcium-binding domains, preventing the conformational shift required for activation.	This prevents enzyme activation without direct active site interaction, offering precise activity control.
Endogenous Inhibition (Calpastatin)	Calpastatin, an endogenous protein, binds calpain in both its active and inactive forms, regulating activity.	Though not used therapeutically, calpastatin's mechanism informs the design of selective inhibitors.

proteases that require calcium for activation, which induces conformational changes that expose the catalytic cysteine residue. Inhibitors such as calpain inhibitor I and MDL28170 interact with this catalytic cysteine, forming either reversible or irreversible covalent bonds that prevent substrate cleavage. This direct competition with the natural substrate is the primary mode of action for most synthetic calpain inhibitors (Etwebi et al., 2018).

7.5.2.2 Allosteric Inhibition

Some inhibitors act by binding to sites outside the active domain of calpain, inducing conformational changes that alter the enzyme's ability to bind calcium or substrate. Allosteric inhibitors can stabilize the inactive form of the enzyme, preventing its activation even in the presence of elevated intracellular calcium levels. This approach offers a more selective modulation of calpain activity, potentially reducing off-target effects associated with direct active site inhibition (Li et al., 2020).

7.5.2.3 Modulation of Calcium Binding

As calpains are activated by binding calcium ions, inhibitors can also function by disrupting the calcium-binding domains of calpain. This prevents the calcium-induced conformational shift necessary for activation. By blocking calcium binding, these inhibitors reduce the enzyme's proteolytic activity without directly interacting with the catalytic site, which may provide a more refined control over calpain function (Tamtaji et al., 2019).

7.6 PRECLINICAL STUDIES OF CALPAIN INHIBITORS IN AD MODELS

Preclinical studies investigating calpain inhibitors as a potential treatment for AD have provided significant insights into their efficiency and protection. These studies, using both *in vitro* and *in vivo* models, have revealed the potential of inhibitors to target key AD pathologies such as tau hyperphosphorylation, Aβ deposition, and synaptic dysfunction (Mughal et al., 2022).

7.6.1 *In Vitro* Studies

In vitro studies on nerve cell cultures and brain tissue samples from AD models give us insights into how stopping calpain affects key AD processes. These tests have revealed a strong connection between calpain activity and both tau and Aβ pathologies. In cultures exposed to Aβ clumps, active calpain leads to tau cutting and phosphate addition, which helps to form tangles in nerve fibers. Drugs that block calpain, like calpain inhibitor I and MDL28170, reduce tau's extra phosphates and breaking, which might prevent nerve cell death linked to tau. Likewise, blocking calpain changes how APP breaks down, making less toxic Aβ bits and fewer Aβ plaques. This suggests that drugs to block calpain could tackle AD from many angles (Mahaman et al., 2019; Ono et al., 2016; Saito et al., 2016).

7.6.2 *In Vivo* Studies

In vivo studies using genetically modified animal models of AD, specifically the APP/PS1 mouse model, have provided valuable insights into the potential of calpain inhibitors as therapeutic agents. These models simulate key AD features such as Aβ plaque formation and tau pathology. Treatment with calpain inhibitors like MDL28170 has shown promising results, including a reduction in Aβ plaque accumulation, improved memory and learning performance (e.g., in the Morris water maze), and decreased tau phosphorylation, suggesting a potential to alleviate tau-related pathology. Additionally, calpain inhibition protects synaptic function by preventing the degradation of synaptic proteins like PSD-95 and spectrin, thereby preserving synaptic structure and plasticity. This could play a critical role in protecting cognitive function in AD, supporting the therapeutic potential of calpain inhibitors (Donkor, 2015; Jantos et al., 2019; Sasaguri et al., 2017).

7.6.3 Efficacy and Safety Findings

Preclinical studies have demonstrated strong evidence supporting the effectiveness of calpain inhibitors in mitigating AD-related pathologies in both *in vitro* and *in vivo* systems (Basheer et al., 2023). Regarding safety, preclinical studies have predominantly demonstrated positive outcomes. Calpain inhibitors like MDL28170 have been shown to be well-tolerated in animal models, with no significant adverse effects reported at therapeutic doses. However, long-term safety studies are still needed to assess the potential for off-target effects, especially given the widespread role of calpains in various physiological processes. Since calpains are involved in essential cellular functions such as cytoskeletal remodeling and cell signaling, the development of highly selective inhibitors that target specific calpain isoforms will be crucial for minimizing the risk of toxicity and ensuring that calpain inhibition does not interfere with normal cellular activities (Yadikar et al., 2020).

7.7 CLINICAL STUDIES AND TRIALS

The translation of calpain inhibitors from preclinical models to human applications necessitates rigorous clinical investigation. Although the majority of research on calpain inhibitors has been confined to *in vitro* and *in vivo* preclinical studies, emerging clinical trials are beginning to explore their therapeutic potential in AD (Lon et al., 2019).

7.7.1 Overview of Clinical Trials Involving Calpain Inhibitors

The transition from preclinical studies to clinical applications involves several phases aimed at assessing the safety, tolerability, pharmacokinetics, and preliminary efficacy of

TABLE 7.5
Overview of Clinical Trials Involving Calpain Inhibitors

Trial Phase	Study Title	Calpain Inhibitor	Study Design	Primary Outcomes	Status
Phase I	Safety and Tolerability of Calpain Inhibitor I	Calpain Inhibitor I	Single-blind, placebo-controlled	Adverse events, pharmacokinetics	Completed
Phase II	Efficacy of MDL28170 in Alzheimer's Disease	MDL28170	Randomized, double-blind	Change in cognitive function (ADAS-Cog)	Ongoing
Phase III	Calpain Inhibition for Cognitive Decline	Calpain Inhibitor II	Multicenter, randomized, controlled	Improvement in clinical global impression	Recruiting
Phase II/III	Calpain Inhibitors in Neurodegenerative Disorders	Calpain Inhibitor III	Open-label, extension study	Long-term safety, efficacy assessments	Completed
Phase I	Evaluation of Safety Profile in Healthy Volunteers	Novel Calpain Inhibitor	Dose-escalation, single-center	Maximum-tolerated dose	Completed
Phase II	Impact of Calpain Inhibition on Biomarkers	Calpain Inhibitor IV	Randomized, placebo-controlled	Changes in biomarker levels (CSF Aβ, tau)	Active, not recruiting

these inhibitors in human subjects (Table 7.5). Noteworthy, calpain inhibitors that have entered clinical trials include compounds such as SNJ-1945 (also known as calpain inhibitor I) and BDA-410, among others (Carragher, 2006).

These clinical trials are primarily designed to evaluate the safety profile of calpain inhibitors in healthy volunteers and AD patients. Phase I clinical trials are primarily designed to establish the optimal dosing regimen and evaluate potential adverse effects of calpain inhibitors. In subsequent phases (Phase II and III), the focus shifts to assessing their efficacy in slowing the progression of AD, enhancing cognitive function, and modulating disease-associated biomarkers (Saatman et al., 2010).

7.7.2 Results and Implications

Early clinical trials of calpain inhibitors show promising early results. Although final efficacy data are still awaited. Safety assessments indicate that these inhibitors are generally well tolerated. For example, a Phase 1 trial of SNJ-1945 reported an acceptable safety profile with no significant dose-limiting toxicity. Preliminary findings on efficacy indicate many advantages, including slight improvements in memory along with decreased levels of tau phosphorylation and tau in the brain. Additionally, imaging studies have shown potential reductions in Aβ plaque accumulation in the brains of treated individuals (Dhir et al., 2020; Opsahl-Sorteberg et al., 2024).

7.8 FUTURE DIRECTIONS AND PERSPECTIVES

The ongoing research and development efforts surrounding calpain inhibitors for AD are paving the way for innovative therapeutic strategies. Future directions focus on enhancing the efficacy and safety of calpain inhibitors, exploring combination therapies, embracing personalized medicine approaches, and investigating the broader roles of calpains in other diseases (Table 7.6).

Advancements in medicinal chemistry and drug design are critical for overcoming the current limitations of calpain inhibitors. Structure-based drug design, leveraging detailed knowledge of calpain isoform structures, can facilitate the development of highly selective and potent inhibitors. Additionally, the incorporation of novel pharmacophores and the optimization of pharmacokinetic properties will enhance the drug-like characteristics of calpain inhibitors.

TABLE 7.6
Future Directions and Perspectives in Calpain Inhibitor Research

Future Directions	Description
Advances in Calpain Inhibitor Development	• Development of novel inhibitors and implementation of high-throughput screening techniques. • Utilization of structure-based drug design for optimizing calpain inhibitors.
Combination Therapies	• Exploration of synergistic treatment approaches alongside existing Alzheimer's disease therapies. • Adoption of multi-target strategies to effectively tackle the complex pathology of Alzheimer's disease.
Personalized Medicine Approaches	• Implementation of genetic and biomarker-driven patient stratification for tailored treatment. • Customization of calpain inhibition strategies based on individual patient profiles.
Exploring Calpain's Role Beyond AD	• Investigation of calpain's potential therapeutic applications in other neurodegenerative disorders. • Examination of the broader implications of calpain biology for therapeutic development.

High-throughput screening and fragment-based drug discovery are emerging as valuable tools for identifying new calpain inhibitors with improved specificity and efficacy. Given the multifactorial nature of AD, combination therapies targeting multiple pathological pathways, such as calpain inhibitors paired with anti-amyloid, anti-tau, neuroprotective, antioxidant, or anti-inflammatory agents, offer promise for enhanced therapeutic outcomes by addressing toxic protein accumulation and preserving synaptic function. Personalized medicine approaches, leveraging genetic, epigenetic, and biomarker profiles, can optimize calpain inhibitor use by identifying patients most likely to benefit and tailoring dosing regimens to individual pharmacokinetic and pharmacodynamic profiles. Furthermore, the role of calpains extends beyond AD, with implications in other neurodegenerative conditions like Parkinson's and Huntington's diseases, as well as non-neurological disorders such as cancer and cardiovascular diseases, highlighting the broad-spectrum potential of calpain inhibitors as therapeutic agents across diverse disease contexts.

7.9 CONCLUSION

The pursuit of calpain inhibitors as a therapeutic strategy for AD represents a promising frontier in neurodegenerative disease research. Calpains play a central role in the pathological mechanisms of AD, making them a promising target for therapeutic intervention. Although the development of calpain inhibitors presents challenges, advancements in drug design, combination treatment strategies, and personalized medicine offer encouraging prospects for addressing these hurdles. Incorporating calpain inhibitors into the therapeutic framework for AD has the potential to enhance existing treatments and provide new opportunities for managing this devastating disease. Achieving clinical success, however, demands overcoming obstacles such as ensuring target specificity, improving brain delivery mechanisms, and reducing potential side effects. Ongoing research, innovative approaches in drug development, and thorough clinical investigations will be essential to establish the effectiveness of calpain inhibitors as a treatment option for AD.

REFERENCES

Basheer, N., Smolek, T., Hassan, I., Liu, F., Iqbal, K., Zilka, N., & Novak, P. (2023). Does modulation of tau hyperphosphorylation represent a reasonable therapeutic strategy for Alzheimer's disease? From preclinical studies to the clinical trials. *Molecular Psychiatry*, *28*(6), 2197–2214.

Breijyeh, Z., & Karaman, R. (2020). Comprehensive review on Alzheimer's disease: Causes and treatment. *Molecules*, *25*(24), 5789.

Briz, V., & Baudry, M. (2017). Calpains: Master regulators of synaptic plasticity. *The Neuroscientist*, *23*(3), 221–231.

Cagmat, E. B., Guingab-Cagmat, J. D., Vakulenko, A. V., Hayes, R. L., & Anagli, J. (2015). *Potential use of calpain inhibitors as brain injury therapy.*

Carragher, N. O. (2006). Calpain inhibition: A therapeutic strategy targeting multiple disease states. *Current Pharmaceutical Design*, *12*(5), 615–638.

Chen, H.-H., Liu, P., Auger, P., Lee, S.-H., Adolfsson, O., Rey-Bellet, L., …, & Muhs, A. (2018). Calpain-mediated tau fragmentation is altered in Alzheimer's disease progression. *Scientific Reports*, *8*(1), 16725.

Cummings, J., Lee, G., Ritter, A., Sabbagh, M., & Zhong, K. (2020). Alzheimer's disease drug development pipeline: 2020. *Alzheimer's & Dementia: Translational Research & Clinical Interventions*, *6*(1), e12050.

Dewitt, S., & Hallett, M. B. (2020). Calpain activation by Ca^{2+} and its role in phagocytosis. In *Molecular and cellular biology of phagocytosis* (pp. 129–151). Springer.

Dhani, S., Zhao, Y., & Zhivotovsky, B. (2021). A long way to go: Caspase inhibitors in clinical use. *Cell Death & Disease*, *12*(10), 949.

Dhir, N., Medhi, B., Prakash, A., Goyal, M. K., Modi, M., & Mohindra, S. (2020). Pre-clinical to clinical translational failures and current status of clinical trials in stroke therapy: A brief review. *Current Neuropharmacology*, *18*(7), 596–612.

Donkor, I. O. (2015). An updated patent review of calpain inhibitors (2012–2014). *Expert Opinion on Therapeutic Patents*, *25*(1), 17–31.

Donkor, I. O. (2020). An update on the therapeutic potential of calpain inhibitors: A patent review. *Expert Opinion on Therapeutic Patents*, *30*(9), 659–675.

Esteras, N., & Abramov, A. Y. (2020). Mitochondrial calcium deregulation in the mechanism of beta-amyloid and tau pathology. *Cells*, *9*(9), 2135.

Etwebi, Z., Landesberg, G., Preston, K., Eguchi, S., & Scalia, R. (2018). Mechanistic role of the calcium-dependent protease calpain in the endothelial dysfunction induced by MPO (myeloperoxidase). *Hypertension*, *71*(4), 761–770.

Fa, M., Zhang, H., Staniszewski, A., Saeed, F., Shen, L. W., Schiefer, I. T., …, & Libien, J. (2016). Novel selective calpain 1 inhibitors as potential therapeutics in Alzheimer's disease. *Journal of Alzheimer's Disease*, *49*(3), 707–721.

Hassen, G. W., Kesner, L., Stracher, A., Shulman, A., Rockenstein, E., Mante, M., …, & Masliah, E. (2018). Effects of novel calpain inhibitors in transgenic animal model of Parkinson's disease/dementia with Lewy bodies. *Scientific Reports*, *8*(1), 18083.

Hyatt, H. W., & Powers, S. K. (2020). The role of calpains in skeletal muscle remodeling with exercise and inactivity-induced atrophy. *International Journal of Sports Medicine*, *41*(14), 994–1008.

Jantos, K., Kling, A., Mack, H., Hornberger, W., Moeller, A., Nimmrich, V., …, & Nijsen, M. (2019). Discovery of ABT-957: 1-Benzyl-5-oxopyrrolidine-2-carboxamides as selective calpain inhibitors with enhanced metabolic stability. *Bioorganic & Medicinal Chemistry Letters*, *29*(15), 1968–1973.

Kalash, L., Cresser-Brown, J., Habchi, J., Morgan, C., Miller, D. J., Glen, R. C., …, & Bender, A. (2018). Structure-based design of allosteric calpain-1 inhibitors populating a novel bioactivity space. *European Journal of Medicinal Chemistry*, *157*, 1264–1275.

Khan, H., Garg, N., Singh, T. G., Kaur, A., & Thapa, K. (2022). Calpain inhibitors as potential therapeutic modulators in neurodegenerative diseases. *Neurochemical Research*, *47*(5), 1125–1149.

Knopman, D. S., Amieva, H., Petersen, R. C., Chételat, G., Holtzman, D. M., Hyman, B. T., …, & Jones, D. T. (2021). Alzheimer disease. *Nature Reviews Disease Primers*, *7*(1), 33.

Kotova, I. M., Pestereva, N. S., Traktirov, D. S., Absalyamova, M. T., & Karpenko, M. N. (2023). Functions and distribution of calpain-calpastatin system components in brain during mammal ontogeny. *Biochimica et Biophysica Acta (BBA)-General Subjects*, *1867*(5), 130345.

Lee, J. W., Ko, J., Ju, C., & Eltzschig, H. K. (2019). Hypoxia signaling in human diseases and therapeutic targets. *Experimental & Molecular Medicine*, *51*(6), 1–13.

Li, M., Zhou, S., Wang, G., Qiao, L., Yi, S., Li, T., …, & Tang, Z. (2022). Calpain inhibitor calpeptin improves Alzheimer's disease–like cognitive impairments and pathologies in a diabetes mellitus rat model. *Neurotoxicity Research*, *40*(5), 1248–1260.

Li, Y., He, Z., Lv, H., Chen, W., & Chen, J. (2020). Calpain-2 plays a pivotal role in the inhibitory effects of propofol against TNF-α-induced autophagy in mouse hippocampal neurons. *Journal of Cellular and Molecular Medicine*, *24*(16), 9287–9299.

Li, Y. H., Yu, C. Y., Li, X. X., Zhang, P., Tang, J., Yang, Q., …, & Tu, G. (2018). Therapeutic target database update 2018: Enriched resource for facilitating bench-to-clinic research of targeted therapeutics. *Nucleic Acids Research*, *46*(D1), D1121–D1127.

Lloret, A., Fuchsberger, T., Giraldo, E., & Viña, J. (2015). Molecular mechanisms linking amyloid β toxicity and Tau hyperphosphorylation in Alzheimer's disease. *Free Radical Biology and Medicine*, *83*, 186–191.

Lon, H. K., Mendonca, N., Goss, S., Othman, A. A., Locke, C., Jin, Z., & Rendenbach-Mueller, B. (2019). Pharmacokinetics, safety, tolerability, and pharmacodynamics of alicapistat, a selective inhibitor of human calpains 1 and 2 for the treatment of Alzheimer disease: An overview of phase 1 studies. *Clinical Pharmacology in Drug Development*, *8*(3), 290–303.

Magi, S., Castaldo, P., Macrì, M. L., Maiolino, M., Matteucci, A., Bastioli, G., …, & Lariccia, V. (2016). Intracellular calcium dysregulation: Implications for Alzheimer's disease. *BioMed Research International*, *2016*(1), 6701324.

Mahaman, Y. A. R., Huang, F., Kessete Afewerky, H., Maibouge, T. M. S., Ghose, B., & Wang, X. (2019). Involvement of calpain in the neuropathogenesis of Alzheimer's disease. *Medicinal Research Reviews*, *39*(2), 608–630.

Metwally, E., Al-Abbadi, H. A., Hussain, T., Murtaza, G., Abdellatif, A. M., & Ahmed, M. F. (2023). Calpain signaling: From biology to therapeutic opportunities in neurodegenerative disorders. *Frontiers in Veterinary Science*, *10*, 1235163.

Miyazaki, T., Akasu, R., & Miyazaki, A. (2020). Calpain proteolytic systems counteract endothelial cell adaptation to inflammatory environments. *Inflammation and Regeneration*, *40*, 1–8.

Mughal, E. U., Obaid, R. J., Sadiq, A., Alsharif, M. A., Naeem, N., Kausar, S., …, & Alsantali, R. I. (2022). Chalcone- and flavone-based novel terpyridine metal complexes: Synthesis, electrochemical, photophysical, photovoltaic and computational studies. *Dyes and Pigments*, *201*, 110248.

Murachi, T. (2019). Introduction: Physiological significance of the calpain-calpastatin system. In: *Intracellular calcium-dependent proteolysis* (pp. 157–162). CRC Press.

Obaid, R. J., Mughal, E. U., Naeem, N., Al-Rooqi, M. M., Sadiq, A., Jassas, R. S., …, & Ahmed, S. A. (2022). Pharmacological significance of nitrogen-containing five and six-membered heterocyclic scaffolds as potent cholinesterase inhibitors for drug discovery. *Process Biochemistry*, *120*, 250–259.

Obaid, R. J., Naeem, N., Mughal, E. U., Al-Rooqi, M. M., Sadiq, A., Jassas, R. S., …, & Ahmed, S. A. (2022). Inhibitory potential of nitrogen, oxygen and sulfur containing heterocyclic scaffolds against acetylcholinesterase and butyrylcholinesterase. *RSC Advances*, *12*(31), 19764–19855.

Ono, Y., Saido, T. C., & Sorimachi, H. (2016). Calpain research for drug discovery: Challenges and potential. *Nature Reviews Drug Discovery*, *15*(12), 854–876.

Opsahl-Sorteberg, H. G., Evju, E., Liang, Z., & Fletcher, J. C. (2024). Plant and animal calpain functions, association with microtubules and possible medical applications. *Medical Research Archives*, *12*(9).

Popugaeva, E., Pchitskaya, E., & Bezprozvanny, I. (2017). Dysregulation of neuronal calcium homeostasis in Alzheimer's disease–A therapeutic opportunity? *Biochemical and Biophysical Research Communications*, *483*(4), 998–1004.

Prajapati, N., & Singh, J. (2024). Calpain: An emerging therapeutic target. *Research Journal of Pharmacology and Pharmacodynamics*, *16*(1), 25–29.

Saatman, K. E., Creed, J., & Raghupathi, R. (2010). Calpain as a therapeutic target in traumatic brain injury. *Neurotherapeutics*, *7*, 31–42.

Saito, T., Matsuba, Y., Yamazaki, N., Hashimoto, S., & Saido, T. C. (2016). Calpain activation in Alzheimer's model mice is an artifact of APP and presenilin overexpression. *Journal of Neuroscience*, *36*(38), 9933–9936.

Sasaguri, H., Nilsson, P., Hashimoto, S., Nagata, K., Saito, T., De Strooper, B., …, & Saido, T. C. (2017). APP mouse models for Alzheimer's disease preclinical studies. *The EMBO Journal*, *36*(17), 2473–2487.

Scheltens, P., De Strooper, B., Kivipelto, M., Holstege, H., Chételat, G., Teunissen, C. E., …, & van der Flier, W. M. (2021). Alzheimer's disease. *The Lancet*, *397*(10284), 1577–1590.

Schultz, B., Taday, J., Menezes, L., Cigerce, A., Leite, M. C., & Gonçalves, C.-A. (2021). Calpain-mediated alterations in astrocytes before and during amyloid chaos in Alzheimer's disease. *Journal of Alzheimer's Disease*, *84*(4), 1415–1430.

Shapovalov, I., Harper, D., & Greer, P. A. (2022). Calpain as a therapeutic target in cancer. *Expert Opinion on Therapeutic Targets*, *26*(3), 217–231.

Shoeman, R. L., & Traub, P. (2019) Calpains and the cytoskeleton. In: *Intracellular calcium-dependent proteolysis* (pp. 191–209). CRC Press.

Sorimachi, H., & Suzuki, K. (2001). The structure of calpain. *The Journal of Biochemistry*, *129*(5), 653–664.

Tahami Monfared, A. A., Byrnes, M. J., White, L. A., & Zhang, Q. (2022). Alzheimer's disease: Epidemiology and clinical progression. *Neurology and Therapy*, *11*(2), 553–569.

Tamtaji, O. R., Mirhosseini, N., Reiter, R. J., Azami, A., & Asemi, Z. (2019). Melatonin, a calpain inhibitor in the central nervous system: Current status and future perspectives. *Journal of Cellular Physiology*, *234*(2), 1001–1007.

Tušar, L., Usenik, A., Turk, B., & Turk, D. (2021). Mechanisms applied by protein inhibitors to inhibit cysteine proteases. *International Journal of Molecular Sciences*, *22*(3), 997.

Uddin, M. S., Tewari, D., Sharma, G., Kabir, M. T., Barreto, G. E., Bin-Jumah, M. N., …, & Ashraf, G. M. (2020). Molecular mechanisms of ER stress and UPR in the pathogenesis of Alzheimer's disease. *Molecular Neurobiology*, *57*, 2902–2919.

Veitch, D. P., Weiner, M. W., Aisen, P. S., Beckett, L. A., Cairns, N. J., Green, R. C., …, & Morris, J. C. (2019). Understanding disease progression and improving Alzheimer's disease clinical trials: Recent highlights from the Alzheimer's disease neuroimaging initiative. *Alzheimer's & Dementia*, *15*(1), 106–152.

Wang, W., Zhao, F., Ma, X., Perry, G., & Zhu, X. (2020). Mitochondria dysfunction in the pathogenesis of Alzheimer's disease: Recent advances. *Molecular Neurodegeneration*, *15*, 1–22.

Wang, Y., Bi, X., & Baudry, M. (2018). Calpain-2 as a therapeutic target for acute neuronal injury. *Expert Opinion on Therapeutic Targets*, *22*(1), 19–29.

Wang, Y., Liu, Y., Bi, X., & Baudry, M. (2020). Calpain-1 and calpain-2 in the brain: New evidence for a critical role of calpain-2 in neuronal death. *Cells*, *9*(12), 2698.

Wang, Y., Shi, Y., & Wei, H. (2017). Calcium dysregulation in Alzheimer's disease: A target for new drug development. *Journal of Alzheimer's Disease & Parkinsonism*, *7*(5).

Weiner, M. W., Veitch, D. P., Aisen, P. S., Beckett, L. A., Cairns, N. J., Green, R. C., …, & Morris, J. C. (2017). Recent publications from the Alzheimer's disease neuroimaging initiative: Reviewing progress toward improved AD clinical trials. *Alzheimer's & Dementia*, *13*(4), e1–e85.

Wu, C., Liu, Y., Yang, Y., Zhang, P., Zhong, W., Wang, Y., …, & Li, X. (2020). Analysis of therapeutic targets for SARS-CoV-2 and discovery of potential drugs by computational methods. *Acta Pharmaceutica Sinica B*, *10*(5), 766–788.

Yadikar, H., Torres, I., Aiello, G., Kurup, M., Yang, Z., Lin, F., …, & Wang, K. K. (2020). Screening of tau protein kinase inhibitors in a tauopathy-relevant cell-based model of tau hyperphosphorylation and oligomerization. *PLoS One*, *15*(7), e0224952.

Zhang, H., Wei, W., Zhao, M., Ma, L., Jiang, X., Pei, H., …, & Li, H. (2021). Interaction between Aβ and tau in the pathogenesis of Alzheimer's disease. *International Journal of Biological Sciences*, *17*(9), 2181.

Zhang, X., Zheng, Y., Wang, Z., Zhang, G., Yang, L., Gan, J., & Jiang, X. (2024). Calpain: The regulatory point of cardiovascular and cerebrovascular diseases. *Biomedicine & Pharmacotherapy*, *179*, 117272.

8 Acetylcholinesterase Inhibitors

A Viable Therapeutic Strategy for Alzheimer's Disease

Kalpana Pandya, Devendra Kumar, Shraddha Manish Gupta, and Sushil Kumar Chaudhary

8.1 INTRODUCTION

Alzheimer's disease (AD) is a complicated neurological disease that clues to death by impairing memory and cognitive function (Hippius & Neundörfer, 2003). According to data from the Centers for Disease Control and Prevention (CDC), in 2022, AD was the seventh most communal reason of death in the US, while COVID-19 was the fourth. After stroke, AD was the sixth most common cause of death before the COVID-19 pandemic. Late-onset AD (LOAD) is the term for AD that usually appears after the age of 65 (Drouin & Drouin, 2017). Early-onset AD (EOAD), which is less common and appears before the age of 65, affects about 5% of AD patients. EOAD is often characterized by typical symptoms and delayed diagnosis, which leads to a more aggressive course of the disease. AD has become a fast-expanding public health concern (Mendez, 2017). AD is a major public health problem because of its steadily rising pervasiveness in the US. It is estimated that roughly 57 million Americans are existing with AD, the majority of them being elderly people. The number of Americans affected by Alzheimer's and other dementias is expected to rise to 13.8 million by 2050. The prevalence of Alzheimer's dementia represents the proportion of people in a given population (Dejanovic et al., 2024). LOAD is more common than EOAD, with statistics differing by age group. In 2018, the prevalence of AD among Medicare enrollers was 2.3%, with a significant rise in cases of moderate cognitive impairment (MCI). According to a study by the Health and Retirement Study (HRS), 30% of older persons had AD, which is greater according to the cognitive performance-based approach than the diagnosis-based methodology (Tahami Monfared et al., 2023). Women, older persons, and those with greater education had higher prevalence rates. The frequency was 1.81% for men and 2.75% for women in 2018. There were also clear racial differences, with White people having greater rates of AD than Black and Asian people (Srivastava & Muhammad, 2020). The Northeast had the highest incidence rates, especially in states like New York and Maryland, where rates were over 12%. The relevance of regional healthcare planning is highlighted by the fact that California has the highest number of people diagnosed with AD (Qiu et al., 2009). Although the rising incidence of AD presents difficulties, it also emphasizes the need for improved diagnosis procedures and initiatives to promote health to alleviate the condition's mounting burden. Between 2014 and 2018, 30% of older persons in India were diagnosed with AD at varying severity levels, according to the HRS, which shows a significant incidence of both AD and MCI (Rasmussen & Langerman, 2019). By talking about important gaps in diagnosis and care, the National Alzheimer's Project Act (NAPA) attempts to increase research funding and strengthen public health responses to AD. NAPA has played a key role in influencing laws to boost the financing of research and public health initiatives, highlighting the necessity of a unified national approach (Vinze et al., 2023). Funding for public health programs targeted at enhancing AD early identification and care have been made possible by recent federal legislation (Vinze et al., 2023). As far as individuals are concerned, diagnosis with AD has serious repercussions on them. This affects their cognitive abilities, memory, and daily routines, decreasing their quality of life. Society bears a huge economic cost of AD. The price of caring for dementia patients, including both state-run and informal care. It has been approximated that the unpaid value of care given by relatives is over $232 billion (2023 Alzheimer's disease facts and figures). Equally, health institutions, nursing homes, and the public health sector have immensely impacted their overall operation as a society. To overcome the challenges associated with Alzheimer's, there is a need to comprehensively deal with early detection, quality care provision, caregiver-supportive services, and further studies that can enhance treatment opportunities and possible remedies (Gaugler et al., 2014). Meanwhile, for the treatment of such conditions, understanding therapeutic targets is crucial in developing effective treatments for diseases like Alzheimer's. Identifying specific molecular targets allows researchers to design drugs that can modulate these targets, potentially altering the disease progression. In the situation of AD, targeting mechanisms like amyloid beta accumulation, tau protein pathology, or cholinergic dysfunction can lead to the development of disease-modifying remedies that may sluggish down or pause the progression of the disease (Grill & Cummings, 2010). As plant-based natural inhibitors are

DOI: 10.1201/9781003602477-8

necessary for disease management, they are crucial in managing target-specific pathways involved in disease progression. These inhibitors work as possible targets for disease management. The main hypotheses proposed as causes for AD are the cholinergic hypothesis, linking AD to cholinergic dysfunction; the amyloid hypothesis, focusing on abnormal production and processing of amyloid β-protein; the tauopathies, hyperphosphorylation of tau pointers to the development of neurofibrillary tangles (NFTs); and the N-methyl-D-aspartate receptors (NMDAR) pathology tangled in synaptic dysfunction, excitotoxicity, and neuronal cell death (Majdi et al., 2020). All these hypotheses provide insights into potential mechanisms underlying AD development, highlighting the importance of neurotransmitter dysfunction and protein aggregation in the disease process. Among them, the cholinergic hypothesis involves the role of acetylcholinesterase (AChE) as a crucial enzyme for AD by breaking down acetylcholine (ACh), a neurotransmitter essential for memory and cognition. In AD, the reduction of ACh levels is linked to cognitive decline, making AChE a significant target for therapeutic intervention. The more detailed mechanism is described in Figure 8.1.

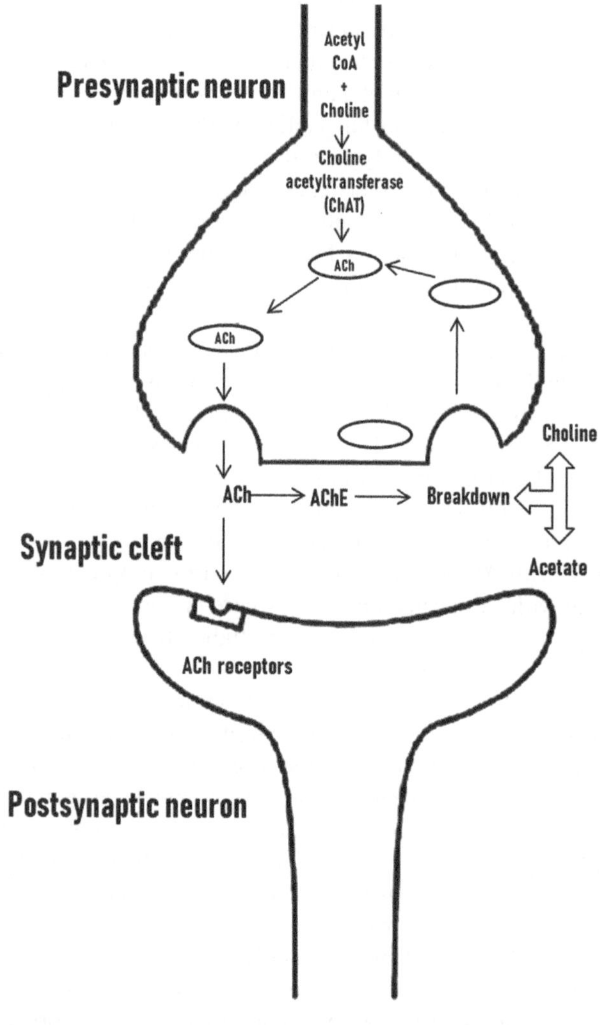

FIGURE 8.1 Release of acetylcholine from receptors and the breaking down of acetylcholinesterase.

AChE is a vital enzyme of the cholinergic scheme mainly accountable for hydrolysis of ACh in the synaptic cleft. The hydrolysis of ACh is an important step in the regulation of cholinergic transmission, which is involved in regulating many neurophysiological processes. AChE is primarily synthesized in neurons of both the central and peripheral nervous systems. Before being delivered to the synaptic cleft, it acts in the endoplasmic reticulum of neuronal cell bodies. AChE biosynthesis requires several key enzymes, including choline acetyltransferase, which is responsible for synthesizing ACh from its building blocks: acetyl-CoA and choline. AChE is produced as a precursor protein that endures a variety of post-translational changes before activation. Stimulation of neuronal activity triggers the expression of this gene, which is facilitated by some of the factors such as neurotrophic factors. With heightened neuronal activity, there may also be heightened expression of AChE and greater levels of this enzyme to keep pace with the heightened demand for ACh breakdown (Chaudhary et al., 2023a). The critical alteration of AChE is subjected to glycosylation, which will make AChE remain stable and active after being synthesized. This will allow the enzyme to fold correctly and work properly within the synaptic cleft. AChE is vital in cholinergic signal transmission through ACh hydrolysis to acetate and choline. Excess stimulation of the ACh receptor is avoided by hydrolysis, which is critical for healthy neurons and proper synaptic transmission (Ali et al., 2024).

Empirical data indicate the cholinergic hypothesis, particularly for AD. Reduced biosynthetic enzyme choline acetyltransferase has been found in AD patients in cognitive regions of the brain, i.e., the cerebral cortex and hippocampus. This observation points to the core function of cholinergic neurotransmission in impairing the cognitive function of AD and other dementia diseases that occur in most. Although numerous treatment strategies aimed at reducing amyloid-β deposition and tau hyperphosphorylation all have been of restricted utility in clinical trials (Gulisano et al., 2018). Enhancing cholinergic neurotransmission remains the foremost symptomatic approach for managing the cognitive and social signs of mild-to-modest AD. Therapeutic agents that enhance the release of ACh in the hippocampus, like linopirdine, muscarinic ACh receptor agonists like xanomeline, and AChEIs like physostigmine and tacrine, have been used to provide better outcomes for patients (Stanciu et al., 2019).

In the brain, ACh is hydrolytically cleaved by two major cholinesterases, AChE and butyrylcholinesterase (BuChE). In AD patients, AChE has been reported to be more abundant than BuChE. It is said to contribute greatly to ACh degradation in regions such as the hippocampus and cerebral cortex. Studies have demonstrated that in AD progression, AChE activity can be condensed by as much as 67% of usual capacity in the temporal lobe and hippocampus. Nevertheless, the activity of BuChE can increase as much as 165% of normal capacity in the course of such progression. Interestingly, reduced activity of BuChE in the medial temporal cortex has been associated with less cognitive decline. Early therapeutic approaches to the increase of defective cholinergic neurotransmission were through the empathy of AChE inhibitors. Later studies focused on the involvement of AChE and BuChE in AD pathology and created a priority for the reticence of both enzymes (Lane et al., 2006). Cholinesterase inhibitors act by inhibiting neurotransmitter breakdown, such as ACh, raising its concentration in the brain and augmenting deficient cholinergic neurotransmission.

In the US and Europe, these ChEIs were some of the first medications permitted for the particular management of AD symptoms. If they inhibit both AChE and BuChE, they are categorized as nonspecific; if they just target AChE, they are categorized as specific. They can also be divided into irreversible, pseudo-irreversible, and reversible inhibitors according to how they work. In summary, improving cholinergic neurotransmission remains a crucial tactic for treating the behavioral and cognitive symptoms linked to mild-to-severe AD (Marucci et al., 2021). Within this chapter, a variety of plant-based therapeutic inhibitors are studied to target important cholinergic signaling pathway components and significantly enhance patient outcomes.

8.2 ROLE OF MEDICINAL PLANTS AS AChE INHIBITORS

Because of their potential therapeutic advantages in treating AD, medicinal plants are becoming acknowledged for their role as AChE inhibitors. Because they Inhibits the break of ACh, a neurotransmitter necessary for memory and cognition, AChE inhibitors are critical for treating the symptoms of AD. Numerous medicinal plants with strong AChE inhibitory action have been found in recent research, opening up a potential therapy possibility. Publications about natural compounds' inhibition of AChE and BChE have significantly increased over the last 20 years. The increasing need for multi-target medications, improvements in screening techniques, and a growing number of molecular docking studies are all contributing factors to this growing interest. Few compounds have been evaluated on animal models; that is essential to confirm their capacity to pass through the blood-brain barrier and have positive effects on the brain. The majority of compounds under investigation have shown in vitro activity.

From thousands of years, humans have used medicinal plants to progress human health and treat a variety of diseases. Systems of traditional medicine are widely acknowledged for their ability to effectively treat neurological conditions. Significantly, substances originating from plants are thought to be a viable source of novel inhibitors, which has resulted in the identification of multiple secondary metabolites and plant extracts that can inhibit AChE. These substances may improve cholinergic functions in AD patients and reduce symptoms related to this neurological condition by raising ACh levels in the brain. (Yuan et al., 2016).

Many studies have highlighted the potential of natural substances like alkaloids, phenolics, flavonoids, and coumarin derivatives as ChEIs. Yokukansan, a traditional Chinese herbal treatment, for instance, has demonstrated efficacy in treating a variety of neurological conditions without any side effects (de Caires & Steenkamp, 2010). Furthermore, the FDA has approved galanthamine, an alkaloid derived from snowdrops, for the treatment of AD. Using medicinal plants as AChE inhibitors has several benefits, fewer adverse effects than manufactured medications and neuroprotective qualities because antioxidants protect neurons from oxidative stress. Even though these plants have potential, further investigation is compulsory to govern the effectiveness and bioavailability of their active ingredients to develop standardized doses and formulations for clinical use.

Recent studies emphasize the promising role of bioactive chemicals from plants, fungi, and algae in pharmacological development aimed at AD (Murray et al., 2024). These natural products, more so AChE inhibitors, serve as important modifiers of AD treatment by enhancing levels of ACh, a key contributor to cognitive functioning. Plants have also become an important source of AChE inhibitors, where different phytochemicals possess promising activity against the enzyme (Vladimir-Knežević et al., 2014). This natural approach offers a compelling alternative to synthetic drugs, which can have some side effects. Phytomedicines are usually presumed to have lesser side effects and hence are very appealing for the long-term therapy of AD (Gajendra et al., 2024). Nevertheless, although the prospect of such plant-derived chemicals holds promise, there is a need for additional experimental confirmation of their efficacy and safety in clinical trials. The combination of traditional knowledge and innovative research methodologies may aid in the finding of effective treatments for neurodegenerative illnesses. Several researchers have reported the practice of several plants as a basis of AChE inhibitors, which has led to the development of phytomedicines for the treatment of AD (Islam et al., 2024). Researchers approximately 260 active compounds from plants that inhibit cholinesterase enzymes, highlighting the importance of bioprospecting in this field. Plants that have been traditionally used as AChE inhibitors is shown below and summarized in Table 8.1.

8.2.1 Medicinal Plants for the Management of AChE

8.2.1.1 *Aconitum anthoroideum* DC

Nine new diterpenoid alkaloids, such as anthoroidine A, a diterpenoid alkaloid with a diterpenoid moiety, and anthoroidine B, a bis-diterpenoid alkaloid joined by a carbon-carbon single bond, were isolated from *Aconitum anthoroideum* DC. The compounds obtained, together with five known diterpenoid alkaloids, were tested for their neuroprotective and AChE inhibitory activities. Rotundifosine F among them showed excellent AChE inhibitory activity with an IC_{50} value of 0.3 μM. In addition, nominine showed strong

TABLE 8.1
Plants That Have Been Conventionally Used as AChE Inhibitors

Sr. No.	Plants Name	Family	Parts Used	Type of Extract	IC_{50} Value or % Inhibition	References
[1].	*Aconitum anthoroideum*	Ranunculaceae	Leaves	Ethanol	0.3 mM	Huang et al. (2020)
[2].	*Allium sativum*	Amaryllidaceae	Bulb (cloves)	Ethanol	0.01 mg/mL or 61.62 μM	Kumar (2015)
[3].	*Alpinia galanga*	Zingiberaceae	Rhizomes	Ethanol	24.6 ± 9.6 μg/mL	Chaiyana et al. (2022)
[4].	*Amaryllis belladonna*	Amaryllidaceae	Bulb	1.0 M H_2SO_4	14.3 ± 2.6 μg/mL	Sibanyoni et al. (2020)
[5].	*Andrographis paniculata*	Acanthaceae	Aerial part	Methanol	2.14 ± 0.04 μM	Panche et al. (2019) and Sani et al. (2019)
[6].	*Aquilaria sinensis*	Thymelaeaceae	Roots and rhizomes	95% ethanol	324 μM	Arya et al. (2021)
[7].	*Asteriscus maritimus*	Asteraceae	Different parts of the Plant	95% ethanol	95 μg/mL	Arya et al. (2021)
[8].	*Callicarpa candicans*	Lamiaceae	Leaves	–	45.67 and 58.38 g/mL	Hung et al. (2022)
[9].	*Centaurium umbellatum*	Gentianaceae	Leaves and stems	Ethanolic extract	94.24% at concentration 3 mg/mL	Neagu et al. (2018)
[10].	*Centella asiatica*	Apiaceae	Leaves	Ethanolic extract	15.05 ± 0.05 and 59.13 ± 0.18 μg/mL	Jusril et al. (2020)
[11].	*Corydalis dubia*	Papaveraceae	Whole plant	Methanol	0.0015 nmol	Wangchuk et al. (2016)
[12].	*Cotinus coggygria*	Anacardiaceae	Leaves and bark	Ethyl acetate and methanol	1.44 ± 0.04 mg/mL	Ozsoy et al. (2017)
[13].	*Ginkgo biloba*	Ginkgoaceae	Bark	Methanol	45.25 μg/mL	Phull et al. (2021)
[14].	*Mentha piperita*	Lamiaceae	Leaves	Hexane and chloroform extract	10.66 ± 0.12 μg/mL	Srief et al. (2023)
[15].	*Moringa oleifera*	Moringaceae	Root	Methanolic extract	0.00845 μg/mL	Nwidu et al. (2018)
[16].	*Marrubium rotundifolium*	Lamiaceae	Aerial part	Methanol, ethyl acetate, and water	11.44 ± 0.40 μg/mL	Ozsoy et al. (2017)
[17].	*Ocimum sanctum*	Lamiaceae	leaves	Methanol	94.22 ± 0.26	Toppo et al. (2019)
[18].	*Polygonum multiflorum*	Polygonaceae	Root	Ethanol	9.11 ± 0.82 μg/mL	Li et al. (2017)
[19].	*Salvia officinalis*	Lamiaceae	Leaves	Methanol	24.04 ± 0.55 and 8.57 ± 0.82 mg/mL	Sharma et al. (2020)
[20].	*Spatholobus suberectus*	Leguminosae	Stem	Ethanol	9.27 μg/mL	Li et al. (2017)
[21].	*Spiranthes sinensis*	Orchidaceae	Whole plant	Ethanolic extract	8.63±0.37 to 19.97 μg/mL	Zou et al. (2021)
[22].	*Stephania venosa*	Menispermaceae	Tuber	Ethanol	93.5, 86.6, and 168.6 μg/mL	Kongkiatpaiboon et al. (2016)
[23].	*Terminalia chebula*	Combretaceae	Fruits	Methanol	29.9 μM	Sancheti et al. (2010)
[24].	*Thymus vulgaris*	Lamiaceae	Leaves	Ethanol	30.23 ± 0.24 μg/mL	Sammi et al. (2017)
[25].	*Zingiber officinale*	Zingiberaceae	Rhizomes	Methanol	51.12 ± 0.37 μg/mL	Li et al. (2017)

neuroprotective effects against MPP^+-induced apoptosis in SH-SY5Y cells, with a rescue rate of 34.4% at 50 μM. These emphasize the therapeutic possibilities of diterpenoid alkaloids of *A. anthoroideum* as potential neuroprotective and anti-Alzheimer drugs (Huang et al., 2020).

8.2.1.2 *Allium sativum* L.

Allicin, the primary ingredient in garlic, is an AChE inhibitor. AChE and BuChE enzymes are inhibited by allicin in a concentration-dependent manner. For AChE, allicin's IC_{50} value is 0.01 mg/mL (61.62 μM) (Kumar, 2015).

8.2.1.3 *Alpinia galanga* (L.) Willd.

Essential oil utilizing gas chromatography-mass spectrometry (GC-MS), the chemical composition was examined. The inhibitory activities on AChE and BChE were evaluated using Ellman's method. The primary constituents were determined to be 1,8-cineole and methyl cinnamonate. IC_{50}

values for AChE and BChE in *Alpinia galanga* oil were 24.6 ± 9.6 and 825.4 ± 340.1 μg/mL, respectively (Chaiyana et al., 2022).

8.2.1.4 *Amaryllis belladonna* L.

Acetylcaranine and undulatine, alkaloids found in *Amaryllis belladonna* of the Amaryllidaceae family, were formerly isolated from buphanidrine and reported. *A. belladonna* was found to be the most active with an IC_{50} value of 14.3 ± 2.6 μg/mL (Sibanyoni et al., 2020).

8.2.1.5 *Andrographis paniculata* Wall

It is popularly called the "king of bitters" or Kalmegh and belongs to the Acanthaceae family. 3,4-di-o-caffeoylquinic acid, one of its active components, inhibits both AChE and BChE; for AChE, its IC_{50} value was found to be 2.14 ± 0.04 μM (Panche et al., 2019; Sani et al., 2019).

8.2.1.6 *Aquilaria sinensis* (Lour.)

Aquilaria sinensis belongs to the Thymelaeaceae family and is usually identified as agarwood or eaglewood. These plants are extracted using methanol by maceration and percolation techniques. It has reasonable inhibitory activity against AChE at 50 μg/mL, with an IC_{50} of 324 μM (Arya et al., 2021).

8.2.1.7 *Asteriscus maritimus* (Linn.)

Asteriscus maritimus belongs to the Asteraceae family and has considerable bioactive characteristics throughout its numerous components, especially when extracted with 95% ethanol. This flower essential oil has strong inhibition properties, with an IC_{50} of 95 μg/mL against AChE (Arya et al., 2021).

8.2.1.8 *Callicarpa candicans* (Burm.f.) Hochr.

The essential oils of different *Callicarpa* species have strong biological activities with great potential in traditional medicine and potential therapeutic applications. The volatile oils of *Callicarpa candicans* had encouraging potential, with IC_{50} ranging from 45.67 to 58.38 μg/mL. The essential oils of *C. sinuata*, *C. petelotii*, *C. nudiflora*, and *C. erioclona* had good activity with IC_{50} ranging from 28.71 to 54.69 μg/mL. In particular, the essential oils of *C. candicans* (Da Nang), *C. candicans* (Quang Nam), *C. erioclona*, *C. sinuata*, *C. petelotii*, and *C. nudiflora* were found to be active, with IC_{50} values ranging from 28.71 ± 3.85 to 54.69 ± 3.05 μg/mL, while *C. candicans* (Nghe An) possessed IC_{50} values of 58.38 ± 2.95 and 73.35 ± 4.61 μg/mL. The major components found in this essential oil are atractylone (37.7%), caryophyllene (19.0%), and selinene (6.2%) (Hung et al., 2022).

8.2.1.9 *Centella asiatica* (L.) Urb.

It was demonstrated to include active substances that enhance memory. Using high-performance liquid chromatography, three CA accession extracts SECA-K017, SECA-K018, and SECA-K019 were chemically profiled. Triterpene compounds inhibited AChE, according to the in vitro investigation, with IC_{50} values ranging from 15.05 ± 0.05 to 59.13 ± 0.18 μg/mL. Madecassic acid was shown to have a higher AChE inhibitory effect than Asiatic acid (Jusril et al., 2020).

8.2.1.10 *Centaurium umbellatum*

Centaurium umbellatum commonly known as centaury, a Gentianaceae family medicinal plant, has been employed for more than 2000 years for its bitterness as an amarum and digestive and for the treatment of febrile disorders, diabetes, hepatitis, and gout. The ethanolic extract of *C. umbellatum* exhibited strong AChE and tyrosinase inhibitory activity ranging from 72.24 to 94.24% (at the maximum dose applied of 3 mg/mL) and 66.96% and 94.03% (at 3 mg/mL), respectively, which corresponded to a strong inhibition of the DPPH radical by 70.29–84.9% (at 3 mg/mL). The results demonstrate the potential of *C. umbellatum* as a natural source of compounds possessing excellent antioxidant and enzyme inhibitory activity, which would be of therapeutic value in curing several ailments and neurodegenerative diseases (Neagu et al., 2018).

8.2.1.11 *Cotinus coggygria* Scop.

Cotinus coggygria exhibited remarkable inhibitory activity against AChE, establishing it as a plant of potential therapeutic value in the treatment of neurodegenerative diseases. The plant had an IC_{50} value of 1.44 ± 0.04 mg/mL, which is an indication of its potency as an AChE inhibitor. *C. coggygria* extracts were emphasized for containing high contents of phenolic compounds responsible for their bioactivity. Among the tested plants, *C. coggygria* was one of the most potent, with great potential for therapeutic use in neurodegenerative disorders like AD (Ozsoy et al., 2017).

8.2.1.12 *Corydalis dubia*

Corydalis plants, which are rich in isoquinoline alkaloids, show potential pharmacological activities for the management of significant ailments like cancer, inflammation, AD, and microbial infections. Eight isoquinoline alkaloids were isolated from *Corydalis dubia*, of which scoulerine and cheilanthifoline were found to be significant compounds. Scoulerine exhibited exceptionally strong anti-AChE activity, strongly inhibiting AChE with a minimum inhibitory requirement (MIR) value of 0.0015 nmol, which is twice that of the standard drug galanthamine (MIR value of 0.003 nmol). With few viable anti-Alzheimer's chemotherapies available, the scoreline is of high potential for research interest. This encompasses lead optimization, structure-activity relationship studies, analog development, pharmacodynamics evaluation, and in vivo animal testing to further its therapeutic use against AD (Wangchuk et al., 2016).

8.2.1.13 *Ginkgo biloba* L.

Ginkgo biloba, which belongs to the family Ginkgoaceae, showed maximum inhibitory activity of AChE with IC_{50}, 45.25 μg/mL. This exceptional enzyme inhibition activity was observed when *G. biloba* was tested against both urease and AChE enzymes. The findings reflect great promise for pharmaceutical uses, indicating the potential of *G. biloba* as a rich source of natural compounds that can be utilized in the synthesis of therapeutic agents (Phull et al., 2021).

8.2.1.14 *Mentha piperita* L.

The main constituents of the chloroform extract were 17-N-acetyl-oroidine (5.363%) and imidazoquinoline (7.767%). In the hexane extract, l-menthone (6.336%) and linolenic acid (19.397%) were the most prevalent ingredients. Moderate antioxidant activity was shown by the non-polar extracts and essential oil of *Mentha piperita.* With IC_{50} values for AChE of 10.66 ± 0.12 μg/mL and 16.33 ± 0.03 μg/mL (Srief et al., 2023).

8.2.1.15 *Marrubium rotundifolium*

Marrubium rotundifolium is a member of the Lamiaceae. The aerial parts of plant are employed for extraction. Methanol extract was often utilized. The phenolic components were more effectively extracted using a water-based extraction process. It inhibits AChE substantially, as shown by an IC_{50} value of 11.44 ± 0.40 μg/mL (Ozsoy et al., 2017).

8.2.1.16 *Moringa oleifera* Lam.

Moringa oleifera shows that the active components were responsible for BChE and an inhibitor of AChE. Its ethanolic extracts contain phenolic and flavonoid chemicals, which reduce the glycemic index, total cholesterol, triglycerides, and low-density lipoprotein cholesterol levels while growing high-density lipoprotein cholesterol in plasma (Nwidu et al., 2018).

8.2.1.17 *Ocimum sanctum* L.

Ocimum sanctum's methanolic leaf extract had the uppermost total phenol contented and antioxidant activities when compared to other plants, The anti-AChE assay's inhibition percentage was 94.22 ± 0.26% (Toppo et al., 2019).

8.2.1.18 *Polygonum multiflorum* Thunb.

Polygonum multiflorum, a traditional Chinese medicine, inhibits AChE and BChE with IC_{50} values of 9.11 and 4.83 μg/mL, respectively, making it one of the most active compounds against these enzymes. *P. multiflorum* inhibited Aβ1-42 peptide aggregation by 87.41%, suggesting a regulatory role in APP processing. The phytoconstituents 2,3,5,4′-tetrahydroxystilbene-2-O-β-D-glucoside (TSG), a principal active molecule, has shown great potential in the treatment of AD and Parkinson's disease via mechanisms, including AChE inhibition, neuroprotection, antioxidant activity, and cognitive improvement (Li et al., 2017).

8.2.1.19 *Salvia officinalis* L.

Commonly known as Sage, has many polyphenolic components that have antioxidant and cognitive effects, and it is frequently used as a spice and food preservative and has been shown to have neuroprotective advantages against AD by inhibiting the AChE enzyme. In Sage, methanol extract strongly inhibited AChE at 24.04 ± 0.55 and 8.57 ± 0.82 mg/mL, in contrast to a positive control of galantamine (6.33 μg/mL) (Sharma et al., 2020).

8.2.1.20 *Stephania venosa* (Blume) Spreng.

Blume Spreng, also called "Sa-Bu-Leud," is a Thai medicinal plant that is utilized as an aphrodisiac, blood tonic, and treatment for diabetes and cancer. The capacity of this alkaloid-rich plant to inhibit AChE has been investigated. HPLC analysis was used to perform a phytochemical screening of *Stephania venosa*, revealing chemical variations between the same species from various provenances. Three alkaloids were obtained by fractionating the extract of *S. venosa*: tetrahydropalmatine, crebanine, and dicentrine. Ellman's AChE inhibition test was used to assess the isolated alkaloids' AChE inhibitory capacity. AChE activity was inhibited by dicentrine, creatinine, and tetrahydropalmatine at IC_{50} values of 93.5, 86.6, and 168.6 μg/mL (Kongkiatpaiboon et al., 2016).

8.2.1.21 *Spatholobus suberectus* Dunn

Spatholobus suberectus, against AChE (IC_{50} = 9.27 μg/mL) and BuChE (IC_{50} = 5.37 μg/mL), belongs to the family Leguminosae. The stem of *S. suberectus* has been used in traditional medicine to enhance blood circulation, reduce inflammation, inhibit tyrosinase, and for conditions such as rheumatism, anemia, and menoxenia (Li et al., 2017).

8.2.1.22 *Spiranthes sinensis* (Pers.) Ames

Spiranthes sinensis (Pers.) Ames, a medicinal orchid, has been shown to have numerous active ingredients with possible AChE inhibitory action, which is important for the treatment of AD. Several substances, including quercetin and kaempferol, have shown noteworthy anti-AChE activity, with IC_{50} values ranging from 8.63 to 19.97 μg/mL (Zou et al., 2021).

8.2.1.23 *Thymus vulgaris* L.

Thyme oil, an essential oil, has been studied for its potential involvement in decreasing AChE, which increases ACh levels in the synaptic cleft, so increasing neurotransmission. Thyme oil contains para-cymene (1-methyl-4-propan-2-ylbenzene), which has minor AChE inhibitory action at lower concentrations (40 and 100 ppm). Significant inhibition was reported at these levels (30.23 ± 0.24 and 29.81 ± 0.40, respectively). However, at a greater dose (500 ppm), thyme oil displayed the opposite effect, demonstrating a complicated interaction with AChE (Sammi et al., 2017).

8.2.1.24 *Terminalia chebula* Retz.

Fruits of *Terminalia chebula* contain 1,2,3,4,6-penta-O-galloyl-β-D-glucose, a potent AChE inhibitor (IC_{50} = 29.9 μM). This gallotanin, which has been obtained from different sources and is known for its various biological functions, was found to have good BChE inhibition and significant antioxidant activity (FRAP assay) (Sancheti et al., 2010).

8.2.1.25 *Zingiber officinale* Roscoe

To ascertain the AChE inhibiting effect of *Zingiber officinale*, commonly referred to as ginger, essential ginger oils were isolated, standardized, and tested for the primary component α-citral. The resulting IC_{50} value was 51.12 ± 0.37 μg/mL. The most prevalent phytoconstituent in ginger oils, according to GC-MS analysis, is α-citral, which ranges from 13.92 to 30.46%. α-Citral, camphene, geraniol, α-pinene, linalool, and α-citronellol were all available in each of the eight ginger oils (Chaudhary et al., 2023b).

8.2.2 Phytoconstituents as a Source of Acetylcholinesterase Inhibitors

Phytoconstituents, particularly alkaloids and flavonoids, have been studied for their capability to inhibit AChE, which is essential for modulating cholinergic transmission in the brain. AChE is a key enzyme that degrades ACh, a neurotransmitter essential for cognitive function. These can inhibit AChE, increasing ACh levels at synapses and boosting cholinergic neurotransmission, which is notably beneficial AD (Balkrishna et al., 2019). Polyphenolic compounds, such as flavonoids, quercetin, apigenin, and kaempferol, have demonstrated substantial AChE inhibitory action. These compounds bind to AChE and inhibit its action, leading to increased ACh levels and stable cholinergic transmission (Cichon et al., 2024). Their mode of action extends beyond enzyme inhibition. They also have neuroprotective properties such as antioxidants and anti-inflammatories, making them promising candidates for the treatment of cognitive and neurodegenerative disorders (Cichon et al., 2024). Apart from flavonoids, alkaloids have become important sources of AChE inhibitors. For example, alkaloids from *Catunaregam spinosa* showed potential AChE inhibition in silico research, revealing numerous compounds with high binding affinity (Thandivel et al., 2024). Furthermore, *Horsfieldia polyspherula* extracts demonstrated significant AChE inhibition, with several compounds exhibiting dual inhibitory action (Idris et al., 2024). Investigation of phytomedicines is crucial, as they have fewer side effects than traditional synthetic pharmaceuticals and so are extremely promising alternatives to control AD (Gajendra et al., 2024; Singh et al., 2020). The therapeutic potential of phytoconstituents as AChE inhibitors appears promising, yet concerns with bioavailability and clinical validation are critical for their effective usage in therapeutic applications. Molecular docking evaluations have revealed that several phytochemicals have higher binding affinities to AChE than typical medicines like donepezil (Azmal et al., 2024). Flavonoids found in extracts of *Phyllanthus emblica* fruits, have substantial AChE inhibitory action, while compounds such as chebulagic acid have a binding affinity (Dawa & Chen, 2024). Phytochemicals such as flavonoids and polyphenols not only block AChE but also have neuroprotective properties that may allow them to pass the blood-brain barrier. These drugs can prevent low levels of ACh in the brain by inhibiting its early breakdown (Hasan et al., 2023). Finally, among the phytoconstituents, flavonoids and alkaloids offer a promising opportunity for the synthesis of natural AChE inhibitors. Their mechanism of action involves cholinergic activation together with additional neuroprotective effects, making them desirable choices for alternative therapies targeting neurodegeneration-related cognitive impairment. Future research should focus on increasing the bioavailability of such compounds and confirming their clinical usefulness in human models to properly utilize their therapeutic worth as medications against diseases such as AD. The detailed study of different phytoconstituents (Figure 8.2) and their role in inhibiting AChEs is described in Table 8.2.

8.2.2.1 Acetylcaranine

It was isolated from the bulbs of the Amaryllidaceae family member, *A. belladonna*. Using Ellman's method, it was discovered that acetylcaranine was the most active of the isolates toward AChE, with an IC_{50} of 11.7 ± 0.7 μM, which was comparable to that of galanthamine (IC_{50} = 6.19 ± 2.60 μM). The docking method was able to successfully predict the binding modes of ligands within receptor binding sites, and the docking workflow predicted that acetylcaranine would have the highest activity, which is in vitro results (Sibanyoni et al., 2020).

8.2.2.2 Apigenin

It is a flavonoid found in a wide variety of plants, including chamomile. The molecular structure is dictated by hydroxyl groups located at positions 4, 5, and 7. Column chromatography and HPLC are prominent isolation methods. AChE inhibition was observed at IC_{50} values of 40.02 ± 0.16 μg/mL and 57.76 ± 0.37 μg/mL (Islam et al., 2021).

8.2.2.3 Berberine

An isoquinoline alkaloid is found in plants of the *Berberis* genus. Usually, ethanol or methanol is used for extraction, while MS and HPLC are used for identification of compounds derived from the *Berberis* species in the Berberidaceae family. The results indicate that the IC_{50} value, or percentage inhibition, on AChE is 2.30 ± 0.29 μM and 7.8 ± 0.8 μM (Murray et al., 2013).

8.2.2.4 Cheilanthifoline

It is a derived compound purified from *C. dubia*, a crop used in indigenous Bhutanese medicine. Cheilanthifoline is one more isoquinoline alkaloid that exhibited poor AChE inhibition with an MIR of 0.31 nmol but possessed analogous poor antibacterial and anti-inflammatory activities: the AChE inhibitory activity, tumor necrosis factor-alpha activity (anti-TNF-α), and activity against one bacterial strain, *Helicobacter pylori* (Wangchuk et al., 2016).

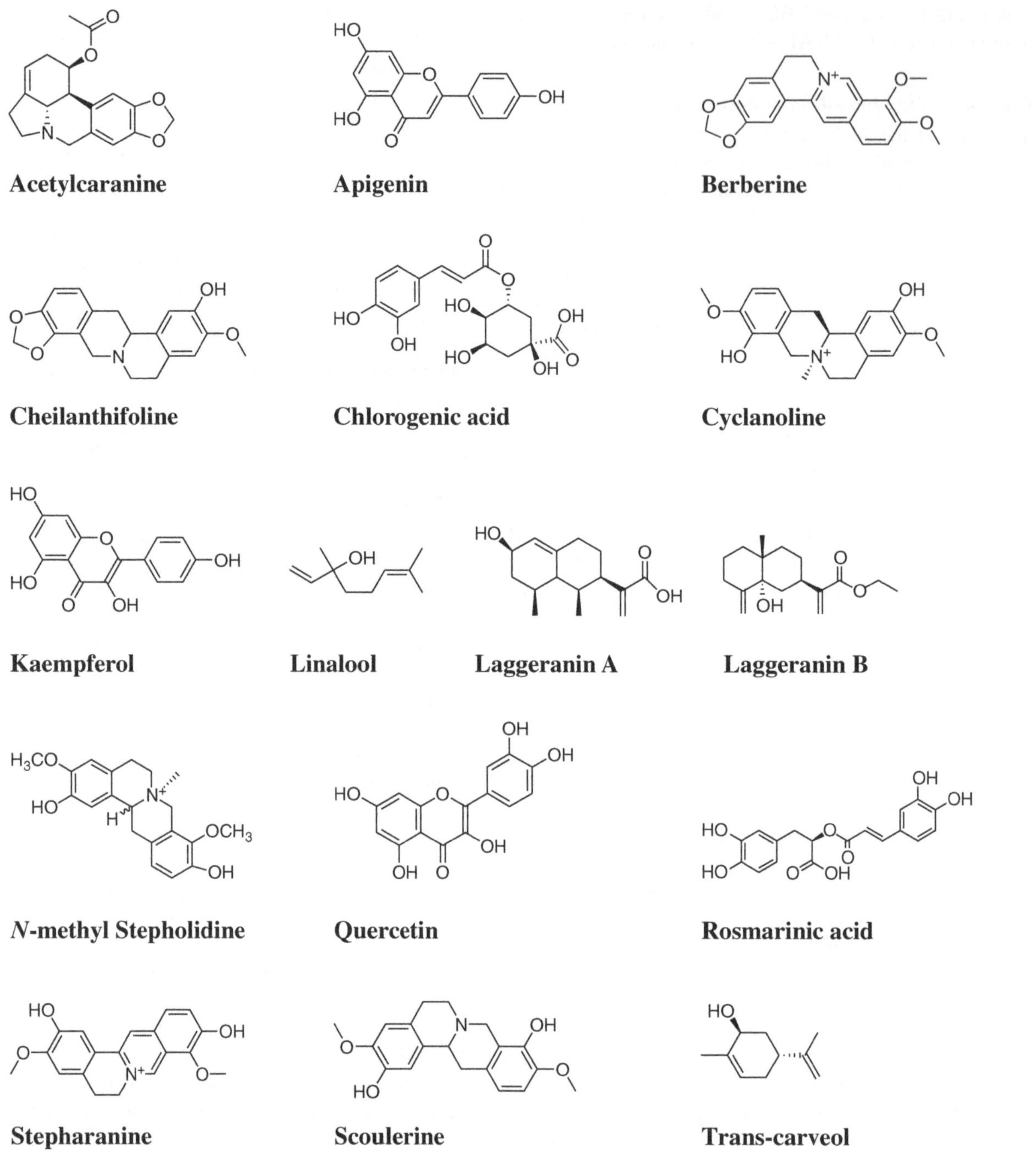

FIGURE 8.2 Phytoconstituents as a source of acetylcholinesterase inhibitors.

8.2.2.5 Cyclanoline

As quaternary protoberberine alkaloids, stepharanine, cyclanoline, and N-methyl stepholidine are obtained from *S. venosa*, a member of the Menispermaceae family. Demonstrated a moderate level of AChE inhibitory activity. *Stephania elegans* and *Stephania cepharantha* have also been reported to contain cyclolanoline (Ingkaninan et al., 2006).

8.2.2.6 Chlorogenic Acid

The highest antioxidative capabilities were shown by chlorogenic acids, which also had IC_{50} values of 1.69 μg/mL for AChE inhibition. The chlorogenic acid content of *Acinos arvensis* and the Lamiaceae family was high (26.06 mg/g) (Vladimir-Knežević et al., 2014).

8.2.2.7 Kaempferol

Kaempferol, a natural flavonoid, inhibited AChE activity in a reversible mixed-inhibition with an IC_{50} of 12.43 ± 0.19 μM. Fluorescence spectra revealed that kaempferol's fluorescence burst mechanism on AChE was static burst in nature, and it was linked with AChE to form a complex with a moderate binding affinity, mostly driven by hydrogen bonding and hydrophobic interactions (Shi et al., 2023).

TABLE 8.2
Phytoconstituents Are Reported as AChE Inhibitors

Sr. No.	Name of the Phytoconstituents	Source (Plant Name with Family)	Type of Extracting Solvent	Isolation Method/Identification Method	IC_{50} Value or % Inhibition on ACHE	References
1	Acetylcaranine	*Ammocharis coranica* (Amaryllidaceae)	Methanol	HPLC	11.7 ± 0.7 µM	Sibanyoni et al. (2020)
2	Apigenin	*Matricaria chamomilla* (Asteraceae)	Ethanol or methanol	Column chromatography and HPLC	40.02 ± 0.16 µg/mL	Islam et al. (2021)
3	Berberine	*Berberis* species (Berberidaceae)	Ethanol	HPLC, MS	2.30 ± 0.29 µM and 7.8 ± 0.8 µM	Murray et al. (2013)
4	Cheilanthifoline	*Corydalis dubia* (Papaveraceae)	Methanol	PTLC	0.31 nmol	Wangchuk et al. (2016)
5	Chlorogenic Acid	*Acinos arvensis* (Lamiaceae)	Ethanol	RP-HPLC	1.69 µg/mL	Vladimir-Knežević et al. (2014)
6	Cyclanoline	*Stephania venosa* (Menispermaceae)	Ethanol	Bioassay-guided fractionation	9.23 ± 3.47 µM	Ingkaninan et al. (2006)
7	Kaempferol	*Glycine max* (Fabaceae), *Hypericum perforatum* (Hypericaceae), *Moringa oleifera* (Moringaceae), and *Rosmarinus officinalis* (Lamiaceae)	Ethanol or Methanol	Column chromatography and HPLC	12.43 ± 0.19 µM	Shi et al. (2023)
8	Linalool	*Zingiber officinale* (ginger) (Zingiberaceae)	Ethyl acetate	GC-MS	55.10 ± 0.191 µg/mL	Chaudhary et al. (2023b)
9	Laggeranin A & B	*Laggera periodontal* (Asteraceae)	Ethyl acetate	Chromatography	437.33 ± 8.33 mM	Li et al. (2023)
10	N-methyl Stepholidine	*S. venosa* (Menispermaceae)	Ethanolic extract	Bioassay-guided fractionation	31.30 µM	Murray et al. (2013)
11	Quercetin	Apples, red onions, and green leafy vegetables (Rosaceae, Alliaceae, Brassicaceae)	Ethanol or methanol	GC-MS, HPLC	0.181 mM and 0.203 mM	Orhan (2021)
12	Rosmarinic acid	*Rosmarinus officinalis* (Lamiaceae)	Ethanol and methanol	HPLC	1.01 µg/mL	Vladimir-Knežević et al. (2014)
13	Stepharanine	*S. venosa* (Menispermaceae)	Ethanol	Bioassay-guided fractionation	14.10 µM	Murray et al. (2013)
14	Scoulerine	*C. dubia* (Papaveraceae)	Methanol	Preparative thin layer chromatography (PTLC)	0.0015 nmol	Wangchuk et al. (2016)
15	Trans-carveol	*Callicarpa candicans* (Lamiaceae)	Hexane, ethanol, or DCM	GC-MS	102.88 g/mL	Hung et al. (2022)

8.2.2.8 Laggeranin A & B

These chemicals, which are sesquiterpenes derived from *Laggera pterodonta*, had specific inhibitory actions on AChE in a dose-dependent manner (Li et al., 2023).

8.2.2.9 Linalool

With an IC_{50} value of 55.10 ± 0.191 µg/mL, linalool, one of the important chemical compounds in ginger (*Z. officinale*) essential oils, has been noted for its AChE inhibiting action. The AChE inhibition activity declined in the evaluated ginger oils in the following order of efficacy: This suggests that linalool was the most efficient of these chemicals in suppressing AChE: linalool > camphene > α-citronellol > α-citral > geraniol > α-pinene. This suggests that linalool has a strong ability to block AChE, which may help it be used to treat neurodegenerative illnesses (Chaudhary et al., 2023b).

8.2.2.10 N-methyl Stepholidine

The ethanolic extract of the Thai medicinal plant *S. venosa* has been reported to possess considerable AChE inhibitory activity using bioassay-guided fractionation. Among the isolated quaternary protoberberine alkaloids, N-methyl stepholidine showed moderate AChE inhibition with an IC_{50} value of 31.30 μM. Although less effective than other alkaloids like cyclanoline (IC_{50} = 9.23 μM) and stepharanine (IC_{50} = 14.10 μM), N-methyl stepholidine was still active, showing the effect of its quaternary nitrogen structure in inhibiting AChE. These results indicate that the quaternary structural characteristics of protoberberine alkaloids are important in contributing to their bioactivity and validate further investigation of N-methyl stepholidine as a therapeutic lead compound for neurodegenerative disease treatment (Murray et al., 2013).

8.2.2.11 Quercetin

It is a flavonoid present in a variety of plants, including apples, red onions, and green leafy vegetables. It is typically extracted with ethanol or methanol and identified using GC-MS and HPLC. Quercetin, an antioxidant, is one of the phenolic chemicals found in many colored fruits and vegetables (Orhan, 2021).

8.2.2.12 Rosmarinic Acid

Rosmarinic acid, which is a polyphenolic phytochemical prevalent in Lamiaceae medicinal herbs, is considered one of the predominant bioactive constituents of bioactive potential as well as a substantial antioxidant and neuroprotective property. Rosmarinic acid occurs most prevalently in plants, including *Mentha piperita* (peppermint), *Satureja subspicata* (savory), and *Melissa officinalis* (lemon balm). Quantitative analyses with RP-HPLC have validated its occurrence in such plants, and it has an IC_{50} value of 1.01 μg/mL of AChE inhibition; its antioxidant activity is greater than the antioxidative potential of Trolox. These biological activities, including its high radical-scavenging potential and neuroprotection, have been credited to the presence of polyphenols such as rosmarinic acid. The compound is also involved in diverse pharmacological activities owing to its anticancer, antimicrobial, and anti-inflammatory properties, thereby rendering these plants important sources of therapeutic and industrial purposes (Vladimir-Knežević et al., 2014).

8.2.2.13 Stepharanine

It was discovered that a Thai medicinal plant, *S. venosa*, belonging to the Menispermaceae family, had a high level of AChE inhibitory action. To find AChEi, bioassay-guided fractionation was used for the ethanolic extract of *S. venosa*. It was possible to isolate the moderately active quaternary protoberberine alkaloids with IC_{50} values of 14.10 μM (Murray et al., 2013).

8.2.2.14 Scoulerine

Scoulerine, an isoquinoline alkaloid, has shown highly significant AChE inhibitory activity and is a lead candidate for anti-Alzheimer's drug discovery. In experiments, the courier showed an MIR value of 0.0015 nmol, two-fold better than the reference drug galanthamine, with an MIR value of 0.003 nmol. It showed very active activity at a concentration as low as 1 ng. In addition to its robust AChE inhibition, scoulerine further demonstrated weak anti-inflammatory and antibacterial activities. It is a compound isolated from *C. dubia*, a plant employed in traditional Bhutanese medicine. Their wide-spectrum biological activities justify the traditional uses of the plant and demonstrate scoulerine's promise as a lead compound for further investigation by structure-activity relationship studies, analog design, and in vivo pharmacodynamic assessments (Wangchuk et al., 2016).

8.2.2.15 Trans-Carveol

Trans-carveol is a monoterpenoid alcohol that has been isolated from the essential oil of the *C. candicans* plant from the Lamiaceae family. The IC_{50} of trans-carveol is 102.88 μg/mL. Trans-carveol has also been investigated in another study, and it has an IC_{50} value for AChE inhibition of 676 μM. AChE inhibitory activity is usually greater in monoterpene ketones than in alcohols, as indicated by comparative studies on monoterpenoids containing the ap-menthane structure. The inhibitory activity of these compounds is further enhanced by the presence of an isopropenyl group. These findings suggest that structural features play a decisive role in the bioactivity of monoterpenoids like trans-carveol (Hung et al., 2022).

8.3 CONCLUSION AND FUTURE PERSPECTIVES

AD is a major and increasing public health issue, with cognitive diminishing and dementia. The cholinergic hypothesis, involving decreased ACh levels, points to AChE as a prime target for therapy. Although existing treatments provide symptomatic relief, the quest for disease-modifying agents goes on. Medicinal plants, with their AChE inhibitory activity, are a potential area of investigation. More studies are vital to establish their efficiency, bioavailability, and safety for clinical use. The future of AD therapy is in a multi-faceted strategy. Ongoing research on plant-based AChE inhibitors, with emphasis on finding and purifying active constituents, performing thorough preclinical and clinical trials to establish efficacy and safety, and creating standardized preparations for consistent dosing. Early detection and diagnosis, enabling early intervention and potentially halting disease progression. Adjusting treatment plans based on the unique traits of each patient, such as their genetic makeup and disease condition. Looking into potential advancements in combining AChE inhibitors with other therapy options that target tau tangles or amyloid plaques are two more aspects of AD pathophysiology. Prevention to lower risk of getting AD entails implementing public health initiatives that promote good lifestyle choices, such as diet, exercise, and cognitive stimulation. Acknowledging the heavy load on caregivers and offering all-encompassing support services to enhance their health and capacity to deliver high-quality care. Increased funding for AD research will hasten the discovery of new pharmaceutical targets, diagnostic tools, and preventative measures. Creation of multi-target drugs that can address several pathogenic features of AD, including tau phosphorylation, amyloid-beta aggregation, and AChE inhibition. Drug delivery systems are optimized to help medications reach their target brain regions and cross the blood-brain barrier.

REFERENCES

2023 Alzheimer's disease facts and figures. (2023). *Alzheimer's & Dementia*, *19*(4), 1598–1695. https://doi.org/10.1002/alz.13016

Ali, A. M., Mohamed, A. A., Ibrahim, A. N., & Elfiky, A. A. (2024). Acetylcholinesterase – Glucose-regulated protein 78 binding site prediction, a hope to cure neurological disorders such as Alzheimer's disease. *Journal of Receptor and Signal Transduction Research*, *44*(3), 122–128. https://doi.org/10.1080/10799893.2024.2426523

Arya, A., Chahal, R., Rao, R., Rahman, M. H., Kaushik, D., Akhtar, M. F., Saleem, A., Khalifa, S. M. A., El-Seedi, H. R., Kamel, M., Albadran, G. M., Abdel-Daim, M. M., & Mittal, V. (2021). Acetylcholinesterase inhibitory potential of various sesquiterpene analogues for Alzheimer's disease therapy. *Biomolecules*, *11*(3). https://doi.org/10.3390/biom11030350

Azmal, M., Hossen, M. S., Shohan, M. N. H., Taqui, R., Malik, A., & Ghosh, A. (2024). A computational approach to identify phytochemicals as potential inhibitor of acetylcholinesterase: Molecular docking, ADME profiling and molecular dynamics simulations. *PLoS ONE*, *19*(6), e0304490.

Balkrishna, A., Pokhrel, S., Tomer, M., Verma, S., Kumar, A., Nain, P., Gupta, A., & Varshney, A. (2019). Anti-acetylcholinesterase activities of mono-herbal extracts and exhibited synergistic effects of the phytoconstituents: A biochemical and computational study. *Molecules*, *24*(22). https://doi.org/10.3390/molecules24224175

Chaiyana, W., Sriyab, S., & Okonogi, S. (2022). Enhancement of cholinesterase inhibition of *Alpinia galanga* (L.) Willd. Essential oil by microemulsions. *Molecules*, *27*(10). https://doi.org/10.3390/molecules27103275

Chaudhary, S. K., Kar, A., Bhardwaj, P. K., Sharma, N., & Mukherjee, P. K. (2023a). High-performance thin layer chromatography method development, validation, and assessment of enzymatic inhibition potential of *Potentilla fulgens* hydroalcoholic roots extract. *Separation Science Plus*, *6*(7), 2300023. https://doi.org/10.1002/sscp.202300023

Chaudhary, S. K., Keithellakpam, O. S., Lalvenhimi, S., Chanda, J., Bhowmick, S., Kar, A., Nameirakpam, B., Bhardwak, P. K., Sharma, N., Rajashekar, Y., Sarangthem, I. D., & Mukherjee, P. K. (2023b). Chemo diversity of ginger – A potent crop in Manipur and its acetylcholinesterase (AChE) inhibitory potential. *Biochemical Systematics and Ecology*, *106*, 104560. https://doi.org/10.1016/j.bse.2022.104560

Cichon, N., Grabowska, W., Gorniak, L., Stela, M., Harmata, P., Ceremuga, M., & Bijak, M. (2024). Mechanistic and therapeutic insights into flavonoid-based inhibition of acetylcholinesterase: Implications for neurodegenerative diseases. *Nutrients*, *17*(1). https://doi.org/10.3390/nu17010078

Dawa, Y., & Chen, J. (2024). Screening and characterization of potential acetylcholinesterase inhibitors from *Phyllanthus emblica* L. fruits using affinity ultrafiltration combined with ultra-high-performance liquid chromatography–mass spectrometry. *Journal of Food Biochemistry*, *2024*(1), 6643716. https://doi.org/10.1155/jfbc/6643716

de Caires, S., & Steenkamp, V. (2010). Use of yokukansan (TJ-54) in the treatment of neurological disorders: A review. *Phytotherapy Research*, *24*(9), 1265–1270. https://doi.org/10.1002/ptr.3146

Dejanovic, B., Sheng, M., & Hanson, J. E. (2024). Targeting synapse function and loss for treatment of neurodegenerative diseases. *Nature Reviews Drug Discovery*, *23*(1), 23–42. https://doi.org/10.1038/s41573-023-00823-1

Drouin, E., & Drouin, G. (2017). The first report of Alzheimer's disease. *The Lancet Neurology*, *16*(9), 687. https://doi.org/10.1016/S1474-4422(17)30258-2

Gajendra, K., Pratap, G. K., Poornima, D. V., Shantaram, M., & Ranjita, G. (2024). Natural acetylcholinesterase inhibitors: A multi-targeted therapeutic potential in Alzheimer's disease. *European Journal of Medicinal Chemistry Reports*, *11*, 100154. https://doi.org/10.1016/j.ejmcr.2024.100154

Gaugler, J. E., Yu, F., Davila, H. W., & Shippee, T. (2014). Alzheimer's disease and nursing homes. *Health Affairs (Millwood)*, *33*(4), 650–657. https://doi.org/10.1377/hlthaff.2013.1268

Grill, J. D., & Cummings, J. L. (2010). Current therapeutic targets for the treatment of Alzheimer's disease. *Expert Review of Neurotherapeutics*, *10*(5), 711–728. https://doi.org/10.1586/ern.10.29

Gulisano, W., Maugeri, D., Baltrons, M. A., Fà, M., Amato, A., Palmeri, A., D'Adamio, L., Grassi, C., Devanand, D. P., Honig, L. S., Puzzo, D., & Arancio, O. (2018). Role of amyloid-β and tau proteins in Alzheimer's disease: Confuting the amyloid cascade. *Journal of Alzheimer's Disease*, *64*(s1), S611–s631. https://doi.org/10.3233/jad-179935

Hasan, S., Khatri, N., Rahman, Z. N., Menezes, A. A., Martini, J., Shehjar, F., Mujeeb, N., & Shah, Z. A. (2023). Neuroprotective potential of flavonoids in brain disorders. *Brain Sciences*, *13*(9). https://doi.org/10.3390/brainsci13091258

Hippius, H., & Neundörfer, G. (2003). The discovery of Alzheimer's disease. *Dialogues in Clinical Neuroscience*, *5*(1), 101–108. https://doi.org/10.31887/DCNS.2003.5.1/hhippius

Huang, S., Zhang, J.-F., Chen, L., Gao, F., & Zhou, X.-L. (2020). Diterpenoid alkaloids from *Aconitum anthoroideum* that offer protection against MPP^+–induced apoptosis of SH-SY5Y cells and acetylcholinesterase inhibitory activity. *Phytochemistry*, *178*, 112459. https://doi.org/10.1016/j.phytochem.2020.112459

Hung, N. H., Quan, P. M., Satyal, P., Dai, D. N., Hoa, V. V., Huy, N. G., Giang, L. D., Ha, N. T., Houng, L. T., Hien, V. T., & Setzer, W. N. (2022). Acetylcholinesterase inhibitory activities of essential oils from Vietnamese traditional medicinal plants. *Molecules*, *27*(20). https://doi.org/10.3390/molecules27207092

Idris, M., Azmi, M. N., Parmusivam, T., Supratman, U., Litaudon, M., & Awang, K. (2024). Phytochemical analysis, antioxidant activity and bioassay-guided isolation of acetylcholinesterase and butyrylcholinesterase inhibitors from *Horsfieldia polyspherula* bark (Myristicaceae). *Tropical Life Sciences Research*, *35*(3), 165–183. https://doi.org/10.21315/tlsr2024.35.3.8

Ingkaninan, K., Phengpa, P., Yuenyongsawad, S., & Khorana, N. (2006). Acetylcholinesterase inhibitors from *Stephania venosa* tuber. *Journal of Pharmacy and Pharmacology, 58*(5), 695–700. https://doi.org/10.1211/jpp.58.5.0015

Islam, M. A., Zaman, S., Biswas, K., Al-Amin, M. Y., Hasan, M. K., Alam, A. H. M. K., Tanaka, T., & Sadik, G. (2021). Evaluation of cholinesterase inhibitory and antioxidant activity of *Wedelia chinensis* and isolation of apigenin as an active compound. *BMC Complementary Medicine and Therapies, 21*(1), 204. https://doi.org/10.1186/s12906-021-03373-4

Islam, M. T., Aktaruzzaman, M., Saif, A., Hasan, A. R., Sourov, M. M. H., Sikdar, B., Rehman, S., Tabassum, A., Haque, S. A. U., Sakib, M. H., Muhib, M. M. A., Setu, M. A. A., Tasnim, F., Rayhan, R., Abdel-Daim, M. M., & Raihan, M. O. (2024). Identification of acetylcholinesterase inhibitors from traditional medicinal plants for Alzheimer's disease using in silico and machine learning approaches. *RSC Advances, 14*(47), 34620–34636. https://doi.org/10.1039/d4ra05073h

Jusril, N. A., Muhamad Juhari, A. N. N., Abu Bakar, S. I., Md Saad, W. M., & Adenan, M. I. (2020). Combining in silico and in vitro studies to evaluate the acetylcholinesterase inhibitory profile of different accessions and the biomarker triterpenes of *Centella asiatica*. *Molecules, 25*(15). https://doi.org/10.3390/molecules25153353

Kongkiatpaiboon, S., Duangdee, N., Prateeptongkum, S., & Chaijaroenkul, W. (2016). Acetylcholinesterase inhibitory activity of alkaloids isolated from *Stephania venosa*. *Natural Product Communications, 11*(12), 1805–1806.

Kumar, S. (2015). Dual inhibition of acetylcholinesterase and butyrylcholinesterase enzymes by allicin. *Indian Journal of Pharmacology, 47*(4), 444–446. https://doi.org/10.4103/0253-7613.161274

Lane, R. M., Potkin, S. G., & Enz, A. (2006). Targeting acetylcholinesterase and butyrylcholinesterase in dementia. *International Journal of Neuropsychopharmacology, 9*(1), 101–124. https://doi.org/10.1017/s1461145705005833

Li, J., Li, F., Wu, G., Gui, F., Li, H., Xu, L., Hao, X., Zhao, Y., Ding, X., & Qin, X. (2023). Acetylcholinesterase inhibitory activity of sesquiterpenoids isolated from *Laggera pterodonta*. *Frontiers in Plant Science, 14*, 1074184. https://doi.org/10.3389/fpls.2023.1074184

Li, Q., Tu, Y., Zhu, C., Luo, W., Huang, W., Liu, W., & Li, Y. (2017). Cholinesterase, β-amyloid aggregation inhibitory and antioxidant capacities of Chinese medicinal plants. *Industrial Crops and Products, 108*, 512–519. https://doi.org/10.1016/j.indcrop.2017.07.001

Majdi, A., Sadigh-Eteghad, S., Rahigh Aghsan, S., Farajdokht, F., Vatandoust, S. M., Namvaran, A., & Mahmoudi, J. (2020). Amyloid-β, tau, and the cholinergic system in Alzheimer disease: Seeking direction in a tangle of clues. *Reviews in the Neurosciences, 31*(4), 391–413. https://doi.org/10.1515/revneuro-2019-0089

Marucci, G., Buccioni, M., Ben, D. D., Lambertucci, C., Volpini, R., & Amenta, F. (2021). Efficacy of acetylcholinesterase inhibitors in Alzheimer's disease. *Neuropharmacology, 190*, 108352. https://doi.org/10.1016/j.neuropharm.2020.108352

Mendez, M. F. (2017). Early-onset Alzheimer disease. *Neurologic Clinics, 35*(2), 263–281. https://doi.org/10.1016/j.ncl.2017.01.005

Murray, A. P., Biscussi, B., Cavallaro, V., Donozo, M., & Rodriguez, S. A. (2024). Naturally occurring cholinesterase inhibitors from plants, fungi, algae, and animals: A review of the most effective inhibitors reported in 2012–2022. *Current Neuropharmacology, 22*(10), 1621–1649. https://doi.org/10.2174/1570159x21666230623105929

Murray, A. P., Faraoni, M. B., Castro, M. J., Alza, N. P., & Cavallaro, V. (2013). Natural AChE inhibitors from plants and their contribution to Alzheimer's disease therapy. *Current Neuropharmacology, 11*(4), 388–413. https://doi.org/10.2174/1570159x11311040004

Neagu, E., Radu, G. L., Albu, C., & Paun, G. (2018). Antioxidant activity, acetylcholinesterase and tyrosinase inhibitory potential of *Pulmonaria officinalis* and *Centaurium umbellatum* extracts. *Saudi Journal of Biological Sciences, 25*(3), 578–585. https://doi.org/10.1016/j.sjbs.2016.02.016

Nwidu, L. L., Elmorsy, E., Aprioku, J. S., Siminialayi, I., & Carter, W. G. (2018). In vitro anti-cholinesterase and antioxidant activity of extracts of *Moringa oleifera* plants from Rivers State, Niger Delta, Nigeria. *Medicines (Basel), 5*(3). https://doi.org/10.3390/medicines5030071

Orhan, I. E. (2021). Cholinesterase inhibitory potential of quercetin towards Alzheimer's disease – A promising natural molecule or fashion of the day? – A narrowed review. *Current Neuropharmacology, 19*(12), 2205–2213. https://doi.org/10.2174/1570159x18666201119153807

Ozsoy, N., Yilmaz-Ozden, T., Serbetci, T., Kultur, S., & Akalin, E. (2017). Antioxidant, anti-inflammatory, acetylcholinesterase and thioredoxin reductase inhibitory activities of nine selected Turkish medicinal plants. *Indian Journal of Traditional Knowledge, 16*(4), 553–561.

Panche, A. N., Chandra, S., & Diwan, A. D. (2019). Multi-target β-protease inhibitors from *Andrographis paniculata:* In silico and in vitro studies. *Plants (Basel), 8*(7). https://doi.org/10.3390/plants8070231

Phull, A.-R., Ali, A., Rafiq, M., Tahir, T., Majid, A., Seo, S.-Y., & Park, H.-J. (2021). Antioxidant potential, urease and acetylcholine esterase inhibitory activity and phytochemical analysis of selected medicinal plants from the Republic of Korea. *Exploratory Research and Hypothesis in Medicine, 6*(2), 51–59. https://doi.org/10.14218/erhm.2020.00069

Qiu, C., Kivipelto, M., & von Strauss, E. (2009). Epidemiology of Alzheimer's disease: Occurrence, determinants, and strategies toward intervention. *Dialogues in Clinical Neuroscience, 11*(2), 111–128. https://doi.org/10.31887/DCNS.2009.11.2/cqiu

Rasmussen, J., & Langerman, H. (2019). Alzheimer's disease – Why we need early diagnosis. *Degenerative Neurological and Neuromuscular Disease, 9*, 123–130. https://doi.org/10.2147/dnnd.s228939

Sammi, S. R., Trivedi, S., Rath, S. K., Nagar, A., Tandon, S., Kalra, A., & Pandey, R. (2017). 1-Methyl-4-propan-2-ylbenzene from *Thymus vulgaris* attenuates cholinergic dysfunction. *Molecular Neurobiology, 54*(7), 5468–5481. https://doi.org/10.1007/s12035-016-0083-0

Sancheti, S., Sancheti, S., Um, B. H., & Seo, S. Y. (2010). 1,2,3,4,6-penta-O-galloyl-β-d-glucose: A cholinesterase inhibitor from *Terminalia chebula*. *South African Journal of Botany, 76*(2), 285–288. https://doi.org/10.1016/j.sajb.2009.11.006

Sani, D., Khatab, N. I. O., Kirby, B. P., Yong, A., Hasan, S., Basri, H., & Stanslas, J. (2019). A standardised *Andrographis paniculata* Burm. Nees aqueous extract prevents lipopolysaccharide-induced cognitive deficits through suppression of inflammatory cytokines and oxidative stress mediators. *Journal of Advanced Research, 16*, 87–97. https://doi.org/10.1016/j.jare.2018.11.005

Sharma, Y., Fagan, J., & Schaefer, J. (2020). In vitro screening for acetylcholinesterase inhibition and antioxidant potential in different extracts of sage (*Salvia officinalis* L.) and rosemary (*Rosmarinus officinalis* L.). *Journal of Biologically Active Products from Nature*, *10*(1), 59–69. https://doi.org/10.1080/22311866.2020.1729239

Shi, W., Zhang, G., Liao, Y., Fei, X., Gong, D., & Hu, X. (2023). Anti-acetylcholinesterase mechanism of kaempferol and its synergistic effect with galanthamine hydrobromide. *Food Bioscience*, *56*, 103174. https://doi.org/10.1016/j.fbio.2023.103174

Sibanyoni, M. N., Chaudhary, S. K., Chen, W., Adhami, H.-R., Combrinck, S., Maharaj, V. Schuster, D., & Viljoen, A. (2020). Isolation, in vitro evaluation and molecular docking of acetylcholinesterase inhibitors from South African Amaryllidaceae. *Fitoterapia*, *146*, 104650. https://doi.org/10.1016/j.fitote.2020.104650

Singh, A. K., Rai, S. N., Maurya, A., Mishra, G., Awasthi, R., Shakya, A., Chellappan, D. K., Dua, K., Vamanu, E., Chaudhary, S. K., & Singh, M. P. (2020). Therapeutic potential of phytoconstituents in management of Alzheimer's disease. *Evidence-Based Complementary and Alternative Medicine*, *2021*(1), 5578574. https://doi.org/10.1155/2021/5578574

Srief, M., Bani, M., Mokrani, E. H., Mennai, I., Hamdi, M., Boumechhour, A., Mustapha, A. M., Derdour, M., Kerkatou, M., El-Shazly, M., Bensouici, C., Nieto, G., & Akkal, S. (2023). Evaluation of in vitro and in silico anti-Alzheimer potential of nonpolar extracts and essential oil from *Mentha piperita*. *Foods*, *12*(1), 190. https://doi.org/10.3390/foods12010190

Srivastava, S., & Muhammad, T. (2020). Violence and associated health outcomes among older adults in India: A gendered perspective. *SSM – Population Health*, *12*, 100702. https://doi.org/10.1016/j.ssmph.2020.100702

Stanciu, G. D., Luca, A., Rusu, R. N., Bild, V., Beschea Chiriac, S. I., Solcan, C., Bild, W., & Ababei, D. C. (2019). Alzheimer's disease pharmacotherapy in relation to cholinergic system involvement. *Biomolecules*, *10*(1). https://doi.org/10.3390/biom10010040

Tahami Monfared, A. A., Ye, W., Sardesai, A., Folse, H., Chavan, A., Aruffo, E., & Zhang, Q. (2023). A path to improved Alzheimer's care: Simulating long-term health outcomes of lecanemab in early Alzheimer's disease from the CLARITY AD trial. *Neurology and Therapy*, *12*(3), 863–881. https://doi.org/10.1007/s40120-023-00473-w

Thandivel, S., Rajan, P., Gunasekar, T., Arjunan, A., Khute, S., Kareti, S. R., & Paranthaman, S. (2024). In silico molecular docking and dynamic simulation of anti-cholinesterase compounds from the extract of *Catunaregam spinosa* for possible treatment of Alzheimer's disease. *Heliyon*, *10*(7), e27880. https://doi.org/10.1016/j.heliyon.2024.e27880

Toppo, A. J., Chandra, S., Jha, D., & Mazumder, P. M. (2019). In vitro evaluation of anti-acetylcholinesterase and free radical scavenging potential of leaf extracts of some selected medicinal plants. *Asian Pacific Journal of Tropical Biomedicine*, *9*(2), 60–65. https://doi.org/10.4103/2221-1691.250856

Vinze, S., Chodosh, J., Lee, M., Wright, J., & Borson, S. (2023). The national public health response to Alzheimer's disease and related dementias: Origins, evolution, and recommendations to improve early detection. *Alzheimer's & Dementia*, *19*(9), 4276–4286. https://doi.org/10.1002/alz.13376

Vladimir-Knežević, S., Blažeković, B., Kindl, M., Vladić, J., Lower-Nedza, A. D., & Brantner, A. H. (2014). Acetylcholinesterase inhibitory, antioxidant and phytochemical properties of selected medicinal plants of the Lamiaceae family. *Molecules*, *19*(1), 767–782. https://doi.org/10.3390/molecules19010767

Wangchuk, P., Sastraruji, T., Taweechotipatr, M., Keller, P. A., & Pyne, S. G. (2016). Anti-inflammatory, anti-bacterial and anti-acetylcholinesterase activities of two isoquinoline alkaloids-scoulerine and cheilanthifoline. *Natural Product Communications*, *11*(12), 1801–1804.

Yuan, H., Ma, Q., Ye, L., & Piao, G. (2016). The traditional medicine and modern medicine from natural products. *Molecules*, *21*(5). https://doi.org/10.3390/molecules21050559

Zou, M., Wang, R., Yin, Q., & Liu, L. (2021). Bioassay-guided isolation and identification of anti-Alzheimer's active compounds from *Spiranthes sinensis* (Pers.) Ames. *Medicinal Chemistry Research*, *30*(10), 1849–1855.

9 Non-Competitive N-Methyl-D-Aspartate Antagonist for Prospective Alzheimer's Disease Treatments

Amit Sharma, Ripunjay Kalita, Hemant R. Jadhav, and Amritansh Bhanot

9.1 INTRODUCTION

Neurodegenerative diseases are a collection of progressive and debilitating conditions characterized by the gradual loss of structure, function, and ultimately cell death in the central nervous system (CNS) (Sharma et al., 2025). The diseases generally interfere with certain populations of neurons that cause impaired cognitive functions, motor dysfunction, and behavioral changes. The most common examples include Alzheimer's disease (AD), Parkinson's disease (PD), Huntington's disease (HD), and amyotrophic lateral sclerosis (ALS) (Sharma et al., 2024a, 2024b).

Their root causes are multifactorial, with mechanisms involving protein misfolding and aggregation, oxidative stress, excitotoxicity, mitochondrial dysfunction, neuroinflammation, and cellular clearance impairment (Joshi et al., 2023; Maselli et al., 2023; Sharma et al., 2024). Many of the different causes may relate to genetic mutation, environmental factors, and many others, yet several neurodegenerative diseases have common pathways of molecular signaling that might represent shared therapeutic targets. However, the specific mechanisms of degeneration in each disease are still not fully understood, with Alzheimer's and Parkinson's being the most prevalent (Arbo et al., 2022).

Currently, around 55 million people worldwide suffer from dementia, with the number expected to rise to 78 million by 2030 and 139 million by 2050 (Agid, 2022; Liu et al., 2023). PD affects approximately 3% of individuals over the age of 65 globally. Given that the majority of neurodegenerative diseases are sporadic, excitotoxicity has emerged as a prominent factor in the degenerative process (Awad et al., 2023; Dash et al., 2024; Mendes et al., 2022; Ontario et al., 2022). Despite the development of several therapeutic molecules targeting these mechanisms, many have failed, largely due to their inability to cross the blood-brain barrier (BBB), which complicates treatment strategies (Rose et al., 2018).

NMDA receptors, integral to excitatory neurotransmission and synaptic plasticity, play a key role in neurodegenerative diseases. Dysregulation of these receptors, often due to overactivation, contributes to excitotoxicity—leading to excessive calcium influx, oxidative stress, mitochondrial dysfunction, and neuronal death. NMDA receptors are ion channels composed of multiple subunits, including GluN1, GluN2 (subtypes GluN2A-D), and GluN3 (subtypes GluN3A-B). Their activation requires both glutamate and a co-agonist, such as glycine or D-serine (van Onselen & Downing, 2024). When activated, NMDA receptors allow calcium ions to flow into neurons, playing a critical role in processes like long-term potentiation (LTP), which strengthens synaptic connections. However, excessive activation, particularly through GluN2B subunits, can exacerbate excitotoxic damage, contributing to neurodegenerative disease progression (Luo et al., 2023; Neves et al., 2023).

NMDA receptor antagonists block the excessive activation of NMDA receptors, helping prevent excitotoxicity (Wang et al., 2024). These antagonists bind to specific sites, such as the PCP (phencyclidine) binding site, to reduce calcium influx and protect neurons from damage. Some NMDA antagonists also interact with the dopaminergic system, enhancing dopamine release, which can both improve therapeutic outcomes and cause unwanted side effects, such as those seen in schizophrenia (Khesmakhi et al., 2024; Shen et al., 2022; Velu et al., 2024). Clinically, these antagonists have diverse uses: amantadine is used to treat dyskinesia in PD, memantine offers neuroprotective benefits in AD by slowing cognitive decline, and ketamine provides rapid antidepressant effects for treatment-resistant depression (Iovino et al., 2020). NMDA receptor antagonists are also being explored in modeling schizophrenia symptoms, providing neuroprotection after strokes or traumatic brain injuries, and showing anticonvulsant potential in epilepsy, though results vary across different conditions.

NMDA receptor antagonists have shown potential in various neurological disorders by mitigating excitotoxicity and improving outcomes (Zhang et al., 2016) In AD, memantine helps reduce cognitive decline, while in PD, amantadine manages symptoms and reduces dyskinesia. Ketamine has shown rapid antidepressant effects for treatment-resistant depression, and new agents such as fluoroethylnormemantine are being researched for the treatment of maladaptive behaviors in post-traumatic stress disorder. However, NMDA antagonists have psychotomimetic side effects, which is one reason they are associated with

DOI: 10.1201/9781003602477-9

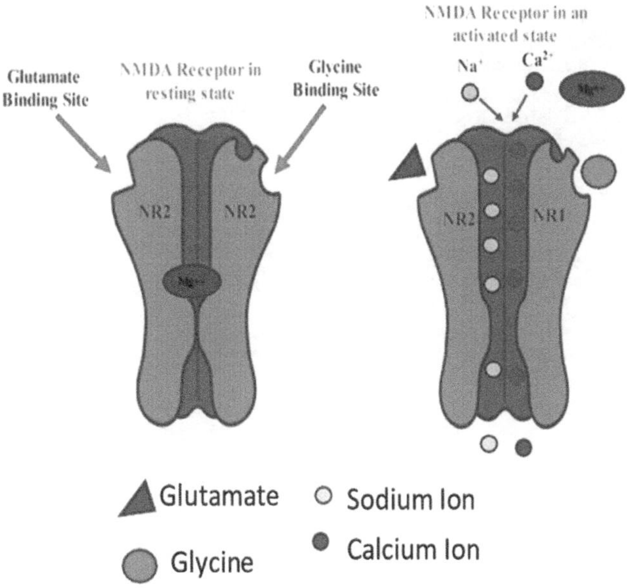

FIGURE 9.1 External magnesium (Mg permeant Ca^{2+}) ions strongly bind to the NMDAR pore at resting membrane potentials and block further ion permeation (left), in contrast to the ions. After postsynaptic membrane depolarization and glutamate release at the presynaptic site, ionic influx across the open channel pore alleviates voltage-dependent Mg block (right). For glutamate or NMDA to activate the channel, glycine must bind to NR1 and function as a co-agonist at NMDA receptors.

schizophrenia research. They are under study for the possibility of preventing or reducing excitotoxicity and thus holding therapeutic potential for ALS, and epilepsy, besides neuroprotection in models of stroke. This broader potential highlights their value for slowing down neurodegenerative processes and the improvement of the condition, placing them as valuable therapy in the practice of neurology (Chen & Lipton, 2006).

This book chapter aims to thoroughly examine the molecular mechanisms of NMDA receptor antagonists at their specific binding sites. The chapter will begin with an overview of the structure and function of NMDA receptors to provide foundational knowledge. Antagonists and modulators will be categorized according to their binding sites, with a focus on side effects that have hindered their clinical development (Kamat et al., 2013). Furthermore, new research directions will be suggested for exploring NMDA receptor antagonists as potential therapies for neurodegenerative disorders. Understanding these mechanisms and overcoming current challenges may lead to innovative strategies for treating these conditions (Figure 9.1) (Bezprozvanny & Mattson, 2008).

9.2 STRUCTURE AND FUNCTION OF NMDA RECEPTOR

The NMDA receptor (N-methyl-D-aspartate receptor) is a crucial ionotropic glutamate receptor in the CNS that plays a vital role in mediating excitatory neurotransmission (Burada et al., 2022). Found primarily at postsynaptic sites of neurons, NMDA receptors are essential for a variety of neural processes, such as synaptic plasticity, memory formation, and learning (Li et al., 2020; Murthy, 2022; Olivero et al., 2023; Yi et al., 2016). Their distinct structure and activation mechanism differentiate them from other glutamate receptors, allowing them to act as central hubs for intracellular signaling and neuronal communication (Zhou & Tajima, 2023).

NMDA receptors are heterotetrameric complexes made up of four subunits symmetrically arranged around a central ion channel (Gawande et al., 2023). The core functional component is the GluN1 subunit, which has to be present for the receptor itself to be formed and operate. This subunit forms the structural backbone of the receptor and is expressed at a consistent level throughout the CNS. Besides GluN1, the receptor consists of GluN2 subunits, which exist in four different forms—GluN2A, GluN2B, GluN2C, and GluN2D. These GluN2 subunits contribute to the functional diversity of the receptor; they influence several key properties such as the opening and closing rates of the channel, as well as its agonist and modulator sensitivity (Mony & Paoletti, 2023).

Apart from GluN1 and GluN2, NMDA receptors can be formed with GluN3 subunits, such as GluN3A and GluN3B. In the presence of GluN1 and GluN2 subunits, these reduce the calcium permeability of the receptor. The particular combination and pattern of subunits determine the physiological and pharmacological properties of the receptor. Besides this, different regions of the brain and various stages of development present differences in the subunit composition of the receptor, and so, it has different specific functions in diverse neural circuits (Wang & Shuaib, 2005).

The ion channel of the NMDA receptor is also uniquely regulated. Mg^{2+} ions block the pore of the channel at resting membrane potential and prevent the ions from flowing across. This is voltage-dependent; when the membrane undergoes depolarization, it releases the block. For the NMDA receptor to be activated, two conditions have to be met simultaneously: glutamate, the primary excitatory neurotransmitter, has to bind to its designated site on the receptor, and a co-agonist, either glycine or D-serine, has to bind to a separate site. This mechanism of dual gating tightly controls receptor activation both in time and space (Bi et al., 1998).

NMDA receptors are involved in so many central functions of the CNS as follows:

1. *Calcium ion permeability*: Opening of NMDA receptor channels facilitates the movement of calcium ions (Ca^{2+}), with lesser amounts of sodium (Na^+) and potassium (K^+) ions. The crucial role is that the permeating calcium ions are an intracellular secondary messenger involved in a variety of pathways for altering gene expression, synaptic efficacy, and even survival of the neurons (Jiang et al., 2024).

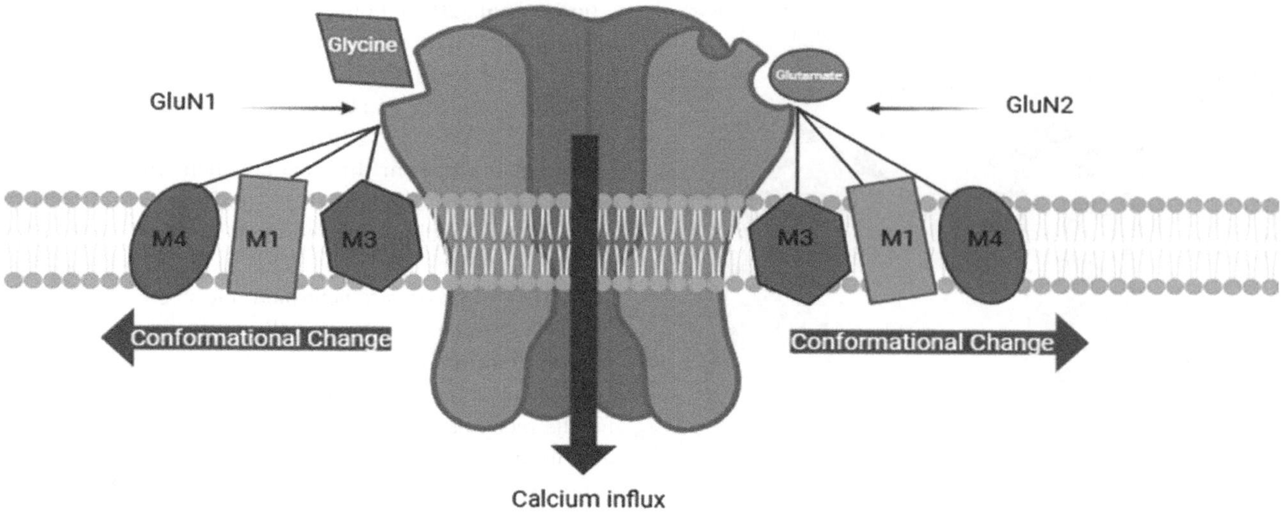

FIGURE 9.2 Schematic illustration of NMDA receptor activation.

2. *Synaptic plasticity*: Synaptic plasticity refers to the capability of synapses to be either strengthened or weakened depending on the level of activity. NMDA receptors play an essentially important role in two principal forms of synaptic plasticity: LTP and long-term depression (LTD). These two phenomena both depend on calcium influx through NMDA receptors. LTP and LTD are fundamental processes for learning and memory; they make it possible for the neurons to adapt and modify their connections (Wu et al., 2023).
3. *Neurodevelopment*: NMDA receptors play a critical role in the formation and maturation of synaptic connections during brain development. They enable the building of functional neural circuits by regulating the processes of synaptogenesis, growth of new dendrites, and activity-dependent selective elimination of redundant connections (Nikam & Meltzer, 2002).
4. *Neurotransmitter release and neural communication*: NMDA receptors modulate the release of other neurotransmitters, such as dopamine and GABA, and therefore influence broad neural network activity and communication. This regulatory role is important for the balance between excitatory and inhibitory signals in the brain (Zhang et al., 2019).
5. *Excitotoxicity and neurodegeneration*: Whereas modest activation of the NMDA receptors is indispensable for normal neuronal operation, severe stimulation may be lethal by itself as a mechanism called excitotoxicity. Pathology by Excess: Such high calcium overload generates inappropriate activations of enzymes, causing disastrous breakdown of cellular structures; ultimately neurons die in a phenomenon considered pathogenetically to underlie most neurodegenerative diseases—including Alzheimer's, HDs, ALS-plus acute diseases, stroke, and head trauma.
6. *Pathophysiology in neurological disorders*: Dysregulated activity of NMDA receptors has been implicated in a wide variety of neurological and psychiatric disorders. Overactivation results in excitotoxic damage associated with neurodegenerative diseases, whereas hypoactivation is correlated with cognitive impairment in schizophrenia. NMDA receptor dysfunction plays a role in conditions such as epilepsy, chronic pain, and mood disorders (Tsai et al., 2024).

Being involved in the most important CNS function and in the pathological processes, NMDA receptors are very critical for therapeutic intervention. Studies involving NMDA receptor modulators and antagonists attempt to reconstitute normal receptor activity in such conditions as AD, PD, depression, and epilepsy. Drugs such as memantine and amantadine have been developed to modulate NMDA receptor activity, while others like ketamine offer rapid therapeutic effects for treatment-resistant depression. Gaining insight into the complex architecture and multifaceted roles of NMDA receptors remains a cornerstone in advancing novel approaches to address neurodegenerative and mental health disorders (Figure 9.2) (Tamborini et al., 2013).

9.3 TYPES AND MOLECULAR MECHANISMS OF NMDA RECEPTOR ANTAGONIST

NMDA receptor antagonists are a group of compounds that modulate the activity of the NMDA receptor, a key ionotropic glutamate receptor involved in mediating excitatory neurotransmission in the brain (Geter-Douglass & Witkin, 1997). The role of excitotoxicity as a major contributor to the development of various neurodegenerative diseases has

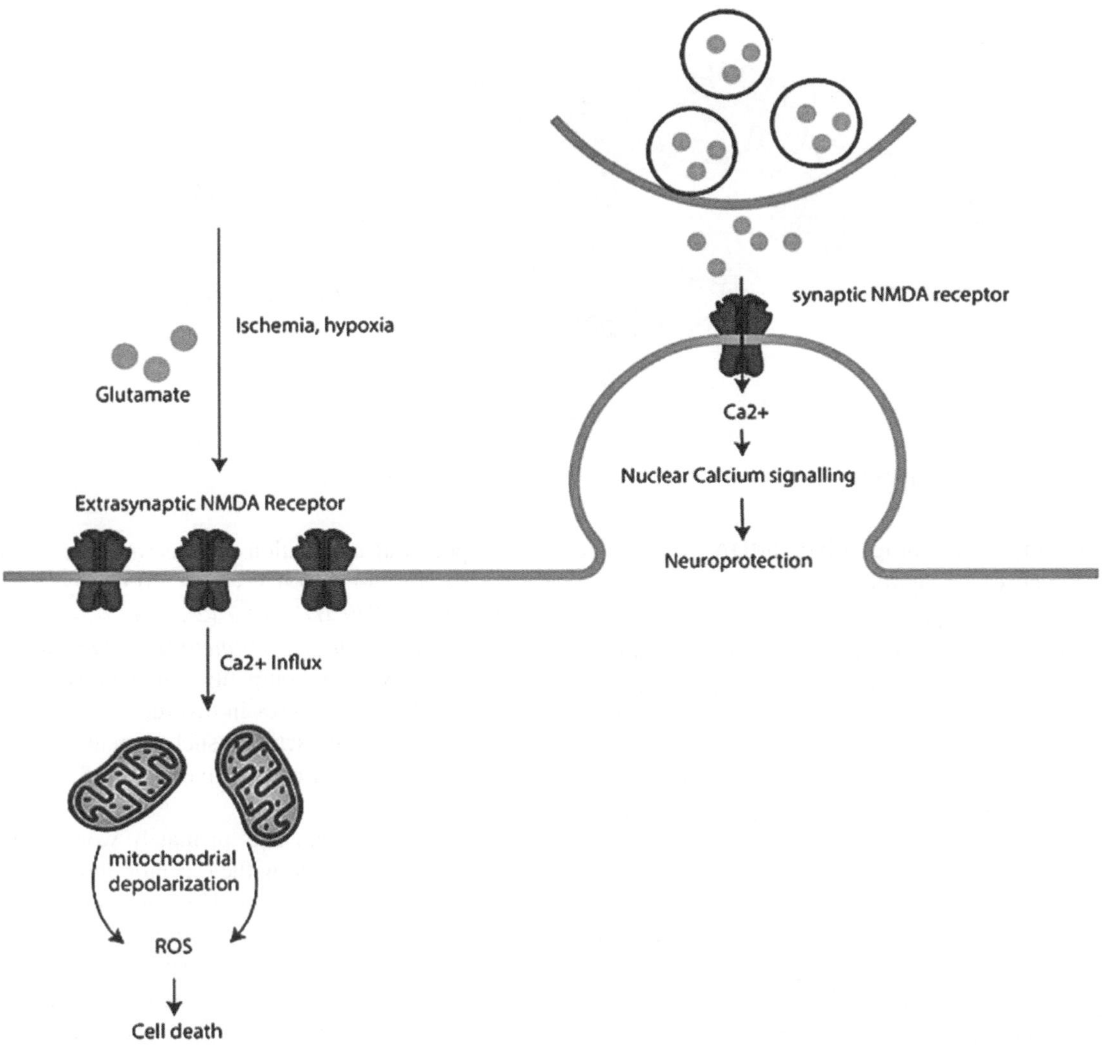

FIGURE 9.3 Role of synaptic and extrasynaptic NMDA receptors in neuronal survival and death.

highlighted NMDA receptor antagonists as a potentially viable avenue for therapeutic intervention. A wide range of NMDA receptor antagonists has been explored for their neuroprotective potential (Jespersen et al., 2014; Kvist et al., 2013; Wlaź et al., 1998). However, despite extensive efforts, their development into successful therapeutic options has encountered substantial difficulties. A primary challenge lies in disrupting the receptors' natural functions, which can trigger severe side effects, including cognitive decline, hallucinations, and psychosis (Figure 9.3) (Driver et al., 2022; Furuya et al., 1999; Lind et al., 2017; Takizawa et al., 1991; Willmore et al., 2001).

Nonetheless, the presence of distinct NMDA receptor subunits, particularly the GluN2 subunit, which is distributed in specific brain regions, offers hope for the design and development of subunit-specific antagonists. These targeted antagonists have the potential to provide therapeutic benefits with reduced side effects, paving the way for more effective treatments (Bubser et al., 1992; Erreger et al., 2005).

These antagonists are classified based on their binding sites and mechanisms of action, which influence their physiological effects and therapeutic potential. The three main categories of NMDA receptor antagonists are competitive antagonists, non-competitive antagonists, and negative allosteric antagonists (Marcus et al., 2001).

9.3.1 Competitive NMDA Receptor Antagonists

Competitive NMDA receptor antagonists directly bind to the glutamate-binding site on the receptor, competing with glutamate for access. By blocking glutamate from activating the receptor, these antagonists prevent the opening of the ion channel, inhibiting the influx of calcium (Ca^{2+}) and sodium (Na^{+}) ions into the neuron. This reduction in ion flow decreases excitatory neurotransmission, which can be beneficial in conditions characterized by excessive glutamate activity (Figure 9.4) (Bespalov et al., 1998; Carmack et al., 2013; Mony et al., 2009; Pérez-Pinzón & Steinberg, 2006; Sharkey et al., 1996).

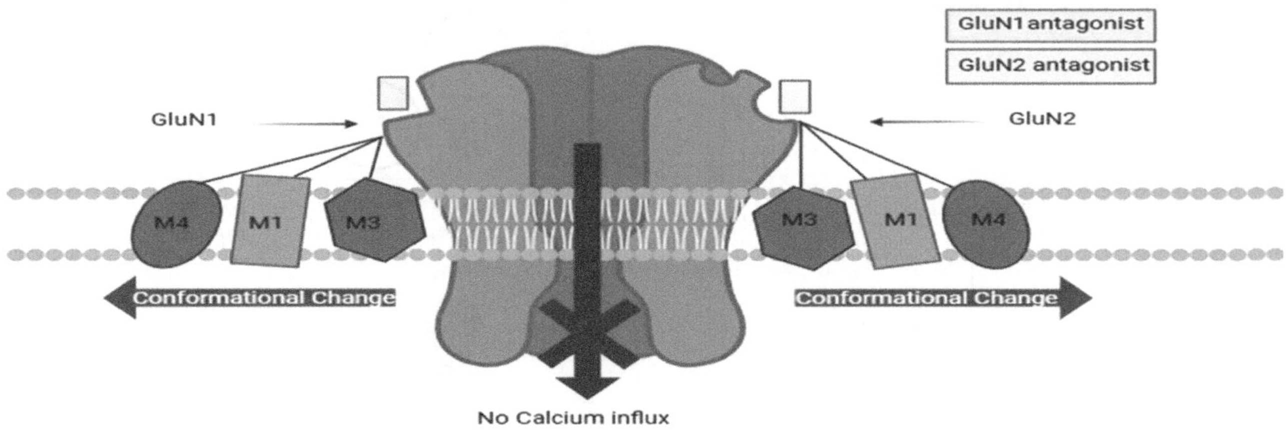

FIGURE 9.4 Schematic representation of competitive antagonism at NMDA receptor.

9.3.1.1 Examples of Competitive NMDA Receptor Antagonists

- *D-CPP (D(-)3-(2-carboxypiperazin-4-yl)-propyl-l-phosphonic acid)*: A selective competitive antagonist that blocks the glutamate binding site on the NMDA receptor, preventing receptor activation. D-CPP has been used in research to explore the role of NMDA receptor-mediated neurotransmission in various brain functions (Bermudo-Soriano et al., 2012; Liu et al., 2021).
- *D-CPP-ene (D(-)3-(2-carboxypiperazin-4-yl)-propenyl-l-phosphonic acid, also known as Midafotel)*: Similar to D-CPP, but with a propenyl group that provides a slightly different pharmacological profile. Midafotel has been studied for its neuroprotective effects, particularly in traumatic brain injury and ischemic stroke (Jullienne et al., 2011).
- *D-AP5 (2-amino-5-phosphonopentanoic acid)*: A widely used competitive antagonist that blocks the NMDA receptor's glutamate-binding site. D-AP5 is frequently used in research to study synaptic plasticity, LTP, and LTD by blocking the receptor's involvement in these processes (Mabrouk et al., 2013).
- *D-AP7 (2-amino-7-phosphonoheptanoic acid)*: A competitive antagonist that selectively inhibits the NMDA receptor's glutamate binding site. D-AP7 is used in studies related to excitotoxicity and neurodegenerative diseases (Mikics et al., 2017).
- *DCKA (5,7-dichlorokynurenic acid)*: A compound acting as a competitive antagonist at the NMDA receptor's glutamate-binding site. DCKA is often employed in research to examine the effects of blocking NMDA receptor activation in brain function and disease models (Papp & Moryl, 1994).
- *CGP-78608 (2-amino-4-methyl-5-phosphono-3-pentenoate-l-ethyl ester)*: A competitive antagonist that selectively inhibits the NMDA receptor by targeting the glutamate-binding site, with potential applications in treating neurodegenerative diseases (Szakács et al., 2003).
- *CGP-37849 (DL-(E)-2-(2-carboxy-2-(4-(trifluoromethyl)phenyl)ethyl)phenyl)-2-phosphonoacetic acid)*: Another competitive antagonist with potential therapeutic uses in disorders involving excessive glutamate activity, such as neurodegenerative diseases and stroke (Tóth et al., 2018).

Competitive antagonists primarily work by preventing glutamate from activating the receptor, thus reducing excitatory signaling in the brain. This mechanism is especially beneficial in mitigating excitotoxicity, a pathological process that contributes to neurodegenerative diseases, stroke, and trauma. (Figure 9.5) (Berger & Rebernik, 2015; Bubser et al., 1992; Flores et al., 2014; Sun & Wessinger, 2004).

9.3.2 Non-Competitive NMDA Receptor Antagonists

Non-competitive NMDA receptor antagonists bind to sites within the ion channel of the receptor, rather than competing with glutamate for its binding site. These antagonists typically interact with the PCP binding site. When bound, they block the ion channel, preventing the flow of calcium and sodium ions into the neuron. This inhibition of ion influx reduces excitatory neurotransmission and can provide neuroprotective effects, especially in conditions where excessive NMDA receptor activation leads to neuronal damage (Blot et al., 2013; Lipton, 2004; Marcus et al., 2001; Rao et al., 2001; Stuchlík & Vales, 2005).

9.3.2.1 Examples of Non-Competitive NMDA Receptor Antagonists

- *PCP*: PCP is one of the first discovered non-competitive NMDA receptor antagonists. It works by binding the PCP site within the ion channel, stopping the flow of ions through the receptor. Since PCP can cause psychotomimetic effects, its use in medical practice is infrequent; however, PCP

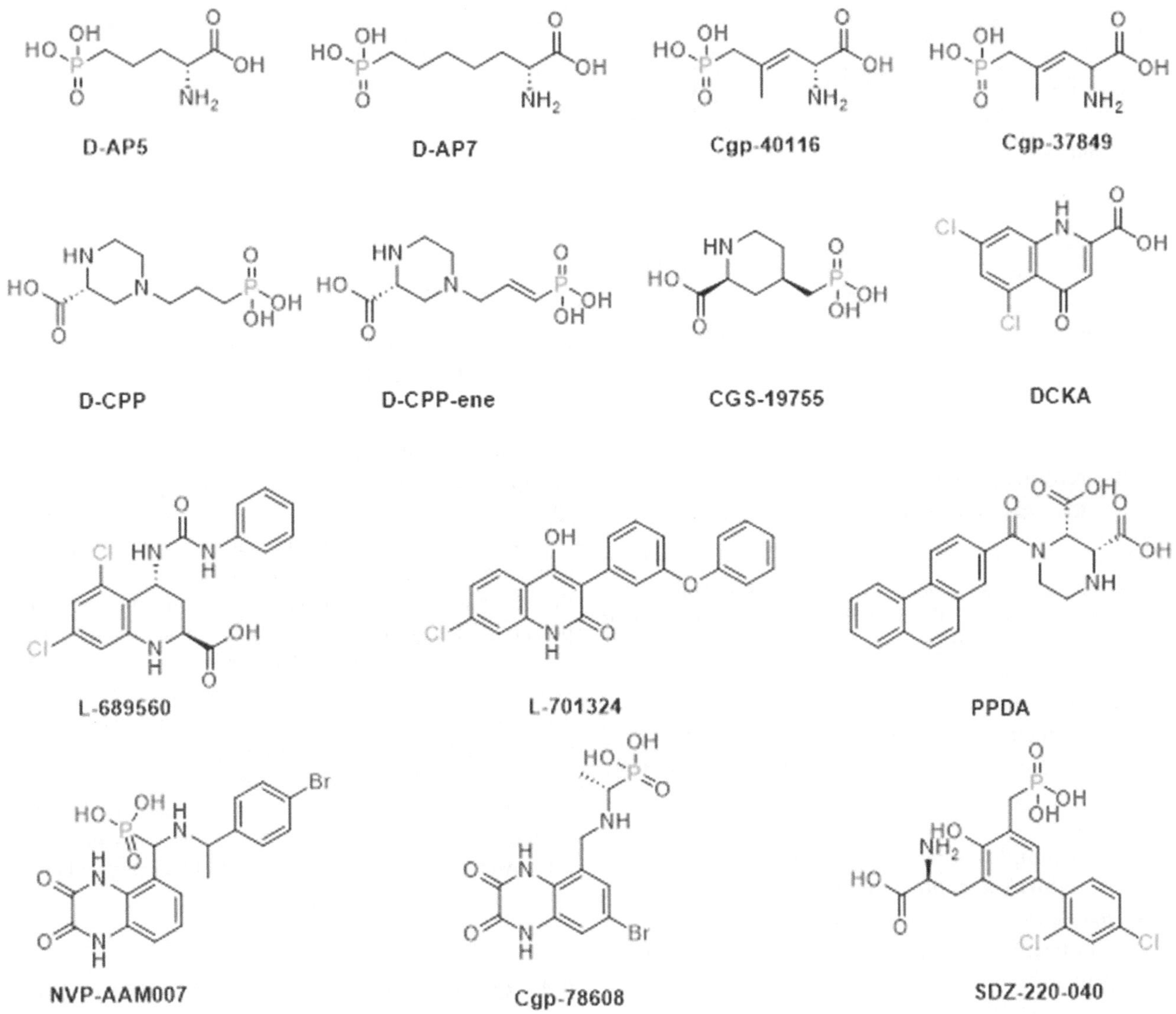

FIGURE 9.5 Competitive NMDA receptor antagonist.

is greatly used in various researches to determine NMDA receptor functions and mechanisms for neurotoxicity (Losi et al., 2006).

- *Dizocilpine (MK-801)*: MK-801 is a potent non-competitive NMDA receptor antagonist that binds to the ion channel and blocks ion flow. MK-801 is used mainly in research for modeling excitotoxicity and the study of neurodegenerative diseases. It has also been examined for its neuroprotective potential in ischemic brain injury.
- *Ketamine*: One of the best known non-competitive NMDA receptor antagonists, ketamine recently gained a new reputation with its rapid-acting anesthetic effect, and even more so with reports of efficacy in treatment-resistant depression. It selectively binds to the PCP site in the ion channel, where it blocks the influx of calcium and sodium and decreases excitatory signaling. Its use has extended beyond anesthesia to include potential therapeutic applications in depression, anxiety, and chronic pain management (Shafiei-Irannejad et al., 2021).
- *Tiletamine*: Tiletamine is another non-competitive NMDA receptor antagonist that is commonly used as part of an anesthetic combination. It works similarly to ketamine by blocking the ion channel, which induces a dissociative anesthetic state by inhibiting excitatory neurotransmission (Lin et al., 2016).

Non-competitive antagonists are particularly useful in conditions where there is overactivation of NMDA receptors, such as neurodegenerative diseases, acute brain injury, and chronic pain. Their neuroprotective effects help prevent neuronal damage due to excitotoxicity (Figure 9.6) (Costa et al., 2010; Sirrieh et al., 2015; Subramanian et al., 2022; Suzuki et al., 2019).

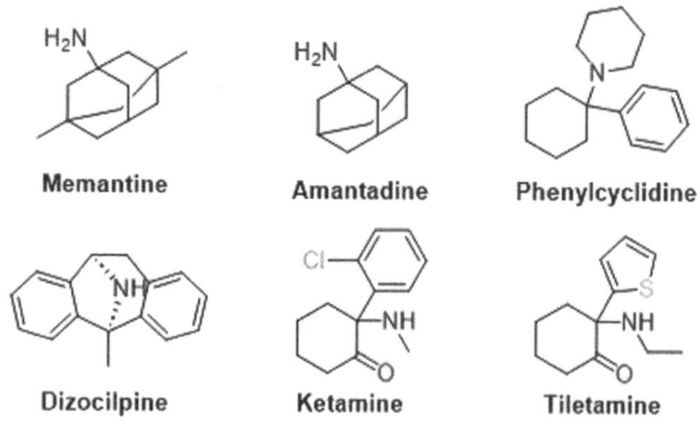

FIGURE 9.6 Non-competitive NMDA receptor antagonists.

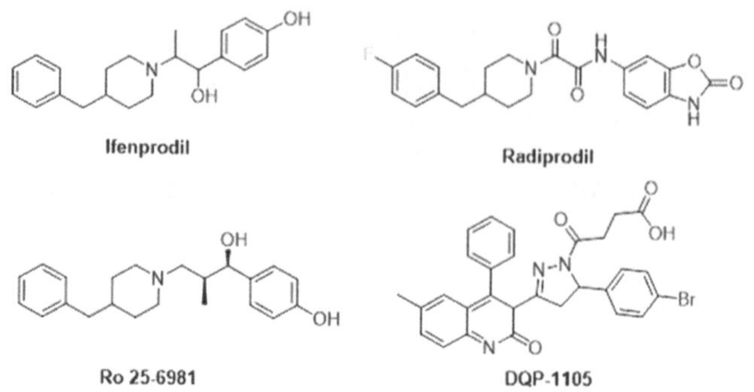

FIGURE 9.7 Compounds showing negative allosteric modulation of NMDA receptors.

9.3.3 Negative Allosteric NMDA Receptor Antagonists

Negative allosteric antagonists bind to a site distinct from the glutamate-binding site, referred to as the allosteric site. Rather than directly blocking glutamate binding or ion channel activity, these antagonists alter the receptor's conformation, reducing its responsiveness to glutamate. By modulating the receptor's structure, negative allosteric antagonists reduce its overall activity and decrease excitability without fully blocking the receptor (Zhang et al., 2018).

9.3.3.1 Examples of Negative Allosteric NMDA Receptor Antagonists

- *Ifenprodil*: Ifenprodil is a selective negative allosteric antagonist of the NMDA receptor and primarily targets the GluN2B subunit. Its binding to the subunit diminishes the response of the receptor to glutamate, leading to a decrease in excitatory neurotransmission. It is very intriguing for treating certain neurological disorders characterized by significant participation of GluN2B-containing receptors, such as forms of pain, depression, and ischemic brain injury (Zhu & Paoletti, 2015).
- *Ro-25698*: It is one of the most selective negative allosteric antagonists. Ro-25698 primarily acts on NMDA receptors GluN2B subunits. It also has low cross-reactivity against other receptor subtypes, providing a potential target in therapeutic interventions without causing widespread blockade at the receptor sites in diseases of excitotoxicity (Zhu et al., 2016).

Negative allosteric antagonists represent a more nuanced approach to the modulation of NMDA receptor activity, as they decrease the sensitivity of receptors without completely eliminating the function of receptors. Such a mechanism has the potential for therapeutic intervention in conditions related to NMDA receptor dysregulation, including neurodegenerative diseases, chronic pain, and psychiatric disorders (Mony et al., 2009; Quan et al., 2023; Zhang et al., 2013).

NMDA receptor antagonists include competitive, non-competitive, and negative allosteric modulators. This diverse class of compounds is of significant therapeutic potential. Their value in the treatment of neurological disorders, management of neurodegenerative diseases, and prevention of excitotoxic damage is their ability to modulate excitatory neurotransmission via different mechanisms of action. Continued research into the pharmacology of these antagonists, including specific subunit selectivity and side effect profiles, will advance our understanding of NMDA receptor function and will guide the development of more effective treatments for a wide range of neurological conditions (Figure 9.7) (Gerber et al., 2013).

9.4 OVERACTIVATION OF NMDA RECEPTOR AND NEUROLOGICAL DISEASES

The overactivation of NMDA receptors plays a pivotal role in the pathogenesis of several neurodegenerative diseases, primarily due to excessive extracellular glutamate. Normally, glutamate, the main excitatory neurotransmitter in the CNS, is tightly controlled by a balance between its release and uptake (Lepeintre et al., 2004). However, in pathological conditions, this balance is disrupted, leading to the accumulation of glutamate and, consequently, overstimulation of NMDA receptors. This excessive activation results in an influx of calcium ions (Ca^{2+}) into the postsynaptic neuron, a central mechanism in excitotoxicity (Bechthold et al., 2021).

Excessive NMDA receptor activation triggers a cascade of detrimental cellular responses, most notably neuronal cell death. The increased concentration of calcium activates the intracellular signaling pathways and activation of enzymes such as calpains and caspases and will tend to induce apoptotic or necrotic death of the neurons. This involves the cell death within individual neurons and contributes to wider neurodegeneration, progressively impairing brain function (Kotajima-Murakami et al., 2022).

Other than neuronal death, excess calcium also deranges synaptic function, mainly synaptic plasticity that facilitates learning and memory. The disruption leads to cognitive impairment as well as diminished memory. In addition, overactivation of NMDA receptors may alter synaptic composition as well as its function. Indeed, modifications in the expression of NMDA receptor subunits, especially GluN2B, have been related to the formation of extrasynaptic NMDA receptors. Such receptors tend to increase excitotoxic damage even more and lead to greater neurodegeneration (Buchan et al., 1992; Janus et al., 2023; Yi et al., 2019).

Another downstream effect of overactivated NMDA receptors is glutamate homeostasis. The glutamate level is kept very low through release from presynaptic and glial uptake. However, when this control fails in the diseased state, an excess of glutamate builds up within the synaptic cleft and its surrounding extracellular space, making the vicious cycle of excitotoxicity more deleterious and hastening neurodegenerative disease progression (Henter et al., 2018).

Overactivation of the NMDA receptor is one reason why targeting this pathway has therapeutic potential in neurodegenerative diseases. Excitotoxicity is implicated significantly in neuronal loss and cognitive impairment in neurodegenerative diseases, such as AD, PD, ALS, and HD. Therapeutic strategies that reduce NMDA receptor activity are likely to arrest or even retard the course of these diseases. Potential therapies may include augmentation of glutamate removal from the synaptic cleft through either increasing the activity of glutamate transporters or through blockade of overactive glutamate release (Yan et al., 2022).

One more promising avenue focuses on the employment of NMDA receptor antagonists that minimize overactivation either by direct block of the receptor or its response to glutamate. Other antagonists of the NMDA receptors include competitive and non-competitive antagonists; the former works by blocking the glutamate binding site, while the latter blocks channel function.

Glial cells are also very important in excitotoxicity, as they maintain glutamate homeostasis by removing excess glutamate. Dysfunctional glial cells can worsen glutamate accumulation and contribute to neurodegeneration. Therefore, strategies aimed at restoring or enhancing glial cell function could offer additional therapeutic benefits (Becker et al., 2019).

NMDA receptor overactivation and the resulting excitotoxicity are central to the development and progression of various neurodegenerative diseases. The excessive calcium influx triggers neuronal death, disrupts synaptic function, and disturbs glutamate homeostasis, all contributing to cognitive decline and neurodegeneration. Targeting NMDA receptor activity or modulating glutamate transmission presents a promising strategy to slow disease progression, while improving glial cell function offers additional potential for therapeutic intervention. Addressing these mechanisms is essential for developing effective treatments for neurodegenerative disorders (Table 9.1) (Kew et al., 1998).

TABLE 9.1
Pharmacological Profiles of NMDA Receptor Antagonists/Modulators

S. No	Compounds	Receptor	Developmental Stage	Therapeutic Effect
		Negative Allosteric Modulators		
1.	Ifenprodil	NMDAR GluN2B	Phase III	a. Antidepressant effect. b. Neuroprotective effect cerebral ischemia. c. Anticonvulsant effect.
2.	Radiprodi	NMDAR GluN2B	Terminated	a. Anticonvulsant effect. b. Antiepileptic effect.
3.	Ro25-6981	NMDAR GluN2B	Preclinical	a. Antidepressant effect. b. Neuroprotective effect. c. Antiparkinsonian effect.
4.	DQP-1105	NMDAR GluN2C and GluN2D	Preclinical	Neuroprotective effect.
		Competitive Antagonists		
5.	D CPP	NMDAR GluN2A	Terminated	a. Antiepileptic effect. b. Neuroprotective effects. c. Alteration of acute behavioral response to cocaine.
6.	D-AP5/D-AP7	Non-subunit selective	Preclinical	a. Block fear acquisition. b. Interrupts acute response to psychostimulants.

(Continued)

TABLE 9.1 ***(Continued)***
Pharmacological Profiles of NMDA Receptor Antagonists/Modulators

S. No	Compounds	Receptor	Developmental Stage	Therapeutic Effect
7.	DCKA	NMDAR GluN1	Preclinical	a. Anxiolytic effect. b. Neuroprotective against NMDA/glycine-induced toxicity.
8.	CGP-78608, CGP-37849, and CGP-40116	NMDAR GluN1	Preclinical	Anticonvulsant effect.
9.	CGS-19755 (Selfotel)	NMDAR GluN2A	Terminated	Neuroprotective effect against global and focal ischemia, trauma and stroke.
10.	L689-560 and L701-324	NMDAR GluN1	Preclinical	a. Anticonvulsant effects. b. Anxiolytic effect in mice.
11.	PPDA	NMDAR GluN2A, GluN2C, and GluN2D	Preclinical	a. Prevents NMDA-induced neuronal death. b. Anti-allynic and anti-hyperalgesic effects.
12.	NVP-AAMO77 (PEAQX)	NMDAR GluN2A, GluN2C, and GluN2B	Preclinical	Impairment of contextual and temporal fear responses.
13.	SDZ-220-040	NMDAR GluN2B	Preclinical	a. Anticonvulsant effect. b. Protection against focal ischemia. c. Attenuate neuropathic pain.
		Non-Competitive Antagonist		
14.	MK-801/DIZOCILPINE	Open-channel blocker	Preclinical	a. Neuroprotective effect in several animal models of cerebral ischemia. b. Anticonvulsant effect.
15.	Memantine	Open-channel blocker	Approved for Alzheimer	a. Neuroprotective effect in AD and prodromal stages of psychosis. b. Antidepressant-like effect. c. Anticonvulsant effect.
16.	Amantadine	Open-channel blocker	Approved for Parkinson	a. Effectively reduce L-Dopa-induced abnormal involuntary movement. b. Anticonvulsant effect. c. Neuroprotective effect.
17.	PCP	Open-channel blocker	Preclinical	a. Anticonvulsant effect. b. Anesthetic effect.
18.	Ketamine	Open-channel blocker	Approved	a. Anticonvulsant effect in NMDA-induced seizure model. b. Anesthetic effect. c. Antidepressant effect.
19.	Tiletamine	Channel blocker	Approved for veterinary use	a. Anesthetic effect. b. Anticonvulsant effects in NMDA-induced seizure model.

9.5 CURRENT CHALLENGES AND FUTURE PERSPECTIVE

The development of NMDA receptor antagonists as therapeutic agents for neurodegenerative diseases faces several significant challenges. One of the key difficulties is the complexity of the NMDA receptor itself. This receptor plays a crucial role in a variety of brain functions, and any disruption—whether hyperactivation or hypoactivation—can lead to adverse health effects (Volgraf et al., 2016). Another challenge is the gap between preclinical studies and clinical trials. While promising therapeutic effects were seen in models of animals, corresponding human studies often fail to replicate the results, indicating a poor understanding of NMDA receptor physiology in the human CNS. Furthermore, heterogeneity in NMDA receptor subtypes makes targeted therapy more cumbersome. The different subunit composition complicates the study of roles and selection of the most appropriate target for treatment through specific subtypes. This complexity, combined with the difficulties in designing selective antagonists, makes it very challenging to achieve the best therapeutic outcome without unintended side effects (Jimenez, 2022).

Another major issue is the side effects for most NMDA receptor antagonists. Most of these drugs are associated with psychotomimetic effects and disturbances in dopaminergic transmission, thus mimicking schizophrenia symptoms—a huge obstacle to their clinical application. Failure in clinical trials has also hindered their implementation. Many antagonists were unable to proceed as they failed due to poor bioavailability or failure to adequately engage target receptors. In addition, the structure and function of the NMDA receptor in humans are not entirely clear, as opposed to the animal model, which confounds drug development and limits further efficacy. It is also quite complex due to subunit variability, which affects the development of selective antagonists that are highly effective and safe (Löschmann et al., 2004).

Altogether, neurodegenerative diseases are very complex by themselves, and they impose another challenge on how to design an effective treatment. As many might be aware, these disorders are the result of numerous interrelated factors: excitotoxicity, oxidative stress, neuroinflammation, protein aggregation, and mitochondrial dysfunction. Most NMDA receptor antagonists were created with a focus primarily on targeting excitotoxicity in a single-target manner. Moreover, some of those antagonists failed to permeate the BBB because they are hydrophilic. Recent studies have evolved to the creation of multifunctional agents that would be able to address glutamate-induced toxicity along with other therapeutic targets through a multi-target approach. These agents, similar to amantadine and memantine, are characterized by polycyclic cage structures that improve lipophilicity, thus increasing their permeability through the BBB (Stan et al., 2014).

Triazole-bridged aryl adamantane compounds, for instance, have been investigated as multipurpose treatments for AD. In neuronal cell lines (SH-SY5Y neuroblastoma cells), these substances have a good safety profile, BBB penetration, and potent inhibition of acetylcholinesterase, amyloid aggregation, and NMDA receptors. The multifunctional properties of polycyclic propargylamine and acetylene derivatives, such as neuroprotection, inhibition of monoamine oxidase, anti-apoptotic actions, and inhibition of NMDA receptors and voltage-gated calcium channels, have also been investigated (Lesuis et al., 2019). Structurally comparable to acetylcholine, carbamate-based cholinesterase inhibitors have also demonstrated promise, displaying a variety of actions, such as metal chelation, antioxidant qualities, anti-neuroinflammation, neuroprotection, and a decrease in amyloid formation. To corroborate these inhibitors' therapeutic effectiveness, more in vivo and clinical research is required, as the majority of the evidence originates from preclinical investigations (Gao et al., 2023).

Namzaric® and Auvelity®, two FDA-approved multifunctional medications, combine an NMDA receptor antagonist with an additional therapeutic component. Auvelity® is used to treat agitation in Alzheimer's patients, whereas Namzaric® treats mild-to-severe Alzheimer's. These medications have identical side effect profiles to their parent medications, yet they have improved activity and adherence compared to their separate components. This emphasizes the continued need for multifunctional hybrids that target additional degenerative processes, such as neuroinflammation, oxidative stress, and mitochondrial failure, in addition to blocking NMDA receptors and reducing symptoms. The development of selective GluN2B antagonists or modulators with polycyclic scaffolds and multi-target properties thus offers a realistic hope in taking care of neurodegenerative diseases. This multi-target approach with polycyclic structures and good bioavailability would present much improved therapy for these complex disorders (Acker et al., 2011).

These notwithstanding, promising future research and development avenues do exist. One is targeting the NMDA receptor subunits specifically implicated in excitotoxicity. The selective antagonists for such subunits would result in a more effective and less toxic therapy. A very promising pathway toward novel NMDA receptor antagonists involves virtual screening, whereby small molecules are identified to interact optimally with the drug-binding site based on a known three-dimensional structure of a target protein. Further studies into the molecular mechanisms of NMDA receptors will also open the way for therapeutic exploitation of this class of receptor in neurodegenerative disorders (Wyllie et al., 2013). Selective negative allosteric modulators, in development, may be safer and hold greater therapeutic potential than traditional antagonists.

In order to make the clinical trials more successful, the receptor physiology needs to be better understood, and the preclinical models need to be more correlated with the human biology. Joint research work among the medicinal chemists, bioinformaticians, and structural biologists can result in substantial information regarding NMDA receptor-targeted therapies. Detailed studies on the involvement of NMDA receptors in excitotoxic neuronal death can further inform the process of designing drugs for prevention of cell death in neurodegenerative diseases. Finally, developing strategies to monitor and mitigate side effects during clinical trials will be essential for the successful adoption of NMDA receptor antagonists in clinical settings. These efforts could pave the way for innovative treatment strategies for neurodegenerative disorders, ultimately improving patient outcomes (Pearlstein et al., 2015).

9.6 CONCLUSION

The development of NMDA receptor antagonists offers promising hope for treating clinical neurodegenerative disorders. NMDA receptors play a crucial role in neurons, modulating the balance of the CNS by facilitating neuroplastic changes through calcium influx into the cells. Modulating NMDA receptors effectively shields neurons from damage caused by excessive activation of excitatory postsynaptic proteins. However, translating the clear potential observed in preclinical models into viable clinical treatments has proven difficult. One of the main challenges is understanding the NMDA receptor as a whole entity because it is not a single structure but different subtypes that are distributed throughout various brain regions.

These subtypes, especially GluN2 subunits, play distinct functions in neuronal communication and synaptic plasticity, thereby complicating drug design to target the subtypes responsible for excitotoxicity while preserving normal neural functions. To date, current NMDA receptor antagonists—amantadine and memantine—only offer symptomatic relief without any change in the course of the underlying disease process. These medications further contribute to psychotomimetic side effects, like hallucinations and delusions. As such, use of such medicines in a neurological disorder complicated them. However, these medicines will help only a few of the symptoms instead of curing neurodegeneration per se.

New research has shown the possibility of targeting specific NMDA receptor subunits, especially the GluN2B subtype that is heavily implicated in excitotoxicity, especially on extrasynaptic regions where it plays a role in neuronal damage. Selective negative allosteric modulators fine-tune the NMDA receptor without completely activating its blockade and could be a safer, more targeted one. The selective inhibition of this subunit may also lead to excitotoxic damage reduction without damaging the physiological function of NMDA receptors at synaptic plasticity and memory formation.

A multifaceted treatment of neurodegenerative diseases is also crucial. NMDA receptor antagonists alone will not be a panacea for the treatment because other pathological processes, such as neuroinflammation, oxidative stress, and mitochondrial dysfunction, also play significant roles in neurodegeneration. Hence, multifunctional agents targeting NMDA receptors and the other pathological processes are necessary. Such agents will not only control excitotoxicity but also take care of the other factors involved in neurodegeneration, which will offer a more comprehensive therapeutic benefit.

The future studies must focus on enhancing the selectivity and bioavailability of NMDA receptor antagonists. The greatest challenge in drug development has been the failure to cross the BBB, which is the key for drugs targeting the CNS. Advanced systems such as nanostructured lipid carriers or others are critical to enhance BBB permeability and the therapeutic efficacy of these drugs.

Ultimately, however, a more complete understanding of the molecular and functional characteristics of NMDA receptor subtypes and how to design drugs that target relevant pathways in neurodegeneration will be necessary for the effective development of therapeutic interventions. Selective NMDA receptor modulation, multifunctional drug design, and innovative drug delivery systems represent the future in the treatment of neurodegenerative diseases. This approach promises better, safer treatments of conditions as serious as Alzheimer's and Parkinson's and other neurodegenerative diseases, all in the pursuit of improved patient outcomes and quality of life.

REFERENCES

Acker, T. M., Yuan, H., Hansen, K. B., Vance, K. M., Ogden, K. K., Jensen, H. S., Burger, P. B., Mullasseril, P., Snyder, J. P., Liotta, D. C., & Traynelis, S. F. (2011). Mechanism for noncompetitive inhibition by novel GluN2C/D N-methyl-D-aspartate receptor subunit-selective modulators. *Molecular Pharmacology*, *80*(5), 782–795.

Agid, Y. (2022). Neurodegenerative disorders: Are we wrong? *Revue neurologique*, *178*(5), 407–413.

Arbo, B. D., Schimith, L. E., Dos Santos, M. G., & Hort, M. A. (2022). Repositioning and development of new treatments for neurodegenerative diseases: Focus on neuroinflammation. *European Journal of Pharmacology*, *919*, 174800.

Awad, R., Avital, A., & Sosnik, A. (2023). Polymeric nanocarriers for nose-to-brain drug delivery in neurodegenerative diseases and neurodevelopmental disorders. *Acta Pharmaceutica Sinica B*, *13*(5), 1866–1886.

Bechthold, E., Schreiber, J. A., Lehmkuhl, K., Frehland, B., Schepmann, D., Bernal, F. A., Daniliuc, C., Álvarez, I., Garcia, C. V., Schmidt, T. J., Seebohm, G., & Wünsch, B. (2021). Ifenprodil stereoisomers: Synthesis, absolute configuration, and correlation with biological activity. *Journal of Medicinal Chemistry*, *64*(2), 1170–1179.

Becker, R., Gass, N., Kußmaul, L., Schmid, B., Scheuerer, S., Schnell, D., Dorner-Ciossek, C., Weber-Fahr, W., & Sartorius, A. (2019). NMDA receptor antagonists traxoprodil and lanicemine improve hippocampal-prefrontal coupling and reward-related networks in rats. *Psychopharmacology*, *236*(12), 3451–3463.

Berger, M. L., & Rebernik, P. (2015). Differential influence of 7 cations on 16 non-competitive NMDA receptor blockers. *Bioorganic & Medicinal Chemistry Letters*, *25*(19), 4131–4135.

Bermudo-Soriano, C. R., Perez-Rodriguez, M. M., Vaquero-Lorenzo, C., & Baca-Garcia, E. (2012). New perspectives in glutamate and anxiety. *Pharmacology Biochemistry and Behavior*, *100*(4), 752–774.

Bespalov, A., Kudryashova, M., & Zvartau, E. (1998). Prolongation of morphine analgesia by competitive NMDA receptor antagonist D-CPPene (SDZ EAA 494) in rats. *European Journal of Pharmacology*, *351*(3), 299–305.

Bezprozvanny, I., & Mattson, M. P. (2008). Neuronal calcium mishandling and the pathogenesis of Alzheimer's disease. *Trends in Neurosciences*, *31*(9), 454–463.

Bi, X., Rong, Y., Chen, J., Dang, S., Wang, Z., & Baudry, M. (1998). Calpain-mediated regulation of NMDA receptor structure and function. *Brain Research*, *790*(1–2), 245–253.

Blot, K., Bai, J., & Otani, S. (2013). The effect of non-competitive NMDA receptor antagonist MK-801 on neuronal activity in rodent prefrontal cortex: An animal model for cognitive symptoms of schizophrenia. *Journal of Physiology-Paris*, *107*(6), 448–451.

Bubser, M., Keseberg, U., Notz, P. K., & Schmidt, W. J. (1992). Differential behavioural and neurochemical effects of competitive and non-competitive NMDA receptor antagonists in rats. *European Journal of Pharmacology*, *229*(1), 75–82.

Buchan, A. M., Slivka, A., & Xue, D. (1992). The effect of the NMDA receptor antagonist MK-801 on cerebral blood flow and infarct volume in experimental focal stroke. *Brain Research*, *574*(1–2), 171–177.

Burada, A. P., Vinnakota, R., Bharti, P., Dutta, P., Dubey, N., & Kumar, J. (2022). Emerging insights into the structure and function of ionotropic glutamate delta receptors. *British Journal of Pharmacology*, *179*(14), 3612–3627.

Carmack, S. A., Kim, J. S., Sage, J. R., Thomas, A. W., Skillicorn, K. N., & Anagnostaras, S. G. (2013). The competitive NMDA receptor antagonist CPP disrupts cocaine-induced conditioned place preference, but spares behavioral sensitization. *Behavioural Brain Research*, *239*, 155–163.

Chen, H. S. V., & Lipton, S. A. (2006). The chemical biology of clinically tolerated NMDA receptor antagonists. *Journal of Neurochemistry*, *97*(6), 1611–1626.

Costa, B. M., Irvine, M. W., Fang, G., Eaves, R. J., Mayo-Martin, M. B., Skifter, D. A., Jane, D. E., & Monaghan, D. T. (2010). A novel family of negative and positive allosteric modulators of NMDA receptors. *The Journal of Pharmacology and Experimental Therapeutics*, *335*(3), 614–621.

Dash, R., Samanta, S., Jena, B. R., & Pradhan, S. (2024). Nanotechnology in drug delivery: An overview of developing the blood brain barrier. *Current Neurovascular Research*, *21*(4), 347–358.

Driver, C., Jackson, T. N., Lagopoulos, J., & Hermens, D. F. (2022). Molecular mechanisms underlying the N-methyl-D-aspartate receptor antagonists: Highlighting their potential for transdiagnostic therapeutics. *Progress in Neuro-Psychopharmacology and Biological Psychiatry*, *119*, 110609.

Erreger, K., Dravid, S. M., Banke, T. G., Wyllie, D. J., & Traynelis, S. F. (2005). Subunit-specific gating controls rat NR1/NR2A and NR1/NR2B NMDA channel kinetics and synaptic signalling profiles. *The Journal of physiology*, *563*(2), 345–358.

Flores, A. J., Bartlett, M. J., So, L. Y., Laude, N. D., Parent, K. L., Heien, M. L., Sherman, S. J., & Falk, T. (2014). Differential effects of the NMDA receptor antagonist MK-801 on dopamine receptor D1-and D2-induced abnormal involuntary movements in a preclinical model. *Neuroscience Letters*, *564*, 48–52.

Furuya, Y., Kagaya, T., Ogura, H., & Nishizawa, Y. (1999). Competitive NMDA receptor antagonists disrupt prepulse inhibition without reduction of startle amplitude in a dopamine receptor-independent manner in mice. *European Journal of Pharmacology*, *364*(2–3), 133–140.

Gao, X., Chen, F., Xu, X., Liu, J., Dong, F., & Liu, Y. (2023). Ro25-6981 alleviates neuronal damage and improves cognitive deficits by attenuating oxidative stress via the Nrf2/ARE pathway in ischemia/reperfusion rats. *Journal of Stroke and Cerebrovascular Diseases*, *32*(3), 106971.

Gawande, D. Y., Shelkar, G. P., Narasimhan, K. K. S., Liu, J., & Dravid, S. M. (2023). GluN2D subunit-containing NMDA receptors regulate reticular thalamic neuron function and seizure susceptibility. *Neurobiology of Disease*, *181*, 106117.

Gerber, Y. N., Privat, A., & Perrin, F. E. (2013). Gacyclidine improves the survival and reduces motor deficits in a mouse model of amyotrophic lateral sclerosis. *Frontiers in Cellular Neuroscience*, *7*, 280.

Geter-Douglass, B., & Witkin, J. M. (1997). Dizocilpine-like discriminative stimulus effects of competitive NMDA receptor antagonists in mice. *Psychopharmacology*, *133*, 43–50.

Henter, I. D., de Sousa, R. T., & Zarate Jr, C. A. (2018). Glutamatergic modulators in depression. *Harvard Review of Psychiatry*, *26*(6), 307–319.

Iovino, L., Tremblay, M. E., & Civiero, L. (2020). Glutamate-induced excitotoxicity in Parkinson's disease: The role of glial cells. *Journal of Pharmacological Sciences*, *144*(3), 151–164.

Janus, A., Lustyk, K., & Pytka, K. (2023). MK-801 and cognitive functions: Investigating the behavioral effects of a non-competitive NMDA receptor antagonist. *Psychopharmacology*, *240*(12), 2435–2457.

Jespersen, A., Tajima, N., Fernandez-Cuervo, G., Garnier-Amblard, E. C., & Furukawa, H. (2014). Structural insights into competitive antagonism in NMDA receptors. *Neuron*, *81*(2), 366–378.

Jiang, L., Liu, N., Zhao, F., Huang, B., Kang, D., Zhan, P., & Liu, X. (2024). Discovery of glun2a subtype-selective N-methyl-D-aspartate (NMDA) receptor ligands. *Acta Pharmaceutica Sinica B*, *14*(5), 1987–2005.

Jimenez, E. C. (2022). Peptide antagonists of NMDA receptors: Structure-activity relationships for potential therapeutics. *Peptides*, *153*, 170796.

Joshi, R., Missong, H., Mishra, J., Kaur, S., Saini, S., Kandimalla, R., Hemachandra Reddy, P., Babu, A., Bhatti, G. K., & Bhatti, J. S. (2023). Nanotheranostics revolutionizing neurodegenerative diseases: From precision diagnosis to targeted therapies. *Journal of Drug Delivery Science and Technology*, *89*, 105067.

Jullienne, A., Montagne, A., Orset, C., Lesept, F., Jane, D. E., Monaghan, D. T., Maubert, E., Vivien, D., & Ali, C. (2011). Selective inhibition of GluN2D-containing N-methyl-D-aspartate receptors prevents tissue plasminogen activator-promoted neurotoxicity both in vitro and in vivo. *Molecular Neurodegeneration*, *6*, 1–11.

Kamat, P. K., Rai, S., Swarnkar, S., Shukla, R., Ali, S., Najmi, A. K., & Nath, C. (2013). Okadaic acid-induced Tau phosphorylation in rat brain: Role of NMDA receptor. *Neuroscience*, *238*, 97–113.

Kew, J. N. C., Trube, G., & Kemp, J. A. (1998). State-dependent NMDA receptor antagonism by Ro 8-4304, a novel NR2B selective, non-competitive, voltage-independent antagonist. *British Journal of Pharmacology*, *123*(3), 463–472.

Khesmakhi, M. V., Salimi, Z., Pourmotabbed, A., Moradpour, F., Rezayof, A., & Nedaei, S. E. (2024). The role of glutamate NMDA receptors of the mediodorsal thalamus in scopolamine-induced amnesia in rats. *Neuroscience Letters*, *820*, 137595.

Kotajima-Murakami, H., Takano, A., Hirakawa, S., Ogai, Y., Funada, D., Tanibuchi, Y., Ban, E., Kikuchi, M., Tachimori, H., Maruo, K., Kawashima, T., Tomo, Y., Sasaki, T., Oi, H., Matsumoto, T., & Ikeda, K. (2022). Ifenprodil for the treatment of methamphetamine use disorder: An exploratory, randomized, double-blind, placebo-controlled trial. *Neuropsychopharmacology Reports*, *42*(1), 92–104.

Kvist, T., Steffensen, T. B., Greenwood, J. R., Tabrizi, F. M., Hansen, K. B., Gajhede, M., Pickering, D. S., Traynelis, S. F., Kastrup, J. S., & Bräuner-Osborne, H. (2013). Crystal structure and pharmacological characterization of a novel N-methyl-D-aspartate (NMDA) receptor antagonist at the GluN1 glycine binding site. *Journal of Biological Chemistry*, *288*(46), 33124–33135.

Lepeintre, J. F., D'Arbigny, P., Mathé, J. F., Vigue, B., Loubert, G., Delcour, J., Kempf, C., & Tadie, M. (2004). Neuroprotective effect of gacyclidine: A multicenter double-blind pilot trial in patients with acute traumatic brain injury. *Neurochirurgie*, *50*(2–3), 83–95.

Lesuis, S. L., Lucassen, P. J., & Krugers, H. J. (2019). Early life stress impairs fear memory and synaptic plasticity; a potential role for GluN2B. *Neuropharmacology*, *149*, 195–203.

Li, W., Kutas, M., Gray, J. A., Hagerman, R. H., & Olichney, J. M. (2020). The role of glutamate in language and language disorders-evidence from ERP and pharmacologic studies. *Neuroscience & Biobehavioral Reviews*, *119*, 217–241.

Lin, J. C., Chan, M. H., Lee, M. Y., Chen, Y. C., & Chen, H. H. (2016). N, N-dimethylglycine differentially modulates psychotomimetic and antidepressant-like effects of ketamine in mice. *Progress in Neuro-Psychopharmacology and Biological Psychiatry, 71*, 7–13.

Lind, G. E., Mou, T. C., Tamborini, L., Pomper, M. G., De Micheli, C., Conti, P., Pinto, A., & Hansen, K. B. (2017). Structural basis of subunit selectivity for competitive NMDA receptor antagonists with preference for GluN2A over GluN2B subunits. *Proceedings of the National Academy of Sciences, 114*(33), E6942–E6951.

Lipton, S. A. (2004). Failures and successes of NMDA receptor antagonists: Molecular basis for the use of open-channel blockers like memantine in the treatment of acute and chronic neurologic insults. *NeuroRx, 1*, 101–110.

Liu, P., Zhang, T., Chen, Q., Li, C., Chu, Y., Guo, Q., Zhang, Y., Zhou, W., Chen, H., Zhou, Z., Wang, Y., Zhao, Z., Luo, Y., Li, X., Song, H., Su, B., Li, C., Sun, T., & Jiang, C. (2021). Biomimetic dendrimer-peptide conjugates for early multi-target therapy of Alzheimer's disease by inflammatory microenvironment modulation. *Advanced Materials, 33*(26), 2100746.

Liu, W., Li, Y., Zhao, T., Gong, M., Wang, X., Zhang, Y., Xu, L., Li, W., Li, Y., & Jia, J. (2023). The role of N-methyl-D-aspartate glutamate receptors in Alzheimer's disease: From pathophysiology to therapeutic approaches. *Progress in Neurobiology, 231*, 102534.

Löschmann, P. A., De Groote, C., Smith, L., Wüllner, U., Fischer, G., Kemp, J. A., Jenner, P., & Klockgether, T. (2004). Antiparkinsonian activity of Ro 25-6981, a NR2B subunit specific NMDA receptor antagonist, in animal models of Parkinson's disease. *Experimental Neurology, 187*(1), 86–93.

Losi, G., Lanza, M., Makovec, F., Artusi, R., Caselli, G., & Puia, G. (2006). Functional in vitro characterization of CR 3394: A novel voltage dependent N-methyl-D-aspartate (NMDA) receptor antagonist. *Neuropharmacology, 50*(3), 277–285.

Luo, L., Li, D., Xu, X., Jia, Q., Li, Z., Xu, R., Chen, Z., & Zhao, Y. (2023). Synthesis and neuroprotective effects of new genipin derivatives against glutamate-induced oxidative damage. *Fitoterapia, 169*, 105616.

Mabrouk, O. S., Mela, F., Calcagno, M., Budri, M., Viaro, R., Dekundy, A., Parsons, C. G., Auberson, Y. P., & Morari, M. (2013). GluN2A and GluN2B NMDA receptor subunits differentially modulate striatal output pathways and contribute to levodopa-induced abnormal involuntary movements in dyskinetic rats. *ACS Chemical Neuroscience, 4*(5), 808–816.

Marcus, M. M., Mathe, J. M., Nomikos, G. G., & Svensson, T. H. (2001). Effects of competitive and non-competitive NMDA receptor antagonists on dopamine output in the shell and core subdivisions of the nucleus accumbens. *Neuropharmacology, 40*(4), 482–490.

Maselli, F., D'Antona, S., Utichi, M., Arnaudi, M., Castiglioni, I., Porro, D., Papaleo, E., Gandellini, P., & Cava, C. (2023). Computational analysis of five neurodegenerative diseases reveals shared and specific genetic loci. *Computational and Structural Biotechnology Journal, 21*, 5395–5407.

Mendes, D., Peixoto, F., Oliveira, M. M., Andrade, P. B., & Videira, R. A. (2022). Mitochondria research and neurodegenerative diseases: On the track to understanding the biological world of high complexity. *Mitochondrion, 65*, 67–79.

Mikics, E., Toth, M., Biro, L., Bruzsik, B., Nagy, B., & Haller, J. (2017). The role of GluN2B-containing NMDA receptors in short-and long-term fear recall. *Physiology & Behavior, 177*, 44–48.

Mony, L., Kew, J. N., Gunthorpe, M. J., & Paoletti, P. (2009). Allosteric modulators of NR2B-containing NMDA receptors: Molecular mechanisms and therapeutic potential. *British Journal of Pharmacology, 157*(8), 1301–1317.

Mony, L., & Paoletti, P. (2023). Mechanisms of NMDA receptor regulation. *Current Opinion in Neurobiology, 83*, 102815.

Murthy, S. E. (2022). Glycine-bound NMDA receptors are stretch-activated. *Trends in Neurosciences, 45*(11), 794–795.

Neves, D., Salazar, I. L., Almeida, R. D., & Silva, R. M. (2023). Molecular mechanisms of ischemia and glutamate excitotoxicity. *Life Sciences, 328*, 121814.

Nikam, S. S., & Meltzer, L. T. (2002). NR2B selective NMDA receptor antagonists. *Current Pharmaceutical Design, 8*(10), 845–855.

Olivero, G., Grilli, M., Marchi, M., & Pittaluga, A. (2023). Metamodulation of presynaptic NMDA receptors: New perspectives for pharmacological interventions. *Neuropharmacology, 234*, 109570.

Ontario, M. L., Siracusa, R., Modafferi, S., Scuto, M., Sciuto, S., Greco, V., Bertuccio, M. P., Trovato Salinaro, A., Crea, R., Calabrese, E. J., Di Paola, R., & Calabrese, V. (2022). Potential prevention and treatment of neurodegenerative disorders by olive polyphenols and hidrox. *Mechanisms of Ageing and Development, 203*, 111637.

Papp, M., & Moryl, E. (1994). Antidepressant activity of non-competitive and competitive NMDA receptor antagonists in a chronic mild stress model of depression. *European Journal of Pharmacology, 263*(1–2), 1–7.

Pearlstein, E., Gouty-Colomer, L. A., Michel, F. J., Cloarec, R., & Hammond, C. (2015). Glutamatergic synaptic currents of nigral dopaminergic neurons follow a postnatal developmental sequence. *Frontiers in Cellular Neuroscience, 9*, 210.

Pérez-Pinzón, M. A., & Steinberg, G. K. (2006). CGS 19755 (Selfotel): A novel neuroprotective agent against CNS injury. *CNS Drug Reviews, 2*(3), 257.

Quan, J., Yang, H., Qin, F., He, Y., Liu, J., Zhao, Y., Ma, C., & Cheng, M. (2023). Discovery of novel tryptamine derivatives as GluN2B subunit-containing NMDA receptor antagonists via pharmacophore-merging strategy with orally available therapeutic effect of cerebral ischemia. *European Journal of Medicinal Chemistry, 253*, 115318.

Rao, V. L. R., Dogan, A., Todd, K. G., Bowen, K. K., & Dempsey, R. J. (2001). Neuroprotection by memantine, a non-competitive NMDA receptor antagonist after traumatic brain injury in rats. *Brain Research, 911*(1), 96–100.

Rose, C. R., Ziemens, D., Untiet, V., & Fahlke, C. (2018). Molecular and cellular physiology of sodium-dependent glutamate transporters. *Brain Research Bulletin, 136*, 3–16.

Shafiei-Irannejad, V., Abbaszadeh, S., Janssen, P. M., & Soraya, H. (2021). Memantine and its benefits for cancer, cardiovascular and neurological disorders. *European Journal of Pharmacology, 910*, 174455.

Sharkey, J., Ritchie, I. M., Butcher, S. P., & Kelly, J. S. (1996). Comparison of the patterns of altered cerebral glucose utilisation produced by competitive and non-competitive NMDA receptor antagonists. *Brain Research, 735*(1), 67–82.

Sharma, A., Rudrawar, S., Bharate, S. B., & Jadhav, H. R. (2025). Recent advancements in the therapeutic approaches for Alzheimer's disease treatment: Current and future perspective. *RSC Medicinal Chemistry, 16*(2), 652–693.

Sharma, A., Rudrawar, S., Sharma, A., Bharate, S. B., & Jadhav, H. R. (2024a). Design, synthesis, in silico, and in vitro evaluation of pyrrol-2-yl-phenyl allylidene hydrazine carboximidamide derivatives as AChE/BACE 1 dual inhibitors. *RSC Advances, 14*(37), 26703–26722.

Sharma, A., Rudrawar, S., Sharma, A., Bharate, S. B., & Jadhav, H. R. (2024b). Unveiling the potential of novel indol-3-yl-phenyl allylidene hydrazine carboximidamide derivatives as AChE/BACE 1 dual inhibitors: A combined in silico, synthesis and in vitro study. *RSC Advances*, *14*(33), 23853–23872.

Shen, Z., Xiang, M., Chen, C., Ding, F., Wang, Y., Shang, C., Xin, L., Zhang, Y., Cui, X. (2022). Glutamate excitotoxicity: Potential therapeutic target for ischemic stroke. *Biomedicine & Pharmacotherapy*, *151*, 113125.

Sirrieh, R. E., MacLean, D. M., & Jayaraman, V. (2015). A conserved structural mechanism of NMDA receptor inhibition: A comparison of ifenprodil and zinc. *Journal of General Physiology*, *146*(2), 173–181.

Stan, T. L., Alvarsson, A., Branzell, N., Sousa, V. C., & Svenningsson, P. (2014). NMDA receptor antagonists ketamine and Ro25-6981 inhibit evoked release of glutamate in vivo in the subiculum. *Translational Psychiatry*, *4*(6), e395.

Stuchlík, A., & Vales, K. (2005). Systemic administration of MK-801, a non-competitive NMDA-receptor antagonist, elicits a behavioural deficit of rats in the Active Allothetic Place Avoidance (AAPA) task irrespectively of their intact spatial pretraining. *Behavioural Brain Research*, *159*(1), 163–171.

Subramanian, S., Haroutounian, S., Palanca, B. J. A., & Lenze, E. J. (2022). Ketamine as a therapeutic agent for depression and pain: Mechanisms and evidence. *Journal of the Neurological Sciences*, *434*, 120152.

Sun, W., & Wessinger, W. D. (2004). Characterization of the non-competitive antagonist binding site of the NMDA receptor in dark Agouti rats. *Life Sciences*, *75*(12), 1405–1415.

Suzuki, A., Murakami, K., Tajima, Y., Hara, H., Kunugi, A., & Kimura, H. (2019). TAK-137, an AMPA receptor potentiator with little agonistic effect, produces antidepressant-like effect without causing psychotomimetic effects in rats. *Pharmacology Biochemistry and Behavior*, *183*, 80–86.

Szakács, R., Weiczner, R., Mihály, A., Krisztin-Péva, B., Zádor, Z., & Zádor, E. (2003). Non-competitive NMDA receptor antagonists moderate seizure-induced c-fos expression in the rat cerebral cortex. *Brain Research Bulletin*, *59*(6), 485–493.

Takizawa, S., Hogan, M., & Hakim, A. M. (1991). The effects of a competitive NMDA receptor antagonist (CGS-19755) on cerebral blood flow and pH in focal ischemia. *Journal of Cerebral Blood Flow & Metabolism*, *11*(5), 786–793.

Tamborini, L., Pinto, A., Mastronardi, F., Iannuzzi, M. C., Cullia, G., Nielsen, B., De Micheli, C., & Conti, P. (2013). 3-Carboxy-pyrazolinalanine as a new scaffold for developing potent and selective NMDA receptor antagonists. *European Journal of Medicinal Chemistry*, *68*, 33–37.

Tóth, Z., Mihály, A., Mátyás, A., & Krisztin-Péva, B. (2018). Non-competitive antagonists of NMDA and AMPA receptors decrease seizure-induced c-fos protein expression in the cerebellum and protect against seizure symptoms in adult rats. *Acta Histochemica*, *120*(3), 236–241.

Tsai, Y. C., Huang, S. M., Peng, H. H., Lin, S. W., Lin, S. R., Chin, T. Y., & Huang, S. M. (2024). Imbalance of synaptic and extrasynaptic NMDA receptors induced by the deletion of CRMP1 accelerates age-related cognitive decline in mice. *Neurobiology of Aging*, *135*, 48–59.

van Onselen, R., & Downing, T. G. (2024). Uptake of β-N-methylamino-L-alanine (BMAA) into glutamate-specific synaptic vesicles: Exploring the validity of the excitotoxicity mechanism of BMAA. *Neuroscience Letters*, *821*, 137593.

Velu, L., Pellerin, L., Julian, A., Paccalin, M., Giraud, C., Fayolle, P., Guillevin, R., & Guillevin, C. (2024). Early rise of glutamate-glutamine levels in mild cognitive impairment: Evidence for emerging excitotoxicity. *Journal of Neuroradiology*, *51*(2), 168–175.

Volgraf, M., Sellers, B. D., Jiang, Y., Wu, G., Ly, C. Q., Villemure, E., Pastor, R. M., Yuen, P. W., Lu, A., Luo, X., Liu, M., Zhang, S., Sun, L., Fu, Y., Lupardus, P. J., Wallweber, H. J., Liederer, B. M., Deshmukh, G., Plise, E., … Schwarz, J. B. (2016). Discovery of GluN2A-selective NMDA receptor positive allosteric modulators (PAMs): Tuning deactivation kinetics via structure-based design. *Journal of Medicinal Chemistry*, *59*(6), 2760–2779.

Wang, C. X., & Shuaib, A. (2005). NMDA/NR2B selective antagonists in the treatment of ischemic brain injury. *Current Drug Targets-CNS & Neurological Disorders*, *4*(2), 143–151.

Wang, X., Zhang, F., Niu, L., Yan, J., Liu, H., Wang, D., Hui, J., Dai, H., Song, J., & Zhang, Z. (2024). High-frequency repetitive transcranial magnetic stimulation improves depressive-like behaviors in CUMS-induced rats by modulating astrocyte GLT-1 to reduce glutamate toxicity. *Journal of Affective Disorders*, *348*, 265–274.

Willmore, C. B., Bespalov, A. Y., & Beardsley, P. M. (2001). Competitive and noncompetitive NMDA antagonist effects in rats trained to discriminate lever-press counts. *Pharmacology Biochemistry and Behavior*, *69*(3–4), 493–502.

Wlaź, P., Ebert, U., Potschka, H., & Löscher, W. (1998). Electrical but not chemical kindling increases sensitivity to some phencyclidine-like behavioral effects induced by the competitive NMDA receptor antagonist D-CPPene in rats. *European Journal of Pharmacology*, *353*(2–3), 177–189.

Wu, E., Zhang, J., Zhang, J., & Zhu, S. (2023). Structural insights into gating mechanism and allosteric regulation of NMDA receptors. *Current Opinion in Neurobiology*, *83*, 102806.

Wyllie, D. J. A., Livesey, M. R., & Hardingham, G. E. (2013). Influence of GluN2 subunit identity on NMDA receptor function. *Neuropharmacology*, *74*, 4–17.

Yan, D., Liu, B., Wei, X., Ou, W., Liao, M., Ji, S., Peng, Y., Liu, J., Wu, S., Wang, M., Ju, Y., Zhang, L., Li, Z., Li, L., & Zhang, Y. (2022). Efficacy and safety of nitrous oxide for patients with treatment-resistant depression, a randomized controlled trial. *Psychiatry Research*, *317*, 114867.

Yi, F., Mou, T. C., Dorsett, K. N., Volkmann, R. A., Menniti, F. S., Sprang, S. R., & Hansen, K. B. (2016). Structural basis for negative allosteric modulation of GluN2A-containing NMDA receptors. *Neuron*, *91*(6), 1316–1329.

Yi, N. X., Zhou, L. Y., Wang, X. Y., Song, Y. J., Han, H. H., Zhang, T. S., Wang, Y. J., Shi, Q., Xu, H., Liang, Q. Q., & Zhang, T. (2019). MK-801 attenuates lesion expansion following acute brain injury in rats: A meta-analysis. *Neural Regeneration Research*, *14*(11), 1919–1931.

Zhang, L., Huang, L., Chen, L., Hao, D., & Chen, J. (2013). Neuroprotection by tetrahydroxystilbene glucoside in the MPTP mouse model of Parkinson's disease. *Toxicology Letters*, *222*(2), 155–163.

Zhang, Y., Li, P., Feng, J., & Wu, M. (2016). Dysfunction of NMDA receptors in Alzheimer's disease. *Neurological Sciences*, *37*, 1039–1047.

Zhang, Z., Liu, J., Fan, C., Mao, L., Xie, R., Wang, S., Yang, M., Yuan, H., Yang, X., Sun, J., Wang, J., Kong, J., Huang, S., & Sun, B. (2018). The GluN1/GluN2B NMDA receptor and metabotropic glutamate receptor 1 negative allosteric modulator has enhanced neuroprotection in a rat subarachnoid hemorrhage model. *Experimental Neurology, 301*, 13–25.

Zhang, Z., Zhang, S., Fu, P., Zhang, Z., Lin, K., Ko, J. K. S., & Yung, K. K. L. (2019). Roles of glutamate receptors in Parkinson's disease. *International Journal of Molecular Sciences, 20*(18), 4391.

Zhou, C., & Tajima, N. (2023). Structural insights into NMDA receptor pharmacology. *Biochemical Society Transactions, 51*(4), 1713–1731.

Zhu, S., & Paoletti, P. (2015). Allosteric modulators of NMDA receptors: Multiple sites and mechanisms. *Current Opinion in Pharmacology, 20*, 14–23.

Zhu, S., Stein, R. A., Yoshioka, C., Lee, C. H., Goehring, A., Mchaourab, H. S., & Gouaux, E. (2016). Mechanism of NMDA receptor inhibition and activation. *Cell, 165*(3), 704–714.

10 Phosphodiesterase Inhibitors
A Promising Avenue for Modulating Cognitive Dysfunction in Alzheimer's Disease

Sylvia F. Fawzi, Daniel E. Mansour, Pierre Mansour, and Mina Y. George

10.1 INTRODUCTION

10.1.1 Alzheimer's Disease

Alzheimer's disease (AD) is a chronic neurological illness impacting cognition and affecting individuals globally. Currently, around 36 million people worldwide are diagnosed with AD, a number anticipated to increase fourfold by 2050 (Ribaudo et al., 2020). Moreover, in the United States alone, the prevalence of AD among individuals aged 65 and older is approximately 6.7 million, with projections estimating an elevation to 13.8 million by 2060 (Alzheimer's Association, 2023).

The clinical presentation of AD patients includes partial or full memory dysfunction, difficulty in learning, limited daily communications, prevention of daily duties, and accordingly reduced patient quality of life (Sanders & Rajagopal, 2020). AD was found to be associated with the deposition of amyloid-β (Aβ) plaques extracellularly, as well as neurofibrillary tau tangles intracellularly. In addition, cholinergic transmission dysfunction, oxidative stress, and neuroinflammation have been involved in AD pathogenesis (Ayoub et al., 2022; Ribaudo et al., 2020).

Currently, available medications for AD do not possess a complete cure for AD but prevent the disease progression over time and cease the threshold of symptoms (Yiannopoulou & Papageorgiou, 2013). Among them, cholinesterase inhibitors are used as a first line for mild-to-moderate AD management. N-methyl-D-aspartate receptor antagonists are another group used for moderate to severe cases of AD (Geerts & Grossberg, 2006). Recently, there has been increasing interest in "disease-modifying" drugs designed to target and disrupt pathogenic factors underlying clinical symptoms, including extracellular amyloid plaque formation, intracellular neurofibrillary tangles, oxidative stress, inflammation, and cholesterol metabolism. These therapeutic approaches aim to prevent the AD progression (Munoz-Torrero, 2008; Parihar & Hemnani, 2004). Clinical trials have primarily focused on acetylcholinesterase (AChE) inhibitors, neurotransmitter receptor agonists and antagonists (intepirdine), β/γ-secretase inhibitors (E2609, verubecestat, AZD3293), monoclonal antibodies targeting Aβ clearance or reduction (gantenerumab, crenezumab, aducanumab), anti-inflammatory drugs, and agents addressing tau tangles aggregation (TRx0237). However, there is growing interest in non-amyloid-based mechanisms, the use of natural products, and biomarker-driven therapeutic approaches for AD in the drug discovery landscape (Cummings et al., 2018; Gad et al., 2023; Hung & Fu, 2017). Thus, the exploration of new therapeutic agents or repurposing of available medications for the treatment of AD becomes a challenge facing medicinal research.

10.1.2 Phosphodiesterase Enzymes

Phosphodiesterases (PDEs) represent a family of endogenous enzymes primarily responsible for producing 5′-cyclic nucleotides *via* hydrolyzing the 3-phosphodiester bond in cyclic adenosine monophosphate (cAMP) and cyclic guanosine monophosphate (cGMP), both of which have functions in numerous cellular pathways. Consequently, PDEs play an essential role in various pharmacological processes through the modulation of cAMP and cGMP levels, thereby affecting cell function (Sanders & Rajagopal, 2020).

As a diverse enzyme group, PDEs encompass 11 isoenzyme families (PDE1–PDE11), each distinguished by differences in substrate specificity and affinity, regulatory mechanisms, kinetic properties, tissue and cellular distribution, and sensitivity to drugs and mediators (Abusnina & Lugnier, 2017; Amin et al., 2018). Collectively, these families include over 40 PDE isoforms that are encoded by 21 genes. They may be specific to either cAMP, such as PDE4, PDE7, and PDE8, or cGMP, such as PDE5, PDE6, and PDE9, while other isoenzymes are specific for both cAMP and cGMP, such as PDE1, PDE2, PDE3, PDE10, and PDE11 (Bollen & Prickaerts, 2012; Prickaerts et al., 2017). Gene families in humans with their targets and specific molecules have been listed in Table 10.1.

PDEis prevent the activity of PDE enzymes and the breakdown of cGMP or cAMP, increasing their levels in smooth muscle cells, which can induce relaxation and vasodilation (Padda & Tripp, 2024). These cyclic nucleotides play vital roles in cellular processes, including neurotransmission and signal transduction (Gorshkov & Zhang, 2014). Key enzymes targeted by cAMP and cGMP include protein

DOI: 10.1201/9781003602477-10

TABLE 10.1
Different Isoforms of Phosphodiesterase Inhibitors in Humans and Their Relevant Specificity to cAMP or cGMP

PDE	Target	Function/Gene	Reference
PDE1	cAMP/cGMP	Calcium/Calmodulin	Omori and Kotera (2007)
PDE2	cAMP/cGMP	PDE2A	Pavlaki and Nikolaev (2018) Omori and Kotera (2007)
PDE3	cAMP/cGMP	PDE3A PDE3B (44 amino acids in their catalytic domain)	Omori and Kotera (2007)
PDE4	cAMP	PDE4A PDE4B PDE4C	Omori and Kotera (2007)
PDE5	cGMP	PDE5A (two GAF domains)	Ahmed et al. (2021)
PDE6	cGMP	Retinal rod and cone cells PDE6A PDE6B PDE6C	Omori and Kotera (2007)
PDE7	cAMP	PDE7A PDE7B	Omori and Kotera (2007)
PDE8	cAMP	PDE8A PDE8B	Omori and Kotera (2007)
PDE9	cGMP	PDE9A	Omori and Kotera (2007)
PDE10	cAMP/cGMP	PDE10A (two GAF domains)	Schultz and Natarajan (2013)
PDE11	cAMP/cGMP	PDE11A4 (two GAF domains and one catalytic domain)	Schultz and Natarajan (2013)

cAMP, cyclic adenosine monophosphate; cGMP, cyclic guanosine monophosphate; GAF, cGMP-specific phosphodiesterase, adenylate cyclase, and FhlA; PDE, phosphodiesterase

kinase A (PKA) and protein kinase G (PKG), which help activate pathways influencing cognitive functions through cAMP response element-binding (CREB) protein and brain-derived neurotrophic factor (BDNF) (Desouky et al., 2024; Sheng et al., 2022). Furthermore, adenylyl cyclase (AC) or guanylyl cyclase (GC) produce cAMP or cGMP, respectively, inside the cell in return for extracellular signals. These molecules can diffuse and interact with target proteins to modulate cellular activities (Bolger, 2021). The inhibition of PDEs thus promotes the AC/cAMP/PKA and GC/cGMP/PKG pathways, enhancing CREB levels, synaptic transmission, and neuronal growth, ultimately reducing cognitive deficits (Sanders & Rajagopal, 2020). The therapeutic advantages of enhancing cAMP or cGMP axis via several PDEi can be linked to their effects on key signaling pathways involving CREB, nuclear factor erythroid 2-related factor 2 (Nrf2), nuclear factor kappa B (NF-κB), and glycogen synthase kinase 3 (GSK3) (Sanders & Rajagopal, 2020). Moreover, Sirtuin-1 (SIRT1) and peroxisome proliferator-activated receptor coactivator-1 (PGC1)-related axes have been reported in AD and other neurological disorders (George et al., 2023; Mohammed et al., 2024).

10.1.3 Normal cAMP and cGMP Signaling

cAMP is synthesized when G protein-coupled complexes activate AC, which converts adenosine triphosphate (ATP) into cAMP. This cAMP then stimulates the exchange proteins and directly activates PKA. PKA is a holoenzyme comprising two catalytic and two regulatory subunits; upon binding with cAMP, the regulatory subunits produce the catalytic ones, allowing them to phosphorylate various proteins (Soberg & Skalhegg, 2018). Regarding the cGMP pathway, nitric oxide synthases (NOSs) generate nitric oxide (NO), which then activates soluble guanylyl cyclase (sGC) in the cytosol. This activation results in the production of cGMP, which subsequently activates PKG (Hasebe & Yoshino, 2016).

10.1.4 Phosphodiesterase Inhibitors and Alzheimer's Disease

There is a significant relationship between PDEi and AD. PDEs have recently emerged as potential therapeutic options for addressing cognitive decline associated with AD (Heckman et al., 2017). The potential of PDEs to improve cognitive function was demonstrated first in a transgenic fly model (Tibbo et al., 2019). It is worth mentioning that several PDE isoforms are abundantly present in the human brain, and PDEis are capable of crossing the blood-brain barrier, reaching therapeutic levels in the central nervous system (CNS) to induce their activities (Ribaudo et al., 2020). Dysregulated cAMP and/or cGMP

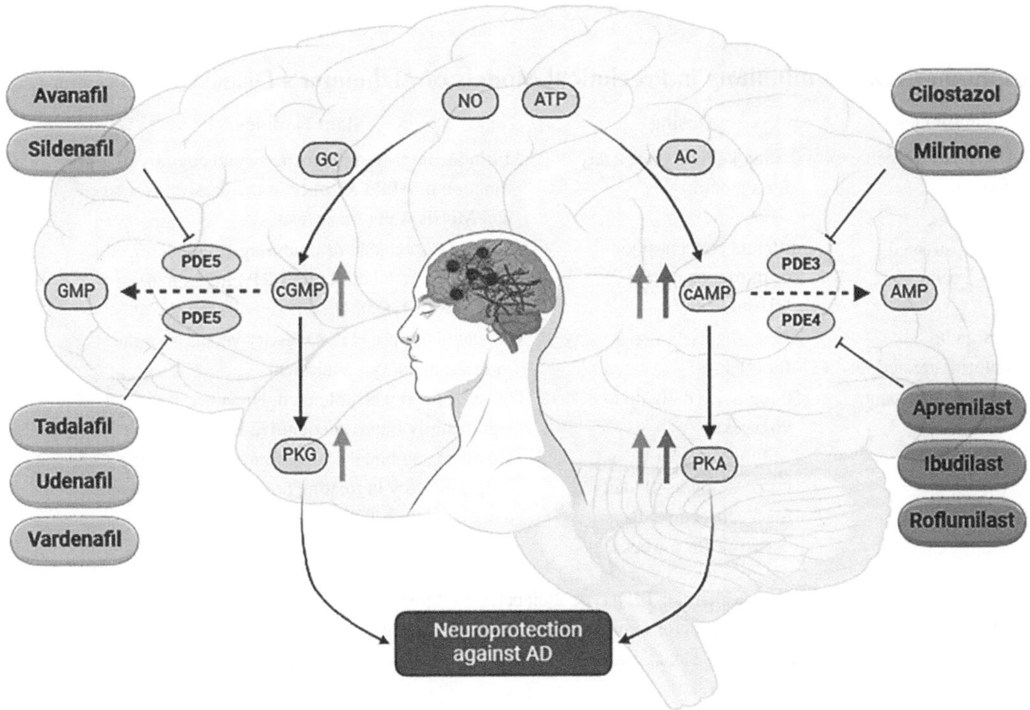

FIGURE 10.1 Targeting phosphodiesterase enzymes: A new frontier in mitigating Alzheimer's disease. The effect of phosphodiesterase inhibitors on cognitive decline associated with AD through affecting cAMP and cGMP. Adenylate cyclase promotes the conversion of ATP to cAMP. The latter activates PKA to stimulate the upregulation of underlying pathways. Guanylate cyclase induces the conversion of NO to cGMP, which allows the activation of PKG. PDEi inhibits PDE enzyme, leading to enhanced levels of cAMP or cGMP. Purple arrows represent PDE3i, green arrows represent PDE4i, and pink arrows represent PDE5i. Created by BioRender. PDE: phosphodiesterase, NO: Nitric oxide, GC: Guanylate cyclase, PKG: Protein kinase G, AC: Adenylate cyclase, PKA: Protein kinase A.

signaling is linked to cognitive dysfunction in neurodegenerative disorders (Di Benedetto et al., 2021). Abnormal PDE function disrupts signal transduction, leading to abnormal cAMP and cGMP levels in memory-related brain areas, increased Aβ production, and impaired memory formation (Nabavi et al., 2019). PDEis may improve AD symptoms through the restoration of synaptic plasticity through the reactivation of the CREB circuit, which plays a crucial role in neuronal synaptic activity (Morid et al., 2023; Nabavi et al., 2019).

In this book chapter, we are going to shed light on the three main classes that were reported for their potential beneficial activity in AD, which are PDE3i, PDE4i, and PDE5i, as well as contemplate the possible associated molecular pathways in preclinical models. Figure 10.1 summarizes all the main points mentioned in this book chapter.

10.2 PHOSPHODIESTERASE-3 INHIBITORS

The phosphodiesterases-3 (PDE3) family members have been recognized to express most of the enzymes of the cAMP-degrading enzyme PDE family. They catalyze the hydrolysis of cAMP to AMP, a potent second messenger important for many events in the cardiovascular, CNS, and immune systems. Regulation or enhancement of cAMP concentration can increase PKA phosphorylating activity and subsequently the synthesis of the regenerative cytokines to enhance cardiac output, inflammation, and neural activity. Transport of cAMP present within the cells, even in its minimal concentration, enhances cell communication through cAMP signaling. In addition, PDE3i such as cilostazol can enhance learning and memory formation by increasing the efficiency of neuroplasticity and memory consolidation in the cortex (Yanai & Endo, 2019). cAMP also increases synaptic plasticity through PKA activation, leading to CaMKII activation pathways as part of its effector pathways. Research using a mouse model has shown that BDNF is expressed shortly after in region 2 of CA1, creating the optimal conditions for synapse formation and activity. It is understood that, among all isoforms, only mature BDNF promotes synapse maturation and is essential for higher cognitive functions (Amidfar et al., 2020). The effect of BDNF is mediated through its interaction with tropomyosin receptor kinase B (TrkB), which induces various intracellular signaling-mediating cascades such as the PI3K/Akt, MAPK/Erk, and PLC-γ pathways. Such pathways can promote neuronal survival and development and synaptic functioning, which is crucial for encoding and retrieving information. Dysregulation of these signaling pathways has been implicated in neurodegenerative diseases such as AD, where there is a loss of cAMP signaling that will in turn result in the deterioration of memory and cognition. The main findings of previous preclinical studies involving PDE3i have been summarized in Table 10.2.

TABLE 10.2
Role of Phosphodiesterase-3 Inhibitors in Preclinical Models of Alzheimer's Disease

Drug	Species	Dosing	Main Findings	References
Milrinone	APP/PS1 transgenic mouse	0.5 mg/kg, i.p., once a day for a month.	Milrinone improved the memory and cognitive function of APP/PS1 mice in the passive avoidance and Morris water maze tests.	Chen et al. (2021)
	LPS/Aβ-treated BV-2 cells	different concentrations of milrinone (0, 10, 25, 50 μmol/L).	Milrinone is capable of inhibiting IL-1β, IL-6, and TNF-α via regulation of NLRP3 inflammasomes and TLR4/MyD88/NF-κB signaling pathway.	Chen et al. (2021)
Cilostazol	Aluminum chloride-treated rats	(50 mg/kg, po) once per day for 60 days.	Cilostazol has protective activity versus aluminum-induced memory impairment.	Khalifa et al. (2022)
	Human participants	50 mg twice daily up to 96 weeks.	Cilostazol was well-tolerated. However, it did not significantly improve cognitive function.	Saito et al. (2023)
	T2D rats	Caffeine 5 mg/kg IP and CSZ 7.5 mg/kg IP daily for 5 consecutive days per week for 8 weeks.	Cilostazol combined with caffeine enhances the drug's efficacy in treating cognitive impairment associated with T2D and potentially AD.	Gomaa et al. (2022)

APP, amyloid precursor protein; LPS, lipopolysaccharide; T2D, type 2 diabetes mellitus; CSZ, cilostazol; TNF-α, tumor necrosis factor-α; IL, interleukin; TLR4, Toll-like receptor 4; NLRP3, Nod-like receptor family pyrin domain-containing 3

10.2.1 Milrinone

Milrinone is a PDE3i that is mostly indicated clinically as an inotrope and vasodilator in cases of acute decompensated heart failure. The neuroprotective property of milrinone on spinal cord injury has been established, but additional evidence should clarify what role milrinone has in AD pathogeny.

A study conducted by Chen et al. (2021) also used APP/PS1 transgenic mouse models and lipopolysaccharide (LPS)/Aβ-treated BV-2 cells for the determination of the efficacy of milrinone in the bi-direction of histopathological changes of AD in vivo and anti-inflammatory mechanism in vitro, respectively. They demonstrated that milrinone helped improve the memory and cognitive functions of APP/PS1 mice in the passive avoidance and Morris water maze tests in the *in vivo* study. Milrinone improved memory retention, as APP/PS1 mice tested without treatment performed with longer latency and greater frequency of entries into the dark compartment that was reduced after milrinone treatment. The escape latency that APP/PS1 mice exhibited was improved with milrinone treatment, and the number of platform crossings was also increased, demonstrating a more developed spatial memory. Furthermore, milrinone reduced Aβ deposits and lowered microglial activation as well as tau protein hyperphosphorylation and reduced oxidative stress and inflammatory cytokine levels. According to *in vitro* study results, milrinone can block interleukin (IL)-1β, IL-6, and tumor necrosis factor-α (TNF-α) production by targeting Nod-like receptor family pyrin domain-containing 3 (NLRP3) inflammasomes and TLR4/MyD88/NF-κB signaling axis. This inflammatory axis was found to play a crucial role in different neurodegenerative disorders and cognitive impairments (Fawzi et al., 2023; Ibrahim et al., 2021b). This study, therefore, revealed that milrinone has a promising curing option for AD since it can attenuate memory loss through inhibiting the multiple pathological events associated with AD (Chen et al., 2021).

10.2.2 Cilostazol

Cilostazol is a PDE3i mainly used as an antiplatelet drug. It is most often recommended for intermittent claudication treatment as a result of poor circulation in the legs and generally peripheral artery disease (PAD).

As mentioned in the study presented by Khalifa et al. (2022), cilostazol was examined for its neuroprotective effects in an aluminum chloride-induced AD rat model. Given cilostazol's cAMP/CREB signaling pathway activation, it significantly counters cognitive decline related to aging. In this regard, rats were subjected to aluminum chloride for two months to induce memory impairment. Then, the rats' cognitive performances were determined by behavioral tasks employing the Morris water maze and y-maze and assessing histological and biochemical contents of the hippocampus. Moreover, behavioral, biochemical, and histological observations made on these aluminum rats were suggestive of having AD. The experiment showed that cilostazol was an effective treatment by increasing neprilysin levels, an enzyme that infiltrates membranes allowing Aβ clearance while decreasing Aβ and phosphorylated tau in the brain. It also reduced inflammatory mediators such as TNF-α, NF-κB, and FAS ligand and ameliorated oxidative stress markers evidenced by opposing lipid peroxidation levels. Furthermore, cilostazol was found to inhibit AChE activity which is characteristically active among enduring dementia of AD patients (Khalifa et al., 2022).

Another study called the COMCID trial (Cilostazol for Prevention of Conversion from Mild Cognitive Impairment to Dementia) was concerned with the safety of cilostazol and its effectiveness in patients with MCI. There were

159 adults with mild cognitive impairment with a mean age of around 75 years enrolled in this active placebo-controlled study, who were assigned either cilostazol (50 mg twice daily) or placebo arms for up to 96 weeks. Change in the score of the Mini-Mental State Examination (MMSE), a well-known test of cognitive reserve, was the main endpoint of the study. The clinical trial results showed that cilostazol was well tolerated and the medication did not worsen patients' conditions. Nevertheless, there was no notable difference in the improvement of cognitive functions or the rate of further cognitive degeneration in MCI patients receiving cilostazol when matched against the placebo group (Saito et al., 2023).

A third study done by Gomaa et al. (2022) focused on the primary effects of cilostazol in the treatment of cognitive impairment associated with type 2 diabetes (T2D) in combination with caffeine in a rat model. A study indicates that combining cilostazol with caffeine enhances the potency of the drug against T2D and possibly AD-related cognitive impairment. Factors responsible for this enhancement include reduction of oxidative stress, inhibition of GSK-3β activity, and enhancement of insulin sensitivity, as well as reduction in levels of pro-inflammatory cytokines. Several studies reported the involvement of Akt/GSK-3β axis in neurological, pulmonary, and renal dysfunction models (George et al., 2020; Habib et al., 2023). These findings indicate that combination therapy of low-dose caffeine and CSZ may provide an effective treatment strategy for AD and T2D-related cognitive decline, offering a promising direction for future therapeutic approaches (Gomaa et al., 2022).

10.3 PHOSPHODIESTERASE-4 INHIBITORS

Phosphodiesterase-4 (PDE4) is the most abundant isoform in immune cells. Through its hydrolytic activity, it converts cAMP to AMP, which promotes the formation of pro-inflammatory cytokines, such as TNF-α, cytokines, and chemokines, as well as the reduction of the release of anti-inflammatory cytokines (Cada et al., 2014; Desouky et al., 2023).

While this storm of cytokines and ILs is advantageous for the proper functioning of immune cells, interrupting this cascade of reactions offers anti-inflammatory properties that might mediate desirable immunomodulatory and neuroprotective effects. That said, PDE4i represents a potential target for the mitigation of neuroinflammatory conditions underlying AD by preventing the hydrolysis of cAMP into AMP and subsequently stunting the downstream inflammatory cytokine production. PDE4 also proved to have neuroplasticity-stimulating effects through the activation of cAMP/CREB signaling, which derives the endogenous rise in BDNF, a robust neuroplasticity modulator (Wang et al., 2018). From that end, it is fairly inferable that PDE4i are promising options for AD treatment. A list of prototypical and novel PDE4Is, as well as their neuroprotective effects, is listed below. The main findings of previous preclinical studies involving PDE4i have been summarized in Table 10.3.

TABLE 10.3
Role of Phosphodiesterase-4 Inhibitors in Preclinical Models of Alzheimer's Disease

Drug	Species	Dosing	Main Findings	References
Apremilast	Male Wistar rats	5 mg/kg/day, i.p., for 3 consecutive days per week for 8 weeks.	Apremilast enhances cognitive function by decreasing pro-inflammatory cytokines, oxidative stress, insulin resistance, and GSK-3β.	Gomaa et al. (2023)
Ibudilast	Fisher transgenic 344-AD rats.	10 mg/kg body weight for 6 months.	Ibudilast has the potential to be repurposed as a drug that targets multiple pathways, including TLR signaling and the ubiquitin/proteasome pathway, to mitigate cognitive deficits and pathology associated with Alzheimer's disease.	Oliveros et al. (2023)
Roflumilast	APP/PS1transgenic	0.1, 0.2, and 0.4 mg/kg. Drugs were continued to be administered to mice 1 h before the test each day during the observation period.	Roflumilast can improve cognitive function in APP/PS1 transgenic mice, potentially due to its activation of the cAMP/CREB/BDNF pathway and its anti-neuroinflammatory effects.	Efentakis et al. (2019)
	APPswe/PS1dE9 mice	5 and 10 mg/kg orally once per day.	Roflumilast was found to improve cognitive function as well as depression-like behavior in AD mice through PDE4B/PDE4D-mediated cAMP/cAMP response element-binding protein/brain-derived neurotrophic factor signaling.	Wang et al. (2020)
	Adult male Swiss albino mice	0.4 mg/kg with or without the PKA inhibitor, H-89, for 21 days.	Roflumilast regulated PXR/ABCB1 expression and arginase/NOS activities, through a PKA-dependent manner, to limit key post-translational modifications of tau, Aβ accumulation, and consequently, cognitive decline in sporadic Alzheimer's disease.	Ashour et al. (2021)

PKA, protein kinase A; GSK-3β, glycogen synthase kinase-3β; TLR, Toll-like receptor; NOS, nitric oxide synthase

10.3.1 Apremilast

Apremilast is an orally available PDE4i whose primary indication is for plaque psoriasis (Cada et al., 2014). In a study by Gomaa et al. (2023), the impact of apremilast on the different characteristics of AD induced by a high-fat/high-fructose diet and streptozotocin was assessed. Behavioral functionality was operationalized using the novel object recognition task, step-through passive avoidance, and Morris water maze tests. In the passive avoidance task, the diabetic rats treated with Apremilast showed step-through latency to enter the aversive-stimulus-associated dark compartment that is similar to the normal control group and had a markedly longer step-through latency relative to the diabetic control group of rats that received DMSO, proving that apremilast reversed learning function. In the novel object recognition task, the exploration time spent by rats for the novel object and the discrimination index were greater in the diabetic apremilast-treated group than those found in the diabetic control group, indicating that apremilast at least partially reversed cognitive dysfunction. In the Morris water maze test, diabetic rats treated with saline or DMSO took considerably longer to locate or search for the platform compared to non-diabetic control rats. However, eight weeks of treatment with apremilast significantly shortened this extended escape latency in diabetic rats compared to those receiving DMSO. Taken together, the results of the three aforementioned tests reveal that apremilast rescued impaired learning and memory.

According to the same study, the pathophysiology underlying the apremilast-induced amelioration of behavioral and learning functions appears to be complex. Treatment with apremilast significantly reduced the number of degenerated cells and restored the expression of AMPA and NMDA receptor subunits, which had been suppressed in the cortex and hippocampus of an AD rat model, in comparison to vehicle-treated rats. Additionally, apremilast treatment led to a marked reduction in hippocampal Aβ levels, tau-positive cell counts, cholinesterase activity, and hippocampal caspase-3 level, a critical marker of neurodegeneration, relative to the placebo group. In aged AD rats, apremilast also notably decreased levels of pro-inflammatory cytokines, GSK-3 levels, oxidative stress, and insulin resistance (Gomaa et al., 2023).

10.3.2 Ibudilast

Ibudilast has multi-targets for its action, being a drug that is a PDEi and TLR4 antagonist at the same time. Owing to its anti-inflammatory activities, it has been used off-label to mitigate a myriad of inflammatory conditions, such as asthma and stroke, in some Asian countries. In a study by Oliveros et al. (2023), it was hypothesized that ibudilast might have mitigatory effects on inflammatory processes relevant to AD. That said, ibudilast was tested on Fisher transgenic 344-AD in hopes of repurposing ibudilast for AD. More specifically, the short-term working memory of both Tg-AD and wild-type rats was assessed using a hippocampal-dependent active place avoidance task. In summary, ibudilast proved to improve spatial learning and short-term memory, as Tg-AD rats that received treatment revealed marked improvement through the memory acquisition phase relative to their counterparts who did not receive treatment. Other findings of the same study that might explain the mechanisms underpinning the ibudilast-induced desirable cognitive outcome include a marked decrease in Aβ accumulation burden, amoeboid/ramified microglia, and tau PHF levels in the dentate gyrus region of the hippocampus. RNA sequencing of the hippocampal tissue demonstrated that ibudilast has distinct effects on the mRNA expression levels of the TLR and ubiquitin-proteasome pathways in transgenic rats. Ibudilast seems to suppress IRAK1 activity by enhancing the expression of its negative regulator, IRAK3, and/or by modulating the levels of TRAF6 and other TLR-associated ubiquitin ligases and conjugates within the TLR4 signaling pathway.

From a microcellular perspective, ibudilast treatment was found to mitigate inflammation by reducing the levels of IL-6, macrophage migration inhibitory factor, interferon receptor 1, and microglial-associated proteins, including arachidonate 5-lipoxygenase activating protein (Alox5ap), transmembrane protein 119 (TMEM119), lysosomal protein transmembrane 5 (LAPTM5), and Toll-like receptor 7 (TLR7). Additionally, the expression of synaptotagmin 2, a synaptic protein, was found to be elevated following ibudilast treatment. Synaptotagmin 2 modulates the vesicle transporter (syntaxins) and cadherin (CDH1) that were found to be involved in neurotransmitter release and synaptic structure maintenance, respectively. Furthermore, ibudilast alleviated the BDNF expression in both male and female rats. Taken together, those results offer various explanations regarding the ibudilast protective effect on AD (Oliveros et al., 2023).

10.3.3 Roflumilast

Roflumilast is a prototypical PDE4i that is FDA-approved for chronic obstructive pulmonary disease treatment. The role of roflumilast as an off-label drug for the reversal of cognitive dysfunction and AD symptoms has been a topic of extensive studies. In a study conducted by Feng et al. (2019), the impact of roflumilast on the cognitive function of APP/PS1 transgenic mice was studied via behavioral testing. Lingering administration of the drug was found to significantly improve cognitive functions. That is, in a novel object recognition task, chronic treatment with roflumilast was shown to reverse the low RI observed in APP/PSI mice compared to the wild-type control. Similarly, roflumilast was found to reverse the less distance and exploratory time compared to the wild-type control in MWM, as well as increase the latency of APP/PS1 mice compared with their counterparts who received no treatment. Roflumilast was also found to augment the cAMP, phosphorylated-CREB, and BDNF while decreasing nuclear translocation of NF-κB p65 subunit and levels of pro-inflammatory cytokines, including IL-6 and IL-1β, in APP/PS1 transgenic mice. It is possible that the stimulation of this molecular pathway might be responsible for the desirable neurocognitive enhancements of roflumilast (Efentakis et al., 2019).

Another study had a similar aim of examining the impact of roflumilast on memory and learning dysfunction. Specifically, the study assessed memory and depression-like behavior in APPswe/PS1dE9 mice. Similarly to the previous study, transgenic mice showed decreased performances on the cognitive tests, which were reversed by roflumilast. Furthermore, AD mice showed increased immobility times in the tail-suspension test and forced swimming test – a finding that was also corrected using roflumilast. Upon neurochemical assays, roflumilast was found to alleviate neuronal cell damage while reducing cell apoptosis in AD mice. Additionally, roflumilast restored the B-cell lymphoma-2 (Bcl-2)/Bcl-2-associated X protein (Bax) ratio and reduced the elevated expression of PDE4B and PDE4D in the cerebral cortex and hippocampus of AD mice. Additionally, roflumilast reinstated cAMP level and enhanced the expression of phosphorylated CREB and BDNF in AD mice, consistent with the results of prior studies (Wang et al., 2020).

More insights about the molecular axes for the neuroprotective activity of roflumilast were offered by another study that aimed to confirm the neurocognitive enhancing effect of roflumilast in mice, as well as decipher the role of downstream mediators of the cAMP effector, PKA, in the desirable effects of roflumilast. The study's findings match those of the aforementioned studies. Roflumilast proved to enhance the capacity of the mice to develop short-term memory in Y-maze and retrieve spatial memories in the Morris water maze relative to the control. Importantly, the co-administration of H-89 – a PKA inhibitor – totally reversed roflumilast's desirable effects, hinting at the cruciality of PKA in the pathways that mediate Roflumilast function. More findings regarding the cellular roflumilast-induced changes included an enhancement in ABCB1 transporters and pregnane X receptors (PXR) expression and a decline in Aβ deposition in the hippocampus. Roflumilast also linked to tau phosphorylation, which was related to increased hippocampal arginase activity and a reduction in the activities of GSK-3β, NOS, and the expression of inducible NOS (Ashour et al., 2021).

10.4 PHOSPHODIESTERASE-5 INHIBITORS

PDE5 enzyme catalyzes the cGMP hydrolysis. Given the powerful vasoactive effects of PDE5i, which induce vasodilation, they are currently approved for the treatment of erectile dysfunction and pulmonary arterial hypertension (Crescioli & Paronetto, 2024).

PDE5 is a metallo-hydrolase that converts cGMP to 5′ GMP, controlling various physiological activities. The PDE5A gene, located on human chromosome 4q26, has 23 exons and three variants. Its structure involves two parts, GAF-A and GAF-B, controlling PDE5 dimerization. The GAF-A domain expresses cGMP binding sites, converting cGMP into inactive 5′ GMP. Given that cGMP buildup is recognized as a repressive inflammatory signal, it is clear that this mechanism serves as a viable strategy to mitigate and regulate the onset and progression of several inflammation-related disorders (Khoshakhlagh et al., 2007; Peixoto & Gomes, 2015). The following section aims to explore various PDE5i, including avanafil, sildenafil, tadalafil, udenafil, and vardenafil, and evaluate their impact on cognitive function based on recent studies. By examining the mechanisms of action and the therapeutic effects observed in preclinical and clinical research, we seek to elucidate their potential role in neuroprotection and cognitive enhancement.

10.4.1 Avanafil

Chowdari Gurram et al. (2023) demonstrated the impact of avanafil, PDE5i, in an experimental model of AD through both *in silico* and *in vivo* approaches. The study focused on avanafil's dual inhibition of IκB kinase-β (IKKβ) and tumor necrosis factor receptor-1 (TNFR1), which are key players in neuroinflammation associated with cognitive decline. Avanafil was shown to bind effectively to IKKβ and TNFR1, inhibiting their activity. This dual inhibition is significant as it targets neuroinflammatory pathways that contribute to Alzheimer's pathology, particularly the NF-κB signaling pathway. In *in silico* studies, the results indicated strong binding affinity and stability, suggesting that avanafil could effectively modulate these targets. Furthermore, in *in vivo* studies, in a mouse model subjected to LPS-induced neuroinflammation, avanafil was given at two different doses (3 and 6 mg/kg) for a duration of 28 days. The higher dose significantly improved cognitive performance in behavioral tests. It also led to decrease in Aβ levels and inflammatory markers. As a PDE5i, avanafil increases levels of cGMP in the brain. Elevated cGMP is associated with neuroprotective effects, including enhanced neuronal survival and reduced neuroinflammation.

10.4.2 Sildenafil

Table 10.4 presents a comprehensive overview of various studies investigating the effects of sildenafil on AD across different models and dosages. The findings consistently highlight sildenafil's potential neuroprotective effects, particularly through its role in enhancing cGMP signaling. In cholinergic-like neurons with the PSEN1 E280A mutation, sildenafil was shown to reverse neuropathological features associated with AD, specifically reducing levels of Aβ and phosphorylated tau while improving neuronal viability and function. This suggests that sildenafil may play a crucial role in mitigating the cellular damage typically seen in Alzheimer's pathology (Giraldo-Berrio et al., 2024).

In rat models, a daily oral administration of 15 mg/kg for six weeks led to significant improvements in cognitive deficits and histopathological changes. Notably, this treatment increased vascular endothelial growth factor (VEGF-A) levels and decreased inflammatory markers

TABLE 10.4
Role of Phosphodiesterase-5 Inhibitors in Preclinical/Clinical Models of Alzheimer's Disease

Drug	Species	Dosing	Main Findings	Reference
Avanafil	Mice	3 and 6 mg/kg, oral administration.	Exhibited dual inhibition of IKKβ and TNFR1, leading to reduced neuroinflammation and improved cognitive function in an Alzheimer's disease model; enhanced memory retention and spatial learning abilities were observed.	Chowdari Gurram et al. (2023)
Sildenafil	Cholinergic-like neurons	Not specified, in vitro study.	Reversed neuropathological features associated with Alzheimer's disease in cholinergic-like neurons with the PSEN1 E280A mutation; reduced levels of Aβ and phosphorylated tau; improved neuronal viability and function.	Giraldo-Berrio et al. (2024)
	Rats	15 mg/kg/day, oral administration for 6 weeks.	Ameliorated cognitive deficits and histopathological changes in a rat model of Alzheimer's disease; increased levels of VEGF-A and decreased VCAM-1 and TNF-α; increased cGMP levels; and improved memory performance in behavioral tests.	Ibrahim et al. (2021a)
	HT-22 Neuronal Cells	Not specified, in vitro study.	Protects neuronal cells from mitochondrial toxicity induced by Aβ peptide; inhibits mitochondrial Ca^{2+} overload, reduces oxidative stress, and prevents cell death through activation of ATP-sensitive K^{+} channels.	Son et al. (2018)
	Patients	50 mg, oral administration.	Improved cerebral hemodynamic function and increased cerebral oxygen metabolism; significant increases in CBF and CMRO2 observed.	Sheng et al. (2017)
	APP/PS1 Mice	10 mg/kg, oral administration.	Prevented neuroinflammation, lowered Aβ levels, and improved cognitive performance in memory tasks.	Zhang et al. (2013)
	SAMP8 mice	7.5 mg/kg, daily for 4 weeks.	Ameliorated cognitive deficits and reduced tau pathology in senescence-accelerated mice; improved performance in memory tasks and decreased levels of β-secretase and cathepsin B, indicating a potential mechanism for cognitive enhancement.	Orejana et al. (2012)
	Tg2576 Mice	15 mg/kg, intraperitoneal injection.	Restored cognitive function in aged Tg2576 mice without altering Aβ burden; reduced tau hyperphosphorylation and increased levels of BDNF and Arc in the hippocampus.	Cuadrado-Tejedor et al. (2011)
	Double transgenic mice with APP/PS1 mutations compared with wild-type littermates	3 mg/kg for 3 weeks, intraperitoneal injection.	Amelioration of synaptic function, CREB phosphorylation, and memory. Increases cGMP and promotes the proliferation of neuronal progenitor cells in the hippocampus.	Puzzo et al. (2009)
	Cynomolgus macaque	0.3, 1, and 3 mg/kg twice weekly, intramuscular injection.	Improved object retrieval performance. cGMP triggers cellular cascades, leading to CREB phosphorylation, causing long-term neuronal signaling changes and performance enhancement	Rutten et al. (2008)
	Tg2576 Mice	10 mg/kg, oral administration.	Restored cognitive function without affecting Aβ burden; improved memory performance in cognitive tests and reduced tau hyperphosphorylation.	Shafiei et al. (2006)
	6-month-old Swiss mice	0.3, 1, and 3 mg/kg, oral administration.	Improved object discrimination performance at 1 mg/kg; increased cGMP levels in hippocampal slices (10 μM)	Rutten et al. (2005)
	Rats	3 mg/kg, oral administration.	Improved memory performance in object recognition after 24 h showed an inverted U-shaped dose-effect curve.	Prickaerts et al. (2005)

(Continued)

TABLE 10.4 *(Continued)*
Role of Phosphodiesterase-5 Inhibitors in Preclinical/Clinical Models of Alzheimer's Disease

Drug	Species	Dosing	Main Findings	Reference
	Rats	1.5, 3.0, and 4.5 mg/kg, intraperitoneal injection.	Attenuated maze learning impairment induced by nitric oxide synthase inhibition; significant improvement in performance observed at 3.0 mg/kg.	Devan et al. (2004)
	Swiss mice – 3 months and aged mice	0.5 and 1 mg/kg.	Improved cognitive function in aging and lipopolysaccharide-induced cognitive dysfunction Enhanced memory retention.	Patil et al. (2004)
	Mice	2, 4, and 8 mg/kg, intraperitoneal injection.	Improved acquisition and retention of memory; administered immediately after training, it showed dose-dependent effects.	Singh & Parle (2003)
	Healthy adults	100 mg, oral administration.	Enhanced auditory selective attention and verbal recognition memory as indicated by event-related brain potentials; no overt behavioral effects observed.	Schultheiss et al. (2001)
	2-month-old Swiss mice	3 mg/kg, intraperitoneal injection.	Improved long-term retention of an inhibitory avoidance response in mice.	Baratti and Boccia (1999)
Tadalafil	Mice	20 mg/kg/day, intraperitoneally for 21 days.	Mitigated cognitive impairment induced by streptozotocin; improved performance in Morris water maze and object recognition tests; reduced Aβ expression by 79% and tau hyperphosphorylation by 60%.	Salem et al. (2021)
	Male rats	20 mg/kg, intraperitoneally.	Reversed spatial learning impairments induced by H-89 and scopolamine in the Morris water maze; improved memory performance and cognitive function through modulation of cGMP and cholinergic signaling pathways.	Tabrizian et al. (2021)
	J20 Mice	10 mg/kg, oral administration.	Crossed the blood-brain barrier and reversed cognitive dysfunction; improved performance in the Morris water maze without altering amyloid burden.	García-Barroso et al. (2013)
Tadalafil derivative 3c	Mice	10 and 30 mg/kg, intragastrically.	Ameliorated cognitive impairment induced by scopolamine via inhibition of AChE and PDE5; demonstrated improved cognitive function.	Ni et al. (2018)
Udenafil	Patients	100 mg (at 3-day intervals for 2 months).	Improved cognitive function, erectile function, and somatization in patients with erectile dysfunction; significant increases in cognitive test scores.	Shim et al. (2011)
Vardenafil	Rats	0.3 and 1 mg/kg, oral administration.	Mitigated cognitive impairment induced by aluminum chloride and d-galactose; improved memory performance in behavioral tests; reduced amyloid-beta and phosphorylated tau levels; enhanced cGMP signaling and reduced endoplasmic reticulum stress.	Awad et al. (2023)
	Mice	0.3 mg/kg, intraperitoneally.	Administration during memory acquisition or early consolidation increased surface levels of AMPA receptors, enhancing spatial memory performance; effects persisted for 24 h post-learning.	Argyrousi et al. (2020)
	Aged rats	0.5 mg/kg, intraperitoneally.	Improved long-term memory performance in aged rats; showed less error in heading angle and increased time spent in the goal quadrant during probe tests after drug administration ceased.	Devan et al. (2014)
	Patients	100 mg, at 3-day intervals for 2 months.	Improved cognitive function, erectile function, and somatization in patients with erectile dysfunction; significant increases in cognitive test scores.	Shim et al. (2011)

AChE, acetylcholinesterase; Aβ, amyloid-beta; CREB, cAMP response element binding protein; BDNF, brain-derived neurotrophic factor; CBF, cerebral blood flow; CMRO2, cerebral metabolic rate of oxygen; TNF-α, tumor necrosis factor-α such as VCAM-1 and TNF-α. The increase in cGMP levels was also associated with improved memory performance in behavioral tests, indicating a potential mechanism through which sildenafil exerts its cognitive benefits (Ibrahim et al., 2021a).

Sildenafil has protected HT-22 neuronal cells from mitochondrial toxicity induced by Aβ peptides in vitro. This protection is attributed to the inhibition of mitochondrial calcium overload and oxidative stress, further underscoring the drug's neuroprotective properties (Son et al., 2018).

In clinical studies involving patients, sildenafil administered at 50 mg orally improved cerebral hemodynamic function, leading to significant increases in cerebral blood flow and oxygen metabolism. These physiological changes are likely mediated by enhanced cGMP signaling, which plays an essential part in vascular function and neuroprotection (Sheng et al., 2017).

Animal studies involving APP/PS1 mice demonstrated that a dosage of 10 mg/kg could prevent neuroinflammation, lower beta-amyloid levels, and enhance cognitive performance in memory tasks. This reinforces the idea that sildenafil's modulation of cGMP contributes to its protective effects against Alzheimer's pathology (Zhang et al., 2013).

Additionally, studies on SAMP8 mice showed that administering 7.5 mg/kg daily for four weeks ameliorated cognitive deficits and reduced tau pathology, again linking these improvements to the modulation of cGMP levels (Orejana et al., 2012). Similarly, in Tg2576 mice, sildenafil restored cognitive function without affecting beta-amyloid burden, further supporting its role in enhancing neuroplasticity and cognitive performance through cGMP signaling pathways (Cuadrado-Tejedor et al., 2011).

Overall, the evidence suggests that sildenafil's ability to increase cGMP levels is central to its neuroprotective effects across various models of AD. The drug's mechanisms appear to involve reducing neuroinflammation, enhancing neuronal survival, and improving cognitive outcomes. These findings collectively support the repurposing of sildenafil as a treatment option for AD, warranting other clinical investigations to confirm its efficacy and safety in human populations (Table 10.4).

10.4.3 Tadalafil

The table below outlines various studies examining the effects of tadalafil and its derivatives on cognitive function, particularly in relation to AD. The findings indicate that tadalafil has potential neuroprotective properties, primarily through the modulation of cGMP signaling. In research by Salem et al. (2021), tadalafil was administered at a dosage of 20 mg/kg/day via intraperitoneal injection for 21 days. Tadalafil mitigated cognitive impairment induced by streptozotocin, a compound known to induce Alzheimer's-like pathology. Notably, the treatment improved performance in both the object recognition and Morris water maze tests while significantly reducing Aβ expression and tau hyperphosphorylation. These conclusions imply that tadalafil may enhance cognitive function by reducing key pathological features associated with AD (Salem et al., 2021).

Further research by Tabrizian et al. (2021) also highlighted the cognitive benefits of tadalafil in male rats, where a dosage of 20 mg/kg was used to reverse spatial learning impairments caused by H-89 and scopolamine. The study reported improved memory performance and cognitive function, attributing these effects to the modulation of cGMP levels and cholinergic signaling pathways (Tabrizian et al., 2021).

A derivative of tadalafil, referred to as tadalafil derivative 3c, was tested in mice at doses of 10 and 30 mg/kg intragastrically. This compound ameliorated scopolamine-induced memory impairment through the inhibition of AChE and PDE5, demonstrating enhanced cognitive function (Ni et al., 2018). In another research involving J20 mice, a specific strain used to model AD, tadalafil was administered at a dose of 10 mg/kg orally. The drug successfully crossed the blood-brain barrier and reversed cognitive dysfunction without altering Aβ burden, indicating its potential for therapeutic application in AD (García-Barroso et al., 2013).

Overall, these studies underscore the potential of tadalafil and its derivatives in improving cognitive function through mechanisms involving cGMP signaling. This modulation appears to play a crucial role in mitigating cognitive deficits associated with AD and related pathologies, warranting further investigation into their therapeutic applications (Table 10.4).

10.4.4 Udenafil

The table below summarizes findings from a study on the effects of udenafil on cognitive function in patients with erectile dysfunction. The results indicate the potential benefits of udenafil in enhancing cognitive performance alongside improvements in erectile function and somatization. In a clinical trial conducted by Shim et al. (2011), patients received 100 mg of udenafil at three-day intervals for two months. The treatment led to significant improvements in cognitive function, as evidenced by increased scores on cognitive tests. This suggests that udenafil may enhance cognitive capabilities, particularly in individuals experiencing erectile dysfunction. The study highlights the dual benefits of udenafil, not only addressing erectile dysfunction but also positively impacting cognitive performance. This aligns with the broader understanding of PDE5i, which are known to enhance vasodilation and may improve cerebral blood flow and neuroprotection through increased levels of cGMP (Table 10.4).

10.4.5 Vardenafil

The table below provides a summary of various studies investigating the effects of vardenafil on cognitive function across different species and dosages. The findings suggest that vardenafil has potential neuroprotective effects, particularly in enhancing cognitive performance and modulating signaling pathways related to memory. In a study conducted by Awad et al. (2023), vardenafil was administered to rats at doses of 0.3 and 1 mg/kg via oral administration. The treatment mitigated cognitive impairment induced by aluminum chloride and d-galactose, leading to improved memory performance in behavioral tests. Notably, vardenafil reduced levels of Aβ and phosphorylated tau while enhancing cGMP signaling and reducing endoplasmic reticulum stress. ER stress becomes activated in case of inhibited proteostasis, which is always associated with a deposition of misfolded/unfolded proteins in the endoplasmic reticulum. Endoplasmic reticulum stress has been reported for its role in neurodegenerative disorders (Desouky et al., 2023), *in vivo* cardiotoxicity (George et al., 2024), and *in vitro* models (Dabour et al., 2024). These results indicate that vardenafil may help alleviate cognitive deficits associated with neurodegenerative processes (Awad et al., 2023).

In another study involving mice, Argyrousi et al. (2020) found that administering vardenafil at a dose of 0.3 mg/kg intraperitoneally during memory acquisition or early consolidation significantly increased the surface levels of AMPA receptors, which are crucial for synaptic plasticity. This enhancement led to improved spatial memory performance, with effects persisting for 24 h post-learning. This suggests that vardenafil may facilitate memory consolidation through mechanisms linked to cGMP signaling (Argyrousi et al., 2020). Further research by Devan et al. (2014) demonstrated that aged rats receiving 0.5 mg/kg of vardenafil intraperitoneally exhibited improved long-term memory performance. The treated rats showed fewer errors in heading angle and spent more time in the goal quadrant during probe tests after drug administration ceased, indicating sustained cognitive benefits from vardenafil treatment (Devan et al., 2014).

In clinical settings, a study by Shim et al. (2011) reported that patients taking 100 mg of vardenafil at three-day intervals for two months experienced improvements not only in erectile function but also in cognitive function and somatization. Significant increases in cognitive test scores were observed, suggesting that vardenafil may have broader applications beyond treating erectile dysfunction (Shim et al., 2011).

Overall, the evidence from these studies supports the potential of vardenafil as a therapeutic agent for enhancing cognitive function through mechanisms involving cGMP signaling and synaptic plasticity. These findings highlight the need for further research to explore the therapeutic implications of PDE5i like vardenafil in neurodegenerative diseases such as AD and other forms of cognitive impairment (Table 10.4).

REFERENCES

(2023). 2023 Alzheimer's disease facts and figures. *Alzheimers Dement*, *19*, 1598–1695.

Abusnina, A., & Lugnier, C. (2017). Therapeutic potentials of natural compounds acting on cyclic nucleotide phosphodiesterase families. *Cell Signal*, *39*, 55–65.

Ahmed, W. S., Geethakumari, A. M., & Biswas, K. H. (2021). Phosphodiesterase 5 (PDE5): Structure-function regulation and therapeutic applications of inhibitors. *Biomedicine & Pharmacotherapy*, *134*, 111128.

Amidfar, M., de Oliveira, J., Kucharska, E., Budni, J., & Kim, Y. K. (2020). The role of CREB and BDNF in neurobiology and treatment of Alzheimer's disease. *Life Sciences*, *257*, 118020.

Amin, S. A., Bhargava, S., Adhikari, N., Gayen, S., & Jha, T. (2018). Exploring pyrazolo[3,4-d]pyrimidine phosphodiesterase 1 (PDE1) inhibitors: A predictive approach combining comparative validated multiple molecular modelling techniques. *Journal of Biomolecular Structure and Dynamics*, *36*, 590–608.

Argyrousi, E. K., Heckman, P. R., van Hagen, B. T., Muysers, H., van Goethem, N. P., & Prickaerts, J. (2020). Pro-cognitive effect of upregulating cyclic guanosine monophosphate signalling during memory acquisition or early consolidation is mediated by increased AMPA receptor trafficking. *Journal of Psychopharmacology*, *34*, 103–114.

Ashour, N. H., El-Tanbouly, D. M., El Sayed, N. S., & Khattab, M. M. (2021). Roflumilast ameliorates cognitive deficits in a mouse model of amyloidogenesis and tauopathy: Involvement of nitric oxide status, Aβ extrusion transporter ABCB1, and reversal by PKA inhibitor H89. *Progress in Neuro-Psychopharmacology & Biological Psychiatry*, *111*, 110366.

Awad, H. H., Desouky, M. A., Zidan, A., Bassem, M., Qasem, A., Farouk, M., AlDeab, H., Fouad, M., Hany, C., & Basem, N. (2023). Neuromodulatory effect of vardenafil on aluminium chloride/d-galactose induced Alzheimer's disease in rats: Emphasis on amyloid-beta, p-tau, PI3K/Akt/p53 pathway, endoplasmic reticulum stress, and cellular senescence. *Inflammopharmacology*, *31*, 2653–2673.

Ayoub, I. M., George, M. Y., Menze, E. T., Mahmoud, M., Botros, M., Essam, M., Ashmawy, I., Shendi, P., Hany, A., Galal, M., Ayman, M., & Labib, R. M. (2022). Insights into the neuroprotective effects of *Salvia officinalis* L. and *Salvia microphylla* Kunth in the memory impairment rat model. *Food & Function*, *13*(4), 2253–2268. https://doi.org/10.1039/d1fo02988f

Baratti, C. M., & Boccia, M. M. (1999). Effects of sildenafil on long-term retention of an inhibitory avoidance response in mice. *Behavioural Pharmacology*, *10*.

Bolger, G. B. (2021). The PDE-opathies: Diverse phenotypes produced by a functionally related multigene family. *Trends in Genetics*, *37*, 669–681.

Bollen, E., & Prickaerts, J. (2012). Phosphodiesterases in neurodegenerative disorders. *IUBMB Life*, *64*, 965–970.

Cada, D. J., Ingram, K., & Baker, D. E. (2014). Apremilast. *Hospital Pharmacy*, *49*, 752–762.

Chen, Q., Yin, Y., Li, L., Zhang, Y., He, W., & Shi, Y. (2021). Milrinone ameliorates the neuroinflammation and memory function of Alzheimer's disease in an APP/PS1 mouse model. *Neuropsychiatric Disease and Treatment*, *17*, 2129–2139.

Chowdari Gurram, P., Satarker, S., Kumar, G., Begum, F., Mehta, C., Nayak, U., Mudgal, J., Arora, D., & Nampoothiri, M. (2023). Avanafil mediated dual inhibition of IKKβ and TNFR1 in an experimental paradigm of Alzheimer's disease: In silico and in vivo approach. *Journal of Biomolecular Structure and Dynamics*, *41*, 10659–10677.

Crescioli, C., & Paronetto, M. P. (2024). The emerging role of phosphodiesterase 5 inhibition in neurological disorders: The state of the art. *Cells*, *13*, 1720.

Cuadrado-Tejedor, M., Hervias, I., Ricobaraza, A., Puerta, E., Pérez-Roldán, J., García-Barroso, C., Franco, R., Aguirre, N., & García-Osta, A. (2011). Sildenafil restores cognitive function without affecting β-amyloid burden in a mouse model of Alzheimer's disease. *British Journal of Pharmacology*, *164*, 2029–2041.

Cummings, J., Lee, G., Ritter, A., & Zhong, K. (2018). Alzheimer's disease drug development pipeline: 2018. *Alzheimers Dement (N Y)*, *4*, 195–214.

Dabour, M. S., George, M. Y., Grant, M. K. O., & Zordoky, B. N. (2024). Canagliflozin differentially modulates carfilzomib-induced endoplasmic reticulum stress in multiple myeloma and endothelial cells. *Archives of Toxicology*. https://doi.org/10.1007/s00204-024-03913-0

Desouky, M. A., George, M. Y., Michel, H. E., & Elsherbiny, D. A. (2023). Roflumilast escalates α-synuclein aggregate degradation in rotenone-induced Parkinson's disease in rats: Modulation of the ubiquitin-proteasome system and endoplasmic reticulum stress. *Chemico-Biological Interactions*, *379*, 110491.

Desouky, M. A., Michel, H. E., Elsherbiny, D. A., & George, M. Y. (2024). Recent pharmacological insights on abating toxic protein species burden in neurological disorders: Emphasis on 26S proteasome activation. *Life Sciences*, *359*, 123206. https://doi.org/10.1016/j.lfs.2024.123206

Devan, B. D., Pistell, P. J., Duffy, K. B., Kelley-Bell, B., Spangler, E. L., & Ingram, D. K. (2014). Phosphodiesterase inhibition facilitates cognitive restoration in rodent models of age-related memory decline. *NeuroRehabilitation*, *34*, 101–111.

Devan, B. D., Sierra-Mercado, D. Jr., Jimenez, M., Bowker, J. L., Duffy, K. B., Spangler, E. L., & Ingram, D. K. (2004). Phosphodiesterase inhibition by sildenafil citrate attenuates the learning impairment induced by blockade of cholinergic muscarinic receptors in rats. *Pharmacology Biochemistry and Behavior*, *79*, 691–699.

Di Benedetto, G., Iannucci, L. F., Surdo, N. C., Zanin, S., Conca, F., Grisan, F., Gerbino, A., & Lefkimmiatis, K. (2021). Compartmentalized signaling in aging and neurodegeneration. *Cells*, *10*.

Efentakis, P., Andreadou, I., Nikolaou, P.-E., Witzler, C., Siemer, S., Kastritis, E., Wenzel, P., Dimopoulos, M. A., & Terpos, E. (2019). Synergistic effect of carfilzomib and metformin in vascular plasticity; the emerging role of autophagy. *Blood*, *134*, 3797.

Fawzi, S. F., Michel, H. E., Menze, E. T., Tadros, M. G., & George, M. Y. (2023). Clotrimazole ameliorates chronic mild stress-induced depressive-like behavior in rats; crosstalk between the HPA, NLRP3 inflammasome, and Wnt/β-catenin pathways. *International Immunopharmacology*, *127*, 111354. https://doi.org/10.1016/j.intimp.2023.111354

Feng, H., Wang, C., He, W., Wu, X., Li, S., Zeng, Z., Wei, M., & He, B. (2019). Roflumilast ameliorates cognitive impairment in APP/PS1 mice via cAMP/CREB/BDNF signaling and anti-neuroinflammatory effects. *Metabolic Brain Disease*, *34*, 583–591.

Gad, S. R., El-Gogary, R. I., George, M. Y., & Hathout, R. M. (2023). Nose-to-brain delivery of 18β-Glycyrrhetinic acid using optimized lipid nanocapsules: A novel alternative treatment for Alzheimer's disease. *International Journal of Pharmaceutics*, *645*, 123387. https://doi.org/10.1016/j.ijpharm.2023.123387

García-Barroso, C., Ricobaraza, A., Pascual-Lucas, M., Unceta, N., Rico, A. J., Goicolea, M. A., Sallés, J., Lanciego, J. L., Oyarzabal, J., & Franco, R. (2013). Tadalafil crosses the blood–brain barrier and reverses cognitive dysfunction in a mouse model of AD. *Neuropharmacology*, *64*, 114–123.

Geerts, H., & Grossberg, G. T. (2006). Pharmacology of acetylcholinesterase inhibitors and N-methyl-D-aspartate receptors for combination therapy in the treatment of Alzheimer's disease. *The Journal of Clinical Pharmacology*, *46*, 8S–16S.

George, M. Y., Dabour, M. S., Rashad, E., & Zordoky, B. N. (2024). Empagliflozin alleviates carfilzomib-induced cardiotoxicity in mice by modulating oxidative stress, inflammatory response, endoplasmic reticulum stress, and autophagy. *Antioxidants (Basel)*, *13*(6), 671. https://doi.org/10.3390/antiox13060671

George, M. Y., El-Derany, M. O., Ahmed, Y., Zaher, M., Ibrahim, C., Waleed, H., Khaled, H., Khaled, G., Saleh, A., Alshafei, H., Alshafei, R., Ahmed, N., Ezz, S., Ashraf, N., & Ibrahim, S. S. (2023, January). Design and evaluation of chrysin-loaded nanoemulsion against lithium/pilocarpine-induced status epilepticus in rats; emphasis on formulation, neuronal excitotoxicity, oxidative stress, microglia polarization, and AMPK/SIRT-1/PGC-1α pathway. *Expert Opinion on Drug Delivery*, *20*(1), 159–174. https://doi.org/10.1080/17425247.2023.2153831

George, M. Y., Menze, E. T., Esmat, A., Tadros, M. G., & El-Demerdash, E. (2020). Potential therapeutic antipsychotic effects of Naringin against ketamine-induced deficits in rats: Involvement of Akt/GSK-3β and Wnt/β-catenin signaling pathways. *Life Sciences*, *249*, 117535. https://doi.org/10.1016/j.lfs.2020.117535

Giraldo-Berrio, D., Jimenez-Del-Rio, M., & Velez-Pardo, C. (2024). Sildenafil reverses the neuropathological Alzheimer's disease phenotype in cholinergic-like neurons carrying the presenilin 1 E280A mutation. *Journal of Alzheimer's Disease*, *99*, 1–18.

Gomaa, A. A., Farghaly, H. S. M., Ahmed, A. M., El-Mokhtar, M. A., & Hemida, F. K. (2022). Advancing combination treatment with cilostazol and caffeine for Alzheimer's disease in high fat-high fructose-STZ induced model of amnesia. *European Journal of Pharmacology*, *921*, 174873.

Gomaa, A. A., Farghaly, H. S. M., Ahmed, A. M., & Hemida, F. K. (2023). Intermittent treatment with Apremilast, a phosphodiesterase-4 inhibitor, ameliorates Alzheimer's-like pathology and symptoms through multiple targeting actions in aged T2D rats. *International Immunopharmacology*, *117*, 109927.

Gorshkov, K., & Zhang, J. (2014). Visualization of cyclic nucleotide dynamics in neurons. *Frontiers in Cellular Neuroscience*, *8*, 395.

Habib, C. N., Ali, A. E., Anber, N. H., & George, M. Y. (2023). Lactoferrin ameliorates carfilzomib-induced renal and pulmonary deficits: Insights to the inflammasome NLRP3/NF-κB and PI3K/Akt/GSK-3β/MAPK axes. *Life Sciences*, *335*, 122245. https://doi.org/10.1016/j.lfs.2023.122245

Hasebe, M., & Yoshino, M. (2016). Nitric oxide/cGMP/PKG signaling pathway activated by M1-type muscarinic acetylcholine receptor cascade inhibits Na+-activated K+ currents in Kenyon cells. *Journal of Neurophysiology*, *115*, 3174–3185.

Heckman, P. R. A., Blokland, A., & Prickaerts, J. (2017). From age-related cognitive decline to Alzheimer's disease: A translational overview of the potential role for phosphodiesterases. *Advances in Neurobiology*, *17*, 135–168.

Hung, S. Y., & Fu, W. M. (2017). Drug candidates in clinical trials for Alzheimer's disease. *Journal of Biomedical Science*, *24*, 47.

Ibrahim, M., Haleem, M., AbdelWahab, S., & Abdel-Aziz, A. (2021a). Sildenafil ameliorates Alzheimer disease via the modulation of vascular endothelial growth factor and vascular cell adhesion molecule-1 in rats. *Human & Experimental Toxicology*, *40*, 596–607.

Ibrahim, S. S., Abo Elseoud, O. G., Mohamedy, M. H., Amer, M. M., Mohamed, Y. Y., Elmansy, S. A., Kadry, M. M., Attia, A. A., Fanous, R. A., Kamel, M. S., Solyman, Y. A., Shehata, M. S., & George, M. Y. (2021b). Nose-to-brain delivery of chrysin transfersomal and composite vesicles in doxorubicin-induced cognitive impairment in rats: Insights on formulation, oxidative stress and TLR4/NF-kB/NLRP3 pathways. *Neuropharmacology, 197*, 108738. https://doi.org/10.1016/j.neuropharm.2021.108738

Khalifa, M., Abdelsalam, R. M., Safar, M. M., & Zaki, H. F. (2022). Phosphodiesterase (PDE) III inhibitor, Cilostazol, improved memory impairment in aluminum chloride-treated rats: Modulation of cAMP/CREB pathway. *Inflammopharmacology, 30*, 2477–2488.

Khoshakhlagh, P., Bahrololoumi-Shapourabadi, M., Mohammadirad, A., Ashtaral-Nakhai, L., Minaie, B., & Abdollahi, M. (2007). Beneficial effect of phosphodiesterase-5 inhibitor in experimental inflammatory bowel disease; Molecular evidence for involvement of oxidative stress. *Toxicology Mechanisms and Methods, 17*, 281–288.

Mohammed, N. N., Tadros, M. G., & George, M. Y. (2024). Empagliflozin repurposing in Parkinson's disease; modulation of oxidative stress, neuroinflammation, AMPK/SIRT-1/PGC-1α, and wnt/β-catenin pathways. *Inflammopharmacology, 32*(1), 777–794. https://doi.org/10.1007/s10787-023-01384-w

Morid, O. F., Menze, E. T., Tadros, M. G., & George, M. Y. (2023). L-carnitine modulates cognitive impairment induced by doxorubicin and cyclophosphamide in rats; insights to oxidative stress, inflammation, synaptic plasticity, liver/brain, and kidney/brain axes. *Journal of Neuroimmune Pharmacology, 18*(3), 310–326. https://doi.org/10.1007/s11481-023-10062-1

Munoz-Torrero, D. (2008). Acetylcholinesterase inhibitors as disease-modifying therapies for Alzheimer's disease. *Current Medicinal Chemistry, 15*, 2433–2455.

Nabavi, S. M., Talarek, S., Listos, J., Nabavi, S. F., Devi, K. P., Roberto de Oliveira, M., Tewari, D., Arguelles, S., Mehrzadi, S., Hosseinzadeh, A., D'Onofrio, G., Orhan, I. E., Sureda, A., Xu, S., Momtaz, S., & Farzaei, M. H. (2019). Phosphodiesterase inhibitors say NO to Alzheimer's disease. *Food and Chemical Toxicology, 134*, 110822.

Ni, W., Wang, H., Li, X., Zheng, X., Wang, M., Zhang, J., Gong, Q., Ling, D., Mao, F., & Zhang, H. (2018). Novel tadalafil derivatives ameliorates scopolamine-induced cognitive impairment in mice via inhibition of acetylcholinesterase (AChE) and phosphodiesterase 5 (PDE5). *ACS Chemical Neuroscience, 9*, 1625–1636.

Oliveros, G., Wallace, C. H., Chaudry, O., Liu, Q., Qiu, Y., Xie, L., Rockwell, P., Figueiredo-Pereira, M. E., & Serrano, P. A. (2023). Repurposing ibudilast to mitigate Alzheimer's disease by targeting inflammation. *Brain, 146*, 898–911.

Omori, K., & Kotera, J. (2007). Overview of PDEs and their regulation. *Circulation Research, 100*, 309–327.

Orejana, L., Barros-Miñones, L., Jordán, J., Puerta, E., & Aguirre, N. (2012). Sildenafil ameliorates cognitive deficits and tau pathology in a senescence-accelerated mouse model. *Neurobiology of Aging, 33*, 625.e11–20.

Padda, I. S., & Tripp, J. (2024). 'Phosphodiesterase Inhibitors.' in, *StatPearls* (Treasure Island (FL) ineligible companies. Disclosure: Jayson Tripp declares no relevant financial relationships with ineligible companies).

Parihar, M. S., & Hemnani, T. (2004). Alzheimer's disease pathogenesis and therapeutic interventions. *Journal of Clinical Neuroscience, 11*, 456–467.

Patil, C. S., Jain, N. K., Singh, V. P., & Kulkarni, S. K. (2004). Differential effect of the PDE5 inhibitors, sildenafil and zaprinast, in aging- and lipopolysaccharide-induced cognitive dysfunction in mice. *Drug Development Research, 63*, 66–75.

Pavlaki, N., & Nikolaev, V. O. (2018). Imaging of PDE2- and PDE3-mediated cGMP-to-cAMP cross-talk in cardiomyocytes. *Journal of Cardiovascular Development and Disease, 5*.

Peixoto, C. A., & dos Santos Gomes, F. O. (2015). The role of phosphodiesterase-5 inhibitors in prostatic inflammation: A review. *Journal of Inflammation, 12*, 54.

Prickaerts, J., Heckman, P. R. A., & Blokland, A. (2017). Investigational phosphodiesterase inhibitors in phase I and phase II clinical trials for Alzheimer's disease. *Expert Opinion on Investigational Drugs, 26*, 1033–1048.

Prickaerts, J., Sik, A., van der Staay, F. J., de Vente, J., & Blokland, A. (2005). Dissociable effects of acetylcholinesterase inhibitors and phosphodiesterase type 5 inhibitors on object recognition memory: Acquisition versus consolidation. *Psychopharmacology (Berl), 177*, 381–390.

Puzzo, D., Staniszewski, A., Deng, S. X., Privitera, L., Leznik, E., Liu, S., Zhang, H., Feng, Y., Palmeri, A., Landry, D. W., & Arancio, O. (2009). Phosphodiesterase 5 inhibition improves synaptic function, memory, and amyloid-beta load in an Alzheimer's disease mouse model. *The Journal of Neuroscience, 29*, 8075–8086.

Ribaudo, G., Ongaro, A., Zagotto, G., Memo, M., & Gianoncelli, A. (2020). Therapeutic potential of phosphodiesterase inhibitors against neurodegeneration: The perspective of the medicinal chemist. *ACS Chemical Neuroscience, 11*, 1726–1739.

Rutten, K., Basile, J., Prickaerts, J., Blokland, A., & Vivian, J. (2008). Selective PDE inhibitors rolipram and sildenafil improve object retrieval performance in adult cynomolgus macaques. *Psychopharmacology, 196*, 643–648.

Rutten, K., Vente, J. D., Şik, A., Ittersum, M. M.-V., Prickaerts, J., & Blokland, A. (2005). The selective PDE5 inhibitor, sildenafil, improves object memory in Swiss mice and increases cGMP levels in hippocampal slices. *Behavioural Brain Research, 164*, 11–16.

Saito, S., Suzuki, K., Ohtani, R., Maki, T., Kowa, H., Tachibana, H., Washida, K., Kawabata, N., Mizuno, T., Kanki, R., Sudoh, S., Kitaguchi, H., Shindo, K., Shindo, A., Oka, N., Yamamoto, K., Yasuno, F., Kakuta, C., Kakuta, R., … & Ihara, M. (2023). Efficacy and safety of cilostazol in mild cognitive impairment: A randomized clinical trial. *JAMA Network Open, 6*, e2344938.

Salem, M. A., Budzyńska, B., Kowalczyk, J., El Sayed, N. S., & Mansour, S. M. (2021). Tadalafil and bergapten mitigate streptozotocin-induced sporadic Alzheimer's disease in mice via modulating neuroinflammation, PI3K/Akt, Wnt/β-catenin, AMPK/mTOR signaling pathways. *Toxicology and Applied Pharmacology, 429*, 115697.

Sanders, O., & Rajagopal, L. (2020). Phosphodiesterase inhibitors for Alzheimer's disease: A systematic review of clinical trials and epidemiology with a mechanistic rationale. *The Journal of Alzheimer's Disease Reports, 4*, 185–215.

Schultheiss, D., Müller, S., Nager, W., Stief, C., Schlote, N., Jonas, U., Asvestis, C., Johannes, S., & Münte, T. (2001). Central effects of sildenafil (Viagra) on auditory selective attention and verbal recognition memory in humans: A study with event-related brain potentials. *World Journal of Urology, 19*, 46–50.

Schultz, J. E., & Natarajan, J. (2013). Regulated unfolding: A basic principle of intraprotein signaling in modular proteins. *Trends in Biochemical Sciences, 38*, 538–545.

Shafiei, M., Mahmoudian, M., Rostami, P., & Nemati, F. (2006). Effect of sildenafil (viagra) on memory retention of a passive avoidance response in rats. *Acta Physiologica Hungarica*, *93*, 53–59.

Sheng, J., Zhang, S., Wu, L., Kumar, G., Liao, Y., Gk, P., & Fan, H. (2022). Inhibition of phosphodiesterase: A novel therapeutic target for the treatment of mild cognitive impairment and Alzheimer's disease. *Frontiers in Aging Neuroscience*, *14*, 1019187.

Sheng, M., Lu, H., Liu, P., Li, Y., Ravi, H., Peng, S.-L., Diaz-Arrastia, R., Devous, M. D., & Womack, K. B. (2017). Sildenafil improves vascular and metabolic function in patients with Alzheimer's disease. *Journal of Alzheimer's Disease*, *60*, 1351–1364.

Shim, Y. S., Pae, C. U., Kim, S. W., Kim, H. W., Kim, J. C., & Koh, J. S. (2011). Effects of repeated dosing with Udenafil (Zydena) on cognition, somatization and erection in patients with erectile dysfunction: A pilot study. *International Journal of Impotence Research*, *23*, 109–114.

Singh, N., & Parle, M. (2003). Sildenafil improves acquisition and retention of memory in mice. *Indian Journal of Physiology and Pharmacology*, *47*, 318–324.

Soberg, K., & Skalhegg, B. S. (2018). The molecular basis for specificity at the level of the protein kinase a catalytic subunit. *Frontiers in Endocrinology (Lausanne)*, *9*, 538.

Son, Y., Kim, K., & Cho, H.-R (2018). Sildenafil protects neuronal cells from mitochondrial toxicity induced by β-amyloid peptide via ATP-sensitive K+ channels. *Biochemical and Biophysical Research Communications*, *500*, 504–510.

Tabrizian, K., Amelinia, F., Belaran, M., Pourheidar, S., Mirzaei, H., & Fanoudi, S. (2021). Tadalafil Reversed H-89–and Scopolamine–Induced spatial learning impairments in male rats. *Drug Research*, *71*, 275–283.

Tibbo, A. J., Tejeda, G. S., & Baillie, G. S. (2019). Understanding PDE4's function in Alzheimer's disease; a target for novel therapeutic approaches. *Biochemical Society Transactions*, *47*, 1557–1565.

Wang, H., Gaur, U., Xiao, J., Xu, B., Xu, J., & Zheng, W. (2018). Targeting phosphodiesterase 4 as a potential therapeutic strategy for enhancing neuroplasticity following ischemic stroke. *International Journal of Biological Sciences*, *14*, 1745–1754.

Wang, H., Zhang, F.-F, Xu, Y., Fu, H.-R, Wang, X.-D, Wang, L., Chen, W., Xu, X.-Y, Gao, Y.-F, Zhang, J.-G, & Zhang, H.-T. (2020). The phosphodiesterase-4 inhibitor roflumilast, a potential treatment for the comorbidity of memory loss and depression in Alzheimer's disease: A preclinical study in APP/PS1 transgenic mice. *International Journal of Neuropsychopharmacology*, *23*, 700–711.

Yanai, S., & Endo, S. (2019). PDE3 inhibitors repurposed as treatments for age-related cognitive impairment. *Molecular Neurobiology*, *56*, 4306–4316.

Yiannopoulou, K. G., & Papageorgiou, S. G. (2013). Current and future treatments for Alzheimer's disease. *Therapeutic Advances in Neurological Disorders*, *6*, 19–33.

Zhang, J., Guo, J., Zhao, X., Chen, Z., Wang, G., Liu, A., Wang, Q., Zhou, W., Xu, Y., & Wang, C. (2013). Phosphodiesterase-5 inhibitor sildenafil prevents neuroinflammation, lowers beta-amyloid levels and improves cognitive performance in APP/PS1 transgenic mice. *Behavioural Brain Research*, *250*, 230–237.

11 Autophagy Enhancer a Promising Therapeutic Approach for Alzheimer's Treatment

Prakash Haloi and Havagiray R. Chitme

11.1 INTRODUCTION

Alzheimer's disease (AD) is a brain condition leading to advanced impairment of memory, cognition, learning, and organizational abilities (Wilson & Geetha, 2020). It is among the primary causes that contribute to age-related dementia. According to current estimates, 15% of people over the age of 60 have mild cognitive impairment (MCI), and 10%–15% of these people develop clinical dementia, primarily diagnosed as AD. It is also estimated that by 2050, there will be about 100 million dementia sufferers worldwide (Lopez & Kuller, 2019). Currently, MCI does not have any approved pharmacological treatments (Ballard et al., 2020). Neurofibrillary tangles (NFTs) and extracellular senile plaques are AD's two primary pathogenic characteristics. The short amyloid-β (Aβ) peptides are found in senile plaques and the hyperphosphorylated tau microtubule-associated protein (MAPT) that forms NFTs. In addition to amyloid plaques, other pathological hallmarks in patients with AD and animal models include synaptic degeneration, impaired mitochondria, inflammation, and imbalanced calcium (Ca) homeostasis (Querfurth & LaFerla, 2010).

AD etiology is very complicated. Some of the risk aspects, including age, genetic history, genetic makeup, education, and acquired brain damage, are considered to have contributed to the development of AD (Collier, et al., 2021). According to the age of onset, AD is broadly classified into two types: (i) Late-onset AD (LOAD), which manifests at age 65 or above, and (ii) Early-onset AD (EOAD), which manifests before the age of 65. APP, PSEN1, and PSEN2 genes have pathogenic mutations associated with many EOAD cases (Hardy, 2017). The inheritance pattern for these mutations is autosomal dominant. EOAD is more aggressive but reduces 3% of all AD cases. Conversely, around 90% of all instances of AD are LOAD (Sun et al., 2017). Although the genetic mechanisms for LOAD are much more complicated, the APOE ε4 gene is considered the primary risk factor for LOAD. Furthermore, GWAS have recently revealed more than 20 additional risk genes linked to LOAD, such as ADAM10, CR1, BIN1, and CD33. The involvement of these risk genes in various AD pathways provides a better understanding of AD pathogenesis (Zhang et al., 2021).

Over recent years, numerous studies suggested that there will be a potential link between defective autophagy and AD. Researchers initially used electron microscopy to examine whether AD patients' brains had structurally damaged neurites and aberrant autophagosomes. Likewise, in the AD animal models, the autophagosomes were localized mainly in the neuronal dendrites and not in the soma. Increasingly, it is being realized that upregulation of the processes that lead to the breakdown of proteins and damaged organelles via autophagy is a viable and efficacious approach for treating AD (Zhang et al., 2021).

The primary aim of this chapter is to generate new ideas on autophagy regulation and the feasibility of its use in the management of AD. In addition, a comprehensive overview of autophagy and its role in AD pathogenesis and currently available drugs that act as autophagy modulators in AD management is provided.

11.2 AUTOPHAGY IN AD PATHOGENESIS

11.2.1 Autophagy in Aβ Metabolism

Multiple mammalian targets of rapamycin inhibitors (mTORs) have been proven to be more effective as autophagy modulators than rapamycin. These inhibitors were potentially diminishing the level of insoluble tau and tau phosphorylation in tau-mutant neurons derived from human neural progenitor cells, indicating that tauopathies are ameliorated by accelerated autophagy (Silva et al., 2020). Several reports have suggested that up-regulating Transcription Factor EB (TFEB), a key regulator of autophagy, enhanced cognitive function and significantly decreased the level of phosphorylated soluble tau and insoluble tau clumps in mouse models (Wang et al., 2016). Moreover, the chaperon-mediated autophagy (CMA) has a significant impact on the tauopathies. Caballero et al. (2021) reported that tau acetylation is resistant to CMA degradation, and tauopathies worsen when CMA is suppressed *in vivo* (Caballero et al., 2021). Prior investigation has indicated that the AD mice treated with rapamycin (a known mTOR inhibitor) resulted in improved autophagy. The mice's brains demonstrated a notable improvement in cognitive function and a significant reduction in intracellular and extracellular Aβ deposits as a result of this treatment. The removal of extracellular Aβ is believed to be facilitated by microglial endocytosis (Huang et al., 2021; Majumder et al., 2011).

DOI: 10.1201/9781003602477-11

Aβ clearance appears to be impacted by autophagy at several stages. The crucial lysosomal protease cathepsin B is involved in the breakdown of autophagic substrates. In animal models with elevated levels of Aβ42 and amyloid deposition, studies of gene knockout models have revealed that the knockdown of cathepsin B worsens the disease score in AD model mice. However, in aged AD mice, the overexpression of cathepsin B through lentiviral transduction reduced amyloid plaques. Such studies have indicated that increasing the autophagy flux at any point in the disease process is favorable for slowing the course of AD (Mueller-Steiner et al., 2006).

Apart from bulk autophagy, Aβ metabolism has been shown to involve selective autophagy. Aggrephagy is a type of selective autophagy that breaks down aggregated proteins (Kirkin & Rogov, 2019). Several selective autophagy receptors unique to aggrephagy have been found in earlier studies, such as p62/SQSTM1, NBR1, and OPTN (Malampati et al., 2020).

11.2.2 Autophagy in Tauopathies

Apart from Aβ metabolism, autophagy is also important in tau pathology. Initially, *in vitro* findings demonstrated that tau clearance was impaired, and unsolvable tau clumps were markedly accumulated upon blocking autophagic flow (Hamano et al., 2008). In the brains of patients with familial AD, hyperphosphorylated tau is shown to colocalize with the autophagy receptor p62/SQSTM1 and the autophagosome marker LC3. Still, control participants without neurodegenerative disorders did not exhibit this overlap (Piras et al., 2016). According to Guo et al. (2016), tau aggregates in a tau model cell line were consistently linked to LC3 and p62/SQSTM1 immunoreactivity. These findings suggest that autophagy also likely affects tauopathies (Guo et al., 2016).

11.2.3 Autophagy in Synaptic Function

Along with tau aggregation and Aβ deposition, another hallmark of AD pathogenesis is synaptic disruption. Synapses are unique structures found in neurons that act as the fundamental building blocks for presynaptic to postsynaptic neuronal communication. During the early stages of AD pathogenesis, fewer synapses have been found (Chen et al., 2019). According to a growing body of research, synaptic processes like neurotransmission and synaptic plasticity depend on functional autophagy (Lieberman & Sulzer, 2020). Synaptojanin-1 is a presynaptic lipid phosphatase that aids in synaptic vesicle endocytosis. Studies have revealed that after Synaptojanin-1 is genetically abated, late endosomes and autophagosomes increase in zebrafish cone photoreceptors (George et al., 2016), indicating that Synaptojanin-1 is essential for preserving regular autophagy at synapses.

Furthermore, autophagy has been shown to play a role in dopamine presynaptic release. According to Hernandez et al. (2012), ATG7 cKO dopaminergic neurons have elevated dopamine release, which can be restored by administering the mTOR inhibitor rapamycin. There are very few studies on autophagy in postsynaptic locations. Autophagy seems to play a part in the breakdown of neurotransmitter receptors, such as GABAA receptors and AMPA receptors (Lieberman & Sulzer, 2020). Additionally, autophagy is required for synaptic plasticity, which denotes synaptic characteristics' ability to alter in number, structure, and function to either enhance or decrease the contact. Synaptic plasticity is widely believed to be a requisite for important brain processes such as memory and knowledge. Long-term potentiation (LTP) and long-term depression (LTD) are cellular forms of learning and memory and are types of synaptic plasticity. However, current research indicated that stimulation of autophagy in mice's hippocampus area was required to generate new memory and that pharmacological suppression of autophagy blocked LTP (Glatigny et al., 2019).

11.2.4 Autophagy in Mitochondrial Dysfunction

Mitochondria are the membrane-bound organelles that produce energy. Neurons require functional mitochondria to preserve cellular survival, synaptic plasticity, and calcium homeostasis (Mattson et al., 2008). Nevertheless, reactive oxygen species (ROS) are produced as byproducts of energy production, and their accumulation affects mitochondria. In the pathophysiology of AD, accumulating Aβ substantially generates excessive ROS and damages mitochondria greatly. The most common method for removing damaged mitochondria is mitophagy. One well-researched mechanism that mediates mitochondrial membrane potential (MMP)-dependent mitophagy is the PINK1-Parkin pathway (Chu, 2019).

As mentioned earlier, one of the simple hallmarks of AD is the impairment of mitochondria's function. Autopsy studies have demonstrated that AD patients detected relatively fewer levels of hippocampus mitophagy (Di Meco, et al., 2020). However, it has also been proposed that sirt-PGC1α dysfunction's role in mitophagy may be a landmark of AD pathology; mitochondrial dysfunction is the first pathogenic process in AD, and the first area of the brain affected by AD is the entorhinal cortex (Kobro-Flatmoen et al., 2021). The same observations were made in neurons derived artificially from iPSCs originating from AD-caught patients and AD mice (Fang et al., 2019; Lou et al., 2020), which means that mitophagy activation can alleviate AD development. As for the therapy, the process of mitophagy, the removal of defective mitochondria, appears to be a potential AD-targeted therapeutic approach, although there is still some ambiguity regarding the connection between mitophagy and AD pathogenesis.

11.3 REGULATION OF AUTOPHAGY-LYSOSOME PATHWAY IN AD

11.3.1 PI3K/Akt/mTOR Pathway

One important regulatory autophagic signaling pathway, PI3K/Akt/mTOR, mostly occurs in response to oxidative stress (OS), infection, and oncostasis (Yin et al., 2018).

Tang et al. (2015) state that aberrant mTOR activation in neuronal cells significantly affects both the initiation of tau secretion and the generation of p-tau. Studies revealed that age-associated cognitive decline is highly influenced by this pathway (Di Domenico et al., 2018). Furthermore, both the hypoactivation and hyperactivation of this pathway are also linked to AD disease conditions. In a clinical study, the data from AD patients, including moderate and late-stage, showed a decline in autophagy and hyperactivation of the PI3K/Akt/mTOR pathway, indicating that autophagy induction may protect neurons and alleviate AD (Tramutola et al., 2015).

11.3.2 AMPK/ULK1 Pathway

The AMPK/ULK1 pathway is one of the important regulatory pathways in which ULK1 serves as a critical initiation complex that integrates nutrient sensing with the induction of autophagy in mammalian cells (Wong et al., 2013). Upon phosphorylation of the downstream ULK1 complex, activated AMPK initiates the autophagic process. Additionally, ULK1 mitigates the inhibitory role of mTORC1, thereby facilitating mTOR signaling that is dependent on autophagy. Current research using APP/PS1 transgenic mice has shown that ALP-related proteins are upregulated, mTOR/S6K signaling is suppressed, and AMPK is activated. On the other hand, transient receptor potential mucolipin-1 (TRPML1) overexpression lowers the amounts of proteins linked to ALP. Additionally, the alteration of the autophagic process via the AMPK/mTOR signaling pathway improves cognition and prevents neuronal death (Zhang, Fang et al., 2017). Some drugs have been reported to be useful in reducing cognitive impairments and removing protein aggregates by increasing AMPK/ULK1-dependent autophagy activation (Fan et al., 2017). Selenomethionine is reported to enhance autophagic flux through AMPK/mTOR signaling in primary 3×Tg neurons and 3×Tg-AD animal models, resulting in the reduction of phosphorylated tau accumulation and cognitive decline (Zhang, Wu et al., 2017).

11.3.3 Bcl-2/Beclin-1 Pathway

Targeting the Bcl-2/Beclin-1 signaling pathway is another key strategy to regulate the autophagic process in AD. A variety of cellular functions, such as phagocytosis, endocytosis, apoptosis, and autophagy, depend on this pathway (McKnight & Zhenyu, 2013). According to recent research, Beclin-1 overexpression may reduce neurological etiology by encouraging autophagy. According to Swaminathan et al. (2016), Beclin-1 affected surface APP internalization and sorting mostly into endolysosomes, decreased the expulsion of APP metabolites, and aided in the lysosomal destruction of surface APP. The BH3 domain of Beclin-1 can be bound by the Bcl-2, which prevents the Beclin-1/PIK3C3 complex from forming and hence inhibits autophagy. It has been demonstrated that Bcl-2 overexpression inhibits autophagy brought on by several pro-autophagic cues, while Bcl-2 knockdown enhances autophagic cell death and heightened cellular vulnerability to amino acid deprivation (Robert et al., 2012).

According to a recent study, animals with the point mutation F121A knocked in the Beclin-1 showed intrinsic induction of autophagy as a result of the disruption of the interaction between Beclin-1 and Bcl-2. Beclin-1 F121A-mediated autophagy reduced cognitive decline and the progression of AD by dramatically suppressing Aβ accumulation (Rocchi et al., 2017). Thus, a key checkpoint for regulating autophagy in AD is the equilibrium between Bcl-2 and Beclin-1.

11.3.4 Transcription Factor EB (TFEB)

Settembre et al. (2011) reported that TFEB is a transcription factor that controls the expression of genes linked to autophagy, including those involved in phagophore initiation, autophagosome formation, autolysosome fusion, and substrate degradation. S211A mutation promotes TFEB nuclear localization, whereas mTORC1 controls TFEB phosphorylation (Vega-Rubin-de-Celis et al., 2017). It has been shown that mTORC1 phosphorylates TFEB and that stress or mTOR inhibition through genetic or pharmacological means causes TFEB to dephosphorylate and go to the nucleus, where it stimulates autophagy and gene expression (Eun et al., 2018).

TFEB is a novel therapeutic target for regulating autophagy. Extensive evidence suggested that TFEB activity induction could facilitate intracellular clearance, a popular treatment strategy for neurodegenerative diseases. In another study, Song et al. (2016) reported that a synthetic curcumin analogue selectively binds to TFEB at its N-terminal and stimulates TFEB nuclear translocation without preventing mTOR activity, which leads to autophagy and lysosome biogenesis. Another study showed that flubendazole inhibited mTOR activation and promoted autophagy induction and TFEB nuclear translocation (Chauhan et al., 2015).

11.4 CURRENT THERAPEUTICS FOR AD PATHOGENESIS

There are currently no effective medications that are capable of preventing or delaying the onset of AD. Donepezil, rivastigmine, galantamine, and memantine are the currently approved medications that the FDA has authorized for clinical use. Among these, Donepezil, Rivastigmine, and Galantamine are acetylcholinesterase inhibitors (AChEIs), preventing acetylcholine from being broken down in the synaptic gap by blocking the action of acetylcholinesterase (AChE). An NMDA receptor antagonist, memantine, could inhibit neuronal apoptosis and neurotoxicity of excitatory amino acids involved in synaptic cleavage (Long & Holtzman, 2019; Peng et al., 2023).

These medications have very limited effectiveness and range from patient to patient. Notably, the US FDA has declared that Aducanumab (brand name 'Aduhelm'), a monoclonal antibody that targets accumulated amyloid β protein, has been approved as a novel treatment for AD. Additionally, the FDA recently approved Lecanemab (BAN2401), an IgG1 monoclonal antibody that demonstrated reduced amyloid markers in patients with early AD and was well tolerated during the trial (van Dyck et al., 2023). Another medication, Donanemab (Kisunla™), was approved by the FDA in 2024. It is an immunotherapy medication that trains the body's immune system to recognize and eliminate amyloid, a protein that accumulates in AD patients' brains (Sims et al., 2023).

There are several medications in clinical trials for AD, but regrettably, many of them have been discontinued due to potential adverse effects or poor efficacy. Ongoing clinical trials so far have primarily focused on two pathological characteristics of AD: tau protein, targeting the mitochondria, amyloid plaques (Aβ), and multiple targets. The new perspectives offer feasible strategies for treating AD in the future.

11.5 AUTOPHAGY-STIMULATING STRATEGIES FOR AD THERAPEUTICS

Since drugs that target tau and amyloid have failed so badly, researchers are looking at alternative strategies to treat AD (Long & Holtzman, 2019). There is a compromise being developed on interventions associated with autophagy. Growing research suggests that autophagy induction may be the best way to improve the breakdown of misfolded proteins and damaged organelles in AD treatment. Since autophagy is a highly dynamic, conserved, and intricately controlled cellular event, it may be possible to stimulate it at different levels, which would enable the development of agonists or antagonists that target different pharmacological targets.

Although gene therapy and immunotherapy are promising approaches for treating AD, small molecules are still preferred because of their ease of blood-brain barrier crossing. There have been hundreds of drugs evaluated clinically against AD in recent years (Cummings et al., 2021). Multiple strategies that target the induction of autophagy through small molecules demonstrated to have neuroprotective effects in animal models of AD. These include activating TFEB, inducing autophagy via modifying the upstream kinases such as mTOR and AMPK, and targeting autophagy elements, lysosomes, and other targets to enhance autophagy. Here we selected multiple drugs that increase autophagy, which have been tested in animal models of AD or are not yet approved for use in different phase clinical trials.

11.5.1 Small Molecule Autophagy Enhancers Targeting the Autophagy

11.5.1.1 mTOR Inhibitors

One of the most proven strategies for preventing neurodegeneration, including AD, is to suppress mTORC1 to trigger autophagy initiated by rapamycin. Rapamycin and its derivatives are well-known mTOR inhibitors as well as macrolide drugs. It has been shown that rapamycin treatment improves cognitive performance and reduces tauopathies and Aβ aggregation in AD mice models. Rapamycin activates autophagy via binding the cytosolic protein FKBP1A/FKBP12. It has been demonstrated that rapamycin treatment improved memory problems in a variety of AD animal models, such as 3×Tg, P301S MAPT, and hAPP (J20) mice, and decreased Ab and MAPT levels. However, glucose intolerance and hyperlipidemia are among the adverse consequences of long-term rapamycin treatment with rapamycin, which may be brought on by the drug's inhibition of mTORC2 (Zhang et al., 2022).

FDA-approved mTOR inhibitors such as temsirolimus and everolimus have shown potential in animal models of AD. In the 3×Tg transgenic AD mouse model, everolimus improves cognitive deficit by lowering the brain mTOR and tau and Aβ levels (Cassano et al., 2019). On the other hand, temsirolimus modulates autophagy in mice's brains by inhibiting mTOR (Frederick et al., 2015). Latrepirdine, an antihistamine medication, can ameliorate Aβ neuropathology, cognitive impairment, and autophagy deficit in the TgCRND8 transgenic AD mouse model (Steele & Gandy, 2013). A potent natural compound, curcumin, modulates the autophagy process via downregulation of PI3K/Akt/mTOR and blocking the Aβ formation (Wang et al., 2014).

11.5.1.2 AMPK Activators

Metformin, a potent AMPK activator, exerts neuroprotective effects in different AD mouse models (Boccardi et al., 2019) and ameliorates memory decline by reducing the levels of the Aβ plaque load and tau (Farr et al., 2019; Ou et al., 2018). Resveratrol, a potent natural compound, induces autophagy by activating AMPK and reducing Aβ levels (Vingtdeux et al., 2010). Berberine, another potent natural compound, increases the breakdown of Aβ, APP, and aggregated MAPT levels in 3×Tg mice and modulates autophagy in AD (Chen et al., 2020; Huang et al., 2017). In a clinical study, *Radix Polygalae* induces autophagy via activating the AMPK/mTOR pathway (Zhao et al., 2015). The aforementioned findings indicated that AMPK activators can act as potent autophagy enhancers in AD pathogenesis.

11.5.1.3 TFEB Activators

The natural disaccharide trehalose stimulates autophagy through mobilization of TFEB (Jeong et al., 2021) and ameliorates AD pathogenesis by decreasing Aβ (Portbury et al., 2017). A curcumin analogue, C1, a synthetic curcuminoid, activates lysosome formation and TFEB-mediated autophagy without mTOR inhibition (Song et al., 2016).

Several small molecules (natural and synthetic), such as cinnamic acid, gemfibrozil, aspirin, and Wy14643, are identified to be effective TFEB activators that promote autophagy and lysosome biogenesis through the stimulation of PPARα. These molecules also demonstrated memory-enhancing effects by reducing Aβ pathology in AD animal models (Chandra et al., 2018, 2019; Luo et al., 2020). Fisetin and flubendazole are two potent small-molecule TFEB activators that exert a protective effect in AD cell lines (Chauhan et al., 2015; Kim et al., 2016).

11.5.1.4 Restoration of Lysosomal Functions

As AD pathogenesis has been associated with lysosome dysfunction, recent interest in actively targeting lysosomal restoration has emerged as a promising therapy strategy for this condition. Based on PLGA-aNP's ability to effectively reduce the lysosomal pH and increase lysosomal activity in multiple AD cell lines, Bourdenx et al. (2016) observed that PLGA-aNP has a potential downregulatory effect on cathepsin B activation in AD. In another study, PLGA-aNP upregulates lysosomal function and protects against Aβ-induced neuronal toxicity, thus underscoring its potential therapeutic role in AD (Wang et al., 2020).

A recent study suggested that β2-adrenergic agonists (e.g., isoproterenol) restore CLCN7/ClC-7 expression and acidify lysosomes in PSEN1-null cells. These findings suggest that restoring lysosomal acidification deficiencies might be a therapeutic approach for AD and other autophagy-related neurodegenerative diseases (Lee et al., 2020).

11.5.1.5 Other Autophagy Enhancers

Besides key regulating signaling cascades such as mTOR and AMPK for autophagy activation, some other small molecules have been identified as autophagy inducers independent of mTOR and AMPK. Among them, lithium chloride and carbamazepine are small molecules that effectively induce autophagy independently in AD animal models (Li et al., 2013; Pan et al., 2018).

11.5.2 Compounds Targeting Mitophagy

There have been considerable efforts toward discovering compounds that are capable of inducing mitophagy. Here, we focus on one such molecule known as nicotinamide, which is used as an anti-aging agent and has been reported to promote mitophagy. Many publications mentioned that the NMU's precursor nicotinamide and its analogues – nicotinamide riboside (NR) and nicotinamide mononucleotide (NMN) – cause mitophagy and protect against aging (Zhang et al., 2021). Recently, several plant-derived compounds have shown promising results in regulating mitophagy and potentially improving cognitive function in preclinical studies (Table 11.1).

TABLE 11.1
Modulatory Effects of Various Plant-Derived Compounds on Mitophagy in the AD Treatment

Natural Compound	AD Model	Daily Dose, Treatment Period	Finding of the Study	References
Resveratrol	Aβ1–42-stimulated PC12 Cells – *in vitro*	3 μM, 24 h	• Lowering oxidative state and apoptosis, • Improve mitophagy	Wang et al. (2018)
Palmatine	PS2APP (mouse) – *in vivo*	10 mg/kg (daily), 28 days, i.p.	• Induce mitophagy, • Stimulate mitochondrial biogenesis	Lee et al. (2023)
Berberine	APP/PS1 (mouse) – *in vivo*	10 mg/kg (daily), 30 days, orally	• Prevent mitophagy dysfunction	Wang et al. (2023)
Salidroside	5×FAD (transgenic mouse) – *in vivo*	1.5 mg/kg (daily), 12 days, i.p.	• Ameliorate mitochondrial dysfunction • Facilitate mitophagy	Yao et al. (2022)
Loganin	3×FAD (transgenicmouse) – *in vivo*	40 mg/kg (daily), 21 days, i.p.	• Modulate mitophagy	Zhou et al. (2023)
Honokiol	APP/PS1 (mouse) – *in vivo*	20 mg/kg (daily), 60 days, i.p.	• Trigger mitophagy	Hou et al. (2022)
PNS	SAMP8 (mouse) – *In vivo*	100 mg/kg (daily), 28 days, i.p.	• Improve oxidative damage • Facilitate mitophagy	Yang et al. (2024)
TSG	LPS stimulated (i) BV2, (ii) N2a, and (iii) SH-SY5Y cells – *in vitro*	10 nM–10 μM, 24 h	• Enhance mitophagy, • Antiinflammation	Gao et al. (2020)
Isoquercitrin	Yeast model – (i) K6001 and (ii) YOM36 – *in vitro*	0.3–100 μM	• Induce mitophagy	Liu et al. (2023)
Luteolin	Glutamate-induced toxicity in neuronal cells (HT-22) – *in vitro*	5–50 μM, 24 h	• Modulate mitophagy	Vongthip et al. (2024)
ALA	Aβ1–42-stimulated SHSY5Y cells – *in vitro*	10–250 nM, 24 h	• Modulate mitophagy, • Induce antiinflammation	Litwiniuk et al. (2020)
ADA	APP/PS1 (mouse) – *in vivo*	1–5 mg/kg (daily), 21 days, i.p.	• Repair mitophagy defects	Zhou et al. (2024)

PNS: *Panax notoginseng* saponins; TSG: Tetrahydroxy stilbene glycoside; ALA: Alpha-linolenic acid; ADA: 3, 14, 19-Triacetyl Andrographolide

11.5.3 Combination Therapy

Theoretically, combining two pharmacological autophagy inducers that operate through distinct regulatory pathways is a powerful strategy for maximizing therapeutic benefits. For instance, mTOR-independent autophagy enhancers like 'trehalose' or 'lithium' combined with the mTOR-dependent enhancer rapamycin greatly increase autophagy and accelerate the clearance of protein aggregates compared to using each treatment alone. Furthermore, a combination of two medications allows for lower dosages of each, enhancing effectiveness while significantly reducing the risk of adverse effects. In this scenario, a strategy that boosts autophagy induction while ensuring effective autophagic degradation could be a highly effective intervention (Stacchiotti & Corsetti, 2020).

11.6 CLINICAL EVALUATIONS OF AUTOPHAGY STIMULATORS IN AD MANAGEMENT

Many small-molecule autophagy activators are being tested for their effectiveness in AD patients. For instance, lithium, a well-researched autophagy enhancer, shows promise in improving cognitive deficiency in patients with amnestic MCI (NCT01055392) (Forlenza et al., 2019). Metformin, another classical anti-diabetic drug, is a safe and well-tolerated agent that mitigates certain cognitive declines in AD patients (NCT04098666) (Luchsinger et al., 2016). Rapamycin has been lengthily studied for its capacity to induce autophagy and its neuroprotective effects in various animal models of AD. Recently, the clinical trial has begun to evaluate its safety and feasibility for patients in the early stages of AD (NCT04629495). A list of clinical trials of autophagy enhancers in AD patients is highlighted in Table 11.2.

11.7 CONCLUDING REMARKS AND FUTURE PERSPECTIVES

AD is a complicated multifactorial disease, and the development of AD therapeutics has presented many frustrating challenges for researchers. Autophagy is an important process in the pathogenesis of AD and has essential functions regulating the generation and degradation of core histopathologic proteins, the Aβ and p-tau proteins. The concept is that more knowledge regarding how the autophagy pathway is changed in AD should allow investigators to develop specific and effective therapies. Various AD-associated genes, such as PSEN1, PICALM, CLU, and TREM2, modulate autophagic flux. Also, more than 30 ATG genes are directly involved in the formation and maturation of the autophagosome.

Indeed, it has been found that certain autophagy enhancers can improve cognition and decrease the progression of AD, at least in animal models. Nevertheless, there was little or no proof of the beneficial effects of autophagy stimulators in AD patients. Established upstream regulators of autophagic signaling such as mTORC1, TFEB, and AMPK may act at multiple levels of the whole autophagy pathway and, therefore, may serve as a target to enhance flux and, to some extent, overall protein degradation. While many substances targeting these pathways, like rapamycin, are used in human treatment, the role of the targets of pleiotropic pathways, like mTOR and AMPK, may have undesired effects when inhibited.

Some extremely safe and/or FDA-approved candidates have been found through clinical trials to manage conditions except for AD. Hence, it should be possible to enroll more participants in clinical trials and to try to determine the efficacy of potentially useful interventions on participants at different AD stages among the recruited population. Furthermore, given the complications of AD, a

TABLE 11.2
List of the Completed/Ongoing Clinical Trials on Autophagy Enhancers for the Management of AD

Drug Candidate	NCT Number	Phase	Year	Sample Size	Results
Rapamycin	NCT04629495	Phase II	2021 and 2023	40	Not applicable
Metformin	NCT01965756	Phase II	2013 and 2015	20	• Safe and well-tolerated • Improve learning, memory, and attention.
Lithium	NCT01055392	Phase II	2007 and 2011	61	• Improve cognitive function • Prevent functional decline (duration 24 months)
Lithium	NCT02129348	Phase II	2014 and 2020	77	Not applicable
Trehalose	NCT04663854	Phase I	2020 and 2022	20	Not applicable
Lithium	NCT03185208	Phase IV	2017 and 2023	80	Not applicable
Resveratrol	NCT01504854	Phase III	2008 and 2010	39	Safe and well tolerated by AD patients
Latrepirdine	NCT00838110	Phase III	2009 and 2010	742	Not effective in AD patients
Metformin	NCT00620191	Phase II	2008 and 2012	80	Results are not promising
Metformin	NCT04098666	Phase II and Phase III	2021 and 2027	326	Not applicable

Source: Zhang et al. (2022).

single drug might not be sufficient to reduce its symptoms. Consequently, future clinical trials should consider combining multiple autophagy activators. In the future, autophagy-directed approaches might offer a novel and fascinating alternative for developing anti-AD drugs.

REFERENCES

Ballard, C., Aarsland, D., Cummings, J., O'Brien, J., Mills, R., Molinuevo, J. L., ..., & Sultana, J. (2020). Drug repositioning and repurposing for Alzheimer disease. *Nature Reviews Neurology*, *16*, 661–673.

Boccardi, V., Murasecco, I., & Mecocci, P. (2019). Diabetes drugs in the fight against Alzheimer's disease. *Ageing Research Reviews*, *54*, 100936.

Bourdenx, M., Daniel, J., Genin, E., Soria, F. N., Blanchard-Desce, M., Bezard, E., & Dehay, B. (2016). Nanoparticles restore lysosomal acidification defects: Implications for parkinson and other lysosomal-related diseases. *Autophagy*, *12*, 472e83.

Caballero, B., Bourdenx, M., Luengo, E., Diaz, A., Sohn, P. D., Chen, X., Wang, C., Juste, Y. R., Wegmann, S., Patel, B., Young, Z. T., Kuo, S. Y., Rodriguez-Navarro, J. A., Shao, H., Lopez, M. G., Karch, C. M., Goate, A. M., Gestwicki, J. E., Hyman, B. T., Gan, L., ... Cuervo, A. M. (2021). Acetylated tau inhibits chaperone-mediated autophagy and promotes tau pathology propagation in mice. *Nature Communications*, *12*, 2238.

Cassano, T., Magini, A., Giovagnoli, S., Polchi, A., Calcagnini, S., Pace, L., Lavecchia, M. A., Scuderi, C., Bronzuoli, M. R., Ruggeri, L., Gentileschi, M. P., Romano, A., Gaetani, S., De Marco, F., Emiliani, C., & Dolcetta, D. (2019). Early intrathecal infusion of everolimus restores cognitive function and mood in a murine model of Alzheimer's disease. *Experimental Neurology*, *311*, 88e105.

Chandra, S., Jana, M., & Pahan, K. (2018). Aspirin induces lysosomal biogenesis and attenuates amyloid plaque pathology in a mouse model of Alzheimer's disease via PPARa. *Journal of Neuroscience*, *38*, 6682e99.

Chandra, S., Roy, A., Jana, M., & Pahan, K. (2019). Cinnamic acid activates PPARa to stimulate lysosomal biogenesis and lower amyloid plaque pathology in an Alzheimer's disease mouse model. *Neurobiology of Disease*, *124*, 379e95.

Chauhan, S., Ahmed, Z., Bradfute, S. B., Arko-Mensah, J., Mandell, M. A., Won Choi, S., Kimura, T., Blanchet, F., Waller, A., Mudd, M. H., Jiang, S., Sklar, L., Timmins, G. S., Maphis, N., Bhaskar, K., Piguet, V., & Deretic, V. (2015). Pharmaceutical screen identifies novel target processes for activation of autophagy with a broad translational potential. *Nature Communications*, *6*, 8620.

Chen, Y., Chen, Y., Liang, Y., Chen, H., Ji, X., & Huang, M. (2020). Berberine mitigates cognitive decline in an Alzheimer's disease mouse model by targeting both tau hyperphosphorylation and autophagic clearance. *Biomedicine & Pharmacotherapy*, *121*, 109670.

Chen, Y., Fu, A. K. Y., & Ip, N. Y. (2019). Synaptic dysfunction in Alzheimer's disease: Mechanisms and therapeutic strategies. *Pharmacology & Therapeutics*, *195*, 186–198.

Chu, C. T. (2019). Mechanisms of selective autophagy and mitophagy: Implications for neurodegenerative diseases. *Neurobiology of Disease*, *122*, 23–34.

Collier, J. J., Guissart, C., Oláhová, M., Sasorith, S., Piron-Prunier, F., Suomi, F., Zhang, D., Martinez-Lopez, N., Leboucq, N., Bahr, A., Azzarello-Burri, S., Reich, S., Schöls, L., Polvikoski, T. M., Meyer, P., Larrieu, L., Schaefer, A. M., Alsaif, H. S., Alyamani, S., Zuchner, S., ... Taylor, R. W. (2021). Developmental consequences of defective ATG7-mediated autophagy in humans. *The New England Journal of Medicine*, *384*, 2406–2417.

Cummings, J., Lee, G., Zhong, K., Fonseca, J., & Taghva, K. (2021). Alzheimer's disease drug development pipeline: 2021. *Alzheimer's & Dementia*, *7*, 12179.

Di Domenico, F., Tramutola, A., Foppoli, C., Head, E., Perluigi, M., & Butterfield, D. A. (2018). mTOR in down syndrome: Role in ass and tau neuropathology and transition to Alzheimer disease-like dementia. *Free Radical Biology and Medicine*, *114*, 94–101.

Di Meco, A., Curtis, M. E., Lauretti, E., & Pratico, D. (2020). Autophagy dysfunction in Alzheimer's disease: Mechanistic insights and new therapeutic opportunities. *Biological Psychiatry*, *87*, 797–807.

Eun, S. Y., Lee, J. N., Nam, I. K., Liu, Z. Q., So, H. S., Choe, S. K., & Park, R. (2018). PEX5 regulates autophagy via the mTORC1-TFEB axis during starvation. *Experimental & Molecular Medicine*, *50*, 4.

Fan, Y., Wang, N., Rocchi, A., Zhang, W., Vassar, R., Zhou, Y., & He, C. (2017). Identification of natural products with neuronal and metabolic benefits through autophagy induction. *Autophagy*, *13*, 41–56.

Fang, E. F., Hou, Y., Palikaras, K., Adriaanse, B. A., Kerr, J. S., Yang, B., Lautrup, S., Hasan-Olive, M. M., Caponio, D., Dan, X., Rocktäschel, P., Croteau, D. L., Akbari, M., Greig, N. H., Fladby, T., Nilsen, H., Cader, M. Z., Mattson, M. P., Tavernarakis, N., & Bohr, V. A (2019). Mitophagy inhibits amyloid-beta and tau pathology and reverses cognitive deficits in models of Alzheimer's disease. *Nature Neuroscience*, *22*, 401–412.

Farr, S. A., Roesler, E., Niehoff, M. L., Roby, D. A., McKee, A., & Morley, J. E. (2019). Metformin improves learning and memory in the SAMP8 mouse model of Alzheimer's disease. *Journal of Alzheimer's Disease*, *68*, 1699e710.

Forlenza, O. V., Radanovic, M., Talib, L. L., & Gattaz, W. F. (2019). Clinical and biological effects of long-term lithium treatment in older adults with amnestic mild cognitive impairment: Randomized clinical trial. *British Journal of Psychiatry*, *215*, 668e74.

Frederick, C., Ando, K., Leroy, K., Héraud, C., Suain, V., Buée, L., & Brion, J. P. (2015). Rapamycin ester analog CCI-779/Temsirolimus alleviates tau pathology and improves motor deficit in mutant tau transgenic mice. *Journal of Alzheimer's Disease*, *44*, 1145e56.

Gao, Y., Li, J., Li, J., Hu, C., Zhang, L., Yan, J., Li, L., & Zhang, L. (2020). Tetrahydroxy stilbene glycoside alleviated inflammatory damage by mitophagy via AMPK related PINK1/Parkin signaling pathway. *Biochemical Pharmacology*, *177*, 113997.

George, A. A., Hayden, S., Stanton, G. R., & Brockerhoff, S. E. (2016). Arf6 and the 5'phosphatase of Synaptojanin 1 regulate autophagy in cone photoreceptors. *Inside Cell*, *1*, 117–133.

Glatigny, M., Moriceau, S., Rivagorda, M., Ramos-Brossier, M., Nascimbeni, A. C., Lante, F., Shanley, M. R., Boudarene, N., Rousseaud, A., Friedman, A. K., Settembre, C., Kuperwasser, N., Friedlander, G., Buisson, A., Morel, E., Codogno, P., & Oury, F. (2019). Autophagy is required for memory formation and reverses age-related memory decline. *Current Biology*, *29*(3), 435–448.e8.

Hamano, T., Gendron, T. F., Causevic, E., Yen, S. H., Lin, W. L., Isidoro, C., Deture, M., & Ko, L. W. (2016). The dynamics and turnover of tau aggregates in cultured cells: Insights into therapies for tauopathies. *Journal of Biological Chemistry, 291*, 13175–13193.

Hernandez, D., Torres, C. A., Setlik, W., Cebrián, C., Mosharov, E. V., Tang, G., Cheng, H. C., Kholodilov, N., Yarygina, O., Burke, R. E., Gershon, M., & Sulzer, D. (2008). Autophagic-lysosomal perturbation enhances tau aggregation in transfectants with induced wild-type tau expression. *European Journal of Neuroscience, 27*, 1119–1130.

Hardy, J. (2017). The discovery of Alzheimer-causing mutations in the APP gene and the formulation of the "amyloid cascade hypothesis". *The FEBS Journal, 284*, 1040–1044.

Hernandez, D., Torres, C. A., Setlik, W., Cebrián, C., Mosharov, E. V., Tang, G., Cheng, H. C., Kholodilov, N., Yarygina, O., Burke, R. E., Gershon, M., & Sulzer, D. (2012). Regulation of presynaptic neurotransmission by macroautophagy. *Neuron, 74*, 277–284.

Hou, M., Bao, W., Gao, Y., Chen, J., & Song, G. (2022). Honokiol improves cognitive impairment in APP/PS1 mice through activating mitophagy and mitochondrial unfolded protein response. *Chemico-Biological Interactions, 351*, 109741.

Huang, M., Jiang, X., Liang, Y., Liu, Q., Chen, S., & Guo, Y. (2017). Berberine improves cognitive impairment by promoting autophagic clearance and inhibiting production of b-amyloid in APP/tau/PS1 mouse model of Alzheimer's disease. *Experimental Gerontology, 91*, 25e33.

Huang, Y., Happonen, K. E., Burrola, P. G., O'Connor, C., Hah, N., Huang, L., Nimmerjahn, A., & Lemke, G. (2021). Microglia use TAM receptors to detect and engulf amyloid beta plaques. *Nature Immunology, 22*, 586–594.

Jeong, S. J., Stitham, J., Evans, T. D., Zhang, X., Rodriguez-Velez, A., Yeh, Y. S., Tao, J., Takabatake, K., Epelman, S., Lodhi, I. J., Schilling, J. D., DeBosch, B. J., Diwan, A., & Razani, B. (2021). Trehalose causes low-grade lysosomal stress to activate TFEB and the autophagy-lysosome biogenesis response. *Autophagy, 17*, 3740e52.

Kim, S., Choi, K. J., Cho, S. J., Yun, S. M., Jeon, J. P., Koh, Y. H., Song, J., Johnson, G. V., & Jo, C. (2016). Fisetin stimulates autophagic degradation of phosphorylated tau via the activation of TFEB and Nrf2 transcription factors. *Scientific Reports, 6*, 24933.

Kirkin, V., & Rogov, V. V. (2019). A diversity of selective autophagy receptors determines the specificity of the autophagy pathway. *Molecular Cell, 76*, 268–285.

Kobro-Flatmoen, A., Lagartos-Donate, M. J., Aman, Y., Edison, P., Witter, M. P., & Fang, E. F. (2021). Re-emphasizing early Alzheimer's disease pathology starting in select entorhinal neurons, with a special focus on mitophagy. *Ageing Research Reviews, 67*, 101307.

Lee, D. Y., Lee, K. M., Um, J. H., Kim, Y. Y., Kim, D. H., & Yun, J. (2023). The natural alkaloid palmatine selectively induces mitophagy and restores mitochondrial function in an Alzheimer's disease mouse model. *International Journal of Molecular Sciences, 24*(22), 16542.

Lee, J. H., Wolfe, D. M., Darji, S., McBrayer, M. K., Colacurcio, D. J., Kumar, A., Stavrides, P., Mohan, P. S., & Nixon, R. A. (2020). β2-Adrenergic agonists rescue lysosome acidification and function in PSEN1 deficiency by reversing defective ER-to-lysosome delivery of ClC-7. *Journal of Molecular Biology, 432*, 2633e50.

Li, L., Zhang, S., Zhang, X., Li, T., Tang, Y., Liu, H., Yang, W., & Le, W. (2013). Autophagy enhancer carbamazepine alleviates memory deficits and cerebral amyloid-b pathology in a mouse model of Alzheimer's disease. *Current Alzheimer Research, 10*, 433e41.

Lieberman, O. J., & Sulzer, D. (2020). The synaptic autophagy cycle. *Journal of Molecular Biology, 432*, 2589–2604.

Litwiniuk, A., Domańska, A., Chmielowska, M., Martyńska, L., Bik, W., & Kalisz, M. (2020). The effects of alpha-linolenic acid on the secretory activity of astrocytes and β amyloid-associated neurodegeneration in differentiated SH-SY5Y cells: Alpha-linolenic acid protects the SH-SY5Y cells against β amyloid toxicity. *Oxidative Medicine and Cellular Longevity, 2020*, 8908901.

Liu, Y., Shen, L., Matsuura, A., Xiang, L., & Qi, J. (2023). Isoquercitrin from *Apocynumvenetum* L. exerts antiaging effects on yeasts via stress resistance improvement and mitophagy induction through the Sch9/Rim15/Msn signaling pathway. *Antioxidants (Basel), 12*(11), 1939.

Long, J. M., & Holtzman, D. M. (2019). Alzheimer disease: an update on pathobiology and treatment strategies. *Cell, 179*, 312–339.

Lopez, O. L., & Kuller, L. H. (2019). Epidemiology of aging and associated cognitive disorders: Prevalence and incidence of Alzheimer's disease and other dementias. *Handbook of Clinical Neurology, 167*, 139–148.

Lou, G., Palikaras, K., Lautrup, S., Scheibye-Knudsen, M., Tavernarakis, N., & Fang, E. F. (2020). Mitophagy and neuroprotection. *Trends in Molecular Medicine, 26*, 8–20.

Luchsinger, J. A., Perez, T., Chang, H., Mehta, P., Steffener, J., Pradabhan, G., Ichise, M., Manly, J., Devanand, D. P., & Bagiella, E. (2016). Metformin in amnestic mild cognitive impairment: Results of a pilot randomized placebo-controlled clinical trial. *Journal of Alzheimer's Disease, 51*, 501e14.

Luo, R., Su, L. Y., Li, G., Yang, J., Liu, Q., Yang, L. X., Zhang, D. F., Zhou, H., Xu, M., Fan, Y., Li, J., & Yao, Y. G. (2020). Activation of PPARA-mediated autophagy reduces Alzheimer disease-like pathology and cognitive decline in a murine model. *Autophagy, 16*, 52e69.

Majumder, S., Richardson, A., Strong, R., & Oddo, S. (2011). Inducing autophagy by rapamycin before, but not after, the formation of plaques and tangles ameliorates cognitive deficits. *PloS One, 6*, 25416.

Malampati, S., Song, J. X., Chun-Kit Tong, B., Nalluri, A., Yang, C. B., Wang, Z., Gopalkrishnashetty Sreenivasmurthy, S., Zhu, Z., Liu, J., Su, C., Krishnamoorthi, S., Iyaswamy, A., Cheung, K. H., Lu, J. H., & Li, A. M. (2020). Targeting aggrephagy for the treatment of Alzheimer's disease. *Cells, 9*, 9.

Mattson, M. P., Gleichmann, M., & Cheng, A. (2008). Mitochondria in neuroplasticity and neurological disorders. *Neuron, 60*, 748–766.

McKnight, N. C., & Zhenyu, Y. (2013). Beclin 1, an essential component and master regulator of PI3K-III in health and disease. *Current Pathobiology Reports, 1*, 231–238.

Mueller-Steiner, S., Zhou, Y., Arai, H., Roberson, E. D., Sun, B., Chen, J., Wang, X., Yu, G., Esposito, L., Mucke, L., & Gan, L. (2006). Antiamyloidogenic and neuroprotective functions of cathepsin B: Implications for Alzheimer's disease. *Neuron, 51*, 703–714.

Ou, Z., Kong, X., Sun, X., He, X., Zhang, L., Gong, Z., Huang, J., Xu, B., Long, D., Li, J., Li, Q., Xu, L., & Xuan, A. (2018). Metformin treatment prevents amyloid plaque deposition and memory impairment in APP/PS1 mice. *Brain, Behavior, and Immunity, 69*, 351e63.

Pan, Y., Short, J. L., Newman, S. A., Choy, K. H. C., Tiwari, D., Yap, C., Senyschyn, D., Banks, W. A., & Nicolazzo, J. A. (2018). Cognitive benefits of lithium chloride in APP/PS1 mice are associated with enhanced brain clearance of b-amyloid. *Brain, Behavior, and Immunity*, *70*, 36e47.

Peng, Y., Jin, H., Xue, Y. H., Chen, Q., Yao, S. Y., Du, M. Q., & Liu, S. (2023). Current and future therapeutic strategies for Alzheimer's disease: An overview of drug development bottlenecks. *Frontiers in Aging Neuroscience*, *15*, 1206572.

Piras, A., Collin, L., Gruninger, F., Graff, C., & Ronnback, A. (2016). Autophagic and lysosomal defects in human tauopathies: Analysis of post-mortem brain from patients with familial Alzheimer disease, corticobasal degeneration and progressive supranuclear palsy. *Acta Neuropathologica Communications*, *4*, 22.

Portbury, S. D., Hare, D. J., Sgambelloni, C., Perronnes, K., Portbury, A. J., Finkelstein, D. I., & Adlard, P. A. (2017). Trehalose improves cognition in the transgenic Tg2576 mouse model of Alzheimer's disease. *Journal of Alzheimer's Disease*, *60*, 549e60.

Querfurth, H. W., & LaFerla, F. M. (2010). Alzheimer's disease. *The New England Journal of Medicine*, *362*, 329–344.

Robert, G., Gastaldi, C., Puissant, A., Hamouda, A., Jacquel, A., Dufies, M., Belhacene, N., Colosetti, P., Reed, J. C., Auberger, P., & Luciano, F. (2012). The anti-apoptotic Bcl-B protein inhibits BECN1-dependent autophagic cell death. *Autophagy*, *8*, 637–649.

Rocchi, A., Yamamoto, S., Ting, T., Fan, Y., Sadleir, K., Wang, Y., Zhang, W., Huang, S., Levine, B., Vassar, R., & He, C. (2017). A Becn1 mutation mediates hyperactive autophagic sequestration of amyloid oligomers and improved cognition in Alzheimer's disease. *PLoS Genetics*, *13*, e1006962.

Settembre, C., Di Malta, C., Polito, V. A., Garcia Arencibia, M., Vetrini, F., Erdin, S., Erdin, S. U., Huynh, T., Medina, D., Colella, P., Sardiello, M., Rubinsztein, D. C., & Ballabio, A. (2011). TFEB links autophagy to lysosomal biogenesis. *Science*, *332*, 1429–1433.

Silva, M. C., Nandi, G. A., Tentarelli, S., Gurrell, I. K., Jamier, T., Lucente, D., Dickerson, B. C., Brown, D. G., Brandon, N. J., & Haggarty, S. J. (2020). Prolonged tau clearance and stress vulnerability rescue by pharmacological activation of autophagy in tauopathy neurons. *Nature Communications*, *11*, 3258.

Sims, J. R., Zimmer, J. A., Evans, C. D., Lu, M., Ardayfio, P., Sparks, J., Wessels, A. M., Shcherbinin, S., Wang, H., Monkul Nery, E. S., Collins, E. C., Solomon, P., Salloway, S., Apostolova, L. G., Hansson, O., Ritchie, C., Brooks, D. A., Mintun, M., Skovronsky, D. M., & TRAILBLAZER-ALZ 2 Investigators (2023). Donanemab in early symptomatic Alzheimer disease: The TRAILBLAZER-ALZ 2 randomized clinical trial. *JAMA*, *330*(6), 512–527.

Song, J. X., Sun, Y. R., Peluso, I., Zeng, Y., Yu, X., Lu, J. H., Xu, Z., Wang, M. Z., Liu, L. F., Huang, Y. Y., Chen, L. L., Durairajan, S. S., Zhang, H. J., Zhou, B., Zhang, H. Q., Lu, A., Ballabio, A., Medina, D. L., Guo, Z., & Li, M. (2016). A novel curcumin analog binds to and activates TFEB *in vitro* and *in vivo* independent of mTOR inhibition. *Autophagy*, *12*, 1372c89.

Stacchiotti, A.,& Corsetti, G. (2020). Natural Compounds and Autophagy: Allies Against Neurodegeneration. *Frontiers in Cell and Developmental Biology*, 8, 555409. doi: 10.3389/fcell.2020.555409.

Steele, J. W., & Gandy, S. (2013). Latrepirdine (Dimebon), a potential Alzheimer therapeutic, regulates autophagy and neuropathology in an Alzheimer mouse model. *Autophagy*, *9*, 617e8.

Sun, Q., Xie, N., Tang, B., Li, R., & Shen, Y. (2017). Alzheimer's disease: From genetic variants to the distinct pathological mechanisms. *Frontiers in Molecular Neuroscience*, *10*, 319.

Swaminathan, G., Zhu, W., & Plowey, E. D. (2016). BECN1/Beclin 1 sorts cell-surface APP/amyloid beta precursor protein for lysosomal degradation. *Autophagy*, *12*, 2404–2419.

Tang, Z., Ioja, E., Bereczki, E., Hultenby, K., Li, C., Guan, Z., Winblad, B., & Pei, J. J. (2015). mTOR mediates tau localization and secretion: implication for Alzheimer's disease. *Biochimica et Biophysica Acta*, *1853*, 1646–1657.

Tramutola, A., Triplett, J. C., Di Domenico, F., Niedowicz, D. M., Murphy, M. P., Coccia, R., Perluigi, M., & Butterfield, D. A. (2015). Alteration of mTOR signaling occurs early in the progression of Alzheimer's disease (AD): Analysis of brain from subjects with preclinical AD, amnestic mild cognitive impairment, and late-stage AD. *Journal of Neurochemistry*, *133*, 739–749.

van Dyck, C. H., Swanson, C. J., Aisen, P., Bateman, R. J., Chen, C., Gee, M., Kanekiyo, M., Li, D., Reyderman, L., Cohen, S., Froelich, L., Katayama, S., Sabbagh, M., Vellas, B., Watson, D., Dhadda, S., Irizarry, M., Kramer, L. D., & Iwatsubo, T. (2023). Lecanemab in early Alzheimer's disease. *The New England Journal of Medicine*, *388*, 9–21.

Vega-Rubin-de-Celis, S., Pena-Llopis, S., Konda, M., & Brugarolas, J. (2017). Multistep regulation of TFEB by mTORC1. *Autophagy*, *13*, 464–472.

Vingtdeux, V., Giliberto, L., Zhao, H., Chandakkar, P., Wu, Q., Simon, J. E., Janle, E. M., Lobo, J., Ferruzzi, M. G., Davies, P., & Marambaud, P. (2010). AMP-activated protein kinase signaling activation by resveratrol modulates amyloid-beta peptide metabolism. *Journal of Biological Chemistry*, *285*, 9100e13.

Vongthip, W., Nilkhet, S., Boonruang, K., Sukprasansap, M., Tencomnao, T., & Baek, S. J. (2024). Neuroprotective mechanisms of luteolin in glutamate-induced oxidative stress and autophagy-mediated neuronal cell death. *Scientific Reports*, *14*(1), 7707.

Wang, C., Zhang, X., Teng, Z., Zhang, T., & Li, Y. (2014). Downregulation of PI3K/Akt/mTOR signaling pathway in curcumin-induced autophagy in APP/PS1 double transgenic mice. *European Journal of Pharmacology*, *740*, 312–20.

Wang, C., Zou, Q., Pu, Y., Cai, Z., & Tang, Y. (2023). Berberine rescues D-ribose-induced Alzheimer's pathology via promoting mitophagy. *International Journal of Molecular Sciences*, *24*(6), 5896.

Wang, H., Jiang, T., Li, W., Gao, N., & Zhang, T. (2018). Resveratrol attenuates oxidative damage through activating mitophagy in an *in vitro* model of Alzheimer's disease. *Toxicology Letters*, *282*, 100–108.

Wang, H., Wang, R., Carrera, I., Xu, S., & Lakshmana, M. K. (2016). TFEB overexpression in the P301S model of tauopathy mitigates increased PHF1 levels and lipofuscin puncta and rescues memory deficits. *eNeuro*, *3*(2), ENEURO.0042-16.2016.

Wang, Y., Wu, Q., Anand, B. G., Karthivashan, G., Phukan, G., Yang, J., Thinakaran, G., Westaway, D., & Kar, S. (2020). Significance of cytosolic cathepsin D in Alzheimer's disease pathology: protective cellular effects of PLGA nanoparticles against beta-amyloid toxicity. *Neuropathology and Applied Neurobiology*, *46*, 686e706.

Wilson, B., & Geetha, K. M. (2020). Neurotherapeutic applications of nanomedicine for treating Alzheimer's disease. *Journal of Controlled Release*, *325*, 25–37.

Wong, P. M., Puente, C., Ganley, I. G., & Jiang, X. (2013). The ULK1 complex: Sensing nutrient signals for autophagy activation. *Autophagy*, *9*, 124–137.

Yang, J., Zhao, H., & Qu, S. (2024). Phytochemicals targeting mitophagy: Therapeutic opportunities and prospects for treating Alzheimer's disease. *Biomedicine & Pharmacotherapy, 177*, 117144.

Yang, Y., Chen, W., Lin, Z., Wu, Y., Li, Y., & Xia, X. (2024). *Panax notoginseng* saponins prevent dementia and oxidative stress in brains of SAMP8 mice by enhancing mitophagy. *BMC Complementary Medicine and Therapies, 24*(1), 144.

Yao, Y., Ren, Z., Yang, R., Mei, Y., Dai, Y., Cheng, Q., Xu, C., Xu, X., Wang, S., Kim, K. M., Noh, J. H., Zhu, J., Zhao, N., Liu, Y. U., Mao, G., & Sima, J. (2022). Salidroside reduces neuropathology in Alzheimer's disease models by targeting NRF2/SIRT3 pathway. *Cell & Bioscience, 12*(1), 180.

Yin, Y., Dang, W., Zhou, X., Xu, L., Wang, W., Cao, W., Chen, S., Su, J., Cai, X., Xiao, S., Peppelenbosch, M. P., & Pan, Q. (2018). PI3K-Akt-mTOR axis sustains rotavirus infection via the 4E-BP1 mediated autophagy pathway and represents an antiviral target. *Virulence, 9*, 83–98.

Zhang, L., Fang, Y., Cheng, X., Lian, Y., Xu, H., Zeng, Z., & Zhu, H. (2017). TRPML1 participates in the progression of Alzheimer's disease by regulating the PPARgamma/AMPK/mTOR signalling pathway. *Cellular Physiology and Biochemistry, 43*, 2446–2456.

Zhang, W., Xu, C., Sun, J., Shen, H. M., Wang, J., & Yang, C. (2022). Impairment of the autophagy-lysosomal pathway in Alzheimer's diseases: Pathogenic mechanisms and therapeutic potential. *Acta Pharmaceutica Sinica B, 12*(3), 1019–1040.

Zhang, Z. H., Wu, Q. Y., Zheng, R., Chen, C., Chen, Y., Liu, Q., Hoffmann, P. R., Ni, J. Z., & Song, G. L (2017). Selenomethionine mitigates cognitive decline by targeting both tau hyperphosphorylation and autophagic clearance in an Alzheimer's disease mouse model. *Journal of Neuroscience, 37*, 2449–2462.

Zhang, Z., Yang, X., Song, Y. Q., & Tu, J. (2021). Autophagy in Alzheimer's disease pathogenesis: Therapeutic potential and future perspectives. *Ageing Research Reviews, 72*, 101464.

Zhao, H., Wang, Z. C., Wang, K. F., & Chen, X. Y. (2015). Aβ peptide secretion is reduced by Radix Polygalae-induced autophagy via activation of the AMPK/mTOR pathway. *Molecular Medicine Reports, 12*(2), 2771–2776.

Zhou, Y., Luo, D., Shi, J., Yang, X., Xu, W., Gao, W., Guo, Y., Zhao, Q., Xie, X., He, Y., Du, G., & Pang, X. (2023). Loganin alleviated cognitive impairment in 3×Tg-AD mice through promoting mitophagy mediated by optineurin. *Journal of Ethnopharmacology, 312*, 116455.

Zhou, Y., Zhao, Q., Zhang, Y., Di, L., Xue, F., Xu, W., Gao, W., Guo, Y., He, Y., Kou, J., Qin, Y., Xie, X., Du, L., Han, G., & Pang, X. (2024). A new andrographolide derivative ADA targeting SIRT3-FOXO3a signaling mitigates cognitive impairment by activating mitophagy and inhibiting neuroinflammation in Apoe4 mice. *Phytomedicine, 124*, 155298. doi: 10.1016/j.phymed.2023.155298.

12 Therapeutic Potential of Receptor for Advanced Glycation End-Products Inhibitors in the Treatment of Alzheimer's Disease

Saikat Dewanjee and Hiranmoy Bhattacharya

12.1 INTRODUCTION

Alzheimer's disease (AD) is a neurodegenerative disease that is characterized by progressive loss of cognitive function and loss of memory, accounting for about 80% of dementia cases globally (Dewanjee et al., 2022; Rayathala et al., 2022). AD is best described as a multifactorial disorder rather than just a downstream effect of deposition of amyloid beta (Aβ) plaques and phosphorylation of tau proteins that leads to the formation of neurofibrillary tangles (NFTs), even though extracellular amyloid plaques and intracellular NFTs are the two main factors behind AD onset (Dewanjee et al., 2022). Senile plaques, NFTs, a malfunctioning cholinergic system, reduced synaptic transmission and plasticity, oxidative stress, and neuroinflammation are some of the defining pathomechanisms of the AD brain.

Advanced glycation end-products (AGEs) are compounds that are created through nonenzymatic interactions between proteins, phospholipoproteins, or nucleic acids and carbonyl compounds (such as glyoxal, methylglyoxal, glyceraldehyde, 3-deoxyglucosone, and sugars) (Dewanjee et al., 2022). Several studies have demonstrated the neurotoxic and neurodegenerative roles of different AGEs (DeTure & Dickson, 2019; Kuzan, 2021; Rungratanawanich et al., 2021; Takeuchi et al., 2017). AGEs interact with the receptor for AGEs (RAGE), which is one of the cascades through which AGEs cause neurodegeneration in AD (Batkulwar et al., 2018). The AGE-RAGE axis's altered proteome caused by increased expression of cathepsin B and asparagine endopeptidase led to an increase in Aβ synthesis and tau phosphorylation. These effects were observed in the AD brain at the same time that AGE levels rose, revealing a novel mechanistic connection between the AGE-RAGE cascade and AD pathogenesis (Batkulwar et al., 2018). However, the role of RAGE in AD is not limited to AGE-RAGE interactions. Downstream consequences of RAGE stimulation that are related to the etiology of AD can also be triggered by other substrates of RAGE, such as high-mobility group box1 proteins (HMGB1), calgranulin proteins (S100), and Aβ (Singh & Agrawal, 2022). Thus, RAGE plays a major role in the onset and progression of AD; hence, it possesses the ability to become a significant target for therapy in the treatment of AD. This chapter delves into the complex structural and functional features of RAGE that contribute to the pathophysiology of AD and identifies the possible points and mechanisms of therapeutic intervention using RAGE inhibitors to ameliorate AD through modification of RAGE and its ligands.

12.2 STRUCTURAL PROPERTIES OF RAGE

Human RAGE proteins are receptors of the immunoglobulin superfamily and are 45-kDa proteins with three major domains: extracellular domain with 23–342 amino acid residues, a transmembrane domain which is hydrophobic in nature with 343–363 amino acid residues, and an intracellular domain suspended in cytoplasm (404–464 residues) (Singh& Agrawal, 2022). The extracellular domain is additionally segmented into three immunoglobulin-like regions. They are called variable domain (V-domain), which has 23–116 amino acids and is coupled with two domains of constant compositions, C1 with 124–221 amino acid residues and C2 with 227–317 amino acid residues (Singh & Agrawal, 2022). A sizable electropositive region and a cavity of hydrophobic nature cover the surface surrounding the V-domain and C1 domain (Twarda-Clapa et al., 2022). In terms of structural details, the Protein Data Bank offers a number of experimental models produced through X-ray crystallography or nuclear magnetic resonance (NMR). 2010 saw the publication of the two earliest crystal structures of the VC1 part of RAGE, which offered light on a potential AGE binding mechanism (Koch et al., 2010; Park & Boyington, 2010). In 2013, Yatime and Andersen presented a more comprehensive picture of the structure that encompassed the C2 region. Three years later, the template of a receptor-substrate complex of RAGE with human S100-A6 was published (Yatime et al., 2016). A structure of the complex of N^{ε}-carboxyethyllysine (CEL)-containing peptide with the V domain was disclosed by Xue et al. in 2011. It showed that the carboxyethyl compound interacts with the electropositive region of the V domain (Xue et al., 2011).

The existence of a sizable, electropositive region on the RAGE surface was revealed by structural analyses, which indicated a mechanism of the ligand interaction. Lysines as

DOI: 10.1201/9781003602477-12

well as arginines that are primarily located in the V domain and partially in the C1 region, line the pocket (Twarda-Clapa et al., 2022). This patch's presence indicates that RAGE interacts with electronegative portions of the altered polypeptide surface where addition of negative charges or elimination of positive charges occurred during AGE modification (Park & Boyington, 2010). Mutagenesis and NMR studies that revealed the role of amino acids lining the electropositive patch for binding AGE-modified bovine serum albumin were the first steps in confirming this theory (Park & Boyington, 2010). This big, electropositive region is an obvious candidate for DNA binding as well. However, the structural analysis of RAGE using an oligonucleotide showed that the contact between the polypeptide and DNA only partly intersected with the electropositive zone (Sirois et al., 2013). The DNA moiety made polarized interactions among the phosphate groups of the nucleic acid mainstay and the side chains of lysine and arginine, the majority of which constituted the positively charged region. The rest of the polar amino acid moieties that are present in that region made no contact with the nucleic acids. As a portion of the electropositive region stayed vacant, it can attach an electronegative portion of AGE and a DNA moiety at the same time, as shown by a CEL compound that touched the RAGE surface around Lys110 and Arg98 (Xue et al., 2011).

Ligand-driven dimerisation and oligomerisation are often observed in RAGE. The stability of these dimers and oligomers could offer a justification for the selectivity or affinity of RAGE regarding the ligands and the resultant signaling consequences (Xie et al., 2007, 2008; Xu et al., 2013; Zong et al., 2010). The C2 region of RAGE is a separate structural component, which is coupled with the C1 region by a connection of 12 amino acid molecules (Dattilo et al., 2007). C2, like the VC1 component, consists of a big electronegative surface surrounded by low-pH amino acid molecules oriented superficially on the VC1 oligomer (Yatime & Andersen, 2013).

Little could be identified about RAGE's transmembrane region (residues 343–363) because there is little data on it. However, its role in signal transmission is most likely caused by the helix-helix homodimerization of RAGE (Sturchler et al., 2008).

RAGE's cytoplasmic region (ct-RAGE) is further segregated into three regions: a domain containing 17-amino acids that is known as the membrane-proximal domain, which is connected to the central domain, which again contains 17-amino acids, and the terminal domain, namely the C-terminal domain. This region also performs critical interplays with ligand-molecules and causes several downstream consequences.

12.3 RAGE AND ITS LIGANDS: INTRICATE INTERACTIONS LEADING TO AD-RELATED COMPLICATIONS

Schmidt and coworkers originally discovered RAGE and their interaction with AGEs in 1992 (Schmidt et al., 1992). Schmidt and colleagues used radio-labeled ^{125}I-AGE albumin as an AGE substrate to show that AGEs bind to RAGE-expressing endothelial cells (Schmidt et al., 1992). More contemporary studies indicate that RAGE attaches to a variety of ligands, which include Aβ, S100, and HMGB1, leading to the generation of pro-inflammatory cytokines (Ramasamy et al., 2005; Wautier & Wautier, 2020; Wuren et al., 2021; Yang et al., 2022). Although HMGB1 functions as a DNA-escort, mending DNA impairment in the nucleus, in the extracellular region, it causes inflammatory effects by attaching itself to the extracellular region of RAGE (Dong et al., 2022). HMGB1-targeted monoclonal antibody has been proven to alleviate the inflammatory consequences in traumatic brain injury (Okuma et al., 2012). The interactions between AGEs and RAGE and the attachment of RAGE with numerous additional ligands not only exacerbate oxidative stress but also lead to RAGE overexpression. The signaling pathways resulting from AGE-RAGE interactions have been linked to a number of clinical conditions, including neurodegeneration that leads to AD, cardiovascular disease, cancer, and diabetes related complications (Adeshara et al., 2022; Faruqui et al., 2022; Hofmann et al., 2002; Jangde et al., 2020; Monu et al., 2022; Muhammad et al., 2008; Patil et al., 2022; Vulichi et al., 2022). Figure 12.1 illustrates the downstream consequences of RAGE activation through ligand binding.

12.3.1 Role of AGEs as a RAGE-Ligand in AD

Glycation of proteins occurs in hyperglycaemic settings as well as in normal conditions, resulting in the formation of irreversible AGEs by means of complex chemical processes such as Schiff base production, the Amadori rearrangement, and the Maillard reaction. These stages consequently cause the creation of hazardous protein aggregates known as AGEs (Salahuddin et al., 2014). Initially, the proteins containing carbonyl groups that come from reducing sugars interact with terminal amino groups or sidechain amino groups, a reaction occurs that is nonenzymatic in nature, resulting in the generation of Schiff bases (Reddy et al., 2022). Under intense oxidative stress, Schiff bases undergo glycoxidation reactions, resulting in highly reactive dicarbonyl molecules, which get rearranged into stable and reversible ketoamines. This rearrangement reaction is known as the Amadori rearrangement. The amino groups in the mentioned proteins interact with dicarbonyl moieties to generate crosslinkages between proteins and intra-protein crosslinkages (by dehydration and rearrangement). This causes the deactivation of AGE-crosslinked proteins,which include receptors and enzymes, which are necessary for regular bodily activities (Salahuddin et al., 2014). In the brain, abnormalities in mitochondrial metabolism of glucose are the main reason behind the production of AGEs (Dewanjee et al., 2022). N^{ε}-(carboxymethyl)lysine (CML) is a significant AGE seen in diabetes and Alzheimer's patients, serving as the primary ligand for RAGE and a biological marker for these disorders. CEL and CML are generated through the Maillard

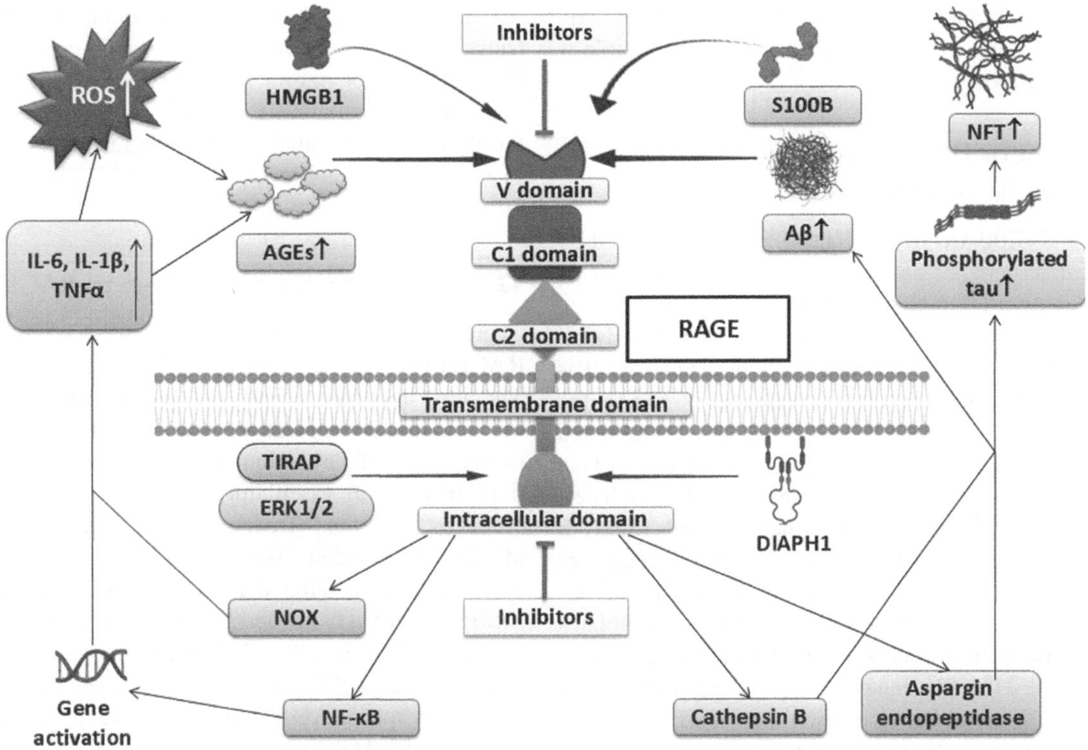

FIGURE 12.1 Downstream consequences of RAGE activation. RAGE is activated by extracellular ligands like AGEs, S100 proteins, Aβ, HMGB1 and intracellular ligands such as TIRAP and DIAPH1. Activation of RAGE leads to the generation of NF-κB and NOX, which produce inflammatory cytokines such as TNFα, IL-6, and IL-1β, which again generate ROS and consequently AGEs, giving rise to a vicious cycle that contributes majorly to AD. Cathepsin B and asparagin endopeptidase are also generated as a downstream consequence of RAGE activation, which contributes to the build-up of Aβ and phosphorylation of tau protein, subsequently increasing intracellular formation of NFTs. AGEs, advanced glycation end-products; RAGE, receptor for AGEs; HMGB1, high mobility group box 1 protein; Aβ, amyloid β; TIRAP, toll-interleukin 1 receptor domain adaptor protein; ROS, reactive oxygen species; DIAPH1, diaphanous related formin-1; ERK, extracellular signal-regulated kinases; NF-κB, nuclear factor κB; IL, interleukin; NOX, NADPH oxidase; TNFα, tumor necrosis factor α; NFTs, neurofibrillary tangles.

reaction, which involves the interaction of lysine with terminal amino groups, typically glyoxal or methylglyoxal molecules (Reddy et al., 2022).

Increased amounts of AGEs were found in the AD brain, indicating an emerging connection between AGEs and the disease's pathophysiology (Batkulwar et al., 2018; Dewanjee et al., 2022). Many of the AGEs have been detected in NFTs, aged plaques, glial cells, and pick bodies (Lüth et al., 2005; Ooi et al., 2022; Yamagishi et al., 2015). AGE-RAGE interaction causes neurodegeneration in a manner that contributes to the etiology of AD (Batkulwar et al., 2018). AGEs attach to the extracellular region of RAGE, triggering a cascade of signaling pathways. This includes stimulation of NFκB and gene activation for pro-inflammatory cytokines like interleukin-6 (IL-6), tumor necrosis factor α (TNFα), and interleukin-1β (IL-1β) (Reddy et al., 2023). This, in turn, exacerbates oxidative stress and AGE overproduction, creating a chain of circumstances of increased AGE-RAGE binding. AGE-RAGE interaction also activates NADPH oxidase (NOX), resulting in the increment of pro-inflammatory cytokines, which are responsible for the abundance of reactive oxygen species (ROS), worsening oxidative stress conditions, and increasing AGE levels (Reddy et al., 2023). This cascade also enhances extracellular signal-regulated kinases 1 and 2 (ERK1/2), and JNK-AP-1 signaling in human endothelial cells, impacting cell division and endothelial cohesion (Adamopoulos et al., 2016). This consequence contributes to one of the main risk factors of AD, that is the disruption of the blood-brain barrier (BBB), which is primarily made of endothelial cells (Cai et al., 2018). AGE-RAGE attachment also causes transportation of beta-amyloid protein across the disrupted BBB, contributing to the development of AD in multitudes (Deane, 2012). Increased expression of cathepsin B and asparagine endopeptidase in the AGE-RAGE axis modifies the proteome, leading to increased Aβ synthesis, mitochondrial malfunction, plaque accumulation, and tau protein-hyperphosphorylation, leading to neuronal demise (Dewanjee et al., 2022; Nowotny et al., 2015). Glyceraldehyde-derived AGEs promote the synthesis of the precursor protein of amyloids (APP) by generating ROS, resulting in neuronal death as well (Ko et al., 2010). Glycated proteins activate microglia and macrophages, leading to oxidative stress and cytokine production, potentially advancing the disease (Azizi et al., 2015). Subsequently activated macrophages also generate cytokines and chemokines, which can produce several

RAGE ligands, including AGEs, HMGB1, S100B, and Aβ. This again creates a vicious cycle in AD pathophysiology (Byun et al., 2017).

12.3.2 Role of HMGB1 and S100B as a RAGE-Ligand in AD

HMGB1 and S100 are RAGE ligands that are released by activated macrophages and injured or stressed cells. This can trigger an immunological response and is detected by RAGE receptors. HMGB1 is present in practically all cells, with the exception of those without a nucleus. It is made up of chromatin-binding proteins that modify the architecture of chromosomes and manage transcription and perform physiological as well as pathological roles (Tian et al., 2007). It is released by numerous cells like dendrites, macrophages, T cells, astrocytes, or necrotic cells (Scaffidi et al., 2002). After release, it connects to RAGE receptors or toll-like receptors (TLRs), activating the inflammatory response (Bianchi & Manfredi, 2007). HMGB1 stimulates NOX, increasing the production of ROS (Tadié et al., 2012). Serum or CSF HMGB1 contents have not been examined in case of AD; however, in a few other neurological illnesses, such as neurological refractory inflammation and Rasmussen's encephalitis, enhanced serum and CSF concentration of HMGB1 has been observed (Walker et al., 2017). HMGB1 is found to react with Aβ peptides, leading to increased deposition and neurotoxicity, and it also stimulates Aβ secretion from microglial cells, causing increased Aβ buildup; these findings can bind HMGB1 with the pathophysiology of AD (Paudel et al., 2020). HMGB1 has also been linked to BBB malfunction, whereas the disrupted BBB permits peripherally present immune cells and inflammatory chemicals to get into the CNS, leading to neuroinflammation and likely contributing to AD pathogenesis (Byun et al., 2017; Paudel et al., 2020). HMGB1 may also contribute to AD pathology by interacting with RAGE, and this interaction activates NF-κB and upregulates proinflammatory cytokines, resulting in increased ROS production and cell death (Gao et al., 2011).

S100 proteins are calcium-binding proteins. S100B, a member of this family, is linked to AD pathology. S100B is a broad collection of calcium-binding proteins with 20 members that possess a regulatory role in a vast range of cellular activities. At healthy quantities, S100B has neurotrophic benefits, while higher levels cause neurotoxicity. Its expression related to the immune system and serum concentrations has been evaluated in a variety of illnesses. S100B is a hallmark of astrocytic stimulation and binds with the RAGE receptors (Sedaghat & Notopoulos, 2008). S100B has been linked to neurodegeneration in AD, and increased levels of S100B have been observed in the brains of AD patients (Chaves et al., 2010; Petzold et al., 2003; Shepherd et al., 2006). Another observation reveals that the expression of physiologically active S100B is higher in brain tissue from Alzheimer's patients (collected post-mortem), and that this enhancement is associated with neuritic disease (Cristóvão & Gomes, 2019). S100B binds with the RAGE, triggering inflammation and oxidative damage by activating NF-κB and NOX in glial cells to release pro-inflammatory cytokines, including TNF-α and IL-1β, which contribute to the inflammation found in AD; this mechanism is similar to AGE-RAGE interactions (Dong et al., 2022; Mosquera, 2010).

12.3.3 Other Ligands that Bind to RAGE

There are a few other ligands that successfully bind with RAGE to contribute to certain pathways implicated in AD. The most important among the rest of them is perhaps the most significant biomarker of AD, i.e., Aβ, which attaches to the extracellular region of RAGE and leads to disruptions in cerebral vasculature, neurones, and microglia in AD-affected brains (Deane et al., 2012). Several proteins of S100 family other than S100B has also been reported to bind with extracellular domains of RAGE such as S100P, S100A6, S100A11, S100A12 (Chiou et al., 2016; Filipek & Leśniak, 2020; Huang et al., 2016; Penumutchu et al., 2014); however, except S100A6, none of them has been evidenced to play any role in AD pathology. Several downstream molecules in the AGE-RAGE axis bind with the intracellular domain of RAGE as well; they are diaphanous related formin-1 (DIAPH1), toll-interleukin 1 receptor domain adaptor protein (TIRAP), and ERK1/2, which eventually lead to the stimulation of the mitogen-activated protein kinase (MAPK) pathway; which is involved in AD pathophysiology (Singh & Agrawal, 2022).

12.4 SOLUBLE RAGE (sRAGE): AN INTERESTING BIOMARKER IN AD

Soluble RAGE (sRAGE) is RAGE that is C-terminal truncated and does not possess the transmembrane region; hence, binding AGEs to sRAGE does not cause a signaling phenomenon (Reddy et al., 2023). The production of sRAGE somewhat mitigates the AGE/RAGE-caused cell destruction. sRAGE plays a competitive role with RAGE in attaching to AGEs, reducing AGE/RAGE interplays and subsequent signaling events (Reddy et al., 2023). Furthermore, sRAGE sequesters AGEs, mitigating the AGE-activated cytotoxic consequences. Delivery of sRAGE to RAGE−/− mice dramatically reduced infarct size in ischaemic brain injury (Muhammad et al., 2008). Administration of recombinant forms of sRAGE has been found to prevent the stimulation of RAGE present at the surface of cells (Metz et al., 2012). sRAGE levels in the serum of AD patients are relatively low as compared to normal levels and hence may act as a biological marker for monitoring AD progression, albeit this biomarker-based assumption is not useful for the distinction of AD from other types of dementia, such as vascular dementia (Xu et al., 2017). Metalloproteinase-activated denaturation of RAGE to the sRAGE has significance for the therapy of neurological ailments and diabetes related complications (Reddy et al., 2023). It is worth mentioning that AD has been called type 3 diabetes mellitus due to profound pathological similarities between the diseases. The metalloproteinases ADAM10 and MMP9 are implicated in the destruction of RAGE to form sRAGE (Liu et al., 2023; Rosenbaum & Saftig, 2024). G-protein-coupled receptors (GPCR), namely the PACAP, induce the production of metalloproteinase by several methods, including stimulation of Ca^{2+} signaling and MAPKs, and are thus associated with the denaturation of the GPCR present in cell surfaces to sRAGE (Metz et al., 2012). This promotion of RAGE to sRAGE conversion has implications for the development of GPCR agonists as treatments for AGE-related disorders such as AD, diabetes, and heart diseases.

sRAGE is also produced by alternative splicing of the human RAGE mRNA. The sRAGE formed due to spliced mRNA is called endogenous secretory RAGE (esRAGE). esRAGE shows high levels in human vascular endothelial cells and pericytes and has been found to impart a cytoprotective effect as opposed to AGEs and mitigate vascular damage in diabetes (Genova et al., 2022). The esRAGE has a similar amino acid sequence compared to sRAGE and does a similar task of AGE trapping, reducing AGE toxicity (Reddy et al., 2023). The esRAGE and sRAGE, since they are without the transmembrane domain, operate as decoy receptors for AGEs and numerous other substrates of RAGE, attenuating the harmful implications of AGEs (Bierhaus et al., 2005).

Thus, in spite of being analogous to RAGE, sRAGE competitively hinders AGE-RAGE interaction by binding with AGE itself. As binding of AGE with sRAGE does not result in triggering of downstream cascades of AGE-RAGE, elevated levels of sRAGE would lead to mitigation of AD pathology. Hence, sRAGE can act as a biomarker in AD, where a depleted level of sRAGE can mean worsening of AD conditions. sRAGE can also be a target for developing novel therapeutic strategies, which would accelerate the transformation of RAGE into sRAGE.

12.5 POTENTIAL OF RAGE INHIBITORS AGAINST AD

Due to the profound impact of ligand-based RAGE activation on RAGE in AD pathophysiology, RAGE presents itself as a prospective therapeutic target for the therapy of AD. RAGE inhibitors can potentially hinder the pathophysiological consequences stemming from the excitation of RAGE. Table 12.1 enlists the RAGE inhibitors acting through different mechanisms that can be useful in the treatment of AD.

12.5.1 Inhibitors of RAGE Binding to Extracellular Domain

Inhibiting ligands that bind to RAGE is the most effective way to reduce the pathophysiology of RAGE-induced inflammation in a variety of diseases. FPS-ZM1 (4-chloro-*N*-cyclohexyl-*N*-(phenylmethyl)benzamide), an inhibitor of RAGE discovered by screening 5000 molecules, effectively blocks attachment of RAGE V-domain to Aβ and alleviates AD symptoms (Deane et al., 2012). This inhibits RAGE's extracellular domain from attaching to HMGB1 and S100B (Deane et al., 2012). FPS-ZM1 suppresses the stimulation of primary microglia by AGEs, resulting in a reduction in RAGE expression, oxidative stress, and thereby lowering the amount of inflammation (Shen et al., 2017). FPS-ZM1 possesses ideal structural characteristics that are mostly related to RAGE interaction. The benzyl group of this compound, which is immediately linked with the nitrogen atom of the amide group, supplies the terminal group with an electron-rich environment (Singh & Agrawal, 2022). Its tertiary amide central core allows it to accept and donate hydrogen bonds with amino acid residues found in RAGE's V domain. The electropositive benzene ring linked to the carbon atom of the carbonyl group, as well as FPS-ZM1's six-membered acyclic chain, gives stability within the RAGE V-domain (Singh & Agrawal, 2022). α-Synuclein, a highly charged protein, interacts with positively charged RAGE residues at the V-domain by means of acidic residues of the C-terminal, triggering neuroinflammatory signaling (Reddy et al., 2023). FPS-ZM1 selectively binds to electropositive residues at the V-region of RAGE, blocking the association of α-synuclein fibrils to it and decreasing the inflammatory signaling at microglia (Reddy et al., 2023). Its fluorine-18 analogue was created for bio-imaging for in vivo research to detect RAGE, capitalizing on its great binding affinity for the V-region of RAGE (Cary et al., 2016). When examined using immunohistochemistry, this chemical co-localized with RAGE in AD brain tissues. It has greater CNS internalization and absorption in numerous zones of the brain that are revealed to express RAGE than its parent analogue, FPS-ZM1 (Cary et al., 2016).

Another exemplar RAGE inhibitor is azeliragon, codenamed PF-04494700. This compound has been found to block RAGE binding of extracellular substrates such as AGEs, S100B, HMGB1, and Aβ (Sabbagh et al., 2011). Azeliragon blocked RAGE's extracellular domain binding to Aβ with an IC50 value of 500 nM, which has been assessed by fluorescence polarisation assays (Bongarzone et al., 2017). Azeliragon also decreased neuroinflammation by lowering Aβ plaque levels and inflammatory cytokines (Burstein et al., 2018). The mechanism of azeliragon binding to RAGE remains confusing because no structural information is available to determine modes and sites of binding. However, molecular docking experiments reveal a few fundamental conceptions of its attachment mechanism with the V-domain of RAGE (Xie et al., 2021). Azeliragon is the only drug candidate that has been considered for the clinical trial to date. Urolithin and its many structural analogues have been shown to suppress AGE-BSA/sRAGE binding, with RAGE blocking comparable to that of Azeliragon (Guérin et al., 2021). Tranilast is a pharmaceutical compound mainly used as an anti-allergic, which substantially blocks the interactions of the RAGE V-domain with S100 ligands, as assessed by H^+ and nitrogen NMR, fluorescence

TABLE 12.1
RAGE Inhibitors Having Potential Anti-AD Therapeutic Effects

Sl. No.	Name of the Inhibitor	Domain of Binding	Model of Experiment	Observed Results	Reference
1.	FPS-ZM1 (4-chloro-N-cyclohexyl-N-(phenylmethyl)benzamide)	Extracellular domain	Primary microglia culture	RAGE expression↓; oxidative stress↓; inflammation↓	Shen et al. (2017)
2.	Fluorine-18 analog of FPS-ZM1	Extracellular domain	Immuno-histochemistry	Binding affinity compared to FPS-ZM1↑; CNS penetration↑	Cary et al. (2016)
3.	Azeliragon	Extracellular domain	Fluorescence polarisation assay	Binding of Aβ and RAGE↓	Burstein et al. (2018)
			Molecular docking	Binding with V-domain of RAGE	Xie et al. (2021)
			Clinical trial phase II	Well-tolerated dose, no significant activity or side effect in high and low doses	Sabbagh et al. (2011)
			Clinical trial phase II	High dose led to cognitive decline and confusion, hence withdrawn; low dose led to slowing down of cognitive decline in mild AD without any toxicity or side effect	Galasko et al. (2014), Burstein et al. (2014)
			Clinical trial phase III	No reported toxicity in the specific dose, terminated due to lack of reported therapeutic benefit	NCT02080364 (Clinicaltrials.gov, 2024)
4.	Urolithin	Extracellular domain	ELISA-assay	AGE-BSA/sRAGE interactions↓; RAGE inhibition↑	Guérin et al. (2021)
5.	Tranilast	Extracellular domain	H⁺ and nitrogen NMR, fluorescence studies, WST-1 assay	RAGE inhibition↑	Chiou et al. (2016), Huang et al. (2016)
6.	Papaverine	Extracellular domain	In vitro cell line assays	RAGE-dependent NF-κB activation via HMGB1↓	El-Far et al. (2018)
7.	Emetine	Extracellular domain	Computational studies	RAGE-Aβ interaction↓	Ahmad et al. (2019)
8.	Aminopyrimidine and phenyl benzoxazole derivatives	Extracellular domain	APP/PS1 mouse model of AD	RAGE-Aβ interaction↓	Han et al. (2012), Choi et al. (2015)
9.	Thirteen small molecules screened from the library of 58,000 molecules by Manigrasso et al. (2016)	Intracellular domain	Fluorescence titration, NMR spectroscopy, and Western blotting	ct-RAGE-DIAPH1 interaction↓; CML-AGE-mediated IL-6 and TNF-α induction↓	Manigrasso et al. (2016)
10.	Pioglitazone	Unknown	High glucose-treated vascular smooth muscle cells	RAGE protein and mRNA expression↓	Gao et al. (2017)
11.	Gliclazide	Unknown	In vitro diabetic atherosclerosis model	IL-1β↓; IL-6↓; TNF-α↓ Anti-inflammatory cytokines↑	Jahan and Choudhary (2021)

studies, and the WST-1 assay (Chiou et al., 2016; Huang et al., 2016). Various amino acids that interact with tranilast in RAGE's active domain (V-domain) were also discovered utilising high ambiguity-driven biomolecular docking (HADDOCK). The mentioned researches reveal molecular data about tranilast, which can then be exploited to design novel RAGE inhibitors (Singh & Agrawal, 2022).

Papaverine (an isoquinoline derivative substituted with a dimethoxy group) strongly inhibited RAGE-dependent NF-κB activation via blocking HMGB1-RAGE interaction (El-Far et al., 2018). Papaverine's influence on HMGB1-inspired NF-κB stimulation was investigated using C6 glioma cells (El-Far et al., 2018). Papaverine at 10 and 20 μM concentrations reduced the upregulation (El-Far et al., 2018). To assess the mode of attachment of papaverine with RAGE, an AGE-RAGE interaction experiment was executed, which demonstrated that papaverine may interact with the binding location of AGE (El-Far et al., 2018).

According to research, papaverine has the ability to reduce chronic inflammatory discomfort in mice models, but it does not demonstrate analgesic effects in oxidative stress at the location of inflammation (Yoshizawa et al., 2021). As a result, papaverine and the derivatives of papaverine could serve as pioneer compounds in the discovery of new and effective RAGE antagonists capable of treating a variety of RAGE-associated disorders. Emetine, a nature-derived small molecule, has also been identified as an efficient blocker of RAGE and its interaction with Aβ through computational studies (Ahmad et al., 2019). Emetine interacts with the exact amino acid fragments of Aβ that are liable to bind with the external domain of RAGE (Ahmad et al., 2019). This implicates that it can be used in the treatment of AD as a potential RAGE-substrate inhibitor molecule. There are a few studies that have been performed for experimental validation of the anti-AD effects of emetine (Ahmad et al., 2021; Han et al., 2021; Manakkadan et al., 2024), however, the perspective of interactions with the RAGE cascade has remained majorly unexplored. Hence, there is a huge research gap in understanding the interaction of emetine with RAGE and its substrate, which might open a new avenue in AD treatment. Similarly, aminopyrimidine and phenyl benzoxazole derivatives identified some years ago showed excellent suppression in binding between RAGE and Aβ, with the therapeutic potential against AD (Choi et al., 2015; Han et al., 2012). However, the pinpoint mode of interaction of these compounds with RAGE has not yet been explored through further research.

In recent years, Kozlyuk et al. used a fragment-oriented method to generate novel RAGE antagonists that can be selectively projected to their ligand-binding region (Kozlyuk et al., 2021). They first tested the attachment of almost 14,000 tiny fragments with the RAGE V-domain using chemical shift perturbations (CSPs) in NMR spectra. The leading fragments were then employed to create an X-ray co-crystallized structure with the V and C1 regions of RAGE to determine precise placement and orientation. Through this method, researchers discovered three optimal binders that can get attached to three different locations on the RAGE (Kozlyuk et al., 2021). Prior to this investigation, no crystallographic evidence was available for inhibitor-bound RAGE. The study's key weakness was a scarcity of data on the direct antagonism of RAGE signaling. However, these findings, along with the crystallographic structures of the three drugs, can be leveraged to develop novel and specific inhibitors of RAGE (Kozlyuk et al., 2021).

12.5.2 Inhibitors of RAGE Binding to Intracellular Domain

The cytoplasmic part of RAGE is equally essential for many forms of RAGE signaling and downstream consequences. Inhibiting the attachment between RAGE and mammalian DIAPH1 has relevance in treating RAGE-induced chronic inflammatory diseases. Thirteen excellent small molecules from a list of 58,000 molecules were selected by Manigrasso et al. that can interact with the intracellular region and prevent the attachment between DIAPH1 and RAGE (Manigrasso et al., 2016). Fluorescence titration tests revealed that these lead compounds have a considerable affinity for ct-RAGE. Among these 13 lead molecules, 12 compounds effectively reduced CML-AGE-induced IL-6 and TNF-α induction (Manigrasso et al., 2016). These small-molecule RAGE antagonists supplement the extracellular RAGE blockers in the development of candidate drugs for neurological diseases and other AGE-related ailments. It is worth noting that a few of these ct-RAGE-targeted lead compounds share structural similarities with FPS-ZM1. That is, they possess a common moiety, benzamide (Manigrasso et al., 2016). ct-RAGE inhibitors derived from quinoline, RAGE203, RAGE208, and RAGE229, were identified as promising treatments for RAGE-induced diabetic problems (Manigrasso et al., 2021). Diabetic mice have considerably higher plasma levels of TNF-α compared to non-diabetic animals. RAGE229 reduced TNF-α and other inflammatory cytokines in diabetes of both types. The mentioned findings reveal that inhibiting ct-RAGE has an immediate effect on the signaling system, preventing the generation of inflammatory cytokines (Manigrasso et al., 2021).

12.5.3 Other RAGE Inhibitors

Pioglitazone, a thiazolidinone-based medication, is a powerful PPAR-γ agonist and RAGE antagonist. In high glucose-treated VSMCs, pioglitazone reduced RAGE protein and mRNA expression (Gao et al., 2017). However, the mechanism of pioglitazone's binding to RAGE domains is uncertain. Gliclazide (a recognised antidiabetic medication) has also been demonstrated to inhibit the AGE-RAGE cascade for the treatment of diabetic atherosclerosis (Jahan & Choudhary, 2021). Research found that gliclazide (1–100 μM) reduced cytokine generation in RAW 264.7 macrophages, including IL-1β, IL-6, and TNF-α, while also enhancing the generation of anti-inflammatory cytokines in a concentration-dependent way (Jahan & Choudhary, 2021). The study, however, did not identify the rudimentary modes of blocking AGE-RAGE signaling by gliclazide.

12.6 CLINICAL PERSPECTIVES

Azeliragon has undergone the first two phases of clinical trials with humans to assess its safety and effectiveness. In a Phase II experiment, three groups of individuals were subjected to be tested for ten weeks with two dosages of azeliragon and placebo (Sabbagh et al., 2011). The study included individuals with mild or moderate AD, with the average age of the individuals being around 75 years. The study included 27 people who were given a small dose of azeliragon (30 mg/day for six days, and then 10 mg/day for the rest). There were 28 individuals who were given a relatively larger dose (80 mg/day for six days, then 20 mg/day for the rest), and 12 who received a placebo treatment

(Sabbagh et al., 2011). Treatment was completed by 88.9% of the subjected individuals at the smaller dose, 85.7% at the high dose, and 66.7% in the placebo treatment group. All the doses were found to be safe, with no major distinction in adverse reactions among the individuals receiving high and low doses (Sabbagh et al., 2011). No clinically significant variations in vital signs, laboratory test findings, or mean ECG parameters in individuals receiving azeliragon were observed. Azeliragon did not consistently affect plasma concentration of Aβ, inflammatory biological markers, or secondary cognitive results (Sabbagh et al., 2011).

Another multinational, 18-month Phase II trial of azeliragon included 399 patients to assess its effectiveness in delaying cognitive deterioration (Burstein et al., 2014; Galasko et al., 2014). Participants in this trial were randomly distributed to three groups: high dose (60 mg/day for six days, and then 20 mg/day till six months), low dose (20 mg/day for six days, and then 5 mg/day till 18 months), and placebo (Burstein et al., 2014; Galasko et al., 2014). The ADAS-cog test served as the key cognitive assessment criterion. After six months, the high-dose medication was stopped because of considerable cognitive decline, confusion, and falls. The adverse reactions were absent in the individuals of the small dose and placebo groups (Burstein et al., 2014; Galasko et al., 2014). The low dose group sustained their ADAS-cog values better than the individuals placed in the placebo group, which favors further exploration of this dose in the next stage of clinical trials. A 12-month sample of participants revealed no significant alterations in biomarkers of CSF, such as Aβ, tau, and phospho-tau, among therapy groups (Galasko et al., 2014; Burstein et al., 2014). The high dose group experienced a reversible increase in cognitive deterioration, which might be attributed to either non-specific RAGE blocking or off-target activity. To further investigate the small dose therapy group, plasma concentrations of azeliragon were measured at three-month intervals (Burstein et al., 2014). Azeliragon plasma concentrations had varying effects on ADAS-cog. Subjects with azeliragon concentrations in the lowest 40% showed a considerable difference in ADAS-cog score compared to the placebo group, indicating a favourable effect of ameliorating cognitive deterioration (Galasko et al., 2014). Subjects with concentrations ranging from 46.8 to 167.7 ng/ml exhibited ADAS-cog levels comparable to the individuals receiving placebo treatment. Patients with mild AD benefited more on low dosage azeliragon than those with moderate AD. This investigation found that taking a lower dose of azeliragon (5 mg/day) can delay cognitive decline more effectively than a placebo (Galasko et al., 2014).

A continuation with a phase III trial of azeliragon containing 880 individuals was terminated in 2021 due to a lack of registered efficacy of the 5 mg dose of azeliragon (NCT02080364). The only takeaway from this clinical study is that the molecule showed no possible toxicity in this dose. However, hopes for clinical translation have not died down, and research into further understanding of the drawbacks needs to be set in motion for azeliragon to make a comeback in the clinical setting. The other potential compounds, such as FPS-ZM1, tranilast, and papaverine, are yet to be tried in a clinical setting and have the potential to be translated as useful drugs against AD.

12.7 DISCUSSION AND FUTURE PERSPECTIVES

AD, having a multifactorial etiology, is a disease that is difficult to manage. Conventional therapy options, i.e., anticholinesterase inhibitors, NMDA receptor blockers, antipsychotic, and anti-amyloid drugs only provides symptomatic management and mitigation of dementia and improvement of cognitive function, only upto a certain level and do not present a long-term and stable solution to aggravating and worsening disease conditions. Similarly, nature-derived small antioxidant-based therapy has only provided limited results due to its limitations, such as non-specificityand less bioavailability. Under these circumstances, novel approaches in the treatment of AD that would impact specific cascades in the causative mechanism of AD have become the need of the hour.

AGE-RAGE axis, RAGE-based ligand binding, and its downstream pathways provide an interesting mechanistic insight into the development of AD. AGEs, which are primary substrates to the RAGE, connect the cascade to impaired glucose metabolism in the brain. AGEs are formed due to oxidative stress caused by mitochondrial dysfunctions that lead to abnormal glucose metabolism, such as the formation of Schiff base and the Malliard reaction (Dewanjee et al., 2022). Whereas other ligands that bind to RAGE link it to other significant mechanisms of AD, such as amyloid plaque production (Aβ), Ca^{2+} homeostasis disruption (S100B). Downstream pathways generated due to RAGE excitation, such as NF-κB, ERK/2, and MAPK pathways, are the ones responsible for neurodegeneration, compromise of the BBB, build-up of Aβ plaques in the brain, and formation of NFTs. Therefore, the AGE-RAGE cascade is intricately linked with all the components of the pathophysiology of AD; hence, it possesses the ability to serve as a pivotal target of therapy in the treatment of AD.

In this chapter, we have discussed several compounds and their mechanism of inhibiting RAGE and the possibilities of their therapeutic impact against AD pathophysiology. We have identified the forerunner in such as therapeutic approach, i.e., azeliragon, which has already been subjected to clinical trials upto phase III. Even though the drug formulated at a certain dose failed in the third stage of clinical trial, the potential of the drug candidate has not died down, and possibilities of new formulation are yet to be tested. Other drug candidates such as FPS-ZM1, tranilast, papaverine, and emetine have also shown overwhelming success in preclinical settings and are waiting to be translated into medicine through clinical studies. Many other compounds are also in development, which have not been completely explored; however, the future holds promise in this treatment axis of AD.

Another interesting aspect of RAGE's involvement in AD pathophysiology is the presence of a RAGE mimicking protein, i.e., sRAGE. sRAGE is structurally similar to RAGE with few small twigs; however, the main interest lies in its similar affinity toward the ligands of RAGE. In spite of binding the ligands of RAGE, sRAGE does not give rise to downstream pathways of RAGE, and thus does not participate in AD pathophysiology. This leads us to a hypothesis that sRAGE competitively scavenges the substrates of RAGE by binding them itself. sRAGE also traps AGEs, which also aggravate AD conditions through other pathways as well. The hypothesis is supported by the finding that sRAGE is underexpressed in AD brains (Xu et al., 2017). Thus, conversion of RAGE into sRAGE may pose as an approach to treatment for the future of AD treatments.

12.8 CONCLUSION

Numerous experimental models, particularly those used to study AD, have demonstrated the detrimental effects of ligand-induced RAGE activation. RAGE activation often has detrimental effects, leading to increased oxidative damage, inflammation, and vascular disruption, as well as altered Aβ transport—all of which are implicated in AD. Blocking RAGE activation has demonstrated a substantial therapeutic advantage in animal models of AD, making it a plausible target for AD medication development in humans. Azeliragon, the first-generation drug that showed promise in preclinical studies, has been subjected to three phases of clinical trials. Even though the trials failed to bear fruitful results, the safety of use of the drug has been established, and further scope remains with optimization, modification, and new formulation of azeliragon, which might have prospects in coming back as a potential agent against AD through RAGE inhibition. Other drugs that showed promise in preclinical setup, such as FPS-ZM1 and tranilast, are yet to be put into clinical tests. A few compounds isolated from natural sources, such as papaverine and emetine, also showed strong anti-AD effects through RAGE inhibition in preclinical tests. Translational research with such molecules is need of the hour to pave a new way in the treatment of the most prevalent dementia related disease in the world.

REFERENCES

Adamopoulos, C., Piperi, C., Gargalionis, A. N., Dalagiorgou, G., Spilioti, E., Korkolopoulou, P., Diamanti-Kandarakis, E., & Papavassiliou, A. G. (2016). Advanced glycation end products upregulate lysyl oxidase and endothelin-1 in human aortic endothelial cells via parallel activation of ERK1/2–NF-κB and JNK–AP-1 signaling pathways. *Cellular and Molecular Life Sciences*, *73*, 1685–1698.

Adeshara, K. A., Bangar, N., Diwan, A. G., & Tupe, R. S. (2022). Plasma glycation adducts and various RAGE isoforms are intricately associated with oxidative stress and inflammatory markers in type 2 diabetes patients with vascular complications. *Diabetes & Metabolic Syndrome: Clinical Research & Reviews*, *16*(3), 102441.

Ahmad, S. S., Khan, H., Danish Rizvi, S. M., Ansari, S. A., Ullah, R., Rastrelli, L., Mahmood, H. M., & Siddiqui, M. H. (2019). Computational study of natural compounds for the clearance of amyloid-beta: A potential therapeutic management strategy for Alzheimer's disease. *Molecules*, *24*(18), 3233.

Ahmad, S. S., Khan, H., Khalid, M., & Almalki, A. S. (2021). Emetine and Indirubin-3-monoxime interaction with human brain acetylcholinesterase: A computational and statistical analysis. *Cellular and Molecular Biology*, *67*(4), 106–114.

Azizi, G., Navabi, S. S., Al-Shukaili, A., Seyedzadeh, M. H., Yazdani, R., & Mirshafiey, A. (2015). The role of inflammatory mediators in the pathogenesis of Alzheimer's disease. *Sultan Qaboos University Medical Journal*, *15*(3), e305.

Batkulwar, K., Godbole, R., Banarjee, R., Kassaar, O., Williams, R. J., & Kulkarni, M. J.(2018). Advanced glycation end products modulate amyloidogenic APP processing and tau phosphorylation: A mechanistic link between glycation and the development of Alzheimer's disease. *ACS ChemicalNeuroscience*,*9*(5),988–1000.

Bianchi, M. E., & Manfredi, A. A. (2007). High-mobility group box 1 (HMGB1) protein at the crossroads between innate and adaptive immunity. *Immunological Reviews*, *220*(1), 35–46.

Bierhaus, A., Humpert, P. M., Morcos, M., Wendt, T., Chavakis, T., Arnold, B., Stern, D. M., & Nawroth, P. P. (2005). Understanding RAGE, the receptor for advanced glycation end products. *Journal of MolecularMedicine*, *83*, 876–886.

Bongarzone, S., Savickas, V., Luzi, F., & Gee, A. D. (2017). Targeting the receptor for advanced glycation endproducts (RAGE): A medicinal chemistry perspective. *Journal of MedicinalChemistry*, *60*(17), 7213–7232.

Burstein, A. H., Grimes, I., Galasko, D. R., Aisen, P. S., Sabbagh, M., & Mjalli, A. M. (2014). Effect of TTP488 in patients with mild to moderate Alzheimer's disease. *BMC Neurology*, *14*, 1–8.

Burstein, A. H., Sabbagh, M., Andrews, R., Valcarce, C., Dunn, I., & Altstiel, L. (2018). Development of Azeliragon, an oral small molecule antagonist of the receptor for advanced glycation endproducts, for the potential slowing of loss of cognition in mild Alzheimer's disease. *The Journal of Prevention of Alzheimer's Disease*, *5*, 149–154.

Byun, K., Yoo, Y., Son, M., Lee, J., Jeong, G. B., Park, Y. M., Salekdeh, G. H., & Lee, B. (2017). Advanced glycation end-products produced systemically and by macrophages: A common contributor to inflammation and degenerative diseases. *Pharmacology & Therapeutics*, *177*, 44–55.

Cai, Z., Qiao, P. F., Wan, C. Q., Cai, M., Zhou, N. K., & Li, Q. (2018). Role of blood-brain barrier in Alzheimer's disease. *Journal of Alzheimer's Disease*, *63*(4), 1223–1234.

Cary, B. P., Brooks, A. F., Fawaz, M. V., Drake, L. R., Desmond, T. J., Sherman, P., Quesada, C. A., & Scott, P. J. (2016). Synthesis and evaluation of [(18)F]RAGER: A first generation small-molecule PET radioligand targeting the receptor for advanced glycation endproducts. *ACS ChemicalNeuroscience*, *7*(3), 391–398.

Chaves, M. L., Camozzato, A. L., Ferreira, E. D., Piazenski, I., Kochhann, R., Dall'Igna, O., Mazzini, G. S., Souza, D. O., & Portela, L. V. (2010). Serum levels of S100B and NSE proteins in Alzheimer's disease patients. *Journal of Neuroinflammation*, *7*, 1–7.

Chiou, J. W., Fu, B., Chou, R. H., & Yu, C. (2016). Blocking the interactions between calcium-bound S100A12 protein and the V domain of RAGE using tranilast. *PLoS One*, *11*(9), e0162000.

Choi, K., Lim, K. S., Shin, J., Kim, S. H., Suh, Y. G., Hong, H. S., Kim, H., Ha, H. J., Kim, Y. H., Lee, J., & Lee, J. (2015). 6-phenoxy-2-phenylbenzoxazoles, novel inhibitors of receptor for advanced glycation end products (RAGE). *Bioorganic & Medicinal Chemistry*, *23*(15), 4919–4935.

"Evaluation of the Efficacy and Safety of Azeliragon (TTP488) in Patients with Mild Alzheimer's disease (STEADFAST)." Clinicaltrials.gov, https://clinicaltrials.gov/study/NCT02080364?term=NCT02080364&rank=1&tab=table (Accessed 8 Dec. 2024).

Cristóvão, J. S., & Gomes, C. M. (2019). S100 proteins in Alzheimer's disease. *Frontiers in Neuroscience*, *13*, 463.

Dattilo, B. M., Fritz, G., Leclerc, E., Vander Kooi, C. W., Heizmann, C. W., & Chazin, W. J. (2007). The extracellular region of the receptor for advanced glycation end products is composed of two independent structural units. *Biochemistry*, *46*(23), 6957–6970.

Deane, R. J. (2012). Is RAGE still a therapeutic target for Alzheimer's disease?*Future Medicinal Chemistry*, *4*(7), 915–925.

Deane, R., Singh, I., Sagare, A. P., Bell, R. D., Ross, N. T., LaRue, B., Love, R., Perry, S., Paquette, N., Deane, R. J., & Thiyagarajan, M. (2012). A multimodal RAGE-specific inhibitor reduces amyloid β–mediated brain disorder in a mouse model of Alzheimer disease. *The Journal of Clinical Investigation*, *122*(4), 1377–1392.

DeTure, M. A., & Dickson, D. W. (2019). The neuropathological diagnosis of Alzheimer's disease. *Molecular Neurodegeneration*, *14*(1), 32.

Dewanjee, S., Chakraborty, P., Bhattacharya, H., Chacko, L., Singh, B., Chaudhary, A., Javvaji, K., Pradhan, S. R., Vallamkondu, J., Dey, A., & Kalra, R. S. (2022). Altered glucose metabolism in Alzheimer's disease: Role of mitochondrial dysfunction and oxidative stress. *Free Radical Biology and Medicine*, *193*, 134–157.

Dong, H., Zhang, Y., Huang, Y., & Deng, H.(2022). Pathophysiology of RAGE in inflammatory diseases. *Frontiers in Immunology*, *13*, 931473.

El-Far, A. H. A. M., Munesue, S., Harashima, A., Sato, A., Shindo, M., Nakajima, S., Inada, M., Tanaka, M., Takeuchi, A., Tsuchiya, H., & Yamamoto, H. (2018). In vitro anticancer effects of a RAGE inhibitor discovered using a structure-based drug design system. *Oncology Letters*, *15*(4), 4627–4634.

Faruqui, T., Khan, M. S., Akhter, Y., Khan, S., Rafi, Z., Saeed, M., Han, I., Choi, E. H., & Yadav, D. K. (2022). RAGE inhibitors for targeted therapy of cancer: A comprehensive review. *International Journal of Molecular Sciences*, *24*(1), 266.

Filipek, A., & Leśniak, W. (2020). S100A6 and its brain ligands in neurodegenerative disorders. *International Journal of Molecular Sciences*, *21*(11), 3979.

Galasko, D., Bell, J., Mancuso, J. Y., Kupiec, J. W., Sabbagh, M. N., Van Dyck, C., Thomas, R. G.,Aisen, P. S.,.(2014). Clinical trial of an inhibitor of RAGE-Aβ interactions in Alzheimer disease. *Neurology*,*82*(17),1536–1542.

Gao, H., Li, H., Li, W., Shen, X., & Di, B. (2017). Pioglitazone attenuates atherosclerosis in diabetic mice by inhibition of receptor for advanced glycation end-product (RAGE) signaling. *Medical Science Monitor: International Medical Journal of Experimental and Clinical Research*, *23*, 6121.

Gao, H. M., Zhou, H., Zhang, F., Wilson, B. C., Kam, W., & Hong, J. S. (2011). HMGB1 acts on microglia Mac1 to mediate chronic neuroinflammation that drives progressive neurodegeneration. *Journal of Neuroscience*, *31*(3), 1081–1092.

Genova, V. M., Gambero, A., de Souza Freitas Campos, P., & Macedo, G. A. (2022). Polyphenolic compounds mechanisms as inhibitors of advanced glycation end products and their relationship to health and disease. In R. Campos-Vega & B. D. Oomah (Eds.), *Molecular Mechanisms of Functional Food* (pp. 1–27). John Wiley & Sons, Inc.

Guérin, T., Waterlot, C., Lipka, E., Gervois, P., Bulteel, D., Betrancourt, D., Moignard, C., Nica, A. S., Furman, C., & Ghinet, A. (2021). Ecocatalysed Hurtley reaction: Synthesis of urolithin derivatives as new potential RAGE antagonists with anti-ageing properties. *Sustainable Chemistry and Pharmacy*, *23*, 100518.

Han, J. L. T., Pang, K. K. L., Ang, S. R. X., Sharma, M., & Sajikumar, S. (2021). Inhibition of lysine methyltransferase G9a/GLP reinstates long-term synaptic plasticity and synaptic tagging/capture by facilitating protein synthesis in the hippocampal CA1 area of APP/PS1 mouse model of Alzheimer's disease. *Translational Neurodegeneration*, *10*, 1–3.

Han, Y. T., Choi, G. I., Son, D., Kim, N. J., Yun, H., Lee, S., Chang, D. J., Hong, H. S., Kim, H., Ha, H. J., & Kim, Y. H. (2012). Ligand-based design, synthesis, and biological evaluation of 2-aminopyrimidines, a novel series of receptor for advanced glycation end products (RAGE) inhibitors. *Journal of Medicinal Chemistry*, *55*(21), 9120–9135.

Hofmann, M. A., Drury, S., Hudson, B. I., Gleason, M. R., Qu, W., Lu, Y., Lalla, E., Chitnis, S., Monteiro, J., Stickland, M. H., & Bucciarelli, L. G. (2002). RAGE and arthritis: The G82S polymorphism amplifies the inflammatory response. *Genes & Immunity*, *3*(3), 123–135.

Huang, Y. K., Chou, R. H., & Yu, C. (2016). Tranilast blocks the interaction between the protein S100A11 and receptor for advanced glycation end products (RAGE) V domain and inhibits cell proliferation. *Journal of Biological Chemistry*, *291*(27), 14300–14310.

Jahan, H., & Choudhary, M. I. (2021). Gliclazide alters macrophages polarization state in diabetic atherosclerosis in vitro via blocking AGE-RAGE/TLR4-reactive oxygen species-activated NF-kβ nexus. *European Journal of Pharmacology*, *894*, 173874.

Jangde, N., Ray, R., & Rai, V. (2020). RAGE and its ligands: from pathogenesis to therapeutics. *Critical Reviews in Biochemistry and Molecular Biology*, *55*(6), 555–575.

Ko, S. Y., Lin, Y. P., Lin, Y. S., & Chang, S. S. (2010). Advanced glycation end products enhance amyloid precursor protein expression by inducing reactive oxygen species. *Free Radical Biology and Medicine*, *49*(3), 474–480.

Koch, M., Chitayat, S., Dattilo, B. M., Schiefner, A., Diez, J., Chazin, W. J., & Fritz, G. (2010). Structural basis for ligand recognition and activation of RAGE. *Structure*, *18*(10), 1342–1352.

Kozlyuk, N., Gilston, B. A., Salay, L. E., Gogliotti, R. D., Christov, P. P., Kim, K., Ovee, M., Waterson, A. G., & Chazin, W. J. (2021). A fragment-based approach to discovery of receptor for advanced glycation end products inhibitors. *Proteins: Structure, Function, and Bioinformatics*, *89*(11), 1399–1412.

Kuzan, A. (2021). Toxicity of advanced glycation end products. *Biomedical Reports*, *14*(5), 1–8.

Liu, J., Jin, Z., Wang, X., Jakoš, T., Zhu, J., & Yuan, Y. (2023). RAGE pathways play an important role in regulation of organ fibrosis. *Life Sciences*, *323*, 121713.

Lüth, H. J., Ogunlade, V., Kuhla, B., Kientsch-Engel, R., Stahl, P., Webster, J., Arendt, T., & Münch, G. (2005). Age-and stage-dependent accumulation of advanced glycation end products in intracellular deposits in normal and Alzheimer's disease brains. *Cerebral Cortex*, *15*(2), 211–220.

Manakkadan, A., Krishnan, D., Rui Xia Ang, S., & Sajikumar, S. (2024). Slow release of hydrogen sulfide in CA1 hippocampal neurons rescues long-term synaptic plasticity and associativity in an amyloid-β induced model of Alzheimer's disease. *Journal of Alzheimer's Disease*, *101*(3), 913–921.

Manigrasso, M. B., Pan, J., Rai, V., Zhang, J., Reverdatto, S., Quadri, N., DeVita, R. J., Ramasamy, R., Shekhtman, A., & Schmidt, A. M. (2016). Small molecule inhibition of ligand-stimulated RAGE-DIAPH1 signal transduction. *Scientific Reports*, *6*(1), 22450.

Manigrasso, M. B., Rabbani, P., Egaña-Gorroño, L., Quadri, N., Frye, L., Zhou, B., Reverdatto, S., Ramirez, L. S., Dansereau, S., Pan, J., & Li, H. (2021). Small-molecule antagonism of the interaction of the RAGE cytoplasmic domain with DIAPH1 reduces diabetic complications in mice. *Science Translational Medicine*, *13*(621), eabf7084.

Metz, V. V., Kojro, E., Rat, D., & Postina, R. (2012). Induction of RAGE shedding by activation of G protein-coupled receptors. *PLoS One*, *7*(7), e41823.

Monu, Agnihotri, P., & Biswas, S.(2022). AGE/non-AGE glycation: An important event in rheumatoid arthritis pathophysiology. *Inflammation*,*45*(2),477–496.

Mosquera, J. A. (2010). Role of the receptor for advanced glycation end products (RAGE) in inflammation. *InvestigacionClinica*, *51*(2), 257–268.

Muhammad, S., Barakat, W., Stoyanov, S., Murikinati, S., Yang, H., Tracey, K. J., Bendszus, M., Rossetti, G., Nawroth, P. P., Bierhaus, A., & Schwaninger, M. (2008). The HMGB1 receptor RAGE mediates ischemic brain damage. *Journal of Neuroscience*, *28*(46), 12023–12031.

Nowotny, K., Jung, T., Höhn, A., Weber, D., & Grune, T. (2015). Advanced glycation end products and oxidative stress in type 2 diabetes mellitus. *Biomolecules*, *5*(1), 194–222.

Okuma, Y., Liu, K., Wake, H., Zhang, J., Maruo, T., Date, I., Yoshino, T., Ohtsuka, A., Otani, N., Tomura, S., & Shima, K. (2012). Anti–high mobility group box-1 antibody therapy for traumatic brain injury. *Annals of Neurology*, *72*(3), 373–384.

Ooi, H., Nasu, R., Furukawa, A., Takeuchi, M., & Koriyama, Y. (2022). Pyridoxamine and aminoguanidine attenuate the abnormal aggregation of β-tubulin and suppression of neurite outgrowth by glyceraldehyde-derived toxic advanced glycation end-products. *Frontiers in Pharmacology*, *13*, 921611.

Park, H., & Boyington, J. C. (2010). The 1.5 Å crystal structure of human receptor for advanced glycation endproducts (RAGE) ectodomains reveals unique features determining ligand binding. *Journal of Biological Chemistry*, *285*(52), 40762–40770.

Patil, G., Kulsange, S., Kazi, R., Chirmade, T., Kale, V., Mote, C., Aswar, M., Koratkar, S., Agawane, S., & Kulkarni, M. (2022). Behavioral and proteomic studies reveal methylglyoxal activate pathways associated with Alzheimer's disease. *ACS Pharmacology & Translational Science*, *6*(1), 65–75.

Paudel, Y. N., Angelopoulou, E., Piperi, C., Othman, I., Aamir, K., & Shaikh, M. F. (2020). Impact of HMGB1, RAGE, and TLR4 in Alzheimer's disease (AD): from risk factors to therapeutic targeting. *Cells*, *9*(2), 383.

Penumutchu, S. R., Chou, R. H., & Yu, C. (2014). Structural insights into calcium-bound S100P and the V domain of the RAGE complex. *PloS One*, *9*(8), e103947.

Petzold, A., Jenkins, R., Watt, H. C., Green, A. J. E., Thompson, E. J., Keir, G., Fox, N. C., & Rossor, M. N. (2003). Cerebrospinal fluid S100B correlates with brain atrophy in Alzheimer's disease. *Neuroscience Letters*, *336*(3), 167–170.

Ramasamy, R., Vannucci, S. J., Yan, S. S. D., Herold, K., Yan, S. F., & Schmidt, A. M. (2005). Advanced glycation end products and RAGE: A common thread in aging, diabetes, neurodegeneration, and inflammation. *Glycobiology*, *15*(7), 16R–28R.

Rayathala, J., Kumar, K., & Venkatesh, P. (2022). Review on Alzheimer's disease: Past, present and future. *Journal of Innovations in Applied Pharmaceutical Science (JIAPS)*, *7*(1), 28–31.

Reddy, V. P., Aryal, P., & Darkwah, E. K. (2022). Advanced glycation end products in health and disease. *Microorganisms*, *10*(9), 1848.

Reddy, V. P., Aryal, P., & Soni, P. (2023). RAGE inhibitors in neurodegenerative diseases. *Biomedicines*, *11*(4), 1131.

Rosenbaum, D., & Saftig, P. (2024). New insights into the function and pathophysiology of the ectodomain sheddase A Disintegrin and Metalloproteinase 10 (ADAM10). *The FEBS Journal*, *291*(13), 2733–2766.

Rungratanawanich, W., Qu, Y., Wang, X., Essa, M. M., & Song, B. J. (2021). Advanced glycation end products (AGEs) and other adducts in aging-related diseases and alcohol-mediated tissue injury. *Experimental & Molecular Medicine*, *53*(2), 168–188.

Sabbagh, M. N., Agro, A., Bell, J., Aisen, P. S., Schweizer, E., & Galasko, D. (2011). PF-04494700, an oral inhibitor of receptor for advanced glycation end products (RAGE), in Alzheimer disease. *Alzheimer Disease & Associated Disorders*, *25*(3), 206–212.

Salahuddin, P., Rabbani, G., & Khan, R. (2014). The role of advanced glycation end products in various types of neurodegenerative disease: A therapeutic approach. *Cellular and Molecular Biology Letters*, *19*(3), 407–437.

Scaffidi, P., Misteli, T., & Bianchi, M. E. (2002). Release of chromatin protein HMGB1 by necrotic cells triggers inflammation. *Nature*, *418*(6894), 191–195.

Schmidt, A. M., Vianna, M., Gerlach, M., Brett, J., Ryan, J., Kao, J., Esposito, C., Hegarty, H., Hurley, W., & Clauss, M. (1992). Isolation and characterization of two binding proteins for advanced glycosylation end products from bovine lung which are present on the endothelial cell surface. *Journal of Biological Chemistry*, *267*(21), 14987–14997.

Sedaghat, F., & Notopoulos, A. (2008). S100 protein family and its application in clinical practice. *Hippokratia*, *12*(4), 198.

Shen, C., Ma, Y., Zeng, Z., Yin, Q., Hong, Y., Hou, X., & Liu, X. (2017). RAGE-specific inhibitor FPS-ZM1 attenuates AGEs-induced neuroinflammation and oxidative stress in rat primary microglia. *Neurochemical Research*, *42*, 2902–2911.

Shepherd, C. E., Goyette, J., Utter, V., Rahimi, F., Yang, Z., Geczy, C. L., & Halliday, G. M. (2006). Inflammatory S100A9 and S100A12 proteins in Alzheimer's disease. *Neurobiology of Aging*, *27*(11), 1554–1563.

Singh, H., & Agrawal, D. K. (2022). Therapeutic potential of targeting the receptor for advanced glycation end products (RAGE) by small molecule inhibitors. *Drug Development Research*, *83*(6), 1257–1269.

Sirois, C. M., Jin, T., Miller, A. L., Bertheloot, D., Nakamura, H., Horvath, G. L., Mian, A., Jiang, J., Schrum, J., Bossaller, L., & Pelka, K. (2013). RAGE is a nucleic acid receptor that promotes inflammatory responses to DNA. *Journal of Experimental Medicine*, *210*(11), 2447–2463.

Sturchler, E., Galichet, A., Weibel, M., Leclerc, E., & Heizmann, C. W. (2008). Site-specific blockade of RAGE-Vd prevents amyloid-β oligomer neurotoxicity. *Journal of Neuroscience*, *28*(20), 5149–5158.

Tadić, J. M., Bae, H. B., Banerjee, S., Zmijewski, J. W., & Abraham, E. (2012). Differential activation of RAGE by HMGB1 modulates neutrophil-associated NADPH oxidase activity and bacterial killing. *American Journal of Physiology-Cell Physiology*, *302*(1), C249–C256.

Takeuchi, M., Takino, J. I., Sakasai-Sakai, A., Takata, T., & Tsutsumi, M. (2017). Toxic AGE (TAGE) theory for the pathophysiology of the onset/progression of NAFLD and ALD. *Nutrients*, *9*(6), 634.

Tian, J., Avalos, A. M., Mao, S. Y., Chen, B., Senthil, K., Wu, H., Parroche, P., Drabic, S., Golenbock, D., Sirois, C., & Hua, J. (2007). Toll-like receptor 9–dependent activation by DNA-containing immune complexes is mediated by HMGB1 and RAGE. *Nature Immunology*, *8*(5), 487–496.

Twarda-Clapa, A., Olczak, A., Białkowska, A. M., & Koziołkiewicz, M. (2022). Advanced glycation end-products (AGEs): Formation, chemistry, classification, receptors, and diseases related to AGEs. *Cells*, *11*(8), 1312.

Vulichi, S. R., Runthala, A., Begari, N., Rupak, K., Chunduri, V. R., Kapur, S., Chippada, A. R., & Sistla, D. S. M. (2022). Type-2 diabetes mellitus-associated cancer risk: In pursuit of understanding the possible link. *Diabetes & Metabolic Syndrome: Clinical Research & Reviews*, *16*(9), 102591.

Walker, L. E., Griffiths, M. J., McGill, F., Lewthwaite, P., Sills, G. J., Jorgensen, A., Antoine, D. J., Solomon, T., Marson, A. G., & Pirmohamed, M. (2017). A comparison of HMGB1 concentrations between cerebrospinal fluid and blood in patients with neurological disease. *Biomarkers*, *22*(7), 635–642.

Wautier, J. L., & Wautier, M. P. (2020). Cellular and molecular aspects of blood cell–endothelium interactions in vascular disorders. *International Journal of Molecular Sciences*, *21*(15), 5315.

Wuren, T., Huecksteadt, T., Beck, E., Warren, K., Hoidal, J., Ostrand-Rosenberg, S., & Sanders, K. (2021). The receptor for advanced glycation endproducts (RAGE) decreases survival of tumor-bearing mice by enhancing the generation of lung metastasis-associated myeloid-derived suppressor cells. *Cellular Immunology*, *365*, 104379.

Xie, J., Burz, D. S., He, W., Bronstein, I. B., Lednev, I., & Shekhtman, A. (2007). Hexameric calgranulin C (S100A12) binds to the receptor for advanced glycated end products (RAGE) using symmetric hydrophobic target-binding patches. *Journal of Biological Chemistry*, *282*(6), 4218–4231.

Xie, J., Reverdatto, S., Frolov, A., Hoffmann, R., Burz, D. S., & Shekhtman, A. (2008). Structural basis for pattern recognition by the receptor for advanced glycation end products (RAGE). *Journal of Biological Chemistry*, *283*(40), 27255–27269.

Xie, J., Xu, H., Wu, X., Xie, Y., Lu, X., & Wang, L. (2021). Design, synthesis and anti-TNBC activity of Azeliragon triazole analogues. *Bioorganic & Medicinal Chemistry Letters*, *54*, 128444.

Xu, D., Young, J. H., Krahn, J. M., Song, D., Corbett, K. D., Chazin, W. J., Pedersen, L. C., & Esko, J. D. (2013). Stable RAGE-heparan sulfate complexes are essential for signal transduction. *ACS Chemical Biology*, *8*(7), 1611–1620.

Xu, X. Y., Deng, C. Q., Wang, J., Deng, X. J., Xiao, Q., Li, Y., He, Q., Fan, W. H., Quan, F. Y., Zhu, Y. P., & Cheng, P. (2017). Plasma levels of soluble receptor for advanced glycation end products in Alzheimer's disease. *International Journal of Neuroscience*, *127*(5), 454–458.

Xue, J., Rai, V., Singer, D., Chabierski, S., Xie, J., Reverdatto, S., Burz, D. S., Schmidt, A. M., Hoffmann, R., & Shekhtman, A. (2011). Advanced glycation end product recognition by the receptor for AGEs. *Structure*, *19*(5), 722–732.

Yamagishi, S. I., Nakamura, N., Suematsu, M., Kaseda, K., & Matsui, T. (2015). Advanced glycation end products: A molecular target for vascular complications in diabetes. *Molecular Medicine*, *21*, S32–S40.

Yang, Y., Liu, G., Li, F., Carey, L. B., Sun, C., Ling, K., Tachikawa, H., Fujita, M., Gao, X. D., & Nakanishi, H. (2022). Receptor for advanced glycation end-products (RAGE) mediates phagocytosis in nonprofessional phagocytes. *Communications Biology*, *5*(1), 824.

Yatime, L., & Andersen, G. R. (2013). Structural insights into the oligomerization mode of the human receptor for advanced glycation end-products. *The FEBS Journal*, *280*(24), 6556–6568.

Yatime, L., Betzer, C., Jensen, R. K., Mortensen, S., Jensen, P. H., & Andersen, G. R. (2016). The structure of the RAGE: S100A6 complex reveals a unique mode of homodimerization for S100 proteins. *Structure*, *24*(12), 2043–2052.

Yoshizawa, K., Takeuchi, K., Nakamura, T., Ukai, S., Takahashi, Y., Sato, A., Takasawa, R., & Tanuma, S. I. (2021). Antinociceptive activity of the novel RAGE inhibitor, papaverine, in a mouse model of chronic inflammatory pain. *Synapse*, *75*(3), e22188.

Zong, H., Madden, A., Ward, M., Mooney, M. H., Elliott, C. T., & Stitt, A. W. (2010). Homodimerization is essential for the receptor for advanced glycation end products (RAGE)-mediated signal transduction. *Journal of Biological Chemistry*, *285*(30), 23137–23146.

13 Targeting Alzheimer's Disease Using Neurotrophic Agents and Mimetics
Current Strategies and Challenges

Shubham Nilkanth Rahmatkar, Amit Kumar, and Damanpreet Singh

13.1 INTRODUCTION

Alzheimer's disease (AD) is an incurable, progressive neurodegenerative disease that gradually affects learning and memory abilities and, ultimately, the capacity to perform the most basic daily task. Initially reported by Alois Alzheimer in 1906, AD is a widely prevalent form of dementia characterized by numerous neuropathological hallmarks, including extracellular beta-amyloid (Aβ) plaques, intracellular accumulation of hyperphosphorylated tau protein leading to neurofibrillary tangles (NFTs), neuronal inflammation, synaptic dysfunction, and neuronal death. There are about 57 million individuals who have dementia, and it is likely to increase to 152 million by 2050 (Nichols et al., 2022). Alterations in dietary habits, lack of exercise, a history of hypertension, depression, obesity, tobacco use, and poor educational background are among the risk factors for AD (Silva et al., 2019). The symptoms of AD typically appear later in life. The primary clinical indications of AD include weakened memory, loss of directionality, language problems, and anxious behaviour. Apart from decreased cognition functions, loss of physical activities and aberrant mental activity, including emotion and behaviour, are considered late-stage signs of AD that finally cause disability and death (Bature et al., 2017). The existing drug regimen is primarily symptomatic, with no long-term therapeutic advantages for functional, behavioural, or cognitive symptoms in AD. In the present scenario, there is an urgent need to find novel approaches to therapy with disease-modifying abilities for AD. Consequently, there is great interest in preventive actions and interventions that promote neuroprotection.

Neurotrophins are naturally occurring proteins essential for the regular functioning of the central nervous system (CNS) and restoration from injury in an adult organism. The brain-derived neurotrophic factor (BDNF), nerve growth factor (NGF), neurotrophin-3 (NT-3), and neurotrophin-4/5 (NT-4–5) are the prominent members of the mammalian neurotrophin family (Rahmatkar & Singh, 2025). These neurotrophins regulate multiple biological processes in the CNS, comprising proliferation, differentiation, migration, cell metabolism, synaptic plasticity, the cell cycle, phenotypic equilibrium, and neuroplasticity support (Gudasheva et al., 2021). Alterations in the levels of neurotrophins have been linked to neurodegeneration and cognitive decline in AD. Among the potential treatment paths, neurotrophins have attracted much interest. This chapter emphasizes the role of neurotrophic agents and mimetics in combating AD as well as their potential benefits, drawbacks, and future directions.

13.2 OVERVIEW OF NEUROTROPHINS AND ALZHEIMER'S DISEASE

Dementia results in the gradual deterioration of mental capabilities, including memory, language, and thinking skills, affecting patients alongside their caregivers. AD is the most prevalent form of dementia, constituting 60% of cases internationally, and is the sixth leading cause of mortality globally (Gadhave et al., 2024). Symptoms of AD in its early stages involve memory loss, confusion, decreased judgement, and trouble doing routine activities. As the disease continues, it impacts more of the brain, resulting in a progressive loss of identity. In its latter stages, AD might cause mortality due to complications such as pneumonia or heart disease.

The major risk factor for AD progression is age and the global increase in population results in spread of the disease drastically within the next decade. A recent study suggested that conditions like diabetes, high blood pressure, obesity, hearing loss, lack of physical activity, depression, smoking, low educational attainment (Livingston et al., 2017), alongside of excessive alcohol consumption, brain injury, and air pollution and social withdrawal are some of the risk factors for the progression of dementia (Livingston et al., 2020). As there is no exact cure for dementia, approximately one-third of the cases could be avoided or postponed by controlling modifiable risk factors. Various diagnosis tools have been developed to identify AD-associated dementia that are primarily based on the cerebrospinal fluid (CSF) biomarkers cognitive impairment measures, and indicators of neuronal death, including medial temporal lobe atrophy or parietotemporal hypoperfusion (Jack Jr et al., 2018, Vos et al., 2015).

AD is known to decline cerebral grey matter, shrinkage of the brain, the presence of amyloid plaques, expansion of ventricles as a consequence of neuronal loss, NFTs, and amyloid deposits (Wu et al., 2021). Amyloid plaques and NFTs propagate distinctly with an imbalance between

DOI: 10.1201/9781003602477-13

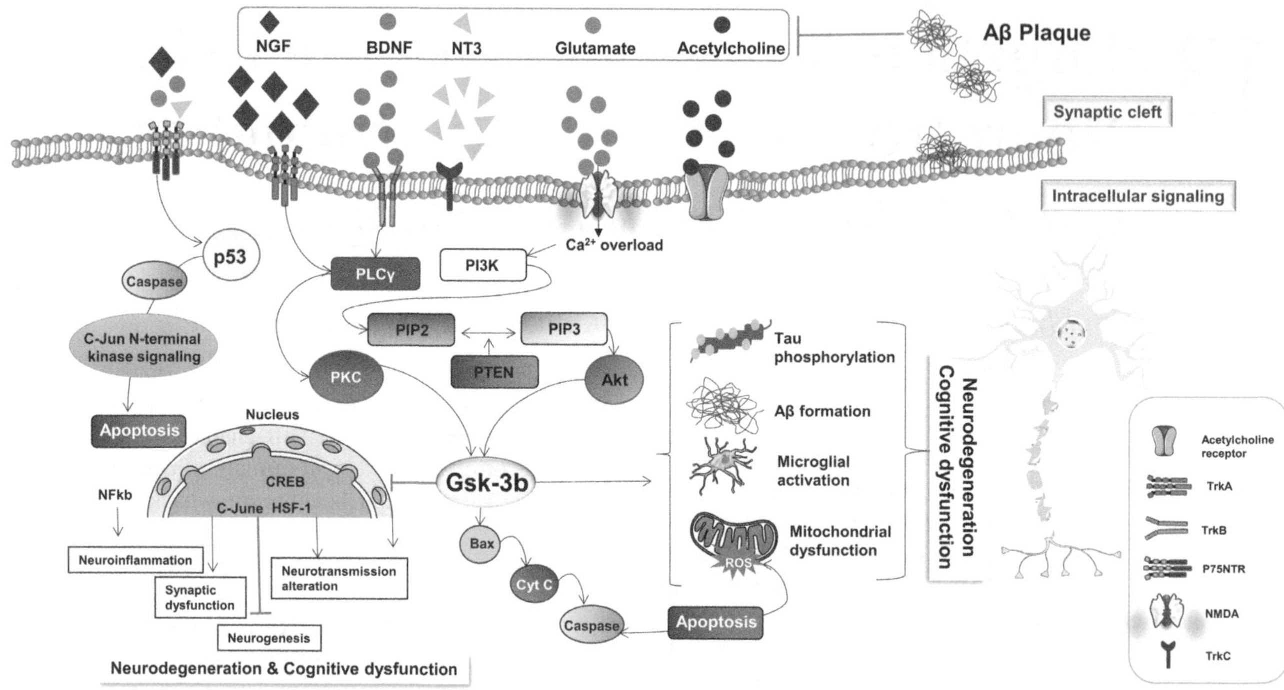

FIGURE 13.1 Schematic representation of neurotrophins BDNF, NGF, and NT3 signalling via Trk receptors, activating PI3K/AKT and PLCγ/PKC pathways. These cascades regulate neuronal survival, plasticity, and stress responses through effectors like CREB, c-Jun, and HSF1. In Alzheimer's disease, dysregulation of p53, GSK, PTEN, and PIP2 contributes to synaptic dysfunction, tau phosphorylation, and neurodegeneration. BDNF – brain-derived neurotrophic factor; NGF – nerve growth factor; NT3 – neurotrophin-3; PI3K – phosphoinositide 3-kinase; AKT – protein kinase B; PLCγ – phospholipase C gamma; PKC – protein kinase C; p53 – tumor protein p53; GSK – glycogen synthase kinase; PTEN – phosphatase and tensin homolog; PIP2 – phosphatidylinositol 4,5-bisphosphate; CREB – cAMP response element-binding protein; c-Jun – jun proto-oncogene; HSF1 – heat shock factor 1.

neurogenesis and neurodegeneration, early synaptic loss, resulting in cognitive decline (Figure 13.1). The disease manifests and debilitates gradually, starting with memory loss and subsequently damaging multiple regions of the brain. The pathological features primarily manifest in the frontal, temporal, and parietal cortices and hippocampus, along with the cholinergic nuclei of the basal forebrain (Auld et al., 2002). Hence, it is essential to comprehend the in-depth mechanisms of AD and design effective therapies. In recent years, numerous therapeutic candidates have been tested for AD, but none have consistently shown favourable effects on disease progression. Modern research is now focused on identifying new targets, with dysregulation of the neurotrophic signalling system believed to be present in the early stages of AD, as evident by research on various experimental models and human post-mortem samples.

The hippocampal neurogenesis is essential for learning and memory. Impaired neurogenesis can result in learning deficits. In adults, the survival and integration of new neurons in the hippocampus rely on neurotrophic factors. Neurotrophins are the most well-studied subset of the neurotrophic factors, including NGF, BDNF, NT-4/5, and NT-3, that support cell maintenance, growth, and survival in the central and peripheral nervous system. Neurotrophins are initially produced in immature forms and then processed into their mature forms through proteolytic cleavage (Teng et al., 2010). The mature molecules exert their effects via selective, rather than specific, mechanisms through high-affinity receptors (TrkA, TrkB, and TrkC) and a single low-affinity receptor (p75NTR) to promote neuronal functions (Figure 13.1). The regulation of neurotrophin levels is crucial for connecting neurons throughout development. Neurotrophins are of great interest now a days due to their therapeutic potential and their mechanistic role in various diseases. However, despite extensive research, their medical use has not been fully explained. Dysregulation of neurotrophic signalling is implicated in the development and progression of neurodegenerative diseases (Miranda-Lourenço et al., 2020).

Neurotrophin-based therapies have effectively shown beneficial effects in neurodegenerative diseases in *in-vivo* models. Among these, NGF has shown a crucial role in modulating depression, schizophrenia, and bipolar disorders and improving cognitive deficits in AD (Mondal & Fatima, 2019). Similarly, BDNF promoted neuronal activity and protected stroke and brain damage. BDNF during a stroke reduced brain damage and neurological issues in rats (Alfonsetti et al., 2023; Rahman et al., 2023). Abnormal build-up of Aβ and tau tangles disrupted communication between brain cells, triggered inflammation, and gradually led to the loss of neurons (Scheltens et al., 2021). Neurotrophins are crucial in supporting neuron survival, growth, and memory function, and activating key intracellular signalling pathways such as PI3K/Akt, MAPK/ERK, and PLCγ (Gao et al., 2022). Aβ disrupted the glutamate signalling, leading to abnormal activation of NMDA receptors. The overstimulation causes excessive influx of calcium ions into neurons, which triggers oxidative stress, mitochondrial dysfunction, and cell damage, accelerating cognitive decline and neurodegeneration (Zhang et al., 2016). One of the key downstream signalling cascades is PI3K/Akt activation, which regulates Glycogen Synthase Kinase-3 beta (Gsk-3β), involved in multiple cellular processes. Normally, Akt phosphorylates and inhibits Gsk-3β, thereby promoting cell survival and preventing abnormal phosphorylation of tau protein. However, in AD, impaired BDNF-TrkB signalling leads to reduced activation of these protective pathways, resulting in overactivation of Gsk-3β (Allen et al., 2011). Hyperactive Gsk-3β contributes to tau hyperphosphorylation, leading to the formation of NFTs, a pathological hallmark of AD. Gsk-3β is involved in promoting the production and aggregation of Aβ peptides by increasing β- and γ-secretase activity (Chauhan et al., 2022), resulting in Aβ plaque formation that disrupts neuronal communication through direct inhibition of neurotransmitters and their receptors (Figure 13.1). The two after effects of tau pathology and Aβ build-up also cause microglial activation, which ultimately results in a chronic pro-inflammatory state. Neuroinflammation further causes neuronal damage and synaptic loss. Aβ and activated Gsk-3β both affect mitochondrial function by increasing oxidative stress, decreasing ATP synthesis, and changing mitochondrial dynamics. Cognitive decline is further affected by the combined effects of tau tangles, Aβ plaques, and inflammation (Garbuz et al., 2021). Another key neurotrophin, NT-3, influences mood and anxiety disorders by affecting neurotransmitters, synaptic plasticity, and neurogenesis (de Miranda et al., 2020). Additionally, NT-4/5 activates TrkB to enhance synaptic activity and plasticity. Although NT-4/5 and BDNF have different effects on TrkB signalling, NT-4/5 causes slower and prolonged signalling (Loch et al., 2015; Xue et al., 2021). In the BDNF-TrkB signalling pathway, key transcription factors such as CREB (cAMP response element-binding protein), c-Jun, and HSF-1 (heat shock factor 1) play crucial roles in maintaining neuronal health and adaptive stress responses. Activated Gsk-3β results in inhibiting these key transcription factors, causing synaptic dysfunction, neurotransmitter alteration. and inhibition of neurogenesis (Figure 13.1) (Rahmatkar & Singh, 2025).

13.3 SMALL-MOLECULE MIMETICS AND THEIR THERAPEUTIC POTENTIAL

The compounds called neurotrophin mimetics replicate the structure and action of neurotrophic factors. Most mimetics bind to neurotrophin receptors in a non-competitive way to control cellular protein expression and encourage neurogenesis. These mimetics may also operate as receptor agonists or antagonists. Several experimental studies are investigating the possibility of neurotrophin mimetics for treating neurodegenerative illnesses because of their enhanced pharmacokinetic characteristics, tissue penetration capacity, and therapeutic effects on neuroprotection, neurogenesis, and synaptic plasticity. Unlike natural neurotrophins, these mimetics may trigger signalling pathways (Madhubala et al., 2024). AD progression is a complicated process with several potential therapeutic interventions. Small-molecule neurotrophic mimetics (Table 13.1) can target specific disease features, including tau hyperphosphorylation, Aβ aggregation, and neurotoxicity.

13.3.1 Peptide Analogues

Peptides are shorter sequences of amino acids that emerged as promising therapeutic agents due to their high target specificity, strong efficacy, and low toxicity. Peptide-based therapeutics have better brain accessibility than large monoclonal antibodies used in immunization strategies (Baazaoui & Iqbal, 2022). Peptide drug delivery faces challenges such as hindrance by the endothelial cell wall, plasma protein binding, degradation by the enzymes of the CNS etc. Overcoming these challenges is crucial for a successful drug design, with proper chemical modifications, such as conjugation with small molecules or incorporating non-natural amino acids (Lien & Lowman, 2003). Peptide therapies provide more delivery alternatives than intravenous injections, improving patient compliance and lowering treatment costs. Synthetic and natural peptides are being tested to treat AD-like disorders.

The small-cyclized dimeric hydrophilic, conserved domain C5 peptide of NGF (KGKE) exerts mimetic neurotrophic activity and protects from neuronal death in dorsal root ganglion neurons. The inhibitory activity of neuronal outgrowth was reversed by increasing the NGF concentration (Longo et al., 1990). Another study on a cyclic peptide (CATDIKGAEC) targeted p75NTR for its potential to ameliorate neuronal damage caused by Aβ 1–40. The peptide strongly protected neurons *in-vitro*, prevented Aβ and NGF binding to p75NTR, and suppressed JNK activation and c-Jun transcription (Yaar et al., 2007). An *in-vivo* study in mouse using CATDIKGAEC that mimics the hazardous KGA motif of Aβ inhibit its binding to p75NTR. Injecting the peptide and Aβ into the mouse cortex markedly reduced neuronal inflammation, thus potentially rescuing it from neurodegeneration via interacting with Aβ-p75NTR (Yaar et al., 2008).

Peptide 021 (P021) is a small water-soluble ciliary neurotrophic factor (CNTF) peptide that can easily cross the blood-brain barrier (BBB) and has favorable pharmacokinetics. P021 improves learning and memory deficits in ageing and AD animal models (Kazim & Iqbal, 2016). Oral P021 administration enhanced neurogenesis and neurotrophic effects by blocking the leukaemia inhibitory factor signalling pathway and boosting BDNF transcription and expression. Treatment with P021 in 3×Tg-AD mice remarkably improves cognitive abilities and synaptic plasticity,

TABLE 13.1
Neuroprotective Mechanisms of Peptide and Non-Peptide Neurotrophic Modulators for Targeting Alzheimer's Disease

Sr. No.	Molecule	Neurotoxicity Model	Structural Features	Mechanisms and Outcomes	Reference
1.	KGKE	NE	Conserved domain C5 peptide of NGF	NGF mimetic neurotrophic activity and protects from neuronal death.	Longo et al. (1990)
2.	Cyclic peptide-CATDIKGAEC	*In-vitro* Aβ 1–40-induced neurotoxicity; *In-vivo* Aβ 1–42-induced neurotoxicity	NGF domain homologous to the mutated NGF-β hairpin loop	Ameliorate *in-vitro* neuronal damage, prevented Aβ binding to p75NTR, and suppressed JNK activation, transcription, and reduced neuronal inflammation.	Yaar et al. (2007), Yaar et al. (2008)
3.	Peptide 021	*In-vivo-transgenic* 3×Tg-AD mouse model	Ciliary neurotrophic factor mimetic	Improved cognitive abilities and synaptic plasticity, reduces abnormal tau hyperphosphorylation, and decreases Aβ generation via elevated BDNF expression and GSK-3β inhibitory activity.	Kazim et al. (2014), Wei et al. (2021)
4.	Tavilermide (MIM-D3)	*In-vivo-transgenic* TgCRND8 mouse model	Partial agonist of the TrkA receptor	Restored neurotrophin signalling, enhanced BBB permeability, activated TrkA-related signalling pathways, and improved cholinergic neurotransmission.	Xhima et al. (2020)
5.	GK-2	*In-vivo-streptozotocin-induced rat* model	Dimeric dipeptide mimetic of NGF loop 4	Improved habituation and spatial memory by causing the phosphorylation of TrkA receptors and Akt-kinases.	Povarnina et al. (2013)
6.	GSB-214	*In-vivo*-scopolamine and streptozotocin-induced AD models	Dimeric dipeptide mimicking the BDNF loop 1	Decreased memory deficits via activation of TrkB, PI3K/AKT, and PLC γ1.	Yu Povarnina et al. (2022)
7.	TAT-TrkB	*In-vivo-transgenic 5XFAD mouse model*	BDNF mimetic on the TrkB	Prevented TrkB-FL cleavage and rescues Aβ-induced synaptic deficits, improved cognitive abilities, restored synaptic plasticity, and slows Tau pathology without causing liver or kidney damage.	Fonseca-Gomes et al. (2024)
8.	cBDNF (1–12) and cNT-3 (1–13)	*In-vitro* SH-SY5Y cell culture	Cyclic mimetics of BDNF and NT-3	Boosted CREB phosphorylation with BDNF and VEGF production, bind Cu^{2+} with great affinity, acting as ionophores, activated Trk/CREB signalling, and improved neuron differentiation.	Magrì et al. (2024)
9.	7,8-DHF	*In-vivo-transgenic 5XFAD mouse model*	TrkB receptor agonist	Restored TrkB signalling and reduced Aβ levels, represses BACE1 expression, improved spatial memory, and increased spine density.	Jang et al. (2010), Andero et al. (2011), Zhang et al. (2014)
10.	R13	*In-vivo-transgenic 5XFAD mouse model*	prodrug of 7,8-DHF-TrkB agonist	Improved cognitive ability, reduced Aβ deposition, senile plaques, and Aβ levels in the different brain regions.	Chen et al. (2018)
11.	CF3CN	*In-vivo-transgenic* 3×Tg-AD *mouse model*	TrkB receptor agonist	Cognitive improvement by separately and synergistically blocking δ-secretase, promoting TrkB/Akt signalling, suppressing δ-secretase activity, decline in APP and Tau cleavage, lowering Aβ load.	Liao et al. (2021)

(Continued)

TABLE 13.1 ***(Continued)***
Neuroprotective Mechanisms of Peptide and Non-Peptide Neurotrophic Modulators for Targeting Alzheimer's Disease

Sr. No.	Molecule	Neurotoxicity Model	Structural Features	Mechanisms and Outcomes	Reference
12.	ACD856	*In-vitro* Aβ 1–42 induced neurotoxicity, *In-vivo* model	Pan-Trk positive allosteric modulator	Improved neurorestoration, resilience, and cognition, boosted SNAP25 and phospho-TrkB levels, promoted BDNF expression, improved NGF-induced neurite outgrowth, and protected cortical neurons.	Parrado Fernandez et al. (2023)
13.	ENT-A013	*In-vitro* primary Dorsal Root Ganglionic neuronal cell culture	Selective TrkA agonist	Selective activation of TrkA receptor and protection of primary hippocampus neurons against oligomeric Aβ toxicity to maintain APP processing, protection of synapse degradation and LTP impairment.	Rogdakis et al. (2022)
14.	MT2	*In-vitro* primary neuronal cell culture	NGF mimetic	Activated neuronal survival-promoted pathways via TrkA autophosphorylation, modulated MAPK and Akt signalling, and increased MKP-1.	Scarpi et al. (2012)
15.	LM11A-31 and LM11A-24	*In-vitro* Aβ 1–42 induced neurotoxicity; *In-vivo* APP, tauP301S transgenic mice models	p75NTR modulators	Inhibited of Aβ-induced Fyn activation, modulated tau phosphorylation, suppressed Aβ-driven synaptotoxic signalling, reduced excessive activation of hippocampal cdk5 and JNK kinases, calpain, and phosphorylation of coflin.	Yang, Tran et al. (2020), Yangv, Liu et al. (2020)
16.	LMDS1	*In-vitro* primary neuronal culture; *In-vivo* Aβ 25–35 induced neurotoxicity mouse model	TrkB agonist	Increased activity on dendritic arborization of neurons.	Fan et al. (2020)
17.	LM22B-10	*In-vitro* Hippocampal neuronal cultures	TrkB and TrkC mimetic	Improved hippocampal cell survival and neurite outgrowth with better brain availability to exert neuroprotective effects.	Yang et al. (2008)
18.	PTX-BD10-2	*In-vivo* transgenic $APP^{L/S}$ mouse model	Analogue of LM22B-10	Reversed BFCN loss, retained cholinergic signalling, and diminished abnormal tau phosphorylation in the brain, protected human cholinergic neurons from Aβ via targeting TrkB and TrkC.	Gonzalez et al. (2022)

Aβ: amyloid-beta, **NGF**: nerve growth factor, **BDNF**: brain-derived neurotrophic factor, **TrkA**: tropomyosin receptor kinase A, **TrkB**: tropomyosin receptor kinase B, **Akt**: protein kinase B, **JNK**: c-Jun N-terminal kinase, **PLCγ1**: phospholipase C gamma 1, **CREB**: cAMP response element binding protein, **7,8-DHF**: 7,8-dihydroxyflavone, **TgCRND8**: transgenic chronic amyloid precursor protein mutant mouse model, **3×Tg-AD**: triple-transgenic Alzheimer's disease model, **5XFAD**: five familial Alzheimer's disease mutations model, **APP**: amyloid precursor protein, **Fyn**: Fyn kinase, **CDK5**: cyclin-dependent kinase 5.

reduces abnormal hyperphosphorylation of tau in the hippocampus, and decreases Aβ generation. The observed effect is likely due to increased BDNF expression, leading to reduced Gsk-3β activity in P021-treated 3×Tg-AD mice (Kazim et al., 2014) (Figure 13.2). Furthermore, Wei et al. (2021) observed that early postnatal treatment with neurotrophic P021 prevents AD-like behaviour and synaptic dysfunction (Table 13.1). The observed neuroprotective effect was further confirmed by the involvement of the BDNF/TrkB pathway. The hippocampal BDNF mRNA and protein levels were significantly higher in 3×Tg-P021 mouse compared to in 3×Tg-Vh mice. These results indicate that initiation of P021 treatment at birth can increase the level of BDNF in mice.

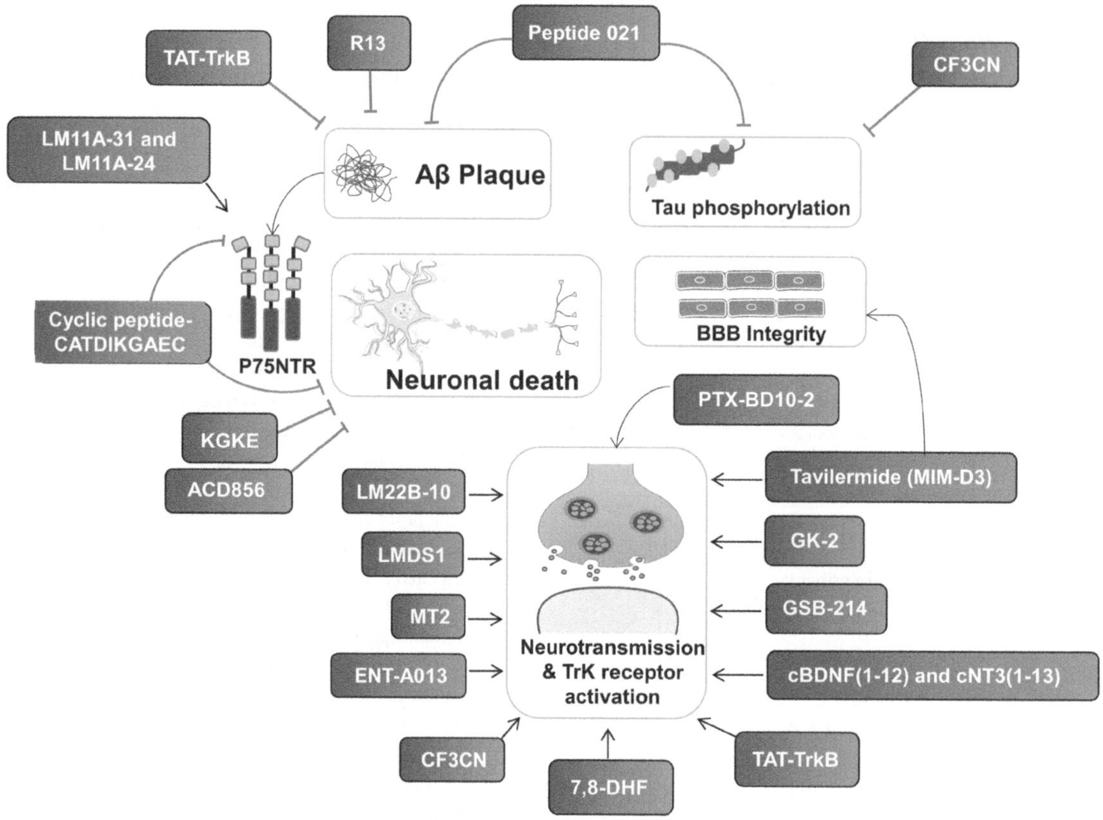

FIGURE 13.2 Schematic illustration of the neuroprotective mechanisms of peptide and non-peptide neurotrophic modulators for targeting Alzheimer's disease. These modulators, including BDNF, NGF, and NT-3 mimetics, exert their effects via diverse molecular pathways to support neuronal health and reduce neurodegeneration.

Cholinergic neuron loss is a key aspect of AD. In studies involving neuronal injury and ageing in animals, NGF has been shown to boost the survival and function of cholinergic neurons, leading to enhanced cognition. Tavilermide, also known as MIM-D3, is a small-molecule mimetic that acts as a partial agonist of the TrkA receptor. TrkA agonist D3 successfully restored neurotrophin signalling in amyloid precursor protein transgenic mice (TgCRND8). Treatment with D3 via intravenous injection and non-invasive administration to the basal forebrain using MRI-guided focused ultrasound enhanced BBB permeability, activated TrkA-related signalling pathways, and improved cholinergic neurotransmission (Xhima et al., 2020). Another dimeric dipeptide mimetic, hexamethylenediamide bis-(N-monosuccinyl- glutamyl-lysine) GK-2, based on NGF loop 4, produced a neuroprotective effect *in-vitro* and animal models of neurodegeneration. In AD models, low doses of systemic treatment of GK-2 improved habituation and spatial memory by causing the phosphorylation of TrkA receptors and Akt-kinases, thus indicating its potential as a therapeutic candidate for AD (Povarnina et al., 2013).

BDNF is recognized to play a neuroprotective role during AD. Whole BDNF still has limited therapeutic use due to its macromolecular protein structure. Bis-(N-monosuccinyl-L-methionyl-L-serine) heptamethylene diamide (GSB-214), a dimeric dipeptide mimicking the BDNF loop 1, has been shown to have neuroprotective properties. GSB-214 stimulates TrkB, PI3K/Akt, and PLCγ1 *in-vitro*. When injected intraperitoneally, GSB-214 demonstrated significant neuroprotective efficacy in a rat brain stroke model and enhanced memory in a novel object recognition test. In scopolamine and streptozotocin-induced AD models, GSB-214 substantially decreased memory deficits, primarily through Trk receptor activity. Its positive effects were counteracted by the Trk inhibitor K252A, (Yu Povarnina et al., 2022).

In AD, Aβ-induced dissociation of TrkB-FL impairs BDNF signalling, affecting nerve survival and synaptic activity. This TrkB-FL cleavage can be detected early in brain and CSF samples. The TAT-TrkB peptide mimics the function of BDNF on the TrkB receptor. It inhibits the breakdown of the full-length TrkB receptor (TrkB-FL) by calpain, maintaining the neuroprotective abilities of TrkB. A study on TAT-TrkB peptide that prevents TrkB-FL cleavage and rescues Aβ-induced synaptic deficits *in-vitro* and *in-vivo* offers evidence of concept for a safe, disease-modifying medication for AD. In 5XFAD mouse, it improved cognitive abilities, restored synaptic plasticity, and slowed Tau pathology without causing liver or kidney damage (Fonseca-Gomes et al., 2024).

Cyclic natural peptides are currently postulated as possible pharmacological molecular entities, combining the benefits of small compounds, such as membrane permeability and oral availability, with the capability to interact with protein surfaces. cBDNF (1–12) and cNT-3 (1–13) have neurotrophin-like actions, acting as both trophic and angiogenic factors and boosting CREB phosphorylation with BDNF and VEGF production (Table 13.1). The cyclic peptides cBDNF (1–12) and cNT-3 (1–13) bind Cu^{2+} with great affinity, acting as ionophores in SH-SY5Y cells. These peptides enhance BDNF, VEGF, and receptor expression, activate Trk/CREB signalling, and improve neuron differentiation in a copper-dependent way (Magrì et al., 2024) (Figure 13.2).

13.3.2 Non-Peptide Analogues

7,8-dihydroxyflavone (7,8-DHF) was found to be a TrkB receptor agonist in cell-based screening, mimicking the effects of BDNF, thus indicating it to be a promising oral therapeutic agent for AD treatment (Jang et al., 2010). Systemic administration of 7,8-DHF activates TrkB receptors in the brain, leading to improved learning, memory, and antidepressant effects in rodents (Andero et al., 2011). Studies have shown that 7,8-DHF ameliorated memory deficits in AD animal models by restoring TrkB signalling and reducing Aβ levels. Additionally, it represses BACE1 expression, improves spatial memory, and increases thin spine density in AD-like neuronal loss models (Zhang et al., 2014). The prodrug approaches were utilized to enhance its bioavailability and stability because of the wide-ranging liver metabolism of 7,8-DHF. The prodrug R13 was developed to improve the reduced oral bioavailability and pharmacokinetic profile 7,8-DHF. A study suggested that the R13 treatment remarkably reduced dose-dependent Aβ deposition, senile plaques and Aβ levels in the brain areas compared to the diseased animals. These findings demonstrated that prolonged oral administration of R13 may stop Aβ deposition and Aβ40 production in 5XFAD mice (Chen et al., 2018).

On the other hand, CF3CN is a small TrkB receptor agonist similar to the actions of BDNF has been studied for its ability to cure neurodegenerative conditions like AD and Parkinson's disease (PD) by activating the BDNF/TrkB signalling pathway. In AD, disruption in BDNF/TrkB signalling increases δ-secretase activation, which causes Aβ and Tau abnormalities. In another study on 3×Tg-AD mouse, a δ-secretase inhibitor (#11a) and a TrkB agonist (CF3CN) improved cognitive function by alone and synergistically blocking δ-secretase. CF3CN promotes TrkB/Akt signalling, suppresses δ-secretase activity, and, together with #11a in 3×Tg-AD mouse leads to a substantial decline in APP and Tau cleavage, lower Aβ load and noteworthy cognitive improvement, (Liao et al., 2021).

The pan-Trk positive allosteric modulator ACD856, a triazinetrione derivative, has favorable pharmacokinetic characteristics and is well tolerated in humans. ACD856 demonstrated a significant pro-cognitive effect, which was responsible for increasing certain neurotransmitter levels and revealed antidepressant-like functions. Reports suggested that ACD856 may improve neurorestoration, resilience, and cognition in models of AD. ACD856 boosted SNAP25 and phospho-TrkB levels, promoted BDNF expression, improved NGF-induced neurite outgrowth, and protected cortical neurons against Aβ and energy-deprivation-induced toxicity *in-vitro*. When administered to aged animals *in-vivo*, ACD856 remarkably improved brain BDNF levels and showed, antidepressant-like effects lasting up to seven days (Parrado Fernandez et al., 2023).

The ENT-A013, a novel and selective TrkA agonist, mimics NGF by activating TrkA and its downstream signalling kinases Akt and Erk1/2. Despite p75NTR-mediated actions, ENT-A013 targets the TrkA receptor and protects primary hippocampus neurons against oligomeric Aβ toxicity while retaining the ability to affect APP processing positively. Lastly, by correcting Aβ-induced synapse degradation and Long term potentiation (LTP) impairment, ENT-A013 improves synaptic functioning (Rogdakis et al., 2022). A small non-peptide substance called MT2 functions as a TrkA receptor agonist, activating the receptor and subsequent signalling pathways to mimic the actions of NGF and encourage cell survival, differentiation, and neurite outgrowth in neuronal cells. The specific signalling of MT2 and its potent neuroprotective characteristics have been recognized as possible therapeutic targets for neurodegenerative diseases as anti-amyloid approaches. Small-molecule MT2 binds preferentially to TrkA, activating survival-promoting pathways via targeted TrkA autophosphorylation and receptor-mediated internalization (Table 13.1). MT2 efficiently controls MAPK and Akt signalling, increases MKP-1, and protects rat hippocampal neurons against metabolic stress and NGF-deficient AD models (Scarpi et al., 2012).

The p75NTR is a potential therapeutic target for AD, as its signalling aligns with pathways that promote Aβ-induced neurite and synapse degradation in disease pathology. p75NTR regulates neuron survival, function, and structure by interacting with various ligands and other receptors. Overexpression of p75NTR was found to be increased in different brain areas, such as basal forebrain cholinergic, hippocampal, cortical, and entorhinal neurons. Aβ aggregates have been identified as potential death-inducing ligands of p75NTR. Aβ toxicity mediated by p75NTR involves c-Jun kinase (JNK) and the transcription factor c-Jun (Knowles et al., 2009). Consequently, modulating p75NTR activity may be an essential disease-modifying treatment approach. The small-molecule LM11A-31, a p75NTR modulator, was reported to inhibit Aβ-induced tau phosphorylation, neuronal inflammation, protein misfolding, neuritic dystrophy, and synaptic loss *in-vitro* and (Nguyen et al., 2014). Exposure of Aβ oligomers in adult hippocampal neurons leads to the loss of drebrin-positive spines and neuritic beading, indicating synaptic degradation associated with AD. Additionally, Aβ exposure in neurons *in-vitro* and APP mutant mouse models led to Fyn kinase activation, enhanced tau hyperphosphorylation and NMDA receptor overactivation, all linked to synaptic dysfunction (Table 13.1). However, LM11A-31 treatment via p75NTR modulation inhibited Aβ-induced Fyn activation, thus modulating tau phosphorylation *in-vitro* and *in-vivo*, which suggests that LM11A-31 suppresses Aβ-driven synaptotoxic signalling through modulation of Fyn kinase activity (Yang, Tran et al., 2020).

The p75NTR protein is reported to activate various tau kinases, calpain, and rhoA-cofilin activities that regulate multiple mechanisms involved in tauopathies. Treatment with the p75NTR modulators LM11A-31 and LM11A-24 in tauP301S mouse for 3 months reduced excessive activation of hippocampal cdk5 and JNK kinases, calpain, and phosphorylation of cofilin. It also decreased tau phosphorylation, insoluble tau aggregates and filaments accumulated, and microglial activation. In addition, LM11A-31 and LM11A-24 treatment reduced tau seeding activity reversed dendritic spine loss in pyramidal neurons and improved dendritic complexity in the hippocampus (Yangv, Liu et al., 2020). A study by Fan et al. (2020) utilized both the Aβ-mediated hippocampal neurons and mouse models to evaluate 12 probable TrkB agonists. 7,8-DHF and LMDS1 were shown to have higher activity on dendritic arborization of neurons compared to other compounds in an *in-vitro* study. Afterwards, LMDS-1 enhanced short-term memory retention by activating TrkB phosphorylation at Y516 in the Aβ-mouse model of AD. An administration of LMDS-1 activated ERK and increased CREB and BDNF levels but decreased Gsk-3β activity and tau phosphorylation.

One more small mimetic, LM22B-10, specifically activated both TrkB and TrkC with their downstream signalling, producing comparable biological effects to BDNF and NT-3 (Yang et al., 2008). *In-vitro*, LM22B-10 improved hippocampus cell survival and neurite outgrowth, whereas *in-vivo* systemic administration has better brain availability to exert neuroprotective effects. PTX-BD10-2, a novel analogue of LM22B-10, is designed to improve BBB permeability when consumed orally. It was found that the PTX-BD10-2 was orally bioavailable and reverse BFCN loss, preserve cholinergic signalling, and reduce abnormal tau phosphorylation in the brain (Gonzalez et al., 2022). Additionally, its analogue protects human cholinergic neurons from Aβ via targeting TrkB and TrkC, suggesting its potential as a treatment option for AD.

13.4 CHALLENGES AND LIMITATIONS IN NEUROTROPHIN THERAPEUTICS

Improving the poor bioavailability and delivery of the neurotrophic factors to degenerating neurons in the CNS could be a powerful way to exert neuroprotective effects. Due to the short half-lives, limited brain availability, and diverse biological effects, neurotrophins face several challenges. One of those challenges is to cross BBB. Poor BBB permeability, uneven plasma levels, unfavourable pharmacokinetics, and systemic side effects limit the therapeutic potential of neurotrophic factors in AD patients. An emerging approach to temporarily open the BBB safely is being utilized to bypass the BBB. A low-frequency focused ultrasound with microbubbles open up the BBB in safe manner to permeate the neurotrophic factors. Clinical trials have confirmed that recombinant BDNF show a low bioavailability and numerous adverse effects (Ochs et al., 2000). A Few clinical studies also support the safety and feasibility of this approach in the patients (Ji et al., 2019). After systemic administration, chemical modification or antibody conjugation with neurotrophins can enhance BBB penetration. A study on rats showed that the covalent modification of NGF with putrescine improved plasma pharmacokinetics and BBB permeability in rats compared to native NGF (Poduslo et al., 1998). Furthermore, CNS delivery of peripherally administered BDNF could be attained by conjugating biotinylated polyethylene glycol-modified-BDNF to a monoclonal antibody via transcytosis through the BBB (Wu & Pardridge, 1999). Apart from BBB integrity and permeability, analysing the probable risks of transplanted cells overexpressing proteins, including tumour formation, host rejection, and inflammation is also important.

Neurotrophins are essential for safeguarding neurons and regulating brain functions. They also impact non-neuronal systems like the immune and vascular systems. This interaction offers treatment possibilities for autoimmune disorders but may pose risks for neurodegenerative diseases. Excessive BDNF levels in the CNS can impair learning and short-term memory, especially in spatial tasks; this indicates that too much BDNF in forebrain circuits could hinder cognitive function (Cunha et al., 2009). Neurotrophins, such as BDNF, regulate neuronal gene activity and impact serotonin pathways. They modulate neurotransmitter dynamics and can rapidly excite central neurons. BDNF and NT-4/5 can depolarize neurons by increasing sodium ion conductance, similar to glutamate, highlighting their excitatory potential in the CNS (Kafitz et al., 1999). Remarkably, NGF can cause cell death in a dose-dependent way by activating the p75NTR receptor, potentially contributing to neurotrophin-induced neurotoxicity (von Bartheld et al., 1994). Infusions of BDNF have been shown to influence behaviour, neuronal activity, and serotonin concentrations. The BDNF administration in a mouse model resulted in hyperactivity and increased food intake, indicating the involvement of hypothalamic nuclei in regulating satiety and locomotor behaviour (Pelleymounter et al., 1995). However, NGF also plays a key role in regulating pain perception by interacting with TrkA and p75NTR receptors, triggering MAPK and NF-κB pathways, which increase pain sensitivity and inflammatory responses.

In both animal and human studies, elevated NGF levels in chronic pain conditions create a feedback loop that aggravates inflammation and hypersensitivity (Seidel et al., 2013). Previously reported pathological investigations have revealed that higher amounts of BDNF and its receptor TrkB are linked to chemotherapy resistance in cancers, including prostate, lung, and breast cancer. These molecular markers are connected to poor prognosis and increased fatalities. By controlling the BDNF/TrkB axis, microRNAs such as miR-101 and miR-107 offer fresh cancer therapy targets (Radin & Patel, 2017). A study suggests that increased neurotrophins like NGF and BDNF can cause enhance cholinergic transmission and sensitize afferent nerve endings in the bladder. The bladder over activity contributes to unusual urinary function in conditions like overactive bladder, as observed in animal models and clinical studies (Kashyap et al., 2018). Further concrete evidences and research are required to control the overproduction of neurotrophic factors to prevent cytotoxic effects and inflammatory responses linked to viral vectors. Therapeutic benefits and potential side effects of neurotrophin activity showed its dual nature. These mimetics selectively target the receptors to modulate specific signalling cascades and replicate the therapeutic effects of neurotrophic factors on neurogenesis, neuronal plasticity, and cognition while having suitable pharmacokinetics and CNS penetration for drug development. These mimetics also avoid unwanted systemic effects and pleiotropic actions by selectively targeting specific receptors (Kazim & Iqbal, 2016) and offer a new therapeutic approach for AD treatment by modulating signalling pathways differently from traditional neurotrophic factors.

13.5 FUTURE DIRECTIONS AND CONCLUSION

Neurotrophic agent-based treatment strategy emerged as an attractive opportunity for AD prevention, particularly by targeting the preliminary synaptic and neuronal dysfunctions that trigger irreversible neuronal degeneration. Few of the major challenges to develop these treatment options are BBB permeability, short half-life, and potential side effects. Future studies should focus on improving the brain permeability, bioavailability, and pharmacokinetic profiling of these mimetics. Additionally, combination treatment could be a more effective approach in the management of AD by targeting various mechanisms or use smaller neurotrophic mimetics than individual neurotrophic factors. Integrating neurotrophic approaches with proteolytic inhibitors can result in synergistic outcomes. Progressive methods of therapy that consider individual patient's genetic and biomarker changes might direct the design of therapeutic approaches. Furthermore, early-phase clinical trials are crucial for confirming safety, efficacy, and long-term cognitive outcomes, paving the way for disease-modifying drugs that understand the heterogeneous AD.

Neurotrophins have diverse effects in the CNS, including playing a key role in neuroprotection and neurogenesis, synaptic plasticity and cognition, potentially offering disease-modifying effects in addressing cognitive dysfunction in AD. A unique and growing strategy for preventing the synaptic loss, tau pathology, and neural degeneration in AD involves the use of neurotrophic factors and their small-molecule mimetics. According to preclinical research, these treatments protect brain circuits and maintain cognitive function by enhancing cellular functions and altering degenerative signalling. A novel therapeutic strategy that differs from existing symptomatic treatments is to target neurotrophic signalling. If neurotrophic approaches are further studied and clinically implemented, there is a great deal of potential to shift the current paradigm in AD from symptomatic therapy to actual disease modification.

ACKNOWLEDGEMENTS

The authors are grateful to the Director, CSIR-Institute of Himalayan Bioresource Technology, Palampur, India, for encouragement and infrastructural support. Shubham N. Rahmatkar is thankful to the CSIR, New Delhi, for providing Senior Research Fellowship vide letter no. 31/GATE/54(06)/2019-EMR-I, respectively. The institute's communication number for this chapter is 5928.

REFERENCES

Alfonsetti, M., d'Angelo, M., & Castelli, V. (2023). Neurotrophic factor-based pharmacological approaches in neurological disorders. *Neural Regeneration Research*, *18*(6), 1220–1228.

Allen, S. J., Watson, J. J., & Dawbarn, D. (2011). The neurotrophins and their role in Alzheimer's disease. *Current Neuropharmacology*, *9*(4), 559–573.

Andero, R., Heldt, S. A., Ye, K., Liu, X., Armario, A., & Ressler, K. J. (2011). Effect of 7, 8-dihydroxyflavone, a small-molecule TrkB agonist, on emotional learning. *American Journal of Psychiatry*, *168*(2), 163–172.

Auld, D. S., Kornecook, T. J., Bastianetto, S., & Quirion, R. (2002). Alzheimer's disease and the basal forebrain cholinergic system: Relations to β-amyloid peptides, cognition, and treatment strategies. *Progress in Neurobiology*, *68*(3), 209–245.

Bature, F., Guinn, B. A., Pang, D., & Pappas, Y. (2017). Signs and symptoms preceding the diagnosis of Alzheimer's disease: A systematic scoping review of literature from 1937 to 2016. *BMJ Open*, *7*(8), e015746.

Chauhan, N., Paliwal, S., Jain, S., Verma, K., Paliwal, S., & Sharma, S. (2022). GSK-3β and its inhibitors in Alzheimer's disease: A recent update. *Mini Reviews in Medicinal Chemistry*, *22*(22), 2881–2895.

Chen, C., Wang, Z., Zhang, Z., Liu, X., Kang, S. S., Zhang, Y., & Ye, K. (2018). The prodrug of 7,8-dihydroxyflavone development and therapeutic efficacy for treating Alzheimer's disease. *Proceedings of the National Academy of Sciences*, *115*(3), 578–583.

Cunha, C., Angelucci, A., D'Antoni, A., Dobrossy, M. D., Dunnett, S. B., Berardi, N., & Brambilla, R. (2009). Brain-derived neurotrophic factor (BDNF) overexpression in the forebrain results in learning and memory impairments. *Neurobiology of Disease*, *33*(3), 358–368.

de Miranda, A. S., de Barros, J. L. V. M., & Teixeira, A. L. (2020). Is neurotrophin-3 (NT-3): A potential therapeutic target for depression and anxiety? *Expert Opinion on Therapeutic Targets*, *24*(12), 1225–1238.

Fan, C. H., Lin, C. W., Huang, H. J., Lee-Chen, G. J., Sun, Y. C., Lin, W., Chen, C. M., Chang, K. H., Su, M. T., & Hsieh-Li, H. M. (2020). LMDS-1, a potential TrkB receptor agonist provides a safe and neurotrophic effect for early-phase Alzheimer's disease. *Psychopharmacology*, *237*, 3173–3190.

Fonseca-Gomes, J., Costa-Coelho, T., Ferreira-Manso, M., Inteiro-Oliveira, S., Vaz, S. H., Alemãn-Serrano, N., Atalaia-Barbacena, H., Ribeiro-Rodrigues, L., Ramalho, R. M., Pinto, R., & Miranda, H. V. (2024). A small TAT-TrkB peptide prevents BDNF receptor cleavage and restores synaptic physiology in Alzheimer's disease. *Molecular Therapy*, *32*(10), 3372–3401.

Gao, L., Zhang, Y., Sterling, K., & Song, W. (2022). Brain-derived neurotrophic factor in Alzheimer's disease and its pharmaceutical potential. *Translational Neurodegeneration*, *11*(1), 4.

Gadhave, D.G., Sugandhi, V.V., Jha, S.K., Nangare, S.N., Gupta, G., Singh, S.K., et al. (2024). Neurodegenerative disorders: Mechanisms of degeneration and therapeutic approaches with their clinical relevance. *Ageing research reviews*, *99*, 102357.

Garbuz, D. G., Zatsepina, O. G., & Evgen'ev, M. B. (2021). Beta amyloid, tau protein, and neuroinflammation: An attempt to integrate different hypotheses of Alzheimer's disease pathogenesis. *Molecular Biology*, *55*, 670–682.

Gonzalez, S., McHugh, T. L., Yang, T., Syriani, W., Massa, S. M., Longo, F. M., & Simmons, D. A. (2022). Small molecule modulation of TrkB and TrkC neurotrophin receptors prevents cholinergic neuron atrophy in an Alzheimer's disease mouse model at an advanced pathological stage. *Neurobiology of Disease*, *162*, 105563.

Gudasheva, T. A., Povarnina, P. Y., Tarasiuk, A. V., & Seredenin, S. B. (2021). Low-molecular mimetics of nerve growth factor and brain-derived neurotrophic factor: Design and pharmacological properties. *Medicinal Research Reviews*, *41*(5), 2746–2774.

Jack Jr, C. R., Bennett, D. A., Blennow, K., Carrillo, M. C., Dunn, B., Haeberlein, S. B., Holtzman, D. M., Jagust, W., Jessen, F., Karlawish, J., & Liu, E. (2018). NIA-AA research framework: Toward a biological definition of Alzheimer's disease. *Alzheimer's & Dementia*, *14*(4), 535–562.

Jang, S. W., Liu, X., Yepes, M., Shepherd, K. R., Miller, G. W., Liu, Y., Wilson, W. D., Xiao, G., Blanchi, B., Sun, Y. E., & Ye, K. (2010). A selective TrkB agonist with potent neurotrophic activities by 7,8-dihydroxyflavone. *Proceedings of the National Academy of Sciences*, *107*(6), 2687–2692.

Ji, R., Smith, M., Niimi, Y., Karakatsani, M. E., Murillo, M. F., Jackson-Lewis, V., Przedborski, S., & Konofagou, E. E. (2019). Focused ultrasound enhanced intranasal delivery of brain derived neurotrophic factor produces neurorestorative effects in a Parkinson's disease mouse model. *Scientific Reports*, *9*(1), 19402.

Kafitz, K. W., Rose, C. R., Thoenen, H., & Konnerth, A. (1999). Neurotrophin-evoked rapid excitation through TrkB receptors. *Nature*, *401*(6756), 918–921.

Kashyap, M. P., Pore, S. K., de Groat, W. C., Chermansky, C. J., Yoshimura, N., & Tyagi, P. (2018). BDNF overexpression in the bladder induces neuronal changes to mediate bladder overactivity. *American Journal of Physiology-Renal Physiology*, *315*(1), F45–F56.

Kazim, S. F., Blanchard, J., Dai, C. L., Tung, Y. C., LaFerla, F. M., Iqbal, I. G., & Iqbal, K. (2014). Disease modifying effect of chronic oral treatment with a neurotrophic peptidergic compound in a triple transgenic mouse model of Alzheimer's disease. *Neurobiology of Disease*, *71*, 110–130.

Kazim, S.F. and Iqbal, K., (2016). Neurotrophic factor small-molecule mimetics mediated neuroregeneration and synaptic repair: emerging therapeutic modality for Alzheimer's disease. *Molecular neurodegeneration*, *11*(1), p.50.

Knowles, J. K., Rajadas, J., Nguyen, T. V. V., Yang, T., LeMieux, M. C., Vander Griend, L., Ishikawa, C., Massa, S. M., Wyss-Coray, T., & Longo, F. M. (2009). The p75 neurotrophin receptor promotes amyloid-β (1–42)-induced neuritic dystrophy in vitro and in vivo. *Journal of Neuroscience*, *29*(34), 10627–10637.

Liao, J., Chen, C., Ahn, E. H., Liu, X., Li, H., Edgington-Mitchell, L. E., Lu, Z., Ming, S., & Ye, K. (2021). Targeting both BDNF/TrkB pathway and delta-secretase for treating Alzheimer's disease. *Neuropharmacology*, *197*, 108737.

Lien, S., & Lowman, H. B. (2003). Therapeutic peptides. *Trends in Biotechnology*, *21*(12), 556–562.

Livingston, G., Huntley, J., Sommerlad, A., Ames, D., Ballard, C., Banerjee, S., Brayne, C., Burns, A., Cohen-Mansfield, J., Cooper, C., & Costafreda, S. G. (2020). Dementia prevention, intervention, and care: 2020 report of the Lancet Commission. *The Lancet*, *396*(10248), 413–446.

Livingston, G., Sommerlad, A., Orgeta, V., Costafreda, S. G., Huntley, J., Ames, D., Ballard, C., Banerjee, S., Burns, A., Cohen-Mansfield, J., & Cooper, C. (2017). Dementia prevention, intervention, and care. *The Lancet*, *390*(10113), 2673–2734.

Loch, A. A., Zanetti, M. V., de Sousa, R. T., Chaim, T. M., Serpa, M. H., Gattaz, W. F., Teixeira, A. L., & Machado-Vieira, R. (2015). Elevated neurotrophin-3 and neurotrophin 4/5 levels in unmedicated bipolar depression and the effects of lithium. *Progress in Neuro-Psychopharmacology and Biological Psychiatry*, *56*, 243–246.

Longo, F. M., Vu, T. K., & Mobley, W. C. (1990). The in vitro biological effect of nerve growth factor is inhibited by synthetic peptides. *Cell Regulation*, *1*(2), 189–195.

Madhubala, D., Mahato, R., Khan, M. R., Bala, A., & Mukherjee, A. K. (2024). Neurotrophin peptidomimetics for the treatment of neurodegenerative diseases. *Drug Discovery Today*, *29*(11), 104156.

Magrì, A., Tomasello, B., Naletova, I., Tabbì, G., Cairns, W. R., Greco, V., Sciuto, S., La Mendola, D., & Rizzarelli, E. (2024). New BDNF and NT-3 cyclic mimetics concur with copper to activate trophic signaling pathways as potential molecular entities to protect old brains from neurodegeneration. *Biomolecules*, *14*(9), 1104.

Miranda-Lourenço, C., Ribeiro-Rodrigues, L., Fonseca-Gomes, J., Tanqueiro, S. R., Belo, R. F., Ferreira, C. B., Rei, N., Ferreira-Manso, M., de Almeida-Borlido, C., Costa-Coelho, T., & Freitas, C. F. (2020). Challenges of BDNF-based therapies: From common to rare diseases. *Pharmacological Research*, *162*, 105281.

Mondal, A. C., & Fatima, M. (2019). Direct and indirect evidences of BDNF and NGF as key modulators in depression: Role of antidepressants treatment. *International Journal of Neuroscience*, *129*(3), 283–296.

Nichols, E., Steinmetz, J.D., Vollset, S.E., Fukutaki, K., Chalek, J., Abd-Allah, F., et al. (2022). Estimation of the global prevalence of dementia in 2019 and forecasted prevalence in 2050: an analysis for the Global Burden of Disease Study 2019. *The Lancet Public Health*, *7*(2), e105–e125.

Nguyen, T. V. V., Shen, L., Vander Griend, L., Quach, L. N., Belichenko, N. P., Saw, N., Yang, T., Shamloo, M., Wyss-Coray, T., Massa, S. M., & Longo, F. M. (2014). Small molecule p75NTR ligands reduce pathological phosphorylation and misfolding of tau, inflammatory changes, cholinergic degeneration, and cognitive deficits in AβPPL/S transgenic mice. *Journal of Alzheimer's Disease*, *42*(2), 459–483.

Ochs, G., Penn, R. D., York, M., Giess, R., Beck, M., Tonn, J., Haigh, J., Malta, E., Traub, M., Sendtner, M., & Toyka, K. V. (2000). A phase I/II trial of recombinant methionyl human brain derived neurotrophic factor administered by intrathecal infusion to patients with amyotrophic lateral sclerosis. *Amyotrophic Lateral Sclerosis and Other Motor Neuron Disorders*, *1*(3), 201–206.

Parrado Fernandez, C., Juric, S., Backlund, M., Dahlström, M., Madjid, N., Lidell, V., Rasti, A., Sandin, J., Nordvall, G., & Forsell, P. (2023). Neuroprotective and disease-modifying effects of the triazinetrione ACD856, a positive allosteric modulator of Trk-receptors for the treatment of cognitive dysfunction in Alzheimer's disease. *International Journal of Molecular Sciences*, *24*(13), 11159.

Pelleymounter, M. A., Cullen, M. J., & Wellman, C. L. (1995). Characteristics of BDNF-induced weight loss. *Experimental Neurology*, *131*(2), 229–238.

Poduslo, J. F., Curran, G. L., & Gill, J. S. (1998). Putrescine-modified nerve growth factor: Bioactivity, plasma pharmacokinetics, blood-brain/nerve barrier permeability, and nervous system biodistribution. *Journal of Neurochemistry*, *71*(4), 1651–1660.

Povarnina, P. Y., Vorontsova, O. N., Gudasheva, T. A., Ostrovskaya, R. U., & Seredenin, S. B. (2013). Original nerve growth factor mimetic dipeptide GK-2 restores impaired cognitive functions in rat models of Alzheimer's disease. *Acta Naturae (англоязычная версия)*, *5*(3(18)), 84–91.

Radin, D. P., & Patel, P. (2017). BDNF: an oncogene or tumor suppressor? *Anticancer Research*, *37*(8), 3983–3990.

Rahman, M. M., Islam, M. R., Supti, F. A., Dhar, P. S., Shohag, S., Ferdous, J., Shuvo, S. K., Akter, A., Hossain, M. S., & Sharma, R. (2023). Exploring the therapeutic effect of neurotrophins and neuropeptides in neurodegenerative diseases: at a glance. *Molecular Neurobiology*, *60*(8), 4206–4231.

Rahmatkar, S. N., & Singh, D. (2025). Decoding the role of neurotrophins in glycogen synthase kinase 3-beta regulation in Alzheimer's disease. *Molecular Neurobiology*, *62*, 1–21.

Rogdakis, T., Charou, D., Latorrata, A., Papadimitriou, E., Tsengenes, A., Athanasiou, C., Papadopoulou, M., Chalikiopoulou, C., Katsila, T., Ramos, I., & Prousis, K. C. (2022). Development and biological characterization of a novel selective TrkA agonist with neuroprotective properties against amyloid toxicity. *Biomedicines*, *10*(3), 614.

Scarpi, D., Cirelli, D., Matrone, C., Castronovo, G., Rosini, P., Occhiato, E. G., Romano, F., Bartali, L., Clemente, A. M., Bottegoni, G., & Cavalli, A. (2012). Low molecular weight, non-peptidic agonists of TrkA receptor with NGF-mimetic activity. *Cell Death & Disease*, *3*(7), e339–e339.

Scheltens, P., De Strooper, B., Kivipelto, M., Holstege, H., Chetelat, G., Teunissen, C. E., Cummings, J., & van der Flier, W. M. (2021). Alzheimer's disease. *The Lancet*, *397*(10284), 1577–1590.

Seidel, M. F., Wise, B. L., & Lane, N. E. (2013). Nerve growth factor: An update on the science and therapy. *Osteoarthritis and Cartilage*, *21*(9), 1223–1228.

Silva, M. V. F., Loures, C. D. M. G., Alves, L. C. V., De Souza, L. C., Borges, K. B. G., & Carvalho, M. D. G. (2019). Alzheimer's disease: Risk factors and potentially protective measures. *Journal of Biomedical Science*, *26*, 1–11.

Teng, K. K., Felice, S., Kim, T., & Hempstead, B. L. (2010). Understanding proneurotrophin actions: Recent advances and challenges. *Developmental Neurobiology*, *70*(5), 350–359.

von Bartheld, C. S., Kinoshita, Y., Prevette, D., Yin, Q. W., Oppenheim, R. W., & Bothwell, M. (1994). Positive and negative effects of neurotrophins on the isthmo-optic nucleus in chick embryos. *Neuron*, *12*(3), 639–654.

Vos, S. J., Verhey, F., Frölich, L., Kornhuber, J., Wiltfang, J., Maier, W., Peters, O., Rüther, E., Nobili, F., Morbelli, S., & Frisoni, G. B. (2015). Prevalence and prognosis of Alzheimer's disease at the mild cognitive impairment stage. *Brain*, *138*(5), 1327–1338.

Wei, W., Liu, Y., Dai, C. L., Baazaoui, N., Tung, Y. C., Liu, F., & Iqbal, K. (2021). Neurotrophic treatment initiated during early postnatal development prevents the Alzheimer-like behavior and synaptic dysfunction. *Journal of Alzheimer's Disease*, *82*(2), 631–646.

Wu, D., & Pardridge, W. M. (1999). Neuroprotection with noninvasive neurotrophin delivery to the brain. *Proceedings of the National Academy of Sciences*, *96*(1), 254–259.

Wu, Z., Peng, Y., Hong, M., & Zhang, Y. (2021). Gray matter deterioration pattern during Alzheimer's disease progression: A regions-of-interest based surface morphometry study. *Frontiers in Aging Neuroscience*, *13*, 593898.

Xhima, K., Markham-Coultes, K., Nedev, H., Heinen, S., Saragovi, H. U., Hynynen, K., & Aubert, I. (2020). Focused ultrasound delivery of a selective TrkA agonist rescues cholinergic function in a mouse model of Alzheimer's disease. *Science advances*, *6*(4), eaax6646.

Xue, Y., Liang, H., Yang, R., Deng, K., Tang, M., & Zhang, M. (2021). The role of pro-and mature neurotrophins in the depression. *Behavioural Brain Research*, *404*, 113162.

Yaar, M., Arble, B. L., Stewart, K. B., Qureshi, N. H., Kowall, N. W., & Gilchrest, B. A. (2008). p75 NTR Antagonistic Cyclic Peptide Decreases the Size of β Amyloid-Induced Brain Inflammation. *Cellular and Molecular Neurobiology*, *28*, 1027–1031.

Yaar, M., Zhai, S., Panova, I., Fine, R. E., Eisenhauer, P. B., Blusztajn, J. K., Lopez-Coviella, I., & Gilchrest, B. A. (2007). A cyclic peptide that binds p75NTR protects neurones from beta amyloid (1–40)-induced cell death. *Neuropathology and Applied Neurobiology*, *33*(5), 533–543.

Yang, T., Knowles, J. K., Lu, Q., Zhang, H., Arancio, O., Moore, L. A., Chang, T., Wang, Q., Andreasson, K., Rajadas, J., & Fuller, G. G. (2008). Small molecule, non-peptide p75NTR ligands inhibit Aβ-induced neurodegeneration and synaptic impairment. *PloS One*, *3*(11), e3604.

Yang, T., Liu, H., Tran, K. C., Leng, A., Massa, S. M., & Longo, F. M. (2020). Small-molecule modulation of the p75 neurotrophin receptor inhibits a wide range of tau molecular pathologies and their sequelae in P301S tauopathy mice. *Acta Neuropathologica Communications*, *8*, 1–23.

Yang, T., Tran, K. C., Zeng, A. Y., Massa, S. M., & Longo, F. M. (2020). Small molecule modulation of the p75 neurotrophin receptor inhibits multiple amyloid beta-induced tau pathologies. *Scientific Reports*, *10*(1), 20322.

Yu Povarnina, P. (2022). A low-molecular-weight BDNF mimetic, dipeptide GSB-214, prevents memory impairment in rat models of Alzheimer's disease. *Acta Naturae (англоязычная версия)*, *14*(4), 94–100.

Zhang, Y., Li, P., Feng, J., et al. (2016). Dysfunction of NMDA receptors in Alzheimer's disease. *Neurological Sciences*, *37*, 1039–1047. https://doi.org/10.1007/s10072-016-2546-5

Zhang, Z., Liu, X., Schroeder, J. P., Chan, C. B., Song, M., Yu, S. P., Weinshenker, D., & Ye, K. (2014). 7,8-dihydroxyflavone prevents synaptic loss and memory deficits in a mouse model of Alzheimer's disease. *Neuropsychopharmacology*, *39*(3), 638–650.

14 Metal Chelators
A Promising Avenue for the Development of Novel Therapeutics for Alzheimer's Disease

Faisal Rehman

14.1 INTRODUCTION

Alzheimer's disease (AD) is a neurological degenerative disorder that is characterized by the impairment of neuronal growth, with the dysregulation in the synaptic networks in various regions of the brain, including the hippocampus and entorhinal cortex. This progressive degeneration over time leads to clinical manifestations such as dementia, depression, and cognitive decline (Jack & Holtzman, 2013). Multiple factors contribute to the onset of AD, including the production of oxidative stress, disruption of the cholinergic neurotransmitter system due to increased acetylcholine levels, development of amyloid-beta (Aβ) plaques, hyperphosphorylation of tau protein, decreased activity of acetylcholine, and dyshomeostasis of biometal ions. However, determining the exact etiology underlying the progression of AD remains a significant challenge due to its multifactorial nature. In recent years, studies have highlighted the roles of Aβ and intraneuronal neurofibrillary tangles in progression of AD, establishing a direct connection with metal ions (Savelief et al., 2013). This correlation has prompted researchers to investigate how metal ions and their chelation contribute to the development of AD. Specifically, reversing Aβ deposition and maintaining metal ion equilibrium may help prevent aggregation and improve cognitive function (Kawada et al., 2015). Moreover, metal ions play an important role in neurodegeneration, particularly in diseases such as Huntington's disease, Alzheimer's, and Parkinson's. Transition metals such as Fe, Zn, Al, and Cu have been involved in the progression and development of numerous neurodegenerative diseases (Filiz et al., 2008; Rao et al., 1999; Zatta et al., 2002). Dysregulation of these transition metal ions can lead to cellular stress such as oxidative stress, which damages neuron cells by promoting the formation of free radicals and disrupting cellular homeostasis. For instance, excess iron can expedite the formation of free radicals through Fenton chemistry, while aberrant copper levels can interfere with protein homeostasis and contribute to the aggregation of neurotoxic species, such as Aβ and tau proteins in AD (Figure 14.1). Furthermore, the imbalance of metal ions can exacerbate neuroinflammation and impair neuroprotective signaling pathways, resulting in neuronal cell death and cognitive decline. Thus, understanding the intricate relationship between metal ions and neurodegenerative processes is essential for developing targeted therapeutic strategies to mitigate their detrimental effects.

Chelators are the molecules that bind to the transition metal ions through the formation of coordinate bonds and remove the excess metal ions in the body fluid, providing effective strategies to combat the dyshomeostasis of metal ions in AD (Santos et al., 2016). Currently reported chelators, such as clioquinol (CQ), epigallocatechin, and curcumin (Table 14.1), have shown promising results, although they also present different assumptions and controversial ideologies regarding their efficacy and mechanisms of action (Di Vaira et al., 2004; Treiber et al., 2004). This chapter delves into a deeper understanding of how chelation therapy helps in the development of AD along with the innovative approaches of targeting the metal complexes.

14.2 METAL IMBALANCE IN AD

The significant increase of the metal ions in the human brain shows evidence of their role in AD-associated pathologies, including Aβ aggregation (Yano et al., 2003) and hyperphosphorylation of tau protein. These metal ions play a role in signal transduction and transcription of various genes through metal-associated regulators. Notably, the accumulation of Cu, Fe, Al, and Zn in various regions of the brain contributes to altered metal transportation and the development of senile plaques (Bishop et al., 2002). Studies revealed that different levels of metal ions influence the progression of AD; for instance, higher levels of Zn (1 mM) and Cu (400 μM) are associated with Aβ plaques, compared to healthy brains where Cu (70 μM) and Zn (350 μM) levels are reported (Huang et al., 1997; Lovell et al., 1998). This accumulation significantly increases during the early onset of AD. Research has demonstrated that divalent ions accumulate in the early stages, while trivalent ions accumulate in later stages, particularly in the hippocampus and frontal cortex. Another study found that the accumulation of Zn (Lee et al., 1999) and Fe (Smith et al., 1998) is closely associated with Aβ in the mouse model (Tg2576) for AD. Various binding sites on Aβ protein have different affinities for Zn^{2+} and Cu^{2+}, which activate the accumulation process. Cu^{2+} and Zn^{2+} bind to Aβ, although Fe^{3+} binds to a lower extent (Atwood et al., 2000). The function of metal ions in

DOI: 10.1201/9781003602477-14

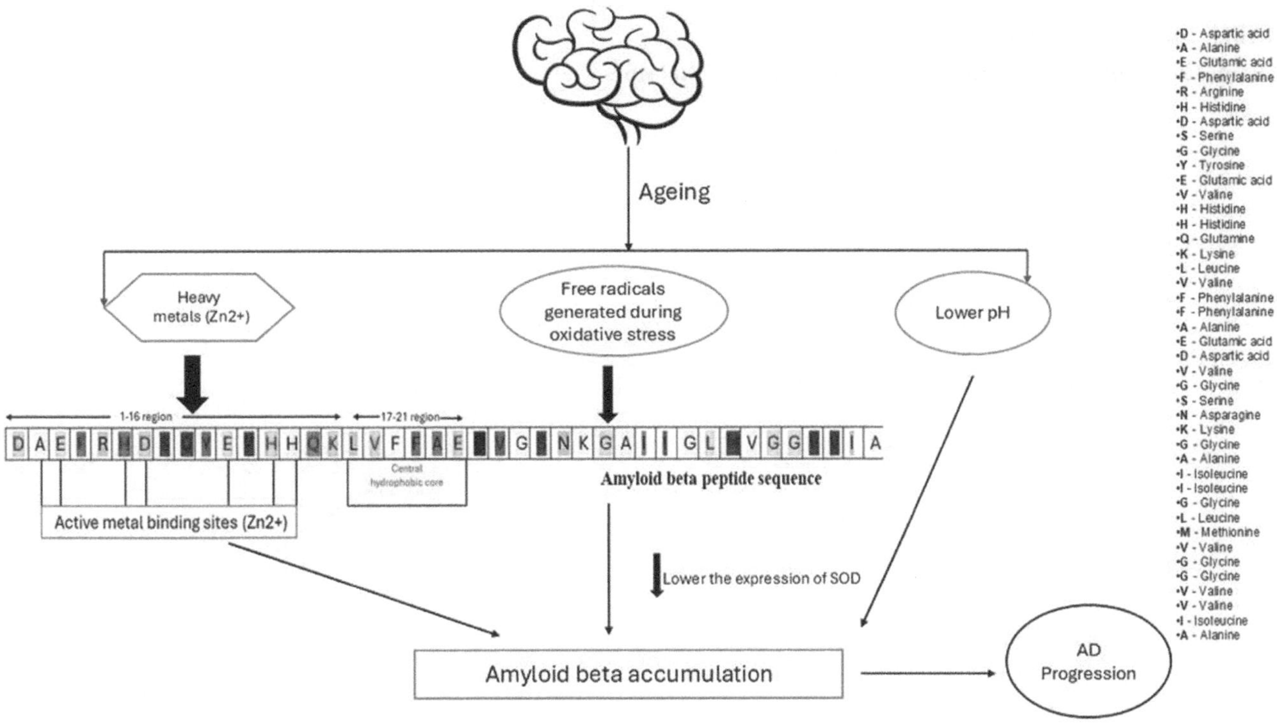

FIGURE 14.1 Schematic representation of the factors contributing to amyloid-beta accumulation and Alzheimer's disease (AD) progression. The diagram illustrates how aging leads to several biochemical changes, including the generation of free radicals during oxidative stress and a decrease in pH, which collectively promote amyloid-beta peptide aggregation. Excess Heavy metals (such as Zn^{2+}) binding to active metal binding sites within the 1–16 region of the amyloid-beta sequence, further facilitating aggregation. The amyloid-beta sequence also contains a central hydrophobic core (amino acids 17–21) critical for its self-association and aggregation. Oxidative stress lowers the expression of superoxide dismutase (SOD), reducing antioxidant defenses and exacerbating amyloid-beta accumulation. The accumulation of amyloid-beta peptides ultimately contributes to the progression of AD.

TABLE 14.1
List of the Common Synthetic and Natural Metal Chelators

S. No.	Category	Compound	Structure
1	Quinolone derivative	8-Hydroxyquinoline	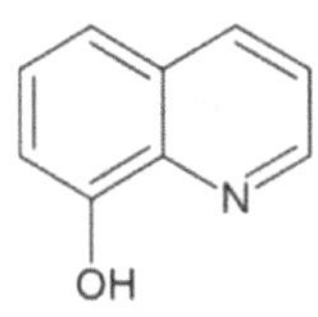
2	Quinolone derivative	Clioquinol	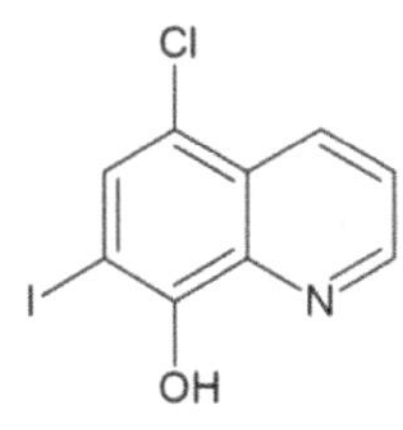
3	Quinolone derivative	PBT2	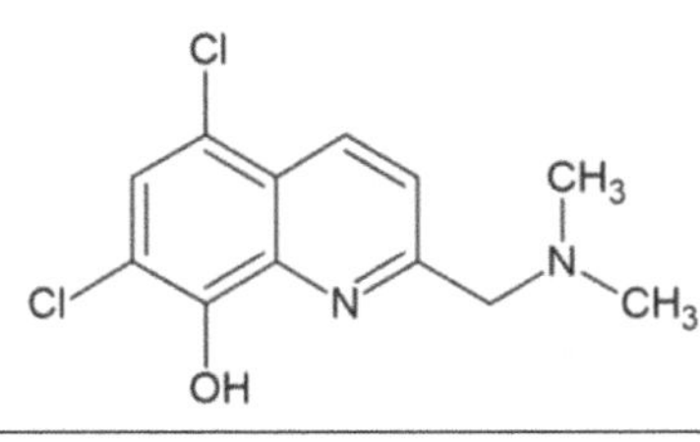

(Continued)

TABLE 14.1 *(Continued)*
List of the Common Synthetic and Natural Metal Chelators

S. No.	Category	Compound	Structure
4	Iron derivative	Deferiprone	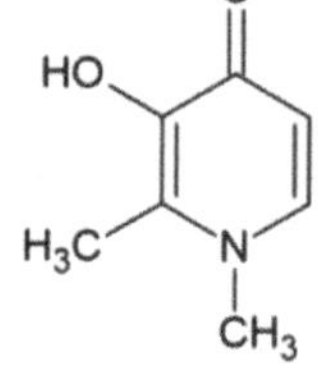
5	Copper chelator	Trientine	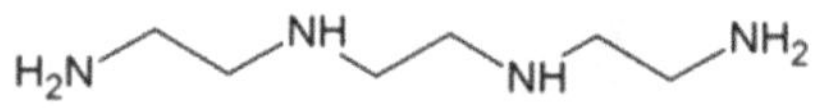
6	Cu/Zn chelators	D-Penicillamine	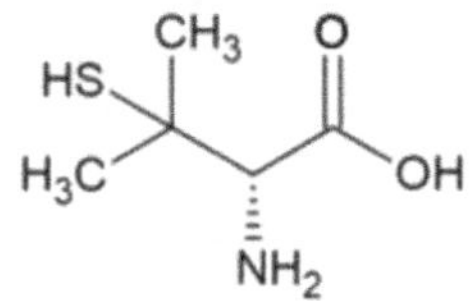
7	Phytotoxin derivative	Tenuazonic acid	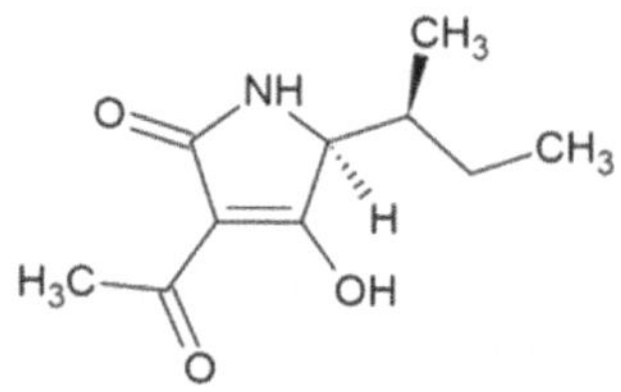
8	Iron chelators	Deferasirox	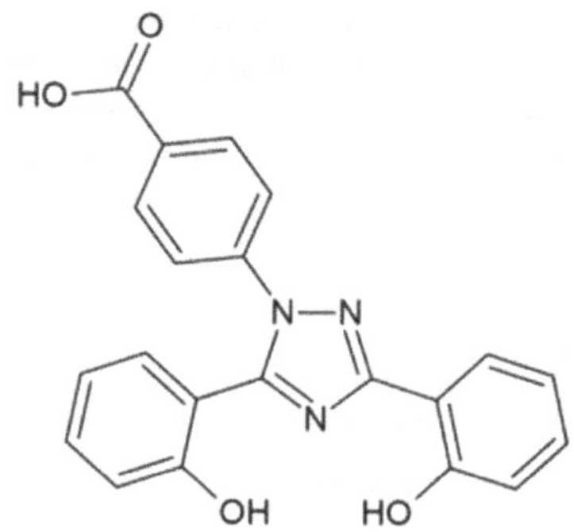
9	Polyphenolic derivative	Curcumin	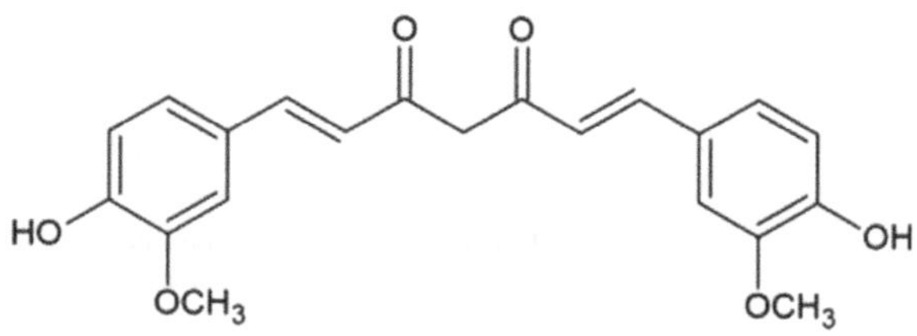
10	Polyphenolic derivative	Epigallocatechin	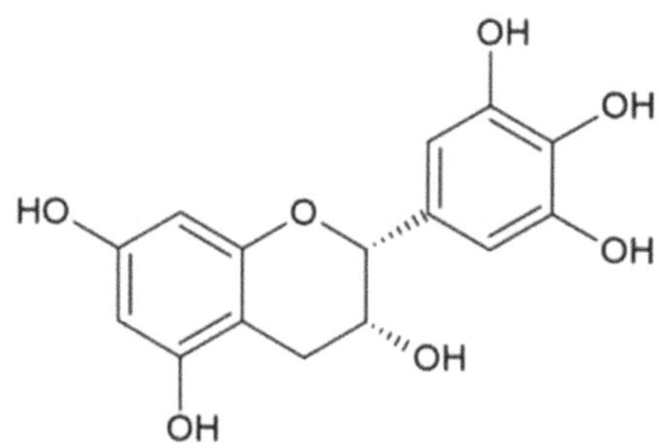

(Continued)

TABLE 14.1 *(Continued)*
List of the Common Synthetic and Natural Metal Chelators

S. No.	Category	Compound	Structure
11	Polyphenolic derivative	Epigallocatechin-3-gallate	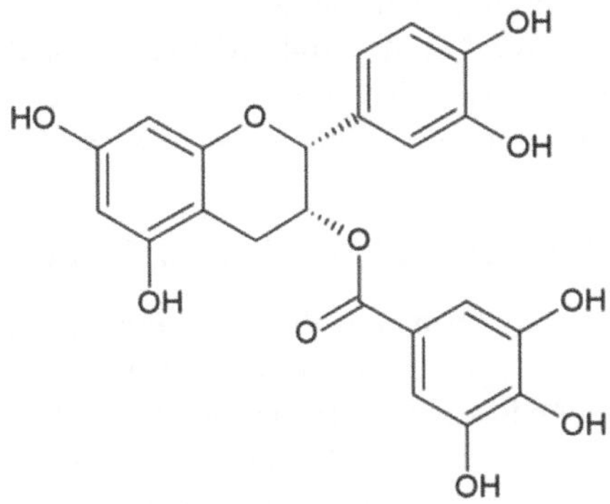
12	Copper chelator	Bathocuproinedisulfonic acid	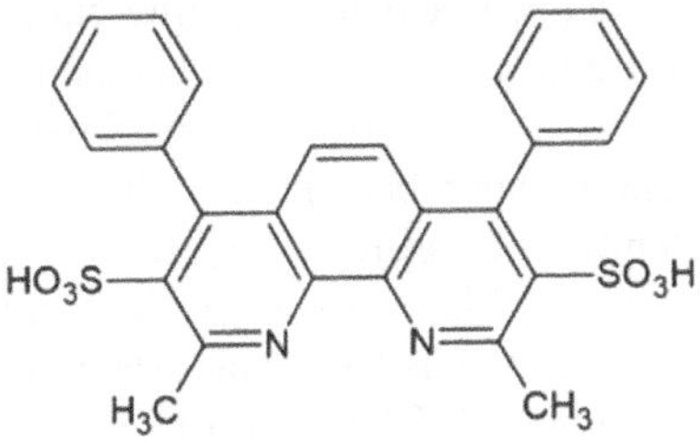
13	Zinc chelator	N,N,N′,N′-tetrakis(2-pyridylmethyl)-ethylenediamine	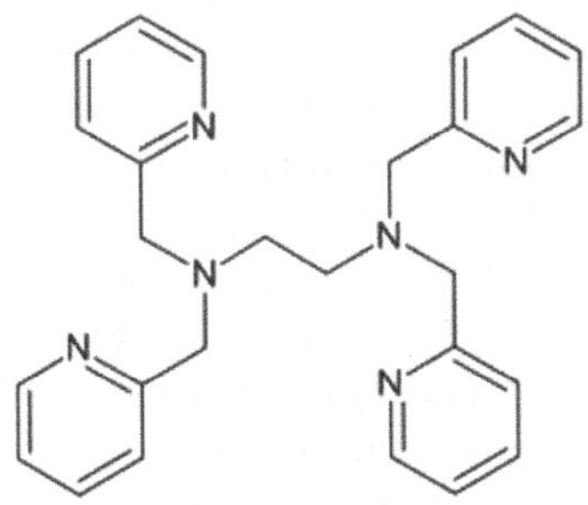
14	Coumarin derivative	Umbelliferone	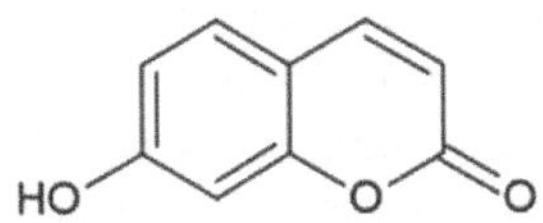
15	Coumarin derivative	Esculetin	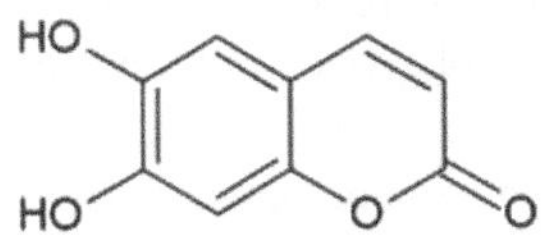
16	Coumarin derivative	Daphnetin	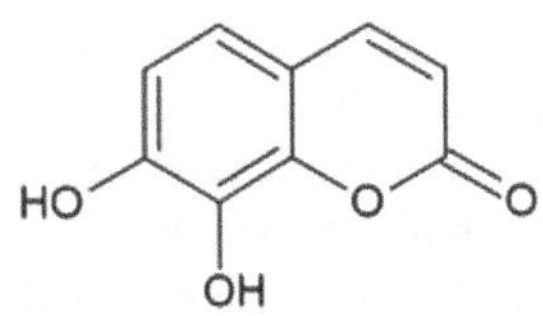

fibril assembly and beta-sheet formation in amyloid peptides remains unclear in vivo. However, studies revealed that Zn^{2+} is more effective aggregator of Aβ than Cu^{2+} at physiological pH. At pH 7, Zn^{2+} forms insoluble complexes, while Cu^{2+} forms soluble ones (Miura et al., 2000). As the human brain ages and becomes slightly acidic, Cu^{2+} is more likely to be associated with the formation of insoluble Aβ aggregates (Atwood et al., 1998). The concentration of metal ions and pH levels are crucial in preventing Aβ accumulation. Recent advancements in metal chelation therapies offer potential therapeutic avenues for Alzheimer's pathology. Oxygen-induced stress is another emerging reason for the pathology of various diseases, including neurological disorders, oncology, and cardiovascular diseases. Normally produced in cells, oxidative stress can become pathological when its levels exceed normal limits, leading to irreversible DNA and protein damage, as well as disruption of redox signaling (Aryal et al., 2014). Metals, particularly Fe and Cu, catalyze oxidative stress, producing reactive oxygen species (Chaturvedi & Shrivastava, 2005).

14.3 MECHANISM OF METAL CHELATORS

Complexes are formed by the interaction of metal ions with electron-rich species (ligands) or they can be described as compounds where cationic metals are bound by molecules that have two or more active sites available for bond formation and held by coordinate bonds. Chelate is a Greek-derived word "Chela" meaning claw, reflecting the molecule's structure. During the chelation process, various organic and inorganic molecules are involved, i.e. metal ions, especially those belonging to transition series, can bind covalently or coordinately with organic species having an active site like oxygen, nitrogen, and sulfur. The use of metal chelators is well-established in various industries. For example, EDTA is used in the textile industry for bleaching, dyeing, and kier boiling. Ar'-disalicylidine-1,2-propanediamine deactivates trace amounts of metals and is used in gasoline to counteract copper oxidation. While metal chelators serve many purposes in industries, their role in clinical applications for human diseases remains limited. One widely used chelator, calcium EDTA, is employed in cases of radioactive poisoning, where it binds to heavy metals in the body, replacing calcium and allowing the complex to be excreted via urine (Knell, 1959). However, its clinical use is far more limited compared to other applications. Numerous metal chelators for clinical purpose have been synthesized in vitro, including iron chelators (deferasirox, deferiprone, deferoxamine), arsenic chelators (zinc, trientine, penicillamine, dimercaprol), lead chelators (succimer, dimercaprol), and mercury chelators (dimercaprol). A few, such as metal porphyrins, are found in nature. The mechanism of action of metal chelators depends on the active sites present on protein surfaces, where the chelators form coordinate bonds with proteins, resulting in complex formation. For example, neuromelanin (NM) exhibits strong chelation properties due to its phenolic groups, which bind to iron (Fe) in a gradual manner (Double et al., 1999). NM poses two linking sites, but only one shows a high affinity for Fe. Fe^{3+} binds to the phenolic group of NM, forming a six-fold high-spin structure of octahedral configuration. In a healthy brain, only 50% of NM is bound to Fe^{2+}, indicating its high potential for chelation during elevated iron levels. Under normal physiological conditions, NM-Fe binding is redox-active, making NM a potent chelating agent with neuroprotective effects. However, in neurological disorders, as the Fe concentration increases over the required threshold levels, the one site with the higher affinity becomes saturated, leading to the binding of Fe on the low-affinity site (Double et al., 2003). In this case, the bound iron becomes redox-active, potentially causing brain damage. An important biomarker for AD is the development of Aβ, which involves the deposition of the Aβ-42 peptide. The Aβ isoform which is about 39–42 residue peptide arises due to the amyloid precursor protein (APP). The Aβ-42 peptide is particularly toxic. Postmortem studies of AD patients have shown amyloid plaque deposition, suggesting that fibril deposition triggers AD progression. However, whether fibrillated plaques or oligomeric forms of Aβ are more toxic remains under debate (Lue et al., 1999). Current studies indicate that the oligomeric form is involved in the early onset of AD (Klein et al., 2001; Wang et al., 2002). Once formed, these oligomers are generated within specialized intracellular vesicles. Inside cells, Aβ monomers form higher-order structures such as dimers and trimers. These water-soluble monomers are found in concentrations six times higher in AD brains compared to healthy ones (McLean et al., 1999). These monomers interact with neurons in vivo, leading to synaptic loss and neurological impairments.

14.4 PRECLINICAL STUDIES ON METAL CHELATION IN ALZHEIMER'S DISEASE

The pharmacological active drugs such as acetylcholinesterase inhibitors (AChE) have been reported to be used in numerous in vivo and in vitro studies (Corey-Bloom et al., 1998) and NMDA receptor partial antagonists (Winblad & Poritis, 1999). However, limited studies exist on the use of metal chelators for treating AD. Aβ is a metalloprotein with a high affinity for Zn^{2+} and Cu^{2+}. The interaction of Aβ with these metal ions leads to the formation of reactive species of oxygen (such as hydroxyl radicals) and the accumulation of plaques, which results to the progression of AD (Curtain et al., 2001). One study, performed on transgenic mice, used oral clioquinol—a Zn/Cu hydrophobic chelator that easily crosses the blood-brain barrier. The results showed a synergetic decrease in Aβ plaques soon after nine weeks of treatment with oral clioquinol (Cherny et al., 2001). Another investigation by Cherny et al. demonstrated that the use of Zn/Cu metal chelators in AD patients dissolved the accumulated plaques (Cherny et al., 1999). A recent study by Ritchie provided solid insight for the efficacy of metal chelators in AD. This randomized, phase two pilot clinical study involved groups of AD patients, some of whom were given a placebo, while others received clioquinol. The results indicated that the placebo group exhibited higher concentrations of metal ions, while the treatment group showed lower concentrations of metal ions and reduced Aβ plaque deposition (Ritchie et al., 2003). Recent work by Treiber and colleagues, however, showed that clioquinol sometimes had the opposite effect, increasing Cu and Zn levels and failing to reduce metal ion accumulation, contrary to the expected action of metal chelators. Notably, two current studies highlight the use of a new ligand for reducing Aβ deposition. This ligand is a derivative of clioquinol, known as hydroxyquinoline. In a 12-week phase II double-blind trial, hydroxyquinoline demonstrated a significant reduction in Aβ concentrations (Adlard et al., 2008; Lannfelt et al., 2008).

14.5 CLINICAL DEVELOPMENT OF METAL CHELATORS FOR AD

Metal ions have been implicating an important role in cellular physiology. An imbalance in metal ion homeostasis can result in numerological diseases such as neurological disorders, cancer, iron-deficiency anemia, and many others. The deposition of excess metals (Zn, Al, Cu, Fe) in organs is a major hallmark of the majority of diseases (Kontoghiorghe et al., 2015). To address metal ion-related issues, numerous chemical agents, including metal chelators, have been developed and progressed to clinical stages. Metal chelators have been applied in various clinical contexts, such as cancer, infectious diseases, aging, free radical pathology, and the removal of xenobiotic metals (Blanusa et al., 2005). At this moment the only FDA-approved drugs which are prescribing for managing AD are the AChE inhibitors (memantine, galantamine, rivastigmine, donepezil) which stimulate cholinergic transmission by preventing the breakdown of acetylcholine. This approach offers limited management of AD, and more advanced treatments are required for comprehensive management. Several well-known anti-inflammatory drugs, anticonvulsants, and antibiotics that possess chelating properties and have undergone clinical trials include aspirin, ibuprofen, indomethacin, and d-penicillamine, which chelate Zn, Fe, and Cu (Perret, 1981; Peterson et al., 1979; Rao & White, 1985; Russanov et al., 1986). Other examples include bleomycin, thalidomide, and ethambutol, which chelate Cu, Fe, and Zn (Dorr, 1992; Sheskin et al., 1981). The Blood Brain Barrier(BBB) is a significant obstacle in brain-targeted therapy and plays a vital role in brain and CNS (Central Nervous System) function, particularly during inflammation (Stolp & Dziegelewska, 2009). The BBB is composed of several types of cells such as astrocytes, pericytes, capillary endothelial cells, and neurons (Bhowmik et al., 2015). These cells separate brain tissue from circulating elements in the vascular system, protecting neuronal tissues from rapid fluctuations in metabolites and external substances. Astrocytes, vital components of the brain, help regulate BBB permeability, neuronal metabolism, and the clearance of waste products (Figley et al., 2011). They also enhance transporter expression and activate specialized enzymes (Abbott, 2002). Inside the BBB the endothelial cells are the main component that acts as shield (Winkler et al., 2011; Wolburg & Lippoldt, 2002). They form tight junctions, also known as blocking junctions, which prevent extracellular fluid transport by closely linking adjacent endothelial cells (Bazzoni & Dejana, 2004). This mechanism is responsible for the presence of specialized transmembrane proteins "Claudins" and "Occludins". Notably, claudins are responsible for the formation of the strong interlocked seal at tight junctions via disulfide (-S-S-) bonds (Cardoso et al., 2010; Singh et al., 2010), while occludins help maintain the barrier's structure and permeability (Yu et al., 2005). Junctional adhesion molecules (endothelial cells, leukocytes) play a critical role in the development of tight junctions and regulate the immunological responses of the BBB during brain pathology (Ebnet et al., 2004; Zlokovic, 2008). Down the tight junctions, adherens junctions, consisting of transmembrane glycoproteins called cadherins, form a polymerization complex that restricts passive diffusion across epithelial and endothelial tissues (Kazmierczak et al., 2007; Lampugnani & Dejana, 2007). These junctions prevent the passive diffusion of compounds and further control the movement of substances from the blood to the brain (Guo et al., 2007; Grimson et al., 2000). Nevertheless, the restriction of the substance via passive diffusion or paracellular transport from blood to the brain and other transports facilitate the movement of these substances across the BBB (Butt et al., 1990; Gao & Jiang, 2013). The BBB is a major obstacle in targeting AD. For instance, deferoxamine (DFO), a well-known chelator, is a charged molecule that cannot cross the BBB and rapidly degrades after administration, but some studies illustrate its ability toward iron chelating. However, it has shown adverse effect and tissue-specific targeting issues. To cross the BBB, a drug's size should be less than 300 nm, and several natural or synthetic polymers have shown the potential to pass the BBB (Liu et al., 2005). Despite the need for metal chelators in AD treatment, understanding their safety, tolerability, and potential side effects is crucial. For example, according to the Centres for Disease Control and Prevention (CDC, 2006), three deaths were reported following cardiac arrest after the administration of Na2EDTA for chelation therapy. While EDTA is a potent chelating agent, it cannot cross the BBB, and its administration has been associated with worsening symptoms of mercurialism or lead poisoning (Aposhian et al., 1996).

14.6 EMERGING CHELATING AGENTS FOR ALZHEIMER'S THERAPY

As we age, the level of free radicals gradually increases, leading to cellular stress. This increase occurs for various reasons, and in AD, one frequently studied factor is the accumulation of metal ion complexes with different cellular proteins, which contributes to numerous neurological conditions. To tackle this issue, chelation therapy has emerged as a novel approach, though it has its strengths and weaknesses. As noted, an ideal chelating agent should demonstrate properties such as lipophilicity, small size, target specificity, and the ability to cross BBB effectively (Shachar et al., 2004). New approaches, like using a prochelator to enhance activity and enable BBB crossing, are under investigation (Galey et al., 1995).

Furthermore, inorganic silicates can form stable complexes with various metal ions and improve blood–brain barrier (BBB) permeability (Fasman & Moore, 1994). However, limited evidence currently exists to confirm the benefits of chelators in AD, indicating the need for advanced targeting strategies for metal ions in AD. Various technologies are being explored to increase the affinity and efficacy of chelators by combining them with nanomaterials. Nanomaterials are evolving as innovative approach for drug delivery due to their structural and size similarity to natural carriers, such as lipoproteins and viruses, facilitating smoother transport across the BBB. Nanoparticles, generally ranging from 10 to 1000 nm in size (Kreuter, 2001), are often coated with polysorbate 80 surfactant (Alyautdin et al., 1998), which enhances biodistribution, reduces toxicity, and improves therapeutic potency (Ravi Kumar, 2000). These nanoparticles use proteins like ApoE and ApoB to mimic LDL(Low density lipoprotein), allowing them to bind to LDL receptors, where they are subsequently taken up by endothelial cells (Dehouck et al., 1997; Muller et al., 2001; Ramge, 1998). When iron chelators are covalently bonded to nanoparticles, the resulting complex could be an ideal option for BBB penetration, overcoming the problem of the chelator being lipophilic and thus reducing toxicity (Alyautdin et al., 1998; Schroeder et al., 1998). An effective iron chelator should cross the BBB, form a complex with its target, and help remove the metal ions from the brain. Non-biodegradable nanoparticles may bind to ApoAI, facilitating the removal of the complex from the brain (Davson & Segal, 1996). However, no well-defined carrier transport system currently exists to eliminate synthetic metal complexes from the brain, though increasing chelator lipophilicity can reduce aqueous solubility, potentially affecting bioavailability (Richardson, 1999). A study conducted by Wong's lab on metal chelators in AD patients with iron overload demonstrated that nanoparticle-bound chelators more effectively removed iron from the brain than traditional options like DFO (Liu et al., 1995).

Postmortem brain studies indicate a high correlation between metal ions and free radicals, which damage DNA, proteins, and lipids (Alam et al., 1997; Dexter et al., 1989; Sanchez-Ramos et al., 1994; Wong, 1991). Various studies—physicochemical, biochemical, and neurochemical—report phenomena such as decreased mitochondrial activity (due to free radical generation), increased hydrogen peroxide (from dopamine metabolism oxidation), and elevated superoxide dismutase production (further increasing hydrogen peroxide) (Mizuno et al., 1989).

A biomarker of AD is the deposition of Aβ plaques from APP in the cortical interstitium and cerebrovasculature. Strong evidence links Aβ with oxidative stress due to its unique physiochemical properties, which include both pro-oxidant and antioxidant activities that initiate redox reactions (Butterfield et al., 2001). Metal homeostasis imbalances allow metals to bind to specific active site residues in neuropil, such as His13/14/6 on the N-terminal, and Met35 on the C-terminal. The N-terminal selectively binds metals like Cu and Fe, while the C-terminal is involved in reducing Cu^{2+} and generating hydrogen peroxide (Atwood et al., 2003).

14.7 CHALLENGES AND LIMITATIONS OF METAL CHELATION THERAPY

The imbalance in metal homeostasis is evidenced in the early progression of AD, supporting the concept of restoring this balance using metal chelation therapy (White et al., 2015). Chelation therapy works by administering metal chelators, which bind to excess endogenous metals, removing them from active binding sites on various proteins, thereby reducing their bioavailability in the extracellular environment. This is achieved either by acting as ionophores or by competing with the metal ions (Buce & Bush, 2010). Notably, a substantial number of metal chelators have advanced through clinical phases, such as an 8-hydroxyquinoline derivative called clioquinol, which chelates Cu, Fe, and Zn, reducing Aβ aggregation but failed in further studies due to toxicity. PBT2, which phosphorylates GSK-3, enhances Aβ degradation and reduces hydrogen peroxide production (Ejaz et al., 2020). Deferiprone is another metal chelator that reduces Aβ and tau protein phosphorylation (Prasanthi et al., 2012). D-penicillamine, a Zn/Cu chelator, helps mitigate free radicals, delaying AD progression (Quitti et al., 2002). Trientine (TETA) inhibits beta-secretase, slowing AD progression in mice (Wang et al., 2013). The conjugated therapy of melatonin and TETA further reduces Aβ and blocks tau protein phosphorylation and gamma-secretase activity (Li et al., 2022). Other chelators designed for different disorders, such as metformin and cyclodipeptides, are being repurposed for AD treatment (Lanza et al., 2018; Yang et al., 2019). In a study by Poliseno et al., tenuazonic acid—a natural chelator and antioxidant—was synthesized into various derivatives that chelate with Cu^{2+}, Zn^{2+}, and Fe^{2+}, demonstrating AChE inhibition properties, making them promising candidates for AD (Poliseno et al., 2021).

14.7.1 Iron Chelation Therapy

Iron is implicated in various neurological conditions, so its chelation presents an effective treatment for AD. Chelators like DFO and deferasirox are currently used in iron overdose and hemochromatosis treatment (Nick, 2007; Piga et al., 2006). Several iron chelators, including CQ, DFP (Deferiprone), DFO (Deferoxamine), and DFX (Deferasirox), are FDA-approved for their capacity to cross the BBB and remove iron from AD-affected brains by suppressing apoptotic pathways (Gal et al., 2010), stabilizing mitochondrial function (Zhu et al., 2007), and restoring protein degradation (Avramovich-Tirosh et al., 2010; Guo et al., 2015; Hanson et al., 2012; Zhu et al., 2022). Additionally, natural Fe chelators, such as epigallocatechin-3-gallate, curcumin, epicatechin, are being studied for chelation therapy (Hegde et al., 2009). Copper Chelation Therapy: Copper (Cu^{2+}) is essential in the body, but its imbalance can activate redox reactions, damaging DNA, proteins, and lipids via free radical production. Metallothionein protects cells from Cu^{2+} toxicity, while chaperones aid in transport to target locations. Cu/Zn superoxide dismutase, utilizing Cu^{2+} as a catalytic center, catalyzes the dismutation of hydroxyl anions, a derivative of free radicals (Tegoni et al., 2014). Copper binds with albumin and ceruloplasmin, reducing Cu^{2+} to Cu^{+} in the presence of ascorbate, though oxygen quickly reoxidizes it to Cu^{2+} (Lovstad, 2002). CQ selectively targets copper toxicity, crossing the BBB and inducing Aβ plaque degradation while increasing metalloproteinase expression (Bareggi & Cornelli, 2012; White et al., 2006). However, prolonged CQ use results in cytotoxicity due to CQ-Cu complexes, similar to 8-hydroxyquinoline (Tardito et al., 2012).

14.7.2 Zinc Chelation Therapy

Zn^{2+} is crucial for many metalloproteins, as well as a "mobile zinc" pool that is necessary for biochemical processes, with imbalances contributing to neurological conditions (Colvin et al., 2010; Sensi et al., 2011). Zn^{2+} appears to facilitate Aβ aggregation, leading researchers to investigate Zn complexation. As both Zn^{2+} and Cu^{2+} promote Aβ aggregation, PBT2 has shown effectiveness in lowering Aβ aggregation and improving cognition in AD patients (Faux et al., 2010). Another chelator, carnosine, displays high affinity in reducing intraneuronal Aβ accumulation, preventing mitochondrial dysfunction in AD mice (Corona et al., 2011). Presenilin, a proteolytic agent of APP, indirectly reduces amyloid plaques by lowering Cu^{2+}/Zn^{2+} levels in cells (Greenough et al., 2011). Chelate design often encounters challenges related to charge, structure, and binding kinetics (Lord & Lippard, 2013). These criteria are optimized using the hard and soft acid-base theory. Apart from metal ligands, CQ also serves as an anti-aggregating agent. Although various Cu chelators are used in neurological conditions, their roles remain controversial. In recent studies, Zatta et al. used cuprizone in mice, which led to spongiosis and edema-like characteristics with neuropil vacuolization in white matter, affecting Cu and Zn homeostasis in the CNS. Dysregulation of metal homeostasis is closely associated with AD progression (Rao et al., 1999). The concentration of metal ions shifts with AD progression, with divalent ions increasing in the initial stages and trivalent ions in later stages. A systematic approach is required to determine the suitability of specific chelators for individualized therapy.

14.8 COMBINATION THERAPIES: METAL CHELATORS AND OTHER ALZHEIMER'S TREATMENTS

Currently, there are only a limited number of FDA-approved drugs for AD treatment. Out of these, four belong to the cholinesterase inhibitor category: galantamine, rivastigmine, donepezil, and tacrine. Most of these are prescribed for acute to chronic AD cases, though donepezil is only recommended for late-stage AD. However, tacrine was lifted from the U.S. market due to its association with liver toxicity. Notably, these drugs only alleviate symptoms rather than address the underlying disease, highlighting the urgent need for new or combination therapies for AD.

According to ALZFORUM (www.alzforum.org, March 2023), 298 AD agents are currently in trials, with 76 of these targeting Aβ aggregates, including immunotherapies and small molecules. These therapeutic agents are categorized as follows:

1. *Reducing Aβ Generation*: Acitretin, lenalidomide, levetiracetam, NIC5-15, and Posiphen (Bakker et al., 2015; Decourt et al., 2020; Endres et al., 2014; Maccecchini et al., 2012).
2. *Enhancing Aβ Clearance*: Bexarotene (Cummings et al., 2016).
3. *Destabilizing or Inhibiting Aβ Aggregates*: ALZ-801, Contraloid, PBT2, and varoglutamstat (Hey et al., 2018; Kutzsche et al., 2020; Lannfelt et al., 2008; Scheltens et al., 2018).
4. *Ameliorating Toxic Effects of Aβ*: ALX-001, CT1812, nasal insulin, and simufilam (Craft et al., 2020; Haas et al., 2017; Izzo et al., 2021; Kellar et al., 2021, 2022; Spurrier et al., 2022; Wang et al., 2017; Wang et al., 2020).

Beyond Aβ targeting, additional types of agents focus on other AD mechanisms. Calcium ions (Ca^{2+}), essential in learning and memory processes, are released within cells when glutamate binds to NMDA receptors. This binding allows the rapid transportation of Ca^{2+} into the cell, supporting long-term plasticity. In AD, however, excessive glutamate activates NMDA receptors excessively, leading to intracellular Ca^{2+} accumulation and neuronal death (Greenamyre & Porter, 1994; Greenamyre & Young, 1989). A partial NMDA receptor antagonist (memantine) works by binding to the receptor and preventing glutamate binding, thereby reducing Ca^{2+} concentration. Notably, memantine got its European approval in 2002 for the AD treatment (Greenamyre & Young, 1989; Winblad & Poritis, 1999). New and innovative approaches are also in clinical phases, such as targeting APP and inhibiting beta- and gamma-secretase. In recent work, Mandel and colleagues proposed a novel AD treatment approach using curcumin and green tea. Both compounds can cross the BBB and exhibit multifunctional activities, including neuroprotective, radical-scavenging, metal chelation, and anti-inflammatory effects. Their mechanism involves targeting Fe ions, and further research is ongoing to provide practical evidence of efficacy and a comprehensive safety profile (Mandel et al., 2007).

14.9 FUTURE DIRECTIONS IN METAL CHELATION RESEARCH

Metal ions are involved in various important biological, physiological, and metabolic activities in the human body. Essential metal transition ions, such as copper, zinc, and iron, are primarily obtained from food sources to maintain health. For instance, Zn is vital for catalytic activity in over 300 zinc-containing proteins and contributes to the transcription of 2000 zinc-dependent transcription factors. An imbalance in metal ions can result in certain diseases, such as iron-deficiency anemia. In some cases, an excess of these ions can cause disorders, including sickle cell anemia, thalassemia, and iron overload in idiopathic hemochromatosis. Understanding the function of metal ions in disease prognosis, there is an urgent need for therapies to remove accumulated metal ions (Kontoghiorghe et al., 2015). Metal ions also support the proliferation and growth of various cells, including cancer cells. Current cancer treatments—such as chemotherapy, surgery, and radiotherapy—aim to destroy cancer cells by directly affecting DNA synthesis and signal transduction pathways. However, a major drawback of these therapies is that they are non-selective, often harming normal cells alongside cancer cells. This underscores the need for new, more targeted therapies, including multidrug approaches that target diverse cellular mechanisms. Since metal ions are involved in cell proliferation, metal chelation therapy is being researched as a potential cancer therapy (Gaur et al., 2018; Lui et al., 2015). For example, iron chelators could inhibit DNA synthesis by binding to iron in ribonucleotide reductase, a key enzyme in DNA synthesis (Chitambar, 2017). Another area where metal ions play a role is in free radical pathology, a significant contributor to many diseases. Free radical production leads to oxidative stress, often mediated by metals like copper and iron. Thus, metal chelators may serve as antioxidant therapy, potentially mitigating free radical damage (Kontoghiorghes & Kontoghiorghes, 2019). For metal chelation therapies, it is essential to understand the efficacy and mechanism of chelators in the context of personalized medicine. Monitoring chelator distribution in organs, including the liver, heart, and brain through MRI scans, helps optimize targeting and improve therapeutic outcomes (Adlard & Bush, 2018; Martin-Bastida et al., 2017; Pennell, 2005). Metal chelators also have a growing application in diagnostic and theranostic therapies. Designed chelating agents bind targeted metals in the body, serving as tracers for diagnostic imaging (Kolnagou et al., 2013). Despite advancements, challenges remain in the field of metal chelation therapy. Continued biomedical research is essential to design metal chelators with optimized physicochemical properties, broadening their diagnostic and therapeutic potential (Parghane & Basu, 2019). Factors such as basic structure, stability, binding affinity, membrane permeability, toxicity, and pharmacology contribute to the mechanism of action of chelators. Additionally, natural competitors in the body can impact the pharmacokinetics (ADME) and are influenced by redoxomic, metallomic, pharmacogenomic, and metabolomic factors (Crisponi et al., 2015; Kontoghiorghes & Kontoghiorghes, 2016). Future directions for chelating agents involve optimization for organ- and cell-specific targeting. For example, novel chelators could incorporate hydroxyurea and platinum complexes for targeted cancer therapies. The use of deferiprone is currently under investigation for managing free radicals and mitochondrial dysfunction (Madaan et al., 2012). The search for early AD biomarkers holds great potential for benefiting patients and advancing the development of disease-modifying therapeutics (Chan et al., 2018). A few established prognostic biomarkers, such as Aβ aggregates and tau protein, are primarily used in AD diagnostics. However, recent research has identified additional biomarkers with promise for early detection. One such biomarker is progranulin, a growth factor expressed in microglia and neurons. Progranulin has neuroprotective effects, helping to prevent astrogliosis and microgliosis, which are inflammatory responses in the brain (Cummings, 2019). It has been studied that cerebrospinal fluid (CSF) levels of progranulin can elevate as soon as ten years before AD symptoms appear, suggesting its potential as a predictive biomarker (Ahmed et al., 2010). Another promising marker is neurogranin, a synaptic protein found in the hippocampus and cortex, which plays a role in synaptic strength and plasticity. Studies have shown that neurogranin levels increase in the CSF of AD patients, particularly in the early stages of the disease (Suarez-Calvet, 2018). Additionally, microRNAs (miRNAs), short RNA sequences of 20–25 nucleotides, are emerging as important AD biomarkers. These miRNAs can alter mRNA function by binding to the 3'UTR, thereby blocking protein translation or increasing mRNA degradation rates, impacting various physiological processes related to AD (Dehghani et al., 2018; Kester et al., 2015; Portelius et al., 2015; Xia et al., 2019).

14.10 CONCLUSION

As AD progresses with aging, early intervention is critical. Metal chelators, among other emerging therapeutics, have shown promising results in slowing AD progression. Two major factors associated with AD are amyloid protein aggregation and oxidative stress. Recently, new research has focused on developing treatments using agents such as bidentate chelators (e.g., hydroxyquinoline, hydroxypyridinones), CQ, curcumin, EGCG, and nanoparticles. However, several challenges limit the potential of chelation therapy for AD, including the lack of specific metal removal, poor absorption, and the inability to cross the BBB. Ongoing research is working to overcome these obstacles by developing multi-targeted molecules that can address multiple AD-related factors. Moreover, previous research has been invaluable in guiding the design of novel therapeutic approaches. Preclinical identification of AD biomarkers is also advancing, aiming for early diagnosis and intervention. While each approach has its unique benefits and limitations, all require careful consideration to optimize new strategies for AD treatment and diagnosis.

REFERENCES

Abbott, N. J. (2002). Astrocyte–endothelial interactions and blood–brain barrier permeability. *Journal of Anatomy, 200*(6), 629–638.

Adlard, P. A., & Bush, A. I. (2018). Metals and Alzheimer's disease: How far have we come in the clinic? *Journal of Alzheimer's Disease, 62*, 1369–1379. https://doi.org/10.3233/JAD-170931

Adlard, P. A., Cherny, R. A., Finkelstein, D. I., Gautier, E., Robb, E., Cortes, M., Volitakis, I., Liu, X., Smith, J. P., Perez, K., Laughton, K., Li, Q. X., Charman, S. A., Nicolazzo, J. A., Wilkins, S., Deleva, K., Lynch, T., Kok, G., Ritchie, C. W., Tanzi, R. E., … Bush, A. I. (2008). Rapid restoration of cognition in Alzheimer's transgenic mice with 8-hydroxy-quinoline analogs is associated with decreased interstitial Aβ. *Neuron, 59*(1), 43–55. https://pubmed.ncbi.nlm.nih.gov/18614028

Ahmed, Z., Sheng, H., Xu, Y. F., Lin, W. L., Innes, A. E., Gass, J., Yu, X., Wuertzer, C. A., Hou, H., Chiba, S., Yamanouchi, K., Leissring, M., Petrucelli, L., Nishihara, M., Hutton, M. L., McGowan, E., Dickson, D. W., & Lewis, J. (2010). Accelerated lipofuscinosis and ubiquitination in granulin knockout mice suggest a role for progranulin in successful aging. *The American Journal of Pathology, 177*, 311–324. https://doi.org/10.2353/ajpath.2010.090915

Alam, Z. I., Daniel, S. E., Lees, A. J., Marsden, D. C., Jenner, P., & Halliwell, B. (1997). A generalised increase in protein carbonyls in the brain of Parkinson's but not incidental Lewy body disease. *Journal of Neurochemistry, 69*, 1326–1329.

Alyautdin, R. N., Tezikov, E. B., Ramge, P., Kharkevich, D. A., Begley, D. J., & Kreuter, J. (1998). Significant entry of tubocurarine into the brain of rats by adsorption to polysorbate 80-coated polybutylcyanoacrylate nanoparticles: An *in situ* brain perfusion study. *Journal of Microencapsulation, 15*, 67–74.

Aposhian, M. M., Maiorino, R. M., Xu, Z., & Aposhian, H. V. (1996). Sodium 2,3-dimercapto-1-propanesulfonate (DMPS) treatment does not redistribute lead or mercury to the brain of rats. *Toxicology, 109*(1), 49–55.

Aryal, B., Jeong, J., & Rao, V. A. (2014). Doxorubicin-induced carbonylation and degradation of cardiac myosin binding protein C promote cardiotoxicity. *Proceedings of the National Academy of Sciences of the United States of America, 111*, 2011–2016. https://doi.org/10.1073/pnas.1321783111

Atwood, C. S., Moir, R. D., Huang, X., Scarpa, R. C., Bacarra, N. M., Romano, D. M., Hartshorn, M. A., Tanzi, R. E., & Bush, A. I. (1998). Dramatic aggregation of Alzheimer Aβ by Cu(II) is induced by conditions representing physiological acidosis. *Journal of Biological Chemistry, 273*, 12817–12826. https://doi.org/10.1074/jbc.273.21.12817

Atwood, C. S., Obrenovich, M. E., Liu, T., Chan, H., Perry, G., Smith, M. A., & Martins, R. N. (2003). Amyloid-beta: A chameleon walking in two worlds: A review of the trophic and toxic properties of amyloid-beta. *Brain Research Reviews, 43*, 1–16.

Atwood, C. S., Scarpa, R. C., Huang, X., Moir, R. D., Jones, W. D., Fairlie, D. P., Tanzi, R. E., & Bush, A. I. (2000). Characterization of copper interactions with Alzheimer amyloid beta peptides: Identification of an attomolar-affinity copper binding site on amyloid beta1–42. *Journal of Neurochemistry, 75*, 1219–1233. https://doi.org/10.1046/j.1471-4159.2000.0751219.x

Avramovich-Tirosh, Y., Bar-Am, O., Amit, T., Youdim, M. B. H., & Weinreb, O. (2010). Up-regulation of hypoxia-inducible factor (HIF)-1α and HIF-target genes in cortical neurons by the novel multifunctional iron chelator anti-Alzheimer drug, M30. *Current Alzheimer Research*. https://doi.org/10.2174/156720510791162403

Bakker, A., Albert, M. S., Krauss, G., Speck, C. L., & Gallagher, M. (2015). Response of the medial temporal lobe network in amnestic mild cognitive impairment to therapeutic intervention assessed by fMRI and memory task performance. *NeuroImage: Clinical, 7*, 688–698. https://doi.org/10.1016/j.nicl.2015.02.009

Bareggi, S. R., & Cornelli, U. (2012). Clioquinol: Review of its mechanisms of action and clinical uses in neurodegenerative disorders. *CNS Neuroscience & Therapeutics*. https://doi.org/10.1111/j.1755-5949.2010.00231.x

Bazzoni, G., & Dejana, E. (2004). Endothelial cell-to-cell junctions: Molecular organization and role in vascular homeostasis. *Physiological Reviews, 84*(3), 869–903. https://doi.org/10.1152/physrev.00035.2003

Bhowmik, A., Khan, R., & Ghosh, M. K. (2015). Blood-brain barrier: A challenge for effectual therapy of brain tumors. *BioMed Research International, 2015*, 320941. https://doi.org/10.1155/2015/320941

Bishop, G. M., Robinson, S. R., Liu, Q., Perry, G., Atwood, C. S., & Smith, M. A. (2002). Iron: A pathological mediator of Alzheimer disease. *Developmental Neuroscience, 24*, 184–187. https://doi.org/10.1159/000067721

Blanusa, M., Varnai, V. M., Piasek, M., & Kostial, K. (2005). Chelators as antidotes of metal toxicity: Therapeutic and experimental aspects. *Current Medicinal Chemistry, 12*, 2771–2794. https://doi.org/10.2174/092986705774462987

Buce, J. A., & Bush, A. I. (2010). Biological metals and Alzheimer's disease: Implications for therapeutics and diagnostics. *Progress in Neurobiology*. https://doi.org/10.1016/j.pneurobio.2010.04.003

Butt, A. M., Jones, H. C., & Abbott, N. J. (1990). Electrical resistance across the blood–brain barrier in anaesthetized rats: A developmental study. *Journal of Physiology, 429*, 47–62.

Butterfield, D. A., Drake, J., Pocernich, C., & Castegna, A. (2001). Evidence of oxidative damage in Alzheimer's disease brain: Central role for amyloid beta-peptide. *Trends in Molecular Medicine, 7*, 548–554. https://doi.org/10.1016/S1471-4914(01)02173-6

Cardoso, F. L., Brites, D., & Brito, M. A. (2010). Looking at the blood–brain barrier: Molecular anatomy and possible investigation approaches. *Brain Research Reviews, 64*(2), 328–363. https://doi.org/10.1016/j.brainresrev.2010.04.002

Centers for Disease Control and Prevention (CDC). (2006). Deaths associated with hypocalcemia from chelation therapy—Texas, Pennsylvania, and Oregon, 2003–2005. *Morbidity and Mortality Weekly Report, 55*, 204–207. https://www.cdc.gov/mmwr/preview/mmwrhtml/mm5508a3.htm

Chan, S., Lian, Q., Chen, M. P., Jiang, D., Ho, J. T. K., Cheung, Y. F., & Chan, G. C. F. (2018). Deferiprone inhibits iron overload-induced tissue factor bearing endothelial microparticle generation by inhibition of oxidative stress-induced mitochondrial injury and apoptosis. *Toxicology and Applied Pharmacology, 338*, 148–158. https://doi.org/10.1016/j.taap.2018.06.014

Chaturvedi, U. C., & Shrivastava, R. (2005). Interaction of viral proteins with metal ions: Role in maintaining the structure and functions of viruses. *FEMS Immunology & Medical Microbiology, 43*, 105–114. https://doi.org/10.1016/j.femsim.2004.11.004

Cherny, R. A., Atwood, C. S., Xilinas, M. E., Gray, D. N., Jones, W. D., McLean, C. A., Barnham, K. J., Volitakis, I., Fraser, F. W., Kim, Y., Huang, X., Goldstein, L. E., Moir, R. D., Lim, J. T., Beyreuther, K., Zheng, H., Tanzi, R. E., Masters, C. L., & Bush, A. I. (2001). Treatment with a copper-zinc chelator markedly and rapidly inhibits beta-amyloid accumulation in Alzheimer's disease transgenic mice. *Neuron*, *30*, 665–676. https://doi.org/10.1016/s0896-6273(01)00317-8

Cherny, R. A., Legg, J. T., McLean, C. A., Fairlie, D. P., Huang, X., Atwood, C. S., Beyreuther, K., Tanzi, R. E., Masters, C. L., & Bush, A. I. (1999). Aqueous dissolution of Alzheimer's disease beta-amyloid deposits by biometal depletion. *Journal of Biological Chemistry*, *274*, 23223–23228. https://doi.org/10.1074/jbc.274.33.23223

Chitambar, C. R. (2017). The therapeutic potential of iron-targeting gallium compounds in human disease: From basic research to clinical application. *Pharmacological Research*, *115*, 56–64. https://doi.org/10.1016/j.phrs.2016.11.009

Colvin, R. A., Holmes, W. R., Fontaine, C. P., & Maret, W. (2010). Cytosolic zinc buffering and muffling: Their role in intracellular zinc homeostasis. *Metallomics*, *2*(5), 306–317. https://doi.org/10.1039/b926662c

Corey-Bloom, J., Anand, R., Veach, J. (1998). A randomized trial evaluating the efficacy and safety of ena 713 (rivastigmine tartrate), a new acetylcholinesterase inhibitor, in patients with mild to moderately severe Alzheimer's disease. *International Journal of Geriatr Psychopharmacol*, *1*, 55–65.

Corona, C., Frazzini, V., Silvestri, E., Lattanzio, R., La Sorda, R., Piantelli, M., Canzoniero, L. M. T., Ciavardelli, D., Rizzarelli, E., & Sensi, S. L. (2011). Effects of dietary supplementation of carnosine on mitochondrial dysfunction, amyloid pathology, and cognitive deficits in 3xTg-AD mice. *PLoS ONE*, *6*(3), e17971. https://doi.org/10.1371/journal.pone.0017971

Craft, S., Raman, R., Chow, T. W., Rafii, M. S., Sun, C. K., Rissman, R. A., Donohue, M. C., Brewer, J. B., Jenkins, C., Harless, K., Gessert, D., & Aisen, P. S. (2020). Safety, efficacy, and feasibility of intranasal insulin for the treatment of mild cognitive impairment and Alzheimer disease dementia: A randomized clinical trial. *JAMA Neurology*, *77*(9), 1099–1109. https://doi.org/10.1001/jamaneurol.2020.1840

Crisponi, G., Nurchi, V. M., Lachowicz, J. I., Crespo-Alonso, M., Zoroddu, M. A., & Peana, M. (2015). Kill or cure: Misuse of chelation therapy for human diseases. *Coordination Chemistry Reviews*, *284*, 278–285. https://doi.org/10.1016/j.ccr.2014.10.007

Cummings, J. (2019). The role of biomarkers in Alzheimer's disease drug development. In *Advances in experimental medicine and biology* (Vol. 1118, pp. 29–61). https://doi.org/10.1007/978-3-030-05542-4_3

Cummings, J. L., Zhong, K., Kinney, J. W., Heaney, C., Moll-Tudla, J., Joshi, A., Pontecorvo, M., Devous, M., Tang, A., & Bena, J. (2016). Double-blind, placebo-controlled, proof-of-concept trial of bexarotene in moderate Alzheimer's disease. *Alzheimer's Research & Therapy*, *8*, 4. https://doi.org/10.1186/s13195-016-0173-2

Curtain, C. C., Ali, F., Volitakis, I., Cherny, R. A., Norton, R. S., Beyreuther, K., Barrow, C. J., Masters, C. L., Bush, A. I., & Barnham, K. J. (2001). Alzheimer's disease amyloid-beta binds copper and zinc to generate an allosterically ordered membrane-penetrating structure containing superoxide dismutase-like subunits. *Journal of Biological Chemistry*, *276*, 20466–20473. https://doi.org/10.1074/jbc.M100556200

Davson, H., & Segal, M. B. (1996). *Physiology of the CSF and blood–brain barriers*. CRC Press.

Decourt, B., Wilson, J., Ritter, A., Dardis, C., DiFilippo, F. P., Zhuang, X., Cordes, D., Lee, G., Fulkerson, N. D., St Rose, T., Hartley, K., & Sabbagh, M. N. (2020). MCLENA-1: A phase II clinical trial for the assessment of safety, tolerability, and efficacy of lenalidomide in patients with mild cognitive impairment due to Alzheimer's disease. *Open Access Journal of Clinical Trials*, *12*, 1–13. https://doi.org/10.2147/oajct.s221914

Dehghani, R., Rahmani, F., & Rezaei, N. (2018). MicroRNA in Alzheimer's disease revisited: Implications for major neuropathological mechanisms. *Reviews in the Neurosciences*, *29*(2), 161–182. https://doi.org/10.1515/revneuro-2017-0033

Dehouck, B., Fenart, L., Dehouck, M. P., Pierce, A., Torpier, G., & Cecchelli, R. (1997). A new function for the LDL receptor: Transcytosis of LDL across the blood–brain barrier. *Journal of Cell Biology*, *138*(4), 877–889. https://doi.org/10.1083/jcb.138.4.877

Dexter, D. T., Wells, F. R., & Lees, A. J. (1989). Increased nigral iron content and alterations in other metal ions occurring in brain in Parkinson's disease. *Journal of Neurochemistry*, *52*(2), 381–389. https://doi.org/10.1111/j.1471-4159.1989.tb09132.x

Di Vaira, M., Bazzicalupi, C., Orioli, P. L., Messori, L., Bruni, B., & Zatta, P. (2004). Clioquinol, a drug for Alzheimer's disease specifically interfering with brain metal metabolism: Structural characterization of its zinc(II) and copper(II) complexes. *Inorganic Chemistry*, *43*(13), 3795–3797. https://doi.org/10.1021/ic0494051

Dorr, R. T. (1992). Bleomycin pharmacology: Mechanism of action and resistance, and clinical pharmacokinetics. *Seminars in Oncology*, *19*(Suppl. 5), 3–8.

Double, K. L., Gerlach, M., Schünemann, V., Trautwein, A. X., Zecca, L., Gallorini, M., Youdim, M. B. H., & Riederer, P. (2003). Iron-binding characteristics of neuromelanin of the human substantia nigra. *Biochemical Pharmacology*, *66*(3), 489–494. https://doi.org/10.1016/s0006-2952(03)00293-4

Double, K. L., Riederer, P. F., & Gerlach, M. (1999). Significance of neuromelanin for neurodegeneration in Parkinson's disease. *Drug News & Perspectives*, *12*, 333–340.

Ebnet, K., Suzuki, A., Ohno, S., & Vestweber, D. (2004). Junctional adhesion molecules (JAMs): Molecules with dual functions. *Journal of Cell Science*, *117*, 19–29. https://doi.org/10.1242/jcs.00925

Ejaz, H. W., Wang, W., & Lang, M. (2020). Copper toxicity links to pathogenesis of Alzheimer's disease and therapeutics approaches. *International Journal of Molecular Sciences*, *21*(20), 7660. https://doi.org/10.3390/ijms21207660

Endres, K., Fahrenholz, F., Lotz, J., Hiemke, C., Teipel, S., Lieb, K., Tüscher, O., & Fellgiebel, A. (2014). Increased CSF APPs-α levels in patients with Alzheimer disease treated with acitretin. *Neurology*, *83*(21), 1930–1935. https://doi.org/10.1212/WNL.0000000000001017

Fasman, G. D., & Moore, C. D. (1994). The solubilization of model Alzheimer tangles: Reversing the beta-sheet conformation induced by aluminum with silicates. *Proceedings of the National Academy of Sciences of the United States of America*, *91*(23), 11232–11235. https://doi.org/10.1073/pnas.91.23.11232

Faux, N. G., Ritchie, C. W., Gunn, A., Rembach, A., Tsatsanis, A., Bedo, J., Harrison, J., Lannfelt, L., Blennow, K., Zetterberg, H., Ingelsson, M., Masters, C. L., Tanzi, R. E., Cummings, J. L., Herd, C. M., & Bush, A. I. (2010). PBT2 rapidly improves cognition in Alzheimer's disease: Additional phase II analyses. *Journal of Alzheimer's Disease*, *20*(2), 509–516. https://doi.org/10.3233/JAD-2010-1390

Figley, C. R., & Stroman, P. W. (2011). The role(s) of astrocytes and astrocyte activity in neurometabolism, neurovascular coupling, and the production of functional neuroimaging signals. *European Journal of Neuroscience*, *33*(4), 577–588. https://doi.org/10.1111/j.1460-9568.2010.07584.x

Filiz, G., Price, K. A., Caragounis, A., Du, T., Crouch, P. J., & White, A. R. (2008). The role of metals in modulating metalloprotease activity in the AD brain. *European Biophysics Journal*, *37*(3), 315–321. https://doi.org/10.1007/s00249-007-0244-1

Gal, S., Zheng, H., Fridkin, M., & Youdim, M. B. H. (2010). Restoration of nigrostriatal dopamine neurons in post-MPTP treatment by the novel multifunctional brain-permeable iron chelator-monoamine oxidase inhibitor drug, M30. *Neurotoxicity Research*. https://doi.org/10.1007/s12640-009-9070-9

Galey, J. B., Dumats, J., Beck, I., Fernandez, B., & Hocquaux, M. (1995). N,N′-bis-dibenzyl ethylenediaminediacetic acid (DBED): A site-specific hydroxyl radical scavenger acting as an 'oxidative stress activatable' iron chelator in vitro. *Free Radical Research*, *22*(1), 67–86. https://doi.org/10.3109/10715769509147529

Gao, H. L., & Jiang, X. G. (2013). Progress on the diagnosis and evaluation of brain tumors. *Cancer Imaging*, *13*(4), 466–481. https://doi.org/10.1102/1470-7330.2013.0037

Gaur, K., Vázquez-Salgado, A. M., Duran-Camacho, G., Dominguez-Martinez, I., Benjamín-Rivera, J. A., Fernández-Vega, L., Carmona Sarabia, L., Cruz García, A., Pérez-Deliz, F., & Méndez Román, J. A. (2018). Iron and copper intracellular chelation as an anticancer drug strategy. *Inorganics*, *6*(4), 126. https://doi.org/10.3390/inorganics6040126

Greenamyre, J. T., & Porter, R. H. (1994). Anatomy and physiology of glutamate in the CNS. *Neurology*, *44*(Suppl 8), S7–S13.

Greenamyre, J. T., & Young, A. B. (1989). Excitatory amino acids and Alzheimer's disease. *Neurobiology of Aging*, *10*(5), 593–602.

Greenough, M. A., Volitakis, I., Li, Q. X., Laughton, K., Evin, G., Ho, M., Dalziel, A. H., Camakaris, J., & Bush, A. I. (2011). Presenilins promote the cellular uptake of copper and zinc and maintain copper chaperone of SOD1-dependent copper/zinc superoxide dismutase activity. *Journal of Biological Chemistry*. https://doi.org/10.1074/jbc.M110.163964

Grimson, M. J., Coates, J. C., Reynolds, J. P., Shipman, M., Blanton, R. L., & Harwood, A. J. (2000). Adherens junctions and beta-catenin-mediated cell signaling in a non-metazoan organism. *Nature*, *408*(6813), 723–731. https://doi.org/10.1038/35047099

Guo, C., Zhang, Y. X., Wang, T., Zhong, M. L., Yang, Z. H., Hao, L. J., Chai, R., & Zhang, S. (2015). Intranasal deferoxamine attenuates synapse loss via up-regulating the P38/HIF-1α pathway on the brain of APP/PS1 transgenic mice. *Frontiers in Aging Neuroscience*, *7*, 104. https://doi.org/10.3389/fnagi.2015.00104

Guo, R., Sakamoto, H., Sugiura, S., & Ogawa, M. (2007). Endothelial cell motility is compatible with junctional integrity. *Journal of Cellular Physiology*, *211*(2), 327–335. https://doi.org/10.1002/jcp.20945

Haas, L. T., Salazar, S. V., Smith, L. M., Zhao, H. R., Cox, T. O., Herber, C. S., Degnan, A. P., Balakrishnan, A., Macor, J. E., Albright, C. F., & Strittmatter, S. M. (2017). Silent allosteric modulation of mGluR5 maintains glutamate signaling while rescuing Alzheimer's mouse phenotypes. *Cell Reports*, *20*(1), 76–88. https://doi.org/10.1016/j.celrep.2017.06.023

Hanson, L. R., Fine, J. M., Renner, D. B., Svitak, A. L., Burns, R. B., Nguyen, T. M., Tuttle, N. J., Marti, D. L., Panter, S. S., & Frey, W. H. (2012). Intranasal delivery of deferoxamine reduces spatial memory loss in APP/PS1 mice. *Drug Delivery and Translational Research*. https://doi.org/10.1007/s13346-011-0050-2

Hegde, M. L., Bharathi, P., Suram, A., Venugopal, C., Jagannathan, R., Poddar, P., Srinivas, P., Sambamurti, K., Rao, K. J., Scancar, J., Messori, L., Zecca, L., & Zatta, P. (2009). Challenges associated with metal chelation therapy in Alzheimer's disease. *Journal of Alzheimer's Disease*. https://doi.org/10.3233/JAD-2009-1068

Hey, J. A., Yu, J. Y., Versavel, M., Abushakra, S., Kocis, P., Power, A., Kaplan, P. L., Amedio, J., & Tolar, M. (2018). Clinical pharmacokinetics and safety of ALZ-801, a novel prodrug of tramiprosate in development for the treatment of Alzheimer's disease. *Clinical Pharmacokinetics*, *57*(3), 315–333. https://doi.org/10.1007/s40262-017-0608-3

Huang, X., Atwood, C. S., Moir, R. D., Hartshorn, M. A., Vonsattel, J. P., Tanzi, R. E., & Bush, A. I. (1997). Zinc-induced Alzheimer's Abeta 1-40 aggregation is mediated by conformational factors. *Journal of Biological Chemistry*, *272*(42), 26464–26470. https://pubmed.ncbi.nlm.nih.gov/9334223/

Izzo, N. J., Yuede, C. M., LaBarbera, K. M., Limegrover, C. S., Rehak, C., Yurko, R., Waybright, L., Look, G., Rishton, G., Safferstein, H., Hamby, M. E., Williams, C., Sadlek, K., Edwards, H. M., Davis, C. S., Grundman, M., Schneider, L. S., DeKosky, S. T., Chelsky, D., … Catalano, S. M. (2021). Preclinical and clinical biomarker studies of CT1812: A novel approach to Alzheimer's disease modification. *Alzheimer's & Dementia*, *17*(8), 1365–1382. https://doi.org/10.1002/alz.12322

Jack, C. R., & Holtzman, D. M. (2013). Biomarker modeling of Alzheimer's disease. *Neuron*, *80*(6), 1347–1358. https://doi.org/10.1016/j.neuron.2013.12.003

Kawada, H., Blessing, K., Kiyota, T., Woolman, T., Winchester, L., & Kador, P. F. (2015). Effects of multifunctional antioxidants on mitochondrial dysfunction and amyloid-β metal dyshomeostasis. *Journal of Alzheimer's Disease*. https://doi.org/10.3233/JAD-132471

Kazmierczak, P., Sakaguchi, H., Tokita, J., Wilson-Kubalek, E. M., Milligan, R. A., Müller, U., & Kachar, B. (2007). Cadherin 23 and protocadherin 15 interact to form tip-link filaments in sensory hair cells. *Nature*, *449*(7158), 87–91. https://doi.org/10.1038/nature06091

Kellar, D., Lockhart, S. N., Aisen, P., Raman, R., Rissman, R. A., Brewer, J., & Craft, S. (2021). Intranasal insulin reduces white matter hyperintensity progression in association with improvements in cognition and CSF biomarker profiles in mild cognitive impairment and Alzheimer's disease. *Journal of Prevention of Alzheimer's Disease*, *8*(3), 240–248. https://doi.org/10.14283/jpad.2021.15

Kellar, D., Register, T., Lockhart, S. N., Aisen, P., Raman, R., Rissman, R. A., Brewer, J., & Craft, S. (2022). Intranasal insulin modulates cerebrospinal fluid markers of neuroinflammation in mild cognitive impairment and Alzheimer's disease: A randomized trial. *Scientific Reports*, *12*, 1346. https://doi.org/10.1038/s41598-022-05165-3

Kester, M. I., Teunissen, C. E., Crimmins, D. L., Herries, E. M., Ladenson, J. H., Scheltens, P., van der Flier, W. M., Morris, J. C., Holtzman, D. M., & Fagan, A. M. (2015). Neurogranin as a cerebrospinal fluid biomarker for synaptic loss in symptomatic Alzheimer disease. *JAMA Neurology*, *72*(11), 1275–1280. https://doi.org/10.1001/jamaneurol.2015.1867

Klein, W. L., Krafft, G. A., & Finch, C. E. (2001). Targeting small Aβ oligomers: The solution to an Alzheimer's disease conundrum? *Trends in Neurosciences*, *24*(4), 219–224. https://doi.org/10.1016/s0166-2236(00)01749-5

Knell, M. (1959). Metal chelates. In *Metal-organic compounds* (pp. 37–45). American Chemical Society. https://doi.org/10.1021/ba-1959-0023.ch003

Kolnagou, A., Natsiopoulos, K., Kleanthous, M., Ioannou, A., & Kontoghiorghes, G. J. (2013). Liver iron and serum ferritin levels are misleading for estimating cardiac, pancreatic, splenic and total body iron load in thalassemia patients: Factors influencing the heterogenic distribution of excess storage iron in organs as identified by MRI T2*. *Toxicology Mechanisms and Methods*, *23*(1), 48–56. https://doi.org/10.3109/15376516.2012.727198

Kontoghiorghe, C. N., Kolnagou, A., & Kontoghiorghes, G. J. (2015). Phytochelators intended for clinical use in iron overload, other diseases of iron imbalance and free radical pathology. *Molecules*, *20*(11), 20841–20872. https://doi.org/10.3390/molecules201119725

Kontoghiorghes, C. N., & Kontoghiorghes, G. J. (2016). Efficacy and safety of iron-chelation therapy with deferoxamine, deferiprone, and deferasirox for the treatment of iron-loaded patients with non-transfusion-dependent thalassemia syndromes. *Drug Design, Development and Therapy*, *10*, 465–481. https://doi.org/10.2147/DDDT.S79458

Kontoghiorghes, G. J., & Kontoghiorghes, C. N. (2019). Prospects for the introduction of targeted antioxidant drugs for the prevention and treatment of diseases related to free radical pathology. *Expert Opinion on Investigational Drugs*, *28*(6), 593–603. https://doi.org/10.1080/13543784.2019.1631284

Kreuter, J. (2001). Nanoparticulate systems for brain delivery of drugs. *Advanced Drug Delivery Reviews*, *47*(1), 65–81. https://doi.org/10.1016/s0169-409x(00)00122-8

Kutzsche, J., Jürgens, D., Willuweit, A., Adermann, K., Fuchs, C., Simons, S., Windisch, M., Hümpel, M., Rossberg, W., Wolzt, M., & Willbold, D. (2020). Safety and pharmacokinetics of the orally available antiprionic compound PRI-002: A single and multiple ascending dose phase I study. *Alzheimer's & Dementia (N.Y.)*, *6*, e12001. https://doi.org/10.1002/trc2.12001

Lampugnani, M. G., & Dejana, E. (2007). Adherens junctions in endothelial cells regulate vessel maintenance and angiogenesis. *Thrombosis Research*, *120*(Suppl 2), S1–S6. https://doi.org/10.1016/S0049-3848(07)70124-X

Lannfelt, L., Blennow, K., Zetterberg, H., Batsman, S., Ames, D., Harrison, J., Masters, C. L., Targum, S., Bush, A. I., Murdoch, R., Wilson, J., & Ritchie, C. W. (2008). Safety, efficacy, and biomarker findings of PBT2 in targeting Aβ as a modifying therapy for Alzheimer's disease: A phase IIa, double-blind, randomized, placebo-controlled trial. *The Lancet Neurology*, *7*(9), 779–786. https://doi.org/10.1016/S1474-4422(08)70150-9

Lanza, V., Milardi, D., Di Natale, G., & Pappalardo, G. (2018). Repurposing copper(II)-chelating drugs for the treatment of neurodegenerative diseases. *Current Medicinal Chemistry*. https://doi.org/10.2174/0929867324666170518094404

Lee, J. Y., Mook-Jung, I., & Koh, J. Y. (1999). Histochemically reactive zinc in plaques of the Swedish mutant beta-amyloid precursor protein transgenic mice. *Journal of Neuroscience*, 19, RC10. https://doi.org/10.1523/JNEUROSCI.19-11-j0002.1999

Li, L. B., Fan, Y. G., Wu, W. X., Bai, C. Y., Jia, M. Y., Hu, J. P., Gao, H. L., Wang, T., Zhong, M. L., Huang, X. S., & Guo, C. (2022). Novel melatonin-trientine conjugate as potential therapeutic agents for Alzheimer's disease. *Bioorganic Chemistry*. https://doi.org/10.1016/j.bioorg.2022.106100

Liu, G., Garrett, M. R., Men, P., Zhu, X., Perry, G., & Smith, M. A. (2005). Nanoparticle and other metal chelation therapeutics in Alzheimer's disease. *Biochimica et Biophysica Acta*, *1741*, 246–252.

Liu, G., Miller, S. C., & Bruenger, F. W. (1995). Synthesis of lipophilic 3-hydroxy-2-methyl-4-pyridinone derivatives. *Synthetic Communications*, *25*, 3247–3253.

Lord, R. J., & Lippard, S. J. (2013). Chelators for investigating zinc metalloneurochemistry. *Current Opinion in Chemical Biology*. https://doi.org/10.1016/j.cbpa.2013.01.009

Lovell, M. A., Robertson, J. D., Teesdale, W. J., Campbell, J. L., & Markesbery, W. R. (1998). Copper, iron and zinc in Alzheimer's disease senile plaques. *Journal of Neurological Sciences*, *158*, 47–52.

Lovstad, R. A. (2002). A kinetic study on the copper-albumin catalyzed oxidation of ascorbate. *Biometals*. https://doi.org/10.1023/a:1020247323914

Lue, L. F., Kuo, Y. M., Roher, A. E., Brachova, L., Shen, Y., Sue, L., Beach, T., Kurth, J. H., Rydel, R. E., & Rogers, J. (1999). Soluble amyloid beta-peptide concentration as a predictor of synaptic change in Alzheimer's disease. *American Journal of Pathology*, *155*, 853–862.

Lui, G. Y., Kovacevic, Z., Richardson, V., Merlot, A. M., Kalinowski, D. S., & Richardson, D. R. (2015). Targeting cancer by binding iron: Dissecting cellular signaling pathways. *Oncotarget*, *6*, 18748–18779. https://doi.org/10.18632/oncotarget.4307

Maccecchini, M. L., Chang, M. Y., Pan, C., John, V., Zetterberg, H., & Greig, N. H. (2012). Posiphen as a candidate drug to lower CSF amyloid precursor protein, amyloid-beta peptide, and tau levels: Target engagement, tolerability, and pharmacokinetics in humans. *Journal of Neurology, Neurosurgery & Psychiatry*, *83*, 894–902.

Madaan, K., Kaushik, D., & Verma, T. (2012). Hydroxyurea: A key player in cancer chemotherapy. *Expert Review of Anticancer Therapy*, *2*, 19–29.

Mandel, S., Amit, T., Bar-Am, O., & Youdim, M. B. (2007). Iron dysregulation in Alzheimer's disease: Multimodal brain permeable iron chelating drugs, possessing neuroprotective-neurorescue and amyloid precursor protein-processing regulatory activities as therapeutic agents. *Progress in Neurobiology*, *82*, 348–360.

Martin-Bastida, A., Ward, R. J., Newbould, R., Piccini, P., Sharp, D., Kabba, C., Patel, M. C., Spino, M., Connelly, J., Tricta, F., Crichton, R. R., & Dexter, D. T. (2017). Brain iron chelation by deferiprone in a phase 2 randomised double-blinded placebo controlled clinical trial in Parkinson's disease. *Scientific Reports*, *7*, 1398. https://doi.org/10.1038/s41598-017-01480-2

McLean, C. A., Cherny, R. A., Fraser, F. W., Fuller, S. J., Smith, M. J., Beyreuther, K., Bush, A. I., & Masters, C. L. (1999). Soluble pool of Aβ amyloid as a determinant of severity of neurodegeneration in Alzheimer's disease. *Annals of Neurology*, *46*, 860–866.

Miura, T., Suzuki, K., Kohata, N., & Takeuchi, H. (2000). Metal binding modes of Alzheimer's amyloid beta-peptide in insoluble aggregates and soluble complexes. *Biochemistry*, *39*, 7024–7031. https://doi.org/10.1021/bi0002479

Mizuno, Y., Ohta, S., Tanaka, M., Takamiya, S., Suzuki, K., Sato, T., Oya, H., Ozawa, T., & Kagawa, Y. (1989). Deficiencies in Complex I subunits of the respiratory chain in Parkinson's disease. *Biochemical and Biophysical Research Communications*, *163*, 1450–1455.

Muller, R. H., Jacobs, C., & Kayser, O. (2001). Nanosuspensions as particulate drug formulations in therapy. Rationale for development and what we can expect for the future. *Advanced Drug Delivery Reviews*, *47*, 3–19.

Nick, H. (2007). Iron chelation, quo vadis? *Current Opinion in Chemical Biology*. https://doi.org/10.1016/j.cbpa.2007.04.025

Parghane, R. V., & Basu, S. (2019). Bilateral orbital soft-tissue metastases from renal neuroendocrine tumor: Successful theranostic application of 68Ga/177Lu-DOTATATE with improvement of vision. *Journal of Nuclear Medicine Technology*, *47*, 171–172. https://doi.org/10.2967/jnmt.118.217455

Pennell, D. J. (2005). T2* magnetic resonance and myocardial iron in thalassemia. *Annals of the New York Academy of Sciences*, *1054*, 373–378. https://doi.org/10.1196/annals.1345.044

Perret, D. (1981). The metabolism and pharmacology of d-penicillamine in man. *Journal of Rheumatology Supplement*, *7*, 41–50.

Peterson, D. A., Gerrard, J. M., Rao, G. H., & White, J. G. (1979). Inhibition of ferrous ion-induced oxidation of arachidonic acid by indomethacin. *Prostaglandins Medicine*, *2*, 97–108.

Piga, A., Galanello, R., Forni, G. L., Cappellini, M. D., Origa, R., & Zappu, A. (2006). Randomized phase II trial of deferasirox (Exjade, ICL670), a once-daily, orally-administered iron chelator, in comparison to deferoxamine in thalassemia patients with transfusional iron overload. *Haematologica*, *91*, 873–880.

Poliseno, V., Chaves, S., Brunetti, L., Loiodice, F., Carrieri, A., Laghezza, A., Tortorella, P., Magalhães, J. D., Cardoso, S. M., Santos, M. A., & Piemontese, L. (2021). Derivatives of tenuazonic acid as potential new multi-target anti-Alzheimer's disease agents. *Biomolecules*. https://doi.org/10.3390/biom11010111

Breakspear, M., Roberts, G., Green, M. J., Nguyen, V. T., Frankland, A., Levy, F., Lenroot, R., & Mitchell, P. B. (2015). Cerebrospinal fluid neurogranin: Relation to cognition and neurodegeneration in Alzheimer's disease. *Brain*, *138*, 3373–3385. https://doi.org/10.1093/brain/awv261

Prasanthi, J. R. P., Schrag, M., Dasari, B., Mararha, G., Dickson, A., Kirsch, W. M., & Ghribi, O. (2012). Deferiprone reduces amyloid-β and tau phosphorylation levels but not reactive oxygen species generation in hippocampus of rabbits fed a cholesterol-enriched diet. *Journal of Alzheimer's Disease*. https://doi.org/10.3233/JAD-2012-111346

Quitti, R., Rossini, P. M., Cassetta, E., Mofa, F., Pasqualetti, P., Cortesi, M., Colloca, A., Rossi, L., & Finazzi-Agro', A. (2002). d-Penicillamine reduces serum oxidative stress in Alzheimer's disease patients. *European Journal of Clinical Investigation*. https://doi.org/10.1046/j.1365-2362.2002.00933.x

Ramge, P. (1998). *Untersuchungen zur Überwindung der Blut-Hirn-Schranke mit Hilfe von Nanopartikeln [Studies on overcoming the blood-brain barrier using nanoparticles]*. Shaker Verlag.

Rao, G. H., & White, J. G. (1985). Comparative pharmacology of cyclooxygenase inhibitors on platelet function. *Prostaglandins Leukotrienes and Medicine*, *18*, 119–131.

Rao, K. S. J., Rao, R. V., Shanmugavelu, P., & Menon, R. B. (1999). Trace elements in Alzheimer's disease brain: A new hypothesis. *Alzheimer's Reports*, *2*, 241–246.

Ravi Kumar, M. N. (2000). Nano and microparticles as controlled drug delivery devices. *Journal of Pharmacy and Pharmaceutical Sciences*, *3*, 234–258.

Richardson, D. R. (1999). The therapeutic potential of iron chelators. *Expert Opinion on Investigational Drugs*, *8*, 2141–2158.

Ritchie, C. W., Bush, A. I., Mackinnon, A., Macfarlane, S., Mastwyk, M., MacGregor, L., Kiers, L., Cherny, R., Li, Q. X., Tammer, A., Carrington, D., Mavros, C., Volitakis, I., Xilinas, M., Ames, D., Davis, S., Beyreuther, K., Tanzi, R. E., & Masters, C. L. (2003). Metal-protein attenuation with iodochlorhydroxyquin (clioquinol) targeting Aβ amyloid deposition and toxicity in Alzheimer disease: A pilot phase 2 clinical trial. *Archives of Neurology*, *60*(12), 1685–1691. https://doi.org/10.1001/archneur.60.12.1685

Russanov, E. M., Dimitrova, D. E., Ivancheva, E. A., & Kirkova, M. D. (1986). The effects of aspirin, indomethacin, and their copper complexes on phospholipase activity and lipid peroxidation in rat liver microsomes. *Acta Physiologica Pharmacologica Bulgarica*, *12*, 36–43.

Sanchez-Ramos, J., Overvick, E., & Ames, B. N. (1994). A marker of oxyradical-mediated DNA damage (8-hydroxy-2′-deoxyguanosine) is increased in nigro-striatum of Parkinson's disease brain. *Neurodegeneration*, *3*, 197–204.

Santos, M. A., Chand, K., & Chaves, S. (2016). Recent progress in multifunctional metal chelators as potential drugs for Alzheimer's disease. *Coordination Chemistry Reviews*. https://doi.org/10.1016/j.ccr.2016.04.013

Savelief, M. G., Lee, S., Liu, Y., & Lim, M. H. (2013). Untangling amyloid-beta, Tau, and metals in Alzheimer's disease. *ACS Chemical Biology*. https://doi.org/10.1021/cb400080f

Scheltens, P., Hallikainen, M., Grimmer, T., Duning, T., Gouw, A. A., Teunissen, C. E., Wink, A. M., Maruff, P., Harrison, J., van Baal, C. M., Bruins, S., Lues, I., & Prins, N. D. (2018). Safety, tolerability, and efficacy of the glutaminyl cyclase inhibitor PQ912 in Alzheimer's disease: Results of a randomized, double-blind, placebo-controlled phase 2a study. *Alzheimer's Research & Therapy*, *10*, 107. https://doi.org/10.1186/s13195-018-0431-6

Schroeder, U., Sommerfeld, P., Ulrich, S., & Sabel, B. A. (1998). Nanoparticle technology for delivery of drugs across the blood–brain barrier. *Journal of Pharmaceutical Sciences*, *87*, 1305–1307.

Sensi, S. L., Paoletti, P., Koh, J. Y., Aizenman, E., Bush, A. I., & Hershfinkel, M. (2011). The neurophysiology and pathology of brain zinc. *Journal of Neuroscience*. https://doi.org/10.1523/JNEUROSCI.3454-11.2011

Shachar, D. B., Kahana, N., Kampel, V., Warshawsky, A., & Youdim, M. B. (2004). Neuroprotection by a novel brain permeable iron chelator, VK-28, against 6-hydroxydopamine lesion in rats. *Neuropharmacology*, *46*, 254–263.

Sheskin, J., Gorodetzky, R., Loewinger, E., & Weinreb, A. (1981). In vivo measurements of iron, copper, and zinc in the skin of prurigo nodularis patients treated with thalidomide. *Dermatologica*, *162*, 86–90.

Singh, A. B., Sharma, A., & Dhawan, P. (2010). Claudin protein family and cancer: An overview. *Journal of Oncology*, *2010*, 541957. https://doi.org/10.1155/2010/541957

Smith, M. A., Hirai, K., Hsiao, K., Pappolla, M. A., Harris, P. L., Siedlak, S. L., Tabaton, M., & Perry, G. (1998). Amyloid-beta deposition in Alzheimer transgenic mice is associated with oxidative stress. *Journal of Neurochemistry*, *70*(5), 2212–2215. https://doi.org/10.1046/j.1471-4159.1998.70052212.x

Spurrier, J., Nicholson, L., Fang, X. T., Stoner, A. J., Toyonaga, T., Holden, D., Siegert, T. R., Laird, W., Allnutt, M. A., Chiasseu, M., Brody, A. H., Takahashi, H., Nies, S. H., Pérez-Cañamás, A., Sadasivam, P., Lee, S., Li, S., Zhang, L., Huang, Y. H., … Strittmatter, S. M. (2022). Reversal of synapse loss in Alzheimer mouse models by targeting mGluR5 to prevent synaptic tagging by C1Q. *Science Translational Medicine*, *14*, eabi8593. https://doi.org/10.1126/scitranslmed.abi8593

Stolp, H. B., & Dziegelewska, K. M. (2009). Role of developmental inflammation and blood-brain barrier dysfunction in neurodevelopmental and neurodegenerative diseases. *Neuropathology and Applied Neurobiology*, *35*(2), 132–146. https://doi.org/10.1111/j.1365-2990.2008.00980.x

Suárez-Calvet, M., Capell, A., Araque Caballero, M. Á., Morenas-Rodríguez, E., Fellerer, K., Franzmeier, N., Kleinberger, G., Eren, E., Deming, Y., Piccio, L., Karch, C. M., Cruchaga, C., Paumier, K., Bateman, R. J., Fagan, A. M., Morris, J. C., Levin, J., Danek, A., Jucker, M., … the Dominantly Inherited Alzheimer Network the Alzheimer's Disease Neuroimaging Initiative. (2018). CSF progranulin increases in the course of Alzheimer's disease and is associated with sTREM2, neurodegeneration, and cognitive decline. *EMBO Molecular Medicine*, *10*(12), e9712. https://doi.org/10.15252/emmm.201809712

Tardito, S., Barilli, A., Bassanetti, I., Tegoni, M., Bussolati, O., Franchi-Gazzola, R., Mucchino, C., & Marchiò, L. (2012). Copper-dependent cytotoxicity of 8-hydroxyquinoline derivatives correlates with their hydrophobicity and does not require caspase activation. *Journal of Medicinal Chemistry*. https://doi.org/10.1021/jm301053a

Tegoni, M., Valensin, D., Toso, L., & Remelli, M. (2014). Copper chelators: Chemical properties and biomedical applications. *Current Medicinal Chemistry*. https://doi.org/10.2174/0929867321666140601161939

Treiber, C., Simons, A., Strauss, M., Hafner, M., Cappai, R., Bayer, T. A., & Multhaup, G. (2004). Clioquinol mediates copper uptake and counteracts copper efflux activities of the amyloid precursor protein of Alzheimer's disease. *Journal of Biological Chemistry*, *279*(49), 51958–51964. https://pubmed.ncbi.nlm.nih.gov/15465814/

Wang, C. Y., Xie, J. W., Xu, Y., Wang, T., Cai, J. H., Wang, X., Zhao, B. L., An, L., & Wang, Z. Y. (2013). Trientine reduces BACE-1 activity and mitigates amyloidosis via the AGE/RAGE/NF-κB pathway in a transgenic mouse model of Alzheimer's disease. *Antioxidants & Redox Signaling*. https://doi.org/10.1089/ars.2012.5158

Wang, H. Y., Lee, K. C., Pei, Z., Khan, A., Bakshi, K., & Burns, L. H. (2017). PTI-125 binds and reverses an altered conformation of filamin A to reduce Alzheimer's disease pathogenesis. *Neurobiology of Aging*, *55*, 99–114. https://doi.org/10.1016/j.neurobiolaging.2017.03.013

Wang, H. Y., Pei, Z., Lee, K. C., Lopez-Brignoni, E., Nikolov, B., Crowley, C. A., Marsman, M. R., Barbier, R., Friedmann, N., & Burns, L. H. (2020). PTI-125 reduces biomarkers of Alzheimer's disease in patients. *Journal of Prevention of Alzheimer's Disease*, *7*(4), 256–264. https://doi.org/10.14283/jpad.2020.29

Wang, S. S., Becerra-Arteaga, A., & Good, T. A. (2002). Development of a novel diffusion-based method to estimate the size of the aggregated Aβ species responsible for neurotoxicity. *Biotechnology and Bioengineering*, *80*(1), 50–59. https://doi.org/10.1002/bit.10131

White, A. R., Du, T., Laughton, K. M., Volitakis, I., Sharples, R. A., Xilinas, M. E., Hoke, D. E., Holsinger, R. M. D., Evin, G., Cherny, R. A., Hill, A. F., Barnham, K. J., Li, Q. X., Bush, A. I., & Masters, C. L. (2006). Degradation of the Alzheimer disease amyloid beta-peptide by metal-dependent up-regulation of metalloprotease activity. *Journal of Biological Chemistry*. https://doi.org/10.1074/jbc.M602487200

White, A. R., Kanninen, K., & Crouch, P. (2015). Editorial: Metals and neurodegeneration: Restoring the balance. *Frontiers in Aging Neuroscience*. https://doi.org/10.3389/fnagi.2015.00127

Winblad, B., & Poritis, N. (1999). Memantine in severe dementia: Results of the M-Best Study (Benefit and efficacy in severely demented patients during treatment with memantine). *International Journal of Geriatric Psychiatry*, *14*(2), 135–146. https://doi.org/10.1002/(sici)1099-1166(199902)14:2<135::aid-gps906>3.0.co;2-0

Winkler, E. A., Bell, R. D., & Zlokovic, B. V. (2011). Central nervous system pericytes in health and disease. *Nature Neuroscience*, *14*(11), 1398–1405. https://doi.org/10.1038/nn.2946

Wolburg, H., & Lippoldt, A. (2002). Tight junctions of the blood-brain barrier: Development, composition, and regulation. *Vascular Pharmacology*, *38*(6), 327–337. https://doi.org/10.1016/s1537-1891(02)00200-8

Wong, S. S. (1991). *Chemistry of protein conjugation and cross-linking*. CRC Press.

Xia, X., Wang, Y., Huang, Y., Zhang, H., Lu, H., & Zheng, J. C. (2019). Exosomal miRNAs in central nervous system diseases: Biomarkers, pathological mediators, protective factors, and therapeutic agents. *Progress in Neurobiology*, *183*, 101694. https://doi.org/10.1016/j.pneurobio.2019.101694

Yang, G. J., Liu, H., Ma, D. L., & Leung, C. H. (2019). Rebalancing metal dyshomeostasis for Alzheimer's disease therapy. *Journal of Biological Inorganic Chemistry*. https://doi.org/10.1007/s00775-019-01712-y

Yano, K., Hirosawa, N., Sakamoto, Y., Katayama, H., & Moriguchi, T. (2003). Aggregations of amyloid beta-proteins in the presence of metal ions. *Toxicology Letters*, *144*(2), 134. https://doi.org/10.1016/S0378-4274(03)90499-1

Yu, A. S., McCarthy, K. M., Francis, S. A., McCormack, J. M., Lai, J., Rogers, R. A., Lynch, R. D., & Schneeberger, E. E. (2005). Knockdown of occludin expression leads to diverse phenotypic alterations in epithelial cells. *American Journal of Physiology-Cell Physiology*, *288*(6), C1231–C1241. https://doi.org/10.1152/ajpcell.00414.2004

Zatta, P., Kiss, T., Suwalsky, M., & Berthon, G. (2002). Aluminium(III) as a promoter of cellular oxidation. *Coordination Chemistry Reviews*, *228*(1), 263–270. https://doi.org/10.1016/S0010-8545(02)00074-7

Zhu, D., Liang, R., Liu, Y., Li, Z., Cheng, L., Ren, J., Guo, Y., Wang, M., Chai, H., Niu, Q., Yang, S., Bai, J., Yu, H., Zhang, H., & Qin, X. (2022). Deferoxamine ameliorated Al(mal)3-induced neuronal ferroptosis in adult rats by chelating brain iron to attenuate oxidative damage. *Toxicology Mechanisms and Methods*. https://doi.org/10.1080/15376516.2022.2053254

Zhu, W., Xie, W., Pan, T., Xu, P., Fridkin, M., Zheng, H., Jankovic, J., Youdim, M. B. H., & Le, W. (2007). Prevention and restoration of lactacystin-induced nigrostriatal dopamine neuron degeneration by novel brain-permeable iron chelators. *FASEB Journal*. https://doi.org/10.1096/fj.07-8386com

Zlokovic, B. V. (2008). The blood-brain barrier in health and chronic neurodegenerative disorders. *Neuron*, *57*(2), 178–201. https://doi.org/10.1016/j.neuron.2008.01.003

15 Apolipoprotein E-Based Treatment Approach for Genetically High-Risk Alzheimer's Patients

Chanchal Mittal, Sayani Banerjee, and Sugato Banerjee

15.1 INTRODUCTION

Among 50 million individuals suffering from dementia, 60–80% suffer from Alzheimer's disease (AD)-associated dementia. Currently, over 50 million individuals globally have been diagnosed with various forms of dementia (Andrade-Guerrero et al., 2023). It is characterised by deposits of amyloid-beta (Aβ) peptide aggregation in neurones as well as the presence of neurofibrillary tangles (NFT) due to hyperphosporylated tau accumulation (González et al., 2022). One to 5% of cases of early-onset Alzheimer's disease (EOAD) or familial AD in those under 65 year of age are autosomal dominant. It is characterised by mutations in the amyloid precursor protein gene (APP, 21q21.3), the presenilin 1 gene (PSEN1, 14q24.2), seen in 70% of familial Alzheimer's cases; and the presenilin 2 gene (PSEN2, 1q42.13) for late-onset Alzheimer's disease (LOAD). However, sporadic AD is a complex disorder with various social, environmental and metabolic risk factors along with genetic predisposition, specifically mutations in the ε4 allele of apolipoprotein E (APOE, 19q13.32), with a heritability estimated at 60–80%. The main risk factor is age (Padhy et al., 2023, 2025). Another significant risk factor for LOAD is the gene that codes for triggering receptor expressed on myeloid cells 2 (TREM2) (Gaugler et al., 2022). Along with mutations in the APOE gene, Single Nucleotide Polymorphism (SNP) in the APOE may also facilitate LOAD. Analyses of AD patients and mouse models show that neuronal deterioration begins 15 years before the disease's obvious symptoms appear (Lane-Donovan & Herz, 2017). APOE, a major cholesterol carrier, has three isoforms termed APOE2, E3, and E4, each made up of 299 amino acids only differing at amino acids 112 and 158. Among these three isoforms, APOE4 is more commonly linked to EOAD and LOAD.

Understanding biophysical and structural variations helps us comprehend the mechanics behind the neurological conditions linked to APOE and opens the door to the possibility of structure-guided treatments to treat these conditions. In this chapter, we also describe how APOE function is impacted by amino acid polymorphisms. It also discussed the insights which are currently understood about APOE structure and lipid binding before summarising the unanswered problems for further research.

Current research demonstrates how the brain's physiological homeostasis is determined by the APOE isoform and how several APOE-targeted treatment methods might have preventative or corrective effects in neurodegenerative proteinopathies like AD. It is generally believed that APOE-targeted medicines could be highly effective against neurodegenerative diseases based on animal and cellular studies. In the era of personalised and targeted medicine, improvements in biomarkers along with novel treatment strategies for early diagnosis and prognosis of AD would be crucial for targeted therapeutics, particularly in a population with knowledge of APOE. It is also necessary to ascertain how such therapies impact vascular function and peripheral lipid balance. APOE-targeted treatment approaches continue to be a promising field of study for halting or postponing the onset of AD-type dementias.

This chapter explores the basic function of APOE and its detailed biophysical structures, a vital tool of emerging treatment approaches targeting APOE for genetically high-risk individuals. We also discussed current therapeutic strategies to mitigate amyloid pathology, modify APOE function, and implement lifestyle interventions. Recent advancements in monoclonal antibody therapies and small molecules that enhance the beneficial effects of alternative APOE alleles show promise in early intervention scenarios.

15.2 APOE BIOLOGY – APOLIPOPROTEIN E

15.2.1 Synthesis of APOE

APOE is synthesised both peripherally and centrally. Peripheral APOE cannot cross the blood-brain barrier (BBB). As a result, it creates two separate resources of APOE: one in the periphery and another in the Central Nervous System (CNS). Plasma APOE (~40–70 μg/ml) is predominantly produced by hepatic synthesis (>75%), while APOE is primarily produced by astrocytes in the CNS with a minor contribution from microglia and distressed neurones. The APOE gene is transcribed into mRNA and subsequently translated into APOE. Macrophages and other cell types also produce APOE (Mahley et al., 2009).

APOE, after getting produced at the plasma membrane, undergoes post-translational glycosylation and sialylation

DOI: 10.1201/9781003602477-15

in the Golgi (Fernández-Calle et al., 2022). POE synthesised by astrocytes unergoes more heavy sialylation and glycosylation (Flowers et al., 2019). The process is controlled by DNA-binding proteins liver X receptor (LXR) and retinoid X receptor (RXR) (Schulman, 2017). LDLR-related protein 1, Very Low Density Lipoprotein (VLDL) receptor, and APOE receptor 2, the primary APOE receptors, are involved in multiple functions, including lipid metabolism and uptake, synaptic transmission, synaptic plasticity and normal astrocyte function (Lane-Donovan & Herz, 2017). The number of APOE epitopes and their lipidation state both influence its receptor-binding selectivity. The cellular absorption of Aβ in astrocytes is significantly increased when LDLR is overexpressed, while its deletion reduces its absorption (Basak et al., 2012). When APOE is lipid-free, most of its N-terminal binding domains are usually hidden, but when lipidation occurs, they become accessible for binding to LDLR (Chen et al., 2011).

LXR can be activated by Brain-derived Neurotrophic Factor (BDNF) to increase APOE expression and enhance astrocyte cholesterol export (Spagnuolo et al., 2018). While Axl receptor tyrosine kinase has been shown to regulate APOE in human astrocytes (Zhao et al., 2020). LXR transcriptionally modulates ABCA1 (Fitz et al., 2012; Jacobo-Albavera et al., 2021). Transporter Niemann-Pick type C 1 protein is a major cholesterol transporter to the brain, while cholesterol 24(S)-hydroxylase is involved in cholesterol metabolism (Quan et al., 2003). ABCA1 regulates cholesterol efflux via lipoproteins and facilitates the formation of high-density lipoprotein (HDL). ABCA1 promotes cholesterol efflux to APOE in the CNS, resulting in particles that resemble HDL-like APOE that is additionally rich in phospholipids and unesterified cholesterol. Apolipoprotein A-I, ApoAI facilitates the process by which APOE interacts with ABCA1 in APOE-expressing macrophages to produce unesterified HDL, which is rich in cholesterol (Vitali et al., 2014). Reductions in ABCA1 membrane recycling result in diminished lipidation of APOE4 particles due to an enhanced propensity for self-aggregation of E4 alleles (Rawat et al., 2019).

15.2.2 Function of APOE

Lipid homeostasis is crucial for optimal brain function (Mahley, 2016; O'Brien & Sampson, 1965). APOE primarily functions as the principal transporter of cholesterol and lipids. Moreover, it contributes to vascular integrity, neuronal plasticity, glucose metabolism, signal transduction, damage repair, and function of mitochondria (Belloy et al., 2019; Huang & Mahley, 2014; Raulin et al., 2022; Zhao et al., 2020). APOE acts as a binding substance in lipoprotein endocytosis. While the plasma HDL mostly contains APOA-1, the CNS HDL predominantly features APOE as its principal apolipoprotein. Additional apolipoproteins in the brain include APOA-1, which originates from the periphery, and Apolipoprotein J (APOJ), which is generated centrally and called clusterin. Members of the Low Density Lipoprotein (LDL) receptor family may degrade or recycle APOE to the cell surface after endocytosis via receptors of APOE-containing lipoprotein particles. Cholesterol derived from APOE-containing lipoprotein particles facilitates synapse formation and maintains its normal functions (Kim et al., 2009).

Cholesterol and phospholipids synthesised by glial cells are essential for creating and preserving healthy synapses. Reduced synapse quantity results from a total genetic absence of APOE or a central nervous system malfunction, which is at least partly caused by the loss of astrocyte-derived particles. While VLDL and LDL contain APOE4, APOE3 and APOE2 preferentially correlate with HDLs (Hatters et al., 2006). Under normal conditions, cortical and hippocampal neurones generate little APOE. However, in chemically induced neurodegeneration, mature APOE mRNA and protein were upregulated in damaged hippocampus neurones (Theendakara et al., 2016). Arg 61 and Lys 95 in the monomeric mutant forms a salt bridge in APOE3, while a hydrogen bond is formed between Thr 194 and Glu 255 (Chen et al., 2011). In APOE4, Arg 112 and Glu 109 form a salt bridge, unveiling Arg 61 outside the four-helix bundle (Dong et al., 1994).

15.3 APOLIPOPROTEIN E STRUCTURE AND THE CO-RELATION WITH RISK FACTORS IN AD

The APOE gene, located on chromosome number 19, codes for 34 kDa apolipoprotein, comprising 299 amino acids in humans. These are important for lipid transport peripherally and centrally. Its N-terminal facilitates its binding with the receptor, and the C-terminal (from 165 to 299 amino acids) facilitates lipid interaction (Hazarika et al., 2021). It was found to be crucial for the metabolism of plasma cholesterol (Chen et al., 2021). Three different isoforms of APOE are identified as APOE2, APOE3, and APOE4. These isoforms differ from one another by two amino acid residues at amino acid positions 112 and 158. APOE3 has Cysteine and arginine at 112 and 158 position respectively while Cysteine is present in APOE2 at both position. While arginine is present in positions 112 and 158 in APOE4 (Das et al., 1985; Martens et al., 2022).

Figure 15.1 illustrates the structure of three distinct APOE isoforms. The three alleles exhibit varying frequencies, which results in different homozygous and heterozygous phenotypes. The E4 allele has been shown to increase the risk of developing AD, while the ε2 allele mitigates this risk (Chartier-Hariln et al., 1994; Farrer et al., 1997; Yamazaki et al., 2019). This risk of developing AD with APOE comes with the basic mechanisms of APOE and lipid interaction. Different alleles of APOE interact differently according to their structural dynamics and amino acid arrangements, creating the basic problem in the AD mechanism (Romas et al., 1999). When APOE is lipid-free, it has two structural domains that fold separately, one at either end of the molecule, with the N-terminal domain as an elongated alpha-helical structure. Both APOE's domains

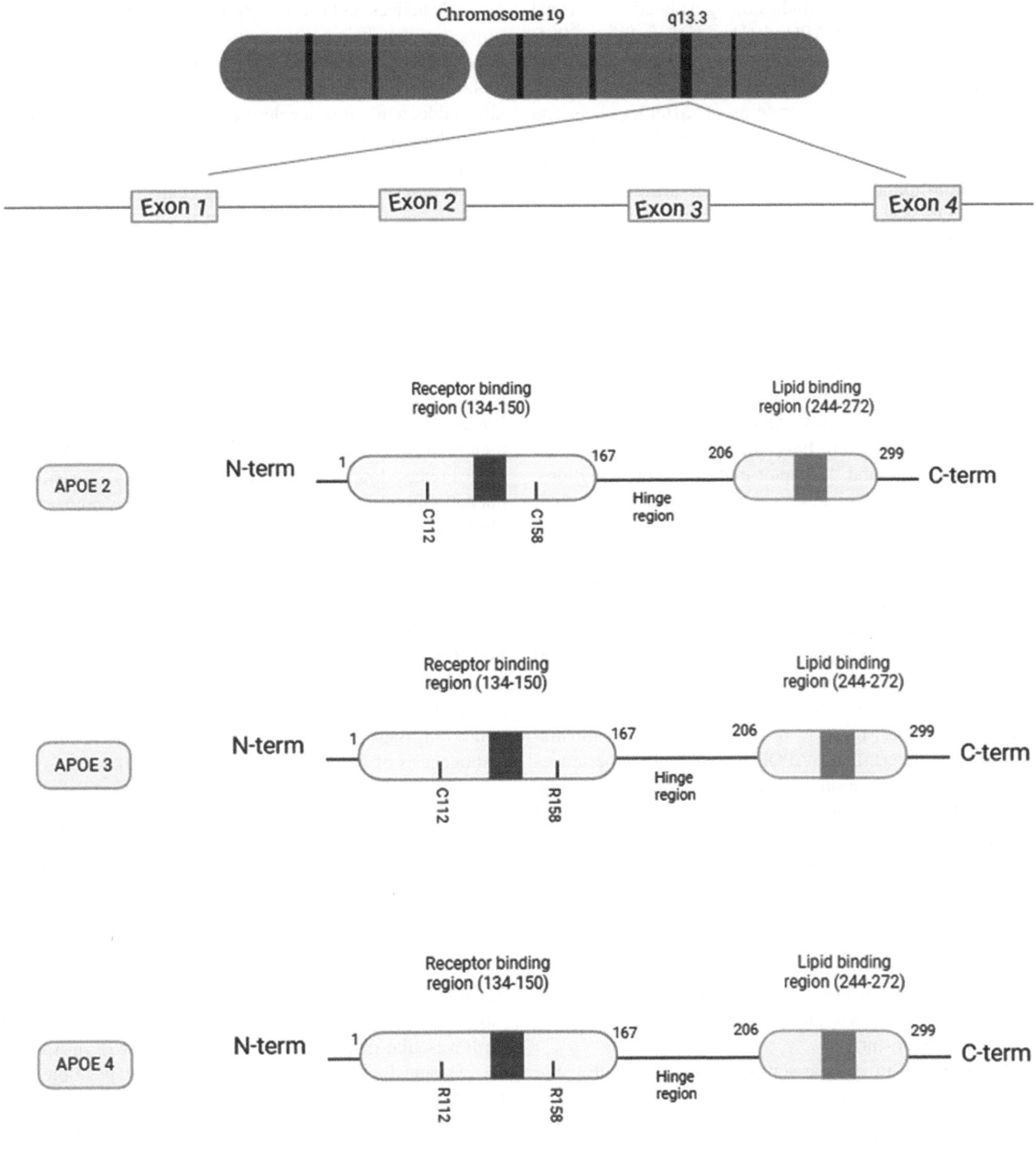

FIGURE 15.1 Different isoforms of apolipoprotein in humans.

fold separately, and their denaturation curves match the full-length proteins exactly. The primary functions of each structural domain are distinct (Weisgraber, 1994).

APOE must be coupled to a lipid (either lipoproteins or phospholipids) to have a high affinity for binding. The primary components of APOE that bind lipoproteins are C-terminal 244–272 AA (Dong et al., 1994). The lipid-bound APOE has been studied using basic synthetic lipoprotein models due to the complexity of plasma lipoprotein composition. When APOE is coupled to phospholipids, there is a significant conformational shift in APOE, forming an α-helical hairpin-like form. Also, helices 1 and 3 appear to separate more upon lipid binding. The capacity of APOE to modify phospholipid is inhibited when the N-terminal helices 1 and 3 are crosslinked together using synthetic disulphides, while in the case of helices 1 and 2 or 3 and 4 cross-linkage, this inhibition is not seen. The α-helical hairpin conformation of APOE is responsible for its high affinity binding to the LDL receptor (Peters-Libeu et al., 2006). The APOE forms a belt-like pattern around the acyl chains based on studies using isolated N-terminal regions of APOE3 and APOE4 (Davidson & Silva, 2005; Fisher & Ryan, 1999).

Numerous pieces of evidence indicate that when APOE forms complexes with lipoproteins of different sizes and shapes, it takes on distinct conformations. Early research demonstrating that the LDL receptor had lower affinity for non-lipidated APOE and high affinity for lipidated APOE was the first to deduce this variation. Later research has demonstrated that the size, lipid content, and presence of additional apolipoproteins inside the lipoprotein particle may affect the receptor-binding for APOE-containing lipoproteins (Kowal et al., 1990; Wisniewski et al., 1995). Numerous investigations have verified that C-apolipoproteins, including apoC-I, apoC-II, and apoC-III, decrease APOE's binding without being displaced from VLDLs. Because of these findings, it is now thought that APOE coupled to lipoproteins possesses different "receptor-active" and "receptor-inactive" conformations. The possible conformational changes of APOE and their biophysical implications must be taken into consideration since APOE can assume a variety of lipid-bound conformations. Specifically, the attached phospholipid can affect the size and stoichiometry of synthetic lipoprotein particles (Newhouse et al., 2005; Pitas et al., 1980).

N-terminal region folding behaviour and conformational stability for APOE may be significant for lipid binding. The three isoforms' varying preferences for lipoprotein binding may be attributed, in part, to variations in the confirmation of their N-terminals. APOE undoubtedly experiences significant conformational changes upon lipid binding, and partly folded states may promote conformational change. Compared to APOE3 or APOE4 mice, APOE with Arg61 folds more easily. It has been demonstrated that the N-terminals of mouse APOE and APOE4 exhibit structural alterations at two amino acid polymorphism sites. APOE3 facilitates the remodelling of the anionic phospholipid dimyristoyl phosphatidyl glycerol into smaller sized components and provides improved protection against LDL aggregation induced via phospholipase C when helix 10 is replaced with a b-turn.

Additionally, in comparison to wild-type APOE3, the created mutant has worse denaturation resistance. The results support the idea that a destabilised global conformation promotes lipid binding, even though this investigation aimed to ascertain if helix 10 contributes to the initiation of binding to lipids (Redmond et al., 2006). In lipid-free contexts, other exchangeable apolipoproteins exhibit characteristics of a destabilised conformation, such as being naturally unfolded, having lipid-binding regions with limited structural stability, or easily forming molten globule states like APOE. All apolipoproteins may bind to lipids due in part to their lack of conformational stability or tendency to produce stable folding intermediates. For instance, apolipoprotein III, which shares structural similarities with APOE, has flexible hinge regions between the bundle's helices and limited conformational stability; these properties may help cause a conformational shift upon lipid attachment. Phospholipid remodelling is slowed down by mutations that make apolipoprotein III more conformationally stable, while it is accelerated by mutations that make it less stable (Wang et al., 2002; Weers et al., 2005).

Although domain interactions are undoubtedly significant, it is uncertain how much they contribute to the stability of the protein. This interaction is especially important for APOE2 and APOE3, as they exhibit less domain interaction than APOE4. Additionally, APOE4's decreased stability in comparison to APOE3 may have a significant pathogenic function in the brain. According to research on Neuro-2a cells, APOE4 and fibrillar Aβ work together to enhance lysosomal activity and leakage more than APOE3. It is believed that APOE4 takes on a less stable form in lysosomes at low pH levels, which causes the lysosomal membrane to rupture. Compared to APOE3, APOE4 in vitro breaks down phospholipid membranes into smaller phospholipid-APOE complexes more quickly and at low pH (Ji et al., 2006). All of these studies collectively demonstrate that the function of APOE4's destabilised conformation merits further study in physiological contexts related to lipid binding and additional processes influenced by variations in stability as shown in Figure 15.2. Partly folded APOE exhibits increased susceptibility to proteolysis in vitro. C-terminal of APOE, specifically around AA 216–299 and AA 193–299 has been identified in amyloid-rich brain tissue fractions, thus may contribute towards the pathogenesis of AD (Wisniewski et al., 1995).

15.3.1 APOE Polymorphism and Its Effect on AD

The polymorphism of APOE results in alterations to its structural and functional characteristics. APOE2, APOE3, and APOE4 represent the three common allelic isoforms. APOE3 is the most prevalent isoform. APOE2 binds to LDL receptors 50–100 times less than APOE3 and APOE4. APOE2 is associated with type III hyperlipoproteinemia, showing high LDL and cholesterol levels with early-onset cardiovascular disease (Weisgraber et al., 1982). In ApoE2, the cysteine is substituted with arginine at position 158. This disrupts the salt bridge between Arg158 and Asp154, thus affecting binding with the LDL receptor (Dong et al., 1996). Since a protein's structure determines its function, it's critical to comprehend how these variations in amino acids impact the APOE isoforms' structure, biophysical characteristics, and, ultimately, their pathophysiological characteristics, including the pathogenesis of AD (Saunders et al., 1993), while also modestly increasing the risk of cardiovascular disease (Luc et al., 1994).

The relationship between APOE4 and illness is caused by two important characteristics of APOE4: domain interactions and decreased stability. The idea of "domain interaction" was proposed to explain the differences in lipoprotein binding preferences between APOE3 and APOE4. Large

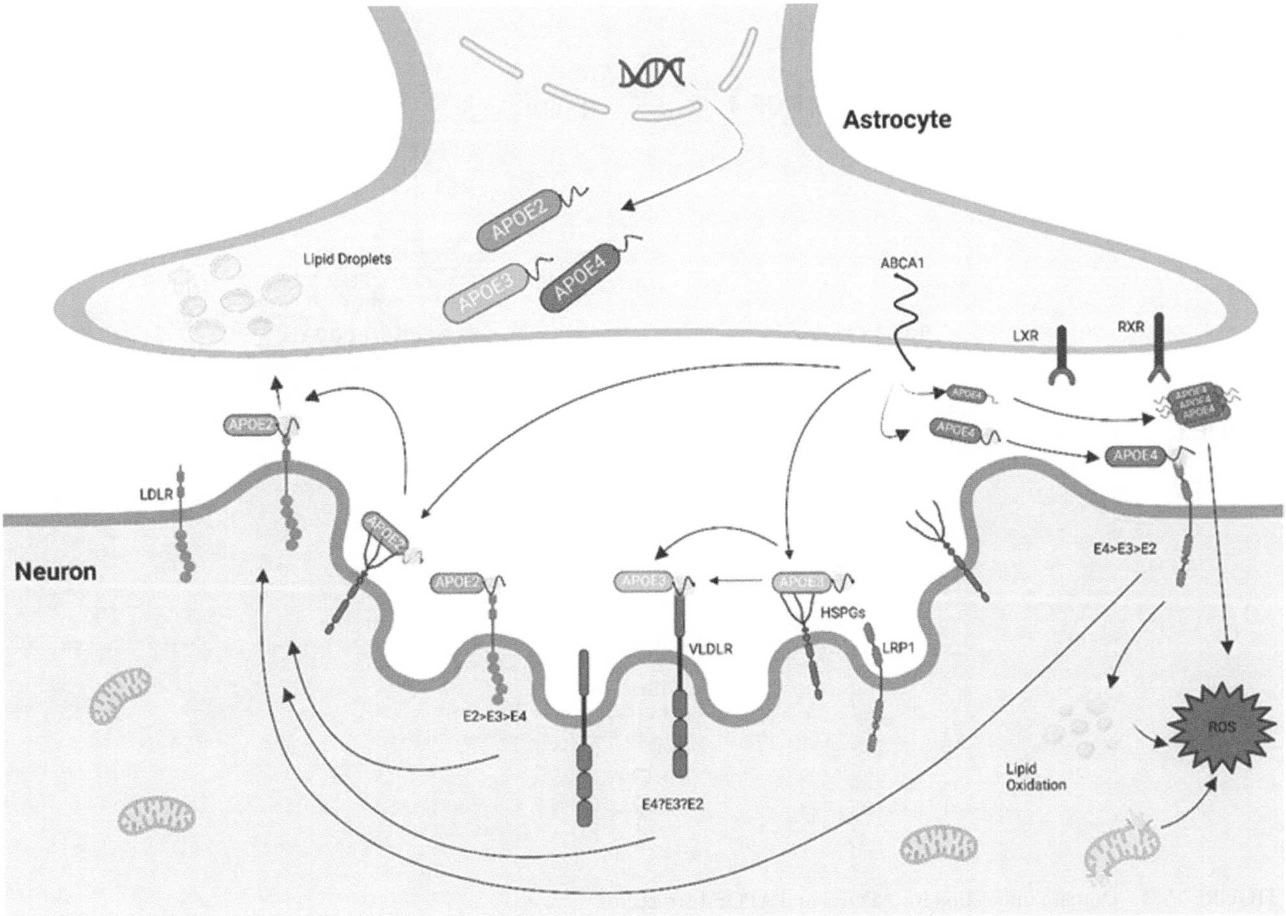

FIGURE 15.2 Interaction of different forms of APOE with their receptor in astrocytes and neurones. The activation of liver X receptor (LXR) and retinoid X receptor (RXR) influences the expression of APOE isoforms in astrocytes. The ATP-binding cassette transporter (ABCA1) promotes the lipidation and secretion of apoE into the extracellular space. ApoE is subsequently recognised by several neuronal lipid receptors, such as the VLDL receptor (VLDLR), LDL receptor (LDLR), low-density lipoprotein receptor-related protein 1 (LRP1), and heparan sulphate proteoglycans (HSPGs). The distinct lipidation states of apoE isoforms affect their affinities for receptors. The apoE4 isoform demonstrates alterations in physiological processes compared to other isoforms, characterised by decreased lipidation status, formation of non-lipidated apoE4 aggregates, and insufficient lipid turnover to astrocytes, leading to the accumulation of neuronal lipid droplets. This promotes mitochondrial dysfunction and the production of reactive oxygen species.

lower density lipoproteins, such as VLDLs and LDLs, are preferred by APOE4, whereas smaller, cholesterol-rich HDLs, like HDLs, are preferred by APOE3 and APOE2. At position 112, arginine or a positive charge determines the binding preference of APOE4. X-ray crystallography of APOE3 and APOE4 N-terminal indicates that position 112 AA can alter the confirmation of side chains (Dong et al., 1994). The side chain of R61 is orientated away from the 4-helix bundle in APOE4 due to the presence of R112, whereas in APOE3, this side chain is obscured between the helices because of cysteine 112 (Figure 15.3). Glu255 was shown to be the likely interacting residue in the C-terminus, suggesting that this change in side chain orientation allows Arg 61 in APOE4 to connect with C-terminal's acidic residue. Forster Resonance Energy Transfer (FRET) experiments imply that N- and C-terminal domain interaction takes place in living neurones that produce APOE fusion proteins (Xu et al., 2004).

All animals, except humans, have APOE with threonine at the same location as Arg61. The N- and C-terminal domain connection is likewise eliminated when human APOE's Arg61 is mutated to threonine. Therefore, APOE in other species functions similarly to APOE3 in domain interaction, even if it lacks arginine at position 112. Mouse APOE, for instance, does not exhibit domain interaction despite having the comparable residues Arg112 and Glu255; nevertheless, in mice, substituting an arginine codon for the threonine codon APOE via gene targeting, also known as "Arg61 mouse APOE," adds domain interaction to a mouse model. The recognised metabolic characteristics of human APOE4 are replicated in plasma by Arg61 mouse APOE, including a propensity for binding VLDL and LDL (Raffaï et al., 2001).

The least heat and chemical denaturation resistance is found in the order APOE2 > APOE3 > APOE4. In comparison to APOE3 and APOE4, the N-terminal domain of APOE2 exhibits a significantly more cooperative unfolding

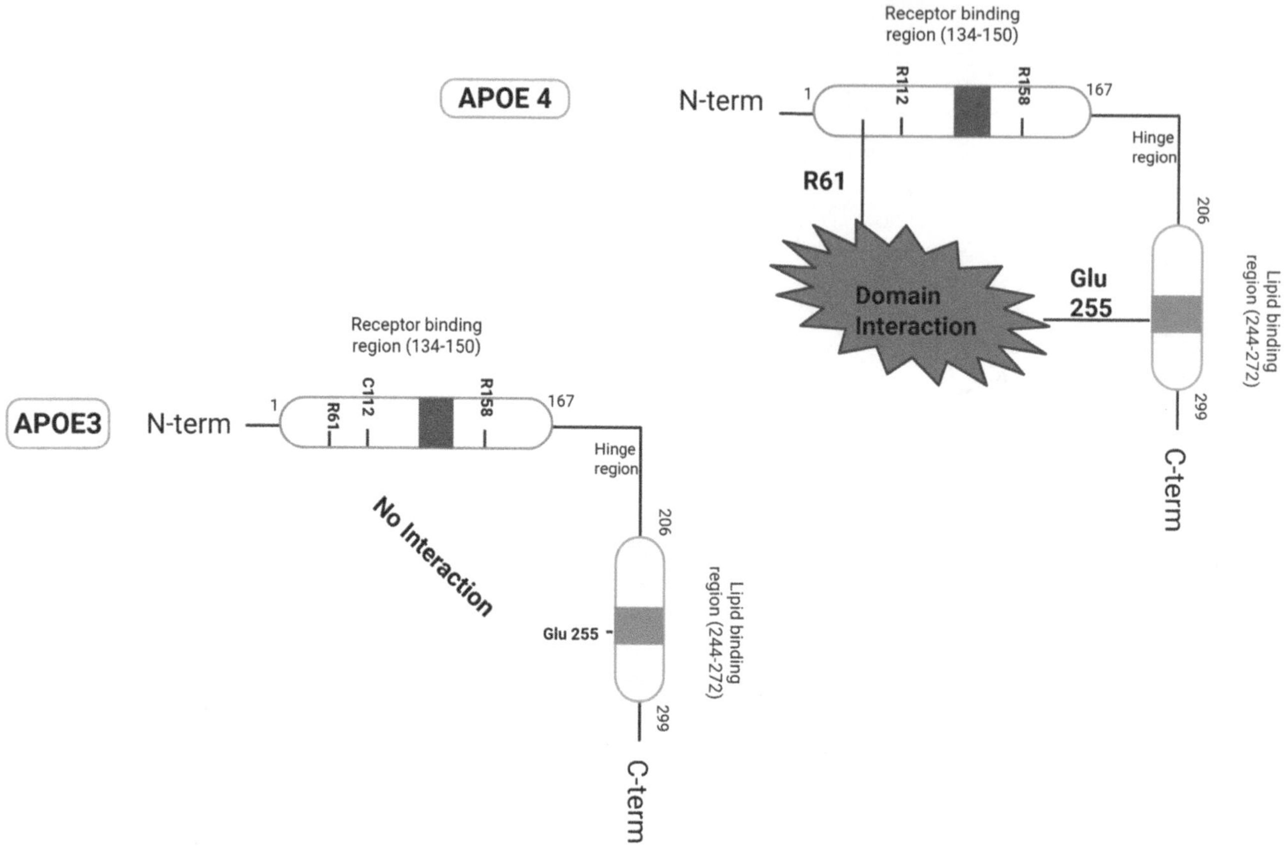

FIGURE 15.3 Domain interaction for APOE3 and APOE4 structures.

transition from the native to the unfolded state. The three-state model can adequately explain the denaturation of APOE3, suggesting that one or more stable folding intermediates are created (Morrow et al., 2002).

Compared to APOE3, APOE4 unfolds even less cooperatively, indicating that APOE4 forms stable intermediate conformations more easily. The folding intermediates of APOE4 display a core alpha-helical structure, increased beta-structure, and an elevated hydrodynamic radius, indicating a configuration that is somewhat more expanded yet compact compared to the folded state. These attributes pertain to the condition of a molten globule. Numerous illnesses have been linked to the destabilisation of protein conformations, which includes the creation of molten globules (Dobson, 2001).

15.4 APOE IN AD PATHOLOGY

15.4.1 APOE in Beta-Amyloid Synthesis and Aggregation

The role of APOE in the formation of Aβ plays a key role in the progression of AD. The Aβ peptides are produced by either the amyloidogenic (α-secretases) or non-amyloidogenic (β-secretases) pathways via the breakdown of the single-pass membrane glycoprotein known as APP. An equilibrium between the formation and removal of Aβ is present under normal conditions; however, when this balance is disrupted, Aβ accumulates in the brain, which is a significant contributor to AD. When the APP's N-terminal is broken down by β-secretases, soluble β-APP and a C-terminal fragment are produced. Successive cleavage of C-terminal fragment by γ-secretases leads to the formation of monomeric peptides predominantly consisting of 40 and 42 amino acid residues. These monomers aggregate to soluble amyloid oligomers and massive, insoluble amyloid fibrils that constitute amyloid plaques. Through a variety of processes, like neuroinflammation, oxidative stress, excitotoxicity, depolarisation, and cytoskeletal disruption, Aβ exhibits neurotoxicity at specific concentrations and conformations. It is well documented that APOE4 hinders Aβ clearance and/or facilitates Aβ accumulation. Amyloid plaque formation, congophilic amyloid angiopathy, and the resulting tau-related disease are all directly impacted by the relationship between APOE and Aβ, which controls Aβ aggregation and clearance. APOE has been postulated to function as a "pathological chaperone" that directly stimulated Aβ fibrillation and aggregation. The APOE (AA 244–272) and Aβ (12–28) interact with each other (Liu et al., 2014). APOE both affects the clearance and aggregation of Aβ. APOE facilitates the aggregation of Aβ in the order APOE4 > APOE3 > APOE2 (Hori et al., 2015). Aβ is eliminated through either an APOE-dependent or APOE-independent mechanism. The APOE-dependent mechanism includes interaction with LRP1 and LDL receptors expressed in astrocytes and pericytes. Astrocytes that express APOE4 exhibit diminished surface expression of LRP1 (Troutwine et al., 2022). According to clinical data from a rare individual who, despite not having APOE, had normal brain MRI results, no cognitive deficits, and normal amounts of tau and Aβ proteins in the CSF (Mak et al., 2014). Preliminary research revealed that APOE is a novel amyloid-associated protein, significantly prevalent in neuritic plaques. It is predicted that APOE served as a "pathological chaperone" that caused Aβ fibrillation and agglomeration.

15.4.2 Role of APOE in Tau Pathology

The APOE4 genotype and amyloid pathology are also associated with tauopathy (Farfel et al., 2016). In females, there is a stronger correlation between CSF tau levels and APOE4 carrier status than in males (Hohman et al., 2018). Increased CSF tau levels in APOE4 carriers are associated with progression of AD and reduced neuroplasticity, emphasising the vital role of APOE4 in tauopathy (Koch et al., 2017). APOE may directly bind to tau and block its phosphorylation sites through an isoform-specific interaction; APOE3 compared to APOE4 has a higher binding affinity to the microtubule-binding region of tau (Strittmatter et al., 1994). By blocking Wnt signalling, the decreased affinity of APOE4 for tau may promote tau hyperphosphorylation mediated by GSK3, producing more NFT (Caruso et al., 2006). It has been shown that tau phosphorylation is enhanced by APOE4 compared to APOE2 and APOE3 in the presence of Aβ, while APOE4 is associated with higher levels of total tau and phospho-tau (Hou et al., 2020). Zhou et al. (2016) demonstrated that, in contrast to E3FAD animals, E4FAD mice exhibited increased site-dependent tau phosphorylation and changed cortical neuronal structure when the calpain-CDK5 signalling pathway was activated. According to the research conducted by Saroja et al., glypican 4 (GPC-4) released by astrocytes has a robust interaction with APOE4, hence intensifying tau hyperphosphorylation induced by APOE4. It exhibits a greater binding affinity for APOE4 than for APOE2. It has been demonstrated that GPC-4 facilitates the formation of NFT. Tau phosphorylation rises in vivo when GPC-4 is activated using Clustered Regularly Interspaced Short Palindromic Repeats CRISPR/dCAS9. The phosphorylation of tau protein caused by APOE4 was significantly inhibited in tau when GPC-4 was not present (Saroja et al., 2022).

15.4.3 Role of APOE in BBB Integrity

According to recent research, an early pathogenic marker implicated in the pathophysiology of AD and cognitive decline is the deterioration of BBB. APOE has been shown to affect the strength of cerebrovascular tight junctions despite cerebral amyloid angiopathy in AD brains (Liu et al., 2013). APOE4 worsens cholesterol dysregulation, induces inflammation, disrupts metabolic homeostasis, and facilitates BBB breakdown, thereby contributing to vascular dementia and AD (Duong et al., 2021). Montagne et al. (2020) showed that APOE4 carriers have BBB damage associated with cognitive impairment, regardless of Aβ or tau pathology. APOE2 and APOE3 facilitate the proliferation of vascular endothelial cells, but APOE4 inhibits this process, potentially resulting in the decrease in BBB integrity (Masuda et al., 2017). APOE4-associated deterioration of the BBB causes synaptic impairment and behavioural impairments (Liu et al., 2022). APOE4-expressing pericytes through MMP9 activation may reduce endothelial cell activity, leading to impaired BBB integrity and cognitive decline (Yamazaki et al., 2020). In amyloid mice, blocking this pathway enhances BBB integrity and lessens behavioural abnormalities linked to APOE4 (Montagne et al., 2021). A recent work employing an advanced Induced Pluripotent stem cell (iPSC)-derived 3D model of BBB revealed that altered calcineurin/NFAT (Nuclear Factor of Activated T-cells) signalling in pericytes results in APOE4-associated cerebral amyloid angiopathy. In contrast, the suppression of calcineurin was observed to reduce APOE expression and vascular amyloid deposition (Blanchard et al., 2020). The studies indicate that APOE is crucial for maintaining BBB integrity.

15.5 APOE AND AD THERAPY

APOE, particularly APOE4, binds to Aβ and is essential for the deposition and removal of Aβ. According to several studies, APP transgenic mice's brain Aβ levels are decreased when APOE4 levels are decreased. Other strategies, like blocking the Aβ-APOE4 interaction, may also reduce the availability of APOE4.

Anti-APOE4 antibodies even with limited BBB permeability can reverse the negative effects of APOE4, much like anti-Aβ antibody-based therapies. APOE4 is present in Aβ deposits along with other amyloid-associated proteins, and it has previously been linked to Aβ deposition. If pathogenic versions of APOE can be sequestered by isoform-specific antibodies, the accumulation of Aβ in the brain can be avoided. Anti-APOE antibodies, when administered before the beginning of the disease, can prevent the development of Aβ plaques in mice (Kim et al., 2012). When administered to animals that already had Aβ deposits, these antibodies could also reduce plaque load, indicating that they may have potential as a therapeutic agent.

15.5.1 APOE Immune Therapeutics

In a subsequent study, anti-APOE antibodies significantly enhanced resting-state functional connectivity and memory in APP transgenic mice, without affecting peripheral cholesterol. Brain administration of anti-APOE antibodies eliminated preexisting plaques and stopped the formation of new Aβ plaques. The antibody " human APOE antibody, HAE-4" reduced Aβ deposition in APP/APOE4 mice. HAE-4 selectively recognises the lipid-free form of APOE4/APOE3 over the lipid-bound substitute and reduces Aβ in FcγR-dependent manner (Liao et al., 2018). After intraperitoneal injection or direct infusion into the brain, HAE-4 decreases the burden of total Aβ plaque. It is more effective in engaging the CNS target when given peripherally than when directly infusing the brain (Borchelt, 2018; Liao et al., 2018).

15.5.2 Genetical Approaches of APOE Expression Modification

Some researchers have employed antisense knockdown techniques based on the theory that APOE4 expression decreases might lessen Aβ build-up and, in turn, ameliorate cognitive problems. Antisense oligonucleotides (ASOs) are considered first-line therapy for a number of neurodegenerative illnesses, such as spinal muscular atrophy, muscular dystrophies, and polyneuropathy. They work as therapeutic agents by interfering with the synthesis of a specific protein.

In AD animal models, ASOs that target APOE receptors have also been explored. ASO specific for corrected splicing of APOER2 improved memory functions in mice. In another study, ASO treatment targeting APOE has also been shown to reduce Aβ aggregation in animals. It is uncertain, nonetheless, if such knockdown techniques will be effective when preexisting Aβ deposits are present, indicating that the window of opportunity for this kind of treatment may be small (Williams et al., 2020).

A family of enzymes known as histone deacetylase (HDAC) silences genes by removing acetyl groups from DNA histones. It has been demonstrated that HDACs are essential for controlling genes related to the lipid metabolic pathway, long-term memory development, and cognition. According to recent research, HDAC inhibition can increase APOE expression in human astrocytes. A phenotypic screening method using multiple chemogenomics libraries revealed that MS275 and CI994, HDAC inhibitors, increase APOE expression. These novel findings provide a fresh strategy for APOE function modulation (Dresselhaus et al., 2018).

Using human PSCs (Pluripotent Stem Cells) and animal cell lines, several studies have tried to change the predominant APOE isoform to address the neurotoxic properties of APOE4. Adeno-associated virus (AAV)-mediated gene transfer has been employed in several investigations of these model systems. However, this technique must overcome ethical and safety challenges before being applied in clinical settings. CRISPR/Cas9 functions similarly to a pair of molecular scissors. To correct a genetic defect, the Cas9 "scissor" is guided to a specific spot on the genome by an editable guide RNA, where it may cut and insert a new nucleotide sequence. Editing of APOE4 by CRISPR decreased tau phosphorylation in iPSCs (Schmid et al., 2019; Wadhwani et al., 2019). Others have modified neuronal APOE to the E3 isoform in neurones, which protects them against cytotoxicity by increasing Aβ clearance. However, there is always a chance of off-target effects due to changes in non-targeted regions of the genome, increasing the risk of cancer (Zhao et al., 2018). Off-target gene editing is an inherent problem with CRISPR/Cas9 that may lead to unsatisfactory therapy with side effects. Even though iPSCs evidence is very encouraging, we must resolve ethical issues before CRISPR/Cas9 gene editing is used as a therapeutic intervention (Kosicki et al., 2018).

15.5.3 Small Molecule-Based APOE Therapy

APOE4 is distinct from the other APOE isoforms in that it has a higher tendency for domain-domain interactions, which decreases interaction of lipid with C-terminus, leading to functional loss (Williams et al., 2020). Therefore, it has been hypothesised that the pathogenicity inherent in APOE4 correlates with the fact that APOE4 is usually lipid-depleted.

ABCA1 helps in the transfer of lipids into apolipoproteins and is recognised for its role in preventing atherosclerosis, which is vital for APOE lipidation. Humans without functioning ABCA1 do, in fact, have low levels of APOE that lead to increased risk of AD and heart disorders. According to the finding, in PDAPP (Platelet-Derived Growth factor promotor-driven amyloid precursor protein) transgenic mice, overexpression of ABCA1 decreased the amyloid burden, but the absence of ABCA1 worsened amyloidogenesis. Subsequent research has shown that ABCA1 is increased by peptides and other small compounds, which supports the idea that ABCA1-mediated lipidation is essential for Aβ elimination. CS-6253, a short peptide, is involved in ABCA1 activation. When CS-6253 is injected intraperitoneally into APOE4 TR mice, it results in increased ABCA1, causes APOE4 to become Lipidation, and decreases Aβ accumulation, tau hyperphosphorylation, and cognitive impairments (Boehm-Cagan et al., 2016; Wahrle et al., 2008).

In APOE4 TR and APOE3 TR mice, CS-6253 was found to normalise plasma APOE4 lipidation and stability similar to APOE3 animals while also partially normalising plasma apoA-I and apoJ levels in humanised APOE4 animals. ASOs can also be used to upregulate ABCA1 by targeting microRNA-33, lowering Aβ levels in APP transgenic mice and cultured neurones. These data provide credence to the idea that activating ABCA1 to stabilise the lipidation profile of APOE4 and provide evidence that APOE4's hypolipidation condition may be sufficiently corrected (Jan et al., 2015; Kim et al., 2015).

As mentioned earlier, Aβ deposits are typically home to APOE, particularly APOE4. Once thought to be unbreakable, small molecules that can destabilise protein-protein interactions are now becoming a masterwork because of significant advancement in our knowledge of PPI scaffold chemistry. This approach has the benefit of frequently using naturally occurring compounds, which can be very selective due to their accurate targeting. A BBB-permeable peptide mimic known as Aβ12-28P is one such inhibitor that prevents APOE from attaching to Aβ (Sadowski et al., 2004).

Nevertheless, astrocyte-produced APOE has the potential to be neurotoxic. Astrocyte APOE is essential for Aβ proteinopathy, which was supported by astrocyte-neurone co-culture studies which accumulated more Aβ plaques than neurones exposed to APOE Knock out astrocytes. Aβ12-28P therapy reduced intraneuronal Aβ in this co-culture system and stopped the loss of synaptic proteins by interfering with the APOE-Aβ interaction (Sadowski et al., 2004; Zheng et al., 2017).

The 6KAPOEp peptide prevents APOE from attaching to APP, thus preventing APOE-Aβ interaction. In the 5XFAD animal model of amyloid disease, 6KAPOEp injection enhanced memory and hippocampus-dependent learning while simultaneously reducing both Aβ and tau pathologies. However, in 5xFAD mice, 6KAPOEp treatment did not affect APOE or cholesterol levels. These findings show that APOE-Aβ interaction inhibitors may be utilised therapeutically to lower the CNS's Aβ and tau load (Sawmiller et al., 2019).

15.6 PRECLINICAL AND CLINICAL STUDIES

Several therapies targeting APOE have reached clinical trials Various studies including both pre-clinical and clinical are listed in Tables 15.1 and 15.2 respectively

15.7 CONCLUSION

The APOE genotype significantly influenced the development of AD and other CNS disorders marked by explicit inflammation, as well as other facets of cell function in the CNS. Interestingly, new APOE genetic variations are continually being found, which sheds more light on how the APOE genotype affects neurodegenerative illnesses. APOE Jacksonville mutant (V236E), which is characterised by a mutation in the lipid-binding area, decreased Aβ accumulation and demonstrated the significance of lipid metabolism during AD. Moreover, Rabinovici and colleagues' R251G variation causes a single amino acid shift at position 251 and was first linked to a lower chance of developing AD (Liu et al., 2021). The APOE genotype will also affect the nearby cells' functioning and microenvironment

TABLE 15.1
List of Drugs Examined in Preclinical Studies

S. No.	Therapeutic Agents	Mode of Action	References
1	Aβ12-28P	• In APP/E4 mice, impairment of memory was averted by blocking the APOE/Aβ association, which leads to decrease in brain Aβ accumulation, apoE co-accumulation within Aβ plaques, and neurite degeneration.	Pankiewicz et al. (2014)
2	CPO_Aβ17–21P	• It reduced the apoE/Aβ association and APOE4 potentiation of Aβ cytotoxicity and pro-fibrillogenic effects on Aβ aggregation *in vitro*. • in an AD APP/PS1 transgenic mouse model, it enhanced cognition and decreased Aβ-related pathologies.	Sadowski et al. (2004) and Liu et al. (2017)
3	AAV-APOE2	• In transgenic mouse models of AD (PDAPP mice & APP.PS1-21/TRE4 mice), it decreases Aβ and amyloid deposition in brain.	Zhao et al. (2016) and Jackson et al. (2024)
4	CS-6253	• Enhance APOE lipidation • Favours Aβ clearance from the brain.	Noveir et al. (2022)
5	COG1410	• In APP/PS1 mice, it increased BDNF levels while decreasing Aβ deposition and reversing A1/A2 reactive astrocytes.	Qiao et al. (2024)
6	6KApoEp	• Prevents apoE from interacting with N-terminal APP • Reducing the phosphorylation of p44/42 mitogen-activated protein kinase and APP cellular membrane trafficking.	Sawmiller et al. (2019)

TABLE 15.2
List of Drugs in Current Clinical Investigations (https://clinicaltrials.gov/)

S. No.	NCT No.	Title of Study	Therapeutic Intervention	Phase of Study/Status
1	NCT03634007	Gene Therapy for APOE4 Homozygote of Alzheimer's Disease	LX1001	Completed
2	NCT04770220	An Efficacy and Safety Study of ALZ-801 in APOE4/4 Early AD Subjects (APOLLOE4)	ALZ-801	Completed
3	NCT05965414	Safety and Pharmacokinetics of Single Ascending Doses and Multiple Ascending Doses of CS6253 in Healthy Volunteers	CS6253	Early phase 1

and the molecular pathways that are subsequently triggered. Furthermore, in addition to APOE's influence on neuroinflammation and neurodegeneration, other genetic and environmental variables will also play a role in the variations in memory task performance between younger and older stages. The last claim alludes to research conducted by Zolakei and associates, which found that elderly APOE4 carriers experienced slower memory loss. It is also pertinent to note that the relationship between the overall concentration of APOE and its isoforms and Aβ, p-tau status, dementia severity, and other biological parameters is unknown. Some studies link structural and biophysical differences between the three isoforms to the varied actions they exhibit in vivo and in cell culture. For instance, APOE is linked to antioxidant activity in vivo and has been demonstrated to have antioxidant activity in cell culture. The brains of people with AD have more oxidative damage than those of non-demented controls. Nonetheless, compared to those without APOE4, Alzheimer's patients with one or more APOE4 alleles exhibit higher levels of lipid peroxidation product 4-hydroxynonenal. Structural variations affecting biological activity are essential to comprehending the disease-causing impacts of APOE. Such initiatives will probably provide two results: first, a reconciliation of the many seemingly unrelated processes of APOE in both diseased and normal conditions; second, the possibility of creating particular medications that can counteract harmful effects of APOE (Hatters et al., 2006). Because of its function as an upstream mediator in intricate networks linked to neurodegeneration and cognitive deficits, APOE is thought to be a prime target for therapy in AD and related dementias, especially as structural insights into the protein continue to advance. Therapeutic measures will be employed to modify APOE levels, structure, lipidation, oligomerisation, and associated consequences to influence the progression of pathology, while enhancing our knowledge of the structure-function relationship of APOE and its mechanistic links to neuropathologies. Future research on cell type-specific activities, novel APOE variations, and systematic alterations through multi-omics analysis may provide important new information for developing AD treatments. Additionally, the creation of innovative AD treatment alternatives will depend heavily on adopting new research and therapeutic delivery methodologies. It is generally believed that existing medicines targeted based on APOE would be more successful at preventing neurodegenerative diseases than treating people currently experiencing them, and many of these experimental techniques have been verified in various preclinical and cell culture models. Early intervention would lower the extent of deterioration (neuropathological or neuropsychiatric), but it is difficult to prevent the progression of the disease completely. Based upon future research indicating that other proteins involved in AD patients, such as tau and α-synuclein, are affected by APOE either directly or by altering Aβ levels. Even certain APOE-targeted therapies may exhibit broader multi-target effects in an APOE isoform-dependent manner, reliant upon future research establishing that APOE influences other AD-related proteins in these patients, such as tau, α-synuclein, or inflammation, either directly or via Aβ. APOE4-targeted medications may be used with other multimodal therapies to address age-related conditions such as neuronal inflammation or BBB permeability. In the developing era of precision medicine, improvements in diagnostics for early detection and prediction of AD would be crucial for targeted therapeutics, particularly in a population with knowledge of APOE subtypes or their modifications. It would also be necessary to ascertain how such therapies might impact vascular function and peripheral lipid balance. Therefore, the overall levels of APOE, lipidation profile, risk factors for vascular disease, inflammatory characteristics, and systemic impacts may be considered while designing a therapeutic strategy against AD. APOE-targeted treatment approaches continue to be a promising field of study for halting or postponing the beginnings of familial AD (Jack et al., 2013).

REFERENCES

Andrade-Guerrero, J., Santiago-Balmaseda, A., Jeronimo-Aguilar, P., Vargas-Rodríguez, I., Cadena-Suárez, A. R., Sánchez-Garibay, C., Pozo-Molina, G., Méndez-Catalá, C. F., Cardenas-Aguayo, M. D., Diaz-Cintra, S., Pacheco-Herrero, M., Luna-Muñoz, J., & Soto-Rojas, L. O. (2023). Alzheimer's disease: An updated overview of its genetics. *International Journal of Molecular Sciences*, *24*(4). https://doi.org/10.3390/ijms24043754

Basak, J. M., Verghese, P. B., Yoon, H., Kim, J., & Holtzman, D. M. (2012). Low-density lipoprotein receptor represents an apolipoprotein E-independent pathway of Aβ uptake and degradation by astrocytes. *Journal of Biological Chemistry*, *287*(17), 13959–13971.

Belloy, M. E., Napolioni, V., & Greicius, M. D. (2019). A quarter century of APOE and Alzheimer's disease: Progress to date and the path forward. *Neuron*, *101*(5), 820–838.

Blanchard, J. W., Bula, M., Davila-Velderrain, J., Akay, L. A., Zhu, L., Frank, A., Victor, M. B., Bonner, J. M., Mathys, H., & Lin, Y.-T. (2020). Reconstruction of the human blood–brain barrier in vitro reveals a pathogenic mechanism of APOE4 in pericytes. *Nature Medicine*, *26*(6), 952–963.

Boehm-Cagan, A., Bar, R., Liraz, O., Bielicki, J. K., Johansson, J. O., & Michaelson, D. M. (2016). ABCA1 agonist reverses the ApoE4-driven cognitive and brain pathologies. *Journal of Alzheimer's Disease*, *54*(3), 1219–1233.

Borchelt, D. R. (2018). Targeting the accomplice to thwart the culprit: A new target for the prevention of amyloid deposition. *The Journal of Clinical Investigation*, *128*(5), 1734–1736.

Caruso, A., Motolese, M., Iacovelli, L., Caraci, F., Copani, A., Nicoletti, F., Terstappen, G. C., Gaviraghi, G., & Caricasole, A. (2006). Inhibition of the canonical Wnt signaling pathway by apolipoprotein E4 in PC12 cells. *Journal of Neurochemistry*, *98*(2), 364–371.

Chartier-Hariln, M.-C., Parfitt, M., Legrain, S., Pérez-Tur, J., Brousseau, T., Evans, A., Berr, C., VIdal, O., Roques, P., & Gourlet, V. (1994). Apolipoprotein E, ε4 allele as a major risk factor for sporadic early and late-onset forms of Alzheimer's disease: Analysis of the 19q13. 2 chromosomal region. *Human Molecular Genetics*, *3*(4), 569–574.

Chen, J., Li, Q., & Wang, J. (2011). Topology of human apolipoprotein E3 uniquely regulates its diverse biological functions. *Proceedings of the National Academy of Sciences*, *108*(36), 14813–14818.

Chen, Y., Strickland, M. R., Soranno, A., & Holtzman, D. M. (2021). Apolipoprotein E: Structural insights and links to Alzheimer disease pathogenesis. *Neuron*, *109*(2), 205–221.

Das, H. K., McPherson, J., Bruns, G. A., Karathanasis, S. K., & Breslow, J. L. (1985). Isolation, characterization, and mapping to chromosome 19 of the human apolipoprotein E gene. *Journal of Biological Chemistry*, *260*(10), 6240–6247. https://doi.org/10.1016/S0021-9258(18)88963-3

Davidson, W. S., & Silva, R. A. G. D. (2005). Apolipoprotein structural organization in high density lipoproteins: Belts, bundles, hinges and hairpins. *Current Opinion in Lipidology*, *16*(3), 295–300.

Dobson, C. M. (2001). The structural basis of protein folding and its links with human disease. *Philosophical Transactions of the Royal Society of London. Series B: Biological Sciences*, *356*(1406), 133–145.

Dong, L.-M., Parkin, S., Trakhanov, S. D., Rupp, B., Simmons, T., Arnold, K. S., Newhouse, Y. M., Innerarity, T. L., & Weisgraber, K. H. (1996). Novel mechanism for defective receptor binding of apolipoprotein E2 in type III hyperlipoproteinemia. *Nature Structural Biology*, *3*(8), 718–722.

Dong, L.-M., Wilson, C., Wardell, M. R., Simmons, T., Mahley, R. W., Weisgraber, K. H., & Agard, D. A. (1994). Human apolipoprotein E. Role of arginine 61 in mediating the lipoprotein preferences of the E3 and E4 isoforms. *Journal of Biological Chemistry*, *269*(35), 22358–22365.

Dresselhaus, E., Duerr, J. M., Vincent, F., Sylvain, E. K., Beyna, M., Lanyon, L. F., LaChapelle, E., Pettersson, M., Bales, K. R., & Ramaswamy, G. (2018). Class I HDAC inhibition is a novel pathway for regulating astrocytic apoE secretion. *PLoS ONE*, *13*(3), e0194661.

Duong, M. T., Nasrallah, I. M., Wolk, D. A., Chang, C. C. Y., & Chang, T.-Y. (2021). Cholesterol, atherosclerosis, and APOE in vascular contributions to cognitive impairment and dementia (VCID): Potential mechanisms and therapy. *Frontiers in Aging Neuroscience*, *13*, 647990.

Farfel, J. M., Yu, L., De Jager, P. L., Schneider, J. A., & Bennett, D. A. (2016). Association of APOE with tau-tangle pathology with and without β-amyloid. *Neurobiology of Aging*, *37*, 19–25.

Farrer, L. A., Cupples, L. A., Haines, J. L., Hyman, B., Kukull, W. A., Mayeux, R., Myers, R. H., Pericak-Vance, M. A., Risch, N., & Van Duijn, C. M. (1997). Effects of age, sex, and ethnicity on the association between apolipoprotein E genotype and Alzheimer disease: A meta-analysis. *JAMA*, *278*(16), 1349–1356.

Fernández-Calle, R., Konings, S. C., Frontiñán-Rubio, J., García-Revilla, J., Camprubí-Ferrer, L., Svensson, M., Martinson, I., Boza-Serrano, A., Venero, J. L., Nielsen, H. M., Gouras, G. K., & Deierborg, T. (2022). APOE in the bullseye of neurodegenerative diseases: Impact of the APOE genotype in Alzheimer's disease pathology and brain diseases. *Molecular Neurodegeneration*, *17*(1), 62. https://doi.org/10.1186/s13024-022-00566-4

Fisher, C. A., & Ryan, R. O. (1999). Lipid binding-induced conformational changes in the N-terminal domain of human apolipoprotein E. *Journal of Lipid Research*, *40*(1), 93–99.

Fitz, N. F., Cronican, A. A., Saleem, M., Fauq, A. H., Chapman, R., Lefterov, I., & Koldamova, R. (2012). Abca1 deficiency affects Alzheimer's disease-like phenotype in human ApoE4 but not in ApoE3-targeted replacement mice. *Journal of Neuroscience*, *32*(38), 13125–13136. https://doi.org/10.1523/jneurosci.1937-12.2012

Flowers, S. A., Grant, O. C., Woods, R. J., & Rebeck, G. W. (2019). O-glycosylation on cerebrospinal fluid and plasma apolipoprotein E differs in the lipid-binding domain. *Glycobiology*, *30*(2), 74–85. https://doi.org/10.1093/glycob/cwz084

Gaugler, J., James, B., Johnson, T., Reimer, J., Solis, M., Weuve, J., Buckley, R. F., & Hohman, T. J. (2022). 2022 Alzheimer's disease facts and figures. *Alzheimers & Dementia*, *18*(4), 700–789.

González, A., Kumar Singh, S., Churruca, M., & Maccioni, R. B. (2022). Alzheimer's disease and tau self-assembly: In the search of the missing link. *International Journal of Molecular Sciences*, *23*(8), 4192.

Hatters, D. M., Peters-Libeu, C. A., & Weisgraber, K. H. (2006). Apolipoprotein E structure: Insights into function. *Trends in Biochemical Sciences*, *31*(8), 445–454.

Hazarika, L., Sen, S., Zawar, A., & Doshi, J. (2021). Identification of APOE4 modulators, targeted therapeutic candidates in coronary artery disease, using molecular docking studies. *bioRxiv*. https://doi.org/10.1101/2021.02.02.429307

Hohman, T. J., Dumitrescu, L., Barnes, L. L., Thambisetty, M., Beecham, G., Kunkle, B., Gifford, K. A., Bush, W. S., Chibnik, L. B., & Mukherjee, S. (2018). Sex-specific association of apolipoprotein E with cerebrospinal fluid levels of tau. *JAMA Neurology*, *75*(8), 989–998.

Hori, Y., Hashimoto, T., Nomoto, H., Hyman, B. T., & Iwatsubo, T. (2015). Role of apolipoprotein E in β-amyloidogenesis: Isoform-specific effects on protofibril to fibril conversion of Aβ in vitro and brain Aβ deposition in vivo. *Journal of Biological Chemistry*, *290*(24), 15163–15174.

Hou, T.-T., Han, Y.-D., Cong, L., Liu, C.-C., Liang, X.-Y., Xue, F.-Z., & Du, Y.-F. (2020). Apolipoprotein E facilitates amyloid-β oligomer-induced tau phosphorylation. *Journal of Alzheimer's Disease*, *74*(2), 521–534.

Huang, Y., & Mahley, R. W. (2014). Apolipoprotein E: Structure and function in lipid metabolism, neurobiology, and Alzheimer's diseases. *Neurobiology of Disease*, *72*, 3–12. https://doi.org/10.1016/j.nbd.2014.08.025

Jack, C. R., Knopman, D. S., Jagust, W. J., Petersen, R. C., Weiner, M. W., Aisen, P. S., Shaw, L. M., Vemuri, P., Wiste, H. J., & Weigand, S. D. (2013). Tracking pathophysiological processes in Alzheimer's disease: An updated hypothetical model of dynamic biomarkers. *The Lancet Neurology*, *12*(2), 207–216.

Jackson, R. J., Keiser, M. S., Meltzer, J. C., Fykstra, D. P., Dierksmeier, S. E., Hajizadeh, S., Kreuzer, J., Morris, R., Melloni, A., & Nakajima, T. (2024). APOE2 gene therapy reduces amyloid deposition and improves markers of neuroinflammation and neurodegeneration in a mouse model of Alzheimer disease. *Molecular Therapy*, *32*(5), 1373–1386.

Jacobo-Albavera, L., Domínguez-Pérez, M., Medina-Leyte, D. J., González-Garrido, A., & Villarreal-Molina, T. (2021). The role of the ATP-binding cassette A1 (ABCA1) in human disease. *International Journal of Molecular Sciences*, *22*(4). https://doi.org/10.3390/ijms22041593

Jan, A., Karasinska, J. M., Kang, M. H., de Haan, W., Ruddle, P., Kaur, A., Connolly, C., Leavitt, B. R., Sorensen, P. H., & Hayden, M. R. (2015). Direct intracerebral delivery of a miR-33 antisense oligonucelotide into mouse brain increases brain ABCA1 expression. *Neuroscience Letters*, *598*, 66–72.

Ji, Z.-S., Müllendorff, K., Cheng, I. H., Miranda, R. D., Huang, Y., & Mahley, R. W. (2006). Reactivity of apolipoprotein E4 and amyloid β peptide: Lysosomal stability and neurodegeneration. *Journal of Biological Chemistry*, *281*(5), 2683–2692.

Kim, J., Basak, J. M., & Holtzman, D. M. (2009). The role of apolipoprotein E in Alzheimer's disease. *Neuron, 63*(3), 287–303. https://doi.org/10.1016/j.neuron.2009.06.026

Kim, J., Eltorai, A. E. M., Jiang, H., Liao, F., Verghese, P. B., Kim, J., Stewart, F. R., Basak, J. M., & Holtzman, D. M. (2012). Anti-apoE immunotherapy inhibits amyloid accumulation in a transgenic mouse model of Aβ amyloidosis. *Journal of Experimental Medicine, 209*(12), 2149–2156.

Kim, J., Yoon, H., Horie, T., Burchett, J. M., Restivo, J. L., Rotllan, N., Ramírez, C. M., Verghese, P. B., Ihara, M., & Hoe, H.-S. (2015). microRNA-33 regulates ApoE lipidation and amyloid-β metabolism in the brain. *Journal of Neuroscience, 35*(44), 14717–14726.

Koch, G., Di Lorenzo, F., Loizzo, S., Motta, C., Travaglione, S., Baiula, M., Rimondini, R., Ponzo, V., Bonnì, S., & Toniolo, S. (2017). CSF tau is associated with impaired cortical plasticity, cognitive decline and astrocyte survival only in APOE4-positive Alzheimer's disease. *Scientific Reports, 7*(1), 13728.

Kosicki, M., Tomberg, K., & Bradley, A. (2018). Repair of double-strand breaks induced by CRISPR–Cas9 leads to large deletions and complex rearrangements. *Nature Biotechnology, 36*(8), 765–771.

Kowal, R. C., Herz, J., Weisgraber, K. H., Mahley, R. W., Brown, M. S., & Goldstein, J. L. (1990). Opposing effects of apolipoproteins E and C on lipoprotein binding to low density lipoprotein receptor-related protein. *Journal of Biological Chemistry, 265*(18), 10771–10779.

Lane-Donovan, C., & Herz, J. (2017). ApoE, ApoE receptors, and the synapse in Alzheimer's disease. *Trends in Endocrinology & Metabolism, 28*(4), 273–284.

Liao, F., Li, A., Xiong, M., Bien-Ly, N., Jiang, H., Zhang, Y., Finn, M. B., Hoyle, R., Keyser, J., & Lefton, K. B. (2018). Targeting of nonlipidated, aggregated apoE with antibodies inhibits amyloid accumulation. *The Journal of Clinical Investigation, 128*(5), 2144–2155.

Liu, C. C., Liu, C. C., Kanekiyo, T., Xu, H., & Bu, G. (2013). Apolipoprotein E and Alzheimer disease: Risk, mechanisms and therapy. *Nature Reviews Neurology, 9*(2), 106–18. https://doi.org/10.1038/nrneurol.2012.263

Liu, C.-C., Murray, M. E., Li, X., Zhao, N., Wang, N., Heckman, M. G., Shue, F., Martens, Y., Li, Y., & Raulin, A.-C. (2021). APOE3-Jacksonville (V236E) variant reduces self-aggregation and risk of dementia. *Science Translational Medicine, 13*(613), eabc9375.

Liu, C.-C., Zhao, J., Fu, Y., Inoue, Y., Ren, Y., Chen, Y., Doss, S. V., Shue, F., Jeevaratnam, S., & Bastea, L. (2022). Peripheral apoE4 enhances Alzheimer's pathology and impairs cognition by compromising cerebrovascular function. *Nature Neuroscience, 25*(8), 1020–1033.

Liu, S., Breitbart, A., Sun, Y., Mehta, P. D., Boutajangout, A., Scholtzova, H., & Wisniewski, T. (2014). Blocking the apolipoprotein E/amyloid β interaction in triple transgenic mice ameliorates Alzheimer's disease related amyloid β and tau pathology. *Journal of Neurochemistry, 128*(4), 577–591.

Liu, S., Park, S., Allington, G., Prelli, F., Sun, Y., Martá-Ariza, M., Scholtzova, H., Biswas, G., Brown, B., & Verghese, P. B. (2017). Targeting apolipoprotein E/amyloid β binding by peptoid CPO_Aβ17-21 P ameliorates Alzheimer's disease related pathology and cognitive decline. *Scientific Reports 7*(1), 8009.

Luc, G., Bard, J.-M., Arveiler, D., Evans, A., Cambou, J.-P., Bingham, A., Amouyel, P., Schaffer, P., Ruidavets, J.-B., & Cambien, F. (1994). Impact of apolipoprotein E polymorphism on lipoproteins and risk of myocardial infarction. The ECTIM study. *Arteriosclerosis Thrombosis: A Journal of Vascular Biology, 14*(9), 1412–1419.

Mahley, R. W. (2016). Central nervous system lipoproteins: ApoE and regulation of cholesterol metabolism. *Arteriosclerosis, Thrombosis, and Vascular Biology, 36*(7), 1305–1315.

Mahley, R. W., Weisgraber, K. H., & Huang, Y. (2009). Apolipoprotein E: Structure determines function, from atherosclerosis to Alzheimer's disease to AIDS. *Journal of Lipid Research, 50*, S183–S188.

Mak, A. C. Y., Pullinger, C. R., Tang, L. F., Wong, J. S., Deo, R. C., Schwarz, J.-M., Gugliucci, A., Movsesyan, I., Ishida, B. Y., & Chu, C. (2014). Effects of the absence of apolipoprotein E on lipoproteins, neurocognitive function, and retinal function. *JAMA Neurology, 71*(10), 1228–1236.

Martens, Y. A., Zhao, N., Liu, C.-C., Kanekiyo, T., Yang, A. J., Goate, A. M., Holtzman, D. M., & Bu, G. (2022). ApoE Cascade hypothesis in the pathogenesis of Alzheimer's disease and related dementias. *Neuron, 110*(8), 1304–1317.

Masuda, T., Shimazawa, M., Hashimoto, Y., Kojima, A., Nakamura, S., Suemori, S., Mochizuki, K., Kawakami, H., Kawase, K., & Hara, H. (2017). Apolipoprotein E2 and E3, but not E4, promote retinal pathologic neovascularization. *Investigative Ophthalmology and Visual Science, 58*(2), 1208–1217.

Montagne, A., Nation, D. A., Sagare, A. P., Barisano, G., Sweeney, M. D., Chakhoyan, A., Pachicano, M., Joe, E., Nelson, A. R., & D'Orazio, L. M. (2020). APOE4 leads to blood–brain barrier dysfunction predicting cognitive decline. *Nature, 581*(7806), 71–76.

Montagne, A., Nikolakopoulou, A. M., Huuskonen, M. T., Sagare, A. P., Lawson, E. J., Lazic, D., Rege, S. V., Grond, A., Zuniga, E., & Barnes, S. R. (2021). APOE4 accelerates advanced-stage vascular and neurodegenerative disorder in old Alzheimer's mice via cyclophilin A independently of amyloid-β. *Nature Aging, 1*(6), 506–520.

Morrow, J. A., Hatters, D. M., Lu, B., Höchtl, P., Oberg, K. A., Rupp, B., & Weisgraber, K. H. (2002). Apolipoprotein E4 forms a molten globule: A potential basis for its association with disease. *Journal of Biological Chemistry, 277*(52), 50380–50385.

Newhouse, Y., Peters-Libeu, C., & Weisgraber, K. H. (2005). Crystallization and preliminary X-ray diffraction analysis of apolipoprotein E-containing lipoprotein particles. *Acta Crystallographica Section F: Structural Biology and Crystallization Communications, 61*(11), 981–984.

Noveir, S. D., Kerman, B. E., Xian, H., Meuret, C., Smadi, S., Martinez, A. E., Johansson, J., Zetterberg, H., Parks, B. A., & Kuklenyik, Z. (2022). Effect of the ABCA1 agonist CS-6253 on amyloid-β and lipoprotein metabolism in cynomolgus monkeys. *Alzheimer's Research & Therapy, 14*(1), 87.

O'Brien, J. S., & Sampson, E. L. (1965). Lipid composition of the normal human brain: Gray matter, white matter, and myelin. *Journal of Lipid Research, 6*(4), 537–544.

Padhy, D. S., Aggarwal, P., Velayutham, R., & Banerjee, S. (2025). Aerobic exercise and metformin attenuate the cognitive impairment in an experimental model of type 2 diabetes mellitus: Focus on neuroinflammation and adult hippocampal neurogenesis. *Metabolic Brain Disease, 40*(1), 1–16.

Padhy, D. S., Palit, P., Ikbal, A. M. A., Das, N., Roy, D. K., & Banerjee, S. (2023). Selective inhibition of peptidyl-arginine deiminase (PAD): Can it control multiple inflammatory disorders as a promising therapeutic strategy? *Inflammopharmacology, 31*(2), 731–744.

Pankiewicz, J. E., Guridi, M., Kim, J., Asuni, A. A., Sanchez, S., Sullivan, P. M., Holtzman, D. M. & Sadowski, M. J. (2014). Blocking the apoE/Aβ interaction ameliorates Aβ-related pathology in APOE ε2 and ε4 targeted replacement Alzheimer model mice. *Acta Neuropathologica Communications 2*(1), 75.

Peters-Libeu, C. A., Newhouse, Y., Hatters, D. M., & Weisgraber, K. H. (2006). Model of biologically active apolipoprotein E bound to dipalmitoylphosphatidylcholine. *Journal of Biological Chemistry, 281*(2), 1073–1079.

Pitas, R. E., Innerarity, T. L., & Mahley, R. W. (1980). Cell surface receptor binding of phospholipid. Protein complexes containing different ratios of receptor-active and-inactive E apoprotein. *Journal of Biological Chemistry, 255*(11), 5454–5460.

Qiao, Y., Liu, H., He, C., & Ma, Y. (2024). ApoE Mimic Peptide COG1410 Reduces Aβ Deposition and Improves Cognitive Function by Inducing the Transformation of A1/A2 Reactive Astrocytes and Increasing the BDNF Concentration in Brain of APP/PS1 Double Transgenic Mice. *Neuroscience, 537*, 116–125.

Quan, G., Xie, C., Dietschy, J. M., & Turley, S. D. (2003). Ontogenesis and regulation of cholesterol metabolism in the central nervous system of the mouse. *Developmental Brain Research, 146*(1-2), 87–98.

Raffaï, R. L., Dong, L.-M., Farese, R. V. Jr, & Weisgraber, K. H. (2001). Introduction of human apolipoprotein E4 "domain interaction" into mouse apolipoprotein E. *Proceedings of the National Academy of Sciences, 98*(20), 11587–11591.

Raulin, A. C., Doss, S. V., Trottier, Z. A., Ikezu, T. C., Bu, G., & Liu, C. C. (2022). ApoE in Alzheimer's disease: Pathophysiology and therapeutic strategies. *Molecular Neurodegeneration, 17*(1), 72. https://doi.org/10.1186/s13024-022-00574-4

Rawat, V., Wang, S., Sima, J., Bar, R., Liraz, O., Gundimeda, U., Parekh, T., Chan, J., Johansson, J. O., Tang, C., Chui, H. C., Harrington, M. G., Michaelson, D. M., & Yassine, H. N. (2019). ApoE4 alters ABCA1 membrane trafficking in astrocytes. *Journal of Neuroscience, 39*(48), 9611–9622. https://doi.org/10.1523/jneurosci.1400-19.2019

Redmond, K. A., Murphy, C., Narayanaswami, V., Kiss, R. S., Hauser, P., Guigard, E., Kay, C. M., & Ryan, R. O. (2006). Replacement of helix 1′ enhances the lipid binding activity of apoE3 n-terminal domain. *The FEBS Journal, 273*(3), 558–567.

Romas, S. N., Tang, M.-X., Berglund, L., & Mayeux, R. (1999). APOE genotype, plasma lipids, lipoproteins, and AD in community elderly. *Neurology, 53*(3), 517–517.

Sadowski, M., Pankiewicz, J., Scholtzova, H., Ripellino, J. A., Li, Y., Schmidt, S. D., Mathews, P. M., Fryer, J. D., Holtzman, D. M., & Sigurdsson, E. M. (2004). A synthetic peptide blocking the apolipoprotein E/β-amyloid binding mitigates β-amyloid toxicity and fibril formation in vitro and reduces β-amyloid plaques in transgenic mice. *The American Journal of Pathology, 165*(3), 937–948.

Saroja, S. R., Gorbachev, K., Julia, T., Goate, A. M., & Pereira, A. C. (2022). Astrocyte-secreted glypican-4 drives APOE4-dependent tau hyperphosphorylation. *Proceedings of the National Academy of Sciences of the United States of America, 119*(34), e2108870119. https://doi.org/10.1073/pnas.2108870119

Saunders, A. M., Strittmatter, W. J., Schmechel, D., St. George-Hyslop, P. H., Pericak-Vance, M. A., Joo, S. H., Rosi, B. L., Gusella, J. F., Crapper-MacLachlan, D. R., & Alberts, M. J. (1993). Association of apolipoprotein E allele ε4 with late-onset familial and sporadic Alzheimer's disease. *Neurology, 43*(8), 1467–1467.

Sawmiller, D., Habib, A., Hou, H., Mori, T., Fan, A., Tian, J., Zeng, J., Giunta, B., Sanberg, P. R., & Mattson, M. P. (2019). A novel apolipoprotein E antagonist functionally blocks apolipoprotein E interaction with N-terminal amyloid precursor protein, reduces β-amyloid-associated pathology, and improves cognition. *Biological Psychiatry, 86*(3), 208–220.

Schmid, B., Prehn, K. R., Nimsanor, N., Garcia, B. I. A., Poulsen, U., Jørring, I., Rasmussen, M. A., Clausen, C., Mau-Holzmann, U. A., & Ramakrishna, S. (2019). Generation of a set of isogenic, gene-edited iPSC lines homozygous for all main APOE variants and an APOE knock-out line. *Stem Cell Research, 34*, 101349.

Schulman, I. G. (2017). Liver X receptors link lipid metabolism and inflammation. *FEBS Letters, 591*(19), 2978–2991. https://doi.org/10.1002/1873-3468.12702

Spagnuolo, M. S., Donizetti, A., Iannotta, L., Aliperti, V., Cupidi, C., Bruni, A. C., & Cigliano, L. (2018). Brain-derived neurotrophic factor modulates cholesterol homeostasis and apolipoprotein E synthesis in human cell models of astrocytes and neurons. *Journal of Cellular Physiology, 233*(9), 6925–6943. https://doi.org/10.1002/jcp.26480

Strittmatter, W. J., Saunders, A. M., Goedert, M., Weisgraber, K. H., Dong, L.-M., Jakes, R., Huang, D. Y., Pericak-Vance, M., Schmechel, D., & Roses, A. D. (1994). Isoform-specific interactions of apolipoprotein E with microtubule-associated protein tau: Implications for Alzheimer disease. *Proceedings of the National Academy of Sciences, 91*(23), 11183–11186.

Theendakara, V., Peters-Libeu, C. A., Spilman, P., Poksay, K. S., Bredesen, D. E., & Rao, R. V. (2016). Direct transcriptional effects of apolipoprotein E. *Journal of Neuroscience, 36*(3), 685–700.

Troutwine, B. R., Hamid, L., Lysaker, C. R., Strope, T. A., & Wilkins, H. M. (2022). Apolipoprotein E and Alzheimer's disease. *Acta Pharmaceutica Sinica B, 12*(2), 496–510. https://doi.org/10.1016/j.apsb.2021.10.002

Vitali, C., Wellington, C. L., & Calabresi, L. (2014). HDL and cholesterol handling in the brain. *Cardiovascular Research, 103*(3), 405–413.

Wadhwani, A. R., Affaneh, A., Van Gulden, S., & Kessler, J. A. (2019). Neuronal apolipoprotein E4 increases cell death and phosphorylated tau release in Alzheimer disease. *Annals of Neurology, 85*(5), 726–739.

Wahrle, S. E., Jiang, H., Parsadanian, M., Kim, J., Li, A., Knoten, A., Jain, S., Hirsch-Reinshagen, V., Wellington, C. L., & Bales, K. R. (2008). Overexpression of ABCA1 reduces amyloid deposition in the PDAPP mouse model of Alzheimer disease. *The Journal of Clinical Investigation, 118*(2), 671–682.

Wang, J., Sykes, B. D., & Ryan, R. O. (2002). Structural basis for the conformational adaptability of apolipophorin III, a helix-bundle exchangeable apolipoprotein. *Proceedings of the National Academy of Sciences, 99*(3), 1188–1193.

Weers, P. M. M., Abdullahi, W. E., Cabrera, J. M., & Hsu, T.-C. (2005). Role of buried polar residues in helix bundle stability and lipid binding of apolipophorin III: Destabilization by threonine 31. *Biochemistry, 44*(24), 8810–8816.

Weisgraber, K. H. (1994). Apolipoprotein E: Structure-function relationships. *Advances in Protein Chemistry, 45*, 249–302.

Weisgraber, K. H., Innerarity, T. L., & Mahley, R. W. (1982). Abnormal lipoprotein receptor-binding activity of the human E apoprotein due to cysteine-arginine interchange at a single site. *Journal of Biological Chemistry, 257*(5), 2518–2521.

Williams, T., Borchelt, D. R., & Chakrabarty, P. (2020). Therapeutic approaches targeting apolipoprotein E function in Alzheimer's disease. *Molecular Neurodegeneration*, *15*, 1–19.

Wisniewski, T., Lalowski, M., Golabek, A., Frangione, B., & Vogel, T. (1995). Is Alzheimer's disease an apolipoprotein E amyloidosis? *The Lancet*, *345*(8955), 956–958.

Xu, Q., Brecht, W. J., Weisgraber, K. H., Mahley, R. W., & Huang, Y. (2004). Apolipoprotein E4 domain interaction occurs in living neuronal cells as determined by fluorescence resonance energy transfer. *Journal of Biological Chemistry*, *279*(24), 25511–25516.

Yamazaki, Y., Shinohara, M., Yamazaki, A., Ren, Y., Asmann, Y. W., Kanekiyo, T., & Bu, G. (2020). ApoE (apolipoprotein E) in brain pericytes regulates endothelial function in an isoform-dependent manner by modulating basement membrane components. *Arteriosclerosis Thrombosis, and Vascular Biology*, *40*(1), 128–144.

Yamazaki, Y., Zhao, N., Caulfield, T. R., Liu, C.-C., & Bu, G. (2019). Apolipoprotein E and Alzheimer disease: Pathobiology and targeting strategies. *Nature Reviews Neurology*, *15*(9), 501–518.

Zhao, L., Gottesdiener, A. J., Parmar, M., Li, M., Kaminsky, S. M., Chiuchiolo, M. J., Sondhi, D., Sullivan, P. M., Holtzman, D. M., & Crystal, R. G. (2016). Intracerebral adeno-associated virus gene delivery of apolipoprotein E2 markedly reduces brain amyloid pathology in Alzheimer's disease mouse models. *Neurobiology of Aging*, *44*, 159–172.

Zhao, W., Fan, J., Kulic, I., Koh, C., Clark, A., Meuller, J., Engkvist, O., Barichievy, S., Raynoschek, C., Hicks, R., Maresca, M., Wang, Q., Brown, D. G., Lok, A., Parro, C., Robert, J., Chou, H. Y., Zuhl, A. M., Wood, M. W., Brandon, N. J., & Wellington, C. L. (2020). Axl receptor tyrosine kinase is a regulator of apolipoprotein E. *Molecular Brain*, *13*(1), 66. https://doi.org/10.1186/s13041-020-00609-1

Zhao, Y., Yang, Y., Xing, R., Cui, X., Xiao, Y., Xie, L., You, P., Wang, T., Zeng, L., & Peng, W. (2018). Hyperlipidemia induces typical atherosclerosis development in Ldlr and Apoe deficient rats. *Atherosclerosis*, *271*, 26–35.

Zheng, J.-Y., Sun, J., Ji, C.-M., Shen, L., Chen, Z.-J., Xie, P., Sun, Y.-Z., & Yu, R.-T. (2017). Selective deletion of apolipoprotein E in astrocytes ameliorates the spatial learning and memory deficits in Alzheimer's disease (APP/PS1) mice by inhibiting TGF-β/Smad2/STAT3 signaling. *Neurobiology of Aging*, *54*, 112–132.

Zhou, M., Huang, T., Collins, N., Zhang, J., Shen, H., Dai, X., Xiao, N., Wu, X., Wei, Z., York, J., Lin, L., Zhu, Y., LaDu, M. J., & Chen, X. (2016). APOE4 induces site-specific tau phosphorylation through calpain-CDK5 signaling pathway in EFAD-Tg mice. *Current Alzheimer Research*, *13*(9), 1048–1055. https://doi.org/10.2174/156720501366616041515 4550

16 Biomarkers
Diagnostic, Prognostic and Therapeutic Monitoring of Alzheimer's Disease

Amit Sharma, Ripunjay Kalita, Hemant R. Jadhav, and Amritansh Bhanot

16.1 INTRODUCTION

16.1.1 Alzheimer's Disease (AD)

Alzheimer's disease (AD) is a progressive and complex neurodegenerative disorder, representing the most common form of dementia and accounting for nearly two-thirds of all dementia cases worldwide (Sharma et al., 2025). As global populations age, the prevalence of AD is projected to rise sharply, creating significant public health and economic challenges (Sharma et al., 2024a, 2024b). Characterized by the gradual loss of neuronal cells, AD leads to cognitive decline, behavioral disturbances, and motor impairments. Despite extensive research, the underlying mechanisms of the disease remain incompletely understood, and current treatments are largely symptomatic, offering limited impact on halting or reversing disease progression (Sharma et al., 2024; Villain & Dubois, 2019). A key factor contributing to the challenges in addressing AD is its heterogeneous nature. The pathobiology of AD includes hallmark neuropathological features such as amyloid-beta (Aβ) plaques and neurofibrillary tangles composed of tau protein (Brookmeyer et al., 2002). These pathological aggregates disrupt normal neuronal function and are accompanied by synaptic loss, neuroinflammation, transmitter imbalances, and reactive astrogliosis, ultimately resulting in cognitive decline (Figure 16.1). The clinical presentation of AD is equally diverse, with typical cases involving amnestic symptoms and atypical cases manifesting as variants like primary progressive aphasia or posterior cortical atrophy. This variability complicates both diagnosis and treatment, particularly when other neurodegenerative or vascular conditions co-occur (Cummings et al., 2022; Dubois et al., 2021; Rasmussen & Langerman, 2019; Scharre, 2019; Snowden et al., 2007).

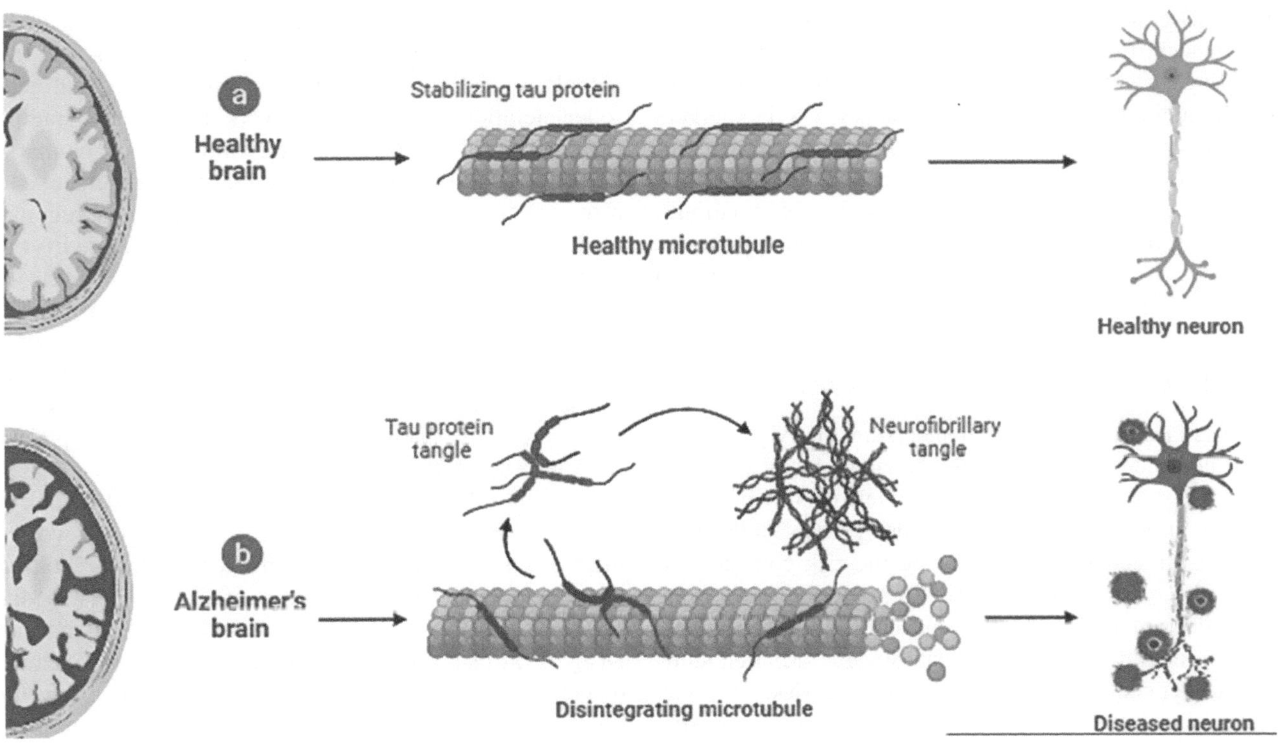

FIGURE 16.1 The pathogenesis and progression of Alzheimer's disease.

DOI: 10.1201/9781003602477-16

In this context, biomarkers have emerged as transformative tools for advancing our understanding and management of AD (Jack et al., 2018). Biomarkers enable early detection, provide insights into disease progression, and offer critical guidance for therapeutic interventions (Hippius & Neundörfer, 2003). In recent years, the development of in vivo biomarkers has shifted the focus of diagnosis from late-stage dementia to earlier, preclinical stages of the disease, where cognitive symptoms are minimal or absent but biomarkers of AD pathology are detectable (Scheltens et al., 1992). These biomarkers have the potential to identify individuals at risk, facilitate early intervention, and enhance the effectiveness of emerging disease-modifying therapies (DMTs), such as aducanumab and other agents targeting Aβ aggregation or tau pathology (Zetterberg & Mattsson, 2014). Biomarkers also play a pivotal role in distinguishing AD from other forms of dementia and neurodegenerative conditions. This is especially important given the overlap of symptoms and the frequent co-occurrence of multiple pathologies in aging populations. Biomarkers, including cerebrospinal fluid (CSF) measures, neuroimaging techniques, and blood-based assays, provide valuable information for tailoring individualized treatment plans and monitoring therapeutic responses. By identifying AD-specific phenotypes and tracking disease dynamics, biomarkers improve diagnostic accuracy, optimize patient management, and support clinical decision-making (Mahaman et al., 2022).

This book chapter provides a comprehensive exploration of the role of biomarkers in the diagnostic, prognostic, and therapeutic monitoring of AD (Blennow et al., 2019; Salloway et al., 2014). It examines the utility of biomarkers in identifying typical and atypical AD phenotypes, enabling early diagnosis, and guiding treatment strategies. Furthermore, the discussion emphasizes how biomarkers enhance patient communication, improve clinical outcomes, and offer hope for addressing the growing societal and economic burden of AD. As the field continues to evolve, the integration of biomarker-based approaches promises to be a cornerstone in the fight against AD (Morris et al., 2016).

16.1.2 Genetic Factors Involved in AD

Genetic factors are crucial in understanding AD, with the Apolipoprotein E (APOE) gene variants being the most extensively studied. The APOE ε4 allele notably increases the risk of developing AD, while the APOE ε2 allele provides a protective effect, reducing the risk compared to other alleles. In addition to APOE, polygenic risk scores, derived from multiple single nucleotide polymorphisms (SNPs) identified through genome-wide association studies, further contribute to understanding AD risk. However, standardized methods for calculating these scores are still lacking (Liu et al., 2018). Numerous SNPs have been associated with AD, enhancing the understanding of genetic predisposition. Genetic testing, particularly for APOE variants, is commonly used to classify individuals into low- and high-risk categories for clinical research trials, though it does not directly assess the neuropathological changes related to the disease. Additionally, genetic factors may inform treatment strategies and dosing, supporting personalized medicine approaches in AD management (Jack et al., 2018).

However, genetic risk factors come with limitations. Variability in how these factors manifest in different individuals makes it difficult to predict the onset and progression of the disease. Moreover, while the APOE genotype is valuable in research, it is not considered a true biomarker of AD according to the ATN criteria, as it does not reflect the actual pathological changes occurring in the brain (Leonenko et al., 2021). Ongoing research is focused on identifying additional genetic markers and better understanding their roles in disease progression, which will enhance risk assessment and therapeutic strategies. While genetic profiling offers valuable insights into AD, the complexity of genetic influences and their implications for personalized treatments remain key areas of exploration (Chandra et al., 2023).

16.2 OVERVIEW OF BIOMARKERS IN AD

16.2.1 Role of Biomarkers in AD Drug Development

Biomarkers play a crucial role in the development of AD therapies, serving several key functions throughout the process (Chandra et al., 2023). They provide objective measures for diagnosing AD by detecting specific pathological changes, such as amyloid plaques and tau pathology, and are central to the amyloid, tau, and neurodegeneration research framework for disease staging (Scheltens et al., 2016). In Phase 2 clinical trials, biomarkers are essential for demonstrating target engagement and providing evidence of disease modification, a requirement for classifying treatments as disease-modifying therapies (Alzheimer's Association, 2016; Prince et al., 2015; Wimo et al., 2017). Additionally, biomarkers help monitor treatment response by evaluating safety, pharmacodynamics, and potential adverse effects. Prognostic biomarkers offer insights into disease progression, while predictive biomarkers assist in forecasting patient responses to treatments, which is valuable for trial design and patient selection (Cummings et al., 2014, 2018; Jack et al., 2018). Biomarkers can also accelerate drug development by enabling shorter, more efficient trials, facilitating the development of companion diagnostics, and expediting the regulatory approval process. The U.S. Food and Drug Administration (FDA) highlights the significance of biomarkers in its AD staging system, particularly for preclinical and prodromal stages. Overall, biomarkers enhance the efficiency and success of AD drug development, supporting better therapeutic outcomes for patients at risk (Cummings, 2019; Guidance, 2018).

16.2.2 Biomarkers of Target Engagement

Biomarkers are essential tools for evaluating normal biological processes, pathological conditions, and responses to treatments, playing a crucial role in the development of AD therapies (Farlow et al., 1992). They are instrumental in diagnosing the disease, assessing target engagement, providing evidence of disease modification, and monitoring safety (Rogers et al., 1998). Target engagement, particularly in clinical trials, is vital for differentiating drug efficacy from trial quality and is key to advancing from Phases 2 to 3 trials. This can be demonstrated through measures such as receptor occupancy, pharmacodynamics, or changes in fluid biomarkers like blood or CSF markers. Key biomarkers for target engagement in AD include amyloid PET, CSF Aβ levels, tau PET, functional MRI, neurodegeneration markers, and blood-based biomarkers, all of which provide valuable insights into the therapeutic effects on amyloid metabolism, tau aggregation, neuronal damage, and symptom improvement (Cook et al., 2014; Cummings, 2008; Morgan et al., 2018; Tan et al., 2014). Despite these advances, challenges remain, including the absence of well-established biomarkers, the "Phase 2 conundrum" where drugs progress without sufficient target engagement data, and discrepancies between biomarker changes and cognitive outcomes. Nevertheless, biomarkers remain indispensable for unraveling the mechanisms of AD and refining therapeutic approaches, highlighting the ongoing need for research and validation (Table 16.1) (Simon et al., 2013).

TABLE 16.1
Clinical Trial Outcomes of Therapeutic Compounds for Alzheimer's Disease

S. No.	Compound	Structure	Mechanism of action	Clinical Trail Status
1.	Solanezumab; Expedition	–	mAb	Expedition I–III Failed
2.	Bapineuzumab; ApoE noncarrier study	–	mAb	Failed
3.	Bapineuzumab; ApoE carrier study	–	mAb	Failed
4.	Semagacestat	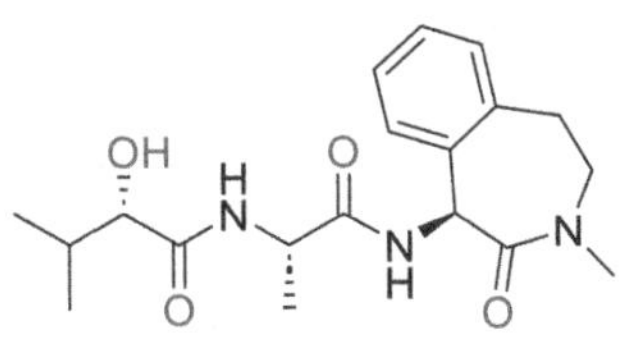	APP proteolysis	Failed
5.	AN1792	Peptide	Reduced Aβ plaque burden	Development terminated
6.	Azeliragon (TTP488)	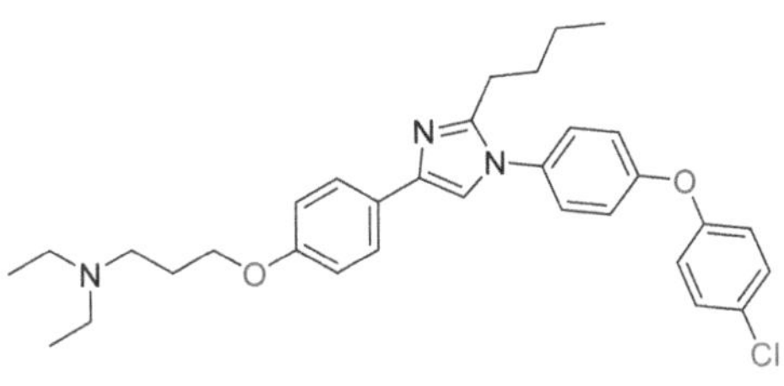	Inhibitor of RAGE	Study aborted
7.	IVIG	–	–	–
8.	Gantenerumab		mAb	Study aborted
9.	Crenezumab	–	mAb	Failed
10.	Verubecestat	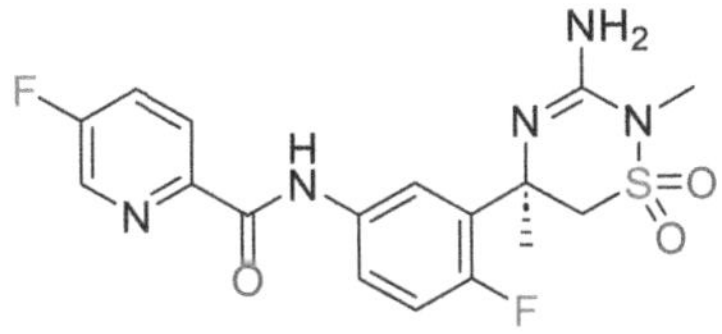	BACE 1 inhibitor	Trials terminated
11.	Aducanumab	–	mAb	Approved
12.	Leqembi	–	Reducing the levels of beta-amyloid from brain	Phase 3
13.	Donanemab/Kisunla	–	Reducing the levels of beta-amyloid from brain	Approved

16.2.3 Biomarkers Evidence of Disease Modification

Biomarkers are measurable characteristics used to evaluate normal biological processes, pathological conditions, or responses to treatments (Mattsson et al., 2017). They are essential for understanding disease baselines, tracking progression, and assessing treatment effectiveness. Biomarkers encompass a wide range of measures, including genetic markers (genomics), protein levels (proteomics), metabolic and lipid profiles (metabolomics and lipidomics), imaging techniques like PET scans, blood tests, electrocardiograms, and organ function assessments (Babić Leko et al., 2016; Chatterjee et al., 2018; Jack et al., 2005; Tarawneh et al., 2016). In AD drug development, biomarkers are critical for diagnosis, verifying target engagement, and providing evidence of disease modification. Diagnostic biomarkers, such as amyloid positivity and low CSF Aβ42 levels, reveal pathological changes in AD (Whitwell et al., 2008). Prognostic biomarkers help predict disease progression and identify patient populations that are most likely to benefit from treatment (Molinuevo et al., 2014). Target engagement biomarkers, crucial in Phase 2 trials, confirm that a drug interacts with its intended target, while pharmacodynamic biomarkers reflect biological responses indicating disease modification (Alexander et al., 2002; Chen et al., 2010; Dupont et al., 2018). Safety biomarkers monitor adverse effects, and their integration into regulatory processes, such as the FDA's accelerated approval pathways, enables early-stage interventions. Companion diagnostics ensure the safe and effective use of new treatments, while biomarkers also help reduce trial sizes and durations by focusing on biological impacts rather than relying solely on clinical outcomes (Mega et al., 2005). In sum, biomarkers are invaluable tools for advancing AD therapies, speeding up drug development, and enhancing patient care (Figure 16.2) (Mercier et al., 2017).

16.2.4 Biomarkers to Support Clinical Decision-Making

Biomarkers play a crucial role in clinical decision-making for AD, supporting diagnosis, treatment planning, and drug development (Dubois et al., 2016). They are classified into diagnostic, prognostic, predictive, response, and safety biomarkers, each serving distinct functions. Diagnostic biomarkers, such as amyloid PET imaging and low CSF Aβ42 levels, help confirm AD pathology, addressing the challenge that clinical criteria can misdiagnose up to 25% of cases (Honig et al., 2023). Prognostic biomarkers, including tau PET imaging and the ApoE-4 genotype, assist in predicting disease progression and identifying patients at higher risk for rapid cognitive decline. Predictive biomarkers are used to anticipate treatment responses, aiding in the selection of patients for clinical trials, while response biomarkers, like changes in amyloid plaques or CSF total tau, evaluate the biological effects of therapies. Safety biomarkers are essential for monitoring adverse effects, ensuring patient safety throughout treatment (Mielke et al., 2018). Overall, biomarkers are integral for accurate diagnosis, monitoring disease progression, optimizing treatment strategies, and

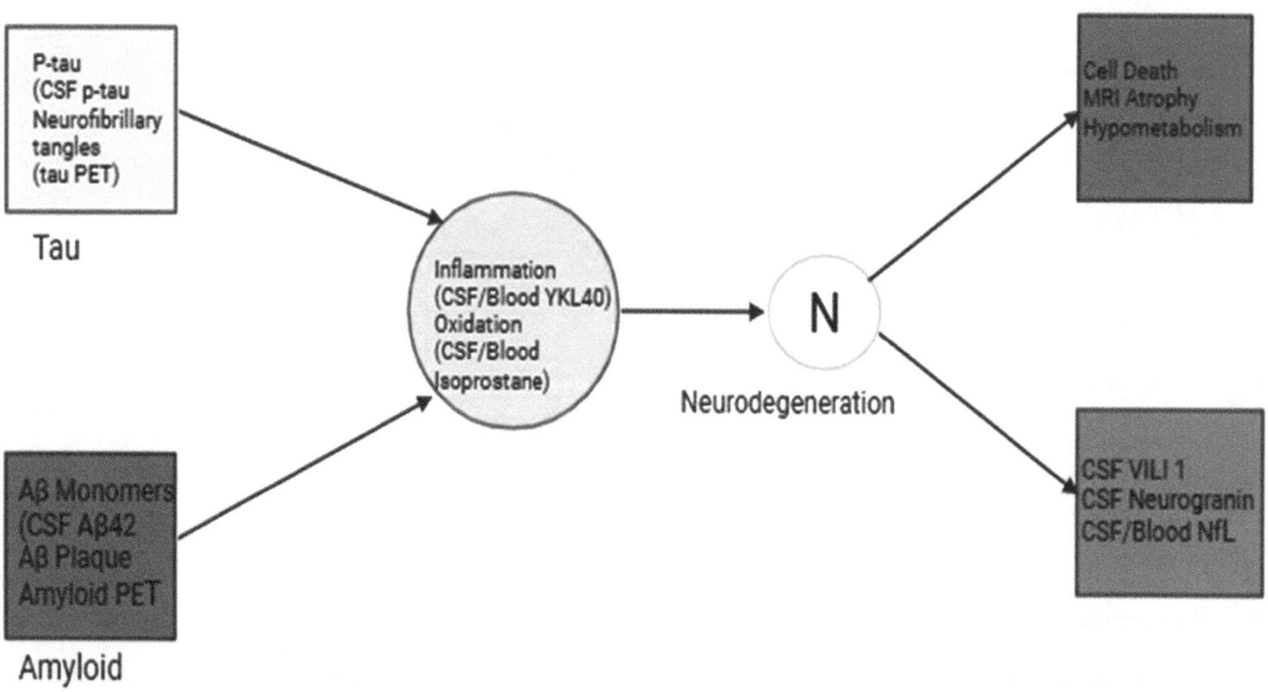

FIGURE 16.2 The amyloid (A), tau (T), and neurodegeneration (N) framework of AD with established and emerging biomarkers.

facilitating the development of disease-modifying therapies (DMTs). Their use in clinical practice improves patient outcomes and drives progress in the management of AD (Mattsson-Carlgren et al., 2020).

16.3 BIOMARKERS FOR SAFETY IN AD DRUG DEVELOPMENT

Biomarkers are crucial for ensuring safety during the development of therapies for AD. Safety biomarkers play an essential role in detecting and monitoring treatment-related adverse events, guiding dosage adjustments, and aiding regulatory approvals (Kleinstreuer et al., 2016). These biomarkers include liver function tests, blood counts, muscle enzyme levels, and electrocardiograms (ECGs), all of which are critical for identifying drug toxicity, a leading cause of drug development failures, accounting for around 30% of program terminations (Arrowsmith & Miller, 2013). For example, liver toxicity has been observed with certain BACE inhibitors and 5-HT6 antagonists like idalopirdine during Phase 2 trials (Wilkinson et al., 2014). Similarly, QTc interval prolongation, a concern for cardiac safety, was noted in a citalopram trial for agitation in AD patients, underscoring the need for ECG monitoring. Off-target adverse effects have also posed challenges, particularly with gamma-secretase inhibitors, which have led to hypopigmentation, skin cancers, and cognitive and functional impairments, compared to placebo. These dermatological effects are attributed to the inhibition of NOTCH proteases, though the causes of cognitive and functional impairments remain unclear. Additionally, amyloid-related imaging abnormalities (ARIA), often linked to certain monoclonal antibodies targeting Aβ, require regular MRI monitoring to ensure patient safety during clinical trials (Porsteinsson et al., 2014).

Biomarkers like amyloid PET scans and CSF Aβ levels are key in confirming AD pathology, while prognostic biomarkers, such as tau PET scans and MRI atrophy measures, help predict disease progression. Predictive biomarkers forecast treatment responses, and pharmacodynamic biomarkers demonstrate biological responses to therapies, especially in Phase 2 trials. Integrating these biomarkers allows clinical trials to be more efficient, with fewer participants and shorter durations, promoting the development of safer and more effective AD therapies (Doody et al., 2013).

Despite these advancements, challenges persist in identifying biomarkers that can definitively predict clinical outcomes, as no universally accepted surrogate markers exist. Nevertheless, the ongoing use of safety biomarkers such as routine monitoring of liver function, cardiac health, and neuroimaging is vital for detecting and addressing treatment-related risks. These measures are essential for enhancing the safety and efficacy of AD treatments, reducing trial failures, and advancing therapeutic options that improve patient outcomes (Figure 16.3) (Sperling et al., 2011).

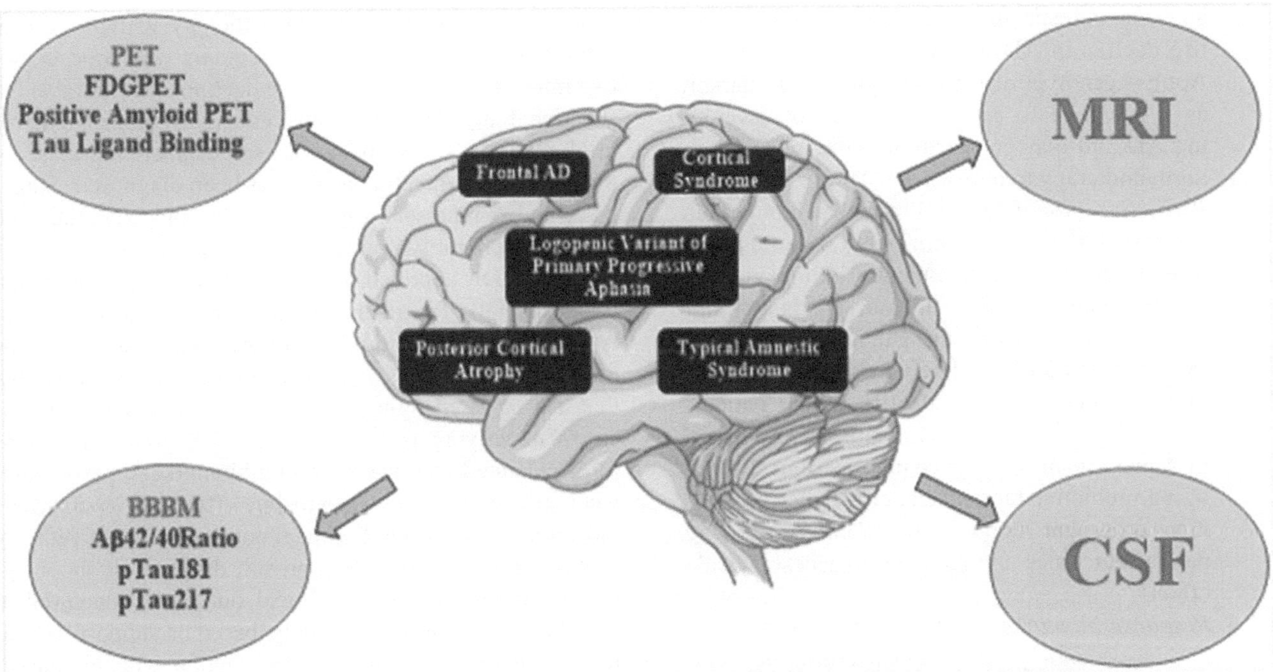

FIGURE 16.3 Diagnostic modalities and clinical syndromes in Alzheimer's disease.

16.4 FDA CLASSIFICATION OF BIOMARKERS AND INTEGRATION INTO STAGES OF AD

The FDA classifies biomarkers into several distinct categories, each serving specific roles in the diagnosis, treatment, and management of AD. These categories help facilitate therapy development and approval by providing clear frameworks for applying biomarkers at different stages of the disease, ultimately enhancing the efficiency and success of clinical trials and drug development.

The FDA categorizes biomarkers based on their role:

1. *Diagnostic biomarkers* – Diagnostic biomarkers are crucial for accurately identifying AD by detecting underlying pathologies. For instance, amyloid PET imaging visualizes amyloid plaques in the brain, and CSF levels of Aβ42 are often reduced in AD patients due to amyloid accumulation. Additionally, changes in the Aβ/tau or Aβ/p-tau ratios serve as diagnostic indicators of AD pathology, offering objective measures to confirm the disease's presence, especially in cases where symptoms overlap with other neurodegenerative conditions.
2. *Prognostic biomarkers* – Prognostic biomarkers are vital for predicting the likely progression of AD, enabling a better understanding of how the disease will unfold in individual patients. These biomarkers are particularly valuable for patient stratification in clinical trials, ensuring that the right patients are selected based on their expected disease trajectory. For example, tau PET scans measure tau protein accumulation in the brain and can indicate the likelihood of rapid cognitive decline in AD patients. The presence of the ApoE-4 genotype also acts as a prognostic marker, as it is linked to an increased risk of developing AD and may contribute to adverse effects in immunotherapy treatments.
3. *Predictive biomarkers* – Predictive biomarkers are essential for forecasting how a patient will respond to specific treatments, aiding clinical decision-making. These biomarkers help tailor treatment strategies, ensuring that patients receive therapies that are most likely to benefit them. For instance, tau PET can identify which patients are more likely to respond positively to anti-tau therapies, enabling targeted interventions. By utilizing predictive biomarkers, clinicians can optimize treatment regimens, potentially improving patient outcomes and reducing unnecessary side effects.
4. *Response biomarkers* – Response biomarkers are key in evaluating the biological effects of therapeutic interventions during clinical trials. These biomarkers reflect how the body is responding to treatment and are crucial in determining whether a drug is effectively modifying the disease. In anti-amyloid treatment programs, for example, a reduction in amyloid plaque burden serves as a response biomarker, indicating that the drug is successfully targeting its intended pathway. Similarly, markers of disease modification, such as changes in CSF total tau levels, can suggest whether a treatment is slowing neurodegeneration. Response biomarkers are particularly important in Phase 2 trials, where the focus is on demonstrating biological activity before advancing to larger Phase 3 trials.
5. *Safety biomarkers* – Safety biomarkers are essential for monitoring the safety of new therapies during clinical development. These biomarkers help detect potential adverse effects or toxicities, ensuring patient safety throughout the trial process. Common safety biomarkers include liver function tests to detect liver toxicity, blood counts to monitor for hematological changes, and ECGs to assess potential cardiac abnormalities. The monitoring of adverse effects is especially critical in Alzheimer's clinical trials, as treatments targeting complex biological pathways may have unintended side effects. The FDA emphasizes the importance of safety biomarkers in supporting the regulatory approval process, ensuring that new therapies are both effective and safe for patients.

The FDA also incorporates biomarkers into its staging system for AD, categorizing the disease into stages based on cognitive and functional impairment. Stage 1 represents asymptomatic individuals with positive biomarkers, who may exhibit signs of AD pathology without clinical symptoms. Stage 2 involves individuals with mild cognitive impairment, where cognitive decline is measurable but functional impairments are not yet evident. Stage 3 features mild cognitive impairment with some functional compromise, though dementia has not yet been diagnosed. Stages 4, 5, and 6 represent various degrees of dementia, with increasing cognitive and functional impairment.

In the FDA approval process, biomarkers play a crucial role, especially for treatments targeting the early stages of Alzheimer's, where cognitive impairment may not yet be apparent. Biomarkers are instrumental in demonstrating that a drug is engaging its target and modifying the disease's underlying pathology. The FDA offers provisions for accelerated approval based on biomarkers, particularly when treatments show promising effects on biomarkers indicative of AD, even if clinical outcomes are not yet fully proven. However, for full approval, drugs must show significant improvements in clinical outcomes compared to placebo. Stage 2 approvals can be based on supportive neuropsychological measures in conjunction with biomarker evidence. This approach helps streamline the approval

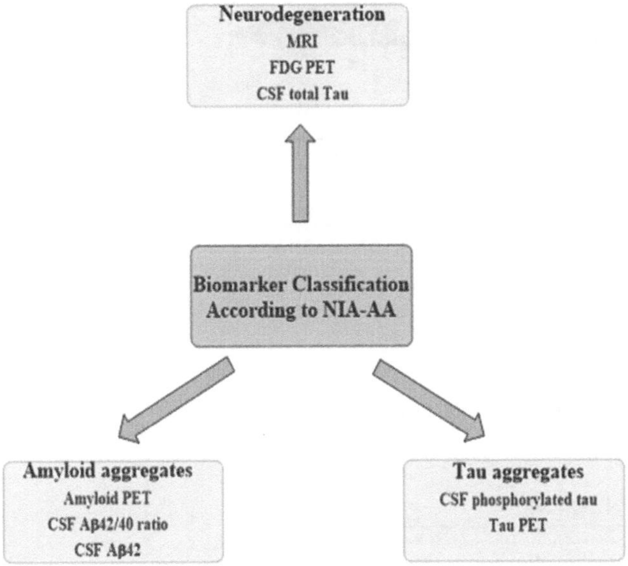

FIGURE 16.4 Biomarker classification according to NIA-AA framework.

process, providing earlier access to potentially beneficial therapies while continuing to collect data on clinical outcomes post-approval (Figure 16.4) (FDA, 2014).

16.5 CURRENT BIOMARKERS FOR DIAGNOSIS OF AD CONTINUUM

Biomarkers play a crucial role in diagnosing AD, offering reliable and objective measures to confirm the presence of the disease and track its progression (Blennow et al., 2015). Given the complexity of AD, clinical diagnoses based solely on cognitive symptoms can be unreliable, particularly in the early stages. Biomarkers provide measurable indicators of the pathological changes associated with AD, enhancing diagnostic accuracy (Alcolea et al., 2014; Palmqvist et al., 2016; Strittmatter, 2013). The current biomarkers used in diagnosing AD, including amyloid, tau, neurodegeneration markers, CSF, blood-based biomarkers, and advanced imaging techniques, are vital in confirming the disease and assessing its progression. They improve diagnostic accuracy, especially in cases where clinical criteria alone might lead to misdiagnosis or delayed treatment. As research into biomarker development continues, integrating these tools into clinical practice holds great promise for enhancing early diagnosis, monitoring treatment response, and advancing our understanding of AD pathology (Forlenza et al., 2015).

Amyloid biomarkers are among the most significant in AD. Amyloid plaques, which are abnormal accumulations of Aβ protein, are a key hallmark of AD pathology (Bloudek et al., 2011). These plaques disrupt normal brain function. To detect amyloid buildup, amyloid PET scans are commonly used, revealing areas of the brain where amyloid deposits are present and providing clear evidence of amyloid-related pathology (Ritchie et al., 2014, 2017). Additionally, CSF biomarkers, such as reduced levels of Aβ42, support the diagnosis of AD (Biscetti et al., 2019).

Tau biomarkers are also essential in AD diagnosis. Tau tangles, which result from the abnormal phosphorylation of tau proteins, are another hallmark of the disease (Golebiowski et al., 1999; Kehoe et al., 2014). Tau biomarkers help detect these neurofibrillary tangles, which are strongly associated with neuronal damage (Juottonen et al., 1999). Tau PET imaging is one method used to visualize tau accumulation in the brain, where radioactive tracers bind to tau tangles, allowing for the assessment of their distribution. This imaging is crucial for differentiating AD from other dementias (Aisen et al., 2010; Anchisi et al., 2005; Ou et al., 2019; Trzepacz et al., 2014).

Neurodegeneration biomarkers are key in evaluating the extent of neuronal damage and loss. Imaging techniques like MRI (magnetic resonance imaging) can detect brain atrophy, especially in regions such as the hippocampus, which is often affected by AD (Yuan et al., 2009). These scans help assess the progression of neurodegeneration and can be used to monitor treatment effectiveness over time (Marcus et al., 2014; O'Brien & Herholz, 2015; Sanchez-Catasus et al., 2017).

Blood-based biomarkers have also gained attention as a non-invasive method for detecting AD-related changes. One such biomarker, neurofilament light (NfL) chain, is a protein released into the bloodstream when nerve cells are damaged. Elevated NfL levels are associated with neurodegeneration and are being investigated as a potential marker for tracking disease progression and treatment response (Zhang et al., 2014).

Imaging biomarkers, particularly PET scans, remain essential in AD diagnosis. In addition to amyloid and tau PET scans, newer imaging technologies are being developed to directly detect neurodegeneration. These techniques offer real-time visual evidence of pathological changes in the brain, making them invaluable for early diagnosis and ongoing monitoring (Martinez et al., 2016).

Genomic and proteomic biomarkers provide deeper insights into the biological mechanisms of AD. Genetic tests, such as those identifying the ApoE genotype, can help determine genetic predispositions to AD, with the ApoE-4 allele being a well-established risk factor. Proteomics, which studies protein expression in biological samples, can detect subtle changes in proteins that may indicate early AD pathology (Mackay et al., 2024).

Emerging fields like metabolomics and lipidomics focus on analyzing metabolic and lipid changes in the body. Abnormalities in metabolism and lipid processing have been linked to AD, and these biomarkers may offer additional insights into the disease process. Metabolic changes can be observed through blood tests or imaging technologies, potentially serving as early indicators of AD, even before clinical symptoms appear (Figure 16.5) (Martínez et al., 1996).

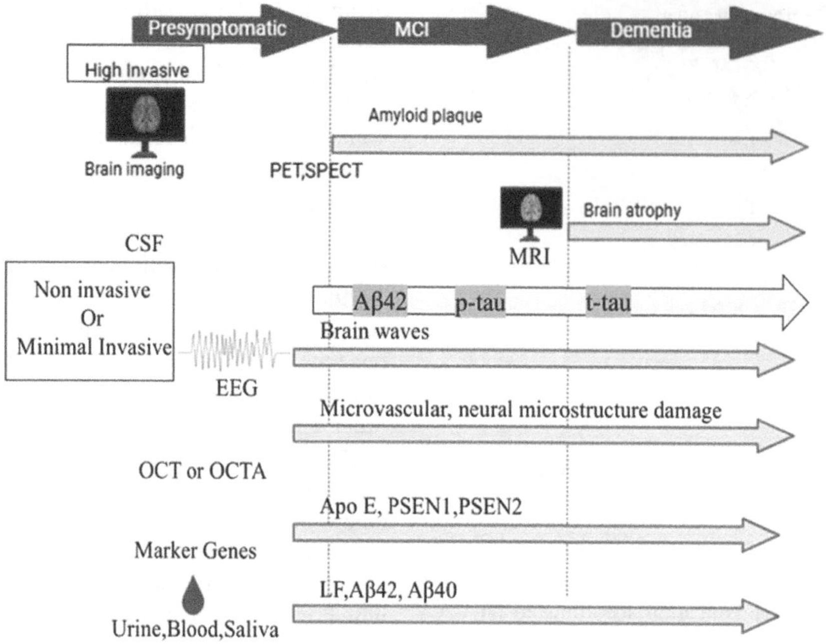

FIGURE 16.5 Biomarker progression and diagnostic modalities across AD stages. This figure illustrates the progression of biomarkers and diagnostic techniques from the presymptomatic stage to mild cognitive impairment (MCI) and dementia. It highlights high-invasive methods like brain imaging (PET, SPECT, MRI) and minimally invasive methods including CSF analysis for Aβ42, p-tau, and t-tau, as well as non-invasive approaches such as electroencephalograms (EEG), optical coherence tomography (OCT/OCTA), genetic markers (ApoE, PSEN1, PSEN2), and biomarkers detectable in urine, blood, and saliva.

16.6 NOVEL BIOMARKERS FOR AD

16.6.1 Non-Invasive and Minimally Invasive Biomarkers

Non-invasive and minimally invasive biomarkers are gaining increasing attention in the field of AD research due to their potential for early detection, monitoring disease progression, and improving diagnostic accuracy (Vecchio et al., 2020). These biomarkers are accessible through less-invasive procedures, making them more patient-friendly compared to traditional methods such as CSF analysis or advanced imaging techniques. Although these biomarkers show promise, further research is required to validate their clinical utility and accuracy (Zhang, Geng, et al., 2021; Zhang, Wang, et al., 2021).

One promising category of non-invasive biomarkers is salivary biomarkers. Saliva, being an easily accessible bodily fluid, has the potential to contain a variety of biomarkers that reflect changes in the brain associated with AD (Izzo et al., 2021). Studies have identified Aβ42, phosphorylated tau (p-tau), total tau (t-tau), and lactoferrin as potential salivary biomarkers for AD. The ability to obtain saliva samples quickly and painlessly makes it an attractive option for large-scale screenings and early detection. However, research into the reliability and sensitivity of these biomarkers is still in its early stages, and further validation is necessary before they can be implemented in routine clinical practice (Zhang et al., 2021).

Plasma biomarkers are also a promising area of research. Plasma levels of Aβ, in particular, have been shown to correlate with both amyloid PET imaging and CSF Aβ levels, providing a less-invasive alternative to traditional diagnostic methods (Gupta et al., 2021). Elevated plasma Aβ concentrations are associated with central nervous system amyloidosis, a key feature of AD. These findings suggest that plasma biomarkers could be useful for identifying individuals at risk for AD, especially in the early stages before clinical symptoms emerge. Ongoing studies are working to refine plasma biomarkers' diagnostic power, focusing on understanding how they correlate with other disease markers and cognitive decline (Colom-Cadena et al., 2020; Jiang et al., 2022).

Another non-invasive biomarker that has shown promise is NfL chain. NfL is a protein released into the bloodstream when neurons are damaged, making it a potential indicator of neurodegeneration. Increased levels of NfL in plasma have been found to correlate with neurodegeneration and cognitive decline, particularly in patients at risk for AD. NfL levels can serve as an early warning sign of neuronal damage, which could help in identifying individuals who are likely to develop AD before the onset of clinical symptoms. This is particularly valuable in monitoring presymptomatic patients who might benefit from early interventions.

Immunomagnetic reduction techniques have also been explored as a method for detecting plasma biomarkers, offering a novel, non-invasive approach to diagnosing AD.

This method has the potential to detect very low concentrations of biomarkers, making it particularly useful for early diagnosis. While this technique is still being refined, it represents an exciting direction for improving diagnostic accuracy in AD and other neurodegenerative diseases (Huan et al., 2018).

In addition to these blood-based biomarkers, CSF biomarkers continue to play an important role in diagnosing AD, although they are more invasive than blood or saliva samples. CSF analysis provides crucial information about Aβ and tau pathology, which are central to AD. Biomarkers like Aβ42, phosphorylated tau, and total tau are measured in CSF to assess the presence of AD-related changes in the brain. While CSF analysis is still considered one of the most reliable methods for confirming an AD diagnosis, its invasive nature makes it less ideal for routine use compared to non-invasive biomarkers.

Genetic risk factors, particularly the presence of the ApoE4 allele, are also considered critical biomarkers in the context of AD. ApoE4 has been shown to influence the progression of AD and is associated with higher levels of amyloid plaques in the brain. Genetic testing for ApoE4, combined with other biomarkers, can help identify individuals at higher risk for developing AD, especially when integrated with blood or imaging biomarkers. The use of genetic testing alongside non-invasive biomarkers may enhance the accuracy of early diagnosis and provide a more comprehensive view of a patient's risk profile (Bekris et al., 2010).

Neuroimaging techniques, including tau PET scans, also provide valuable information about the presence of neuroinflammation and neurodegeneration in the brain. While these imaging techniques are non-invasive, they are more expensive and less accessible than blood-based biomarkers. Nonetheless, neuroimaging plays an important complementary role in diagnosing AD and monitoring disease progression, especially in conjunction with fluid biomarkers.

Non-invasive and minimally invasive biomarkers, such as those found in saliva, blood, and through imaging techniques, hold significant promise for improving the early detection and management of AD. While the potential for these biomarkers is clear, further research and validation through larger studies are necessary to establish their reliability and clinical utility. The integration of these biomarkers with genetic risk factors and advanced imaging techniques may lead to more accurate and efficient diagnostic strategies, ultimately aiding in the development of targeted interventions and improving patient outcomes in Alzheimer's care. Continued exploration of these biomarkers will be essential for their eventual adoption into routine clinical practice (González-Sánchez et al., 2020).

16.6.2 Under Investigation Biomarkers

Several promising biomarkers for AD are currently under investigation, offering the potential for more accurate, non-invasive, and early detection of the disease. These biomarkers could revolutionize AD diagnosis, monitor disease progression, and identify responses to treatment. Although many of these biomarkers are still in the research phase, they show promise and require further validation to confirm their clinical utility (Alber et al., 2020).

One innovative area of research is retinal imaging. Techniques like optical coherence tomography (OCT) have shown potential in detecting structural and pathological changes in the retina that may correlate with AD. Because the retina is part of the central nervous system and shares similarities with the brain's vascular and neural structures, it may be a promising site for detecting early AD-related changes. Studies suggest that retinal thinning and amyloid plaque accumulation in the retina could offer insights into the neurodegenerative process of AD, enabling non-invasive detection. These methods are attractive due to their accessibility, low cost, and ease of use in clinical settings. However, further research is needed to confirm their specificity and sensitivity in relation to AD pathology (Mahaman et al., 2022).

Another exciting area of exploration is peripheral body fluid analysis for AD biomarkers. Saliva, tears, and nasal discharge are being studied as less-invasive alternatives to CSF and blood tests. These fluids contain molecular signatures that may reflect central nervous system pathology, including Aβ, tau, and other potential AD markers. Salivary biomarkers, for example, could be particularly useful for early-stage AD detection, as they can be collected with minimal discomfort. While preliminary studies show promise, larger clinical trials are needed to verify the accuracy and consistency of these biomarkers across different populations.

Small non-coding RNAs (scnRNAs), such as microRNAs, are another promising area of research. These molecules regulate gene expression and may play a role in the molecular processes behind neurodegeneration. Changes in scnRNA expression could serve as early indicators of AD onset, offering insights into the disease's biological processes. However, research in this field is still in its infancy, and more work is needed to understand how scnRNAs can be reliably used as biomarkers in clinical practice (Chidlow et al., 2017).

Plasma biomarkers are also gaining attention due to their non-invasive nature and potential to reflect AD pathology. Biomarkers like Aβ, tau proteins, NfL chain, and glial fibrillary acidic protein (GFAP) are being studied for their ability to diagnose and track disease progression. Plasma levels of Aβ, especially the Aβ42/Aβ40 ratio, have shown potential in reflecting amyloid deposition in the brain, as confirmed by amyloid PET scans. Similarly, tau protein levels, including phosphorylated tau (p-tau), are linked to tau tangles in the brain and may correlate with cognitive decline. NfL, a marker of neurodegeneration, is promising for detecting early neurodegeneration, even in asymptomatic individuals. GFAP, a marker of glial activation, is being studied as glial cells play a key role in neuroinflammation, a hallmark of AD. Plasma biomarkers offer an accessible alternative to CSF, and if validated, they could become crucial for diagnosing AD and monitoring treatment responses (Koronyo et al., 2017).

Genetic profiling, particularly identifying genetic risk factors, also plays a role in AD biomarker research. The APOE ε4 allele is the most well-known genetic risk factor for AD, and its study in combination with other biomarkers can help assess individual risk for the disease. While APOE ε4 is not a definitive biomarker, its presence can aid in stratifying individuals for targeted interventions and clinical trials. Research is also underway to explore other genetic factors and mutations that may contribute to AD, potentially guiding early detection and personalized therapeutic approaches.

Inflammatory markers are another active research area in AD biomarker development. Neuroinflammation is increasingly recognized as a key factor in AD pathophysiology, and markers of inflammation such as cytokines, chemokines, and microglial activation are being studied. These markers are being explored in both blood and CSF to understand the role of immune cells, particularly microglia, in AD progression. Tracking neuroinflammation could lead to biomarkers that predict disease progression and serve as targets for immunomodulatory therapies, although more research is needed to confirm their reliability as indicators of AD-specific inflammation.

Advanced imaging techniques also hold promise for AD diagnosis and monitoring. In addition to amyloid and tau PET scans, newer neuroimaging technologies are being developed to provide more detailed and earlier insights into brain changes related to AD. These techniques may uncover additional biomarkers related to neuroinflammation, neuronal damage, and brain network dysfunction. Functional magnetic resonance imaging (fMRI) and diffusion tensor imaging (DTI) are also being studied to measure brain activity and connectivity, offering potential biomarkers to track disease progression. These imaging methods may complement other biomarkers like blood and CSF markers to improve diagnostic accuracy.

Finally, the development of combination biomarker panels is gaining momentum. Combining multiple biomarkers, such as GFAP with the Aβ 42/40 ratio, could enhance diagnostic accuracy and provide a more comprehensive view of AD pathology. A panel of biomarkers may offer a more reliable way to diagnose AD at different stages, predict disease progression, and assess treatment efficacy. Researchers are also exploring how combining genetic, imaging, and fluid biomarkers could improve clinical decision-making and enable earlier interventions.

Under-investigation biomarkers for AD hold significant promise for early detection, diagnosis, and treatment monitoring. While many of these biomarkers show considerable potential, they require further validation through rigorous clinical trials. Once validated, these biomarkers could transform AD diagnostics, offering non-invasive, accessible methods for early detection, tracking disease progression, and evaluating therapeutic responses, ultimately improving patient outcomes (Figure 16.6) (Mirzaei et al., 2020).

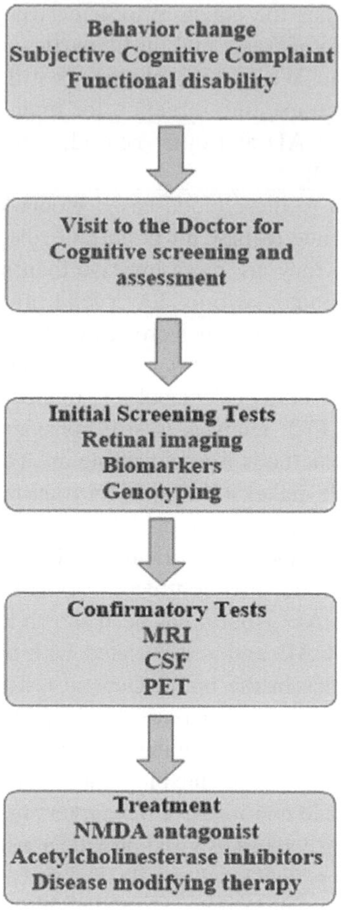

FIGURE 16.6 Flowchart of Alzheimer's disease diagnosis and treatment pathway.

16.7 DISCUSSION/LIMITATIONS

The shift toward a biological definition of AD highlights the importance of diagnostic biomarkers in identifying the disease at an early stage, even before clinical symptoms appear (Khosravi et al., 2019). However, the integration of these biomarkers into routine clinical practice faces several challenges. First, no single biomarker has proven sufficient for an independent AD diagnosis, making it necessary to use a combination of imaging techniques and CSF biomarkers to improve diagnostic accuracy. The high cost and limited accessibility of amyloid PET imaging, which can exceed $4000, complicate the widespread use of these biomarkers. Moreover, the premature use of biomarkers could lead to overdiagnosis, potentially causing unnecessary anxiety and treatments in patients who do not have clinically relevant forms of AD (Weigand et al., 2021). CSF analysis, while valuable, remains an invasive procedure that may not be suitable for all patients.

A key limitation is the lack of standardization in biomarker measurement techniques across different studies, hindering their regulatory approval. Imaging biomarkers are affected by issues such as variability in test results and concerns over radiation exposure from repeated PET scans. While MRI avoids radiation, its use is constrained by patient factors, such as the presence of medical implants (Custodio et al., 2017). CSF biomarkers, despite their significance, face variability between assays and laboratories, along with high rates of false positives and a lack of consensus on appropriate cut-off values, complicating their diagnostic value. Despite these challenges, the future of AD biomarkers remains promising. Ongoing research aims to standardize methods, explore novel biomarkers beyond amyloid and tau, and improve understanding of how biomarkers correlate with cognitive performance and histopathological findings. Future studies must also account for patient factors, including age and genetic predisposition, to enhance diagnostic accuracy (Niemantsverdriet et al., 2017). Overcoming these challenges is essential for unlocking the full potential of biomarkers in the clinical diagnosis and treatment of AD (Jack et al., 2019).

16.8 CONCLUSION

Biomarkers have revolutionized the understanding and management of AD, playing a critical role in diagnosis, prognosis, and therapeutic monitoring. While AD biomarkers have demonstrated their utility in detecting hallmark pathologies, such as Aβ deposition, tau aggregation, and neurodegeneration, they are most effective when used in conjunction with comprehensive clinical assessments. This integrated approach ensures greater diagnostic accuracy, especially in atypical AD presentations that often mimic other forms of dementia. Early diagnosis facilitated by biomarkers enables timely interventions, individualized care, and tailored therapeutic strategies, which are vital as new disease-modifying treatments (DMTs) emerge. Although significant progress has been made, the current reliance on invasive and costly biomarker technologies, such as CSF analysis and advanced imaging, limits their accessibility. Research is increasingly focused on identifying minimally invasive, cost-effective biomarkers, such as blood-based assays and saliva markers, to improve early detection and broaden their application. Additionally, integrating artificial intelligence and machine learning into biomarker analysis holds promise for enhancing diagnostic precision and uncovering patterns in the complex AD continuum.

The A/T/N research framework and related staging systems have provided a structured approach for incorporating biomarkers into clinical practice and drug development. These frameworks emphasize the importance of using biomarkers not only for diagnosis but also for monitoring disease progression, demonstrating therapeutic target engagement, and ensuring treatment safety. By aligning biomarkers with advancements in drug development, such as companion diagnostics, researchers can accelerate the development of effective treatments and better stratify patients for clinical trials.

Looking ahead, the expansion of the biomarker repertoire and deeper understanding of AD's biology are poised to transform patient care. While challenges remain, including the heterogeneity of AD and the need for robust, non-invasive diagnostic tools, the potential of biomarkers to guide early diagnosis, predict disease progression, and optimize therapeutic strategies is undeniable. By bridging the gap between research and clinical application, biomarkers represent a beacon of hope for mitigating the growing global burden of AD.

REFERENCES

Aisen, P. S., Petersen, R. C., Donohue, M. C., Gamst, A., Raman, R., Thomas, R. G., Walter, S., Trojanowski, J. Q., Shaw, L. M., Beckett, L. A., Jack, C. R., Jr, Jagust, W., Toga, A. W., Saykin, A. J., Morris, J. C., Green, R. C., Weiner, M. W., & Alzheimer's Disease Neuroimaging Initiative. (2010). Clinical core of the Alzheimer's disease neuroimaging initiative: Progress and plans. *Alzheimer's & Dementia*, 6(3), 239–246.

Alber, J., Goldfarb, D., Thompson, L. I., Arthur, E., Hernandez, K., Cheng, D., DeBuc, D. C., Cordeiro, F., Provetti-Cunha, L., den Haan, J., Van Stavern, G. P., Salloway, S. P., Sinoff, S., & Snyder, P. J. (2020). Developing retinal biomarkers for the earliest stages of Alzheimer's disease: What we know, what we don't, and how to move forward. *Alzheimer's & Dementia*, *16*(1), 229–243.

Alcolea, D., Martínez-Lage, P., Izagirre, A., Clerigué, M., Carmona-Iragui, M., Alvarez, R. M., Fortea, J., Balasa, M., Morenas-Rodríguez, E., Lladó, A., Grau, O., Blennow, K., Lleó, A., & Molinuevo, J. L. (2014). Feasibility of lumbar puncture in the study of cerebrospinal fluid biomarkers for Alzheimer's disease: A multicenter study in Spain. *Journal of Alzheimer's Disease*, *39*(4), 719–726.

Alexander, G. E., Chen, K., Pietrini, P., Rapoport, S. I., & Reiman, E. M. (2002). Longitudinal PET evaluation of cerebral metabolic decline in dementia: A potential outcome measure in Alzheimer's disease treatment studies. *American Journal of Psychiatry*, *159*(5), 738–745.

Alzheimer's Association (2016). 2016 Alzheimer's disease facts and figures. *Alzheimer's & Dementia*, *12*(4), 459–509.

Anchisi, D., Borroni, B., Franceschi, M., Kerrouche, N., Kalbe, E., Beuthien-Beumann, B., … & Perani, D. (2005). Heterogeneity of brain glucose metabolism in mild cognitive impairment and clinical progression to Alzheimer disease. *Archives of Neurology*, *62*(11), 1728–1733.

Arrowsmith, J., & Miller, P. (2013). Phase II and phase III attrition rates 2011–2012. *Nature Reviews Drug Discovery*, *12*(8), 569–570.

Babić Leko, M., Borovečki, F., Dejanović, N., Hof, P. R., & Šimić, G. (2016). Predictive value of cerebrospinal fluid visinin-like protein-1 levels for Alzheimer's disease early detection and differential diagnosis in patients with mild cognitive impairment. *Journal of Alzheimer's Disease*, 50(3), 765–778.

Bekris, L. M., Yu, C. E., Bird, T. D., & Tsuang, D. W. (2010). Genetics of Alzheimer disease. *Journal of Geriatric Psychiatry and Neurology*, *23*(4), 213–227.

Biscetti, L., Salvadori, N., Farotti, L., Cataldi, S., Eusebi, P., Paciotti, S., & Parnetti, L. (2019). The added value of Aβ42/Aβ40 in the CSF signature for routine diagnostics of Alzheimer's disease. *Clinica Chimica Acta*, *494*, 71–73.

Blennow, K., Dubois, B., Fagan, A. M., Lewczuk, P., De Leon, M. J., & Hampel, H. (2015). Clinical utility of cerebrospinal fluid biomarkers in the diagnosis of early Alzheimer's disease. *Alzheimer's & Dementia*, *11*(1), 58–69.

Blennow, K., Shaw, L. M., Stomrud, E., Mattsson, N., Toledo, J. B., Buck, K., … & Hansson, O. (2019). Predicting clinical decline and conversion to Alzheimer's disease or dementia using novel Elecsys Aβ (1–42), pTau and tTau CSF immunoassays. *Scientific Reports*, *9*(1), 19024.

Bloudek, L. M., Spackman, D. E., Blankenburg, M., & Sullivan, S. D. (2011). Review and meta-analysis of biomarkers and diagnostic imaging in Alzheimer's disease. *Journal of Alzheimer's Disease*, *26*(4), 627–645.

Brookmeyer, R., Corrada, M. M., Curriero, F. C., & Kawas, C. (2002). Survival following a diagnosis of Alzheimer disease. *Archives of Neurology*, *59*(11), 1764–1767.

Chandra, A., Coile, C., & Mommaerts, C. (2023). What can economics say about Alzheimer's disease? *Journal of Economic Literature*, *61*(2), 428–470.

Chatterjee, P., Goozee, K., Sohrabi, H. R., Shen, K., Shah, T., Asih, P. R., Dave, P., ManYan, C., Taddei, K., Chung, R., Zetterberg, H., Blennow, K., & Martins, R. N. (2018). Association of plasma neurofilament light chain with neocortical amyloid-β load and cognitive performance in cognitively normal elderly participants. *Journal of Alzheimer's Disease*, *63*(2), 479–487.

Chen, K., Langbaum, J. B., Fleisher, A. S., Ayutyanont, N., Reschke, C., & Lee, W., Liu, X., Bandy, D., Alexander, G. E., Thompson, P. M., Foster, N. L., Harvey, D. J., de Leon, M. J., Koeppe, R. A., Jagust, W. J., Weiner, M. W., Reiman, E. M., & Alzheimer's Disease Neuroimaging Initiative. (2010). Twelve-month metabolic declines in probable Alzheimer's disease and amnestic mild cognitive impairment assessed using an empirically pre-defined statistical region-of-interest: Findings from the Alzheimer's disease neuroimaging initiative. *Neuroimage*, *51*(2), 654–664.

Chidlow, G., Wood, J. P., Manavis, J., Finnie, J., & Casson, R. J. (2017). Investigations into retinal pathology in the early stages of a mouse model of Alzheimer's disease. *Journal of Alzheimer's Disease*, *56*(2), 655–675.

Colom-Cadena, M., Spires-Jones, T., Zetterberg, H., Blennow, K., Caggiano, A., & DeKosky, S. T., Fillit, H., Harrison, J. E., Schneider, L. S., Scheltens, P., de Haan, W., Grundman, M., van Dyck, C. H., Izzo, N. J., Catalano, S. M., & Synaptic Health Endpoints Working Group. (2020). The clinical promise of biomarkers of synapse damage or loss in Alzheimer's disease. *Alzheimer's Research & Therapy*, *12*, 1–12.

Cook, D., Brown, D., Alexander, R., March, R., Morgan, P., Satterthwaite, G., & Pangalos, M. N. (2014). Lessons learned from the fate of AstraZeneca's drug pipeline: A five-dimensional framework. *Nature Reviews Drug Discovery*, *13*(6), 419–431.

Cummings, J. (2019). The national institute on aging—Alzheimer's association framework on Alzheimer's disease: Application to clinical trials. *Alzheimer's & Dementia*, *15*(1), 172–178.

Cummings, J., Lee, G., Nahed, P., Kambar, M. E. Z. N., Zhong, K., Fonseca, J., & Taghva, K. (2022). Alzheimer's disease drug development pipeline: 2022. *Alzheimer's & Dementia: Translational Research & Clinical Interventions*, *8*(1), e12295.

Cummings, J., Ritter, A., & Zhong, K. (2018). Clinical trials for disease-modifying therapies in Alzheimer's disease: A primer, lessons learned, and A blueprint for the future. *Journal of Alzheimer's Disease*, *64*(s1), S3–S22.

Cummings, J. L. (2008). Optimizing phase II of drug development for disease-modifying compounds. *Alzheimer's & Dementia*, *4*(1), S15–S20.

Cummings, J. L., Morstorf, T., & Zhong, K. (2014). Alzheimer's disease drug-development pipeline: Few candidates, frequent failures. *Alzheimer's Research & Therapy*, *6*, 1–7.

Custodio, N., Montesinos, R., Lira, D., Herrera-Pérez, E., Bardales, Y., & Valeriano-Lorenzo, L. (2017). Mixed dementia: A review of the evidence. *Dementia & Neuropsychologia*, *11*, 364–370.

Doody, R. S., Raman, R., Farlow, M., Iwatsubo, T., Vellas, B., Joffe, S., Kieburtz, K., He, F., Sun, X., Thomas, R. G., Aisen, P. S., Alzheimer's Disease Cooperative Study Steering Committee, Siemers E, Sethuraman G., Mohs, R., & Semagacestat Study Group. (2013). A phase 3 trial of semagacestat for treatment of Alzheimer's disease. *New England Journal of Medicine*, *369*(4), 341–350.

Dubois, B., Hampel, H., Feldman, H. H., Scheltens, P., Aisen, P., Andrieu, S., Bakardjian, H., Benali, H., Bertram, L., Blennow, K., Broich, K., Cavedo, E., Crutch, S., Dartigues, J. F., Duyckaerts, C., Epelbaum, S., Frisoni, G. B., Gauthier, S., Genthon, R., … Proceedings of the Meeting of the International Working Group (IWG) American Alzheimer's Association on "The Preclinical State of AD"; July 23, 2015; Washington DC, USA (2016). Preclinical Alzheimer's disease: Definition, natural history, and diagnostic criteria. *Alzheimer's & Dementia*, *12*(3), 292–323.

Dubois, B., Villain, N., Frisoni, G. B., Rabinovici, G. D., Sabbagh, M., Cappa, S., Bejanin, A., Bombois, S., Epelbaum, S., Teichmann, M., Habert, M. O., Nordberg, A., Blennow, K., Galasko, D., Stern, Y., Rowe, C. C., Salloway, S., Schneider, L. S., Cummings, J. L., & Feldman, H. H. (2021). Clinical diagnosis of Alzheimer's disease: Recommendations of the international working group. *The Lancet Neurology*, *20*(6), 484–496.

Dupont, A. C., Largeau, B., Guilloteau, D., Santiago Ribeiro, M. J., & Arlicot, N. (2018). The place of PET to assess new therapeutic effectiveness in neurodegenerative diseases. *Contrast Media & Molecular Imaging*, *2018*(1), 7043578.

Farlow, M., Gracon, S. I., Hershey, L. A., Lewis, K. W., Sadowsky, C. H., Dolan-Ureno, J., … & Myers, S. (1992). A controlled trial of tacrine in Alzheimer's disease. *JAMA*, *268*(18), 2523–2529.

FDA, U. (2014). Guidance for industry and FDA staff: Qualification process for drug development tools. *Federal Register*, *21*, 83100–83102.

Forlenza, O. V., Radanovic, M., Talib, L. L., Aprahamian, I., Diniz, B. S., Zetterberg, H., & Gattaz, W. F. (2015). Cerebrospinal fluid biomarkers in Alzheimer's disease: Diagnostic accuracy and prediction of dementia. *Alzheimer's & Dementia: Diagnosis, Assessment & Disease Monitoring*, *1*(4), 455–463.

Golebiowski, M., Barcikowska, M., & Pfeffer, A. (1999). Magnetic resonance imaging-based hippocampal volumetry in patients with dementia of the Alzheimer type. *Dementia and Geriatric Cognitive Disorders*, *10*(4), 284–288.

González-Sánchez, M., Bartolome, F., Antequera, D., Puertas-Martín, V., González, P., Gómez-Grande, A., … & Carro, E. (2020). Decreased salivary lactoferrin levels are specific to Alzheimer's disease. *EBioMedicine*, 57.

Guidance, D. (2018). Early Alzheimer's Disease: Developing Drugs for Treatment. *Center for Biologics Evaluation and Research (CBER)*.

Gupta, V. B., Chitranshi, N., den Haan, J., Mirzaei, M., You, Y., Lim, J. K., … & Gupta, V. (2021). Retinal changes in Alzheimer's disease—Integrated prospects of imaging, functional and molecular advances. *Progress in Retinal and Eye Research*, *82*, 100899.

Hippius, H., & Neundörfer, G. (2003). The discovery of Alzheimer's disease. *Dialogues in Clinical Neuroscience*, *5*(1), 101–108.

Honig, L. S., Barakos, J., Dhadda, S., Kanekiyo, M., Reyderman, L., Irizarry, M., Kramer, L. D., Swanson, C. J., & Sabbagh, M. (2023). ARIA in patients treated with lecanemab (BAN2401) in a phase 2 study in early Alzheimer's disease. *Alzheimer's & Dementia: Translational Research & Clinical Interventions*, *9*(1), e12377.

Huan, T., Tran, T., Zheng, J., Sapkota, S., MacDonald, S. W., Camicioli, R., Dixon, R. A., & Li, L. (2018). Metabolomics analyses of saliva detect novel biomarkers of Alzheimer's disease. *Journal of Alzheimer's Disease*, *65*(4), 1401–1416.

Izzo, N. J., Yuede, C. M., LaBarbera, K. M., Limegrover, C. S., Rehak, C., Yurko, R., Waybright, L., Look, G., Rishton, G., Safferstein, H., Hamby, M. E., Williams, C., Sadlek, K., Edwards, H. M., Davis, C. S., Grundman, M., Schneider, L. S., DeKosky, S. T., Chelsky, D., Pike, I., Henstridge, C., Blennow, K., Zetterberg, H., LeVine, H., Spires-Jones, T. L., Cirrito, J. R., & Catalano, S. M. (2021). Preclinical and clinical biomarker studies of CT1812: A novel approach to Alzheimer's disease modification. *Alzheimer's & Dementia*, *17*(8), 1365–1382.

Jack, C. R. Jr, Bennett, D. A., Blennow, K., Carrillo, M. C., Dunn, B., Haeberlein, S. B., Holtzman, D. M., Jagust, W., Jessen, F., Karlawish, J., Liu, E., Molinuevo, J. L., Montine, T., Phelps, C., Rankin, K. P., Rowe, C. C., Scheltens, P., Siemers, E., Snyder, H. M., Sperling, R. (2018). NIA-AA research framework: Toward a biological definition of Alzheimer's disease. *Alzheimer's & Dementia*, *14*(4), 535–562.

Jack, C. R. Jr, Shiung, M. M., Weigand, S. D., O'brien, P. C., Gunter, J. L., Boeve, B. F., Knopman, D. S., Smith, G. E., Ivnik, R. J., Tangalos, E. G., & Petersen, R. C. (2005). Brain atrophy rates predict subsequent clinical conversion in normal elderly and amnestic MCI. *Neurology*, *65*(8), 1227–1231.

Jack, C. R., Therneau, T. M., Weigand, S. D., Wiste, H. J., Knopman, D. S., Vemuri, P., Lowe, V. J., Mielke, M. M., Roberts, R. O., Machulda, M. M., Graff-Radford, J., Jones, D. T., Schwarz, C. G., Gunter, J. L., Senjem, M. L., Rocca, W. A., & Petersen, R. C. (2019). Prevalence of biologically vs clinically defined Alzheimer spectrum entities using the National Institute on Aging–Alzheimer's Association research framework. *JAMA Neurology*, *76*(10), 1174–1183.

Jiang, Y., Zhou, X., Ip, F. C., Chan, P., Chen, Y., Lai, N. C. H, Cheung, K., Lo, R. M. N., Tong, E. P. S., Wong, B. W. Y., Chan, A. L. T., Mok, V. C. T., Kwok, T. C. Y., Mok, K. Y., Hardy, J., Zetterberg, H., Fu, A. K. Y., & Ip, N. Y. (2022). Large-scale plasma proteomic profiling identifies a high-performance biomarker panel for Alzheimer's disease screening and staging. *Alzheimer's & Dementia*, *18*(1), 88–102.

Juottonen, K., Laakso, M. P., Partanen, K., & Soininen, H. (1999). Comparative MR analysis of the entorhinal cortex and hippocampus in diagnosing Alzheimer disease. *American Journal of Neuroradiology*, *20*(1), 139–144.

Kehoe, E. G., McNulty, J. P., Mullins, P. G., & Bokde, A. L. (2014). Advances in MRI biomarkers for the diagnosis of Alzheimer's disease. *Biomarkers in Medicine*, *8*(9), 1151–1169.

Khosravi, M., Peter, J., Wintering, N. A., Serruya, M., Shamchi, S. P., Werner, T. J., Alavi, A., & Newberg, A. B. (2019). 18F-FDG is a superior indicator of cognitive performance compared to 18F-florbetapir in Alzheimer's disease and mild cognitive impairment evaluation: A global quantitative analysis. *Journal of Alzheimer's Disease*, *70*(4), 1197–1207.

Kleinstreuer, N. C., Sullivan, K., Allen, D., Edwards, S., Mendrick, D. L., Embry, M., Matheson, J., Rowlands, J. C., Munn, S., Maull, E., & Casey, W. (2016). Adverse outcome pathways: From research to regulation scientific workshop report. *Regulatory Toxicology and Pharmacology*, *76*, 39–50.

Koronyo, Y., Biggs, D., Barron, E., Boyer, D. S., Pearlman, J. A., Au, W. J., Kile, S. J., Blanco, A., Fuchs, D.T., Ashfaq, A., Frautschy, S., Cole, G. M., Miller, C. A., Hinton, D. R., Verdooner, S. R., Black, K. L., & Koronyo-Hamaoui, M. (2017). Retinal amyloid pathology and proof-of-concept imaging trial in Alzheimer's disease. *JCI Insight*, *2*(16), e93621.

Leonenko, G., Baker, E., Stevenson-Hoare, J., Sierksma, A., Fiers, M., Williams, J., de Strooper, B. & Escott-Price, V. (2021). Identifying individuals with high risk of Alzheimer's disease using polygenic risk scores. *Nature Communications*, *12*(1), 4506.

Liu, Z., Liu, Y., Zhang, J., Zhang, Y., Ma, G., Cui, L., Zhang, L., & Li, T. (2018). Development of a high-sensitivity method for the measurement of human nasal Aβ42, Tau, and phosphorylated Tau. *Journal of Alzheimer's Disease*, *62*(2), 737–744.

Mackay, G. A., Gall, C., Jampana, R., Sleith, C., & Lip, G. (2024). Scottish Intercollegiate Guidelines Network (SIGN) Guidance on Dementia. The investigation of suspected Dementia (SIGN 168) with focus on biomarkers. *Thrombosis and Haemostasis.*

Mahaman, Y. A. R., Embaye, K. S., Huang, F., Li, L., Zhu, F., Wang, J. Z., Liu, R., Feng, J., Wang, X. (2022). Biomarkers used in Alzheimer's disease diagnosis, treatment, and prevention. *Ageing Research Reviews*, *74*, 101544.

Marcus, C., Mena, E., & Subramaniam, R. M. (2014). Brain PET in the diagnosis of Alzheimer's disease. *Clinical Nuclear Medicine*, *39*(10), e413–e426.

Martinez, G., Flicker, L., Vernooij, R. W., Fuentes Padilla, P., Zamora, J., Roque i Figuls, M., Urrutia, G. & Bonfill Cosp, X. (2016). F-18 PET ligands for the early diagnosis of Alzheimer's disease dementia and other dementias in people with mild cognitive impairment (MCI). *Cochrane Database of Systematic Reviews.*

Martínez, G., Vernooij, R. W., Fuentes Padilla, P., Zamora, J., Flicker, L., Bonfill Cosp, X., & Cochrane Dementia and Cognitive Improvement Group. (1996). 18F PET with flutemetamol for the early diagnosis of Alzheimer's disease dementia and other dementias in people with mild cognitive impairment (MCI). *Cochrane Database of Systematic Reviews*, 2017(11).

Mattsson, N., Andreasson, U., Zetterberg, H., Blennow, K., & Alzheimer's Disease Neuroimaging Initiative (2017). Association of plasma neurofilament light with neurodegeneration in patients with Alzheimer disease. *JAMA Neurology*, *74*(5), 557–566.

Mattsson-Carlgren, N., Janelidze, S., Palmqvist, S., Cullen, N., Svenningsson, A. L., Strandberg, O., Mengel, D., Walsh, D. M., Stomrud, E., Dage, J. L., & Hansson, O. (2020). Longitudinal plasma p-tau217 is increased in early stages of Alzheimer's disease. *Brain*, *143*(11), 3234–3241.

Mega, M. S., Dinov, I. D., Porter, V., Chow, G., Reback, E., Davoodi, P., O'Connor, S. M., Carter, M. F., Amezcua, H., & Cummings, J. L. (2005). Metabolic patterns associated with the clinical response to galantamine therapy: A fludeoxyglucose f 18 positron emission tomographic study. *Archives of Neurology*, *62*(5), 721–728.

Mercier, J., Provins, L., & Valade, A. (2017). Discovery and development of SV2A PET tracers: Potential for imaging synaptic density and clinical applications. *Drug Discovery Today: Technologies*, *25*, 45–52.

Mielke, M. M., Hagen, C. E., Xu, J., Chai, X., Vemuri, P., Lowe, V. J., Airey, D. C., Knopman, D. S., Roberts, R. O., Machulda, M. M., Jack, C. R. Jr, Petersen, R. C., & Dage, J. L. (2018). Plasma phospho-tau181 increases with Alzheimer's disease clinical severity and is associated with tau-and amyloid-positron emission tomography. *Alzheimer's & Dementia*, *14*(8), 989–997.

Mirzaei, N., Shi, H., Oviatt, M., Doustar, J., Rentsendorj, A., Fuchs, D. T., Sheyn, J., Black, K. L., Koronyo, Y., & Koronyo-Hamaoui, M. (2020). Alzheimer's retinopathy: Seeing disease in the eyes. *Frontiers in Neuroscience*, *14*, 921.

Molinuevo, J. L., Blennow, K., Dubois, B., Engelborghs, S., Lewczuk, P., Perret-Liaudet, A., Teunissen, C. E., & Parnetti, L. (2014). The clinical use of cerebrospinal fluid biomarker testing for Alzheimer's disease diagnosis: A consensus paper from the Alzheimer's Biomarkers Standardization Initiative. *Alzheimer's & Dementia*, *10*(6), 808–817.

Morgan, P., Brown, D. G., Lennard, S., Anderton, M. J., Barrett, J. C., Eriksson, U., Fidock, M., Hamrén, B., Johnson, A., March, R. E., Matcham, J., Mettetal, J., Nicholls, D. J., Platz, S., Rees, S., Snowden, M. A., & Pangalos, M. N. (2018). Impact of a five-dimensional framework on R&D productivity at AstraZeneca. *Nature Reviews Drug Discovery*, *17*(3), 167–181.

Morris, E., Chalkidou, A., Hammers, A., Peacock, J., Summers, J., & Keevil, S. (2016). Diagnostic accuracy of 18F amyloid PET tracers for the diagnosis of Alzheimer's disease: A systematic review and meta-analysis. *European Journal of Nuclear Medicine and Molecular Imaging*, *43*(2), 374–385.

Niemantsverdriet, E., Valckx, S., Bjerke, M., & Engelborghs, S. (2017). Alzheimer's disease CSF biomarkers: Clinical indications and rational use. *Acta neurologica belgica*, *117*, 591–602.

O'Brien, J. T., & Herholz, K. (2015). Amyloid imaging for dementia in clinical practice. *BMC Medicine*, *13*, 1–3.

Ou, Y.-N., Xu, W., Li, J. Q., Guo, Y., Cui, M., Chen, K. L., Huang, Y. Y., Dong, Q., Tan, L., & Yu, J. T. (2019). FDG-PET as an independent biomarker for Alzheimer's biological diagnosis: A longitudinal study. *Alzheimer's Research & Therapy*, *11*, 1–11.

Palmqvist, S., Mattsson, N., Hansson, O., & Alzheimer's Disease Neuroimaging Initiative (2016). Cerebrospinal fluid analysis detects cerebral amyloid-β accumulation earlier than positron emission tomography. *Brain*, *139*(4), 1226–1236.

Porsteinsson, A. P., Drye, L. T., Pollock, B. G., Devanand, D. P., Frangakis, C., Ismail, Z., Marano, C., Meinert, C. L., Mintzer, J. E., Munro, C. A., Pelton, G., Rabins, P. V., Rosenberg, P. B., Schneider, L. S., Shade, D. M., Weintraub, D., Yesavage, J., Lyketsos, C. G., & CitAD Research Group. (2014). Effect of citalopram on agitation in Alzheimer disease: The CitAD randomized clinical trial. *JAMA*, *311*(7), 682–691.

Prince, M., Wimo, A., Guerchet, M., Ali, G. C., Wu, Y. T., & Prina, M. (2015). World Alzheimer Report 2015. *The Global Impact of Dementia: An analysis of prevalence, incidence, cost and trends.*

Rasmussen, J., & Langerman, H. (2019). Alzheimer's disease—Why we need early diagnosis. *Degenerative Neurological and Neuromuscular Disease*, *9*, 123–130.

Ritchie, C., Noel-Storr, A. H., Ukoumunne, O., Ladds, E. C., & Martin, S. (2017). CSF tau and the CSF tau/ABeta ratio for the diagnosis of Alzheimer's disease dementia and other dementias in people with mild cognitive impairment (MCI). *Cochrane Database of Systematic Reviews*, *3*(3), CD010803.

Ritchie, C., Smailagic, N., Noel-Storr, A. H., Takwoingi, Y., Flicker, L., Mason, S. E., & McShane, R. (2014). Plasma and cerebrospinal fluid amyloid beta for the diagnosis of Alzheimer's disease dementia and other dementias in people with mild cognitive impairment (MCI). *Cochrane Database of Systematic Reviews*, *2014*(6), CD008782.

Rogers, S. L., Farlow, M. R., Doody, R. S., Mohs, R., Friedhoff, L. T., & Donepezil Study Group* (1998). A 24-week, double-blind, placebo-controlled trial of donepezil in patients with Alzheimer's disease. *Neurology*, *50*(1), 136–145.

Salloway, S., Sperling, R., Fox, N. C., Blennow, K., Klunk, W., Raskind, M., Sabbagh, M., Honig, L. S., Porsteinsson, A. P., Ferris, S., Reichert, M., Ketter, N., Nejadnik, B., Guenzler, V., Miloslavsky, M., Wang, D., Lu, Y., Lull, J., Tudor, I. C., … for the Bapineuzumab 301 and 302 Clinical Trial Investigators (2014). Two phase 3 trials of bapineuzumab in mild-to-moderate Alzheimer's disease. *New England Journal of Medicine*, *370*(4), 322–333.

Sanchez-Catasus, C. A., Stormezand, G. N., Jan van Laar, P., De Deyn, P. P., Alvarez Sanchez, M., & Dierckx, R. A. J. O. (2017). FDG-PET for prediction of AD dementia in mild cognitive impairment. A review of the state of the art with particular emphasis on the comparison with other neuroimaging modalities (MRI and perfusion SPECT). *Current Alzheimer Research*, 14(2), 127–142.

Scharre, D. W. (2019). Preclinical, prodromal, and dementia stages of Alzheimer's disease. *Practical Neurology*, *15*, 36–47.

Scheltens, P., Blennow, K., Breteler, M. M., De Strooper, B., Frisoni, G. B., Salloway, S., & Van der Flier, W. M. (2016). Alzheimer's disease. *The Lancet*, *388*(10043), 505–517.

Scheltens, P., Leys, D., Barkhof, F., Huglo, D., Weinstein, H. C., Vermersch, P., Kuiper, M., Steinling, M., Wolters, E. C., & Valk, J. (1992). Atrophy of medial temporal lobes on MRI in "probable" Alzheimer's disease and normal ageing: Diagnostic value and neuropsychological correlates. *Journal of Neurology, Neurosurgery & Psychiatry*, *55*(10), 967–972.

Sharma, A., Rudrawar, S., Bharate, S. B., & Jadhav, H. R. (2025). Recent advancements in the therapeutic approaches for Alzheimer's disease treatment: Current and future perspective. *RSC Medicinal Chemistry, 16(2), 652–693.*

Sharma, A., Rudrawar, S., Sharma, A., Bharate, S. B., & Jadhav, H. R. (2024a). Design, synthesis, in silico, and in vitro evaluation of pyrrol-2-yl-phenyl allylidene hydrazine carboximidamide derivatives as AChE/BACE 1 dual inhibitors. *RSC Advances*, *14*(37), 26703–26722.

Sharma, A., Rudrawar, S., Sharma, A., Bharate, S. B., & Jadhav, H. R. (2024b). Unveiling the potential of novel indol-3-yl-phenyl allylidene hydrazine carboximidamide derivatives as AChE/BACE 1 dual inhibitors: A combined in silico, synthesis and in vitro study. *RSC Advances*, 14(33), 23853–23872.

Simon, G. M., Niphakis, M. J., & Cravatt, B. F. (2013). Determining target engagement in living systems. *Nature Chemical Biology*, *9*(4), 200–205.

Snowden, J. S., Stopford, C. L., Julien, C. L., Thompson, J. C., Davidson, Y., Gibbons, L., Pritchard, A., Lendon, C. L., Richardson, A. M., Varma, A., Neary, D., & Mann, D. M. (2007). Cognitive phenotypes in Alzheimer's disease and genetic risk. *Cortex*, *43*(7), 835–845.

Sperling, R. A., Jack Jr, C. R., Black, S. E., Frosch, M. P., Greenberg, S. M., Hyman, B. T., Scheltens, P., Carrillo, M. C., Thies, W., Bednar, M. M., Black, R. S., Brashear, H. R., Grundman, M., Siemers, E. R., Feldman, H. H., & Schindler, R. J. (2011). Amyloid-related imaging abnormalities in amyloid-modifying therapeutic trials: Recommendations from the Alzheimer's Association Research Roundtable Workgroup. *Alzheimer's & Dementia*, *7*(4), 367–385.

Strittmatter, W. J. (2013). Bathing the brain. *The Journal of Clinical Investigation*, *123*(3), 1013–1015.

Tan, C. C., Yu, J. T., Wang, H. F., Tan, M. S., Meng, X. F., Wang, C., … & Tan, L. (2014). Efficacy and safety of donepezil, galantamine, rivastigmine, and memantine for the treatment of Alzheimer's disease: A systematic review and meta-analysis. *Journal of Alzheimer's Disease*, *41*(2), 615–631.

Tarawneh, R., D'Angelo, G., Crimmins, D., Herries, E., Griest, T., Fagan, A. M., Zipfel, G. J., Ladenson, J. H., Morris, J. C., & Holtzman, D. M. (2016). Diagnostic and prognostic utility of the synaptic marker neurogranin in Alzheimer disease. *JAMA neurology*, *73*(5), 561–571.

Trzepacz, P. T., Yu, P., Sun, J., Schuh, K., Case, M., Witte, M. M., Hochstetler, H., Hake, A., & Alzheimer's Disease Neuroimaging Initiative (2014). Comparison of neuroimaging modalities for the prediction of conversion from mild cognitive impairment to Alzheimer's dementia. *Neurobiology of Aging*, *35*(1), 143–151.

Vecchio, F., Miraglia, F., Alù, F., Menna, M., Judica, E., Cotelli, M., & Rossini, P. M. (2020). Classification of Alzheimer's disease with respect to physiological aging with innovative EEG biomarkers in a machine learning implementation. *Journal of Alzheimer's Disease*, *75*(4), 1253–1261.

Villain, N., & Dubois, B. (2019). Alzheimer's disease including focal presentations. *Seminars in Neurology*, *39*(02), 213–226.

Weigand, A. J., Thomas, K. R., Bangen, K. J., Eglit, G. M., Delano-Wood, L., Gilbert, P. E., Brickman, A. M., Bondi, M. W., & Alzheimer's Disease Neuroimaging Initiative (2021). APOE interacts with tau PET to influence memory independently of amyloid PET in older adults without dementia. *Alzheimer's & Dementia*, *17*(1), 61–69.

Whitwell, J. L., Josephs, K. A., Murray, M. E., Kantarci, K., Przybelski, S. A., Weigand, S. D., Vemuri, P., Senjem, M. L., Parisi, J. E., Knopman, D. S., Boeve, B. F., Petersen, R. C., Dickson, D. W., & Jack, C. R., Jr (2008). MRI correlates of neurofibrillary tangle pathology at autopsy: A voxel-based morphometry study. *Neurology*, *71*(10), 743–749.

Wilkinson, D., Windfeld, K., & Colding-Jørgensen, E. (2014). Safety and efficacy of idalopirdine, A 5-HT6 receptor antagonist, in patients with moderate Alzheimer's disease (LADDER): A randomised, double-blind, placebo-controlled phase 2 trial. *The Lancet Neurology*, *13*(11), 1092–1099.

Wimo, A., Guerchet, M., Ali, G. C., Wu, Y. T., Prina, A. M., Winblad, B., Jönsson, L., Liu, Z., & Prince, M. (2017). The worldwide costs of dementia 2015 and comparisons with 2010. *Alzheimer's & Dementia*, *13*(1), 1–7.

Yuan, Y., Gu, Z. X., & Wei, W. S. (2009). Fluorodeoxyglucose–positron-emission tomography, single-photon emission tomography, and structural MR imaging for prediction of rapid conversion to Alzheimer disease in patients with mild cognitive impairment: A meta-analysis. *American Journal of Neuroradiology*, *30*(2), 404–410.

Zetterberg, H., & Mattsson, N. (2014). Understanding the cause of sporadic Alzheimer's disease. *Expert Review of Neurotherapeutics*, *14*(6), 621–630.

Zhang, H., Geng, X., Wang, Y., Guo, Y., Gao, Y., Zhang, S., Du, W., Liu, L., Sun, M., Jiao, F., Yi, F., Li, X., & Wang, L. (2021). The significance of EEG alpha oscillation spectral power and beta oscillation phase synchronization for diagnosing probable Alzheimer disease. *Frontiers in Aging Neuroscience*, *13*, 631587.

Zhang, S., Smailagic, N., Hyde, C., Noel-Storr, A. H., Takwoingi, Y., McShane, R., & Feng, J. (2014). 11 C-PIB-PET for the early diagnosis of Alzheimer's disease dementia and other dementias in people with mild cognitive impairment (MCI). *Cochrane Database of Systematic Reviews*, *2014*(7), CD010386.

Zhang, Y., Wang, Y., Shi, C., Shen, M., & Lu, F. (2021). Advances in retina imaging as potential biomarkers for early diagnosis of Alzheimer's disease. *Translational Neurodegeneration*, *10*, 1–9.

17 Non-Peptidic Anti-Aggregate Compounds

A Promising Avenue for Alzheimer's Disease Therapy

Akib Ali, Babita Kumari, and Pankaj Bharali

17.1 INTRODUCTION

Alzheimer's disease (AD) is a progressive neurological disorder that slowly weakens memory, thinking capacity, and ability to do the basic tasks of a normal human being. It eventually leads to dementia. While early inception of Alzheimer's occurs between the ages of 30 and mid-60, late and most common Alzheimer's symptoms usually begin in the mid-60s. Eponymously named after Dr. Alois Alzheimer, this disease is characterized by the occurrence of amyloid-β (Aβ) plaques (clumps in brain tissue), neurofibrillary or tau tangles (bundles of tangled fibers), and loss of connections between neurons. It initially affects the entorhinal cortex and hippocampus that also known as memory region followed by moving toward other regions of the brain that control language, logic, and behavior. The accumulation of Aβ, especially oligomeric Aβ, is associated with cascade of pathological processes, including neuroinflammation, oxidative stress, and synaptic dysfunction (Selkoe & Hardy, 2016). Nonetheless, current therapeutic approaches for AD, such as cholinesterase inhibitors and N-methyl-D-aspartate receptor antagonists, are primarily focused on symptom relief. They could offer only a moderate improvement in cognitive function, yet do not address the underlying mechanisms of the disease for complete remedy (Yiannopoulou & Papageorgiou, 2020). This restriction emphasizes the importance of new therapy approaches capable of modifying the course of the disease by targeting the root pathogenic processes, i.e., protein aggregation. Since protein aggregation constitutes a central part of AD pathophysiology, methods aimed at preventing or reversing this process have appeared as a viable option for disease-modifying therapy.

Although researchers like Aggidis et al. (2021) and Sehra et al. (2025) clearly highlighted the peptide inhibitors effectively blocking Aβ aggregation both in vitro and in preclinical models, the clinical translation of peptide-based treatments is marred with harsh challenges. These encompass poor blood-brain barrier (BBB) penetration, proteolytic vulnerability in the complicated biological milieu, along with high manufacturing expenses attributed to peptide synthesis (Ahmed et al., 2020; Sighencea et al., 2024). To overcome these challenges, non-peptidic anti-aggregate compounds emerged as a novel and promising approach offering several key advantages over peptide compounds. These compounds exhibit low immunogenicity coupled with high stability that favors low risk of antagonistic immune reactions and less susceptible to enzymatic degradation within human body, respectively. Notably, physicochemical properties and smaller size further improved BBB penetration, facilitating effective drug targeting and delivery (Díaz-Caballero et al., 2020).

Development of rational drug designing methodology and high-throughput screening procedures further enabled rapid and effective screening of meta-libraries in identifying potent chemical compounds with anti-aggregation activity. Furthermore, rational drug design tools also allow selective modulation of lead compounds to enhance potency, selectivity, and pharmacokinetic profile. Wang et al. (2023) showed that certain classes of non-peptidic compound have promising preclinical activity in AD models such as metal chelators that inhibit metal-catalyzed Aβ aggregation, polyphenols, having both anti-amyloidogenic and antioxidant properties, and small molecule chaperones, having the ability to stabilize soluble species of Aβ and inhibit their aggregation and subsequent formation of toxic oligomers. Among them, naturally occurring bioactive compounds like the polyphenols gained significant attention due to their extensive biological activities. For instance, epigallocatechin gallate (EGCG), the major polyphenol in green tea, has been shown to induce conformational changes in Aβ fibrils, leading to the formation of non-toxic, off-pathway aggregates (Fernandes et al., 2021; Youn et al., 2022). Similarly, Gervais et al., in 2007, reported that even small molecule chaperones, such as tramiprosate, can stabilize soluble forms of Aβ, thereby inhibiting their aggregation and subsequent formation of toxic oligomers. Gomez-Sequeda et al. (2024) reported that the combination of tramiprosate, curcumin, and SP600125 reduces the neuropathological phenotype in cholinergic-like neurons with the PSEN1 I416T mutation associated with familial AD. An overview of non-peptidic anti-aggregate compounds in AD therapy is shown in Figure 17.1.

DOI: 10.1201/9781003602477-17

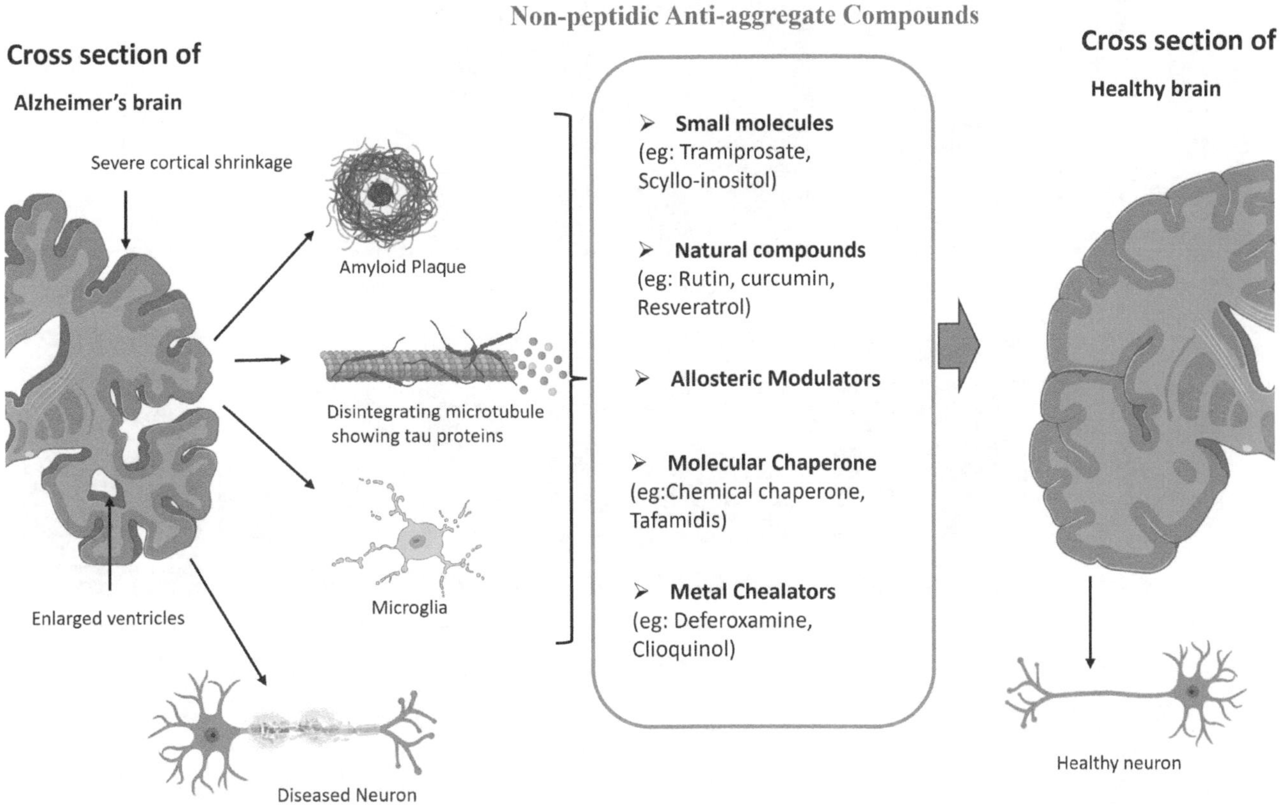

FIGURE 17.1 Illustration summarizing non-peptidic anti-aggregate compounds used in Alzheimer's disease therapy, including small molecules targeting amyloid-β and tau aggregation, such as curcumin, EGCG, methylene blue, and metal chelators like clioquinol.

Despite these encouraging findings, translating non-peptidic anti-aggregate drugs from bench to bedside remains a significant challenge. Further research is crucial to optimize their pharmacokinetic and pharmacodynamic properties, ensuring effective drug delivery to the brain, maximizing their therapeutic efficacy, and minimizing potential side effects. This chapter critically examines the potential of non-peptidic anti-aggregate compounds as a therapeutic approach for AD, focusing on their molecular mechanisms of action, emerging therapeutic candidates, and translational hurdles. By elucidating their strengths and limitations, we aim to provide a comprehensive overview of these compounds' therapeutic potential and pave the way for the development of more effective disease-modifying therapies for AD.

17.2 TRADITIONAL HERBAL EXTRACTS AND ALZHEIMER'S DISEASE

Plants are rich in diverse bioactive compounds, including alkaloids, flavonoids, phenols, glycosides, terpenoids, coumarins, and saponins. With these as key components, herbal and complementary medicines have played a significant role in traditional medical practices for centuries, especially in treating neurological disorders. Around the world, various herbal remedies have been used to manage neurodegenerative conditions effectively (Figure 17.2). For instance, in Europe, *Salvia lavandulaefolia* (Spanish sage) and *Salvia officinalis* (common sage) have been utilized since the 16th century to enhance memory, a practice validated by clinical trials (Akhondzadeh et al., 2003). Similarly, *Bacopa monnieri* (water hyssop), a chief species in the Indian Ayurvedic medicine, has been traditionally used to boost memory and cognitive functions. Another Ayurvedic remedy, *Centella asiatica* (Asiatic pennywort), is often administered with milk to promote memory enhancement. *Withania somnifera* root and *Celastrus paniculatus*, known as a rejuvenative tonic in Ayurveda, has also been employed to improve cognitive abilities. The increasing popularity of herbal medicines is largely due to their perceived efficacy, safety, and cost-effectiveness. Additionally, a randomized, double-blind exploratory trial found that a *Ginkgo biloba* extract, when combined with donepezil, was more effective than donepezil alone for Alzheimer's patients with neuropsychiatric symptoms (Veerendra Kumar et al., 2003). The experimental and clinical evidence for raw extract of other plant includes *Terminalia chebula* fruits, *Ocimum sanctum*, *Clitoria ternatea*, *W. somnifera*, *C. paniculatus*, *Evolvulus alsinoides*, *Desmodium gangeticum*, *Eclipta alba*, *Moringa oleifera*, and *Convolvulus pluricaulis*, showed potential in addressing cognitive impairment. Furthermore, some of these herbal compounds act as non-peptidic anti-aggregate agents that inhibit the aggregation of Aβ proteins, a hallmark of Alzheimer's pathology (Mehla et al., 2020).

FIGURE 17.2 Image showing various medicinal plants known for their potent anti-Alzheimer's properties and commonly used for their neuroprotective and cognitive-enhancing effects.

Notable Indian herbal formulations studied for their potential in managing AD include Trasina, Memorin, Abana, Mentat, and Brahmi Ghrita (Mehla et al., 2020). Traditional Chinese medicine has a rich history of treating dementia, backed by thousands of years of accumulated knowledge. Modern pharmacological research has validated the therapeutic potential of many bioactive compounds from Chinese herbal medicines (Figure 17.2).

For example, ginsenoside Rg1 from *Radix ginseng* acts as a γ-secretase inhibitor, reducing $A\beta$ aggregation and preventing neuronal apoptosis. Similarly, tanshinone IIA from Radix *Salvia miltiorrhiza* and baicalin from *Radix scutellariae* have shown efficacy in mitigating oxidative stress-induced neuronal damage. Icariin from *Epimedium brevicornum* not only lowers $A\beta$ levels and prevents tau protein hyperphosphorylation but also combats oxidative stress and apoptosis. Huperzine A from *Huperzia serrata* serves as a cholinesterase inhibitor, while evodiamine from Fructus evodiae and curcumin from *Curcuma longa* exhibit anti-inflammatory properties. Curcumin also targets both $A\beta$ and tau proteins. The multi-targeted effects and minimal side effects of Chinese herbal medicines make them particularly suitable for long-term dementia management. Recent studies have highlighted the potential of sesquiterpene lactones as therapeutic agents for AD. Methanol extracts from *Adhatoda vasica* and *Ferula assa-foetida* were evaluated *in vitro* for their inhibitory effects on acetylcholinesterase (AChE) and cyclooxygenase-1 (COX-1) enzymes, respectively. The findings revealed significant AChE inhibition by *A. vasica* and notable COX-1 inhibitory activity by *F. assa-foetida*, suggesting their promise as potential AD treatments (Menon et al., 2013).

17.3 NON-PEPTIDIC ANTI-AGGREGATE COMPOUNDS AND MOLECULAR STRATEGIES

Over the past decade, research on non-peptidic anti-aggregate compounds targeting Alzheimer's Aβ aggregation has significantly advanced. Key developments and trends include identification of novel compounds, elucidating the molecular mechanism with special focus on Aβ and tau followed by multiple therapeutic potentials. In 2014, Arai et al. leveraged the structure-activity relationship of cyclic Aβ16–20 (cyclo-[KLVFF]) to identify key pharmacophore motifs. Their findings highlighted the critical involvement of side chains from specific residues (Leu, Val, Phe), while demonstrating that backbone amide bonds were not essential. This discovery enabled the rational design of potent, non-peptidic small molecules, marking a breakthrough as the first successful aggregation inhibitors of amyloids developed through molecular design. Since then, researchers have identified various non-peptidic small molecules with anti-aggregate properties (Table 17.1), including natural products like curcumin and EGCG and synthetic compounds like tramiprosate.

As reported by Eisenberg and Sawaya (2016), tryptanthrin and its analogs have shown potent inhibition of tau

TABLE 17.1
Emerging Therapeutic Compounds for Alzheimer's Disease

Category	Compound Name	Chemical Structure	Mechanism and Therapeutic Explanation	References
Synthetic and natural compounds	Buntanetap (Posiphen)	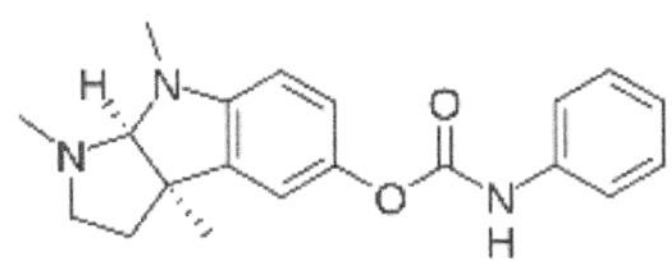	Inhibits neurotoxic protein production by altering β-secretase activity and reducing Aβ peptide synthesis.	Fang et al. (2023)
	Cyclophilin D inhibitors	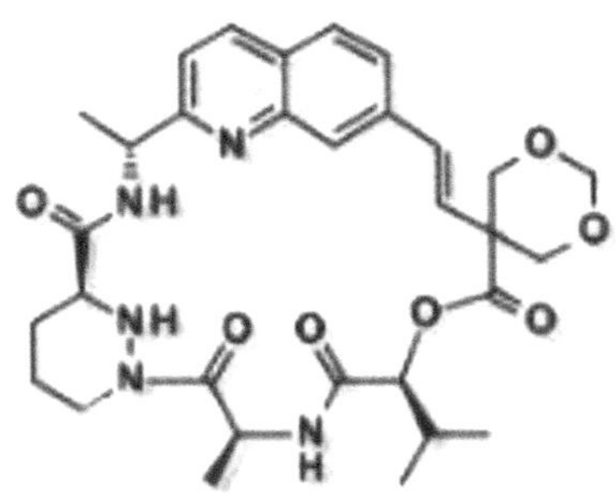	Promote neuroprotection by blocking mPTP synthesis, preventing neuronal death.	Haleckova et al. (2022)
	p75 Neurotrophin receptor (p75NTR) ligands	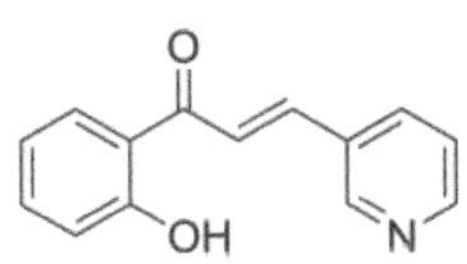	Enhance survival signaling, inhibit Aβ-induced synaptic dysfunction, and reduce neuronal death.	Longo et al. (2007)
	IDO1 inhibitors	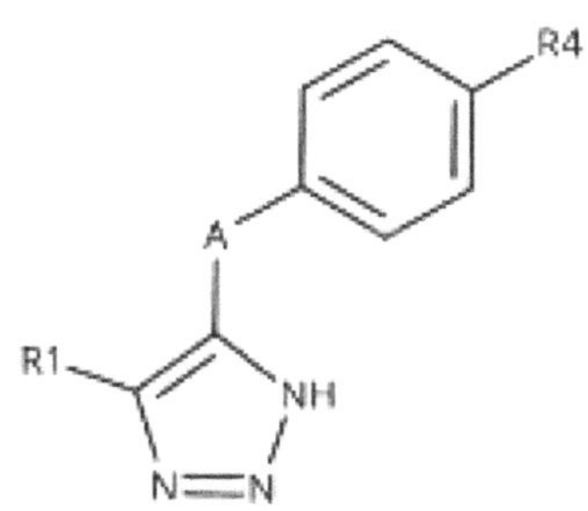	Improve glucose metabolism in astrocytes, restoring memory and brain function; effective for early-stage AD.	Duan et al. (2023)

(Continued)

TABLE 17.1 ***(Continued)***
Emerging Therapeutic Compounds for Alzheimer's Disease

Category	Compound Name	Chemical Structure	Mechanism and Therapeutic Explanation	References
	Methylene blue (tau aggregation inhibitors)	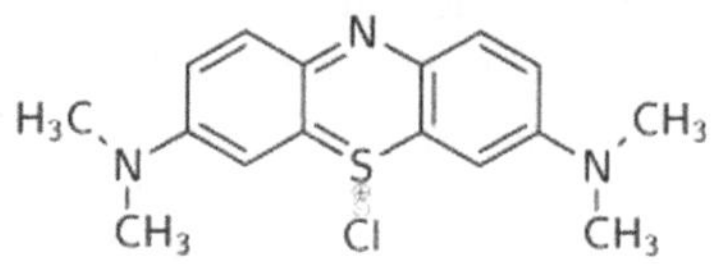	Reduces tau aggregation and neurotoxicity by modulating oxidative stress and mitochondrial function.	Wischik et al. (2014)
	Anthraquinones (tau aggregation inhibitors)	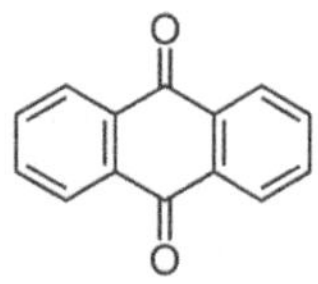	Stabilize tau protein, preventing misfolding and aggregation while promoting clearance.	Pickhardt et al. (2005)
Natural compounds and derivatives	Epigallocatechin gallate (EGCG)	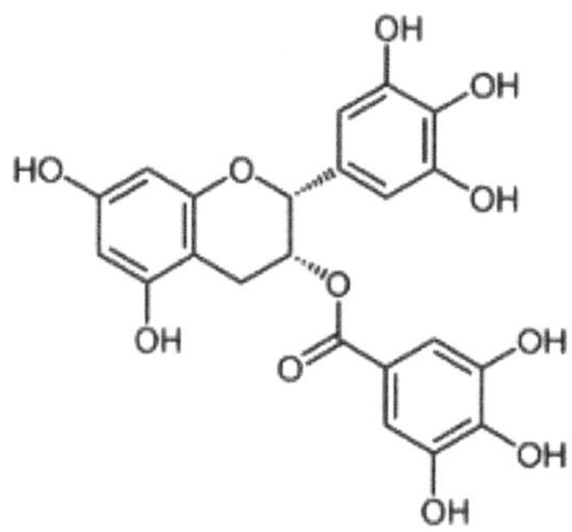	Disaggregates tau and amyloid fibrils; exhibits antioxidant properties but poor bioavailability.	Zhang et al. (2014)
	Baicalein	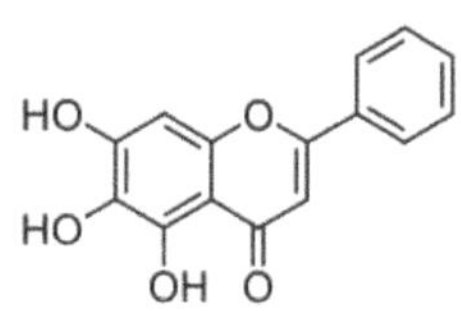	Destabilizes Aβ protofibrils by disrupting hydrophobic core interactions, reducing aggregation.	Sowndhararajan et al. (2017)
	Curcumin	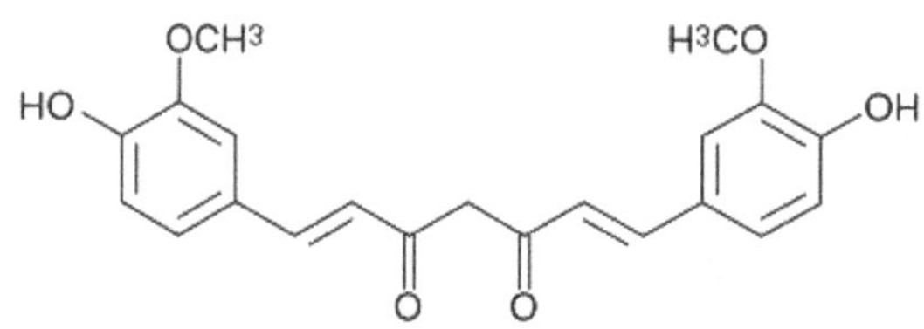	Binds to Aβ plaques, inhibits plaque formation, promotes disaggregation, and has anti-inflammatory effects.	Yang et al. (2005)
	Resveratrol	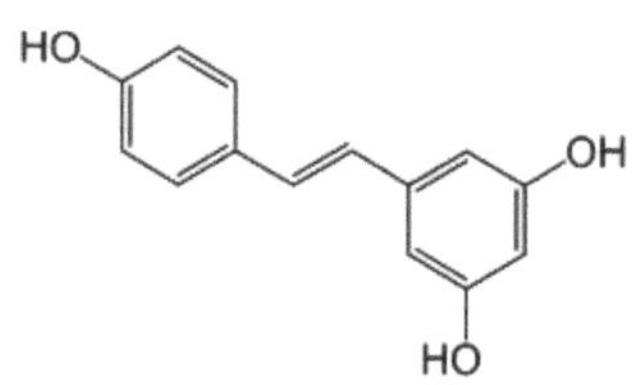	Modulates Aβ aggregation, enhances clearance via autophagy, and provides antioxidant and anti-inflammatory effects.	Baur and Sinclair (2006)
	Huperzine A	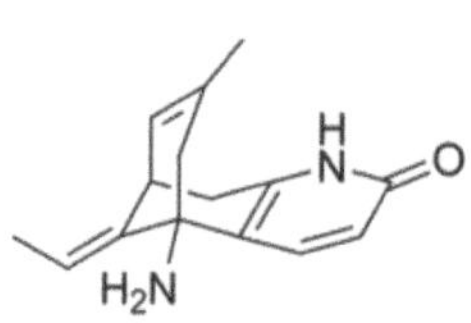	Inhibits acetylcholinesterase, increasing acetylcholine levels to improve cognitive function.	Zangara (2003)

(Continued)

TABLE 17.1 *(Continued)*
Emerging Therapeutic Compounds for Alzheimer's Disease

Category	Compound Name	Chemical Structure	Mechanism and Therapeutic Explanation	References
	Rutin	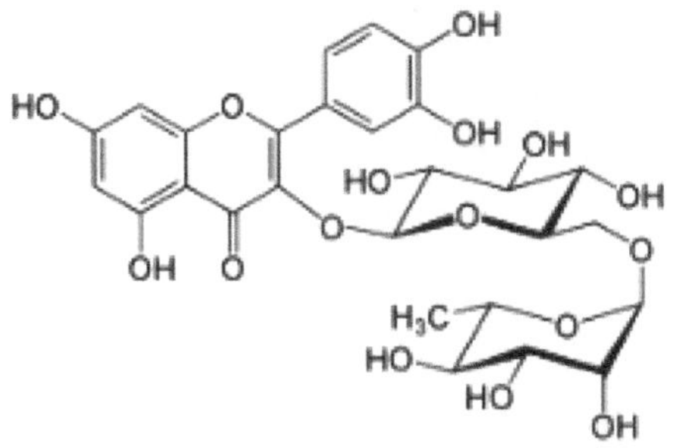	Exhibits antioxidant and anti-inflammatory effects; inhibits Aβ aggregation and tau hyperphosphorylation.	Habtemariam (2016)
	Brazilin	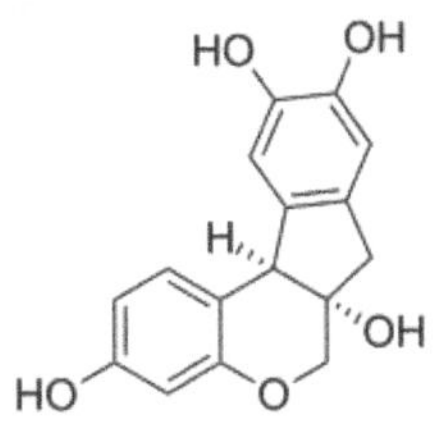	Inhibits Aβ fibrillogenesis, dismantles toxic aggregates, and protects neurons from oxidative stress.	Du et al. (2015)
	Ginnalin A	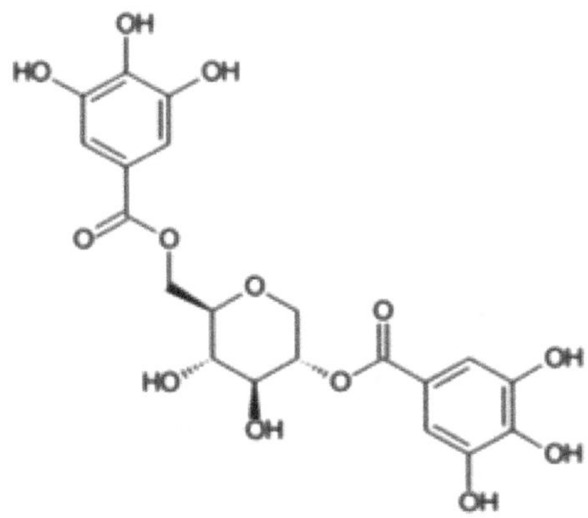	Inhibits Aβ aggregation, disassembles fibrils, and mitigates oxidative stress and membrane damage.	Fan et al. (2020)
	Wgx-50	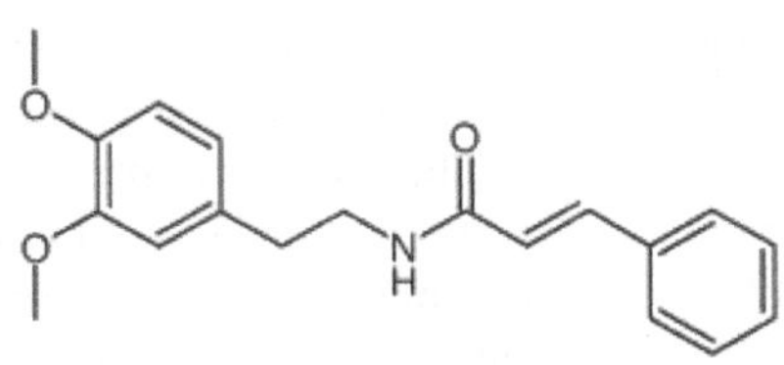	Destabilizes Aβ protofibrils, preventing fibril elongation and promoting disassembly into non-toxic forms.	Fan et al. (2015)
	Myricetin	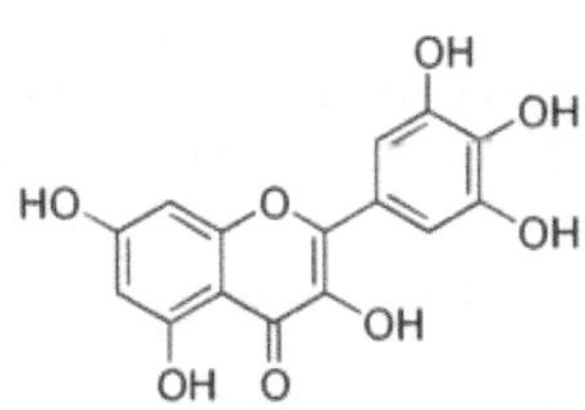	Prevents Aβ oligomerization by stabilizing monomers; has antioxidant properties protecting neurons.	Ono et al. (2012)
	Oleuropein	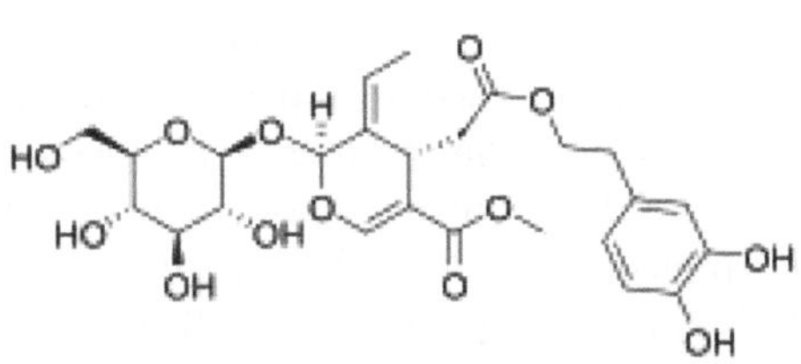	Reduces inflammation, oxidative stress, and Aβ aggregation with anti-neuroinflammatory effects.	Muscat et al. (2020)

(Continued)

TABLE 17.1 ***(Continued)***
Emerging Therapeutic Compounds for Alzheimer's Disease

Category	Compound Name	Chemical Structure	Mechanism and Therapeutic Explanation	References
	Rosmarinic acid	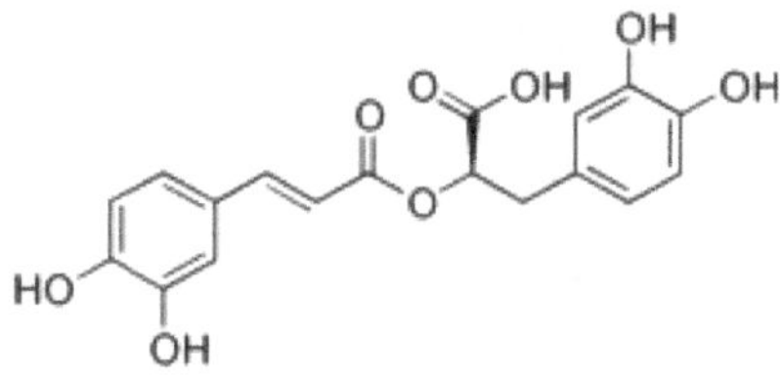	Inhibits fibril elongation, remodels existing fibrils, and reduces cytotoxicity with antioxidant properties.	Muscat et al. (2020)
	Sclerotiorin	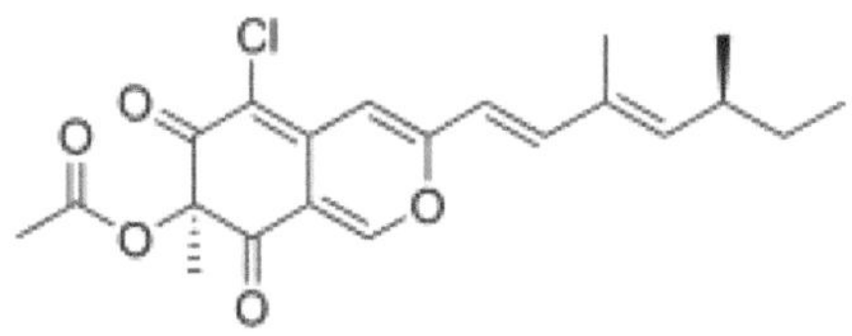	Stabilizes nonfibrillar Aβ oligomers, preventing toxic fibril formation and reducing seeding activity.	Wiglenda et al. (2020)
	Tanshinone	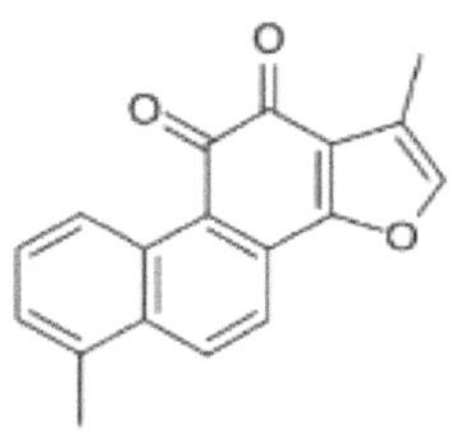	Inhibits amyloid aggregation by stabilizing non-aggregated forms and disrupting existing fibrils.	Dong et al. (2017)
	Uncarinic acid C	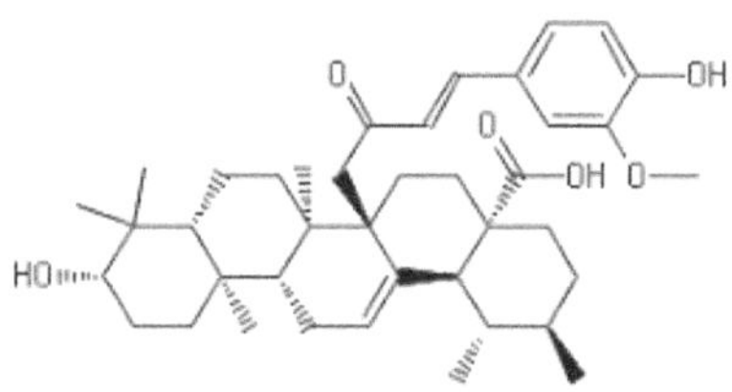	Screened for preventing amyloid aggregation; potential for disrupting toxic aggregates and mitigating damage.	Murakami et al. (2023)

protein aggregation, a hallmark of AD. Further, it also effectively blocks the seeding process critical for the propagation of tau pathogenesis within the brain. This dual mechanism of action validates the potentiality in halting disease progression as well as reverse existing pathology. Likewise, natural products, like ashwagandha (*W. somnifera*), have shown promise in neurodegenerative disease models. Especially, bioactive components such as withanolides may enhance neuronal function and reduce protein aggregation through antioxidant and anti-inflammatory mechanisms (Singh et al., 2011).The development of *in vivo* platforms has significantly accelerated the search for effective non-peptidic inhibitors. Several studies described rapid *in vivo* screening method for identifying small molecule inhibitors of amyloid production (Abelein & Johansson, 2023; Hart et al., 2016). Although clinical validation of peptide-based compounds has elevated since 2018, only a few non-peptide compounds are under clinical trials. While aducanumab, a monoclonal antibody targeting Aβ plaques, was approved by the FDA in 2021 (Huang et al., 2023), non-peptide compounds are still under trial phases. Based on their therapeutic potential and versatility, the key benefits of non-peptidic anti-aggregate compounds over peptide-based inhibitors have been highlighted in Table 17.2.

The molecular mechanism of action involves multiple strategies, including inhibition of aggregation initiation, disruption of aggregation propagation, or modulation of cellular processes. These compounds inhibit protein aggregation by preventing misfolding, disrupting fibril formation, and promoting the clearance of preformed aggregates. The types and examples of each strategy are highlighted below.

17.3.1 Inhibition of Aggregation Initiation

17.3.1.1 Monomer Stabilization

Certain chemicals interact with monomeric tau or Aβ to stabilize their natural conformations, preventing initial misfolding. This involves disrupting key intermolecular interactions, such as hydrogen bonding and hydrophobic interactions, early in the aggregation process. These compounds efficiently stop the synthesis of damaging oligomers by attaching to particular monomers and blocking important

TABLE 17.2
Key Benefits of Non-Peptidic Anti-Aggregate Compounds over Peptide-Based Inhibitors, Emphasizing Their Therapeutic Potential and Versatility

Advantage	Description	References
Enhanced stability	Non-peptidic compounds are more resistant to enzymatic degradation, ensuring longer systemic circulation and sustained therapeutic benefits.	Aguzzi et al. (2023)
Improved bioavailability	Due to their smaller size and hydrophobic nature, non-peptidic molecules are better absorbed and distributed within the body compared to hydrophilic peptides.	Silva et al. (2022)
Superior BBB penetration	Non-peptidic compounds exhibit enhanced ability to cross the blood-brain barrier due to their smaller size and greater lipophilicity, crucial for treating neurodegenerative diseases.	Barisano et al. (2022)
Reduced immunogenicity	These compounds have a lower likelihood of triggering immune responses, making them safer for long-term use compared to peptide-based therapies.	Wang et al. (2023)
Greater design flexibility	Their chemical structures can be tailored to target specific regions of misfolded proteins, such as amyloid-β or tau, with high precision.	Matera et al. (2020)
Potential for oral administration	Many non-peptidic drugs can be administered orally, offering greater convenience and improved patient compliance compared to parenterally administered peptide-based drugs.	Mekala et al. (2020)
Reduced production costs	Non-peptidic compounds are often more cost-effective to produce due to simpler synthesis processes compared to peptide-based drugs.	Abdallah (2024)
Versatility in drug delivery	These compounds can be formulated into advanced drug delivery systems, such as nanoparticles or liposomes, to enhance brain targeting and therapeutic efficacy.	Cerrato and Langel (2022)
Potential for combination therapies	Non-peptidic drugs can be combined with other therapeutic agents to achieve synergistic effects and improve overall outcomes.	Cummings et al. (2023)
Reduced risk of aggregation	Non-peptidic compounds exhibit greater stability and are less likely to aggregate, ensuring better safety and efficacy than peptide-based therapies.	Kucwaj-Brysz et al. (2021)

residues involved in aggregation (Teplow et al., 2012). For instance, it has been discovered that natural substances like azaphilone can deconstruct preexisting tau aggregates in a dose-dependent manner and limit the construction of tau filaments. It can also decrease the quantity and overall length of tau polymers (Paranjape et al., 2015). Similarly, rosmarinic acid and myricetin prevent Aβ aggregation by directly interacting with Aβ peptides and stopping them from oligomerizing and forming fibrils (Ono et al., 2012).

17.3.1.2 Nucleation Inhibition

Non-peptide compounds can also disrupt the onset of initial aggregates, or nuclei, which serve as templates for subsequent aggregation. They prevent the formation of stable nuclei and hence stop the aggregation process by attaching to protein areas important in early oligomer formation (Naiki et al., 2010). It has been demonstrated that tau phosphorylation and aggregation are inhibited by increasing O-GlcNAcylation levels on tau proteins using OGA inhibitors. Sharma et al. (2024) reported that substances such as thiamet-G can elevate tau O-GlcNAcylation, which lowers aggregation in vivo (Sharma et al., 2024).

17.3.2 Disruption of Aggregation Propagation

17.3.2.1 Seeding Inhibition

Some compounds block the seeding process by binding to the surfaces of existing aggregates, thus preventing further monomer addition. For instance, compounds that attach to Aβ fibril surfaces can effectively stop further growth by hindering additional monomer binding (Estrada & Soto, 2006). Natural compounds like brazilin extracted from *Caesalpinia sappan* inhibit $A\beta_{42}$ fibrillogenesis by directly binding to $A\beta_{42}$ species through hydrophobic interactions and hydrogen bonding, effectively remodeling mature fibrils (Du et al., 2015).

17.3.2.2 Fibril Destabilization

According to Ono et al. (2004), certain compounds destabilize preformed aggregates, promoting their breakdown into less toxic species. By targeting specific fibril regions and disrupting stable interactions, these compounds induce conformational changes that drive disassembly. EGCG, a polyphenol found in green tea, has been shown to destabilize amyloid fibrils by binding to their surface, inducing conformational changes, and converting them into non-toxic, amorphous aggregates.

17.3.3 Modulation of Cellular Processes

17.3.3.1 Chaperone Activity

Some compounds function as chemical chaperones, aiding in correct protein folding and preventing misfolding. They stabilize native protein conformations or facilitate the refolding of misfolded proteins. Compounds enhancing molecular chaperones' ability to prevent aggregation can reduce misfolded protein accumulation (Kopito, 2000). While Reddy and Beal (2008) reported potential role of 4-phenylbutyrate (PBA), chemical chaperone alleviates ER stress and reduces Aβ aggregation in preclinical models.

17.3.3.2 Modulation of Cellular Pathways

Rubinsztein et al. (2012) highlighted that targeting cellular pathways, particularly the proteasome and autophagy, plays a pivotal role in enhancing the clearance of misfolded proteins, which is crucial in mitigating the toxic effects of protein aggregates associated with neurodegenerative diseases, including Alzheimer's, Parkinson's, and Huntington's diseases.

17.4 PROCEDURE AND CHALLENGES FOR CLINICAL DEVELOPMENT

The clinical development of non-peptidic anti-aggregate compounds typically follows a phased approach, similar to other drug development programs.

17.4.1 Preclinical Studies

This phase involves in vitro and in vivo studies to evaluate the efficacy and safety of the compound. In vitro studies often utilize cell-based models or purified Aβ peptides to assess the compound's ability to inhibit Aβ aggregation, disaggregate preformed fibrils, and reduce Aβ toxicity. Animal models of AD, such as transgenic mice expressing human Aβ, are commonly used in in vivo investigations. The key goals of these investigations include understanding the pharmacokinetics, pharmacodynamics, and possible adverse effects of the potent chemical compounds.

17.4.2 Clinical Trial

Preclinical studies are followed by clinical trials on volunteers.

Phase-I: In phase-I trial, pharmacokinetic parameters such as absorption, distribution, metabolism, and excretion of the dosed compounds are examined on the healthy volunteers to assess the safety and tolerability of the compound.

Phase-II: Unlike phase-I, phase-II trail is conducted on patients with AD to ascertain the efficacy and safety of the compound. Phase-II trials usually involve tests on a few patients with varying dosages and treatment durations.

Phase-III: Phase-III is an extensive investigation in a broader group of patients intended to validate the compound's safety and effectiveness. These types of trials are frequently carried out across many locations and usually involve hundreds to thousands of individuals.

17.4.3 Post-Clinical Surveillance

After satisfactory completion of phase-III studies, a drug may be submitted for regulatory approval followed by post-marketing monitoring to track any negative effects of the compound in the general population.

Recently, some non-peptide anti-aggregate compounds were subjected to preclinical research. In 2019, China granted provisional clearance for the treatment of mild to moderate Alzheimer's using sodium oligomannate (GV-971) synthesized from the isolated oligosaccharides of marine algae *Ecklonia kurome*. Its proposed mechanisms include Aβ disaggregation and modulation of neuroinflammation. However, due to skepticism regarding its clinical data, further trials were conducted. In 2022, phase-III trials in the United States were terminated early, and development was suspended. Ensuring the selectivity of these compounds is crucial to target disease-related aggregates while preserving the physiological functions of essential proteins. Moreover, optimizing pharmacokinetic properties to achieve effective concentrations in the central nervous system (CNS) remains a major hurdle. Ceperognastat (LY3372689), an O-GlcNAcase inhibitor developed by Eli Lilly, was investigated for its potential to reduce tau protein aggregation in early symptomatic AD. The PROSPECT-ALZ study, which concluded in July 2024, did not demonstrate clinical benefit, as both doses failed to meet the primary endpoint. Azeliragon is another potent small molecule antagonist to the receptor for advanced glycation end-product (RAGE). This drug was prescribed to be taken orally by the moderate Alzheimer's patients to decrease cognitive deterioration in people. However, due to ineffectiveness against the disease, phase-III STEADFAST experiment was canceled (Burstein et al., 2018). Later, potent application of azeliragon was diverted toward oncology focusing on understanding efficacy to treat glioblastoma in conjunction with temozolomide (Sepulveda et al., 2024). Several factors impact the establishment of a new drug. Target specificity, BBB penetration, translational efficacy, and regulatory barriers are among the major issues impeding the successful development of non-peptidic anti-aggregate medicines for AD. To mitigate these obstacles and improve patients' health, individualized strategies along with creative research and sophisticated modeling are necessary. Some of the major challenges are highlighted below:

a. It is still very difficult to get high specificity to pathogenic aggregates, like tau tangles or Aβ plaques, without interfering with vital protein functions. Especially, safety and effectiveness of the targeted chemical may be diminished by off-target interactions. Verubecestat and other BACE1 inhibitors, for instance, showed early promise but failed because of off-target toxicity and patient cognitive deterioration (Egan et al., 2018).
b. Due to inadequate BBB penetration, a large number of non-peptidic substances are unable to reach

appropriate therapeutic concentrations in the brain. Therefore, to ensure efficient distribution to the CNS, pharmacokinetic qualities must be optimized. Despite targeting γ-secretase, compounds like semagacestat failed clinical trials because of their poor CNS delivery and related side effects (Doody et al., 2013).

c. In many cases, improvements in preclinical models often do not translate into clinical advantages due to complex pathophysiology. Animal models may not imitate exact course as in humans which could result in differences between preclinical and clinical results (LaFerla et al., 2007).
d. In order to prevent neurotoxicity or other negative consequences, a thorough safety profiling is highly necessary for the long-term use of non-peptidic drugs that target CNS systems. Green et al. (2009) reported that Tarenflurbil was stopped in phase-III trials because of its lack of effectiveness and possible safety concerns.
e. The development of innovative AD treatments requires a significant financial cost as well as overcoming strict regulatory requirements. For instance, smaller biotech businesses are discouraged from pursuing innovative therapeutics because it is projected that bringing a new drug to market will cost more than $1 billion (Cummings et al., 2019).
f. The intricacy of the disease is frequently not addressed by single-agent treatments. As Imbimbo et al. (2005) point out, this emphasizes the necessity of tailored approaches, including combined treatments that target several pathways, viz. tau, Aβ, and neuroinflammation.
g. Trial failures can be caused by methodological problems, such as utilizing inadequate dosages or starting treatment too late in the course of the disease. Numerous studies concentrate on moderate-to-severe stages of AD, where neurodegeneration has progressed and the effectiveness of treatment is limited (Cummings, 2011) (Figure 17.3).
h. Most clinical trials target patients with moderate-to-severe AD, which may be too late for effective intervention. Early intervention, potentially during the preclinical or prodromal stages, is critical for slowing disease progression (Rasmussen & Langerman, 2019).
i. AD is a heterogeneous disease with diverse underlying mechanisms. A one-size-fits-all approach may not be effective. Identifying biomarkers to stratify patients and predict therapeutic responses remains a significant challenge (Armstrong, 2013).
j. Gaining regulatory approval for novel therapies is a complex and lengthy process. For instance, the Medicines and Healthcare Products Regulatory Agency (MHRA) and NICE evaluate therapies like *HMTM* based on robust clinical evidence, delaying market entry (Coombes, 2010).
k. The AD therapeutic landscape is highly competitive, with multiple compounds targeting similar pathways. For example, Roche's experimental antibody *trontinemab* demonstrates enhanced BBB penetration and efficacy, posing competition for other treatments (Meglio, 2023).
l. Developing affordable therapies is crucial for widespread adoption. Oral medications like *HMTM* may reduce healthcare costs compared to intravenous treatments. However, high pricing of innovative drugs remains a barrier to accessibility (Meglio, 2022).
m. Non-peptidic compounds may exhibit off-target toxicities, limiting their clinical application. For example, some γ-secretase inhibitors were associated with gastrointestinal and immune-related side effects, leading to trial discontinuation (Pagano et al., 2020).
n. Ensuring compounds effectively inhibit protein aggregation without causing adverse effects is crucial. Limited clinical benefits observed with drugs like *aducanumab* caution against overestimating the impact of aggregation inhibitors (Cummings, 2011).

17.5 NANOMATERIALS AS THERAPEUTIC TOOLS IN ALZHEIMER'S DISEASE THERAPY

Nanotechnology, through the development of nanomaterials, offers a promising avenue for addressing the challenges faced in AD therapy. Due to their unique physicochemical properties such as small size, high surface area, and functionalizability, nanomaterials are gradually being explored for targeted drug delivery, imaging, and theranostic applications in AD therapy. While nanoparticles (NPs), nanoliposomes, nanogels, dendrimers, and carbon-based structures have shown significant promise in facilitating the delivery of therapeutic agents across the BBB, which remains a substantial challenge in AD treatment (Faiyaz et al., 2021). Notably, in 2016, Quan et al. successfully designed a quantum dot nano-vehicle capable of targeting surface cells and playing an important role in AD detection. Other magnetic nanoparticles (MNPs) by Pedram et al. (2016) have demonstrated the capacity to traverse biological and physical barriers, including the formidable BBB. They have been effectively utilized in the diagnosis of amyloidosis by facilitating the imaging of the amyloid plaque or fibril deposit. The magnetic nanoparticles (FMNPs) have remarkable biocompatibility, excellent magnetic relaxivities for MRI applications, and high selectivity in targeting Aβ plaques within human Alzheimer transgenic mice (Nasr et al., 2018). They can also detect Aβ plaques and effectively traverse the BBB. For instance, as reported by Sánchez-López et al. (2018), lipid-based nanocarriers and polymeric NPs have demonstrated improved medication bioavailability, solubility, and stability while reducing

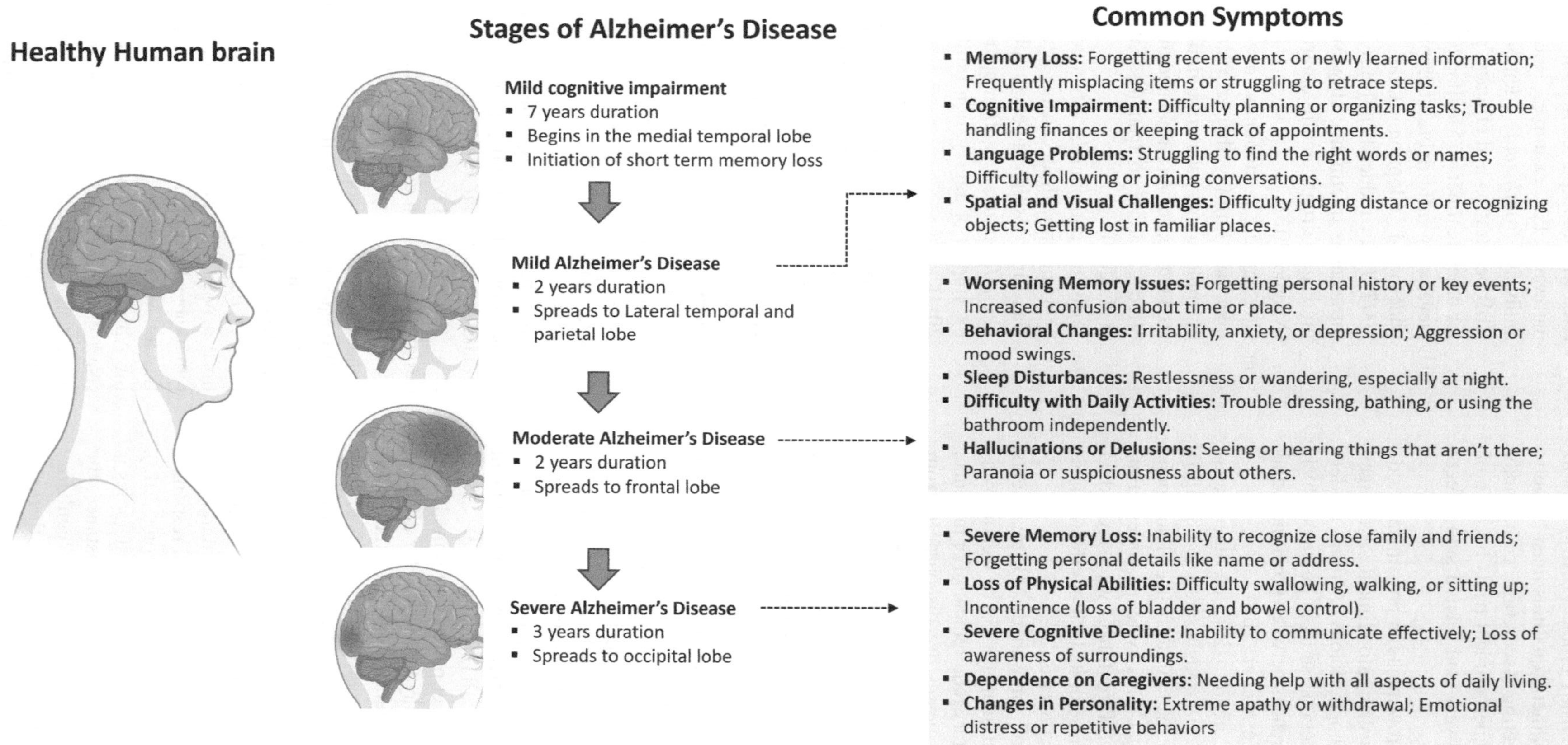

FIGURE 17.3 Diagram showing the different stages of Alzheimer's disease—mild, moderate, and severe—along with associated symptoms such as memory loss, confusion, difficulty with language, behavioral changes, and loss of physical functions.

systemic toxicity (Sanchez-Lopez et al., 2018). These compounds precisely localize the drug to the brain through surface modifications with ligands such as transferrin and lactoferrin; besides, targeted administration minimizes off-target effects. This is especially important in AD therapy as it can be difficult to achieve therapeutic concentrations at the pathological site. Modulating Aβ aggregation, a characteristic of AD, is another significant application for nanomaterials. Metal-based nanomaterials, such as gold and silver NPs, can disintegrate premade plaques and prevent Aβ fibrillation via surface plasmon resonance and catalytic activity (Skaat et al., 2013). Additionally, Aβ plaques have been imaged using graphene oxide and quantum dots, allowing for early detection and disease progression tracking (Huang & Isidoro, 2017). Nevertheless, these materials' potential neurotoxicity and long-term biocompatibility are still issues that need for thorough preclinical testing. The potential neurotoxicity and long-term biocompatibility of these materials are still an issue that too need for a thorough preclinical testing. Meanwhile, the antioxidant and anti-inflammatory qualities of nanomaterials have also been investigated; these qualities are essential for reducing oxidative stress and neuroinflammation, two major factors in the pathophysiology of AD. Cerium oxide NP is a good example that provides neuroprotection by exhibiting regenerating antioxidant activity and scavenging free radicals (Heckman et al., 2020). Furthermore, Moreira et al. (2023) reported that dendrimers coupled with anti-inflammatory medications have also proven effective in lowering pro-inflammatory cytokine production and glial activation in preclinical AD models.

17.6 THE ROLE OF ARTIFICIAL INTELLIGENCE AND MACHINE LEARNING IN ALZHEIMER'S DISEASE THERAPY

Artificial intelligence (AI) and machine learning (ML) indeed are among the most significant inventions of the 20th century. Application of AIML in various fields including health science is diverse; especially it has played a crucial role in Alzheimer's therapy accelerating the discovery and optimization of non-peptidic anti-aggregate compounds. It is evident that Alzheimer's diagnosis research involves analysis of extensive datasets, including molecular libraries, omics-based information, and reports from several data of pre-/post-clinical trials. AIML-driven system can integrate these datasets allowing researchers to uncover complex mechanisms of Aβ and tau aggregation that play central role in pathogenesis (Sarma & Chatterjee, 2024). Additionally, AIML can also enhance procedures for virtual screening by streamlining the selection of non-peptidic inhibitors. We can also identify specific compounds with high binding affinities for amyloidogenic targets using computational tools like molecular docking or dynamics simulations. It can optimize their chemical structures to improve specificity, bioavailability, and BBB permeability (AlMohimeed et al., 2023; Johnson et al., 2018). AI can facilitate predictive modeling by predicting the aggregation-inhibiting capability of candidate drugs through the analysis of molecular descriptors and experimental data. This capability speeds up the process of drug discovery by precisely focusing potent bioactive compounds for further studies (Johnson et al., 2018). AI can further aid in the field of designing precision medicine specifically targeted for a group of sub-populations based on their genetic makeup and medical history (Taherdoost & Ghofrani, 2024). Overall, AIML can speed up target-specific de-novo drug designing as well as repurposing.

17.7 FUTURE PROSPECTS AND CONCLUDING REMARK

Non-peptidic anti-aggregate compounds represent a promising frontier therapeutic avenue for Alzheimer's treatment, potentially surpassing traditional strategies. Their capacity to inhibit both Aβ and tau aggregation, combined with favorable pharmacological properties, makes them strong candidates for clinical development. However, further optimization of design, safety, and delivery is essential in unlocking their therapeutic potential. Collaborative analytical research among medicinal chemists, neuroscientists, and pharmacologists will be instrumental in driving progress in advancing this field. Moreover, integrating biomarker-based diagnosis into the AD model system could facilitate more precise and personalized medicine. The computational biology and high-throughput screening have become crucial for identifying and optimizing non-peptidic inhibitors. For this, by utilizing structure-based drug design, researchers are investigating key molecular interactions among these inhibitors as well as amyloidogenic proteins, paving the way for the development of highly selective as well as effective therapeutic agents. As discussed, advanced tools like molecular docking techniques, dynamic stimulations, and ML with AI-enabled scientists were used to design specific changes that focused on improvement of drug specificity and endurance. High-throughput research on polyphenolic derivatives such as resveratrol, rutin, or curcumin needs further refinement. Incorporating nanotechnology into non-peptidic anti-aggregate therapies can provide better solutions to key challenges, notably BBB permeability and the rapid systematic degradation. Nanocarrier systems, encompassing dendrimers, liposomes, and polymeric NPs, aid in encapsulating these compounds and enhance target-specific drug delivery with optimum dose timing for effective therapeutic impact. Additionally, these advanced technologies are enabling diagnosis, seamlessly merging treatment with real-time disease monitoring, and evaluating how well the body is clearing harmful substances. For successful clinical applications, it is essential to improve how these non-peptidic inhibitors are absorbed and processed by the body (pharmacokinetics) and ensure safety profiles. Further research on personalized medicine which are guided by the help of biomarkers, e.g., cerebrospinal fluid $A\beta_{42}$ and tau levels, can lead to more precision and effective therapeutic. However, expanding the structural diversity among non-peptidic compounds by exploring novel molecular frameworks such as chaperone-mimicking molecules and metal-chelating agents holds significant potential. These innovative approaches offer two prolonged therapeutic benefits: (a) disrupting harmful protein interactions and (b) stabilizing non-toxic forms of amyloid-forming proteins. Overall, given the increasing global prevalence of this disease, the development of non-peptidic anti-aggregate therapies offers hope for affected individuals and their families and holds the potential to significantly impact the treatment of neurodegenerative diseases and public health.

AUTHOR CONTRIBUTION

Conceptualization: PB, AA, and BK; Data acquisition and manuscript writing: AA and BK; Revision and supervision: PB; all authors read and approved the final manuscript.

CONFLICT OF INTEREST

No potential conflict of interest was reported by the authors.

ACKNOWLEDGMENTS

The authors express their sincere gratitude to the Director of CSIR-NEIST, Jorhat, Assam, for his constant support and for providing the necessary resources. AA and BK would like to acknowledge Department of Biotechnology, Govt. of India for DBT-Research Associateship under DBT-RA program.

REFERENCES

Abdallah, A. E. (2024). Review on anti-Alzheimer drug development: Approaches, challenges and perspectives. *RSC Advances*, *14*(16), 11057–11088.

Abelein, A., & Johansson, J. (2023). Amyloid inhibition by molecular chaperones in vitro can be translated to Alzheimer's pathology in vivo. *RSC Medicinal Chemistry*, *14*(5), 848–857.

Aggidis, A., Chatterjee, S., Townsend, D., Fullwood, N. J., Ortega, E. R., Tarutani, A., Hasegawa, M., Lucas, H., Mudher, A., & Allsop, D. (2021). Peptide-based inhibitors of tau aggregation as a potential therapeutic for Alzheimer's disease and other tauopathies. *bioRxiv*, 2021–2006.

Aguzzi, C., Perinelli, D. R., Cespi, M., Zeppa, L., Mazzara, E., Maggi, F., Petrelli, R., Bonacucina, G., & Nabissi, M. (2023). Encapsulation of hemp (*Cannabis sativa* L.) essential oils into nanoemulsions for potential therapeutic applications: Assessment of cytotoxicological profiles. *Molecules*, *28*(18), 6479.

Ahmed, A., Ghallab, E. H., Shehata, M., Zekri, A. R. N., & Ahmed, O. S. (2020). Impact of nano-conjugate on Drosophila for early diagnosis of Alzheimer's disease. *Nanotechnology*, *31*(36), 365102.

Akhondzadeh, S., Noroozian, M., Mohammadi, M., Ohadinia, S., Jamshidi, A. H., & Khani, M. (2003). *Salvia officinalis* extract in the treatment of patients with mild to moderate Alzheimer's disease: A double blind, randomized and placebo-controlled trial. *Journal of Clinical Pharmacy and Therapeutics*, *28*(1), 53–59.

AlMohimeed, A., Saad, R. M., Mostafa, S., El-Rashidy, N. M., Farrag, S., Gaballah, A., Abd Elaziz, M., El-Sappagh, S., & Saleh, H. (2023). Explainable artificial intelligence of multi-level stacking ensemble for detection of Alzheimer's disease based on particle swarm optimization and the sub-scores of cognitive biomarkers. *IEEE Access*, *11*, 123173–123193.

Arai, T., Sasaki, D., Araya, T., Sato, T., Sohma, Y., & Kanai, M. (2014). A cyclic KLVFF-derived peptide aggregation inhibitor induces the formation of less-toxic off-pathway amyloid-β oligomers. *ChemBioChem*, *15*(17), 2577–2583.

Armstrong, R. A. (2013). What causes Alzheimer's disease? *Folia Neuropathologica*, *51*(3), 169–188.

Barisano, G., Montagne, A., Kisler, K., Schneider, J. A., Wardlaw, J. M., & Zlokovic, B. V. (2022). Blood–brain barrier link to human cognitive impairment and Alzheimer's disease. *Nature Cardiovascular Research*, *1*(2), 108–115.

Baur, J. A., & Sinclair, D. A. (2006). Therapeutic potential of resveratrol: The in vivo evidence. *Nature Reviews Drug Discovery*, *5*(6), 493–506.

Burstein, A. H., Sabbagh, M., Andrews, R., Valcarce, C., Dunn, I., & Altstiel, L. (2018). Development of Azeliragon, an oral small molecule antagonist of the receptor for advanced glycation endproducts, for the potential slowing of loss of cognition in mild Alzheimer's disease. *The Journal of Prevention of Alzheimer's Disease*, *5*, 149–154.

Cerrato, C. P., & Langel, Ü (2022). An update on cell-penetrating peptides with intracellular organelle targeting. *Expert Opinion on Drug Delivery*, *19*(2), 133–146.

Coombes, R. (2010). NICE and value based pricing: What do we know? *BMJ*, *341*.

Cummings, J. L. (2011). Alzheimer's disease clinical trials: Changing the paradigm. *Current Psychiatry Reports*, *13*, 437–442.

Cummings, J., Lee, G., Ritter, A., Sabbagh, M., & Zhong, K. (2019). Alzheimer's disease drug development pipeline: 2019. *Alzheimer's & Dementia: Translational Research & Clinical Interventions*, *5*, 272–293.

Cummings, J., Zhou, Y., Lee, G., Zhong, K., Fonseca, J., & Cheng, F. (2023). Alzheimer's disease drug development pipeline: 2023. *Alzheimer's & Dementia: Translational Research & Clinical Interventions*, *9*(2), e12385.

Díaz-Caballero, M., Navarro, S., Nuez-Martínez, M., Peccati, F., Rodríguez-Santiago, L., Sodupe, M., Teixidor, F., & Ventura, S. (2020). pH-responsive self-assembly of amyloid fibrils for dual hydrolase-oxidase reactions. *ACS Catalysis*, *11*(2), 595–607.

Dong, M., Zhao, W., Hu, D., Ai, H., & Kang, B. (2017). N-terminus binding preference for either tanshinone or analogue in both inhibition of amyloid aggregation and disaggregation of preformed amyloid fibrils-toward introducing a kind of novel anti-Alzheimer compounds. *ACS Chemical Neuroscience*, *8*(7), 1577–1588.

Doody, R. S., Raman, R., Farlow, M., Iwatsubo, T., Vellas, B., Joffe, S., Kieburtz, K., He, F., Sun, X., Thomas, R. G., & Aisen, P. S. (2013). A phase 3 trial of semagacestat for treatment of Alzheimer's disease. *New England Journal of Medicine*, *369*(4), 341–350.

Du, W. J., Guo, J. J., Gao, M. T., Hu, S. Q., Dong, X. Y., Han, Y. F., Liu, F. F., Jiang, S., & Sun, Y. (2015). Brazilin inhibits amyloid β-protein fibrillogenesis, remodels amyloid fibrils and reduces amyloid cytotoxicity. *Scientific Reports*, *5*(1), 7992.

Duan, Z., Shi, L., He, Z. N. T., Kuang, C., Han, T., & Yang, Q. (2023). The protective effect of IDO1 inhibition in Aβ-treated neurons and APP/PS1 mice. *American Journal of Alzheimer's Disease & Other Dementias®*, *38*, 15333175231214861.

Egan, M. F., Kost, J., Tariot, P. N., Aisen, P. S., Cummings, J. L., Vellas, B., Sur, C., Mukai, Y., Voss, T., Furtek, C., & Mahoney, E. (2018). Randomized trial of verubecestat for mild-to-moderate Alzheimer's disease. *New England Journal of Medicine*, *378*(18), 1691–1703.

Eisenberg, D. S., & Sawaya, M. R. (2016). Implications for Alzheimer's disease of an atomic resolution structure of amyloid-β (1–42) fibrils. *Proceedings of the National Academy of Sciences*, *113*(34), 9398–9400.

Estrada, L. D., & Soto, C. (2006). Inhibition of protein misfolding and aggregation by small rationally-designed peptides. *Current Pharmaceutical Design*, *12*(20), 2557–2567.

Faiyaz, M., Ganayee, M. A., Akhtar, S., Krishnan, S., Flora, B., Dogra, D., Jha, N. K., Chellappan, D. K., Negi, P., Dua, K., & Kesari, K. K. (2021). Nanomaterials in Alzheimer's disease treatment: A comprehensive review. *Frontiers in Bioscience-Landmark*, *26*(10), 851–865.

Fan, H. M., Gu, R. X., Wang, Y. J., Pi, Y. L., Zhang, Y. H., Xu, Q., & Wei, D. Q. (2015). Destabilization of Alzheimer's Aβ42 protofibrils with a novel drug candidate wgx-50 by molecular dynamics simulations. *The Journal of Physical Chemistry B*, *119*(34), 11196–11202.

Fan, Q., Liu, Y., Wang, X., Zhang, Z., Fu, Y., Liu, L., Wang, P., Ma, H., Ma, H., Seeram, N. P., & Zheng, J. (2020). Ginnalin A inhibits aggregation, reverses fibrillogenesis, and alleviates cytotoxicity of amyloid β (1–42). *ACS Chemical Neuroscience*, *11*(4), 638–647.

Fang, C., Hernandez, P., Liow, K., Damiano, E., Zetterberg, H., Blennow, K., Feng, D., Chen, M., & Maccecchini, M. (2023). Buntanetap, a novel translational inhibitor of multiple neurotoxic proteins, proves to be safe and promising in both Alzheimer's and Parkinson's patients. *The Journal of Prevention of Alzheimer's Disease*, *10*(1), 25–33.

Fernandes, L., Cardim-Pires, T. R., Foguel, D., & Palhano, F. L. (2021). Green tea polyphenol epigallocatechin-gallate in amyloid aggregation and neurodegenerative diseases. *Frontiers in Neuroscience*, *15*, 718188.

Gervais, F., Paquette, J., Morissette, C., Krzywkowski, P., Yu, M., Azzi, M., Lacombe, D., Kong, X., Aman, A., Laurin, J., & Szarek, W. A. (2007). Targeting soluble Aβ peptide with Tramiprosate for the treatment of brain amyloidosis. *Neurobiology of Aging*, *28*(4), 537–547.

Gomez-Sequeda, N., Jimenez-Del-Rio, M., & Velez-Pardo, C. (2024). Combination of tramiprosate, curcumin, and SP600125 reduces the neuropathological phenotype in familial Alzheimer disease PSEN1 I416T cholinergic-like neurons. *International Journal of Molecular Sciences*, *25*(9), 4925.

Green, R. C., Schneider, L. S., Amato, D. A., Beelen, A. P., Wilcock, G., Swabb, E. A., & Zavitz, K. H., & Tarenflurbil Phase 3 Study Group. (2009). Effect of tarenflurbil on cognitive decline and activities of daily living in patients with mild Alzheimer disease: A randomized controlled trial. *JAMA*, *302*(23), 2557–2564.

Habtemariam, S. (2016). Rutin as a natural therapy for Alzheimer's disease: Insights into its mechanisms of action. *Current Medicinal Chemistry*, *23*(9), 860–873.

Haleckova, A., Benek, O., Zemanová, L., Dolezal, R., & Musilek, K. (2022). Small-molecule inhibitors of cyclophilin D as potential therapeutics in mitochondria-related diseases. *Medicinal Research Reviews*, *42*(5), 1822–1855.

Hart, N. J., Koronyo, Y., Black, K. L., & Koronyo-Hamaoui, M. (2016). Ocular indicators of Alzheimer's: Exploring disease in the retina. *Acta Neuropathologica*, *132*, 767–787.

Heckman, K. L., Estevez, A. Y., DeCoteau, W., Vangellow, S., Ribeiro, S., Chiarenzelli, J., Hays-Erlichman, B., & Erlichman, J. S. (2020). Variable in vivo and in vitro biological effects of cerium oxide nanoparticle formulations. *Frontiers in Pharmacology*, *10*, 1599.

Huang, C. C., & Isidoro, C. (2017). Raman spectrometric detection methods for early and non-invasive diagnosis of Alzheimer's disease. *Journal of Alzheimer's Disease*, *57*(4), 1145–1156.

Huang, L. K., Kuan, Y. C., Lin, H. W., & Hu, C. J. (2023). Clinical trials of new drugs for Alzheimer disease: A 2020–2023 update. *Journal of Biomedical Science*, *30*(1), 83.

Imbimbo, B. P., Lombard, J., & Pomara, N. (2005). Pathophysiology of Alzheimer's disease. *Neuroimaging Clinics*, *15*(4), 727–753.

Johnson, K. W., Torres Soto, J., Glicksberg, B. S., Shameer, K., Miotto, R., Ali, M., Ashley, E., & Dudley, J. T. (2018). Artificial intelligence in cardiology. *Journal of the American College of Cardiology*, *71*(23), 2668–2679.

Kopito, R. R. (2000). Aggresomes, inclusion bodies and protein aggregation. *Trends in Cell Biology*, *10*(12), 524–530.

Kucwaj-Brysz, K., Baltrukevich, H., Czarnota, K., & Handzlik, J. (2021). Chemical update on the potential for serotonin 5-HT6 and 5-HT7 receptor agents in the treatment of Alzheimer's disease. *Bioorganic & Medicinal Chemistry Letters*, *49*, 128275.

LaFerla, F. M., Green, K. N., & Oddo, S. (2007). Intracellular amyloid-β in Alzheimer's disease. *Nature Reviews Neuroscience*, *8*(7), 499–509.

Longo, F. M., Yang, T., Knowles, J. K., Xie, Y., Moore, L. A., & Massa, S. M. (2007). Small molecule neurotrophin receptor ligands: Novel strategies for targeting Alzheimer's disease mechanisms. *Current Alzheimer Research*, *4*(5), 503–506.

Matera, C., Camarero, N., Masferrer, C., Coll, E., Vitiello, S., Rus, A., Segovia, C., Rubio, C., Suarez-Cabrera, C., Gomila, A., & Libergoli, M. (2020). *Phototrexate: A novel drug candidate for cancer and psoriasis.*

Meglio, M. (2022). Hydromethylthionine Mesylate's place as a potential tau aggregator of Alzheimer disease. *Neurology Live*, NA–NA.

Meglio, M. (2023). Alzheimer disease pipeline update: Inside look at promising agents. *Neurology Live*, NA–NA.

Mehla, J., Gupta, P., Pahuja, M., Diwan, D., & Diksha, D. (2020). Indian Medicinal herbs and formulations for Alzheimer's disease, from traditional knowledge to scientific assessment. *Brain Sciences*, *10*(12), 964.

Mekala, S., Nelson, G., & Li, Y. M. (2020). Recent developments of small molecule γ-secretase modulators for Alzheimer's disease. *RSC Medicinal Chemistry*, *11*(9), 1003–1022.

Menon, J. S., Krishnakumar, K., Dineshkumar, B., John, A., Paul, D., & Cherian, J. (2013). Herbs for Alzheimer disease-a review. *International Research Journal of Plant Science*, *3*, 54–56.

Moreira, D. A., Santos, S. D., Leiro, V., & Pêgo, A. P. (2023). Dendrimers and derivatives as multifunctional nanotherapeutics for Alzheimer's disease. *Pharmaceutics*, *15*(4), 1054.

Murakami, K. (2023). High-throughput searches for natural products as aggregation modulators of amyloidogenic proteins. *New Tide of Natural Product Chemistry*, 313–332.

Muscat, S., Pallante, L., Stojceski, F., Danani, A., Grasso, G., & Deriu, M. A. (2020). The impact of natural compounds on S-shaped Aβ42 fibril: From molecular docking to biophysical characterization. *International Journal of Molecular Sciences*, *21*(6), 2017.

Naiki, H., Hasegawa, K., Ono, K., & Yamada, M. (2010). Search for antiamyloidogenic compounds based on a nucleation-dependent polymerization model. *Yakugaku Zasshi: Journal of the Pharmaceutical Society of Japan*, *130*(4), 503–509.

Nasr, S. H., Kouyoumdjian, H., Mallett, C., Ramadan, S., Zhu, D. C., Shapiro, E. M., & Huang, X. (2018). Detection of β-Amyloid by sialic acid coated bovine serum albumin magnetic nanoparticles in a mouse model of Alzheimer's disease. *Small*, *14*(3), 1701828.

Ono, K., Hasegawa, K., Naiki, H., & Yamada, M. (2004). Anti-amyloidogenic activity of tannic acid and its activity to destabilize Alzheimer's β-amyloid fibrils in vitro. *Biochimica et Biophysica Acta (BBA)-Molecular Basis of Disease*, *1690*(3), 193–202.

Ono, K., Li, L., Takamura, Y., Yoshiike, Y., Zhu, L., Han, F., Mao, X., Ikeda, T., Takasaki, J. I., Nishijo, H., & Takashima, A. (2012). Phenolic compounds prevent amyloid β-protein oligomerization and synaptic dysfunction by site-specific binding. *Journal of Biological Chemistry*, *287*(18), 14631–14643.

Pagano, K., Tomaselli, S., Molinari, H., & Ragona, L. (2020). Natural compounds as inhibitors of Aβ peptide aggregation: Chemical requirements and molecular mechanisms. *Frontiers in Neuroscience*, *14*, 619667.

Paranjape, S. R., Riley, A. P., Somoza, A. D., Oakley, C. E., Wang, C. C., Prisinzano, T. E., Oakley, B. R., & Gamblin, T. C. (2015). Azaphilones inhibit tau aggregation and dissolve tau aggregates in vitro. *ACS Chemical Neuroscience*, *6*(5), 751–760.

Pedram, M. Z., Shamloo, A., Alasty, A., & Ghafar-Zadeh, E. (2016). Optimal magnetic field for crossing super-paramagnetic nanoparticles through the brain blood barrier: A computational approach. *Biosensors*, *6*(2), 25.

Pickhardt, M., Gazova, Z., von Bergen, M., Khlistunova, I., Wang, Y., Hascher, A., Mandelkow, E. M., Biernat, J., & Mandelkow, E. (2005). Anthraquinones inhibit tau aggregation and dissolve Alzheimer's paired helical filaments in vitro and in cells. *Journal of Biological Chemistry*, *280*(5), 3628–3635.

Quan, L., Wu, J., Lane, L. A., Wang, J., Lu, Q., Gu, Z., & Wang, Y. (2016). Enhanced detection specificity and sensitivity of Alzheimer's disease using amyloid-β-targeted quantum dots. *Bioconjugate Chemistry*, *27*(3), 809–814.

Rasmussen, J., & Langerman, H. (2019). Alzheimer's Disease – Why We Need Early Diagnosis. *Degenerative Neurological and Neuromuscular Disease*, *9*, 123–130.

Reddy, P. H., & Beal, M. F. (2008). Amyloid beta, mitochondrial dysfunction and synaptic damage: Implications for cognitive decline in aging and Alzheimer's disease. *Trends in Molecular Medicine*, *14*(2), 45–53.

Rubinsztein, D. C., Codogno, P., & Levine, B. (2012). Autophagy modulation as a potential therapeutic target for diverse diseases. *Nature Reviews Drug Discovery*, *11*(9), 709–730.

Sánchez-López, E., Ettcheto, M., Egea, M. A., Espina, M., Cano, A., Calpena, A. C., Camins, A., Carmona, N., Silva, A. M., Souto, E. B., & García, M. L. (2018). Memantine loaded PLGA PEGylated nanoparticles for Alzheimer's disease: In vitro and in vivo characterization. *Journal of Nanobiotechnology*, *16*, 1–16.

Sarma, M., & Chatterjee, S. (2024). Etiology of Late-Onset Alzheimer's disease, biomarker efficacy, and the role of machine learning in stage diagnosis. *Diagnostics*, *14*(23), 2640.

Sehra, N., Parmar, R., & Jain, R., (2025). Peptide-based amyloid-beta aggregation inhibitors. *RSC Medicinal Chemistry*, *16*(3), 1083–1104.

Selkoe, D. J., & Hardy, J. (2016). The amyloid hypothesis of Alzheimer's disease at 25 years. *EMBO Molecular Medicine*, *8*(6), 595–608.

Sepulveda, J. M., Valiente, M., Vaz-Salgado, M., Pineda, E., Martínez-García, M., Ruiz Vico, M., Velilla, G., Mazariegos, M. A., Valduvieco, I., Castro-Henriques, M., & Cano, L. (2024). *Azeliragon, a RAGE inhibitor, in combination with temozolomide and radiotherapy in patients with newly diagnosed glioblastoma: Phase Ib/II CAN-201 NDG trial design.*

Sharma, A., Singh, A., Debnath, R., Gupta, G. D., & Sharma, K., (2024). Role of O-GlcNAcylation in Alzheimer's disease: Insights and perspectives. *European Journal of Medicinal Chemistry Reports*, *12*, 100195.

Sighencea, M. G., Popescu, R., & Trifu, S. C. (2024). From fundamentals to innovation in Alzheimer's disease: Molecular findings and revolutionary therapies. *International Journal of Molecular Sciences*, *25*(22), 12311.

Silva, M. C., Nandi, G., Donovan, K. A., Cai, Q., Berry, B. C., Nowak, R. P., Fischer, E. S., Gray, N. S., Ferguson, F. M., & Haggarty, S. J. (2022). Discovery and optimization of tau targeted protein degraders enabled by patient induced pluripotent stem cells-derived neuronal models of tauopathy. *Frontiers in Cellular Neuroscience*, *16*, 801179.

Singh, N., Bhalla, M., de Jager, P., & Gilca, M. (2011). An overview on ashwagandha: A Rasayana (Rejuvenator) of Ayurveda. *African Journal of Traditional, Complementary and Alternative Medicines*, *8*(5S).

Skaat, H., Corem-Slakmon, E., Grinberg, I., Last, D., Goez, D., Mardor, Y., & Margel, S. (2013). Antibody-conjugated, dual-modal, near-infrared fluorescent iron oxide nanoparticles for antiamyloidgenic activity and specific detection of amyloid-β fibrils. *International Journal of Nanomedicine*, *8*, 4063–4076.

Sowndhararajan, K., Deepa, P., Kim, M., Park, S. J., & Kim, S. (2017). Baicalein as a potent neuroprotective agent: A review. *Biomedicine & Pharmacotherapy*, *95*, 1021–1032.

Taherdoost, H., & Ghofrani, A. (2024). AI and the evolution of personalized medicine in pharmacogenomics. *Intelligent Pharmacy*.

Teplow, D. B., Yang, M., Roychaudhuri, R., Pang, E., Huynh, T. P., Chen, M. S., & Beroukhim, S. (2012). The amyloid beta-protein and Alzheimer's disease. In *Alzheimer's disease: Targets for new clinical diagnostic and therapeutic strategies* (pp. 1–47).

Veerendra Kumar, M. H., & Gupta, Y. K. (2003). Effect of *Centella asiatica* on cognition and oxidative stress in an intracerebroventricular streptozotocin model of Alzheimer's disease in rats. *Clinical and Experimental Pharmacology and Physiology*, *30*(5–6), 336–342.

Wang, C., Shao, S., Li, N., Zhang, Z., Zhang, H., & Liu, B. (2023). Advances in Alzheimer's disease-associated Aβ therapy based on peptide. *International Journal of Molecular Sciences*, *24*(17), 13110.

Wiglenda, T., Groenke, N., Hoffmann, W., Manz, C., Diez, L., Buntru, A., Brusendorf, L., Neuendorf, N., Schnoegl, S., Haenig, C., & Schmieder, P. (2020). Sclerotiorin stabilizes the assembly of nonfibrillar Abeta42 oligomers with low toxicity, seeding activity, and beta-sheet content. *Journal of Molecular Biology*, *432*(7), 2080–2098.

Wischik, C. M., Harrington, C. R., & Storey, J. M. (2014). Tau-aggregation inhibitor therapy for Alzheimer's disease. *Biochemical Pharmacology*, *88*(4), 529–539.

Yang, F., Lim, G. P., Begum, A. N., Ubeda, O. J., Simmons, M. R., Ambegaokar, S. S., Chen, P. P., Kayed, R., Glabe, C. G., Frautschy, S. A., & Cole, G. M. (2005). Curcumin inhibits formation of amyloid β oligomers and fibrils, binds plaques, and reduces amyloid in vivo. *Journal of Biological Chemistry*, *280*(7), 5892–5901.

Yiannopoulou, K. G., & Papageorgiou, S. G. (2020). Current and future treatments in Alzheimer disease: An update. *Journal of Central Nervous System Disease*, *12*, 1179573520907397.

Youn, K., Ho, C. T., & Jun, M. (2022). Multifaceted neuroprotective effects of (–)-epigallocatechin-3-gallate (EGCG) in Alzheimer's disease: An overview of pre-clinical studies focused on β-amyloid peptide. *Food Science and Human Wellness*, *11*(3), 483–493.

Zangara, A. (2003). The psychopharmacology of huperzine A: An alkaloid with cognitive enhancing and neuroprotective properties of interest in the treatment of Alzheimer's disease. *Pharmacology Biochemistry and Behavior*, *75*(3), 675–686.

Zhang, X., Wu, Z., & Weng, P. (2014). Antioxidant and hepatoprotective effect of (–)-epigallocatechin 3-O-(3-O-methyl) gallate (EGCG3″Me) from Chinese oolong tea. *Journal of Agricultural and Food Chemistry*, *62*(41), 10046–10054.

18 Translational Potentials of Nanotechnology-Based Interventions for Alzheimer's Disease

Garima Varshney, Giriraj T Kulkarni, Mandeep Kumar Arora, Satish Kumar Gupta, Rajendra Awasthi, and Bhupesh Sharma

18.1 DIAGNOSTIC STRATEGIES AND THERAPEUTIC ADVANCES IN ALZHEIMER'S DISEASE

Diagnosis of Alzheimer's disease (AD) involves clinical and cognitive assessments (Mini-Mental State Examination, Montreal Cognitive Assessment), biomarker analysis (Cerebrospinal Fluid (CSF) levels of Aβ42, tau, and blood-based markers like neurofilament light chain), neuroimaging (positron emission tomography for amyloid and tau detection, magnetic resonance imaging for hippocampal atrophy), and genetic testing (apolipoprotein E ε4, β-amyloid precursor protein, PSEN1/2 mutations). Current pharmacological treatments include N-methyl-D-aspartate (NMDA) antagonist memantine and cholinesterase inhibitors (rivastigmine, donepezil) for symptom management, while anti-amyloid monoclonal antibodies (lecanemab, aducanumab) aim to slow disease progression. Emerging therapies target tau aggregation, neuroinflammation, and synaptic loss through gene therapy, RNA-based approaches, and nanotechnology-driven drug delivery systems, enhancing blood-brain barrier (BBB) penetration. Non-pharmacological strategies, like physical exercise, cognitive rehabilitation, Mediterranean diet, and sleep regulation, contribute to slowing cognitive decline. Despite advancements, no curative treatment exists, emphasizing the need for early diagnosis and novel interventions.

18.1.1 The Need for Innovative Approaches to Address AD Diagnosis and Treatment Challenges

The accumulation of amyloid-beta (Aβ) plaques and neurofibrillary tangles (NFT) represents the hallmark pathology of AD. Despite decades of research, early and accurate diagnosis remains a major challenge due to overlapping clinical symptoms with other neuropsychiatric disorders and the absence of definitive biomarkers in the early stages. This limitation reduces the effectiveness of conventional diagnostic methods such as cerebrospinal fluid analysis, cognitive assessments, and neuroimaging (Kas et al., 2019; Lockhart & Walther, 2009; Trojanowski et al., 2010). Furthermore, standard therapeutic options, including NMDA receptor antagonists and cholinesterase inhibitors, primarily provide symptomatic relief without addressing the underlying disease mechanisms (Mauricio et al., 2019). The integration of artificial intelligence-driven analytics with novel molecular biomarkers, alongside large-scale initiatives such as the "European Prevention of Alzheimer's Dementia Registry," underscores the urgent need for innovative diagnostic and therapeutic strategies aimed at enabling early intervention and improving patient outcomes (Kale et al., 2024; Vermunt et al., 2018).

18.1.2 Nanoparticle-Based Diagnostic Approaches for AD

Nanotechnology presents significant potential to address the limitations of existing diagnostic methods for AD. Recent advancements have facilitated the creation of nanoparticle-based platforms that offer improved sensitivity and specificity for the synthesis of vital AD biomarkers. Silica nanoparticles are gaining considerable interest due to facile surface functionalization and tunable pore properties, chemical stability, high payload, biocompatibility, ease of production, controlled drug release, multimodal imaging potential, and prolonged circulation time (Singh et al., 2024). Silica nanoparticles, carbon nanotubes, quantum dots, and gold nanoparticles, when functionalized with specific aptamers or ligands, have been effectively utilized for the detection of tau proteins and Aβ in biological fluids (Chen et al., 2024; Elsori et al., 2023). Such nanoscale platforms not only offer superior imaging contrast when integrated with techniques like positron emission tomography and magnetic resonance imaging but also allow for the simultaneous detection of NFT and amyloid peptides, thus providing a comprehensive molecular fingerprint of AD pathology (Kong et al., 2023).

In addition, the integration of AI-driven data analytics with these nanodiagnostic tools is anticipated to refine risk stratification and improve disease modeling, thereby enabling real-time monitoring of disease progression and personalized intervention strategies (Babu et al., 2024).

DOI: 10.1201/9781003602477-18

18.1.3 Benefits of Nanotechnology-Based Interventions in Neurodegenerative Disorders

Beyond diagnostics, nanotechnology holds significant promise in the realm of therapeutics. One of the principal challenges in treating AD is the effective delivery of therapeutic agents across the BBB, a hurdle that conventional small-molecule drugs often cannot overcome. Nanocarriers such as polymeric nanoparticles, liposomes, dendrimers, and lipid-based systems have been engineered to facilitate targeted drug delivery, ensuring enhanced bioavailability and reducing systemic side effects (Dhariwal et al., 2025; Kim & Kim, 2024). Nanotechnology-based therapeutic strategies also offer multifunctional capabilities. Engineered nanoparticles can be designed to mitigate oxidative stress, modulate neuroinflammatory responses, and promote neuronal survival. Furthermore, emerging nanotherapeutic approaches are targeting pathological processes associated with both tau protein aggregation and Aβ deposition. Such multifunctional nanocarriers enable the co-delivery of neuroprotective agents, anti-amyloid compounds, and gene therapies, thus offering a comprehensive strategy to counteract AD pathology (Dhariwal et al., 2025; Garcés et al., 2024).

18.2 BBB AND DRUG TRANSPORT TO THE BRAIN

BBB serves as a selective permeability barrier that safeguards the brain from potentially harmful substances, while facilitating the passage of essential nutrients. It has a tight junction composed of different types of cells (Hawkins et al., 2005; Pardridge, 2012; Singh et al., 2021; Stamatovic et al., 2008). The constituents of BBB include endothelial cells, basement membrane, astrocyte feet, and pericytes. The functional components, cellular structure of BBB, are illustrated in Figure 18.1.

BBB core is constituted by endothelial cells that line the cerebral blood capillaries. The cells are densely aggregated, forming tight connections that inhibit the passage of most substances between them. The tight junctions consist of proteins such as claudins and occludins. Basement membrane is situated beneath the endothelial cells, which is a thin, fibrous layer that provides structural support. It comprises extracellular matrix proteins such as laminin, collagen, and fibronectin. Pericytes are contractile cells that cover endothelial cells. Pericytes play a role in the repair and remodeling of BBB. They regulate blood circulation within the capillaries and contribute to the stability and maintenance of the BBB. Astrocyte feet are a type of glial cell in the

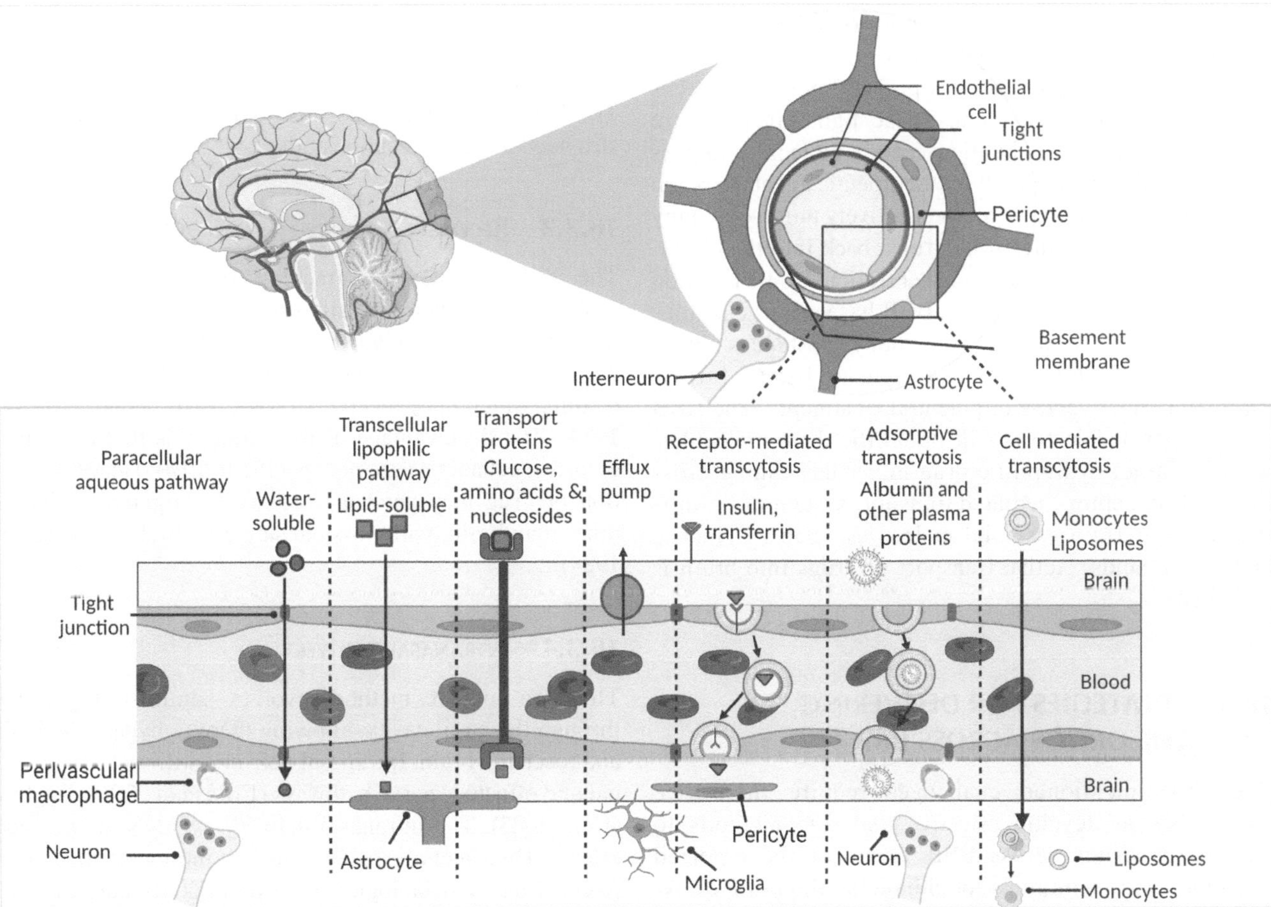

FIGURE 18.1 Mechanisms of nanoparticle transportation across the BBB and cellular structure of BBB (tight junctions, endothelial cells, and astrocytes) (with permission from Raghav et al., 2023). (Copyright Elsevier.)

brain. Their end-foot processes surround the blood vessels and contribute to the integrity of BBB. They release signaling chemicals that assist in preserving the tight junctions among endothelial cells.

BBB has a multi-layered structure, which is highly selective and ensures that the brain remains protected from potentially harmful substances while receiving the nutrients and molecules it needs to function properly (Hawkins et al., 2005; Stamatovic et al., 2008). The BBB has specialized transport systems for exchange of essential nutrients and molecules between blood and brain cells. The transport systems found in BBB for exchange of essential nutrients include carrier-mediated transport, receptor-mediated transport, and efflux transporters. Carrier-mediated transport consists of integral membrane proteins, which transport glucose, amino acids, and other small molecules, in the form of facilitated diffusion. Receptor-mediated transport is an active transport mechanism involving protein receptors that bind to specific molecules such as transferrin or insulin, thereby facilitating their entry into the brain. Efflux transporters are the proteins such as P-glycoprotein that actively pump out toxins and waste products from the brain back into the bloodstream. They effectively inhibit drug absorption into the brain.

The BBB presents significant obstacles in the delivery of therapeutic agents to the brain. These obstacles are tight junctions, efflux transporters, enzymatic degradation, size and polarity, receptor-mediated transport (Banks, 2009). Tight junctions prevent most substances from passing between them. These tight junctions are highly selective and only allow certain molecules to pass through, making it difficult for therapeutic agents to reach the brain. The efflux transporters actively pump out many foreign substances, including drugs, back into the bloodstream. This reduces the concentration of therapeutic agents that can reach the brain. BBB has various enzymes that can metabolize and degrade many substances before they can enter the brain. This enzymatic barrier further limits the effectiveness of potential treatments. The BBB is more permeable to small, lipophilic drug molecules, whereas large, hydrophilic drug molecules cannot cross the BBB. Receptor-mediated transport system of brain is specific to certain essential molecules, such as glucose and amino acids. Active transport of drugs into brain is a challenge.

18.3 STRATEGIES FOR DELIVERING THE DRUGS ACROSS BBB

The above-mentioned challenges require innovative approaches to develop effective and safe methods to deliver the drugs across BBB. Some of the reported strategies and techniques for enhancing the brain delivery of drugs include use of nanoparticles, focused ultrasound, intranasal delivery, receptor-mediated transport, and viral vectors.

18.3.1 Focused Ultrasound

In this technique, ultrasonic waves are used to temporarily disrupt the BBB, allowing drugs to pass through. Microbubbles are injected into the bloodstream and then targeted with ultrasound waves, which create transient and reproducible openings in the BBB. Microbubble-assisted focused ultrasound delivery of mRNA encapsulated in lipid nanoparticles to the brain has been reported (Ogawa et al., 2022). The bioavailability of transferrin-receptor targeting liposomes has been increased by 40% using focused ultrasound and microbubble treatment (Olsman et al., 2021). Similarly, doxorubicin and endogenous antibodies have been delivered successfully to brain in animal models (Cho et al., 2011; Gross et al., 2011; Jordão et al., 2013; Raymond et al., 2008). However, this technique has safety concerns (Tsai et al., 2018).

18.3.2 Nanoformulations

Nanoparticles, such as polymeric, lipid-based, magnetic, and silica nanoparticles, can be engineered to cross the BBB and deliver drugs directly to the brain. These nanoparticles can be designed to target specific receptors or cells, enhancing their effectiveness. Nanocarriers utilize a range of transport mechanisms to successfully navigate the BBB, including exocytosis, transcytosis, simple diffusion, and receptor-mediated endocytosis (Tiwari et al., 2023). This is one of the safe methods to deliver the drugs to the brain. The long-term effects and potential toxicity of nanoparticles need to be thoroughly studied.

18.3.3 Receptor-Mediated Transport

The method involves attaching therapeutic agents to molecules that can naturally cross the BBB via receptor-mediated transport (Gabathuler, 2010; Marcello & Chiono, 2023). For example, drugs can be linked to transferrin or insulin, which have receptors on the BBB (Pond & Tozer, 1984). The disadvantage of this strategy is that the transcytosis mechanisms are not specific to BBB. Hence, a fraction of drug may get non-specifically targeted to tissues other than brain, with possible side effects (Pond & Tozer, 1984).

18.3.4 Intranasal Delivery

This non-invasive method involves administering drugs through the nasal cavity, allowing them to bypass the BBB and reach the brain. Intranasal delivery utilizes the trigeminal and olfactory nerve pathways (Erdő et al., 2018; Raghav et al., 2023). The intranasal delivery avoids systemic side effects (Dhuria et al., 2010). Through intranasal route, different drugs and biologics, such as oligonucleotides, proteins, and peptides, and even cells can be delivered to brain (Dhuria et al., 2010). However, there are certain limitations of intranasal delivery to the brain. They include low permeability of hydrophilic drugs and macromolecules through nasal epithelium, loss of drug through rapid mucociliary clearance, metabolism of drugs in the nasal mucosa by proteolytic enzymes, and entry of non-absorbed drugs in the respiratory and digestive tracts leading to systemic absorption and side effects (Mittal et al., 2014).

18.3.5 Viral Vectors

Certain viruses have naturally evolved the ability to bypass the BBB. Researchers use these viruses as carriers to deliver therapeutic agents, especially genes, to the brain (Ingusci et al., 2024). Many viral vectors that have been successfully used as vectors for gene delivery to the brain include adeno-associated viruses, retroviruses, adenoviruses, and herpesvirus (Lentz et al., 2012). By modifying the viral vectors, they can enhance their ability to cross the BBB and target specific tissues (Ingusci et al., 2024; Walther & Stein, 2000).

18.4 NANOCARRIER SYSTEMS FOR IMPROVED DRUG DELIVERY TO THE BRAIN

To address the challenges encountered by drugs in crossing the BBB, novel strategies in drug development have emerged as a critical necessity (Banks, 2009; Hawkins & Davis, 2005; Stamatovic et al., 2008). Various nanocarrier-based approaches meet this demand. The nanocarriers improve the efficacy of the AD therapy. Nanocarriers can sustain the release of entrapped drug. Many nanocarrier-based brain delivery systems have been explored for the treatment of AD. Majority of such systems/approaches use functionalized/multifunctional nanoparticles. The mechanisms of nanoparticle transportation across the BBB are illustrated in Figure 18.1. The advantages, limitations and safety aspects of various nanocarrier systems for targeting to brain are given in Table 18.1.

18.4.1 Liposomes

Liposomes are the widely accepted choice for brain delivery of drugs due to their properties such as biodegradability, low toxicity, non-immunogenicity, biocompatibility, and flexibility (Bozzuto & Molinari, 2015; Naseri et al., 2015; Zahin et al., 2020). They are amphiphilic in nature and consist of a lipid bilayer encircling an aqueous core (Elkhoury et al., 2019; Hasan et al., 2019). Liposomes can encapsulate hydrophobic drugs in their bilayer and hydrophilic bioactive molecules in their core (Hasan et al., 2020). Multifunctional liposomal vesicles containing a

TABLE 18.1
Advantages, Limitations, and Modifications Affecting Safety Associated with Various Nanoformulations in Brain Drug Delivery

Nanoparticle Type	Advantages	Limitations	Modifications Affecting Safety
Liposomes	• Biocompatible and biodegradable. • Can encapsulate lipophilic and hydrophilic molecules. • Surface can be modified for BBB targeting.	• Prone to leakage and instability. • Short circulation half-life. • Rapid clearance by RES.	• PEGylation improves stability but may cause hypersensitivity reactions (PEG-induced immune response). • Cholesterol addition improves rigidity but may impact clearance rate.
SLNs	• High drug stability. • Biodegradable lipids. • Can be sterilized easily. • Suitable for both hydrophilic and lipophilic drugs.	• Limited drug loading. • Drug expulsion during storage. • Potential aggregation.	• Surfactants like poloxamer improve stability but may cause irritation or toxicity at high concentrations. • Use of specific lipids (tristearin) may impact metabolic pathways.
Polymeric nanoparticles	• Controlled and sustained drug release. • Good biocompatibility. • Can be modified for targeted delivery.	• Complex synthesis process. • Potential polymer toxicity. • Low drug loading in some cases.	• Surface functionalization with ligands (transferrin) enhances BBB targeting but may alter immune recognition. • Use of toxic organic solvents during synthesis can impact biocompatibility.
Dendrimers	• High surface functionalization. • Precise drug targeting and controlled release. • Good solubility and stability.	• Potential cytotoxicity. • Complex synthesis. • Difficult to scale up production.	• Surface charge modification (PEGylation) can reduce toxicity. • Use of amine-terminated dendrimers can increase cytotoxicity due to interaction with cell membranes.
Silica nanoparticles	• High surface area and tunable pore size. • Can be loaded with multiple drugs. • Surface modifiable for BBB targeting.	• Potential silica toxicity. • Slow degradation *in vivo*. • Risk of aggregation.	• Surface coating (PEG, albumin) can enhance safety. • Poorly designed MSNs may induce ROS production and inflammation.

BBB transport mediator (anti-transferrin antibody) and a curcumin have been reported earlier, which were found to be potentially useful in diagnosis and therapy of AD (Mourtas et al., 2014).

18.4.2 Solid Lipid Nanoparticles (SLNs)

Solid Lipid Nanoparticles (SLNs) (10–1000 nm) are composed of lipids or modified lipids such as fatty acids, triglycerides, waxes. SLNs have a solid lipid core, in which lipophilic and hydrophilic drugs can be dispersed (Chen et al., 2001; Kaur et al., 2008). They can cross the reticuloendothelial system (RES) of the BBB easily (Müller et al., 2000; Pardeshi et al., 2012). SLNs provide better protection to the entrapped drug from biochemical degradation due to their solid lipid composition. Since SLNs are physiological solid lipid emulsion systems, they exhibit better biocompatibility and less toxicity when compared to polymeric nanoparticles (Tabatt et al., 2004).

Manjunath and Venkateswarlu (2006) designed SLNs using PEG and Tween 80 to improve the bioavailability of the antipsychotic drugs nitrendipine and clozapine. Wang et al. (2002) have reported good brain targeting efficiency of SLNs containing 5-fluoro-2′-deoxyuridine to the brain, 3′,5′-dioctanoyl-5-fluoro-2′-deoxyuridine. Song et al. (2014) developed high-density lipoprotein nanoparticles reconstituted with apolipoprotein E3 for the removal of Aβ. Following four weeks of daily treatment, there was a reduction in Aβ deposits, microgliosis, neurological changes, and memory deficits.

18.4.3 Polymeric Nanoparticles

Polymeric nanoparticles protect drugs from degradation and thereby improve the bioavailability of the incorporated drugs (Van Giau et al., 2018). A brain targeting nanopolymeric system based on n-butyl-2-cyanoacrylate has been reported (Kulkarni et al., 2010). The nanoparticles encapsulated a radio-labeled amyloid affinity drug, 125I-clioquinol, for diagnosis of AD. The nanoparticles crossed the BBB in mice and increased the retention of the drug in the brain. Chitosan nanoparticles of rivastigmine have been reported by Wilson et al. (2011). The nanoparticles exhibited a biphasic Fickian diffusion release pattern.

18.4.4 Dendrimers

Dendrimers are called wonder polymers and are used to solubilize sparingly soluble drugs. They have been used as drug carriers for AD (Jose & Charyulu, 2016). Dendrimers are nanosized macromolecular branched systems, which have three structural components, namely, core, interior dendritic part, and functionalized surface. Patel et al. (2007) have developed sialic acid-conjugated dendrimers, which mimicked the sialic acid found in the cell surface and reduced the neurotoxic effect of Aβ in plaque.

18.5 TRANSLATIONAL AND REGULATORY CONSIDERATIONS OF NANO-BASED INTERVENTIONS IN THE THERAPY OF AD

The nano-based approaches for diagnosis and treatment of AD are aimed at improving the efficacy of the therapy and decreasing the risks of systemic side effects via reduction of dose of the drug. To bring nano-based systems from bench to bedside, many challenges are to be addressed. Some of the key issues in clinical translation of nano-based systems include scaleup and large-scale manufacturing, safety, biocompatibility and biodegradability, intellectual property, federal regulations, and cost-effectiveness in comparison to the existing therapeutic options.

18.5.1 Scaleup and Large-Scale Manufacturing

An important bottleneck in the clinical translation of nano-based systems is the complex nature of the formulation and the complex processes used in the preparation of the formulation. Platforms that require complex preparation procedures exhibit limited translation potential as large-scale manufacturing can pose problems in manufacturing and quality control (Hua et al., 2018; Sainz et al., 2015). The hurdles in the large-scale manufacturing of nano-based systems include high cost of starting materials, high manufacturing costs, low product yield, complex processes used in the preparation, scalability issues, removal of impurities such as byproducts and starting materials, physical and chemical instability issues with respect to the incorporated drug and the finished product, insufficient batch-to-batch consistency, lack of infrastructure for production (Hafner et al., 2014; Hare et al., 2017; Sainz et al., 2015).

Methods for large-scale production of several nano-based drug delivery systems have been developed successfully. For example, methods to manufacture liposomes on industrial scale are available. Industry-scale manufacturing becomes challenging with increased complexity of the system. For example, modification of surface with coatings and ligands and incorporation of more than one drug in the system increase the number of steps in the manufacturing, which leads to increased production cost and problems in cGMP (Tinkle et al., 2014).

Characterization and validation of nano-based delivery systems involves several parameters, such as particle size and size distribution, zeta potential of the preparation, surface characteristics, purity, drug content, encapsulation efficiency, drug release profile, degree of conjugation with ligand, and many more, which makes the quality control of the final product a challenge. Ascertaining batch-to-batch consistency is another challenge due to several parameters as mentioned above. Batch-to-batch variations can lead to changes in physicochemical properties, physical and chemical stability issues, and pharmacokinetic parameters (Tinkle et al., 2014).

18.5.2 Safety

Safety of drug products is of utmost importance in getting regulatory approval for the marketing of the product. Detailed safety evaluation of the drug product is necessary before it is tested in humans. During clinical trials or after the drug product is introduced in the market, if any changes are made to the processes or formulation which might affect safety, the sponsor needs to assess these changes carefully to ensure the safety of the drug product. The United States Food and Drug Administration (US FDA) enlists the modifications throughout INDA that can affect the safety of the drug product.

Nanoparticles and nanomaterials are toxic. They exhibit various toxicities such as cytotoxicity (Lewinski et al., 2008), immunotoxicity (Jackman et al., 2016), genotoxicity (Accomasso et al., 2018), male reproductive toxicity (Dantas et al., 2022; Habas et al., 2021), developmental toxicity on brain as the nanoparticles get transferred from mother's blood to fetal circulation (Umezawa et al., 2017) and distressing of cells by internal and external interaction with nanoparticles (Sharma et al., 2024).

Rational design of nano-based drug delivery system is of utmost importance in increasing their translational potential. Rational design starts from selection of raw materials, production method, scaleup techniques, optimization of production method, purification techniques and their validation, and finally evaluation and quality control parameters and their optimization (Accomasso et al., 2018). Safety of some of the raw materials such as lipids used in liposomes and SLNs is already studied and proved. But, as nano-based systems become more complex with the use of synthetic materials, surfactants, coatings, ligands, antibody-tagging, etc., the establishment of safety profile also gets more complex (Narang et al., 2013; Tinkle et al., 2014).

Evaluation of nanotoxicity needs validated, standardized, sensitive *in vitro*, *ex vivo*, and *in vivo* methods to assess the toxicity. Such methods need to be developed in the earlier stage of drug development (Accomasso et al., 2018; Jackman et al., 2016) before preclinical testing in animal models. In the earlier stages of drug discovery, toxicity studies can be performed on different cell lines such as liver, brain, lungs, kidneys, and placenta, using standardized protocols (Kulkarni, 2002) to assess the potential risk of genotoxicity and cytotoxicity.

There is also a need to establish safety of nano-based systems in animal models through acute, sub-acute, and chronic toxicity studies. This is to establish the safety of nanomaterials in short-term and long-term use. For conducting such studies also, standard assessment protocols are needed. Along with these, establishment of pharmacokinetics, distribution in the body, localized tissue or organ accumulation, and finally, therapeutic efficacy of nanomaterials is mandatory for regulatory approvals (Tinkle et al., 2014).

18.5.3 Regulatory Requirements

Nano-based drug delivery systems have great potential to improve therapeutic effectiveness and provide great health benefits. But commercialization of nano-based systems needs regulatory approvals. The regulatory approvals need well-established large-scale manufacturing facilities and procedures, quality control protocols, along with establishment of safety and efficacy of the drug product (Tinkle et al., 2014). Currently, there is a large scientific and regulatory gap with respect to nano-based systems. Nano-based systems are currently regulated within conventional regulatory frameworks, which cover conventional immediate release and modified release drug products. However, these regulations are not appropriate for nano-based drug products to confirm the safety, quality, and efficacy of the products for new drug approvals (NDA) (Gaspar, 2007). There is a need to establish an appropriate regulatory framework for nano-based drug delivery systems, which can evaluate them properly. For this, a collaboration between researchers, industry, academia, and regulatory agencies is needed. The regulatory agencies need to approve and include newer analytical tools and standardization methods to evaluate the physical, chemical, and pharmaceutical properties of nano-based drug delivery systems (Gaspar, 2007).

Quality control procedures and good manufacturing practice must be adapted to ensure large-scale production (Operti et al., 2021). Regulatory agencies, including European Medicines Agency and the US FDA, have established guidelines for nanopharmaceuticals; however, the complexity of nanomaterials often makes standard regulatory pathways inadequate (Allan et al., 2021). Unlike conventional drugs, nanoparticles exhibit unique physicochemical properties that can alter pharmacokinetics, biodistribution, and toxicity (Haripriyaa & Suthindhiran, 2023). This necessitates stringent preclinical and clinical evaluation to assess their stability, long-term effects, and biocompatibility (Dobrovolskaia et al., 2008). AD patients may experience cognitive decline, making it difficult for them to provide complete informed consent (Kim, 2011). Ensuring transparent communication about risks and benefits to patients and caregivers is critical. Additionally, equitable access to nanomedicine-based AD treatments must be prioritized. High development costs could lead to limited affordability, exacerbating healthcare disparities (Uskoković, 2021). Regulatory policies should promote affordability and global accessibility while encouraging responsible innovation. In conclusion, while nanotechnology-based interventions offer transformative potential for AD, addressing ethical and regulatory challenges is essential to ensure their safe, effective, and equitable implementation in clinical practice.

18.5.4 Intellectual Property of Nano-Based Drug Delivery Systems

Currently, global standards for grant of patents are based on product patents or process patents or a combination of both. This offers a simple straightforward method to grant patents. But, when the complex nature of nano-based drug delivery systems is considered, there can be many issues

TABLE 18.2
Status of Various Nanotechnology-Based Patents for Brain Drug Delivery

Patent Number	Description	Status/Application Granted Year	Anticipated Expiration	Inventors
CN110859826B	The invention presents brain tumor-targeted biomimetic nanoparticles using doxorubicin-loaded polyglutamic acid, coated with U87 glioma cell membranes, enhancing blood-brain barrier penetration, tumor targeting, sustained drug release, and chemotherapy efficacy for brain tumor treatment.	Active/2022	19-12-2039	Lintao et al.
US10307372B2	PEG-coated, neutral-charged 110-nm non-adhesive nanoparticles diffuse in brain ECS, enhancing CNS drug delivery by enabling higher payload, improved loading efficiency, and prolonged drug release.	Active/2019	24-11-2033	Hanes et al.
US10123969B2	A method enhances therapeutic drug transport to the brain's perivascular space by introducing a drug and an osmolyte (0.5–12.9 M) into cerebrospinal fluid, improving delivery to cerebral blood vessels and CNS parenchyma. The osmolyte facilitates enhanced drug penetration, optimizing treatment efficacy for neurological disorders.	Active/2018	14-10-2036	Thorne et al.
US10842755B2	A nanoparticle with a polymer or inorganic core and a brain metastatic cancer cell membrane shell enables targeted drug delivery across the blood-brain barrier, enhancing brain tissue targeting. Disclosed methods for nanoparticle formation and application.	Active/2020	19-02-2039	Xu

in granting patents. For example, developing nano-delivery systems for existing drugs using a new technology, or existing drug with existing technology for a newer use, can lead to complex and confusing IP situations as either the drug or the nanosystem or both are already commercialized or protected by patent separately. If the nano-delivery system utilizes coating with commercially available material such as Eudragit®, patent is already owned by a company. Hence, to protect the nano-based systems, newer and simplified IP practices and protocols are required (Murday et al., 2009). Table 18.2 presents patent status of various nanotechnology-based patents for brain drug delivery.

18.5.5 Cost-Effectiveness of Nano-Based Drug Delivery Systems

Nanocarriers are very complex and involve multiple-step procedures in their manufacture, along with use of a wide variety of raw materials. As the complexity of nanosystem increases, the yield, purification, impurity profiling, and quality control become complicated. Establishment of methods for large-scale production of nano-based drugs along with control points is another issue. All these aspects together render the nano-based drug delivery systems costly. In case of life-threatening diseases, the increased cost with reduced toxicity and improved therapeutic efficacy is justifiable. In case of other diseases, the increased cost of therapy may not be easily accepted by the patients or by the insurance companies. If nano-delivery systems are developed for biotechnology-based drugs, the cost of the product will be further higher. Hence, the cost-benefit ratio for nano-based systems needs to be carefully drawn for the commercial success of the product.

18.6 FUTURE PROSPECTS

Conventional strategies to treat AD tend to fail due to lack of drug bioavailability in the brain. Nano-based drug delivery systems offer better and innovative means to achieve required drug concentrations in the brain by modulating different pathways in the target tissues. Nanotechnology provides a better means of managing AD. Hence, there is a need to investigate nano-based strategies of treating AD, which may delay disease progression significantly. However, stress also needs to be given to toxicity and safety of nano-based strategies through use of biocompatible and biodegradable materials in the development of nano-drug delivery systems. Even though a plethora of research is already undertaken, there are many hurdles in bringing the nano-based systems from bench to bedside. Majority of the research is currently limited to preclinical level.

The pace of clinical translation of nano-based systems for therapy of AD has remained slow, due to reasons such as time and cost involved in development, complexity of the nanosystems, lack of facilities for scaleup and large-scale production, physical and chemical stability issues with nano-based systems, establishment of safety of these systems, along with regulatory hurdles. Cost-benefit analysis is another limitation to the clinical translation of nano-based systems for AD therapy when compared to the approved existing therapeutic strategies.

18.7 CONCLUSION

The use of nanotechnology in medicine can have a major impact on health. In case of AD, this technology can give better therapeutic intervention to manage the disease more effectively. But the commercialization of nano-based therapies has many hurdles, which might be due to the complex nature of the nano-based systems. If researchers can reduce the complexity of nano-based systems, their translation into clinically applicable therapeutic intervention can be achieved.

REFERENCES

Accomasso, L., Cristallini, C., & Giachino, C. (2018). Risk assessment and risk minimization in nanomedicine: A need for predictive, alternative, and 3Rs strategies. *Frontiers in Pharmacology*, *9*, 228. https://doi.org/10.3389/fphar.2018.00228

Allan, J., Belz, S., Hoeveler, A., Hugas, M., Okuda, H., Patri, A., Rauscher, H., Silva, P., Slikker, W., Sokull-Kluettgen, B., Tong, W., & Anklam, E. (2021). Regulatory landscape of nanotechnology and nanoplastics from a global perspective. *Regulatory Toxicology and Pharmacology*, *122*, 104885. https://doi.org/10.1016/j.yrtph.2021.104885

Babu, B., Parvathy, G., Bawa, F. S. M., Gill, G. S., Patel, J., Sibia, D. S., Sureddi, J., & Newgent, H. (2024). Comparing the artificial intelligence detection models to standard diagnostic methods and alternative models in identifying Alzheimer's disease in at-risk or early symptomatic individuals: A scoping review. *Cureus*, *16*(12): e75389. https://doi.org/10.7759/cureus.75389

Banks, W. A. (2009). Characteristics of compounds that cross the blood-brain barrier. *BMC Neurology*, *9*(Suppl 1), S3.

Bozzuto, G., & Molinari, A. (2015). Liposomes as nanomedical devices. *International Journal of Nanomedicine*, 975–999. https://doi.org/10.2147/IJN.S68861

Chen, D. B., Lu, W. L., & Zhang, Q. (2001). In vitro and in vivo study of two types of long-circulating solid lipid nanoparticles containing paclitaxel. *Chemical and Pharmaceutical Bulletin*, *49*(11), 1444–1447.

Chen, W., Li, J., Guo, J., Li, L., & Wu, H. (2024). Diagnosis and therapy of Alzheimer's disease: Light-driven heterogeneous redox processes. *Advances in Colloid and Interface Science*, 103253. https://doi.org/10.1016/j.cis.2024.103253

Cho, E. E., Drazic, J., Ganguly, M., Stefanovic, B., & Hynynen, K. (2011). Two-photon fluorescence microscopy study of cerebrovascular dynamics in ultrasound-induced blood–brain barrier opening. *Journal of Cerebral Blood Flow & Metabolism*, *31*(9), 1852–1862. https://doi.org/10.1038/jcbfm.2011.59

Dantas, G. P., Ferraz, F. S., Andrade, L. M., & Costa, G. M. (2022). Male reproductive toxicity of inorganic nanoparticles in rodent models: A systematic review. *Chemico-Biological Interactions*, *363*, 110023. https://doi.org/10.1016/j.cbi.2022.110023

Dhariwal, R., Jain, M., Mir, Y. R., Singh, A., Jain, B., Kumar, P., Tariq, M., Verma, D., Deshmukh, K., Yadav, V. K., & Malik, T. (2025). Targeted drug delivery in neurodegenerative diseases: The role of nanotechnology. *Frontiers in Medicine*, *12*, 1522223. https://doi.org/10.3389/fmed.2025.1522223

Dhuria, S. V., Hanson, L. R., & Frey II, W. H. (2010). Intranasal delivery to the central nervous system: Mechanisms and experimental considerations. *Journal of Pharmaceutical Sciences*, *99*(4), 1654–1673.

Dobrovolskaia, M. A., Aggarwal, P., Hall, J. B., & McNeil, S. E. (2008). Preclinical studies to understand nanoparticle interaction with the immune system and its potential effects on nanoparticle biodistribution. *Molecular Pharmaceutics*, *5*(4), 487–495. https://doi.org/10.1021/mp800032f

Elkhoury, K., Russell, C. S., Sanchez-Gonzalez, L., Mostafavi, A., Williams, T. J., Kahn, C., Peppas, N. A., Arab-Tehrany, E., & Tamayol, A. (2019). Soft-nanoparticle functionalization of natural hydrogels for tissue engineering applications. *Advanced Healthcare Materials*, *8*(18), 1900506.

Elsori, D., Rashid, G., Khan, N. A., Sachdeva, P., Jindal, R., Kayenat, F., Sachdeva, B., Kamal, A. Z., Baker, A. M., & Fahmy, S. A. (2023). Nanotube breakthroughs: Unveiling the potential of carbon nanotubes as a dual therapeutic arsenal for Alzheimer's disease and brain tumors. *Frontiers in Oncology*, *13*, 1265347. https://doi.org/10.3389/fonc.2023.1265347

Erdő, F., Bors, L. A., Farkas, D., Bajza, Á, & Gizurarson, S. (2018). Evaluation of intranasal delivery route of drug administration for brain targeting. *Brain Research Bulletin*, *143*, 155–170. https://doi.org/10.1016/j.brainresbull.2018.10.009

Gabathuler, R. (2010). Approaches to transport therapeutic drugs across the blood–brain barrier to treat brain diseases. *Neurobiology of Disease*, *37*(1), 48–57.

Garcés, P., Amaro, A., Montecino, M., & van Zundert, B. (2024). Inorganic polyphosphate: From basic research to diagnostic and therapeutic opportunities in ALS/FTD. *Biochemical Society Transactions*, *52*(1), 123–135. https://doi.org/10.1042/BST20230257

Gaspar, R. (2007). Regulatory issues surrounding nanomedicines: Setting the scene for the next generation of nanopharmaceuticals. *Nanomedicine*, *2*(2), 143–147. https://doi.org/10.2217/17435889.2.2.143

Gross, M. E., Nelson, E. T., Mone, M. C., Hansen, H. J., Sklow, B., Glasgow, R. E., & Scaife, C. L. (2011). A comparison of postoperative outcomes utilizing a continuous preperitoneal infusion versus epidural for midline laparotomy. *The American Journal of Surgery*, *202*(6), 765–770. https://doi.org/10.1016/j.amjsurg.2011.05.016

Habas, K., Demir, E., Guo, C., Brinkworth, M. H., & Anderson, D. (2021). Toxicity mechanisms of nanoparticles in the male reproductive system. *Drug Metabolism Reviews*, *53*(4), 604–617. https://doi.org/10.1080/03602532.2021.1917597

Hafner, A., Lovrić, J., Lakoš, G. P., & Pepić, I. (2014). Nanotherapeutics in the EU: An overview on current state and future directions. *International Journal of Nanomedicine*, 1005–1023. https://doi.org/10.2147/IJN.S55359

Hanes, J., Woodworth, G. F., & Nance, E. A. (2019). *U.S. Patent No. 10,307,372*. Washington, DC: U.S. Patent and Trademark Office.

Hare, J. I., Lammers, T., Ashford, M. B., Puri, S., Storm, G., & Barry, S. T. (2017). Challenges and strategies in anti-cancer nanomedicine development: An industry perspective. *Advanced Drug Delivery Reviews*, *108*, 25–38. https://doi.org/10.1016/j.addr.2016.04.025

Haripriyaa, M., & Suthindhiran, K. (2023). Pharmacokinetics of nanoparticles: Current knowledge, future directions and its implications in drug delivery. *Future Journal of Pharmaceutical Sciences*, *9*(1), 113. https://doi.org/10.1186/s43094-023-00569-y

Hasan, M., Elkhoury, K., Belhaj, N., Kahn, C., Tamayol, A., Barberi-Heyob, M., Arab-Tehrany, E., & Linder, M. (2020). Growth-inhibitory effect of chitosan-coated liposomes encapsulating curcumin on MCF-7 breast cancer cells. *Marine Drugs*, *18*(4), 217.

Hasan, M., Elkhoury, K., Kahn, C. J., Arab-Tehrany, E., & Linder, M. (2019). Preparation, characterization, and release kinetics of chitosan-coated nanoliposomes encapsulating curcumin in simulated environments. *Molecules*, *24*(10), 2023.

Hawkins, B. T., & Davis, T. P. (2005). The blood-brain barrier/neurovascular unit in health and disease. *Pharmacological Reviews*, *57*(2), 173–185.

Hua, S., De Matos, M. B., Metselaar, J. M., & Storm, G. (2018). Current trends and challenges in the clinical translation of nanoparticulate nanomedicines: Pathways for translational development and commercialization. *Frontiers in Pharmacology*, *9*, 790. https://doi.org/10.3389/fphar.2018.00790

Ingusci, S., Hall, B. L., Goins, W. F., Cohen, J. B., & Glorioso, J. C. (2024). Viral vectors for gene delivery to the central nervous system. *Handbook of Clinical Neurology*, *205*, 59–81. https://doi.org/10.1016/B978-0-323-90120-8.00001-0

Jackman, J. A., Mészáros, T., Fülöp, T., Urbanics, R., Szebeni, J., & Cho, N. J. (2016). Comparison of complement activation-related pseudoallergy in miniature and domestic pigs: Foundation of a validatable immune toxicity model. *Nanomedicine: Nanotechnology, Biology and Medicine*, *12*(4), 933–943. https://doi.org/10.1016/j.nano.2015.12.377

Jordão, J. F., Thévenot, E., Markham-Coultes, K., Scarcelli, T., Weng, Y. Q., Xhima, K., O'Reilly, M., Huang, Y., McLaurin, J., Hynynen, K., & Aubert, I. (2013). Amyloid-β plaque reduction, endogenous antibody delivery and glial activation by brain-targeted, transcranial focused ultrasound. *Experimental Neurology*, *248*, 16–29. https://doi.org/10.1016/j.expneurol.2013.05.008

Jose, J., & Charyulu, R. N. (2016). Prolonged drug delivery system of an antifungal drug by association with polyamidoamine dendrimers. *International Journal of Pharmaceutical Investigation*, *6*(2), 123. https://doi.org/10.4103/2230-973X.177833

Kale, M. B., Wankhede, N. L., Pawar, R. S., Ballal, S., Kumawat, R., Goswami, M., Khalid, M., Taksande, B., Upaganlawar, A., Umekar, M., Kopalli, S. R., & Koppula, S. (2024). AI-driven innovations in Alzheimer's disease: Integrating early diagnosis, personalized treatment, and prognostic modelling. *Ageing Research Reviews*, 102497. https://doi.org/10.1016/j.arr.2024.102497

Kas, M. J., Penninx, B., Sommer, B., Serretti, A., Arango, C., & Marston, H. (2019). A quantitative approach to neuropsychiatry: The why and the how. *Neuroscience & Biobehavioral Reviews*, *97*, 3–9. https://doi.org/10.1016/j.neubiorev.2017.12.008

Kaur, I. P., Bhandari, R., Bhandari, S., & Kakkar, V. (2008). Potential of solid lipid nanoparticles in brain targeting. *Journal of Controlled Release*, *127*(2), 97–109.

Kim, H. Y., & Kim, Y. (2024). Chemical-driven amyloid clearance for therapeutics and diagnostics of Alzheimer's disease. *Accounts of Chemical Research*, *57*(22), 3266–3276. https://doi.org/10.1021/acs.accounts.4c00458

Kim, S. Y. (2011). The ethics of informed consent in Alzheimer disease research. *Nature Reviews Neurology*, *7*(7), 410–414. https://doi.org/10.1038/nrneurol.2011.76

Kong, A. H. Y., Wu, A. J., Ho, O. K. Y., Leung, M. M. K., Huang, A. S., Yu, Y., Zhang, G., Lyu, A., Li, M., & Cheung, K. H. (2023). Exploring the potential of aptamers in targeting neuroinflammation and neurodegenerative disorders: Opportunities and challenges. *International Journal of Molecular Sciences*, *24*(14), 11780. https://doi.org/10.3390/ijms241411780

Kulkarni, G. T. R. (2002). *Biotechnology and its applications in pharmacy*. Jaypee Bros Medical Publishers.

Kulkarni, P. V., Roney, C. A., Antich, P. P., Bonte, F. J., Raghu, A. V., & Aminabhavi, T. M. (2010). Quinoline-n-butyl cyanoacrylate-based nanoparticles for brain targeting for the diagnosis of Alzheimer's disease. *Wiley Interdisciplinary Reviews: Nanomedicine and Nanobiotechnology*, *2*(1), 35–47. https://doi.org/10.1002/wnan.59

Lentz, T. B., Gray, S. J., & Samulski, R. J. (2012). Viral vectors for gene delivery to the central nervous system. *Neurobiology of Disease*, *48*(2), 179–188. https://doi.org/10.1016/j.nbd.2011.09.014

Lewinski, N., Colvin, V., & Drezek, R. (2008). Cytotoxicity of nanoparticles. *Small*, *4*(1), 26–49. https://doi.org/10.1002/smll.200700595

Lintao, C., Shengping, Z., Ruijing, L., Huaqing, C., Lanlan, L., & Huamei, H., (2022). Brain tumor-targeted biomimetic drug-loaded nanoparticle, preparation method and use thereof. CN110859826B.

Lockhart, B. P., & Walther, B. (2009). Les biomarqueurs:«Found in translation». *médecine/Sciences*, *25*(4), 423–430. https://doi.org/10.1051/medsci/2009254423

Manjunath, K., & Venkateswarlu, V. (2006). Pharmacokinetics, tissue distribution and bioavailability of nitrendipine solid lipid nanoparticles after intravenous and intraduodenal administration. *Journal of Drug Targeting*, *14*(9), 632–645.

Marcello, E., & Chiono, V. (2023). Biomaterials-enhanced intranasal delivery of drugs as a direct route for brain targeting. *International Journal of Molecular Sciences*, *24*(4), 3390. https://doi.org/10.3390/ijms24043390

Mauricio, R., Benn, C., John Davis, J., Dawson, G., Dawson, L. A., Evans, A., Fox, N., Gallacher, J., Hutton, M., Isaac, J., Jones, D. N. C., Jones, L., Lalli, G., Libri, V., Lovestone, S., Moody, C., Noble, W., Perry, H., Pickett, J., Reynolds, D., Ritchie, C., Rohrer, J. D., Routledge, C., Rowe, J., Snyder, H., Spires-Jones, T., Swartz, J., Truyen, L., Whiting, P., & Therapeutics for Dementia Consortium. (2019). Tackling gaps in developing life-changing treatments for dementia. *Alzheimer's & Dementia: Translational Research & Clinical Interventions*, *5*, 241–253. https://doi.org/10.1016/j.trci.2019.05.001

Mittal, D., Ali, A., Md, S., Baboota, S., Sahni, J. K., & Ali, J. (2014). Insights into direct nose to brain delivery: Current status and future perspective. *Drug Delivery*, *21*(2), 75–86.

Mourtas, S., Lazar, A. N., Markoutsa, E., Duyckaerts, C., & Antimisiaris, S. G. (2014). Multifunctional nanoliposomes with curcumin–lipid derivative and brain targeting functionality with potential applications for Alzheimer disease. *European Journal of Medicinal Chemistry*, *80*, 175–183.

Müller, R. H., Mäder, K., & Gohla, S. (2000). Solid lipid nanoparticles (SLN) for controlled drug delivery–A review of the state of the art. *European Journal of Pharmaceutics and Biopharmaceutics*, *50*(1), 161–177.

Murday, J. S., Siegel, R. W., Stein, J., & Wright, J. F. (2009). Translational nanomedicine: Status assessment and opportunities. *Nanomedicine: Nanotechnology, Biology and Medicine*, *5*(3), 251–273. https://doi.org/10.1016/j.nano.2009.06.001

Narang, A., Chang, R. K., & Hussain, M. A. (2013). Pharmaceutical development and regulatory considerations for nanoparticles and nanoparticulate drug delivery systems. *Journal of Pharmaceutical Sciences*, *102*(11), 3867–3882. https://doi.org/10.1002/jps.23691

Naseri, N., Valizadeh, H., & Zakeri-Milani, P. (2015). Solid lipid nanoparticles and nanostructured lipid carriers: Structure, preparation and application. *Advanced Pharmaceutical Bulletin*, *5*(3), 305. https://doi.org/10.15171/apb.2015.043

Ogawa, K., Kato, N., Yoshida, M., Hiu, T., Matsuo, T., Mizukami, S., Omata, D., Suzuki, R., Maruyama, K., Mukai, H., & Kawakami, S. (2022). Focused ultrasound/microbubbles-assisted BBB opening enhances LNP-mediated mRNA delivery to brain. *Journal of Controlled Release*, *348*, 34–41. https://doi.org/10.1016/j.jconrel.2022.05.042

Olsman, M., Sereti, V., Mühlenpfordt, M., Johnsen, K. B., Andresen, T. L., Urquhart, A. J., & de Lange Davies, C. (2021). Focused ultrasound and microbubble treatment increases delivery of transferrin receptor-targeting liposomes to the brain. *Ultrasound in Medicine & Biology*, *47*(5), 1343–1355. https://doi.org/10.1016/j.ultrasmedbio.2021.01.014

Operti, M. C., Bernhardt, A., Grimm, S., Engel, A., Figdor, C. G., & Tagit, O. (2021). PLGA-based nanomedicines manufacturing: Technologies overview and challenges in industrial scale-up. *International Journal of Pharmaceutics*, *605*, 120807. https://doi.org/10.1016/j.ijpharm.2021.120807

Pardeshi, C., Rajput, P., Belgamwar, V., Tekade, A., Patil, G., Chaudhary, K., & Sonje, A. (2012). Solid lipid based nanocarriers: An overview/Nanonosaci na bazi cvrstih lipida: Pregled. *Acta pharmaceutica*, *62*(4), 433.

Pardridge, W. M. (2012). Drug transport across the blood–brain barrier. *Journal of Cerebral Blood Flow & Metabolism*, *32*(11), 1959–1972.

Patel, D. A., Henry, J. E., & Good, T. A. (2007). Attenuation of β-amyloid-induced toxicity by sialic-acid-conjugated dendrimers: Role of sialic acid attachment. *Brain Research*, *1161*, 95–105. https://doi.org/10.1016/j.brainres.2007.05.055

Pond, S. M., & Tozer, T. N. (1984). First-pass elimination basic concepts and clinical consequences. *Clinical Pharmacokinetics*, *9*, 1–25.

Raghav, M., Gupta, V., Awasthi, R., Singh, A., & Kulkarni, G. T. (2023). Nose-to-brain drug delivery: Challenges and progress towards brain targeting in the treatment of neurological disorders. *Journal of Drug Delivery Science and Technology*, *86*, 104756. https://doi.org/10.1016/j.jddst.2023.104756

Raymond, S. B., Treat, L. H., Dewey, J. D., McDannold, N. J., Hynynen, K., & Bacskai, B. J. (2008). Ultrasound enhanced delivery of molecular imaging and therapeutic agents in Alzheimer's disease mouse models. *PloS One*, *3*(5), e2175. https://doi.org/10.1371/journal.pone.0002175

Sainz, V., Conniot, J., Matos, A. I., Peres, C., Zupanǒiǒ, E., Moura, L., Silva, L. C., Florindo, H. F., & Gaspar, R. S. (2015). Regulatory aspects on nanomedicines. *Biochemical and Biophysical Research Communications*, *468*(3), 504–510. https://doi.org/10.1016/j.bbrc.2015.08.023

Sharma, N., Kurmi, B. D., Singh, D., Mehan, S., Khanna, K., Karwasra, R., Kumar, S., Chaudhary, A., Jhakmola, V., Sharma, A., Singh, S. K., Dua, K., & Kakkar, D. (2024). Nanoparticles toxicity: An overview of its mechanism and plausible mitigation strategies. *Journal of Drug Targeting*, *32*(5), 457–469. https://doi.org/10.1080/1061186X.2024.2316785

Singh, A. K., Singh, S., Minocha, T., Yadav, S. K., Narayan, R., Nayak, U. Y., Singh, S. K., & Awasthi, R. (2024). In vitro profiling and molecular dynamics simulation studies of berberine loaded MCM-41 mesoporous silica nanoparticles to prevent neuronal apoptosis. *Nanoscale Advances*, *6*(9), 2469–2486. https://doi.org/10.1039/D3NA01142A

Singh, A. K., Singh, S. S., Rathore, A. S., Singh, S. P., Mishra, G., Awasthi, R., Mishra, S. K., Gautam, V., & Singh, S. K. (2021). Lipid-coated MCM-41 mesoporous silica nanoparticles loaded with berberine improved inhibition of acetylcholine esterase and amyloid formation. *ACS Biomaterials Science & Engineering*, *7*(8), 3737–3753. https://doi.org/10.1021/acsbiomaterials.1c00514

Song, Q., Huang, M., Yao, L., Wang, X., Gu, X., Chen, J., Chen, J., Huang, J., Hu, Q., Kang, T., Rong, Z., Qi, H., Zheng, G., Chen, H., & Gao, X. (2014). Lipoprotein-based nanoparticles rescue the memory loss of mice with Alzheimer's disease by accelerating the clearance of amyloid-beta. *ACS Nano*, *8*(3), 2345–2359. https://doi.org/10.1021/nn4058215

Stamatovic, S. M., Keep, R. F., & Andjelkovic, A. V. (2008). Brain endothelial cell-cell junctions: How to "open" the blood brain barrier. *Current Neuropharmacology*, *6*(3), 179–192.

Tabatt, K., Kneuer, C., Sameti, M., Olbrich, C., Müller, R. H., Lehr, C. M., & Bakowsky, U. (2004). Transfection with different colloidal systems: Comparison of solid lipid nanoparticles and liposomes. *Journal of Controlled Release*, *97*(2), 321–332.

Thorne, R. G., Pizzo, M. E., & Wolak, D. J. (2018). *U.S. Patent No. 10,123,969*. Washington, DC: U.S. Patent and Trademark Office.

Tinkle, S., McNeil, S. E., Mühlebach, S., Bawa, R., Borchard, G., Barenholz, Y., Tamarkin, L., & Desai, N. (2014). Nanomedicines: Addressing the scientific and regulatory gap. *Annals of the New York Academy of Sciences*, *1313*(1), 35–56. https://doi.org/10.1111/nyas.12403

Tiwari, V., Tiwari, A., Sharma, A., Kumar, M., Kaushik, D., & Sagadevan, S. (2023). An optimistic approach to nanotechnology in Alzheimer's disease management: An overview. *Journal of Drug Delivery Science and Technology*, *86*, 104722.

Trojanowski, J. Q., Arnold, S. E., Karlawish, J. H., Brunden, K., Cary, M., Davatzikos, C., Detre, J., Gaulton, G., Grossman, M., Hurtig, H., Jedrziewski, K., McCluskey, L., Naylor, M., Polsky, D., Schellenberg, G. D., Siderowf, A., Shaw, L. M., Deerlin, V. V., Wang, L. S., Werner, R., Xie,, S. X., & Lee, V. M. Y. (2010). Design of comprehensive Alzheimer's disease centers to address unmet national needs. https://doi.org/10.1016/j.jalz.2009.11.004

Tsai, H. C., Tsai, C. H., Chen, W. S., Inserra, C., Wei, K. C., & Liu, H. L. (2018). Safety evaluation of frequent application of microbubble-enhanced focused ultrasound blood-brain-barrier opening. *Scientific Reports*, *8*(1), 17720. https://doi.org/10.1038/s41598-018-35677-w

Umezawa, M., Onoda, A., & Takeda, K. (2017). Developmental toxicity of nanoparticles on the brain. *Yakugaku zasshi: Journal of the Pharmaceutical Society of Japan*, *137*(1), 73–78. https://doi.org/10.1248/yakushi.16-00214

Usković, V. (2021). Nanomedicine for the poor: A lost cause or an idea whose time has yet to come? *Nanomedicine*, *16*(14), 1203–1218. https://doi.org/10.2217/nnm-2021-0024

Van Giau, V., An, S. S. A., & Hulme, J. P. (2018). Mitochondrial therapeutic interventions in Alzheimer's disease. *Journal of the Neurological Sciences*, *395*, 62–70. https://doi.org/10.1016/j.jns.2018.09.033

Vermunt, L., Veal, C. D., Ter Meulen, L., Chrysostomou, C., Van Der Flier, W., Frisoni, G. B., …, & Visser, P. J. (2018). European Prevention of Alzheimer's dementia registry: Recruitment and prescreening approach for a longitudinal cohort and prevention trials. *Alzheimer's & Dementia*, *14*(6), 837–842. https://doi.org/10.1016/j.jalz.2018.02.010

Walther, W., & Stein, U. (2000). Viral vectors for gene transfer: A review of their use in the treatment of human diseases. *Drugs*, *60*(2), 249–271. https://doi.org/10.2165/00003495-200060020-00002

Wang, J. X., Sun, X., & Zhang, Z. R. (2002). Enhanced brain targeting by synthesis of 3′, 5′-dioctanoyl-5-fluoro-2′-deoxyuridine and incorporation into solid lipid nanoparticles. *European Journal of Pharmaceutics and Biopharmaceutics*, *54*(3), 285–290.

Wilson, B., Samanta, M. K., Muthu, M. S., & Vinothapooshan, G. (2011). Design and evaluation of chitosan nanoparticles as novel drug carrier for the delivery of rivastigmine to treat Alzheimer's disease. *Therapeutic Delivery*, *2*(5), 599–609. https://doi.org/10.4155/tde.11.21

Xu, P. (2020). *U.S. Patent No. 10,842,755*. Washington, DC: U.S. Patent and Trademark Office.

Zahin, N., Anwar, R., Tewari, D., Kabir, M. T., Sajid, A., Mathew, B., Uddin, M. S., Aleya , L., & Abdel-Daim, M. M. (2020). Nanoparticles and its biomedical applications in health and diseases: Special focus on drug delivery. *Environmental Science and Pollution Research*, *27*(16), 19151–19168. https://doi.org/10.1007/s11356-019-05211-0

19 Target of the Neurodegenerative Diseases and Mechanism of the Potent Natural Molecule

Sonima Prasad, Ankit Rathee, Geetanjali Singh, Iram Khan, Charan Singh, and Arti Singh

19.1 INTRODUCTION TO NEURODEGENERATIVE DISEASES

Neurodegenerative diseases are defined by the persistent and accelerated deterioration of neuronal cell structures, which can lead to severe neurological dysfunction and considerable impairment of daily activities that individuals may struggle to perform (Hou et al., 2019). A myriad of different molecules have been studied intensively in the ongoing quest to combat these devastating diseases effectively and to improve the quality of life for those affected. The neuroprotective potential of several natural compounds has been thoroughly explored and tested against various diseases that impact the nervous system (Pohl & Kong Thoo Lin, 2018). The vast majority of these biologically active compounds are fundamentally based on polyphenolic molecules, senile plaque formations, and other relevant peptides that are directly linked to the intricate and multifaceted mechanisms that underlie Alzheimer's disease (AD), one of the most prominent neurodegenerative disorders (Vicente-Zurdo et al., 2024). These compounds frequently contain enzyme inhibitors or activator molecules that have displayed a wide array of therapeutic potentials, some of which include applications in cancer treatment, alleviation from the adverse effects of neurodegenerative diseases, and strategies for combating infections that complicate patient care (Modi et al., 2009). The encouraging and promising results that have emerged from these studies strongly confirm that a diet abundantly rich in flavonoids, phenolics, and other beneficial biocompounds can elicit substantial potential physiological activities in both animals and humans alike, ultimately leading to improved health outcomes and enhanced well-being. In addition to these natural compounds, some powerful derivatives emerging from well-established drugs, such as propargylamines and prilactone monoamine oxidase (MAO) inhibitors, have shown noteworthy potential for application in the treatment of neurodegenerative diseases, further underscoring the significance of ongoing research in this critical area of medical science and the continuous need for innovation to uncover new therapeutic avenues (Gadhave et al., 2024; Hider et al., 2011).

19.1.1 Definition and Types of Neurodegenerative Diseases

Neurodegeneration is a complex process in which neurons, the basic building blocks of the nervous system, gradually lose their shape or function (Hussain et al., 2018). Neurodegenerative diseases comprise a wide range of disorders that collectively affect millions of people globally. These debilitating diseases ultimately lead to either the death of neurons or compromise the protection of healthy neurons (Duchen, 2012). This area of study is particularly complex and remains the focus of ongoing and extensive research. Neurodegenerative disorders can be classified based on several criteria, including the specific brain region that is affected, unique pathological characteristics, predominant motor effects, the mechanisms of pathogenesis, and various genetic traits that may be present (Erkkinen et al., 2018). The following are some of the more well-known neurodegenerative diseases: Huntington's disease (HD), which involves movement disorders and cognitive decline; AD, which primarily affects memory and cognitive functions; Parkinson's disease, which is characterized by motor symptoms and tremors; Amyotrophic Lateral Sclerosis (ALS), also known as Lou Gehrig's disease; Hereditary Spastic Paraplegia; and Spinal and Bulbar Muscular Atrophy (Manoharan et al., 2016; Naumann, 2015).

These diseases broadly fall into three distinct categories that encompass various debilitating conditions. The first category includes motor neuron diseases, which consist of serious conditions such as "ALS" and "spinal muscular atrophy" (SMA) (Abed, 2023). These conditions primarily affect the nerve cells in the brain and spinal cord, leading to severe muscle weakness and atrophy. The second category is focused on extrapyramidal diseases, with well-known examples like Parkinson's disease, which is marked by tremors and movement difficulties, and HD, characterized by progressive motor dysfunction and cognitive decline (Brotini & Gigli, 2004). Lastly, the third category consists of diseases associated with protein accumulation or those related to mental degeneration, which can significantly impact a person's cognitive and emotional well-being (Sanaeifar et al., 2024). The increasing number of patients suffering from these debilitating

DOI: 10.1201/9781003602477-19

neurodegenerative disorders has imposed a heavy burden on families, caregivers, and society as a whole. This situation creates not only emotional distress but also economic challenges due to associated healthcare costs and lost productivity (Goetzel et al., 2004). Understanding the root causes of neurodegenerative diseases remains a complex challenge, as the underlying mechanisms are not yet fully understood. There is currently no established method to achieve perfect cures for these conditions, leaving patients and their families with limited options for effective treatment (Wareham et al., 2022). Research has identified that defects in protein structure or function can be linked to many of these diseases. The cell has evolved specific protein degradation mechanisms that serve to accurately counteract the accumulation of damaged or misfolded proteins, which can lead to cellular dysfunction (Hanna et al., 2019). Among these crucial pathways are the ubiquitin-proteasome system and the autophagy-lysosome pathway, both of which play essential roles in maintaining cellular health. Consequently, it has been recognized that neurodegenerative diseases are often implicated in the disruption of these critical protein degradation pathways, leading to the accumulation of toxic proteins and contributing to the progression of the disease (Rubinsztein, 2006). Understanding and addressing these disruptions could pave the way for potential therapeutic strategies in the battle against neurodegenerative diseases (Toader et al., 2024).

19.1.2 Epidemiology and Impact

Almost one-third of the entire global disease burden can be traced back to non-communicable disorders, and astonishingly, approximately three-quarters of these intricate and perplexing conditions emerge predominantly in developing countries (Saceleanu et al., 2023). Among the multitude of these disorders, neurodegeneration emerges as a notable issue, not only standing out as a severe medical condition by itself but also representing an immensely underappreciated challenge that demands our attention (Hermosillo-Abundis et al., 2024). This widespread neglect is largely due to the fact that neurodegenerative diseases contribute both directly and indirectly to the significant difficulties encountered in managing and curing various associated comorbidities that frequently accompany these conditions (Safiri et al., 2024). The burden imposed by neurodegenerative diseases is increasingly becoming an alarming and weighty reality, with its rise being significantly magnified by the rapid aging of populations around the globe (Manji & Sisodiya, 2024). Alarmingly, the stark truth reveals that there are currently no effective therapies available for many neurodegenerative diseases, which are notoriously known to have invariably fatal consequences (Alraawi et al., 2022). The combination of these twin, concerning factors presents a daunting and substantial challenge to the field of modern medicine, as well as to the socio-economic organization of both developed and underdeveloped societies alike (Silove et al., 2017). Regardless of whether the eventual death stems from the direct consequences of neurological failure or from co-occurring factors that could be indirectly linked to the overarching effects and complications stemming from neurodegenerative diseases, these sobering and unfortunate data strongly emphasize both the significance of the issue and the pressing questions associated with the prevention and management of neurodegenerative diseases on a global scale (Bernheim et al., 2014). As the number of elderly individuals in the world continues to grow rapidly and at an unprecedented rate, it becomes strikingly apparent that there exists an urgent need for effective interventions aimed at improving not only the quality and length of life, but also the economic efficiency of life among aging populations (Prince et al., 2016). The protection, development of cures, and even the implementation of better management practices for neurodegenerative diseases are critically needed if a significant portion of the world's hopeful vision of a healthy, productive, and extended lifespan is ever to be realized (Poovaiah et al., 2018). The importance of prioritizing dedicated research and allocating sufficient resources in our efforts to tackle this ever-escalating public health crisis cannot be overstated. Indeed, the advancement of science and medicine through targeted research could provide invaluable insights and lead to breakthroughs that genuinely change the trajectory of health outcomes for affected populations worldwide (Centola, 2013).

19.2 THE MECHANISMS OF NEURODEGENERATIVE PROCESSES

The primary process of neurodegeneration includes neuronal death through the loss of cellular signaling, ultimately leading to the loss of the signal itself (Chi et al., 2018). This complete downregulation cannot be recovered. Loss of connectivity can lead to neuron death through multiple signaling pathways, including the activation of potent caspases or increased calcium levels leading to endonuclease (Folch et al., 2012). The loss of a connection from the neuron to a target cell, including muscle in the case of motor neurons, leads to target death. Not only do end-stage neurons die, but also neurons lose their functionality through processes such as apoptosis, either through the activation of a particular pathway in the neuron, the action of glial cells such as astrocytes, or through the action of the immune system (Carson et al., 2006). While some argue whether neurodegeneration begins with inflammation itself, there is no argument that the potent immune cells we have in our body that can neutralize and kill bacteria can trigger and activate pathways within neurons that cause them to die (Pickard et al., 2017).

Other factors that contribute to the progression of neurodegenerative diseases include the upregulation of signaling that leads to mitochondrial dysfunction within neurons themselves (Golpich et al., 2017). Given the tide of energy these cells use to maintain this electrical charge, the upregulation of excessive amounts of reactive molecules as a byproduct of the cell's metabolic usage, while maintaining a balance is provided by protecting cells from free radicals (Pamenter, 2014). Little spokes at the end of neurons, filled with potent molecules, help support neurons, and even upregulated signaling of these deadly molecules can lead to neuronal death. Take a heart attack: if one blood vessel to the heart is blocked, cells, including some heart cells, can die within 20 minutes. Cells generally die quickly. However, brain cells are the exception in that if there is such an attack, surrounding cells arrive on the scene and help keep some neurons alive that may not have died, but rather could have died, helping to sequester the infarct, and then help make connections to the neurons of the infarcted area to keep them alive (Klimaschewski, 2022). Brain neurons like their own pulsating rhythm and, while potentially suggested therapies such as stem cell therapies have a powerful potential, the symptoms associated with the neurons from dying in conjunction with potential stem therapies become increasingly clear, suggesting that small protective molecules come to the brain, cross the blood-brain barrier, and keep neurons alive. Such a mechanism may be challenging to find (Wareham et al., 2022).

19.2.1 Underlying Cellular and Molecular Mechanisms in Neurodegeneration

Neurodegeneration is represented as the progressive degeneration of the structure and function of neurons, resulting in the loss of a specific population of neurons related to a given neurodegenerative disorder (Jellinger, 2010). Diverse types of abnormal features come with neurodegeneration, including the accumulation of toxic proteins in the cytoplasm of neural cells as well as numerous aggregated environmental agents; the degeneration of structures such as tangles and plaques, including beta-amyloid and phosphorylated tau protein. Mitochondrial bioenergy failure, oxidative injury, glutamate excitotoxicity, dysfunction of calcium homeostasis, dysfunctional behavior of proteasomes, and the ubiquitin-proteasome system are examples of other factors that can directly lead to neuronal disorders (Jurcau, 2021).

Mitochondrial autophagy is a possible phenomenon contributing to neurodegenerative disorders. Mitophagy can remove damaged mitochondria associated with mitochondrial dysfunction and recycle these for energy, as mitochondrial bioenergy failure and cell death can be prevented (Wang et al., 2019). Especially in Parkinson's disease, excessive accumulations of alpha-synuclein-related fragmented or damaged mitochondria and overreactive oxidative species are the most common features of pathology and a link to the development of Parkinson's disease. In the later stage, alpha-synuclein is released from the damaged mitophagosome and thereby inhibits autophagy-lysosome function, causing an incremental pathological cascade. Furthermore, dysfunction of the ubiquitin-proteasome system is another consistent neurodegenerative aspect. In several neurodegenerative disorders, the function of the ubiquitin-proteasome system is presented as abnormal. Especially, mutations of ubiquitin-related proteases are linked with numerous neurodegenerative diseases, including Parkinson's, Alzheimer's, and HD (Ross and Pickart, 2004) (Table 19.1).

19.2.1.1 Pathophysiology of AD

From a histological perspective, AD is distinguished by the existence of extracellular deposits known as beta amyloid plaques, which are made up of the transmembrane amyloid precursor protein's (APP's) proteolytic byproduct, beta amyloid peptide, which is mostly found on synapses (Penke et al., 2017). AD patients show cognitive impairment with a loss of short-term memory followed by impairments in reasoning and decision making, with increasing neurodegeneration and reduced brain mass. In addition to amyloid-β (Aβ) plaques, AD is characterized by the presence of intracellular neurofibrillary tangles consisting of filaments composed of the abnormally hyperphosphorylated protein tau (DeTure & Dickson, 2019). The hyperphosphorylation of tau leads to microtubule destabilization and formation of paired helical filaments, leading to neuronal dysfunction and cell death. In patients with mild cognitive impairment, hyperphosphorylated tau leads to the conversion to tangles, finally leading to neurodegeneration and cognitive impairment, converting mild cognitive impairment to clinical AD (Bejanin et al., 2017).

The major risk for the occurrence of AD is age, and this is consistent with the observation that elderly patients with neuroinflammation often develop cognitive decline and some form of AD (Calsolaro & Edison, 2016). The neuroinflammatory hypothesis supports the theory that the inflammatory mediators released into the brain lead to oxidative stress and, consequently, to synaptic damage and ultimately the formation of neurofibrillary tangles and Aβ plaques. The induction of a chronic inflammatory response interferes with the lysosomes' degradation pathway mechanism of damaged cellular structures, accumulating altered proteins (Jellinger, 2010). Neurons are not able to withstand the presence of proteins such as Aβ, which then accumulate to form extracellular plaques. These neuroinflammatory and oxidative components represent an interface link between AD and other comorbidities such as cardiovascular diseases. It has been seen that other risk factors are the presence of brain atrophy, the presence of alleles, smoking, insulin resistance/diabetes, severe traumatic brain injury, exposure to high levels of heavy metals or drinking water containing aluminum, and less education (McKenzie et al., 2017) (Figure 19.1).

19.2.1.1.1 Amyloid Cascade Hypothesis

For AD, the most well-known hypothesis is the amyloid cascade hypothesis. The Aβ precursor proteins from neuronal cells are found in extracellular spaces, forming neurotoxic plaques, possibly through the mediation of certain factors that destroy mechanisms, causing an organism to produce immune reactions that weaken the connected biomolecules, ultimately leading to their dysfunction (Hampel et al., 2021). The immune system is activated, which is related to intense

TABLE 19.1
Pathophysiological Mechanisms of Neurodegenerative Disease

Sr. No.	Disease	Pathophysiological Mechanism	Symptoms	Reference
1	AD	Amyloid cascade	Trigger immune responses and inflammation, leading to neurodegeneration	Zhang et al. (2023)
		Tau protein	Under pathological conditions, tau undergoes hyperphosphorylation, leading to misfolding and aggregation into harmful paired helical filaments	Sonawane et al. (2022)
		Neuroinflammation	Prolonged activation due to excessive harmful stimuli leads to chronic inflammation, causing neuron damage and worsening the condition	Zhang et al. (2023)
		Cholinergic	Reducing cognitive decline and neurodegeneration, particularly those that act as bifunctional agents, targeting both oxidative stress and cholinesterase activity	Przybyłowska et al. (2021)
		Oxidative stress	The accumulation of lipid peroxides, damage to proteins, changes in nucleic acids, and alterations in membrane permeability	Lai et al. (2024)
		Excitotoxicity	Increases reactive oxygen species production, lipid peroxidation, DNA fragmentation, and cytokine release	Tripathi et al. (2021)
		Mitochondrial dysfunction	Neurotoxins, hypoxia, injury, or glutamate exposure can further damage mitochondria. This damage causes lipid and DNA/RNA oxidation, leading to membrane dysfunction and impaired apoptosis	Alkhalifa et al. (2024)
2	Parkinson's disease	Dopaminergic neuronal loss	The presynaptic components that damage dopaminergic neurons cause oxidative stress	Imbriani et al. (2022)
		α-Synuclein and Lewy bodies	In normal conditions, vesicle formation helps manage cellular stress, but under oxidative stress, excessive vesicle formation can damage mitochondria	Zeng et al. (2024)
		Neuroinflammation	Natural Nrf2 activators help neural cells and brain tissues resist damage from oxidative stress and proteinopathies linked to neurodegenerative diseases	Kim et al. (2024)
		Oxidative stress	Protein damage includes loss of function, aggregation, and misfolding	Koszła and Sołek (2024)
3	Huntington's disease	Expanded Cytomegalovirus-Associated Glycoprotein (CAG) repeats	Mitochondrial dysfunction, impaired axonal and dendritic transport, and altered gene expression, all contributing to the disease's progression	Dai et al. (2023)
4	"Amyotrophic lateral sclerosis" (ALS)	Degeneration of both upper motor neurons in the brain and lower motor neurons in the spinal cord	ALS includes muscle weakness, atrophy, dysarthria, and dysphagia. Cognitive dysfunction occurs in some patients, but in general, compared to frontotemporal dementia	Van Es (2024)

inflammatory response mechanisms that develop, inducing a harmful cycle of neurodegeneration. A second important element in the appearance of the disease is the formation of Aβ peptide plaques inside the neuron, thus creating a harmful cycle of neurodegeneration. Most treatments developed until now aim to ameliorate symptoms or to stabilize the neurotransmission process of enzymes related to the activities of neurotransmitters (Kandimalla & Reddy, 2017). Therefore, there is a need, on one hand, to inhibit the activity of APPs and, on the other hand, to inhibit all known activities related to the overproduction of amyloid peptides, including the suppression of cleavage from large produced molecules. A possible approach to solving the problem associated with neurodegenerative diseases is the inhibition of the activity of APPs, without destroying the physiological capabilities of these proteins (Khanam et al., 2016). Additionally, it will be necessary to explore ways to use low molecular weight substances that can intervene in this enzymatic mechanism, targeting the active site of the APPs or disrupting pre- and intracellular connections. This will eventually lead to the inhibition of the formation of amyloid peptide oligomers in the cytoplasm and the failure of the production of large molecules, which accumulate in extracellular spaces, ultimately constituting abnormal plaques that affect the cognitive capabilities of the brain (Stefani, 2012).

19.2.1.1.2 Tau Protein and Neurofibrillary Tangles

The microtubule-associated τ-protein, commonly known as tau, is an essential protein that plays a crucial role in the

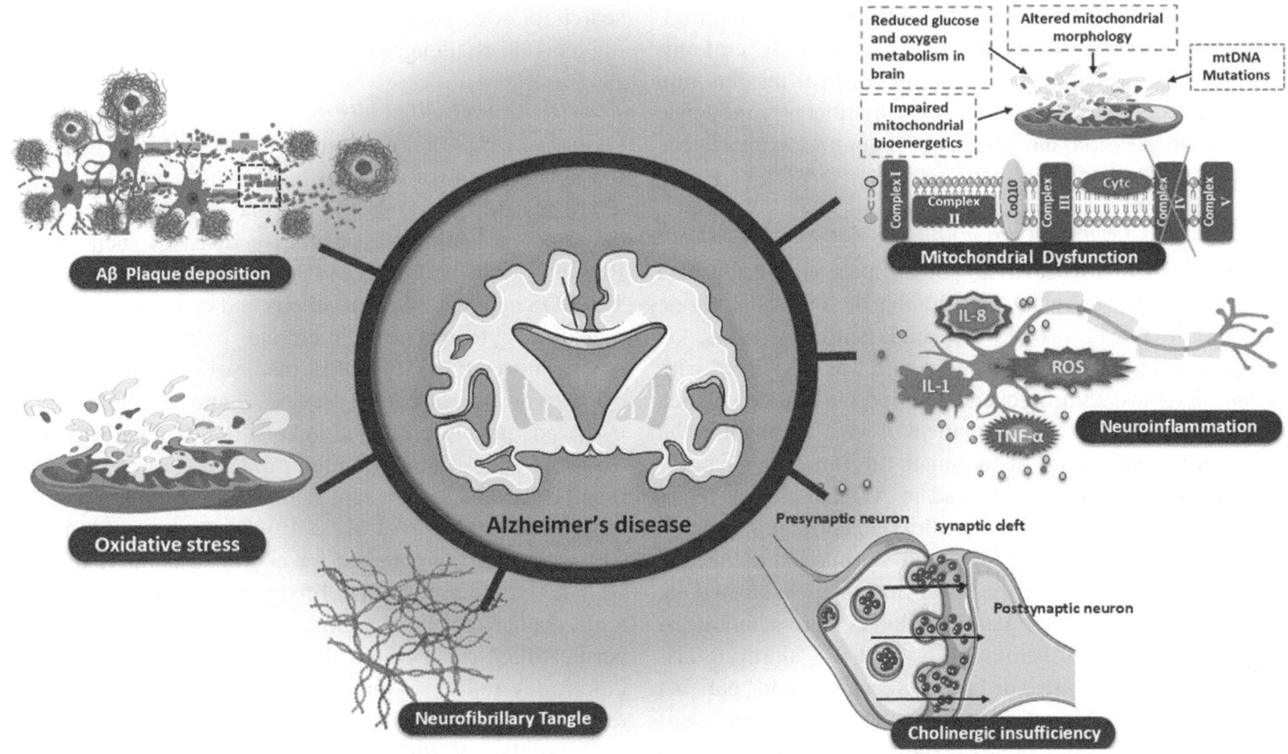

FIGURE 19.1 The pathophysiology of Alzheimer's disease (AD) encompasses the principal molecular and cellular mechanisms contributing to its progression, such as the accumulation of amyloid-β plaques, hyperphosphorylation of tau protein resulting in neurofibrillary tangles, synaptic dysfunction, neuroinflammation, and oxidative stress, culminating in neuronal death and cognitive decline.

majority of neurodegenerative dementias, which are generally classified as tauopathies (Sinsky et al., 2021). Among these, AD stands out as a prominent example where tau is especially significant. This protein is indispensable not only for the proper development but also for the ongoing maintenance of the structural integrity of axonal microtubules, which are vital for neuronal function (Prokop, 2013). It is theorized that a variety of functions performed by tau come about through a complex interplay of conformational states that are either permitted or promoted by specific phosphorylation events that occur over time. Under normal physiological conditions, tau tends to form soluble protein subunits that are highly charged, allowing them to function effectively within the cellular milieu. However, when confronted with pathological conditions, tau undergoes a critical transformation; the protein begins to self-aggregate and assembles into various high molecular weight species. This aggregation includes the characteristic straight filaments that are found in paired helical filaments, which represent the most defining structures encountered in the progression of AD (Goedert, 2015).

Hyperphosphorylation is a primary trait in conformational maladjustment that is crucial to the pathogenesis of AD and other tauopathies (Li, 2016). Truncation, proteolysis, glycosylation, nitration, pyroglutamate formation, ubiquitination, glycation, advanced glycosylation end product formation, and isomerization (including cis-trans isomerization of proline residues) have all been recognized as occurring in parallel to and cooperating with phosphorylation, leading to the formation of the pathological species of tau (Polanco et al., 2018). The accumulation of paired helical filaments has direct harmful impacts on synaptic activity, which is also considered to be causing the memory impairment characteristic of patients with AD, especially in the early stages (Bae & Kim, 2017).

19.2.1.1.3 Neuroinflammation

In response to a broad range of foreign and endogenous pathogenic stimuli, the central nervous system (CNS) experiences a complicated and sophisticated set of inflammatory processes known as neuroinflammation (Kölliker-Frers et al., 2021). In normal circumstances, the early activation of microglia is closely associated with the effective prevention of harmful factors, thus providing crucial and vital neuroprotection. When microglial cells become activated, they produce a diverse and extensive range of inflammatory factors, such as nitric oxide, reactive oxygen species (ROS), cytokines, and chemokines, all of which play significant and essential roles in the immune response (Smith et al., 2012). However, when there is a significant and excessive accumulation of harmful stimuli, the inflammatory response can become protracted and chronic, ultimately leading to severe destruction of the CNS and establishing a vicious and destructive cycle of neuron death and further damage. Over the past five years, a multitude of reports and studies have addressed the potential therapeutic, diagnostic, and pathogenic relationships that exist between neuroinflammation and various neurodegenerative diseases in several animal models, providing important and critical insights into how these complex interactions may significantly influence disease progression and severity (Guzman-Martinez et al., 2019).

19.2.1.1.4 *Cholinergic Hypothesis*

In the year 1982, researchers put forth the significant idea that there was a crucial and notable loss of cholinergic neurons in the brains of individuals suffering from AD (Contestabile, 2011). This phenomenon was associated with a considerable decline in overall cholinergic function, which subsequently led to the development of what is known as the cholinergic hypothesis. According to this intriguing hypothesis, the clinical stages observed in AD can closely relate to a cumulative loss of large cholinergic cortical projection neurons that are essential for cognitive function (Schaeffer & Gattaz, 2008). Based on the principles established by this hypothesis, various AChE inhibitors were synthesized with the intention of helping to prevent this alarming crisis in neurodegenerative disorders, specifically for patients affected by AD. Nevertheless, a major disadvantage associated with these compounds is their high susceptibility to side effects (Trippier et al., 2013). This occurs either through non-selective inhibition or selective inhibition at alternative AChE targets rather than at specific isoforms of AChE. This particular situation certainly heightens the significance of exploring natural AChE inhibitors as a viable and potentially safer alternative (Ramsay et al., 2016).

To date, extensive research has reported that various natural compounds show potential in reversing deficiencies associated with clinical symptoms of mild memory loss primarily by preventing the degenerative processes that contribute to cognitive decline (Baquero & Martín, 2015). Among these promising agents, a select group of AChE inhibitors has been rigorously evaluated in vivo and has demonstrated a significant ability to reduce neurodegeneration in a variety of preclinical studies. The beneficial effects of certain bifunctional compounds have consistently shown neuroprotective properties, in addition to their role as AChE inhibitors (Šebestík et al., 2011). This presents a compelling narrative regarding their relevance in therapeutic interventions. The discovery of multifunctional small-molecule compounds that effectively target both oxidative stress and cholinesterase inhibition represents a particularly promising strategy for the treatment of both symptomatic and neurodegenerative disorders, including but not limited to AD (Walczak-Nowicka & Herbet, 2021). These advancements highlight the potential of these compounds in clinical applications. In particular, natural MBOA alkaloids exhibiting both AChE and botulinum toxin activities are of immense interest, as they can serve as a new and exciting class of lead compounds for treating AD as well as conditions such as myasthenia gravis or organophosphate poisoning, which are closely linked to the inhibition of cholinesterase-related effects. This represents a significant step forward in the search for effective treatments that can address these pressing medical challenges (Pohanka, 2011).

19.2.1.1.5 *Oxidative Stress*

Oxidative stress causes cellular injuries and is considered the key mechanism responsible for both neurodegenerative diseases and aging. "ROS" play a crucial role in the development of neurodegenerative diseases such as AD, Parkinson's disease, HD, and ALS (Rekatsina et al., 2020). The ROS can be both endogenous and exogenous in origin, including cytochrome, glutathione, oxidative phosphorylation, nitric oxide, NADPH oxidases, xanthine oxidase, cyclooxygenase, and lipoxygenase. It appears that the brain is not adequately protected against the damage caused by oxidative stress due to its high production of ROS and the relative lack of antioxidant defense systems compared to other tissues (Cenini et al., 2019). The proposed oxidative stress-mediated hypotheses for neurodegenerative diseases include the accumulation of lipid peroxides, damage to proteins, changes in nucleic acids, and alterations in membrane permeability. As a result, dietary antioxidant supplementation may reduce oxidative damage to brain tissue, thus reducing vulnerability to age-related diseases (Singh, 2015).

A number of studies have also shown the antioxidant capacity of certain compounds, such as reducing nitric oxide induction and suppressing lipopolysaccharide-induced ROS production, as well as mitochondrial hyperpolarization of microglia and cortical astrocytes (Hassan et al., 2022). Diosgenin significantly ameliorates oxidative stress through the activation of Extracellular Signal-Regulated Kinase (ERK) phosphorylation. The therapeutic potential of curcumin against neuropathological events in Parkinson's disease has been demonstrated in a mouse model in which neurobehavioral abnormalities, release of proinflammatory cytokines, apoptotic neuronal loss, and accumulation of Lewy bodies and Lewy neurites were all significantly attenuated, especially when in combination with a vehicle. Resveratrol also protected dopaminergic neurons in a Parkinson's disease model by regulating HIF-1α, miR-34a, and DJ-1 (Batiha et al., 2023).

19.2.1.1.6 *Excitotoxicity*

Excitotoxicity has long been implied to be a major contributor in several neurodegenerative diseases, especially AD, Parkinson's disease, HD, and motoneuron disease (Flores et al., 2012). Excitotoxicity is the result of excessive levels of glutamate in the synaptic cleft and N-methyl-D-aspartate (NMDA) receptor activation. Overactivation of NMDA receptors leads to Ca^{2+} influx into neurons, which increases ROS production, lipid peroxidation, DNA fragmentation, and cytokine release. This condition triggers a series of cellular events such as the activation of several protein kinase cascade signaling and caspase-dependent apoptosis (Papa et al., 2006). In addition, overactivation of these receptors also causes energy depletion, and excitotoxicity disrupts neuronal function or eventually leads to apoptotic and necrotic neuronal loss. Excitotoxicity-induced neurotoxicity directly causes glutamate neurotoxicity and degrades glutamatergic function. Several natural and synthetic molecules are the focus of current research into neuroprotection against glutamate-induced neurotoxicity. The abundant source of cannabinoids and derivatives from *Cannabis sativa* provides opportunities to screen their potential in neuroprotection in the context of AD and Parkinson's disease (Pugazhendhi et al., 2021). Although several studies demonstrate that cannabinoids could trigger neuromodulatory actions and could suppress excitotoxicity and oxidative stress-related toxicity in neuronal cells, the endocannabinoid system still plays a multifactorial and diverse role in neurotransmission, neuroinflammation, preserving antioxidant defense, gene regulation, and the interaction of several cellular receptors in the CNS (Chiurchiù et al., 2018).

19.2.1.1.7 Mitochondrial Dysfunction

Mitochondrial dysfunction is often associated with the onset or progression of neurodegenerative diseases (Johnson et al., 2021). Mitochondria are the main site of oxidative phosphorylation, the primary cellular Adenosine Triphosphate (ATP) generation process in most neurons. ROS generation occurs due to the leakage of electrons from the electron transport system because of the partial reduction of oxygen, and some are considered by-products of the process (Madkour, 2019). Production is essential for normal cellular signaling and homeostasis in mitochondria. However, in mitochondria, oxidative damage, uncontrolled ROS production, or insufficient ROS removal by the mitochondria-associated antioxidant systems in relation to pathophysiological processes could be induced and cause damage (Rehman et al., 2023).

Moreover, mitochondrial damage also occurs due to any harmful factor that targets mitochondria, such as neurotoxins, hypoxia, injury, and exposure to high concentrations of glutamate in neurons (Facecchia et al., 2011). ROS generation can cause a chain reaction of lipid oxidation and DNA/RNA oxidation, leading to membrane dysfunction and reduced apoptosis in neurons. Dysfunctions in the proteins involved in electron transport from complexes I, II, and III and ROS production of these proteins are related to the pathophysiological processes associated with the onset or progression of neurodegenerative diseases (Islam, 2017).

19.2.1.1.8 Synaptic Dysfunction and Neurodegeneration

Since synaptic integrity is crucial for cognitive function, it is not surprising that the compromise of this function defines a group of genetic abnormalities that lead to neurocognitive disorders (Barrantes, 2024). Our nervous system contains many synapses, and it is estimated that there are 10,000 synapses in a vertebrate neuron. If synaptic function is impaired, it will impair the parallel computing power of the nervous system, cause cognitive inability, and lead to a variety of neuropsychiatric disorders. The accumulation of synaptic damage is also a hallmark of various neurodegenerative diseases, such as AD and Parkinson's disease (Mazzucchi et al., 2020). One of the two most prevalent neurodegenerative disorders is Parkinson's disease. Movement becomes functionally impaired due to brain degeneration. Bradykinesia, tremor, and stiffness are examples of release disorders. Parkinson's disease is characterized by a reduction in dopamine in the striatum and a loss of dopaminergic neurons in the substantia nigra. The primary pathogenic characteristics of dopamine degeneration are thought to be the existence of Lewy bodies and Lewy processes (Jellinger, 2012).

In AD patients, early synaptic disorders occur even earlier than Aβ or tau. They start to affect cognitive functions such as learning and memory. Synaptic loss is closely related to cognitive dysfunction both in AD and other neurodegenerative diseases, including Lewy bodies, frontotemporal dementia, and ALS (John & Reddy, 2021). In the underlying neuropathology, synaptic failure has been widely accepted as a key factor mediating cognitive impairment, rather than the total length of neuronal axons. Whether the spine density is normal or abnormal can be correlated with learning ability. The connection between spine density and learning ability is probably the foundation of neurogenetic ability (Batool et al., 2019). However, synaptic and dendritic abnormalities assessed by microscopy do not exclusively contribute to cognition, but at the same time increase other neurological pathophysiological conditions, such as Parkinson's disease, Lewy body dementia, or corticobasal degeneration. In previous studies, there has been fragmentary proof that dendritic deterioration and synaptic dysfunction are the potentiated rudiments for most neurological abnormalities, for instance, cognitive deficit, dystonia, or tremor, among others (Ciurea et al., 2017). Indeed, in some of the most frequent neuropsychiatric disturbances, the more accepted change occurs not in myelinated axons, but rather in the dendrites and their ensembles at the postsynaptic complex. Given these previous studies, in addition to diminished spine density, deficient dendritic spine plasticity, and morphological flaws can directly disturb synaptic biology, producing disproportionate neuronal action (Dejanovic et al., 2024).

Cognitive function requires overall synaptic plasticity. Excitatory and inhibitory synapses must have well-coordinated operation; otherwise, the imbalance between excitatory and inhibitory stability and loss will lead to cognitive and neuropsychiatric disorders, for example, epilepsy, autism, schizophrenia, or bipolar affective disorder (Priego, 2021). Consequently, potential treatments should focus on rectifying the imbalance existing in neuropathology that potentially causes cognitive and neuropsychiatric disorders. Furthermore, this study identifies the main synaptic pathology and the residual unanswered questions, and the study proposes potential future research to aid in surmounting deficits in most neurological illnesses (Nazir et al., 2024).

19.2.1.2 Pathophysiology of Parkinson's Disease

PD is a chronic and progressive neurodegenerative movement disorder whose cardinal pathological features include a gradual destruction of dopaminergic neurons in the

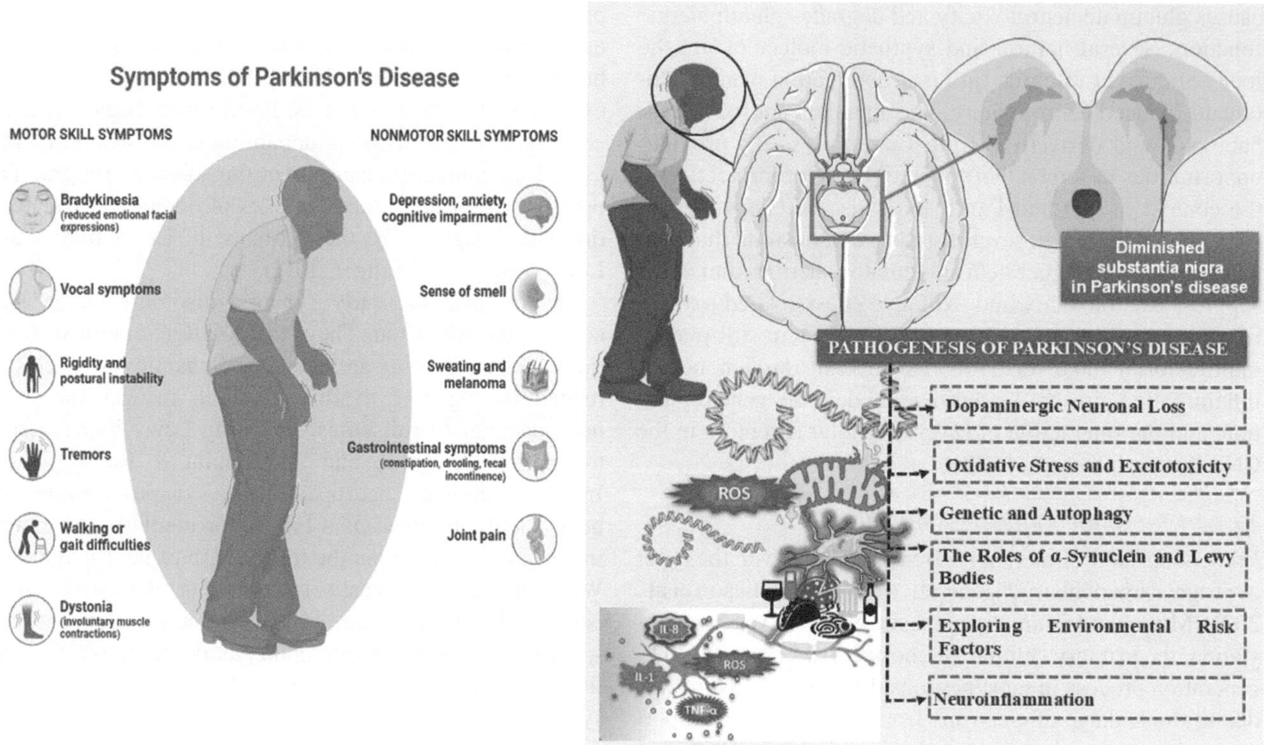

FIGURE 19.2 Pathophysiology of Parkinson's disease (PD)—Illustration of the central mechanisms implicated in PD, encompassing the gradual degeneration of dopaminergic neurons in the substantia nigra, the accumulation of alpha-synuclein aggregates resulting in Lewy bodies, mitochondrial dysfunction, oxidative stress, neuroinflammation, and disrupted dopamine signaling, culminating in the motor and non-motor symptoms typical of the disease.

substantia nigra pars compacta and the presence of cytoplasmic Lewy bodies (Rahul & Siddique, 2022). Although motor symptoms are the primary clinical hallmark of PD, the disease is complex and involves a number of other pathophysiological processes, including oxidative stress, mitochondrial dysfunction, and inflammation. The exact cause of neuronal degeneration is still unknown, but data suggest the involvement of multifactorial pathways, such as oxidative stress, proteasomal dysfunction, mitochondrial impairment, excitotoxicity, inflammation, and apoptosis (Radi et al., 2014).

As oxidative stress accumulates and becomes detrimental, it has the potential to initiate neuronal cell death. RNS and ROS are associated with the oxidation of cellular lipids, proteins, and DNA, leading to a cascade of events culminating in caspase activation and ultimately, the destruction of neurons (Moujalled et al., 2021). The distinctive anatomy, physiology, and biochemistry of the neurons that differentiate them from other cell types in the brain might contribute to the vulnerability of the neurons to oxidative stress. Glutamate-mediated excitotoxicity is a key pathophysiological event in a variety of neurological disorders that consequently damage dopaminergic neurons and promote neuronal cell death (Iovino et al., 2020). In response to excitotoxicity, mitochondrial dysfunction, and oxidative stress, an inflammatory reaction is activated. The microglia-mediated neuroinflammation might induce a decrease in dopamine uptake and release, oxidative stress, inflammation activation, and mitochondrial dysfunction. Besides the neuropathologic processes mentioned in the text, others, like ER stress and decreased neurotrophic factor signaling, might also play a role in the loss of dopaminergic neurons (Karvandi et al., 2023) (Figure 19.2).

19.2.1.2.1 Dopaminergic Neuronal Loss

The loss of dopaminergic neurons in the substantia nigra and their striato-nigral projection causes Parkinson's disease. Parkinson's disease is the major age-related movement disorder (Tong et al., 2017). The dopaminergic neuromelanin-containing substantia nigra neurons in the midbrain are sensitive to neurotoxic damage by oxidants, energy inhibitors, and various environmental and genetic factors. The basis of Parkinson's disease is a mild chronic augmentation of active iron and dopamine, along with their redox incompatibilities that cause age-related vulnerability and neuronal degeneration's oxidative stress (Zhu et al., 2023). Cell culture studies have demonstrated that neuromelanin iron could become redox-active if it were discharged from the organelles. The loss of a regular reduction point, like glutathione, would enhance the production of monoamines' active quinone, along with other protein-linked molecules. Oxidized proteins might catalyze iron into reactive iron, exacerbating cell death. Discharge of dopamine from the presynaptic components that damaged dopaminergic neurons causes oxidative stress (Wise et al., 2022). Consequently, medications that lessen the symptoms of Parkinson's disease could also increase the formation of dopamine quinones within the cell, which is a harmful effect. In addition to aging and toxin-elicited oxidative stress, Lewy bodies are also a sign of heritable and idiopathic Parkinson's disease. Animal studies implicate the ubiquitin-proteasome pathway in various forms of Parkinson's disease, encouraging the formation of drugs that influence the proteasome and enhance protein degeneration (Lim & Tan, 2007).

19.2.1.2.2 Role of α-Synuclein and Lewy Bodies

It is also believed that oxidative stress plays an important role in the development of this disorder due to its prevalent aggregation and rapid production of ROS (Chen et al., 2012). This point of view is logically supported by the fact that its specific inhibitor has a higher content of polyunsaturated fatty acids. Increased oxidative damage to these acids could, in turn, enhance the aggregation and toxicity effects, causing further cell damage and potentially pathogenic events. Thus, the relationship between oxidative stress and Parkinson's disease can be taken a step further by implicating a link that plays between increased oxidative stress and mitochondrial functions (Dias et al., 2013).

These specific mutations are noted to have quite an increased priming effect on the protein. Interestingly, the mutations that enhance the association with the mitochondrial membrane are also those association with the discovery of Parkinson's disease (Borsche et al., 2021). By interacting with the oxidation state of the membrane, it mediates ROS production. These mutations enhance the secretory pathway and, unlike non-mutants, enhance vesication during oxidative stress. This suggests that one mechanism in neurodegeneration is to promote inappropriate vesicle formation on the mitochondria. Preventing vesicle formation by inactivating it may be a way to protect against excessive oxidative stress. However, the precise mechanism by which it causes oxidative damage to enhance mitochondrial defects and neurodegeneration is not known (Chen et al., 2012).

19.2.1.2.3 Neuroinflammation

Overall, many new molecules upregulating Nrf2 were discovered by High-Throughput Screening (HTS) for the protection of neural cells from apoptosis. The Nrf2-activating effect of the phytochemicals paeoniflorin and liquiritin was discovered and studied by HTS, using a recently synthesized hybrid of arylroscorcinol and Cytochrome C Oxidase (CCO) in control and SH-SY5Y cells (dopaminergic neuroblastoma) (Velagapudi et al., 2018). The study of silbinins A and B by Gene Set Analysis (GSA) was performed in BV-2 cells (mouse microglial cell lines) treated with Lipopolysaccharide (LPS) and C2-Exo-Glc. In this biological model, silibinin A and silibinin B targeted Nrf2 with different potencies, demonstrated their neuroprotective properties at a concentration of 3 μM, and reduced both LPS- and C2-Exo-Glc-enhanced ROS production. Here, we comprehensively discuss the role of naturally endowed Nrf2 activators in the acquisition of resistance of neural cells and brain tissues to proteinopathies and oxidative stress, for instance, in excitotoxicity, neuroinflammation, apoptosis, amyloidogenesis, and neurofibrillary degeneration, revealing the impact of the natural molecules on structural and functional features of proteins, pathologically correlated with the development of neurodegenerative diseases (Kim et al., 2024).

19.2.1.2.4 Oxidative Stress

Oxidative stress is central to the progression of many neurodegenerative diseases such as Alzheimer's, Parkinson's, Huntington's, prion diseases, and ALS (Islam, 2017). The brain is highly susceptible to oxidative damage due to excessive use of oxygen, high levels of polyunsaturated fatty acids, high production of superoxide radicals, and low levels of detoxifying enzymes. Oxidative stress arises from an imbalance between the generation and amelioration of toxic free radical species. Generally, ROS are derived from the incomplete reduction of oxygen, such as the superoxide anion radical or hydrogen peroxide (Kehrer & Klotz, 2015). The hydroxyl radical, the most potent toxic ROS, is generated via a reaction involving transition metal ions. On the other hand, damage from oxidative stress mostly implicates lipids, proteins, and nucleic acids. For lipids, oxidation leads to the production of aldehydes, which generate adducts with proteins or DNA that disturb cellular communication and the normal function of cell components. High levels of accumulated oxidative damage may lead to impaired mitochondria, reduced ATP production, and loss of membrane potential (Cui et al., 2012).

In parallel, oxidative stress may lead to protein impairments, including loss of function, aggregation, and the formation of misfolded proteins. In order to repair oxidative damage to cellular components, the rapid and effective action of detoxifying enzymes and non-enzymatic antioxidants is required (Gregersen & Bross, 2010). However, the high reactivity of ROS causes disruption of the cellular defense system, inducing physiological damage. Notably, oxidative stress in the CNS is responsible for declining cognitive abilities during aging, ultimately leading to neuropsychiatric diseases. Thus, scavenging ROS or improving cellular protection against them are critical strategies for preventing neurodegenerative diseases (de Vries et al., 2008).

19.2.1.2.5 Excitotoxicity

The mechanisms leading to neuronal loss are still unknown. It is believed that there is a group of mechanisms that cause degradation or cell death of neurons:—Apoptosis: it is characterized by the loss of normal plasma membrane integrity, chromatin condensation and fragmentation, cell shrinkage, and the formation of membrane vesicles called apoptotic bodies that are easily phagocytized by neighboring cells (Mierke & Mierke, 2020). One of the genes that regulate apoptosis is the Bcl-2 gene that belongs to the same family as the previous gene. This gene has the opposite effect of the previous gene, thereby inhibiting the action of the apoptotic genes. Despite having a similar origin to the sequence, this gene has an antagonistic action to P53. Excitotoxicity: this mechanism leads to neuronal death in situations of over-stimulation by excitatory amino acids. The result is the overload of calcium, which induces the activation of calcium-dependent proteins that will cause irreversible damage to neurons. When this happens, the presynaptic action is increased considerably, implying loss of control in relation to the postsynaptic action. The reuptake system will fail, and the glutamate levels in the synaptic cleft will increase. This increase occurs by attacking two areas of the glutamatergic system: glutamate transporters and second messenger-gated genes (Brasnjo & Otis, 2001).

The result of the excitatory amino acids is the activation of receptors in various neurons of the CNS. Glutamatergic synapses are expressed in most of the neocortex's forebrain, basal ganglia, and pontine nuclei (Hassel & Dingledine, 2012). This system has glutamate receptors, kainate, sigma, and NMDA. These last receptors are involved in most trophic relationships and plasticity of the CNS. Indeed, in NMDA, we find at least four subunits, each with at least seven identified excitotops forming various pentameric complexes. This effect is mediated by an ion flux through the plasma membrane and the connections to intracellular signaling pathways, including the action of specific enzymes (Pandey et al., 2007).

19.2.1.2.6 Genetic Factors

Familial occurrences of ALS account for approximately 10–20% of all diagnosed cases of this neurodegenerative condition. It is important to note that familial and sporadic cases of ALS are clinically and neuropathologically indistinguishable from one another, which adds complexity to diagnosis and research efforts. Clinical presentations commonly involve uni- or bilaterally focal limb weakness, which may manifest in a variety of ways (Longinetti et al., 2017). In contrast, sporadic cases are typically observed in older individuals and often present as generalized disease progression over time. Occasionally, one specific aspect of ALS can be traced back to a mutation in the SOD1 gene, which has garnered significant attention. Although over 100 mutations have been reported regarding this gene, current research has comprehensively examined their relationship with oxidative stress within various animal models (Abokyi et al., 2020). Another important gene implicated in both familial and sporadic cases of ALS is TDP-43. This gene has been the subject of extensive study, particularly concerning its role in the pathophysiology of the disease. Remarkably, approximately 96% of ALS cases have been found to present with TDP-43 aggregates, indicating a strong association between this gene and ALS. Beyond these two prominent genes, research has also documented the involvement of other genetic factors. For instance, mutations in the VAPB and SQSTM1 genes have been identified in approximately 25% of familial ALS cases exhibiting frontotemporal lobar degeneration, highlighting the complexity and genetic variability of the disease (Ranganathan et al., 2020). Furthermore, there are several prosite markers related to TDP-43 that are more prevalent in familial cases of ALS. These include proteins such as valosin-containing ATPase, optineurin, and UBQLN2, which play a role in cellular stress responses and protein homeostasis. Most familial forms of ALS exhibit the presence of these markers, while sporadic cases remain predominantly characterized by their irregular and less predictable patterns. The ongoing research into the genetic underpinnings of both familial and sporadic ALS is crucial for advancing our understanding of the disease and ultimately developing targeted therapies that address both forms effectively. The complexity of ALS genetics continues to open new avenues for exploration and insight in the quest for effective treatment and management strategies (Akçimen et al., 2023).

19.2.1.2.7 Impaired Proteostasis and Autophagy

Neurodegenerative diseases are classically characterized by the abnormal accumulation of pathogenic protein aggregates within the affected neural tissues. The reasons behind the formation of these pathological aggregates can be numerous and complex, including factors such as misfolded proteins, damaged proteins, and excessive synthesis of certain proteins (Gandhi et al., 2019). However, in the context of neurodegenerative diseases, the misfolded proteins and aggregated proteins are not effectively modulated and cleared from the cellular environment due to impaired proteostasis mechanisms. Consequently, if the cellular clearance mechanisms of these accumulated proteins are insufficient or fail altogether, it can lead to toxic effects that instigate a self-propagating cascade, exhibiting characteristics akin to prion-like conformations. Moreover, an alteration in the transcriptional expression of genes that encode molecular chaperones also significantly impacts the tendency and rate at which misfolded proteins aggregate (Kim et al., 2013). These aggregated proteins, noted for being insoluble and intracellular, are commonly found in a variety of neurodegenerative diseases, including Alzheimer's, Parkinson's, and HD, among others. In conclusion, the efficient clearance of these aggregated proteins from specialized cells, such as through proteasomal degradation pathways and autophagy processes, is utterly crucial for maintaining neuronal health and function. The impairment of proteolytic pathways ultimately leads to an alarming accumulation of misfolded proteins, resulting in the aggregation of these disease-associated proteins that contribute to the progression and severity of neurodegenerative conditions (Singh et al., 2024).

19.2.1.3 Pathophysiology of Huntington's Disease

Huntingtin is a large and extraordinarily complex protein that is critically involved in a multiplicity of essential normal functions within the cells. Among its various roles, it participates in the transportation of nutrients, the intricate process of protein trafficking, and, importantly, the activation of a transcription program that is fundamental for cellular activities (Tabrizi et al., 2020). The functions that are

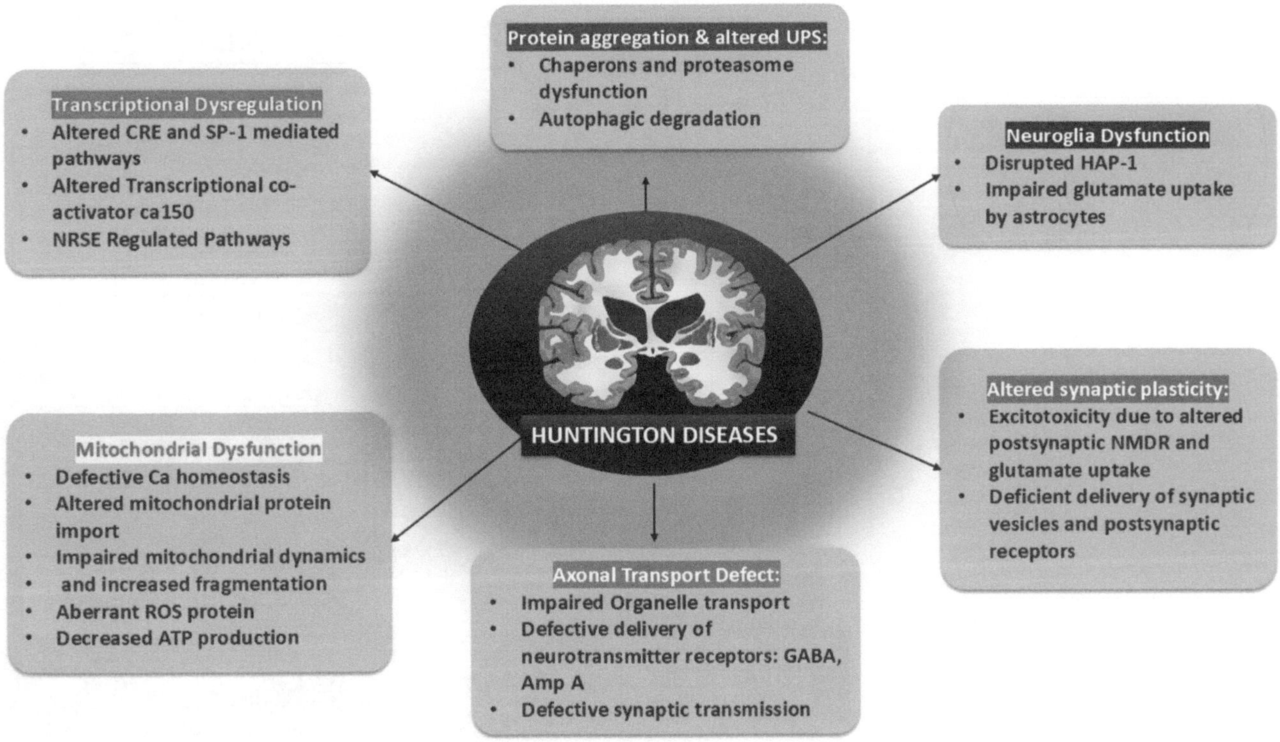

FIGURE 19.3 Pathophysiology of Huntington's disease (HD)—Depiction of the molecular and cellular processes in HD, emphasizing the mutation in the huntingtin (HTT) gene that results in expanded CAG repeats, the formation of toxic mutant huntingtin protein aggregates, neuronal dysfunction, excitotoxicity, mitochondrial impairment, oxidative stress, and progressive neurodegeneration, particularly in the striatum and cortex, culminating in motor, cognitive, and psychiatric manifestations.

predominantly significant to the neurons, especially those that shed light on the potential underlying mechanisms of neurodegeneration, represent new vital clues about the complexities associated with the pathophysiology of HD. Consequently, treating HD is specifically focused on these crucial aspects of Huntingtin's functions. Disruption of the normal function of Huntingtin is caused by the presence of a mutant protein characterized by expanded CAG repeats, which significantly alters its role within the body (Aronin et al., 1995). There is an early and progressive loss of the striatum as observed through functional magnetic resonance imaging in human patients, which correlates with progressively severe phenotypic manifestations of the disease. In various rodent models, numerous mitochondrial problems, dysfunctions in axonal and dendritic trafficking, and alterations in gene transcription have been documented (Calkins et al., 2011). These prevalent findings strongly support their potential importance to the pathogenesis of HD, derived not only from in vitro models but also from information gathered through imaging studies involving patients alongside transgenic animal models. As a result of the identification of remarkable disease-linked molecular functions for N-terminal Huntingtin fragments that are not only present but also toxic in cells affected by HD, recent research studies have begun to focus extensively on N-terminal mutant Huntingtin. This area of investigation is proving to be critical, as these particular structures can significantly influence the therapeutic strategies involving both antibodies and synthetic molecules geared toward the treatment of HD (Denis et al., 2019) (Figure 19.3).

19.2.1.4 Pathophysiology of Amyotrophic Lateral Sclerosis (ALS)

ALS is a fatal neurodegenerative disease characterized by the degeneration of both upper motor neurons in the brain and lower motor neurons in the spinal cord. With progressive muscle atrophy due to muscle weakness and paralysis, it is a chronic, debilitating disease that results in severe disability and death within several years following diagnosis (Mora & Chiò, 2015). The major symptoms of ALS include muscle weakness, atrophy, dysarthria, and dysphagia. Cognitive dysfunction occurs in some patients, but in general, compared to frontotemporal dementia, it occurs at an early stage among the majority of ALS patients. The etiology of the majority of ALS is unknown, but some cases, approximately 10%, are linked to genetic mutations (Majoor-Krakauer et al., 2003). A proportion of such familial cases have been associated with autosomal dominant mutations in copper- and zinc-superoxide dismutase 1, causing reduced capacity to decrease the superoxide levels and neutralize the toxic action of the excess of free radicals. In addition to the familial mutation, mutations in copper and zinc-dependent extracellular copper chaperone for superoxide dismutase have been identified in some patients.

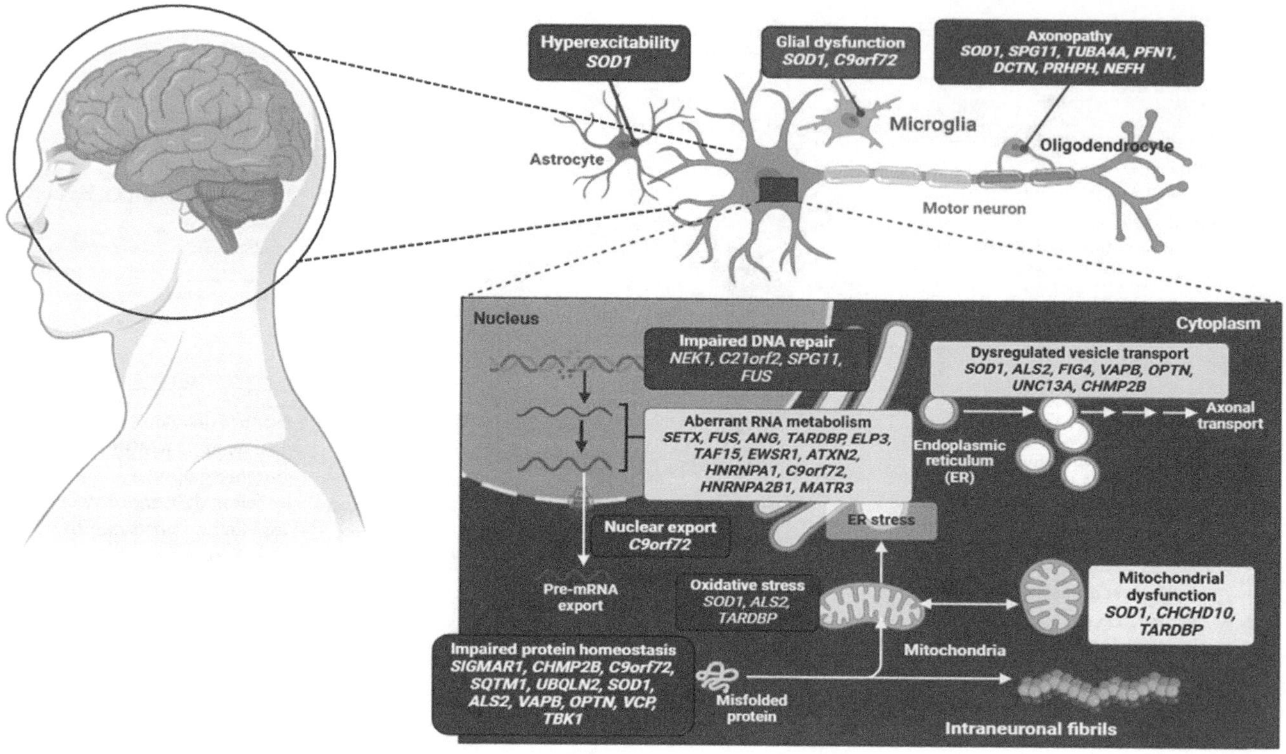

FIGURE 19.4 Pathophysiology of amyotrophic lateral sclerosis (ALS)—Illustration of the principal mechanisms involved in ALS, encompassing the aggregation of misfolded proteins such as TDP-43, SOD1, and FUS, glutamate excitotoxicity, mitochondrial dysfunction, oxidative stress, neuroinflammation, and impairment of axonal transport, culminating in the progressive degeneration of motor neurons in the brain and spinal cord, which leads to muscle weakness and atrophy.

The ALS2 gene, which codes for alsin, the product of which contributes to motor neuron viability and maintenance, is a recessive juvenile ALS gene that exhibits an unusual spectrum, such as the lymphoma-like phenotype (Iżycka-Świeszewska, 2023).

Several pathomechanisms, such as oxidative stress, glutamate toxicity, protein aggregation, abnormal mitochondrial dynamics, neuroinflammation, decreased growth factors, and axonal transport, have been reported to be involved in ALS pathogenesis (Tedeschi et al., 2021). First, regarding oxidative stress that is implicated in the etiology and progression of ALS, several lines of evidence showed that superoxide dismutase, which is decreased in copper, zinc, or its zinc-free status, could play a pivotal role in the pathophysiologic mechanism for familial and sporadic onset of ALS. Second, several extended works have indicated that an excess of glutamate in the synaptic area of motor neurons, pathological transactivation of glutamate receptors, changes in glutamate transporter dynamics, and a dysfunction of glutamate transporters influence the pathophysiologic mechanism of ALS. Glutamate toxicity would be induced by an excessive extracellular concentration of glutamate, leading to the entry of an excessive amount of calcium, which induces mitochondrial swelling, deficit, and cytochrome c release, elevating the expression of the glutamate transporter and causing energy shortfalls (Balkhi et al., 2014). Third, the generation of insoluble SOD1-immunopositive inclusions that have a tendency to accumulate in motor neurons, negative interneurons, and the ventral horn of the spinal cord is a characteristic feature of ALS, accounting for a protein-seeding change, a significant factor of SOD1 conformational abnormalities, and glutathione synthesis, protein aggregation, mitochondrial localization, the expression of TNF-α, NF-κB activation, increased oxidative DNA damage, neuroinflammation, and decreased axonal transport. In particular, an abnormality in the dynamics of several mitochondrial-related proteins, which include mitochondrial uncoupling proteins and SOD1, would operate through the pathomechanism of ALS. Finally, a dysfunction in the axonal transport of several molecular proteins, such as neurofilament, kinesin-1 heavy chain, and dynactin complex, would be linked to the decreased motor axon transport hypothesis of ALS (Grierson & Miller, 2021) (Figure 19.4).

19.2.1.5 Exploring the Pathophysiology of Multiple Sclerosis

Sclerosis means hardening and indicates local fibrosis or hardening of various tissues. It attacks both the gray and white matter of the nerve fibers in the various parts of the CNS. The actual cause and pathogenic process of multiple sclerosis are largely unknown (Chaudhuri, 2013). Multiple sclerosis is considered a type of disease due to autoimmune damage and is also called a systemic disease because it affects a person's entire body. Pathogenetic studies showed that MS is an autoimmune inflammatory disease (Wang et al., 2015).

The reason for the hardening is that the myelin sheath of the nerve fibers is largely reduced due to demyelination, and axon and glial proliferation is caused by hardening or sclerosis (Hampton et al., 2008). Multiple sclerosis occurs after viruses, bacteria, toxins, and other factors injure neurons, but lesions on myelin sheaths provoke autoimmune responses in the body, activate T-lymphocytes, and destroy the normal structure of the nerves by releasing neurotoxic cytokines. Furthermore, CD4+ activation of Th1, Th17, and other pathways leads to macrophage activation and the release of proinflammatory mediators, erosive myelin, and axons. The blood-brain barrier is destroyed, and the antibody and immunoreactive mediators penetrate the blood-brain barrier and damage the nerve tissue and tissue structure. This causes inflammatory demyelination, blockage of nerve conduction, reduction, and disappearance of stable and long-term effects on neurological functions, and a decrease in quality of life (Pirko & Noseworthy, 2009).

19.3 CURRENT TREATMENT APPROACHES

The social-demographic situation of many countries results in the necessity to reduce public expenses in the area of health. Meanwhile, the population gets older, and the number of elderly and longevity population groups increases (Maheshwari & Maheshwari, 2024). This occurs in a global context, where neurodegenerative diseases are a matter of great concern due to several factors such as the lack of a cure until now, the fact that at present there are only symptomatic treatments, the large public expenditure on the maintenance of patients who require daily care and lose the ability to carry out their personal activities as a consequence of the development of the disease (Alzheimer's Association, 2013). Once these diseases have been strongly linked to oxidative stress, the lines of research for the development of possible treatments are mostly focused on antioxidants. It is very important that the therapeutic agents have high activity, low or no toxicity, do not interfere with the clinical progression of the disease, have difficulty in the development of resistance, and do not interfere with drugs used in long-term therapies, since many patients are of advanced age and have more than two pathologies and take several chronic medications (Ingersoll & Cohen, 2008).

The choice of an adequate model is important for the reproducibility of the results. In vitro models, where cells in culture mimic the characteristics of pathologies, use cancer cells, which have high metabolism and growth, with subsequent demands for high concentrations of the compound (Barbosa et al., 2021). Animal models raise questions inherent to the models. The volume of pharmacokinetics and all aspects of bioavailability interfere with the model, since many molecules behave differently between species, even when plants have humans and rodents among the target species. Therefore, the research of bioactive natural compounds with therapeutic potential for neurodegenerative diseases using in vivo, in vitro, and ex vivo models, associated with the understanding of their mechanisms, as well as the development of their nanoformulation with high bioavailability, represents critical points to be considered in the search for possible new therapies. This review seeks to bring this information to light (Agrahari et al., 2019).

19.3.1 Pharmacological Interventions

Currently, the therapeutic criteria for neurodegenerative disorders are neuronal restoration, neuronal replacement, and neuronal protection, including neuroprotection with drug-based therapy. The first step of treatment is generally to control the symptoms through drugs that will modify chemicals within neuronal axons, helping communication (Reddy et al., 2020). An increased understanding of neurodegeneration maps the routes to prevention, repair, and protection of neurodegenerative status, which is a critical service to an aging society. Basically, neuroprotection is a way to promote the survival of neurons by trying to prevent their destruction in the brain. In this context, several functional channels and receptors play a pivotal role in the action mechanism of naturally occurring chemical moieties that have been found to be greatly effective in the suppression of neuronal death due to ischemic events (Jia et al., 2024).

Potent neuroprotective molecules isolated from varied sources, such as marine organisms, various medicinal plants, and other environmental resources like fruits, peppers, tea, and grapes, belong to a different class of molecules, including alkaloids, polyphenols, flavonoids, and terpenoids (Gomes et al., 2024). These chemicals were found to create a neuroprotective effect via altered expression of AMPA and NMDA channels, blocking the function of ionotropic receptor cation channels, or agonizing GABAA receptors with the prevention of glutamate binding, controlling oxidative stress, and augmenting intraneuronal survival molecular pathways. It has also been found that they contribute to the development of neuronal growth and the creation of neurotransmitters (Weng, 2024). With respect to the different modes of action, a number of different molecules have been found to modulate pathological dysregulation by interacting with transmitter-indicated receptors, which are basically G-protein-coupled receptor types of molecules in which the gliotransmitter that activates metabotropic types of glutamate receptors is found (Bradley & Challiss, 2012).

19.3.2 Non-Pharmacological Interventions

Prominent among non-pharmacological strategies is dietary enhancement. It has been reported that significant comorbidities are linked with abnormal or deficient levels of vitamins and essential fatty acids, and some dietary supplements can be used to treat or reduce the progression of certain neurodegenerative disorders. A severe issue related to the use of natural dietary enhancement, such as antioxidants and anti-inflammatory molecules, is the lack of a definite association with neuroprotection (Kidd, 2005). However, in the elderly, particular dietary supplements are linked to lower odds of dementia and enhanced memory function. Additionally, calcium and vitamin D supplementation may help inhibit bone fractures in the elderly, and correction of a vitamin D deficit, alongside muscle repair, is linked to a decreased risk of falls (Bischoff-Ferrari et al., 2018). In addition to dietary modifiers, integrative physical activity boosts life expectancy and accelerates aging, and it is connected with a decreased risk of cognitive decline in older adults. Integrative physical activity and training can enhance cognitive functioning and nerve health in people of any age. Exercise can increase blood supply to the brain, boost brain-derived nerve survival factor, and support nerve connections that are significant for memory and learning. One of the most significant modifiable factors associated with lower cognitive function and quicker progression to dementia is increased oxidative stress and inflammation. Therefore, the current information implies that integrating medical treatment with physical training, a balanced diet, and other potentially preventative programs can enhance the odds of effective neurodegenerative disorder treatment (McGough et al., 2017).

19.4 NATURAL MOLECULES AS POTENTIAL THERAPEUTICS

Neurodegenerative diseases include a diversity of diseases characterized by a progressive loss of structure or function of neurons that ultimately leads to death. The number of diagnosed patients around the world continues to grow, but the target for treating these diseases is still very limited (Kovacs, 2019). With the advance of scientific knowledge and the constant search for disease treatments, plants have been providing new chemotherapeutic possibilities for the treatment of neuronal damage. However, the use of plants is not yet fully explored due to low selectivity, low bioavailability, rapid metabolism, and side effects of substances that penetrate the blood-brain barrier, such as flavonoids and phenolic acids present in plant composition. As a prototype of an oral chemopreventive agent, many studies have proven the role of natural bioactive components in maintaining normal brain function and safeguarding the brain from damage by fighting neurodegenerative processes (Shabir et al., 2022).

Knowledge about the mechanism through which exogenous substances reach the brain after passing through systemic circulation, and how the biological barrier operates as a neurovascular unit, may provide the basis for optimizing system delivery for treating neurodegenerative diseases. Such knowledge can forge new treatments for natural products with poor bioavailability (Pandit et al., 2020). In addition, since many of the secondary metabolites have high lipophilicity and low molecular weight, numerous attempts have been made to develop the first soluble natural products for discovery that protect brain tissue. It is an ancestral pursuit, while unlocking the significant potential of therapeutic natural bioactive components, combined with findings from the application of pharmacogenomics, nanotechnology, and other fields, to advance the treatment of drugs for neurodegenerative diseases in an innovative manner (Viskelis & Surguchov, 2024). This review provides a systematic summary concerning the underlying mechanism of some candidate neuroprotective molecules derived from natural sources. This includes alkaloids, terpenes, biophenolic compounds, and other recurring substances known to have neuromodulatory potential used to treat neurodegenerative disorders (Gomes et al., 2024).

19.4.1 Characteristics of Natural Molecules with Neuroprotective Properties

Despite sharing no direct structural similarity, most natural molecules with neuroprotective properties exhibit the ability to act as potent antioxidants. Noteworthy are ascorbic acid, tocopherols, terpenes, flavonoids, and isoquinol-related alkaloids, chalcones, or coumarins (Mani et al., 2023). Antioxidative properties of a substance not only result from its ability to neutralize ROS, although it certainly is important. The antioxidative activity of a substance includes the ability to inhibit the formation of the Chronic Myelogenous Leukemia (CML) receptor responsible for ROS and to attenuate neurotoxicity or toxicity of other cells. Neuroprotective molecules, apart from scavenging oxidized Low-Density Lipoprotein (LDL) presented on microglial surfaces, should interact with TLR4 and, consequently, inhibit the microglia-to-neuron proinflammatory signaling. Perfectly fitting four characteristics provide remarkable neuroprotective potency (Ciruela et al., 2010).

Neurodegenerative diseases usually involve the loss of neurons as a result of several processes that culminate in the increase of oxidative/nitrative stress and finally lead to the accumulation of toxic protein oligomers or aggregates. Based on available experimental data, the chapter aims to critically review the roles of some natural molecules that are known as potent antioxidants in the prevention of neuronal loss during the development of neurodegenerative diseases (Grimm et al., 2011). There is abundant experimental evidence showing the efficiency of natural antioxidants in the prevention of neurodegenerative diseases, along with the analysis of three herbs with the most promising properties and interesting chemical profiles (Marino et al., 2022) (Table 19.2).

19.4.1.1 Neuroprotective Effects of Natural Molecules in AD

Neurodegenerative diseases, especially AD, Parkinson's disease, and HD, are problems that concern worldwide healthcare services. New molecules to treat these diseases should meet the requirements of stability in biological environments, membrane permeability, specificity, and feasibility (Zahra et al., 2020). However, some of the efficient

TABLE 19.2
Natural Molecules with Neuroprotective Properties and Their Mechanisms of Action

Sr. No.	Natural Molecules	Source	Mechanism of Action	Targeted Disease	Reference
1	Curcumin	Turmeric	Anti-inflammatory, antioxidative, anti-protein misfolding, and anti-apoptosis	Ischemic stroke, multiple sclerosis, AD, and Parkinson's disease	Garodia et al. (2023)
2	Resveratrol	Berries, grapes, and nuts	Antioxidant, anti-apoptotic	Alzheimer's, Huntington's disease	Zhang et al. (2021)
3	Huperzine A	*Huperzia serrate*	Inhibits acetylcholinesterase, neuroprotective	Alzheimer's, disease	Ohba et al. (2020)
4	Ginsenosides	Ginseng	Enhances neuronal survival, anti-inflammatory, antioxidant, anti-inflammatory, neuroprotective, anti-apoptotic, and immune-modulatory properties	Alzheimer's, Parkinson's disease	Gong et al. (2022)
5	Omega-3 fatty acids	Fish oil	Reduces inflammation, protects against neurotoxicity	Alzheimer's, Parkinson's disease	Avallone et al. (2019)
6	Ashwagandha	*Withania somnifera*	Anti-inflammatory, antitumor, antioxidant, anti-arthritic, and immunomodulatory effects	Alzheimer's, Parkinson's, Huntington's disease, amyotrophic lateral sclerosis	Mikulska et al. (2023)
7	Quercetin	Fruits and vegetables	Anti-inflammatory, anticancer, antivirus, anti-aging, antioxidant, neurite growth-promoting, and neuroprotective activities	Alzheimer's, Parkinson's disease	Chiang et al. (2023)
8	Lycopene	Tomato, watermelon	Protective against DNA damage and to reduce the risk of cancer	Alzheimer's, Parkinson's disease	Imran et al. (2020)
9	Sulforaphane	Cruciferous vegetables	Several cancer, cardiovascular, and metabolic disease models, including animal models of type 2 diabetes	Alzheimer's, disease	Baralić et al. (2024)
10	Berberine	Berberis species	Anti-inflammatory, anti-diabetic, anti-microbial, and antitumor	Alzheimer's, disease	Belwal et al. (2020)
11	*Mucuna pruriens*	Velvet bean	Crossing the blood-brain barrier and increasing dopamine levels in the brain	Alzheimer's, Parkinson's, Huntington's disease, amyotrophic lateral sclerosis, multiple sclerosis	Gupta and Sharma (2025)
12	Coenzyme Q10	Ubiquinone	Antioxidants, neuronal degeneration	Parkinson's disease	Rauchova (2021)
13	Baicalein	Mulberry	Degeneration of dopaminergic neurons, alleviate neuroinflammation, and reduce oxidative stress levels	Parkinson's disease	Bhushan et al. (2024)
14	Naringenin	Citrus fruits	Antioxidant, anti-inflammatory, anticancer, and neuropharmacological effects	Parkinson's, AD, and amyotrophic lateral sclerosis	Kar and Panda (2024)
15	Epigallocatechin gallate	Green tea	Gene expression influenced by tumor necrosis factor-α, reduces activity in the mTOR/S6K signaling pathway	Parkinson's disease	Papuc et al. (2021)
16	Cannabinoids	*Cannabis sativa*	Neuroprotective, anti-inflammatory, and antioxidant effects	Parkinson's Disease	Cásedas et al. (2022)
17	*Rhodiola rosea*	Salidroside	Antioxidants reduce oxidative stress and improve mitochondrial function	Parkinson's disease	Bernatoniene et al. (2023)
18	*Ginkgo biloba*	Maidenhair tree	Vasodilation enhances cerebral blood flow, provides antioxidant defense, scavenges free radicals, reduces platelet aggregation, and supplies neuroprotection	Parkinson's disease	Wendo et al. (2024)
19	Hesperidin	Citrus fruits	anti-inflammatory, neuroprotective, antitumor, and antioxidative effects	Alzheimer's, Parkinson's disease	Joshi et al. (2022)

natural substances, such as flavonoids, are already being used as food. Research has indicated that these natural substances should be the first step in developing novel drugs against neurodegenerative diseases (Divino da Rocha et al., 2011). In this chapter, the key molecular biology mechanisms of these diseases and the natural molecules that suppress disease progression, such as ginkgolides, baicalein, quercetin, resveratrol, and curcumin, will be discussed. These natural molecules could be potential drug candidates and future sources of pharmacophores to treat AD (Ayaz et al., 2019).

As the β-amyloid cascade hypothesis remains the most accepted etiology of AD, many of the neuroprotective molecules are reported to target the amyloidogenic processing pathway. Baicalein, which results in the decreased ratio of Aβ42 and Aβ40 (found to be more toxic and more aggregation-prone), is also reported to prevent cognitive impairments and the accumulation of Aβ1–42 (Barage & Sonawane, 2015). In addition to baicalein, ginkgolide B can affect the amyloidogenic processing pathway. By reducing the production of Aβ42 and enhancing the non-amyloidogenic processing through activating α-secretase, ginkgolide B can reduce the level of Aβ1–42 and further rescue the spatial memory of animal models, even inhibiting the c-Jun N-terminal kinase-related stress and inflammatory signals. These results provide molecular-level evidence of the neuroprotective roles of EGb761 (Monteiro et al., 2024).

19.4.1.1.1 Prominent and Significant Natural Molecules in Current Alzheimer's Research and Their Impact

Natural remedies are globally known and have been attracting a lot of attention in recent years. More recently, many researchers around the world have isolated, characterized, and recorded the beneficial nature of natural molecules in health-related biological studies. For many years, these were not discussed and were rejected (Aware et al., 2022). Those beliefs are from ancient medicine times because of their healing capacity. The natural molecules presented in this review, like hybrids of the entity, are approved; natural small molecules have more beneficial homogeneous effects in curing neurodegenerative diseases like AD (Guzior et al., 2015). Recent clinical and pivotal natural molecule studies, such as peptide alteration, precise mechanistic peptides, dual functional and redox-active peptides, and triple functional peptides intended for neurodegenerative disease treatment, mainly AD, are updated in this region (Trapani et al., 2018).

Peptides derived from natural elements and the study of these peptides represent around 10,000 hand-curated studies, involving diverse methodologies, to emerge as potential new therapeutics for various targeted diseases. Peptides are small molecules that can selectively modulate the properties of the Aβ peptides, a degenerative factor for AD (Pereira & Aires-de-Sousa, 2018). Several major causes of AD and age-related neurological diseases are associated with the accumulation of not only Aβ peptides but also phosphorylated tau proteins, which generate toxic multi-nuclear aggregates. Concentrating on the process of converting peptides of natural origin through different approaches into therapeutic entities illustrates the paramount interest in this area (Gan et al., 2021). The rest of the review sums up the prominence and significance of these tiny molecules in current research and various pitfalls that are yet to be resolved. Both traditional and modern, sophisticated methodologies that have revealed the finer details are discussed. Additionally, the possible opportunities for new therapies in many other disease-related fields are also pointed out with this particular need (Lang & Espay, 2018).

19.4.1.1.2 Curcumin

Curcumin, a well-known bioactive compound, exhibits potential effects in a diverse range of neurological diseases, including ischemic stroke, multiple sclerosis, AD, and Parkinson's disease. The primary effects and underlying mechanisms of curcumin consist of anti-inflammatory, antioxidative, anti-protein misfolding, and anti-apoptosis pathways (Cai et al., 2020). Curcumin also exerts a neuroprotective effect in the amelioration of neurodegenerative damage in various brain areas. In addition, it has been indicated that the protective activities of curcumin against DNA damage and apoptosis involve regulating Bcl-2 and Bax, and inhibiting caspase-3 and c-fibril formation, which are associated with loss of motor function in 6-OHDA-activated rats (El-Agnaf et al., 2004). Furthermore, curcumin ameliorated alpha-synuclein-induced substantia nigra dopaminergic neuron apoptotic behaviors and upregulated PTEN-induced putative kinase 1 (PINK1)/Parkin-dependent mitophagy in vitro. The protective effect of curcumin also protects PC12 cells by inhibiting mitochondrial apoptosis, preventing calcium overload, and releasing cytochrome C to activate caspase-9 and caspase-3 in PC12 cells (Fan et al., 2017).

Considering its significant effects on various neurodegenerative diseases, curcumin has become the most popular polyphenol across the globe. Although curcumin has a short half-life and low cell uptake rates, which cause poor bioavailability, these two complicated problems—curcumin distribution and absorption—have a critical impact on the bioavailability and efficacy of curcumin (Bjørklund et al., 2024). In order to solve these two problems, many approaches have been adopted to improve the bioavailability of curcumin, including co-delivery strategies and using smart drug delivery systems. Despite its imperfections, curcumin may be employed as a standard neuroprotective molecule in drug research and development, or it may warrant further investigation in the future (Joshi et al., 2022).

19.4.1.1.3 Resveratrol

Oxidative stress is characterized by an increased production of free radicals, particularly ROS, and a reduced level of antioxidant molecules. This imbalance results in cellular damage, causing significant implications and worsening aging and age-related diseases (Lushchak, 2014).

Resveratrol is a natural polyphenolic compound found in foods such as berries, grapes, and nuts. It is considered a powerful antioxidant that can cleanse ROS, increase the antioxidant defense, and reduce damage to the CNS. The role of resveratrol extends to decreasing oxidative stress and enhancing crucial inhibitors of apoptosis and autophagy in neurodegenerative diseases (Pourhanifeh et al., 2019). Due to its multiple properties, many experimental studies of resveratrol for the treatment and prevention of ischemia, aging, inflammatory diseases, and neurodegenerative disorders have been completed with positive results. Additionally, the clinical trials conducted with resveratrol have shown ameliorative effects in the treatment of neurological diseases (Singh et al., 2019).

Resveratrol has a multifaceted effect on neurodegenerative diseases because of its characterized mechanism. This mechanism includes antioxidant and anti-apoptotic activities, the modulation of neuroinflammation, and the regulation of mitochondrial dynamics. Some researchers recently presented evidence that resveratrol may become a supplement for the therapy of neurodegenerative diseases attributed to its regulation of autophagy (Utpal et al., 2024). The enhanced application of resveratrol to neuropathies has achieved adequate outcomes due to its fast metabolism. Approaches to improve its efficacy are required, for instance, amalgamation with advanced carriers and technologies to develop its bioavailability in the CNS. These approaches could lead to an urgent demand for the application of resveratrol for neuroprotection. In this review, we summarize the progress in therapeutic applications of resveratrol for neurodegenerative diseases. The mechanisms and characteristics that make resveratrol an effective potential therapy for neurodegenerative diseases are discussed for readers (Pangeni et al., 2014).

19.4.1.1.4 Huperzine A

Huperzine A is a lycopodium alkaloid, isolated from the club moss *Huperzia serrata*. Used for centuries in China for the treatment of swelling, fever, and blood disorders, clinical and preclinical studies have revealed a wide range of therapeutic potential for HupA (Wang et al., 2006). In neurodegenerative diseases like AD, the progression of cholinesterase inhibitors slows down cognitive loss, which is associated with dramatic changes in the brain. It seems promising that acetylcholinesterase prevents AChE-induced Aβ peptide aggregation and enhances the release of vascular endothelium. The significant effects of HupA on the NMDAR, free radical mitigation, and calcium modulation suggest a significant role as a survivor (Wang & Tang, 2005). In fact, several studies identified a neuroprotective correlation between the composition of HupA, the BCL-2 family, and the mitogen-activated protein kinase pathways. Although there is no obvious HupA protection mechanism against neurodegeneration, some mechanisms are presented as a combination of anti-apoptotic effects through the activation of cell survival and suppression of pathogenic apoptosis, caspase-dependent cell death prevented through the PI-3K/Akt pathway, inhibition of glutamate-induced dysregulation, or NMDA receptor destruction through excessive calcium delivery and stimulation of the ERK pathway (Xia, 2009). In preventing the decline of neural signaling, HupA also has an important role in the development of neurodegeneration. Even though reversal can be recommended, the focus will now be put on the supplementation of its neuroprotective potential with the provocation of antioxidants, protection, and/or synaptic rejuvenation or apoptosis inhibition. However, because this area is not completely documented, more research, especially looking at pigmented neurodegenerative systems, is necessary to concentrate on proliferation, differentiation, and the design of new therapeutic drug combinations as potential neuroprotective agents alone or in combination as possible protective mechanisms (Lapchak & Araujo, 2007).

19.4.1.1.5 Ginsenosides

Ginsenosides are saponins that exist in the roots, stems, and leaves of the ginseng plant, which has been used extensively in the treatment of a variety of diseases, including cancer, cardiovascular diseases, diabetes, cognitive disorders, neurodegenerative diseases, and depression. There have been many reviews on the therapeutic uses of ginsenosides (Lee & Kim, 2014). Here, lupenol and pseudoginsenoside-F11 are mentioned, but there is a huge amount of potential ginsenosides that interact with neurodegenerative diseases. Peptides have been reported in the caudate and leptomeninges of the ginseng root. Genniposide, ginsenoside Rg (with Rg3) has been isolated and identified, but ginsenoside Rg5, ginsenoside Rb1 may also have greater neuroprotective effects than alpinetin. Ginsenosides Rg3 and Rk1 are ginsenosides that have received much attention from many researchers as potential active compounds. Animal studies have proved that ginsenosides Rb1, Rg1, Rg2, Rg3, and Rg5 can resist the degradation of dopaminergic neurons, which help increase cholinergic activity, decrease activity, and achieve anti-addiction effects (d'Anglemont de Tassigny et al., 2015). The effect of ginsenoside Rd on gerbils with cerebral ischemia can help cure ischemic stroke. Ginsenoside Rg1, ginsenoside Rg3, ginsenoside Rd, ginsenoside Ro, ginsenoside Rf, ginsenoside Cs, ginsenoside F11, and ginsenoside Rb1 have potent nitric oxide-inhibitory activities with IC50 values of 13.72 μM, 30.17 μM, 17.16 μM, 25.58 μM, 28.45 μM, 35.10 μM, 38.89 μM, 30.21 μM, and 34.69 μM, respectively (Zhao et al., 2022).

19.4.1.1.6 Omega-3 Fatty Acids

Omega-3 fatty acids in the diet have also been demonstrated to possess protective effects against neurodegenerative diseases. The typical feature of AD and several other neurodegenerative diseases is the presence of misfolded proteins (Mazza et al., 2007). Abnormal calcification constitutes one of the reasons for the misfolding of proteins. One of the neural protective mechanisms of omega-3 fatty acids is that they can capture calcium through a chemical reaction, which promotes an unfavorable environment for Aβ. In vitro studies have suggested that increased omega-3 intake is inversely associated with Aβ levels in the brain and with the risk of developing AD, although contradictory findings in clinical trials indicate the necessity of further comprehensive pathophysiological research (Jia et al., 2014).

Omega-3 decreases inflammation and demonstrates neuroprotective roles. Germline knockout mice lacking the receptor for advanced glycation end products and APPswe/PS1ΔE9 mice exhibited reduced age-related accumulation of Aβ in their brains, manifesting the role of brain RAGE in Aβ accumulation (Liu et al., 2018). High levels of brain fatty acid inflammation markers suggest that RAGE ligands, such as advanced glycation end products and high mobility group box 1, primarily upregulate the inflammatory burden in the aged brain. The production of these RAGE ligands stimulated the levels of cytokines and other inflammatory markers in COX-2 upregulation and production in the aged mouse brain (Dong et al., 2022). Additionally, brain and plasma fatty acid markers of inflammation were significantly higher in aged RAGE sidecar mice. These brain and peripheral fatty acid changes were selectively reduced in aged RAGE mice treated with the Docosahexaenoic Acid (DHA) diet. Docosahexaenoic acid supplementation exerts anti-inflammatory properties in subjects with mild cognitive impairment and can also reduce plasma levels of the proinflammatory adipokine fatty acid-binding protein 3, which can protect the brain and blood vessels and may be beneficial for ameliorating AD. However, translations of these positive findings are still lacking, and omega-3 supplementation trials for neurodegenerative diseases are challenging due to the complexity of drug administration and lack of an optimal assessment timeframe (Mora et al., 2022).

19.4.1.1.7 Ashwagandha (Withania somnifera)

Neurodegenerative disease is a condition characterized by extreme nerve cell decline in one particular part of the body or in numerous regions in general. The common neurodegenerative diseases include AD, HD, ALS, and Parkinson's disease. *Withania somnifera* possesses multiple therapeutic properties such as anti-inflammatory, antitumor, antioxidant, anti-arthritic, and immunomodulatory effects (Ciurea et al., 2023). There are large numbers of active compounds isolated from *W. somnifera* extract, including withanolide derivatives and sitoindosides. However, *W. somnifera* is often utilized to enhance memory, lower glucocorticoid levels, increase resistance to harmful chemicals, and restore damaged neurons. Among various secondary metabolites in *W. somnifera*, the alkaloid withanine is able to resist the immunosuppressive consequences of various morphines (Halder et al., 2022).

In a recent scientific study, the healing components of *W. somnifera* have been revealed to manage neurodegenerative diseases. The ethanolic extract from *W. somnifera* lifts epithelium growth levels and reduces the impacts of pancreatitis in rats. In order to reveal healing components, information regarding active components from *W. somnifera* was investigated and then tested for healing influence against oxidative damage of CNS cells (Bhat et al., 2022). The obtained experimental results suggest that *W. somnifera* ranges from a herb with swelling healing components and is indicated for the management of human as well as animal neuroinflammatory diseases, especially degenerative disease (Yatoo et al., 2018).

19.4.1.1.8 Quercetin

Quercetin is a polyphenolic flavonol commonly found in fruits and vegetables. It is known for its anti-inflammatory, antioxidant, neurite growth-promoting, and neuroprotective activities. Quercetin has been reported to have the potential to prevent or delay inflammation and chronic diseases related to inflammation, as well as metabolic syndrome (Muratori et al., 2022). Numerous studies have suggested that aging and diseases like AD and Parkinson's disease are conditions resulting from inflammatory responses. Quercetin has been shown to decrease dopamine levels and alleviate memory impairments in Parkinson's disease (Calabrese et al., 2018). It has been reported to be effective in reducing the toxicity caused by 6-hydroxydopamine. It has shown significant improvement in the ability of mitochondrial fusion and fission in Parkinson's disease models. Moreover, Quercetin helps in decreasing the level of Aβ, which is limited due to its effect on amyloid beta production and clearance, and it also has a protective effect on cells that are injured by amyloid beta (Khan et al., 2019).

Thus, Quercetin can be considered a potential compound for the cure and prevention of diseases like AD and Parkinson's disease. The key impacts of Quercetin include decreased inflammation, consistent alteration in gut microbiota, and decreased dopamine and amyloid beta levels (Khan et al., 2019). However, the mechanism behind Quercetin is not completely understood, but the main sources of neurodegeneration in the aforementioned diseases are apoptosis, mitochondrial dysfunction, oxidative stress, and inflammatory responses. Antioxidants and polyphenolic compounds could possibly alleviate both conditions by interfering with signaling enzymes and inhibiting inflammatory mediators (Teleanu et al., 2022). They propose the idea of combining Quercetin with a more potent anti-inflammatory agent for better neuroprotective effects. However, due to its slow absorption and poor bioavailability, the effect of Quercetin is limited. Quercetin may be combined with other compounds to enhance bioavailability. Due to a lack of information on the biodistribution and pharmacokinetics of Quercetin-based nanoparticles, it is necessary to further investigate the biological mechanisms of Quercetin in treating neurodegenerative diseases (Tomou et al., 2023).

19.4.1.1.9 Lycopene

Lycopene is one of the most potent antioxidants in tomato, watermelon, and some other fruits and vegetables. It belongs to the carotenoid family and has long been known to be protective against DNA damage and to reduce the risk of cancer. Many studies have suggested that lycopene also exhibits neuroprotective activities. The pretreatment of lycopene in PC12 cells prevented the decrease in cell viability induced by Aβ (Kim et al., 2014). Furthermore, lycopene also exhibited an effect on suppressing ROS overproduction and Aβ accumulation by the activation of autophagy. The intraperitoneal administration of lycopene in the transgenic mouse model also showed a suppressive effect on autophagic degradation, mitochondrial dysfunction, oxidative stress, and apoptosis. These findings suggested that lycopene-derived modulation of the Aβ clearance pathway might be protective (Li et al., 2024). Meanwhile, in the MPTP mouse model, oral treatment of lycopene also decreased lipid peroxidation and oxidative DNA damage. The markers of neuroinflammation, such as microglia, TNF-α, and iNOS, were also ameliorated by lycopene. In microglia, lycopene suppressed the upregulated IκB and the translocation of p65 to the nucleus to suppress microglia activation. Similarly, lycopene also decreased. These results indicated that lycopene exerted its neuroprotective effects by the reduction of ROS, mitochondrial dysfunction, neuroinflammation, and the protection of mitochondria (Chen et al., 2019).

19.4.1.1.10 Sulforaphane

Sulforaphane (SFN) is one of the isothiocyanates containing the sulfur group, and it is present in cruciferous vegetables in high concentration because it occurs as a glucosinolate precursor. SFN is much more stable than its precursor, glucoraphanin. Upon breakdown of the respective vegetable tissue containing glucosinolate by either the enzyme present in the same tissue called myrosinase, following damage to the vegetable tissue by chewing or by blending for cooking, degradation of glucoraphanin to SFN is possible (Li et al., 2024). Therefore, the cancer prevention properties of SFN have led to an increase in biomedical interest in this molecule and to an expansion of studies over time. SFN has been shown to provide benefits in several cancer, cardiovascular, and metabolic disease models, including animal models of type 2 diabetes (Cefalu, 2006). Prevention of neurodegeneration and enhancement of neuronal function have also been accomplished by treatment with SFN in various settings known to cause oxidative stress, inflammation, and neuronal demise. For example, SFN has been shown to protect neuronal cells against cognitive loss induced by carnosine in AD models. However, the clinical advantages of SFN are characterized by poor retention time and metabolic conversion by the gastrointestinal microbiota unless a mechanism is provided to protect against these outcomes, such as the administration of SFN in the form of stable lipophilic precursors (Zhang et al., 2024).

19.4.1.1.11 Berberine

Berberine (BBR) is an alkaloid extracted from several herbal plants that has been associated with multiple biological and pharmacological activities, including anti-inflammatory, anti-diabetic, anti-microbial, and antitumor applications, among others. Among the various activities of BBR, neuroprotective effects have been widely reported (Gaba et al., 2021). BBR exerts a protective effect on neurons by scavenging free radicals and attenuating the levels of lipid peroxidation and ROS. BBR is a potential agent to combat neurodegenerative diseases and brain injury by maintaining mitochondrial membrane potential, inhibiting cytochrome C release from mitochondria, and inhibiting oxidative stress-induced apoptosis (Tian et al., 2023).

Another important target of BBR activity for the treatment of neurodegenerative diseases is the increase in the levels of neurotrophic factors. BBR significantly increases the expression levels of brain-derived neurotrophic factor and nerve growth factor, which are neurotrophic factors possessing antioxidant properties, as well as the ability to regulate metal ion homeostasis and protect DNA (Tian et al., 2023). Thus, it is suggested that the neuroprotective mechanism of BBR is accomplished by reducing oxidative damage through regulating neurotrophism. Mitochondria are an important cause of cell apoptosis, and mitochondrial damage leads to neurodegenerative diseases (Li et al., 2020). BBR has exerted beneficial effects on mitochondrial dysfunction by regulating the formation of Bcl-2 and Bax, as well as ameliorating mitochondrial function. BBR protects against beta-amyloid-induced neurotoxicity in cultured cortical neurons by suppressing the involvement of mitochondrial apoptotic signaling and restoring Brain-Derived Neurotrophic Factor (BDNF) expression. The studies regarding the neuroprotective mechanism of BBR have gradually matured and are likely to attract considerable attention in the coming years (Fang et al., 2020).

19.4.1.2 Neuroprotective Effects of Natural Molecules in Parkinson's Disease

PD is a common neurodegenerative disorder that is characterized by progressive motor dysfunction due to degenerative lesions in the nigrostriatal dopaminergic pathway. The molecular pathologies for PD, such as oxidative stress, mitochondrial dysfunction, loss of trophic factor function, and protein aggregation, have been well studied and represent major therapeutic targets against this disease (Pavese & Brooks, 2009). Due to the severe side effects of L-DOPA treatment and the lifelong nature of PD, developing novel therapeutics with low toxicity and greater efficacy to relieve symptoms and slow the progression of PD is a prime goal in its treatment (Yadav & Li, 2015). Nowadays, the discovery of natural molecules with neuroprotective properties against many neurodegenerative diseases has received great attention. Most of these natural molecules, such as curcumin, resveratrol, flavonoids, polyphenols, alkaloids, and triterpenoids, are of low toxicity and have been shown to be safe for human consumption. Also, the probable neuroprotective properties of these natural molecules have been shown to target pathologies pertaining to PD (Abushouk et al., 2017).

Although multi-target therapeutics have advantages in treating a multifactorial, progressive disease such as PD, the major prospect of a neuroprotective agent is its ability to enhance the function and survival of dopaminergic neurons in the substantia nigra. Many natural molecules have been proven to protect dopaminergic neurons and/or improve their functions through in vitro and in vivo models of PD (Cavalli et al., 2008). Since various pathways of dopaminergic cell loss have been uncovered in PD, these studies may generate many beneficial strategies for preventing or slowing PD progression through targeting different pathologies. In addition, many natural molecules showing neuroprotective properties may suggest that nutraceuticals and/or food containing certain compounds could be beneficial for the prevention or therapy of PD. The available natural molecules with neuroprotective properties against PD and relevant information are summarized in the following sections (Nebrisi, 2021).

19.4.1.2.1 Mucuna pruriens (Velvet Bean)

It is a specific type of neurodegenerative disorder that primarily affects the neurons in the brain, particularly those responsible for coordinating movement and maintaining balance. It falls within the broader category of conditions that include AD, HD, ALS, multiple sclerosis, and prion disease (Khan et al., 2019). In Parkinson's disease, the death of dopamine-producing neurons in the substantia nigra region of the brain is a hallmark feature, leading to the characteristic symptoms such as tremors, stiffness, and difficulty with movement. Neurons are essential to the function of the nervous system, which consists of the brain and spinal cord. Unfortunately, neurons do not have the ability to replicate or regenerate; therefore, once they are damaged or lost due to Parkinson's disease, the body cannot restore them. This condition is described as progressive and currently has no cure, resulting in a gradual decline in motor control and overall neurological function (Ontaneda et al., 2017).

In the context of PD management, various therapeutic approaches reflect the guidelines of product development. While numerous drug therapies are available, they primarily aim to alleviate symptoms rather than fully address the disease's root causes. Among the drug therapies, several synthetic compounds are commonly prescribed for PD, including levodopa, biperiden, trihexyphenidyl, amantadine, and selegiline. However, these medications can lead to toxic effects and significant adverse reactions (Maetzler et al., 2016). In contrast, traditional herbs and natural products have emerged as promising alternatives in the management of PD, showing efficacy in slowing the progression of neurodegenerative conditions. Notably, *Mucuna pruriens* has gained attention for its beneficial effects on neurological disorders, particularly PD, AD, and other forms of neuronal dysfunction. This is largely due to its potent antioxidant properties. This review seeks to underscore the potential therapeutic benefits of *M. pruriens* in the treatment of neurodegenerative diseases, particularly in the context of PD management (Das et al., 2024).

19.4.1.2.2 Curcumin

Curcumin, the primary active compound of Curcuma longa, exhibits various biological protective effects relevant to PD. Previous research has highlighted curcumin's potential in combating issues such as cancer, oxidative stress, and inflammation, all of which can contribute to neurodegenerative processes. In PD, neurotoxicity linked to glutamate plays a significant role in neuronal damage, with excess glutamate contributing to cellular loss. This neuroexcitotoxicity is a critical factor in the progression of PD, leading to cell death within specific brain areas associated with its characteristic symptoms. In studies focused on curcumin's protective abilities, it has demonstrated potential in safeguarding neuronal health in models relevant to Parkinson's (Islam et al., 2024). Curcumin has been shown to enhance cell survival rates while mitigating neuroinflammation and oxidative stress by acting as a ROS scavenger, thus protecting neurons from free radical damage and apoptosis. Furthermore, curcumin addresses the aggregation of proteins associated with neurodegenerative conditions. Specifically, it has been observed to inhibit the aggregation of Parkinson's-related proteins, suggesting that it may play a role in disrupting pathogenic processes. Overall, curcumin presents a promising therapeutic avenue for protecting neurons, potentially improving outcomes in individuals affected by PD and offering an effective strategy for addressing the cellular challenges posed by this condition (Lang & Espay, 2018).

19.4.1.2.3 Resveratrol

Resveratrol, a naturally occurring polyphenol from the stilbene family, is primarily found in fruits and seeds such as grapes, berries, *Polygonum cuspidatum*, and peanuts. Research indicates that both injectable and oral forms of resveratrol can significantly reduce neuropathological lesions and cognitive deficits in mouse models, which could have implications for conditions like PD. Additionally, resveratrol demonstrates neuroprotective effects against cerebral injury and mitochondrial-dependent apoptosis, with these benefits tied to the increased transcription of antioxidant enzymes. Despite concerns regarding the bioavailability of resveratrol for effective clinical application, several chemical derivatives, micronutrient supplements, and nanoparticles have been developed, showing promising improvements in the absorption and efficacy of resveratrol for potential use in treating neurodegenerative diseases such as Parkinson's (Kaur et al., 2022).

Resveratrol exhibits significant benefits in adjunctive therapy for patients with PD, particularly when combined with cells and/or exosomes. This combination may lead to delayed onset of symptoms and potential neurogenesis following neurodegenerative events. Such properties make resveratrol a promising candidate for future pharmaceutical therapies aimed at regenerative medicine for individuals affected by Parkinson's. Notably, resveratrol has demonstrated strong mitochondria-targeted anti-necroptosis effects, functioning as an autophagy inducer that helps mitigate injuries to dopamine-producing neurons. This action may subsequently reduce neuroinflammation and cell necroptosis commonly observed in PD. Furthermore, the drug's mitochondria-specific binding and enhancement of activity have clarified the role of the mitochondria-CaMKII-PINK1 axis in providing neuroprotection, particularly relevant to the mechanisms underlying PD progression (Duta-Bratu et al., 2023).

19.4.1.2.4 Ginsenosides

Ginsenosides from Panax ginseng show considerable therapeutic potential for Parkinson's disease due to their antioxidant, anti-inflammatory, neuroprotective, anti-apoptotic, and immune-modulatory properties. These benefits may help reduce neuronal damage associated with neurodegenerative diseases. Structurally, ginsenosides are amphiphilic, comprising a hydrophobic aglycone and hydrophilic sugar linked by ether or ester bonds. There are over 86 distinct ginsenosides categorized into four structural types: protopanaxadiol, protopanaxatriol, ocotillol, and oleanolic acid. While they are absorbed in the small intestine, the mechanisms of absorption are not well understood. Some research indicates that facilitated passive diffusion and micelle solubilization may play a role, necessitating further investigation. Their high molecular weights and low water solubility challenge absorption, yet bioconversion metabolites could improve bioavailability, offering a promising approach for managing PD (Rajabian et al., 2019).

The metabolites of protopanaxadiol, which are absorbed from the intestines, have been found in the plasma and are widely distributed in various organs, including the brain, heart, lungs, liver, and kidneys. This indicates that these metabolites can cross the blood-brain barrier and engage in the bidirectional movement of ginsenosides. Within the brain, specific ginsenosides such as RG1 and RB1 have been identified; however, other hydrophilic protopanaxadiols have not been detected, even though RB1 is present in the plasma. This phenomenon can be explained by the active efflux mediated by P-glycoprotein and reduced absorption due to their low lipophilicity. Conversely, the metabolites of ginsenosides derived from protopanaxatriol are excreted in an unchanged form in the bile, where colonic bacteria in the large intestine can convert them back into protopanaxadiol and panaxtriol. Notably, these metabolites do not originate from pancreatic bacteria, marking a significant finding that indicates their potential impact on the variety of metabolites that enter systemic circulation, which may have implications for the management and treatment of PD (Kim et al., 2018).

19.4.1.2.5 Quercetin

Quercetin is a flavonoid predominantly found in various plants, including vegetables, fruits, and tea. It is recognized as one of the most significant polyphenol derivatives, garnering attention for its diverse array of biofunctions, including anticancer, anti-inflammatory, antivirus, anti-aging, and neuroprotective properties. Within the context of PD, quercetin has demonstrated remarkable efficacy in protecting the brain against neurotoxicity. It facilitates neurite outgrowth through specific signaling pathways and exhibits protective effects against oxidative stress in neuronal cells. Furthermore, it reduces multiple neuroinflammatory responses both in vitro and in vivo. In animal studies, quercetin has the ability to traverse the blood-brain barrier, allowing it to concentrate in the brain. Its beneficial effects extend to various CNS disorders, making it a valuable compound for neurological health. The underlying molecular mechanisms involve diminishing neurotoxic inflammation by lowering the expression of specific inflammatory markers in cerebral cells and inhibiting induced post-synaptic signals while preventing microglial activation. These insights suggest that quercetin may serve as a critical nutrient for brain health, with potential therapeutic applications for mitigating neurological conditions such as PD. In research assessing the accumulation of Aβ in the brain, an assay revealed weak immunoreactivity in the negative control group. In contrast, mice treated with quercetin exhibited a marked reduction in Aβ accumulation, particularly observed in staining from the cortex and hippocampus. These findings indicate that quercetin may help alleviate amyloid accumulation through chronic treatments, lending support to its role as a therapeutic candidate for managing symptoms associated with PD (Deepika et al., 2024).

19.4.1.2.6 Coenzyme Q10 (Ubiquinone)

Coenzyme Q10 (CoQ10) plays a crucial role in cellular energy production and function, particularly in the context of Parkinson's disease. Within the mitochondrial respiratory chain, CoQ10 acts as an essential electron carrier, helping to ensure efficient ATP production, which is vital for neuronal health. This compound also recognizes and works alongside antioxidants such as melatonin and vitamin E, contributing significantly to the body's defenses against oxidative stress. In individuals with PD, oxidative stress is a known contributor to neuronal degeneration. CoQ10, as a powerful antioxidant, has the ability to inhibit lipid peroxidation, a harmful process linked to cell death and apoptosis. Its various protective qualities—including promoting mitochondrial biogenesis, enhancing mitochondrial quality control, and exhibiting substantial antioxidant activity—support the overall function and survival of neurons affected by neurodegenerative processes. Moreover, CoQ10 can mediate the effects of oxidative stress and metabolic disturbances associated with aging, including its interaction with vitamin D, which is known to influence the CNS. Importantly, this natural molecule operates through diverse mechanisms, and various formulations have been shown to provide health benefits without causing toxic effects, even at higher doses. It is noteworthy that part of CoQ10's efficacy is attributed to its ability to regenerate vitamin E, further bolstering its antioxidant capacity. For individuals at risk of or living with PD, it is advisable to incorporate CoQ10 supplementation early in life and maintain it throughout, as this may help mitigate disease progression and enhance quality of life (Pallotti et al., 2021).

19.4.1.2.7 Omega-3 Fatty Acids

Omega-3 long-polyunsaturated fatty acids (n-3 LC-PUFAs) exhibit diverse effects on health, particularly in relation to neurological conditions such as PD. Research indicates that these fatty acids may influence the progression of brain diseases and contribute to cognitive decline associated with aging by engaging various mechanisms, including maintaining neuronal lipid integrity, modulating neuro-immune responses, and facilitating autophagy processes. Preclinical studies have highlighted their potential in ameliorating impaired neuronal connectivity, which is often disrupted in PD. Within the CNS, n-3 LC-PUFAs appear to promote the synthesis of pro-resolving lipid mediators derived from their metabolic pathway, which may help mitigate neuroinflammatory processes and enhance the overall activity of the neuroimmune system. The presence of metabolic syndrome, particularly obesity, is associated with increased risks of cognitive decline by diminishing circulating levels of n-3 LC-PUFAs and essential nutrients for brain health. Nonetheless, even low levels of n-3 LC-PUFAs, when integrated into a balanced diet, may exert positive effects, particularly in the context of metabolic syndrome. Low dietary intake of n-3 LC-PUFAs has been identified as a significant risk factor for cognitive decline, including the progression of dementia and conditions such as AD. Although there is no conclusive evidence that n-3 LC-PUFA supplementation prevents cognitive decline or dementia, substantial data support their potential as adjunctive therapy, especially during the prodromal and early stages of PD. Further investigation is necessary, utilizing appropriate biomarkers and advanced neuroimaging techniques, to comprehensively evaluate their effects. There is a need for more interventional clinical trials to definitively establish the cognitive protective role of n-3 LC-PUFAs in PD. Overall, these fatty acids may serve as a critical supplement to combat the prevalent n-3 LC-PUFA deficiency seen in Western diets and the heightened risk of neurodegeneration. Additionally, maternal dietary habits during pregnancy can have long-lasting effects on the prevention of neurodegenerative disorders, and early dietary influences on offspring also appear to play a crucial role in their neurological health (Bousquet et al., 2011; Dyall & Michael-Titus, 2008).

19.4.1.2.8 Hesperidin

Hesperidin, a citrus flavonoid, constitutes approximately 90–95% of the total flavonoid content in citrus fruits, primarily from the seeds, leaves, and peels of oranges, grapefruits, tangerines, lemons, and limes. Numerous studies suggest that hesperidin possesses various beneficial properties, including anti-inflammatory, neuroprotective, antitumor, and antioxidative effects. These properties have been specifically analyzed in connection with neurodegenerative disorders such as PD, as well as other conditions mediated by NMDA excitotoxicity, AD, and oxidative stress. The effects of hesperidin regarding neuroprotection, antioxidant activity, immunomodulation, and anti-inflammatory responses can be attributed to its active metabolite, hesperetin. However, hesperidin is challenging to solubilize due to its low water solubility. To address this issue, several drug delivery systems have been developed to shield hesperidin from gastric acid and the microflora of the small intestine, enhancing its solubility in water. In the context of PD and related CNS disorders, hesperidin's potential can be harnessed through diverse drug delivery methods. Research has focused on its role in inflammation-mediated processes, oxidative stress, amyloid beta-related pathology, and neurodegenerative progression, alongside its influence on nerve growth factor-regulated mitochondrial dynamics and calcium homeostasis. Furthermore, hesperidin is noted to function as a protective factor against chronic stress, regulated by stress-activated protein kinases and poly (ADP-ribose) polymerase 1, modulating various cytokines and growth factors. While much attention has been given to improving hesperidin's solubility, understanding its neuroprotective capabilities is crucial for future clinical applications, especially for the combined treatment of CNS diseases, including PD. Additionally, conducting efficacy studies will support the pharmaceutical industry's efforts to develop advanced formulations of hesperidin, overcoming challenges such as its limited solubility and the high doses that could potentially lead to gastrointestinal discomfort. This chapter discusses the neuroprotective, antioxidant, immunomodulatory, and anti-inflammatory properties of hesperidin, focusing on its relevance to oxidative stress, the immune response, and the activity of microglial cells, particularly in the context of chronic exposure (Atoki et al., 2023; Parhiz et al., 2015).

19.4.1.2.9 Baicalein

Baicalein is a flavonoid known scientifically as 5,6,7-trihydroxy-2-phenyl-4H-chromen-4-one. It is found in mulberry leaves and can also be extracted from the plant *Scutellaria baicalensis* Georgi. Research suggests that baicalein offers protective benefits for individuals with PD. This compound may mitigate the degeneration of dopaminergic neurons, alleviate neuroinflammation, and reduce oxidative stress levels. The protective mechanisms of baicalein include the downregulation of α-synuclein phosphorylation, which is associated with inflammation and astrocyte activation. Moreover, baicalein helps shield neurons from stresses related to the endoplasmic reticulum and calcium imbalances. It has demonstrated the ability to reduce β-amyloid and enhance the activation of ERK1/2, positively influencing amyloidogenesis and tau pathology in transgenic mouse models. Importantly, baicalein not only exhibits antioxidative properties and inhibits amyloid aggregation but also has anti-inflammatory effects and the capability to cross the blood-brain barrier. Baicalein improves LPS-induced neuroinflammatory responses in part through the activation of α-secretase signaling, which is mediated by β-arrestin 2. This multifaceted approach showcases baicalein's potential as a therapeutic agent in managing Parkinson's disease (Song et al., 2021).

19.4.1.2.10 Naringenin

Flavonoids are widely acknowledged for their varied pharmacological effects and their crucial role in preserving the body's homeostasis, particularly in the context of neurodegenerative diseases like Parkinson's. Research has delved into the structure-activity relationship of these compounds to gain insights into the underlying mechanisms that contribute to their therapeutic benefits. Among the flavonoids, naringenin stands out due to its therapeutic properties, including antioxidant, anti-inflammatory, anticancer, and neuropharmacological effects, rendering it particularly significant in addressing Parkinson's disease. The specific mechanisms by which naringenin acts in relation to Parkinson's, as well as other neurodegenerative conditions like Alzheimer's and ALS, are examined. This includes its impact on critical processes such as signal transduction, neuroinflammation, protein aggregation, energy metabolism, and the inhibition of enzymes that may initiate or worsen neurodegeneration. Flavonoids, being naturally occurring polyphenolic compounds, are characterized by their distinct structural composition, which features three phenolic rings and a central pyran ring containing oxygen, underscoring their potential use in the treatment of conditions like Parkinson's disease (Sahoo et al., 2024).

19.4.1.2.11 EGCG (Epigallocatechin Gallate)

Epigallocatechin gallate (EGCG), a key component of green tea, has garnered attention for its potential neuroprotective effects, particularly in the context of Parkinson's disease. Research indicates that EGCG can enhance the activity of various chaperone proteins, which are crucial for maintaining protein stability. It has the ability to bind to intrinsically unstructured proteins, thus preventing their transformation into harmful amyloid fibrils, a process that is often associated with neurodegenerative conditions like Parkinson's. In the context of Parkinson's disease, EGCG has been proposed to safeguard neurons by interacting with several important signaling molecules involved in cellular health, such as p70 ribosomal S6 protein kinase, SirT1, and histone deacetylase. Studies conducted on living organisms suggest that EGCG may effectively inhibit neurodegeneration, thus providing a potential therapeutic avenue for those affected by Parkinson's. Moreover, EGCG plays a role in regulating vital physiological processes such as synaptic transmission, which is critical for maintaining communication between neurons in the brain. It has been observed to stimulate the formation of lipid particles and facilitate the opsonization of Aβ, a protein implicated in the progression of Parkinson's disease. Additionally, EGCG may upregulate gene expression influenced by tumor necrosis factor-α, reduce activity in the mTOR/S6K signaling pathway, and promote the functions of the ubiquitin-proteasome system, all of which can contribute to neuronal health. EGCG also exhibits antioxidant properties, influences APP metabolism, and affects the balance of α/β-secretase activity, which is important for Aβ processing. Furthermore, it modulates the fibrillation process of Aβ and can prevent amyloid fibrils from attaching to neuronal cell surfaces, thereby reducing potential cellular damage. Despite these advantages, it is important to consider that some studies report a toxic effect associated with EGCG. Additionally, its low bioavailability remains a challenge for its clinical application. Nonetheless, the pharmaceutical potential of EGCG, particularly for neurodegenerative disorders like Parkinson's disease, continues to be a subject of investigation, either used independently or in conjunction with other therapeutic agents (Singh et al., 2015; Sutherland et al., 2005).

19.4.1.2.12 Cannabinoids (CBD)

Cannabidiol (CBD) has emerged as a significant compound derived from *C. sativa*, notable for its non-psychotropic effects. Recent research has increasingly focused on its potential neuroprotective properties in the realm of Parkinson's Disease, among other neurodegenerative disorders; however, the exact molecular mechanisms behind these effects remain to be fully elucidated. One area of interest is neurogenesis, which may serve as an important pathway through which CBD exerts its protective actions. It exhibits a unique blend of neuroprotective, anti-inflammatory, and antioxidant effects, functioning effectively in both enzymatic and non-enzymatic contexts. This combination contributes to our understanding of the pathological processes involved in Parkinson's Disease, essential for developing innovative therapeutic strategies. Neurogenesis plays a pivotal role in the cellular processes affected by neurodegenerative diseases such as Parkinson's, where neuronal degeneration is a central concern. High-CBD plant extracts are being explored as promising candidates for new pharmacological interventions aimed at promoting neurogenesis. These extracts hold the potential not only for treating psychiatric conditions but also for specifically targeting the neuronal cells affected in neurodegenerative diseases like Parkinson's. Given the growing interest in the therapeutic applications of CBD, the market for this compound is rapidly expanding. Alongside commercial products boasting its high concentrations, the cultivation of hemp as a natural source is on the rise, increasing the availability of CBD-rich extracts. However, it is important to note that the processing of hemp into CBD products may lead to contamination with other compounds or variability in CBD concentrations across different samples. This variability poses challenges for ensuring the purity and consistency of CBD intended for therapeutic use in patients with Parkinson's Disease and underscores the need for rigorous quality control (Martínez et al., 2020).

19.4.1.2.13 Rhodiola rosca (Salidroside)

Rhodiola rosea is a perennial plant found in various cold regions around the world, such as the Alps, the Pyrenees, the Carpathian Mountains, Scandinavia, Iceland, Great Britain, and the Arctic. It is recognized as a valuable adaptogen, known for its ability to mitigate stress effects. A key compound in *R. rosea* is salidroside, which serves as a potent scavenger of ROS, playing a critical role in the neurodegenerative processes associated with diseases like Parkinson's disease. This unique characteristic positions *R. rosea* as a promising candidate for the prevention, retardation, or even cessation of the progression of neurodegenerative conditions, including Parkinson's, as well as stroke. However, one significant limitation of *R. rosea* is its low permeability across the blood-brain barrier. This poses a challenge, as it can lead to insufficient amounts of salidroside reaching the necessary receptors within the brain. Furthermore, the beneficial effects of *R. rosea* are primarily linked to its salidroside content. To enhance the therapeutic potential and ensure optimal absorption of active compounds in the brain, it is crucial for researchers to work on increasing the bioavailability of *R. rosea* and to explore innovative drug formulation strategies (Nabavi et al., 2016).

19.4.1.2.14 Bacopa monnieri

The Bacopa species is known for its neuroprotective compounds, crucial for Parkinson's disease management. Researchers use chromatography and nuclear magnetic resonance for identification, validating results with mass spectroscopy. Bacopa's key compounds include bacopasaponins, which are cytotoxic but offer neuroprotective benefits, as well as flavonoid structures like bacoparosides and bacopaflavones. It contains various other compounds such as alkanes, alkaloids, alcohols, and more, contributing to its neuroprotective potential. The neuroprotective qualities of *Bacopa monnieri*, particularly relevant for cognitive decline in Parkinson's patients, have gained significant research attention. Studies suggest its memory-enhancing effects stem from key signaling mechanisms like choline acetyltransferase and nitric oxide synthase, important for neuroprotection and cognitive improvement. Bacopaside I and II expressions differ based on Bacopa source, with research showing that learning thresholds can boost cognitive abilities, leading to better memory retention and recall accuracy. Phytochemicals in Bacopa and wheat enhance dendritic spine expression in the hippocampus, vital for memory and learning, supported by findings that spermine activates neuronal nitric oxide synthase for neuronal health. Many older adults consume *B. monnieri* as a remedy for cognitive decline linked to Parkinson's disease. Studies also show its anti-amnestic properties, including butyrylcholinesterase inhibition and synergistic effects, which will be discussed further. Moreover, *B. monnieri* extracts have demonstrated improved learning outcomes in animal models, highlighting its cognitive-enhancing potential in neurological disorders such as Parkinson's disease (Sangeet et al., 2023).

19.4.1.2.15 Lycopene

Lycopene, a potent antioxidant found in tomatoes and tomato products, is essential for managing Parkinson's disease. This natural pigment, prevalent in red fruits, is noted for its nutritional advantages. Research links lycopene to reduced risks of chronic illnesses like cancer and cardiovascular disease. Its neuroprotective role in the CNS is crucial for neurodegenerative conditions. Studies suggest lycopene helps prevent stroke, AD, and other neurodegenerative disorders, which are detrimental to those with Parkinson's. These disorders often experience oxidative stress and inflammation, two factors that exacerbate neurodegeneration. While various dietary compounds have neuroprotective effects, lycopene's benefits present new therapeutic options for managing Parkinson's. Its significant antioxidant capacity makes lycopene a vital nutritional component, especially compared to other carotenoids. Notably, lycopene cannot be produced by the body and must be sourced from the diet. Emerging evidence suggests it may help prevent neurodegenerative diseases linked to oxidative stress, emphasizing dietary factors in Parkinson's management. This review aims to elucidate lycopene's protective effects against neurodegenerative diseases, with a focus on molecular mechanisms and signaling pathways, exploring current treatment possibilities for those affected by these conditions. The findings could lead to new research and nutritional therapies centered on lycopene's advantages (Saini et al., 2020).

19.4.1.2.16 Gingko Biloba

Ginkgo biloba extract is rich in glycosides, flavonoids, terpenoids, and trace elements, offering notable pharmacological benefits for Parkinson's disease management. It promotes vasodilation, enhances cerebral blood flow, provides antioxidant defense, scavenges free radicals, reduces platelet aggregation, and supplies neuroprotection. Research indicates that it inhibits amyloid deposition, preventing senile plaque formation, which is significant in neurodegenerative disease contexts. In rat models of chronic cerebral ischemia-reperfusion injury, *G. biloba* extract reduced proteins linked to cellular stress and neuronal death, such as JNK, Caspase-3, and GFAP. In Parkinson's models, it affects α-synuclein buildup at dopaminergic terminals and encourages the co-synthesis of HSPB1, improving blights associated with α-synuclein accumulation. Oral administration in aged rat models has shown enhancements in histopathological conditions, minimizing astrocyte and microglia activation in the prefrontal cortex, critical for cognitive performance. In a model simulating cerebral small vessel occlusion, learning and memory functions improved, with significant decreases in hippocampal damage. *G. biloba* also appears to alleviate cognitive deficits from conditions like multiple sclerosis and may influence BACE1 activity, crucial for reducing spatial memory loss and Aβ1-42 accumulation, exerting inhibitory effects on TGF-β signaling in Parkinson's cases. By boosting synaptin levels, it mitigates oxidative stress through various mechanisms, including free radical scavenging, lipid peroxidation prevention, and enhancement of antioxidant enzyme activity. Additionally, it diminishes neuroinflammation by inhibiting microglia activation and proinflammatory cytokine release, potentially delaying neurological decline across neurodegenerative disorders. Overall, *G. biloba* extract may offer crucial protective benefits for those with Parkinson's and related neurological conditions by targeting essential pathological mechanisms (Wang et al., 2021).

19.4.1.3 Neuroprotective Potential of Natural Molecules in Huntington's Disease

Natural molecules exhibit potential in providing therapy to common neurodegenerative diseases, including HD. To combat HD, a number of natural agents have been employed. However, targeting only one pathogenic pathway of HD, considering the multifaceted nature of its pathogenesis, is not the most effective method, as prior studies have shown. Rather, a multi-target approach, as is the case with natural molecules, has been shown to be more efficient. A number of pathogenic pathways have been targeted by the potential therapeutic natural molecules, with results that are encouraging (Bhat et al., 2023). Although the safety of such natural molecules has only been confirmed through various in vitro studies and animal experiments, an understanding of their mechanism suggests that many clinical therapeutic drugs perform multiple functions in the body. Most of the natural molecules illustrated herein exhibit safety profiles, making them excellent candidates for clinical application. Furthermore, well-designed clinical trials and future studies in humans will be useful to investigate in-depth signaling pathways to uncover new, promising drugs in neurodegenerative diseases, particularly in HD. Neurodegenerative diseases are prevalent in clinical settings and are a significant cause of fatality, stress, and financial pressure for both patients and families. HD is an autosomal dominant brain-linked hereditary neurodegenerative disease for which no effective treatment is available (Song et al., 2020). Recent studies suggest that its symptoms may be associated with energy metabolism dysfunction, free radical accumulation, and excitatory amino acid toxicity. Therefore, modifying energy dysmetabolism and oxidative stress may represent ideal therapeutic strategies for neurodegenerative diseases, including HD. Natural molecules possess potential neuroprotective effects against HD symptoms by concurrently improving disrupted energy metabolism, excessive oxidative stress, reduced energy levels, redox imbalance, mitochondrial dysfunction, and DNA damage. This review ascertains the potential neuroprotective effects of the aforementioned 13 major natural compounds, mobilizing further preclinical or clinical applications against HD and expanding our knowledge of these natural molecules' activities in HD (Alqahtani et al., 2023).

19.4.1.4 Neuroprotective Effects of Natural Molecules in Amyotrophic Lateral Sclerosis (ALS)

ALS is a progressive neurodegenerative disease characterized by the degeneration of the upper motor neurons in the brain and the lower motor neurons in the spinal cord, leading to muscle atrophy and paralysis. Several drugs are available to treat the symptoms of ALS; however, to date, riluzole and edaravone are the only two drugs used to treat ALS patients (Nijssen et al., 2017). Neuroinflammation and excitotoxicity are two of the disease-causing mechanisms of familial or sporadic ALS. In addition, an abnormal accumulation of the disease-causing proteins, mutations in SOD1, and TAR-DNA-binding protein 43 kDa (TDP-43) are often observed in fALS and sALS patients. Many reports have indicated that natural molecules have beneficial effects on ALS (Silva et al., 2020).

Vitamin B1, a natural molecule, suppresses neuroinflammatory responses through the downregulation of NF-κB and AKT signaling and inhibits excitotoxicity by suppressing AMPK activation. P7C3-like compound inhibits TDP-43 aggregation and clears the protein from the nucleus via enhanced proteasomal degradation and autophagic activity. Additionally, increased levels of nicotinamide adenine dinucleotide (NAD) through nicotinamide riboside and nicotinamide mononucleotide exhibit beneficial effects on ALS via mitochondrial protection, protein respiration inhibition, and SIRT3/Akt/mTOR signaling. Resveratrol increases cell viability through the activation of the DBC1/LKB1/AMPK/Sirt1 axis. In addition, resveratrol inhibits the aggregation of mutated SOD1 by accelerating the clearance of the protein via the AMPK/Sirt1/P62 pathway (Ahmad et al., 2024).

19.4.1.5 Neuroprotective Effects of Natural Molecules in Multiple Sclerosis

Multiple sclerosis is an inflammatory demyelinating disease in the CNS, resulting in axon damage and loss. Studies have shown that the imbalance between apoptosis and autophagy in the disease state can lead to demyelination in the body, whereby apoptosis results in the destruction and loss of oligodendrocytes, while autophagy is essential for the survival and differentiation of oligodendrocyte precursor cells. It has been demonstrated that quercetin can activate protein kinase A and NAD-dependent deacetylase sirtuin-1 to increase the protein expression of GAP-43, enhance the mitochondrial translocation level of the endoplasmic reticulum stress sensor PERK, and the interaction between Drp1, Fis1, and lysosome-associated protein type 2 after injury, and ultimately promote cell autophagy protection assuring better functional and cytoprotective effects (Brück, 2005). Further verification of the protective effect of quercetin on demyelination and the related regulatory mechanism was conducted using cuprizone-induced demyelination in mice. Results revealed that quercetin might play a role in promoting Oligodendrocyte Precursor Cell (OPC) differentiation and remyelination by inhibiting demyelination and promoting immature oligodendrocytes in the demyelinated corpus callosum. In recent years, silymarin has been found to inhibit inflammatory and oxidative pathways, consequently inhibiting the expression of phospho-glycogen synthase kinase 3β, promoting oligodendrocyte differentiation, and finally enhancing the remyelination process against the demyelination condition in the mouse multiple sclerosis brain (Gross et al., 2004). Additionally, it has been found that after quercetin and simvastatin treatment, the expression of Glypican-1 mRNA in the corpus callosum of the Experimental Autoimmune Encephalomyelitis (EAE) group was higher than that of the EAE model group, which showed that both drugs could improve the differentiation ability of oligodendrocytes. The results of behavioral experiments also confirmed that both drugs could improve the clinical score of EAE mice, but the effect was not statistically significant. In response, some scientists suggested the direct neuroprotective effect of resveratrol in establishing a mouse model of multiple sclerosis. The results of animal experiments showed that resveratrol increased the size of nerve axons, the number of nerve axons, and increased the expression of β-tubulin (Wu et al., 2020).

19.5 INVESTIGATING PRECLINICAL MODELS AND FINDINGS

Overall, instrumental neuronal damage is a critical factor in neurodegenerative diseases. Neurodegenerative diseases are characterized by the progressive decline in nervous system function, which affects the body's ability to manage motor control problems. The ND1 gene is associated with mitochondrial cytopathies, which are central mitochondrial proteins. Disruptions in the energy system may cause pathological symptoms and exacerbate issues in nerve cells that have higher energy demands than other cells. Thus, diminished energy for these cells may limit their ability to repair themselves, leading to cell death and highlighting the symptoms of neurodegenerative diseases. Preclinical models are needed for each condition in the search for therapeutic agents that can repair damaged nerves in these disorders (Marcos-Rabal et al., 2021).

The most critical problem faced in the treatment of neurodegenerative diseases is the regeneration of axons, which are one of the most important cell types in the peripheral nervous system. The impaired ability of the nerves to maintain functional synchronization may initiate the progression of vulnerability to axon degeneration and apoptosis. The key factor here is the search for potential factors and signaling pathways that have beneficial effects on progressive neurodegenerative disorders. This approach would yield useful and logical results for testing in the treatment of human neurodegenerative disorders. There is no question that much more work will be required to translate discoveries from preclinical models to the stage at which clinical trials are justified, designed, and conducted. With commitment from academia, government agencies, and the private sector, these challenges should be met (Curcio & Bradke, 2018).

19.6 CASE STUDIES AND CLINICAL TRIALS

AD, Parkinson's disease, HD, and ALS are among the most prominent and concerning neurodegenerative disorders, impacting the lives of millions globally, and rendering them with significant challenges and hardships. Despite the extensive and consistently ongoing endeavors to devise and establish effective and targeted therapies for these profoundly debilitating and devastating conditions, substantial advancements in treatment options continue to be frustratingly limited and elusive. As a direct consequence of this stagnation in progress, the rates of both the incidence and prevalence of these specific neurodegenerative diseases continue to rise at an alarming and concerning rate (Iyer et al., 2024). The complexity that is inherently woven into the fabric of these diseases at a genomic and molecular level, when combined with the myriad of individual genetic variations that exist, alongside the overlapping functions and intricate interactions of various genetic products and biomarkers, poses monumental and significant challenges for researchers and clinicians alike who are striving valiantly to develop effective targeted treatments and interventions. Nevertheless, amidst these challenges, a glimmer of hope has started to emerge recently from innovative research endeavors in progressive fields such as nutrigenomics and proteomics, which have collectively yielded a promising and diverse array of therapeutic candidates that are currently undergoing thorough testing in various clinical trials spanning across the globe. In the specific context of AD, innovative and potentially transformative compounds like aducanumab and lecanemab, which specifically target the harmful accumulation of Aβ plaques within the brain, are currently under detailed and comprehensive investigation. Conversely, in the context of Parkinson's disease, groundbreaking drugs such as nusinersen and tofersen are being employed and investigated with the aim of not only modifying the underlying disease processes but also actively seeking improvements in neuronal function and resilience (Yu et al., 2024). Regarding HD, cutting-edge and promising therapies such as deutetrabenazine and antisense oligonucleotides are demonstrating notable potential and success in clinical settings and trials, thereby offering new hope for those affected. In relation to ALS, ongoing studies are thoroughly evaluating the efficacy and safety, alongside the possible therapeutic benefits, of compounds like edaravone and riluzole, which are aimed at slowing disease progression and enhancing the overall quality of life for patients dealing with these serious conditions. This diverse range of promising candidate drugs, coupled with a host of traditional plant-based medicines and natural compounds, embodies a multifaceted and integrative approach to the ongoing development of treatments specifically tailored for these challenging diseases. Each specific compound is currently being scrutinized for its unique mechanism of action and therapeutic potential, thus providing a renewed and invigorated sense of hope for the development of more effective interventions and strategies aimed at combating these challenging and debilitating diseases. This current document serves to meticulously highlight and underscore these significant neurodegenerative diseases, detailing their intricate and complex mechanisms, recent clinical outcomes, and prospective future developments related to these promising drugs. Furthermore, it brings critical attention to natural compounds and extracts such as 6-methoxy-4-methyl-5,7-dihydroxyisoflavone, morin, ginkgolide A and B, *Hypericum perforatum*, chitosan-coated spherical mesoporous silica nanoparticles, ferulic acid, EGCG, and

Moringa oleifera leaf extract—all of which are actively being evaluated for their potential therapeutic benefits in the context of these challenging and complex disorders (Vinceković et al., 2021).

19.6.1 Efficacy and Safety Profiles

AD is a complex neurodegenerative condition marked by the degeneration of cholinergic neurons and significant deposition of amyloid plaques in the brain, which correlates with neurofibrillary tangle formation. Additionally, key pathological features of AD include neuroinflammation, oxidative stress, mitochondrial dysfunction, decreased neurotrophic factor levels, and compromised integrity of the blood-brain barrier. The progression of AD is driven by Aβ aggregates resulting from excessive production of APP, prolonged degradation inhibition, and inadequate Aβ clearance. Enzymes involved in Aβ production, as well as tau kinases and neurotrophic factors, present potential targets for AD therapies (Majdi et al., 2020). Since effective pharmacological treatments for AD remain unavailable, numerous clinical trials are investigating novel compounds like aducanumab, lecanemab, and gantenerumab, which aim to address the essential etiopathogenic factors underlying AD. Similarly, Parkinson's disease (PD) is characterized by the degeneration of dopaminergic neurons, particularly in the substantia nigra pars compacta, leading to reduced dopamine levels and impaired motor function. PD pathology is often associated with the formation of Lewy bodies within dopaminergic neurons. The underlying mechanisms of PD include protein misfolding and aggregation, mitochondrial dysfunction, microglial activation, neuroinflammation, instability of the blood-brain barrier, oxidative stress, melatonin deficiency, and depletion of trophic factors. Various animal models of PD and known gene mutation models are utilized to elucidate its etiology in vivo. Current clinical trials are ongoing to evaluate the effectiveness of emerging drugs such as dupage, istradefylline, and valtoco, which target key molecular pathways in PD management. The urgent need for effective therapies underscores the importance of identifying innovative drug candidates through comprehensive data mining and systems biology analyses (Nasr, 2024).

19.7 CHALLENGES AND FUTURE DIRECTIONS

In the current scenario, people are suffering from different types of neurodegenerative diseases. This review aims to summarize the most potent and industrially promising natural compounds with neurodegenerative disease-curing capabilities. Although nature contains unlimited resources to diagnose and cure these diseases, there are many limitations when selecting natural molecules because there are no specific guidelines and parameters to assay a natural molecule with different types of disease-curing capabilities. Additionally, due to the large number of molecules, it is very difficult to select a suitable molecular target to develop a potent drug. In research, we also found that very few natural molecules have been recognized for their curing potential and have been in clinical trials. Due to a lack of scientific recognition and limited clinical use, they have not been available in hospitals.

So far, no treatment can completely cure neurodegenerative diseases; often, the only option to suppress them is to treat the symptoms in individual patients. It is a fact that without understanding the underlying causes and mechanisms of neurodegenerative diseases, we cannot develop a treatment strategy. There are many reasons why the development of neurodegenerative disease treatments is very challenging, such as poor penetration of the molecules into the brain, a diverse and ill-defined patient population, symptom assessment, and rapid disease progression in patients. In natural product-based drugs, we found many limitations, such as bioavailability issues, difficulty in evaluating the correct dosage, and low potency. However, these molecules have many advantages over synthetic drugs, such as lower toxicity and multiple targets, making them attractive for modern disease treatment. The combination of nature-derived and other drug molecules with multiple targets and improved specificity may achieve therapeutic benefits in the future. In the near future, the exploration of newer molecules and targets against these diseases and their development into clinical phases to study their effects on real patients is an immediate need. For example, stem cells and gene-editing technology have the potential to facilitate the study of various diseases more easily. Furthermore, the use of advanced technology for high-throughput screening and new drug discovery can help explore newer molecules and accelerate clinical development rapidly at a lower cost and risk. The development of novel molecular targeting compounds and biomarkers can assist scientists and researchers in diagnosing and curing neurodegenerative diseases. Although many advances have shown promising results and have been translated into clinical applications, there are still hurdles to overcome. The use of newly advanced technology in basic and clinical research, along with other prevention strategies and therapeutics, needs to be implemented with ongoing work. Meanwhile, the optimized use of older and newer drugs, procedures, and other options may reduce some of the challenges of end-stage diseases. Although this goal is challenging, the next generation of scientists and researchers will be in a good position to develop evidence-based knowledge for the prevention and alleviation of aging-associated vulnerabilities and behaviors.

19.7.1 Barriers to Natural Molecule Research and Development

In the year 2019, a significant observation was made regarding the quantity of scientific research manuscripts focused on the effects of natural molecules on neurological diseases, sourced from prestigious international academic journals dedicated to natural molecules. It became evident that this body of work was not only limited in scale but also lacked substantial insight. Such findings suggest that the process of researching and developing natural molecules for commercialization continues to encounter numerous obstacles. Typically, research and development efforts concerning natural molecules are viewed as preliminary endeavors. Furthermore, chronic conditions such as arthritis, cardiovascular issues, central neuromuscular disorders, cancer, infectious diseases, and diabetes have been identified as ailments that frequently incorporate the use of natural molecules. The implementation of in vitro and in vivo studies, alongside clinical trials, plays a crucial role in enhancing the comprehension of the advantages provided by these natural compounds. Such research is essential for identifying alternative therapies and can also bolster the efficacy of synthetic medications. In this context, various candidate natural molecules—among them asarone, naringin, nepitrin, rosmarinic acid, gossypol, xanthones, cocoa-derived flavonoids, tectoridin, scolymoside, puerarin, farrerol, alshanol A, and alshanol B—have demonstrated promising outcomes in discouraging clinical trial results and pathogenesis investigations. It is posited that further contemporary scientific research aimed at elucidating how these potent candidates can be more prominently featured may lead to enhanced outcomes.

19.7.2 Emerging Strategies and Technologies

Novel and ever-evolving technologies and innovative strategies are absolutely essential to advance our comprehensive knowledge of cellular, molecular, and environmental events that converge in the intricate mechanisms of neurodegenerative progression and can be effectively exploited for the discovery and development of groundbreaking new therapeutic agents. Some of the most exciting and promising recent progress—particularly studies that provide groundbreaking new mechanistic insights—utilize advanced systems biology and functional genomics approaches, as well as organoids and patient-derived cells, which are crucial for this research. A modern and frontal role of imaging techniques highlights the acute need for such methodologies in preclinical neurodegenerative disease research, where understanding intricacies is vital. But of utmost interest and significant importance appears to be the identification and utilization of natural products and natural extracts that hold the potential to prevent initiation, slow progression, and effectively treat a range of symptoms associated with neurodegenerative diseases. Furthermore, when these approaches are combined with a modern and thorough understanding of the molecular and cellular events underlying the neurodegeneration process, together they provide exciting, continually evolving new paradigms for the treatment of these devastating and increasingly challenging issues facing modern medicine today. This synthesis of knowledge promises to pave the way for innovative solutions in the realm of neurodegenerative diseases.

REFERENCES

Abed, M. (2023). A comprehensive examination of human brain disorders. *Journal of Biomedical and Sustainable Healthcare Applications*, *3*(2), 141–152.

Abokyi, S., To, C.-H., Lam, T. T., & Tse, D. Y. (2020). Central role of oxidative stress in age-related macular degeneration: Evidence from a review of the molecular mechanisms and animal models. *Oxidative Medicine and Cellular Longevity*, *2020*(1), 7901270.

Abushouk, A. I., Negida, A., Ahmed, H., & Abdel-Daim, M. M. (2017). Neuroprotective mechanisms of plant extracts against MPTP induced neurotoxicity: Future applications in Parkinson's disease. *Biomedicine & Pharmacotherapy*, *85*, 635–645.

Agrahari, V., Burnouf, P.-A., Burnouf, T., & Agrahari, V. (2019). Nanoformulation properties, characterization, and behavior in complex biological matrices: Challenges and opportunities for brain-targeted drug delivery applications and enhanced translational potential. *Advanced Drug Delivery Reviews*, *148*, 146–180.

Ahmad, H. A., Wahab, F., Ullah, M., & Khan, M. I. (2024). Exploring the therapeutic oower of flavonoids on chronic disease: Unraveling the mechanisms of action especially by following MAPKs/NF-KB signaling pathways. In *Role of flavonoids in chronic metabolic diseases: From bench to clinic* (pp. 1–49). Wiley- Scrivener Publishing.

Akçimen, F., Lopez, E. R., Landers, J. E., Nath, A., Chiò, A., Chia, R., & Traynor, B. J. (2023). Amyotrophic lateral sclerosis: Translating genetic discoveries into therapies. *Nature Reviews Genetics*, *24*(9), 642–658.

Alkhalifa, A. M. E. R., Alkhalifa, O., Durdanovic, I., & Ibrahim, D. R. (2024). *Oxidative stress and mitochondrial dysfunction in Alzheimer's disease: Insights into pathophysiology and treatment.*

Alqahtani, T., Deore, S. L., Kide, A. A., Shende, B. A., Sharma, R., Dadarao Chakole, R., Nemade, L. S., Kishor Kale, N., Borah, S., & Shrikant Deokar, S. (2023). Mitochondrial dysfunction and oxidative stress in Alzheimer's disease, and Parkinson's disease, Huntington's disease and amyotrophic lateral sclerosis-an updated review. *Mitochondrion*, *71*, 83–92.

Alraawi, Z., Banerjee, N., Mohanty, S., & Kumar, T. K. S. (2022). Amyloidogenesis: What do we know so far? *International Journal of Molecular Sciences*, *23*(22), 13970.

Alzheimer's Association. (2013). 2013 Alzheimer's disease facts and figures. *Alzheimer's & Dementia*, *9*(2), 208–245.

Aronin, N., Chase, K., Young, C., Sapp, E., Schwarz, C., Matta, N., Kornreich, R., Lanwehrmeyer, B., Bird, E., & Beal, M. F. (1995). CAG expansion affects the expression of mutant Huntingtin in the Huntington's disease brain. *Neuron*, *15*(5), 1193–1201.

Atoki, A. V., Aja, P. M., Shinkafi, T. S., Ondari, E. N., & Awuchi, C. G. (2023). Hesperidin plays beneficial roles in disorders associated with the central nervous system: A review. *International Journal of Food Properties*, *26*(1), 1867–1884.

Avallone, R., Vitale, G., & Bertolotti, M. (2019). Omega-3 fatty acids and neurodegenerative diseases: New evidence in clinical trials. *International Journal of Molecular Sciences*, *20*(17). https://doi.org/10.3390/ijms20174256

Aware, C. B., Patil, D. N., Suryawanshi, S. S., Mali, P. R., Rane, M. R., Gurav, R. G., & Jadhav, J. P. (2022). Natural bioactive products as promising therapeutics: A review of natural product-based drug development. *South African Journal of Botany*, *151*, 512–528.

Ayaz, M., Ullah, F., Sadiq, A., Kim, M. O., & Ali, T. (2019). *Natural products-based drugs: potential therapeutics against Alzheimer's disease and other neurological disorders.* Frontiers Media SA.

Bae, J. R., & Kim, S. H. (2017). Synapses in neurodegenerative diseases. *BMB Reports*, *50*(5), 237.

Balkhi, H. M., Gul, T., Banday, M. Z., & Haq, E. (2014). Glutamate excitotoxicity: An insight into the mechanism. *International Journal of Advanced Research*, *2*(7), 361–373.

Baquero, M., & Martín, N. (2015). Depressive symptoms in neurodegenerative diseases. *World Journal of Clinical Cases: WJCC*, *3*(8), 682.

Barage, S. H., & Sonawane, K. D. (2015). Amyloid cascade hypothesis: Pathogenesis and therapeutic strategies in Alzheimer's disease. *Neuropeptides*, *52*, 1–18.

Baralić, K., Živanović, J., Marić, Đ, Bozic, D., Grahovac, L., Antonijević Miljaković, E., Ćurčić, M., Buha Djordjevic, A., Bulat, Z., Antonijević, B., & Đukić-Ćosić, D. (2024). Sulforaphane-A compound with potential health benefits for disease prevention and treatment: Insights from pharmacological and toxicological experimental studies. *Antioxidants (Basel)*, *13*(2). https://doi.org/10.3390/antiox13020147

Barbosa, M. A. G., Xavier, C. P. R., Pereira, R. F., Petrikaitė, V., & Vasconcelos, M. H. (2021). 3D cell culture models as recapitulators of the tumor microenvironment for the screening of anti-cancer drugs. *Cancers*, *14*(1), 190.

Barrantes, F. J. (2024). Cognitive synaptopathy: Synaptic and dendritic spine dysfunction in age-related cognitive disorders. *Frontiers in Aging Neuroscience*, *16*, 1476909.

Batiha, G. E.-S., Al-Kuraishy, H. M., Al-Gareeb, A. I., & Elekhnawy, E. (2023). SIRT1 pathway in Parkinson's disease: A faraway snapshot but so close. *Inflammopharmacology*, *31*(1), 37–56.

Batool, S., Raza, H., Zaidi, J., Riaz, S., Hasan, S., & Syed, N.I. (2019). Synapse formation: From cellular and molecular mechanisms to neurodevelopmental and neurodegenerative disorders. *Journal of Neurophysiology*, *121*. 1381–1397

Bejanin, A., Schonhaut, D. R., La Joie, R., Kramer, J. H., Baker, S. L., Sosa, N., Ayakta, N., Cantwell, A., Janabi, M., & Lauriola, M. (2017). Tau pathology and neurodegeneration contribute to cognitive impairment in Alzheimer's disease. *Brain*, *140*(12), 3286–3300.

Belwal, T., Bisht, A., Devkota, H. P., Ullah, H., Khan, H., Pandey, A., Bhatt, I. D., & Echeverría, J. (2020). Phytopharmacology and clinical updates of Berberis species against diabetes and other metabolic diseases. *Frontiers in Pharmacology*, *11*, 41.

Bernatoniene, J., Jakstas, V., & Kopustinskiene, D. M. (2023). Phenolic compounds of *Rhodiola rosea* L. as the potential alternative therapy in the treatment of chronic diseases. *International Journal of Molecular Sciences*, *24*(15), 12293.

Bernheim, J. L., Distelmans, W., Mullie, A., & Ashby, M. A. (2014). Questions and answers on the Belgian model of integral end-of-life care: Experiment? Prototype? "Eu-Euthanasia": The close historical, and evidently synergistic, relationship between palliative care and euthanasia in Belgium: An interview with a doctor involved in the early development of both and two of his successors. *Journal of Bioethical Inquiry*, *11*, 507–529.

Bhat, J. A., Akther, T., Najar, R. A., Rasool, F., & Hamid, A. (2022). *Withania somnifera* (L.) Dunal (Ashwagandha); current understanding and future prospect as a potential drug candidate. *Frontiers in Pharmacology*, *13*, 1029123.

Bhat, S. A., Ahamad, S., Dar, N. J., Hassan Siddique, Y., & Nazir, A. (2023). The emerging landscape of natural small-molecule therapeutics for Huntington's disease. *Current Neuropharmacology*, *21*(4), 867–889.

Bhushan, B., Singh, N. K., & Singh, R. (2024). Traditional Chinese medicine: Its growing potential in treating neurological disorders. *Pharmacological Research-Modern Chinese Medicine*, *11*, 100422.

Bischoff-Ferrari, H. A., Bhasin, S., & Manson, J. A. E. (2018). Preventing fractures and falls: A limited role for calcium and vitamin D supplements? *JAMA*, *319*(15), 1552–1553.

Bjørklund, G., Oliinyk, P., Khavrona, O., Lozynska, I., Lysiuk, R., Darmohray, R., Antonyak, H., Dub, N., Zayachuk, V., & Antoniv, O. (2024). The effects of fisetin and curcumin on oxidative damage caused by transition metals in neurodegenerative diseases. *Molecular Neurobiology*, *62*, 1–22.

Borsche, M., Pereira, S. L., Klein, C., & Grünewald, A. (2021). Mitochondria and Parkinson's disease: Clinical, molecular, and translational aspects. *Journal of Parkinson's Disease*, *11*(1), 45–60.

Bousquet, M., Calon, F., & Cicchetti, F. (2011). Impact of omega-3 fatty acids in Parkinson's disease. *Ageing Research Reviews*, *10*(4), 453–463.

Bradley, S. J., & Challiss, R. A. J. (2012). G protein-coupled receptor signalling in astrocytes in health and disease: A focus on metabotropic glutamate receptors. *Biochemical Pharmacology*, *84*(3), 249–259.

Brasnjo, G., & Otis, T. S. (2001). Neuronal glutamate transporters control activation of postsynaptic metabotropic glutamate receptors and influence cerebellar long-term depression. *Neuron*, *31*(4), 607–616.

Brotini, S., & Gigli, G. L. (2004). Epidemiology and clinical features of sleep disorders in extrapyramidal disease. *Sleep Medicine*, *5*(2), 169–179.

Brück, W. (2005). The pathology of multiple sclerosis is the result of focal inflammatory demyelination with axonal damage. *Journal of Neurology*, *252*, v3–v9.

Cai, B., Zhang, Y., Wang, Z., Xu, D., Jia, Y., Guan, Y., Liao, A., Liu, G., Chun, C. J., & Li, J. (2020). Therapeutic potential of diosgenin and its major derivatives against neurological diseases: Recent advances. *Oxidative Medicine and Cellular Longevity*, *2020*(1), 3153082.

Calabrese, V., Santoro, A., Monti, D., Crupi, R., Di Paola, R., Latteri, S., Cuzzocrea, S., Zappia, M., Giordano, J., & Calabrese, E. J. (2018). Aging and Parkinson's disease: Inflammaging, neuroinflammation and biological remodeling as key factors in pathogenesis. *Free Radical Biology and Medicine*, *115*, 80–91.

Calkins, M. J., Manczak, M., Mao, P., Shirendeb, U., & Reddy, P. H. (2011). Impaired mitochondrial biogenesis, defective axonal transport of mitochondria, abnormal mitochondrial dynamics and synaptic degeneration in a mouse model of Alzheimer's disease. *Human Molecular Genetics*, *20*(23), 4515–4529.

Calsolaro, V., & Edison, P. (2016). Neuroinflammation in Alzheimer's disease: Current evidence and future directions. *Alzheimer's & Dementia*, *12*(6), 719–732.

Carson, M. J., Thrash, J. C., & Walter, B. (2006). The cellular response in neuroinflammation: The role of leukocytes, microglia and astrocytes in neuronal death and survival. *Clinical Neuroscience Research*, *6*(5), 237–245.

Cásedas, G., Moliner, C., Maggi, F., Mazzara, E., & López, V. (2022). Evaluation of two different *Cannabis sativa* L. extracts as antioxidant and neuroprotective agents. *Frontiers in Pharmacology*, *13*, 1009868.

Cavalli, A., Bolognesi, M. L., Minarini, A., Rosini, M., Tumiatti, V., Recanatini, M., & Melchiorre, C. (2008). Multi-target-directed ligands to combat neurodegenerative diseases. *Journal of Medicinal Chemistry, 51*(3), 347–372.

Cefalu, W. T. (2006). Animal models of type 2 diabetes: Clinical presentation and pathophysiological relevance to the human condition. *ILAR Journal, 47*(3), 186–198.

Cenini, G., Lloret, A., & Cascella, R. (2019). Oxidative stress in neurodegenerative diseases: From a mitochondrial point of view. *Oxidative Medicine and Cellular Longevity, 2019*(1), 2105607.

Centola, D. (2013). Social media and the science of health behavior. *Circulation, 127*(21), 2135–2144.

Chaudhuri, A. (2013). Multiple sclerosis is primarily a neurodegenerative disease. *Journal of Neural Transmission, 120*, 1463–1466.

Chen, D., Huang, C., & Chen, Z. (2019). A review for the pharmacological effect of lycopene in central nervous system disorders. *Biomedicine & Pharmacotherapy, 111*, 791–801.

Chen, X., Guo, C., & Kong, J. (2012). Oxidative stress in neurodegenerative diseases☆. *Neural Regeneration Research, 7*(5), 376–385.

Chi, H., Chang, H.-Y., & Sang, T.-K. (2018). Neuronal cell death mechanisms in major neurodegenerative diseases. *International Journal of Molecular Sciences, 19*(10), 3082.

Chiang, M. C., Tsai, T. Y., & Wang, C. J. (2023). The potential benefits of quercetin for brain health: A review of anti-inflammatory and neuroprotective mechanisms. *International Journal of Molecular Sciences, 24*(7). https://doi.org/10.3390/ijms24076328

Chiurchiù, V., van der Stelt, M., Centonze, D., & Maccarrone, M. (2018). The endocannabinoid system and its therapeutic exploitation in multiple sclerosis: Clues for other neuroinflammatory diseases. *Progress in Neurobiology, 160*, 82–100.

Ciruela, F., Llobet, A., Martín-Satué, M., Solsona, C., Correia-de-Sá Pelegrin, P., Rotllán, P., Borea, P. A., Franco, R., Sen, R., & Cunha, R. (2010). Tarragona-Barcelona 30th May–2nd June Guest editors. *Purinergic Signalling, 6*(1), S1–S162.

Ciurea, A., Ciurea, J., & Opris, I. (2017). Neuromodulation of consciousness disorders. In *Electroceuticals: Advances in electrostimulation therapies* (pp. 317–346). Springer International Publishing.

Ciurea, A. V., Mohan, A. G., Covache-Busuioc, R.-A., Costin, H.-P., Glavan, L.-A., Corlatescu, A.-D., & Saceleanu, V. M. (2023). Unraveling molecular and genetic insights into neurodegenerative diseases: Advances in understanding Alzheimer's, Parkinson's, and Huntington's diseases and amyotrophic lateral sclerosis. *International Journal of Molecular Sciences, 24*(13), 10809.

Contestabile, A. (2011). The history of the cholinergic hypothesis. *Behavioural Brain Research, 221*(2), 334–340.

Cui, H., Kong, Y., & Zhang, H. (2012). Oxidative stress, mitochondrial dysfunction, and aging. *Journal of Signal Transduction, 2012*(1), 646354.

Curcio, M., & Bradke, F. (2018). Axon regeneration in the central nervous system: Facing the challenges from the inside. *Annual Review of Cell and Developmental Biology, 34*(1), 495–521.

d'Anglemont de Tassigny, X., Pascual, A., & López-Barneo, J. (2015). GDNF-based therapies, GDNF-producing interneurons, and trophic support of the dopaminergic nigrostriatal pathway. Implications for Parkinson's disease. *Frontiers in Neuroanatomy, 9*, 10.

Dai, Y., Wang, H., Lian, A., Li, J., Zhao, G., Hu, S., & Li, B. (2023). A comprehensive perspective of Huntington's disease and mitochondrial dysfunction. *Mitochondrion, 70*, 8–19.

Das, S., Rajeswari, V. D., Venkatraman, G., & Ramanathan, G. (2024"). Phytochemicals in Parkinson's disease: A pathway to neuroprotection and personalized medicine.". *Cell Biochemistry and Biophysics, 83*, 1–17.

de Vries, H. E., Witte, M., Hondius, D., Rozemuller, A. J. M., Drukarch, B., Hoozemans, J., & van Horssen, J. (2008). Nrf2-induced antioxidant protection: A promising target to counteract ROS-mediated damage in neurodegenerative disease? *Free Radical Biology and Medicine, 45*(10), 1375–1383.

Deepika, T. C. D., Sharma, N. K., Ranga, V., & Maurya, P. K. (2024). Naringenin orchestrates and regulates the reactive oxygen species-mediated pathways and proinflammatory signaling: Targeting hallmarks of aging-associated disorders. *Rejuvenation Research, 27*(1), 3–16.

Dejanovic, B., Sheng, M., & Hanson, J. E. (2024). Targeting synapse function and loss for treatment of neurodegenerative diseases. *Nature Reviews Drug Discovery, 23*(1), 23–42.

Denis, H. L., David, L. S., & Cicchetti, F. (2019). Antibody-based therapies for Huntington's disease: Current status and future directions. *Neurobiology of Disease, 132*, 104569.

DeTure, M. A., & Dickson, D. W. (2019). The neuropathological diagnosis of Alzheimer's disease. *Molecular Neurodegeneration, 14*(1), 32.

Dias, V., Junn, E., & Mouradian, M. M. (2013). The role of oxidative stress in Parkinson's disease. *Journal of Parkinson's Disease, 3*(4), 461–491.

Divino da Rocha, M., Dias Viegas, F. P., Cristina Campos, H., Carolina Nicastro, P., Calve Fossaluzza, P., Manssour Fraga, C. A., Barreiro, E. J., & Viegas, C. (2011). The role of natural products in the discovery of new drug candidates for the treatment of neurodegenerative disorders II: Alzheimer's disease. *CNS & Neurological Disorders-Drug Targets-CNS & Neurological Disorders), 10*(2), 251–270.

Dong, H., Zhang, Y., Huang, Y., & Deng, H. (2022). Pathophysiology of RAGE in inflammatory diseases. *Frontiers in Immunology, 13*, 931473.

Duchen, M. R. (2012). Mitochondria, calcium-dependent neuronal death and neurodegenerative disease. *Pflügers Archiv-European Journal of Physiology, 464*, 111–121.

Duta-Bratu, C.-G., Mihai Nitulescu, G., Mihai, D. P., & Olaru, O. T. (2023). Resveratrol and other natural oligomeric stilbenoid compounds and their therapeutic applications. *Plants, 12*(16), 2935.

Dyall, S. C., & Michael-Titus, A. T. (2008). Neurological benefits of omega-3 fatty acids. *Neuromolecular Medicine, 10*(4), 219–235.

El-Agnaf, O. M. A., Paleologou, K. E., Greer, B., Abogrein, A. M., King, J. E., Salem, S. A., Fullwood, N. J., Benson, F. E., Hewitt, R., & Ford, K. J. (2004). A strategy for designing inhibitors of α-synuclein aggregation and toxicity as a novel treatment for Parkinson's disease and related disorders. *The FASEB Journal, 18*(11), 1315–1317.

Erkkinen, M. G., Kim, M.-O., & Geschwind, M. D. (2018). Clinical neurology and epidemiology of the major neurodegenerative diseases. *Cold Spring Harbor Perspectives in Biology, 10*(4), a033118.

Facecchia, K., Fochesato, L.-A., Ray, S. D., Stohs, S. J., & Pandey, S. (2011). Oxidative toxicity in neurodegenerative diseases: Role of mitochondrial dysfunction and therapeutic strategies. *Journal of Toxicology, 2011*(1), 683728.

Fan, C.-D., Li, Y., Fu, X.-T., Wu, Q.-J., Hou, Y.-J., Yang, M.-F., Sun, J.-Y., Fu, X.-Y., Zheng, Z.-C., & Sun, B.-L. (2017). Reversal of beta-amyloid-induced neurotoxicity in PC12 cells by curcumin, the important role of ROS-mediated signaling and ERK pathway. *Cellular and Molecular Neurobiology*, *37*, 211–222.

Fang, Z., Tang, Y., Ying, J., Tang, C., & Wang, Q. (2020). Traditional Chinese medicine for anti-Alzheimer's disease: Berberine and evodiamine from *Evodia rutaecarpa*. *Chinese Medicine*, *15*, 1–16.

Flores, G., Negrete-Díaz, J. V., Carrión, M., Andrade-Talavera, Y., Bello, S. A., Sihra, T. S., & Rodríguez-Moreno, A. (2012). Excitatory amino acids in neurological and neurodegenerative disorders. In *Amino acids in human nutrition and health* (pp. 427–453). CABI.

Folch, J., Junyent, F., Verdaguer, E., Auladell, C., Pizarro, J. G., Beas-Zarate, C., Pallas, M., & Camins, A. (2012). Role of cell cycle re-entry in neurons: A common apoptotic mechanism of neuronal cell death. *Neurotoxicity Research*, *22*, 195–207.

Gaba, S., Saini, A., Singh, G., & Monga, V. (2021). An insight into the medicinal attributes of berberine derivatives: A review. *Bioorganic & Medicinal Chemistry*, *38*, 116143.

Gadhave, D. G., Sugandhi, V. V., Jha, S. K., Nangare, S. N., Gupta, G., Singh, S. K., Dua, K., Cho, H., Hansbro, P. M., & Paudel, K. R. (2024). Neurodegenerative disorders: Mechanisms of degeneration and therapeutic approaches with their clinical relevance. *Ageing Research Reviews*, *99*, 102357.

Gan, B. H., Gaynord, J., Rowe, S. M., Deingruber, T., & Spring, D. R. (2021). The multifaceted nature of antimicrobial peptides: Current synthetic chemistry approaches and future directions. *Chemical Society Reviews*, *50*(13), 7820–7880.

Gandhi, J., Antonelli, A. C., Afridi, A., Vatsia, S., Joshi, G., Romanov, V., Murray, I. V. J., & Khan, S. A. (2019). Protein misfolding and aggregation in neurodegenerative diseases: A review of pathogeneses, novel detection strategies, and potential therapeutics. *Reviews in the Neurosciences*, *30*(4), 339–358.

Garodia, P., Hegde, M., Kunnumakkara, A. B., & Aggarwal, B. B. (2023). Curcumin, inflammation, and neurological disorders: How are they linked? *Integrative Medicine Research*, *12*(3), 100968. https://doi.org/10.1016/j.imr.2023.100968

Goedert, M. (2015). Alzheimer's and Parkinson's diseases: The prion concept in relation to assembled Aβ, tau, and α-synuclein. *Science*, *349*(6248), 1255555.

Goetzel, R. Z., Long, S. R., Ozminkowski, R. J., Hawkins, K., Wang, S., & Lynch, W. (2004). Health, absence, disability, and presenteeism cost estimates of certain physical and mental health conditions affecting US employers. *Journal of Occupational and Environmental Medicine*, *46*(4), 398–412.

Golpich, M., Amini, E., Mohamed, Z., Azman Ali, R., Mohamed Ibrahim, N., & Ahmadiani, A. (2017). Mitochondrial dysfunction and biogenesis in neurodegenerative diseases: Pathogenesis and treatment. *CNS Neuroscience & Therapeutics*, *23*(1), 5–22.

Gomes, T. M., Sousa, P., Campos, C., Perestrelo, R., & Câmara, J. S. (2024). Secondary bioactive metabolites from foods of plant origin as theravention agents against neurodegenerative disorders. *Foods*, *13*(14), 2289.

Gong, L., Yin, J., Zhang, Y., Huang, R., Lou, Y., Jiang, H., Sun, L., Jia, J., & Zeng, X. (2022). Neuroprotective mechanisms of ginsenoside Rb1 in central nervous system diseases. *Frontiers in Pharmacology*, *13*, 914352. https://doi.org/10.3389/fphar.2022.914352

Gregersen, N., & Bross, P. (2010). Protein misfolding and cellular stress: An overview. In *Protein misfolding and cellular stress in disease and aging: Concepts and protocols* (pp. 3–23). Humana Press (Springer).

Grierson, A., & Miller, C. (2021). Axonal transport and amyotrophic lateral sclerosis. In *Amyotrophic lateral sclerosis, second edition* (pp. 309–321). CRC Press.

Grimm, S., Hoehn, A., Davies, K. J., & Grune, T. (2011). Protein oxidative modifications in the ageing brain: Consequence for the onset of neurodegenerative disease. *Free Radical Research*, *45*(1), 73–88.

Gross, E. R., Hsu, A. K., & Gross, G. J. (2004). Opioid-induced cardioprotection occurs via glycogen synthase kinase β inhibition during reperfusion in intact rat hearts. *Circulation Research*, *94*(7), 960–966.

Gupta, A., & Sharma, B. (2025). Neurodegenerative diseases (ND): An introduction. In *Synaptic plasticity in neurodegenerative disorders* (pp. 3–20). CRC Press.

Guzior, N., Wieckowska, A., Panek, D., & Malawska, B. (2015). Recent development of multifunctional agents as potential drug candidates for the treatment of Alzheimer's disease. *Current Medicinal Chemistry*, *22*(3), 373–404.

Guzman-Martinez, L., Maccioni, R. B., Andrade, V., Patricio Navarrete, L., Gabriela Pastor, M., & Ramos-Escobar, N. (2019). Neuroinflammation as a common feature of neurodegenerative disorders. *Frontiers in Pharmacology*, *10*, 1008.

Halder, T., Kundu, S., & Ghosh, B. (2022). *Withania somnifera*: A future pharma factory. In *Applications in plant biotechnology* (pp. 3–44). CRC Press.

Hampel, H., Hardy, J., Blennow, K., Chen, C., Perry, G., Kim, S. H., Villemagne, V. L., Aisen, P., Vendruscolo, M., & Iwatsubo, T. (2021). The amyloid-β pathway in Alzheimer's disease. *Molecular Psychiatry*, *26*(10), 5481–5503.

Hampton, D. W., Anderson, J., Pryce, G., Irvine, K.-A., Giovannoni, G., Fawcett, J. W., Compston, A., Franklin, R. J. M., Baker, D., & Chandran, S. (2008). An experimental model of secondary progressive multiple sclerosis that shows regional variation in gliosis, remyelination, axonal and neuronal loss. *Journal of Neuroimmunology*, *201*, 200–211.

Hanna, J., Guerra-Moreno, A., Ang, J., & Micoogullari, Y. (2019). Protein degradation and the pathologic basis of disease. *The American Journal of Pathology*, *189*(1), 94–103.

Hassan, S. S., Samanta, S., Dash, R., Karpiński, T. M., Habibi, E., Sadiq, A., Ahmadi, A., & Bungau, S. (2022). The neuroprotective effects of fisetin, a natural flavonoid in neurodegenerative diseases: Focus on the role of oxidative stress. *Frontiers in Pharmacology*, *13*, 1015835.

Hassel, B., & Dingledine, R. (2012). Glutamate and glutamate receptors. In *Basic neurochemistry* (pp. 342–366). Elsevier.

Hermosillo-Abundis, C., Méndez-Rojas, M. A., & Arias-Carrión, O. (2024). Implications of environmental nanoparticles on neurodegeneration. *Journal of Neuroscience Research*, *102*(5), e25340.

Hider, R. C., Roy, S., Ma, Y. M., Le Kong, X., & Preston, J. (2011) The potential application of iron chelators for the treatment of neurodegenerative diseases. *Metallomics*, *3*(3), 239–249.

Hou, Y., Dan, X., Babbar, M., Wei, Y., Hasselbalch, S. G., Croteau, D. L., & Bohr, V. A. (2019). Ageing as a risk factor for neurodegenerative disease. *Nature Reviews Neurology*, *15*(10), 565–581.

Hussain, R., Zubair, H., Pursell, S., & Shahab, M. (2018). Neurodegenerative diseases: Regenerative mechanisms and novel therapeutic approaches. *Brain Sciences*, *8*(9), 177.

Imbriani, P., Martella, G., Bonsi, P., & Pisani, A. (2022). Oxidative stress and synaptic dysfunction in rodent models of Parkinson's disease. *Neurobiology of Disease*, *173*, 105851.

Imran, M., Ghorat, F., Ul-Haq, I., Ur-Rehman, H., Aslam, F., Heydari, M., Shariati, M. A., Okuskhanova, E., Yessimbekov, Z., Thiruvengadam, M., Hashempur, M. H., & Rebezov, M. (2020). Lycopene as a natural antioxidant used to prevent human health disorders. *Antioxidants (Basel)*, *9*(8). https://doi.org/10.3390/antiox9080706

Ingersoll, K. S., & Cohen, J. (2008). The impact of medication regimen factors on adherence to chronic treatment: A review of literature. *Journal of Behavioral Medicine*, *31*, 213–224.

Iovino, L., Tremblay, M. E., & Civiero, L. (2020). Glutamate-induced excitotoxicity in Parkinson's disease: The role of glial cells. *Journal of Pharmacological Sciences*, *144*(3), 151–164.

Islam, M. R., Rauf, A., Akash, S., Islam Trisha, S., Nasim, A. H., Akter, M., Dhar, P. S., Ogaly, H. A., Hemeg, H. A., & Wilairatana, P. (2024). Targeted therapies of curcumin focus on its therapeutic benefits in cancers and human health: Molecular signaling pathway-based approaches and future perspectives. *Biomedicine & Pharmacotherapy*, *170*, 116034.

Islam, M. T. (2017). Oxidative stress and mitochondrial dysfunction-linked neurodegenerative disorders. *Neurological Research*, *39*(1), 73–82.

Iyer, K., Tenchov, R., Sasso, J., Ralhan, K., Jotshi, J., Polshakov, D., Maind, A., & Zhou, Q. A. (2024). *Rare diseases: Insights from landscape analysis of current research, spotlighting amyotrophic lateral sclerosis, Huntington's disease, and myasthenia gravis.*

Iżycka-Świeszewska, E. (2023). *"Neuropatologia 2023." Polish academy of sciences.*

Jellinger, K. A. (2010). Basic mechanisms of neurodegeneration: A critical update. *Journal of Cellular and Molecular Medicine*, *14*(3), 457–487.

Jellinger, K. A. (2012). Neuropathology of sporadic Parkinson's disease: Evaluation and changes of concepts. *Movement Disorders*, *27*(1), 8–30.

Jia, J., Jiao, W., Wang, G., Wu, J., Huang, Z., & Zhang, Y. (2024). Drugs/agents for the treatment of ischemic stroke: Advances and perspectives. *Medicinal Research Reviews*, *44*(3), 975–1012.

Jia, Q., Deng, Y., & Qing, H. (2014). Potential therapeutic strategies for Alzheimer's disease targeting or beyond β-amyloid: Insights from clinical trials. *BioMed Research International*, *2014*(1), 837157.

John, A., & Reddy, P. H. (2021). Synaptic basis of Alzheimer's disease: Focus on synaptic amyloid beta, P-tau and mitochondria. *Ageing Research Reviews*, *65*, 101208.

Johnson, J., Mercado-Ayon, E., Mercado-Ayon, Y., Dong, Y. N., Halawani, S., Ngaba, L., & Lynch, D. R. (2021). Mitochondrial dysfunction in the development and progression of neurodegenerative diseases. *Archives of Biochemistry and Biophysics*, *702*, 108698.

Joshi, P., Bisht, A., Joshi, S., Semwal, D., Kumar Nema, N., Dwivedi, J., & Sharma, S. (2022). Ameliorating potential of curcumin and its analogue in central nervous system disorders and related conditions: A review of molecular pathways. *Phytotherapy Research*, *36*(8), 3143–3180.

Joshi, S., Dhingra, A. K., Chopra, B., Guarve, K., & Bhateja, D. (2022). Therapeutic potential and clinical evidence of hesperidin as neuroprotective agent. *Central Nervous System Agents in Medicinal Chemistry*, *22*(1), 5–14.

Jurcau, A. (2021). Insights into the pathogenesis of neurodegenerative diseases: Focus on mitochondrial dysfunction and oxidative stress. *International Journal of Molecular Sciences*, *22*(21), 11847.

Kandimalla, R., & Reddy, P. H. (2017). Therapeutics of neurotransmitters in Alzheimer's disease. *Journal of Alzheimer's Disease*, *57*(4), 1049–1069.

Kar, A., & Panda, S. (2024). Phytochemicals: Promising alternatives for metabolic regulation in neurodegenerative diseases. In *Altered metabolism: A major contributor of comorbidities in neurodegenerative diseases* (pp. 417–438). Springer.

Karvandi, M. S., Sheikhzadeh Hesari, F., Aref, A. R., & Mahdavi, M. (2023). The neuroprotective effects of targeting key factors of neuronal cell death in neurodegenerative diseases: The role of ER stress, oxidative stress, and neuroinflammation. *Frontiers in Cellular Neuroscience*, *17*, 1105247.

Kaur, A., Tiwari, R., Tiwari, G., & Ramachandran, V. (2022). Resveratrol: A vital therapeutic agent with multiple health benefits. *Drug Research*, *72*(01), 5–17.

Kehrer, J. P., & Klotz, L.-O. (2015). Free radicals and related reactive species as mediators of tissue injury and disease: Implications for health. *Critical Reviews in Toxicology*, *45*(9), 765–798.

Khan, A. U., Akram, M., Daniyal, M., & Zainab, R. (2019). Awareness and current knowledge of Parkinson's disease: A neurodegenerative disorder. *International Journal of Neuroscience*, *129*(1), 55–93.

Khan, H., Ullah, H., Aschner, M., Cheang, W. S., & Akkol, E. K. (2019). Neuroprotective effects of quercetin in Alzheimer's disease. *Biomolecules*, *10*(1), 59.

Khanam, H., Ali, A., & Asif, M. (2016). Neurodegenerative diseases linked to misfolded proteins and their therapeutic approaches: A review. *European Journal of Medicinal Chemistry*, *124*, 1121–1141.

Kidd, P. M. (2005). Neurodegeneration from mitochondrial insufficiency: Nutrients, stem cells, growth factors, and prospects for brain rebuilding using integrative management. *Alternative Medicine Review*, *10*(4), 268.

Kim, C. H., Park, M. K., Kim, S. K., & Cho, Y. H. (2014). Antioxidant capacity and anti-inflammatory activity of lycopene in watermelon. *International Journal of Food Science & Technology*, *49*(9), 2083–2091.

Kim, K. H., Lee, D., Lee, H. L., Kim, C.-E., Jung, K., & Kang, K. S. (2018). Beneficial effects of Panax ginseng for the treatment and prevention of neurodegenerative diseases: Past findings and future directions. *Journal of Ginseng Research*, *42*(3), 239–247.

Kim, Y., Lim, J., & Oh, J. (2024). Taming neuroinflammation in Alzheimer's disease: The protective role of phytochemicals through the gut–brain axis. *Biomedicine & Pharmacotherapy*, *178*, 117277.

Kim, Y. E., Hipp, M. S., Bracher, A., Hayer-Hartl, M., & Ulrich Hartl, F. (2013). Molecular chaperone functions in protein folding and proteostasis. *Annual Review of Biochemistry*, *82*(1), 323–355.

Klimaschewski, L. P. (2022). Aging and neurodegenerative diseases: Why do nerve cells die? In *Parkinson's and Alzheimer's today: About neurodegeneration and its therapy* (pp. 27–103). Springer.

Kölliker-Frers, R., Udovin, L., Otero-Losada, M., Kobiec, T., Herrera, M. I., Palacios, J., Razzitte, G., & Capani, F. (2021). Neuroinflammation: An integrating overview of reactive-Neuroimmune cell interactions in health and disease. *Mediators of Inflammation*, *2021*(1), 9999146.

Koszła, O., & Sołek, P. (2024). Misfolding and aggregation in neurodegenerative diseases: Protein quality control machinery as potential therapeutic clearance pathways. *Cell Communication and Signaling*, *22*(1), 421.

Kovacs, G. G. (2019). Molecular pathology of neurodegenerative diseases: Principles and practice. *Journal of Clinical Pathology*, *72*(11), 725–735.

Lai, W., Zhang, J., Sun, J., Min, T., Bai, Y., He, J., Cao, H., Che, Q., Guo, J., & Su, Z. (2024). Oxidative stress in alcoholic liver disease, focusing on proteins, nucleic acids, and lipids: A review. *International Journal of Biological Macromolecules*, *278*, 134809.

Lang, A. E., & Espay, A. J. (2018). Disease modification in Parkinson's disease: Current approaches, challenges, and future considerations. *Movement Disorders*, *33*(5), 660–677.

Lapchak, P. A., & Araujo, D. M. (2007). Advances in ischemic stroke treatment: Neuroprotective and combination therapies. *Expert Opinion on Emerging Drugs*, *12*(1), 97–112.

Lee, C. H., & Kim, J.-H. (2014). A review on the medicinal potentials of ginseng and ginsenosides on cardiovascular diseases. *Journal of Ginseng Research*, *38*(3), 161–166.

Li, L., Ma, P., Nirasawa, S., & Liu, H. (2024). Formation, immunomodulatory activities, and enhancement of glucosinolates and sulforaphane in broccoli sprouts: A review for maximizing the health benefits to human. *Critical Reviews in Food Science and Nutrition*, *64*(20), 7118–7148.

Li, M., Tang, S., Peng, X., Sharma, G., Yin, S., Hao, Z., Li, J., Shen, J., & Dai, C. (2024). Lycopene as a therapeutic agent against aflatoxin B1-related toxicity: Mechanistic insights and future directions. *Antioxidants*, *13*(4), 452.

Li, Y. J. (2016). Intra-maxillary molecular releasing and its application in the assistance of neurodegenerative disease therapeutics. *International Journal of Clinical Therapeutics and Diagnosis*, *4*(1), 100–109.

Li, Z., Jiang, T., Lu, Q., Xu, K., He, J., Xie, L., Chen, Z., Zheng, Z., Ye, L., & Xu, K. (2020). Berberine attenuated the cytotoxicity induced by t-BHP via inhibiting oxidative stress and mitochondria dysfunction in PC-12 cells. *Cellular and Molecular Neurobiology*, *40*, 587–602.

Lim, K.-L., & Tan, J. M. M. (2007). Role of the ubiquitin proteasome system in Parkinson's disease. *BMC Biochemistry*, *8*, 1–10.

Liu, X.-A., Das, B., Chen, Y., Chen, Z., Avchalumov, Y., Tian, X., & Puthanveettil, S. V. (2018). Recent advances in Alzheimer's drug discovery research. *Frontiers in Clinical Drug Research-Alzheimer Disorders*, *7*, 26.

Longinetti, E., Mariosa, D., Larsson, H., Ye, W., Ingre, C., Almqvist, C., Lichtenstein, P., Piehl, F., & Fang, F. (2017). Neurodegenerative and psychiatric diseases among families with amyotrophic lateral sclerosis. *Neurology*, *89*(6), 578–585.

Lushchak, V. I. (2014). Free radicals, reactive oxygen species, oxidative stress and its classification. *Chemico-Biological Interactions*, *224*, 164–175.

Madkour, L. H. (2019). Function of reactive oxygen species (ROS) inside the living organisms and sources of oxidants. *Pharmaceutical Sciences & Analytical Research Journal*, *2*, 180023.

Maetzler, W., Klucken, J., & Horne, M. (2016). A clinical view on the development of technology-based tools in managing Parkinson's disease. *Movement Disorders*, *31*(9), 1263–1271.

Maheshwari, A., & Maheshwari, G. (2024). Aging population in Vietnam: Challenges, implications, and policy recommendations. *International Journal of Aging*, *2*(1), e1.

Majdi, A., Sadigh-Eteghad, S., Rahigh Aghsan, S., Farajdokht, F., Vatandoust, S. M., Namvaran, A., & Mahmoudi, J. (2020). Amyloid-β, tau, and the cholinergic system in Alzheimer's disease: Seeking direction in a tangle of clues. *Reviews in the Neurosciences*, *31*(4), 391–413.

Majoor-Krakauer, D., Willems, P. J., & Hofman, A. (2003). Genetic epidemiology of amyotrophic lateral sclerosis. *Clinical Genetics*, *63*(2), 83–101.

Mani, S., Dubey, R., Lai, I.-C., Babu, M. A., Tyagi, S., Swargiary, G., Mody, D., Singh, M., Agarwal, S., & Iqbal, D. (2023). Oxidative stress and natural antioxidants: Back and forth in the neurological mechanisms of Alzheimer's disease. *Journal of Alzheimer's Disease*, *96*(3), 877–912.

Manji, H., & Sisodiya, S. (2024). Global burden of neurological disease and the neurology of climate change. In *Neurology: A queen square textbook* (pp. 1–15). Elsevier.

Manoharan, S., Guillemin, G. J., Abiramasundari, R. S., Essa, M. M., Akbar, M., & Akbar, M. D. (2016). The role of reactive oxygen species in the pathogenesis of Alzheimer's disease, Parkinson's disease, and Huntington's disease: A mini review. *Oxidative Medicine and Cellular Longevity*, *2016*(1), 8590578.

Marcos-Rabal, P., González-Fuentes, J., Castro-Vázquez, L., Lozano, M. V., Rodríguez-Robledo, V., Santander-Ortega, M. J., Selva-Clemente, J., Villaseca-González, N., & del Mar Arroyo-Jiménez, M. (2021). Neurodegenerative diseases: A multidisciplinary approach. *Current Pharmaceutical Design*, *27*(30), 3305–3336.

Marino, A., Battaglini, M., Moles, N., & Ciofani, G. (2022). Natural antioxidant compounds as potential pharmaceutical tools against neurodegenerative diseases. *ACS Omega*, *7*(30), 25974–25990.

Martínez, V., Iriondo De-Hond, A., Borrelli, F., Capasso, R., Del Castillo, M. D., & Abalo, R. (2020). Cannabidiol and other non-psychoactive cannabinoids for prevention and treatment of gastrointestinal disorders: Useful nutraceuticals? *International Journal of Molecular Sciences*, *21*(9), 3067.

Mazza, M., Pomponi, M., Janiri, L., Bria, P., & Mazza, S. (2007). Omega-3 fatty acids and antioxidants in neurological and psychiatric diseases: An overview. *Progress in Neuro-Psychopharmacology and Biological Psychiatry*, *31*(1), 12–26.

Mazzucchi, S., Palermo, G., Campese, N., Galgani, A., Della Vecchia, A., Vergallo, A., Siciliano, G., Ceravolo, R., Hampel, H., & Baldacci, F. (2020). The role of synaptic biomarkers in the spectrum of neurodegenerative diseases. *Expert Review of Proteomics*, *17*(7–8), 543–559.

McGough, E., Kirk-Sanchez, N., & Liu-Ambrose, T. (2017). Integrating health promotion into physical therapy practice to improve brain health and prevent Alzheimer disease. *Journal of Neurologic Physical Therapy*, *41*, S55–S62.

McKenzie, J. A., Spielman, L. J., Pointer, C. B., Lowry, J. R., Bajwa, E., Lee, C. W., & Klegeris, A. (2017). Neuroinflammation as a common mechanism associated with the modifiable risk factors for Alzheimer's and Parkinson's diseases. *Current Aging Science*, *10*(3), 158–176.

Mierke, C. T., & Mierke, C. T. (2020). Cell proliferation, survival, necrosis and apoptosis. In *Cellular mechanics and biophysics: Structure and function of basic cellular components regulating cell mechanics* (pp. 743–824). Springer.

Mikulska, P., Malinowska, M., Ignacyk, M., Szustowski, P., Nowak, J., Pesta, K., Szeląg, M., Szklanny, D., Judasz, E., Kaczmarek, G., Ejiohuo, O. P., Paczkowska-Walendowska, M., Gościniak, A., & Cielecka-Piontek, J. (2023). Ashwagandha (*Withania somnifera*)-current research on the health-promoting activities: A narrative review. *Pharmaceutics*, *15*(4). https://doi.org/10.3390/pharmaceutics15041057

Modi, G., Pillay, V., Choonara, Y. E., Ndesendo, V. M. K., du Toit, L. C., & Naidoo, D. (2009). Nanotechnological applications for the treatment of neurodegenerative disorders. *Progress in Neurobiology*, *88*(4), 272–285.

Monteiro, K. L. C., de Aquino, T. M., & da Silva-Júnior, E. F. (2024). Natural compounds as inhibitors of Aβ peptide and tau aggregation. *CNS & Neurological Disorders-Drug Targets-CNS & Neurological Disorders)*, *23*(10), 1234–1250.

Mora, G., & Chiò, A. (2015). Disorders of upper and lower motor neurons. In *Prognosis of neurological diseases* (pp. 261–272). Springer.

Mora, I., Arola, L., Caimari, A., Escoté, X., & Puiggròs, F. (2022). Structured long-chain omega-3 fatty acids for improvement of cognitive function during aging. *International Journal of Molecular Sciences*, *23*(7), 3472.

Moujalled, D., Strasser, A., & Liddell, J. R. (2021). Molecular mechanisms of cell death in neurological diseases. *Cell Death & Differentiation*, *28*(7), 2029–2044.

Muratori, L., Fregnan, F., Maurina, M., Haastert-Talini, K., & Ronchi, G. (2022). The potential benefits of dietary polyphenols for peripheral nerve regeneration. *International Journal of Molecular Sciences*, *23*(9), 5177.

Nabavi, S. F., Braidy, N., Orhan, I. E., Badiee, A., Daglia, M., & Nabavi, S. M. (2016). *Rhodiola rosea* L. and Alzheimer's disease: From farm to pharmacy. *Phytotherapy Research*, *30*(4), 532–539.

Nasr, F. (2024). The power of translational biology: From genomics big data to concept building and therapeutic design. In *Drug discovery and evaluation: Safety and pharmacokinetic assays* (pp. 1–16). Springer.

Naumann, J. P. (2015). Amyotrophic lateral sclerosis, the primary motor neuron disease. *The Downtown Review*, *1*(2), 5.

Nazir, A., Assad, A., Hussain, A., & Singh, M. (2024). Alzheimer's disease diagnosis using deep learning techniques: Datasets, challenges, research gaps and future directions. *International Journal of System Assurance Engineering and Management*, *12*, 1–35.

Nebrisi, E. E. (2021). Neuroprotective activities of curcumin in Parkinson's disease: A review of the literature. *International Journal of Molecular Sciences*, *22*(20), 11248.

Nijssen, J., Comley, L. H., & Hedlund, E. (2017). Motor neuron vulnerability and resistance in amyotrophic lateral sclerosis. *Acta Neuropathologica*, *133*, 863–885.

Ohba, T., Nakamura, S., Shimazawa, M., Hayashi, Y., Kono, H., & Hara, H. (2020). Protective effects of *Huperzia serrata* and its components against oxidative damage and cognitive dysfunction, *PharmaNutrition*, *13*, 100203, https://doi.org/10.1016/j.phanu.2020.100203

Ontaneda, D., Thompson, A. J., Fox, R. J., & Cohen, J. A. (2017). Progressive multiple sclerosis: Prospects for disease therapy, repair, and restoration of function. *The Lancet*, *389*(10076), 1357–1366.

Pallotti, F., Bergamini, C., Lamperti, C., & Fato, R. (2021). The roles of coenzyme Q in disease: Direct and indirect involvement in cellular functions. *International Journal of Molecular Sciences*, *23*(1), 128.

Pamenter, M. E. (2014). Mitochondria: A multimodal hub of hypoxia tolerance. *Canadian Journal of Zoology*, *92*(7), 569–589.

Pandey, S., Zhang, W., & Assmann, S. M. (2007). Roles of ion channels and transporters in guard cell signal transduction. *FEBS Letters*, *581*(12), 2325–2336.

Pandit, R., Chen, L., & Götz, J. (2020). The blood-brain barrier: Physiology and strategies for drug delivery. *Advanced Drug Delivery Reviews*, *165*, 1–14.

Pangeni, R., Sahni, J. K., Ali, J., Sharma, S., & Baboota, S. (2014). Resveratrol: Review on therapeutic potential and recent advances in drug delivery. *Expert Opinion on Drug Delivery*, *11*(8), 1285–1298.

Papa, S., Bubici, C., Zazzeroni, F., Pham, C. G., Kuntzen, C., Knabb, J. R., Dean, K., & Franzoso, G. (2006). The NF-κB-mediated control of the JNK cascade in the antagonism of programmed cell death in health and disease. *Cell Death & Differentiation*, *13*(5), 712–729.

Papuc, C., Goran, G. V., Predescu, C. N., Tudoreanu, L., & Ştefan, G. (2021). Plant polyphenols mechanisms of action on insulin resistance and against the loss of pancreatic beta cells. *Critical Reviews in Food Science and Nutrition*, *62*(2), 325–352.

Parhiz, H., Roohbakhsh, A., Soltani, F., Rezaee, R., & Iranshahi, M. (2015). Antioxidant and anti-inflammatory properties of the citrus flavonoids hesperidin and hesperetin: An updated review of their molecular mechanisms and experimental models. *Phytotherapy Research*, *29*(3), 323–331.

Pavese, N., & Brooks, D. J. (2009). Imaging neurodegeneration in Parkinson's disease. *Biochimica et Biophysica Acta (BBA)-Molecular Basis of Disease*, *1792*(7), 722–729.

Penke, B., Bogár, F., & Fülöp, L. (2017). β-Amyloid and the pathomechanisms of Alzheimer's disease: A comprehensive view. *Molecules*, *22*(10), 1692.

Pereira, F., & Aires-de-Sousa, J. (2018). Computational methodologies in the exploration of marine natural product leads. *Marine Drugs*, *16*(7), 236.

Pickard, J. M., Zeng, M. Y., Caruso, R., & Núñez, G. (2017). Gut microbiota: Role in pathogen colonization, immune responses, and inflammatory disease. *Immunological Reviews*, *279*(1), 70–89.

Pirko, I., & Noseworthy, J. H. (2009). Demyelinating disorders of the central nervous system. In *Textbook of clinical neurology* (p. 1103). Elsevier Inc.

Pohanka, M. (2011). Cholinesterases, a target of pharmacology and toxicology. *Biomedical Papers of the Medical Faculty of Palacky University in Olomouc*, *155*(3).

Pohl, F., & Kong Thoo Lin, P. (2018). The potential use of plant natural products and plant extracts with antioxidant properties for the prevention/treatment of neurodegenerative diseases: In vitro, in vivo and clinical trials. *Molecules*, *23*(12), 3283.

Polanco, J. C., Li, C., Bodea, L.-G., Martinez-Marmol, R., Meunier, F. A., & Götz, J. (2018). Amyloid-β and tau complexity—Towards improved biomarkers and targeted therapies. *Nature Reviews Neurology*, *14*(1), 22–39.

Poovaiah, N., Davoudi, Z., Peng, H., Schlichtmann, B., Mallapragada, S., Narasimhan, B., & Wang, Q. (2018). Treatment of neurodegenerative disorders through the blood–brain barrier using nanocarriers. *Nanoscale*, *10*(36), 16962–16983.

Pourhanifeh, M. H., Shafabakhsh, R., Reiter, R. J., & Asemi, Z. (2019). The effect of resveratrol on neurodegenerative disorders: Possible protective actions against autophagy, apoptosis, inflammation and oxidative stress. *Current Pharmaceutical Design*, *25*(19), 2178–2191.

Priego, C. S. (2021). Characterization of cell-type-specific and activity-dependent enhancers in human neuronal populations and their impact on neuropsychiatric disorders. *Icahn school of medicine at mount sinai.*

Prince, M., Comas-Herrera, A., Knapp, M., Guerchet, M., & Karagiannidou, M. (2016). World Alzheimer Report 2016. Improving healthcare for people living with dementia: Coverage, quality and costs now and in the future. *Alzheimer's Disease International.*

Prokop, A. (2013). The intricate relationship between microtubules and their associated motor proteins during axon growth and maintenance. *Neural Development*, *8*, 1–10.

Przybyłowska, M., Dzierzbicka, K., Kowalski, S., Chmielewska, K., & Inkielewicz-Stepniak, I. (2021). Therapeutic potential of multifunctional derivatives of cholinesterase inhibitors. *Current Neuropharmacology*, *19*(8), 1323–1344.

Pugazhendhi, A., Suganthy, N., Chau, T. P., Sharma, A., Unpaprom, Y., Ramaraj, R., Karuppusamy, I., & Brindhadevi, K. (2021). Cannabinoids as anticancer and neuroprotective drugs: Structural insights and pharmacological interactions—A review. *Process Biochemistry*, *111*, 9–31.

Radi, E., Formichi, P., Battisti, C., & Federico, A. (2014). Apoptosis and oxidative stress in neurodegenerative diseases. *Journal of Alzheimer's disease*, *42*(s3), S125–S152.

Rahul, & Siddique, Y. (2022). Neurodegenerative disorders and the current state, pathophysiology, and management of Parkinson's disease." *CNS & Neurological Disorders-Drug Targets-CNS & Neurological Disorders)*, *21*(7), 574–595.

Rajabian, A., Rameshrad, M., & Hosseinzadeh, H. (2019). Therapeutic potential of Panax ginseng and its constituents, ginsenosides and gintonin, in neurological and neurodegenerative disorders: a patent review. *Expert Opinion on Therapeutic Patents*, *29*(1), 55–72.

Ramsay, R. R., Majekova, M., Medina, M., & Valoti, M. (2016). Key targets for multi-target ligands designed to combat neurodegeneration. *Frontiers in Neuroscience*, *10*, 375.

Ranganathan, R., Haque, S., Coley, K., Shepheard, S., Cooper-Knock, J., & Kirby, J. (2020). Multifaceted genes in amyotrophic lateral sclerosis-frontotemporal dementia. *Frontiers in Neuroscience*, *14*, 684.

Rauchova, H. (2021). Coenzyme Q10 effects in neurological diseases. *Physiological Research*, *70*(Suppl 4), S683.

Reddy, A. P., Ravichandran, J., & Carkaci-Salli, N. (2020). Neural regeneration therapies for Alzheimer's and Parkinson's disease-related disorders. *Biochimica et Biophysica Acta (BBA)-Molecular Basis of Disease*, *1866*(4), 165506.

Rehman, M. U., Sehar, N., Dar, N. J., Khan, A., Arafah, A., Rashid, S., Rashid, S. M., & Ganaie, M. A. (2023). Mitochondrial dysfunctions, oxidative stress and neuroinflammation as therapeutic targets for neurodegenerative diseases: An update on current advances and impediments. *Neuroscience & Biobehavioral Reviews*, *144*, 104961.

Rekatsina, M., Paladini, A., Piroli, A., Zis, P., Pergolizzi, J. V., & Varrassi, G. (2020). Pathophysiology and therapeutic perspectives of oxidative stress and neurodegenerative diseases: A narrative review. *Advances in Therapy*, *37*, 113–139.

Ross, C. A., & Pickart, C. M. (2004). The ubiquitin–proteasome pathway in Parkinson's disease and other neurodegenerative diseases. *Trends in Cell Biology*, *14*(12), 703–711.

Rubinsztein, D. C. (2006). The roles of intracellular protein-degradation pathways in neurodegeneration. *Nature*, *443*(7113), 780–786.

Saceleanu, V. M., Toader, C., Ples, H., Covache-Busuioc, R.-A., Costin, H. P., Bratu, B.-G., Dumitrascu, D.-I., Bordeianu, A., Corlatescu, A. D., & Ciurea, A. V. (2023). Integrative approaches in acute ischemic stroke: From symptom recognition to future innovations. *Biomedicines*, *11*(10), 2617.

Safiri, S., Ghaffari Jolfayi, A., Fazlollahi, A., Morsali, S., Sarkesh, A., Daei Sorkhabi, A., Golabi, B., Aletaha, R., Motlagh Asghari, K., & Hamidi, S. (2024). Alzheimer's disease: A comprehensive review of epidemiology, risk factors, symptoms diagnosis, management, caregiving, advanced treatments and associated challenges. *Frontiers in Medicine*, *11*, 1474043.

Sahoo, L., Tripathy, N. S., & Dilnawaz, F. (2024). Naringenin Nanoformulations for neurodegenerative diseases. *Current Pharmaceutical Biotechnology*, *25*(16), 2108–2124.

Saini, R. K., Rengasamy, K. R. R., Mahomoodally, F. M., & Keum, Y.-S. (2020). Protective effects of lycopene in cancer, cardiovascular, and neurodegenerative diseases: An update on epidemiological and mechanistic perspectives. *Pharmacological Research*, *155*, 104730.

Sanaeifar, F., Pourranjbar, S., Pourranjbar, M., Ramezani, S., Mehr, S. R., Soliman Wadan, A.-H., & Khazeifard, F. (2024). Beneficial effects of physical exercise on cognitive-behavioral impairments and brain-derived neurotrophic factor alteration in the limbic system induced by neurodegeneration. *Experimental Gerontology*, *195*, 112539.

Sangeet, S., Khan, A., Mahanta, S., Roy, N., Das, S. K., Mohanta, Y. K., Saravanan, M., Tag, H., & Hui, P. K. (2023). Computational analysis of *Bacopa monnieri* (L.) Wettst. Compounds for drug development against neurodegenerative disorders. *Current Computer-Aided Drug Design*, *19*(1), 24–36.

Schaeffer, E. L., & Gattaz, W. F. (2008). Cholinergic and glutamatergic alterations beginning at the early stages of Alzheimer disease: Participation of the phospholipase A2 enzyme. *Psychopharmacology*, *198*, 1–27.

Šebestík, J., Marques, S. M., Falé, P. L., Santos, S., Arduíno, D. M., Cardoso, S. M., Oliveira, C. R., Serralheiro, M. L. M., & Santos, M. A. (2011). Bifunctional phenolic-choline conjugates as anti-oxidants and acetylcholinesterase inhibitors. *Journal of Enzyme Inhibition and Medicinal Chemistry*, *26*(4), 485–497.

Shabir, I., Kumar Pandey, V., Shams, R., Dar, A. H., Dash, K. K., Khan, S. A., Bashir, I., Jeevarathinam, G., Rusu, A. V., & Esatbeyoglu, T. (2022). Promising bioactive properties of quercetin for potential food applications and health benefits: A review. *Frontiers in Nutrition*, *9*, 999752.

Silove, D., Ventevogel, P., & Rees, S. (2017). The contemporary refugee crisis: An overview of mental health challenges. *World Psychiatry*, *16*(2), 130–139.

Silva, J. M., Nobre, M. S. C., Albino, S. L., Lócio, L. L., Nascimento, A. P. S., Scotti, L., Scotti, M. T., Oshiro-Junior, J. A., Lima, M. C. A., & Mendonça-Junior, F. J. B. (2020). Secondary metabolites with antioxidant activities for the putative treatment of amyotrophic lateral sclerosis (ALS): "Experimental evidences". *Oxidative Medicine and Cellular Longevity*, *2020*(1), 5642029.

Singh, A. P., Singh, R., Singh Verma, S., Rai, V., Kaschula, C. H., Maiti, P., & Gupta, S. C. (2019). Health benefits of resveratrol: Evidence from clinical studies. *Medicinal Research Reviews*, *39*(5), 1851–1891.

Singh, J., Sarkar, B., Mukherjee, A., & Mutsuddi, M. (2024). Protein metabolism: Critical factors implicated in neurodegenerative diseases. In *Altered metabolism: A major contributor of comorbidities in neurodegenerative diseases* (pp. 173–205). Springer.

Singh, N. A., Mandal, A. K. A., & Khan, Z. A. (2015). Potential neuroprotective properties of epigallocatechin-3-gallate (EGCG). *Nutrition Journal*, *15*, 1–17.

Singh, S. (2015). Antioxidants as a preventive therapeutic option for age related neurodegenerative diseases. *Therapeutic Targets for Neurological Diseases*, *2*.

Sinsky, J., Pichlerova, K., & Hanes, J. (2021). Tau protein interaction partners and their roles in Alzheimer's disease and other tauopathies. *International Journal of Molecular Sciences*, 22(17), 9207.

Smith, J. A., Das, A., Ray, S. K., & Banik, N. L. (2012). Role of pro-inflammatory cytokines released from microglia in neurodegenerative diseases. *Brain Research Bulletin*, *87*(1), 10–20.

Sonawane, S. K., Dubey, T., Balmik, A. A., Das, R., & Chinnathambi, S. (2022). *Alzheimer's disease pathology: A tau perspective*.

Song, H., Sieurin, J., Wirdefeldt, K., Pedersen, N. L., Almqvist, C., Larsson, H., Valdimarsdóttir, U. A., & Fang, F. (2020). Association of stress-related disorders with subsequent neurodegenerative diseases. *JAMA Neurology*, *77*(6), 700–709.

Song, Q., Peng, S., & Zhu, X. (2021). Baicalein protects against MPP+/MPTP-induced neurotoxicity by ameliorating oxidative stress in SH-SY5Y cells and mouse model of Parkinson's disease. *Neurotoxicology*, *87*, 188–194.

Stefani, M. (2012). Structural features and cytotoxicity of amyloid oligomers: Implications in Alzheimer's disease and other diseases with amyloid deposits. *Progress in Neurobiology*, *99*(3), 226–245.

Sutherland, B. A., Shaw, O. M., Clarkson, A. N., Jackson, D. M., Sammut, I. A., & Appleton, I. (2005). Neuroprotective effects of (−)-epigallocatechin gallate after hypoxia-ischemia-induced brain damage: Novel mechanisms of action. *The FASEB Journal*, *19*(2), 1–22.

Tabrizi, S. J., Flower, M. D., Ross, C. A., & Wild, E. J. (2020). Huntington disease: New insights into molecular pathogenesis and therapeutic opportunities. *Nature Reviews Neurology*, *16*(10), 529–546.

Tedeschi, V., Petrozziello, T., & Secondo, A. (2021). Ca^{2+} dysregulation in the pathogenesis of amyotrophic lateral sclerosis. *International Review of Cell and Molecular Biology*, *363*, 21–47.

Teleanu, D. M., Niculescu, A.-G., Lungu, I. I., Radu, C. I., Vladâcenco, O., Roza, E., Costăchescu, B., Grumezescu, A. M., & Teleanu, R. I. (2022). An overview of oxidative stress, neuroinflammation, and neurodegenerative diseases. *International Journal of Molecular Sciences*, *23*(11), 5938.

Tian, E., Sharma, G., & Dai, C. (2023). Neuroprotective properties of berberine: Molecular mechanisms and clinical implications. *Antioxidants*, *12*(10), 1883.

Toader, C., Tataru, C. P., Munteanu, O., Serban, M., Covache-Busuioc, R.-A., Ciurea, A. V., & Enyedi, M. (2024). Decoding neurodegeneration: A review of molecular mechanisms and therapeutic advances in Alzheimer's, Parkinson's, and ALS. *International Journal of Molecular Sciences*, *25*(23), 12613.

Tomou, E.-M., Papakyriakopoulou, P., Saitani, E.-M., Valsami, G., Pippa, N., & Skaltsa, H. (2023). Recent advances in nanoformulations for quercetin delivery. *Pharmaceutics*, *15*(6), 1656.

Tong, J., Rathitharan, G., Meyer, J. H., Furukawa, Y., Ang, L.-C., Boileau, I., Guttman, M., Hornykiewicz, O., & Kish, S. J. (2017). Brain monoamine oxidase B and A in human parkinsonian dopamine deficiency disorders. *Brain*, *140*(9), 2460–2474.

Trapani, G., Satriano, C., & La Mendola, D. (2018). Peptides and their metal complexes in neurodegenerative diseases: From structural studies to nanomedicine prospects. *Current Medicinal Chemistry*, *25*(6), 715–747.

Tripathi, R., Gupta, R., Sahu, M., Srivastava, D., Das, A., Ambasta, R. K., & Kumar, P. (2021). Free radical biology in neurological manifestations: Mechanisms to therapeutics interventions. *Environmental Science and Pollution Research*, *29*, 1–48.

Trippier, P. C., Jansen Labby, K., Hawker, D. D., Mataka, J. J., & Silverman, R. B. (2013). Target-and mechanism-based therapeutics for neurodegenerative diseases: Strength in numbers. *Journal of Medicinal Chemistry*, *56*(8), 3121–3147.

Utpal, B. K., Mokhfi, F. Z., Zehravi, M., Sweilam, S. H., Gupta, J. K., Kareemulla, S., Rao, A., Kumar, V. V., Krosuri, P., & Prasad, D. (2024). Resveratrol: A natural compound targeting the PI3K/Akt/mTOR pathway in neurological diseases. *Molecular Neurobiology*, *62*, 1–30.

Van Es, M. A. (2024). Chapter one – Amyotrophic lateral sclerosis; clinical features, differential diagnosis and pathology. In P. J. Shaw & J. Kirby (Eds.), *International review of neurobiology* (pp. 1–47). Academic Press.

Velagapudi, R., El-Bakoush, A., & Olajide, O. A. (2018). Activation of Nrf2 pathway contributes to neuroprotection by the dietary flavonoid tiliroside. *Molecular Neurobiology*, *55*, 8103–8123.

Vicente-Zurdo, D., Gómez-Mejía, E., Rosales-Conrado, N., & León-González, M. E. (2024). A comprehensive analytical review of polyphenols: Evaluating neuroprotection in Alzheimer's disease. *International Journal of Molecular Sciences*, *25*(11), 5906.

Vinceković, M., Jurić, S., Marijan, M., Viskić, M., Vlahoviček-Kahlina, K., & Bandić, L. M. (2021). Encapsulation of herb extracts (Aromatic and medicinal herbs). In *Aromatic herbs in food* (pp. 263–322). Elsevier.

Viskelis, J., & Surguchov, A. (2024). *Essential oils: Recent advances, new perspectives and applications*.

Walczak-Nowicka, Ł. J., & Herbet, M. (2021). Acetylcholinesterase inhibitors in the treatment of neurodegenerative diseases and the role of acetylcholinesterase in their pathogenesis. *International Journal of Molecular Sciences*, *22*(17), 9290.

Wang, J., Chen, X., Bai, W., Wang, Z., Xiao, W., & Zhu, J. (2021). Study on mechanism of *Ginkgo biloba* L. leaves for the treatment of neurodegenerative diseases based on network pharmacology. *Neurochemical Research*, *46*(7), 1881–1894.

Wang, L., Wang, F. S., & Gershwin, M. E. (2015). Human autoimmune diseases: A comprehensive update. *Journal of Internal Medicine*, *278*(4), 369–395.

Wang, R., & Tang, X. C. (2005). Neuroprotective effects of huperzine A: A natural cholinesterase inhibitor for the treatment of Alzheimer's disease. *Neurosignals*, *14*(1–2), 71–82.

Wang, R., Yan, H., & Tang, X. (2006). Progress in studies of huperzine A, a natural cholinesterase inhibitor from Chinese herbal medicine 1. *Acta Pharmacologica Sinica*, *27*(1), 1–26.

Wang, Y., Xu, E., Musich, P. R., & Lin, F. (2019). Mitochondrial dysfunction in neurodegenerative diseases and the potential countermeasure. *CNS Neuroscience & Therapeutics*, *25*(7), 816–824.

Wareham, L. K., Liddelow, S. A., Temple, S., Benowitz, L. I., Di Polo, A., Wellington, C., Goldberg, J. L., He, Z., Duan, X., & Bu, G. (2022). Solving neurodegeneration: Common mechanisms and strategies for new treatments. *Molecular Neurodegeneration*, *17*(1), 23.

Wendo, J. K., Mbaria, J. M., Nyariki, J. N., & Isaac, A. O. (2024). *Ginkgo biloba* attenuated detrimental inflammatory and oxidative events due to *Trypanosoma brucei rhodesiense* in mice treated with melarsoprol. *PLOS Neglected Tropical Diseases*, *18*(4), e0012103.

Weng, H.-R. (2024). Emerging molecular and synaptic targets for the management of chronic pain caused by systemic lupus erythematosus. *International Journal of Molecular Sciences*, *25*(7), 3602.

Wise, R. M., Wagener, A., Fietzek, U. M., Klopstock, T., Mosharov, E. V., Zucca, F. A., Sulzer, D., Zecca, L., & Burbulla, L. F. (2022). Interactions of dopamine, iron, and alpha-synuclein linked to dopaminergic neuron vulnerability in Parkinson's disease and Neurodegeneration with Brain Iron Accumulation disorders. *Neurobiology of Disease*, *175*, 105920.

Wu, Y., Pang, Y., Wei, W., Shao, A., Deng, C., Li, X., Chang, H., Hu, P., Liu, X., & Zhang, X. (2020). Resveratrol protects retinal ganglion cell axons through regulation of the SIRT1-JNK pathway. *Experimental Eye Research*, *200*, 108249.

Xia, Y. (2009). *Role of the PI-3K/Akt and ERK pathways in phencyclidine-induced neurotoxicity in neonatal rats and the protection by lithium and BDNF.*

Yadav, H. P., & Li, Y. (2015). The development of treatment for Parkinson's disease. *Advances in Parkinson's Disease*, *4*(3), 59–78.

Yatoo, M. I., Gopalakrishnan, A., Saxena, A., Parray, O. R., Tufani, N. A., Chakraborty, S., Tiwari, R., Dhama, K., & Iqbal, H. M. N. (2018). Anti-inflammatory drugs and herbs with special emphasis on herbal medicines for countering inflammatory diseases and disorders-a review. *Recent Patents on Inflammation & Allergy Drug Discovery*, *12*(1), 39–58.

Yu, Y., Yu, S., Battaglia, G., & Tian, X. (2024). Amyloid-β in Alzheimer's disease: Structure, toxicity, distribution, treatment, and prospects. *Ibrain*, *10*

Zahra, W., Nand Rai, S., Birla, H., Sen Singh, S., Dilnashin, H., Singh Rathore, A., & Singh, S. P. (2020). The global economic impact of neurodegenerative diseases: Opportunities and challenges. In *Bioeconomy for sustainable development* (pp. 333–345). Springer Singapore.

Zeng, H., Liu, Y., Liu, X., Li, J., Lu, L., Xue, C., Wu, X., Zhang, X., Zheng, Z., & Lu, G. (2024). Interplay of α-synuclein oligomers and endoplasmic reticulum stress in Parkinson's disease: Insights into cellular dysfunctions. *Inflammation*, *1*, 1–17.

Zhang, L.-X., Li, C.-X., Kakar, M. U., Khan, M. S., Wu, P.-F., Amir, R. M., Dai, D.-F., Naveed, M., Li, Q.-Y., Saeed, M., Shen, J.-Q., Rajput, S. A., & Li, J.-H. (2021). Resveratrol (RV): A pharmacological review and call for further research, *Biomedicine & Pharmacotherapy*, *143*, 112164, https://doi.org/10.1016/j.biopha.2021.112164

Zhang, W., Xiao, D., Mao, Q., & Xia, H. (2023). Role of neuroinflammation in neurodegeneration development. *Signal Transduction and Targeted Therapy*, *8*(1), 267. https://doi.org/10.1038/s41392-023-01486-5

Zhang, Y., Zhang, W., Zhao, Y., Peng, R., Zhang, Z., Xu, Z., Simal-Gandara, J., Yang, H., & Deng, J. (2024). Bioactive sulforaphane from cruciferous vegetables: Advances in biosynthesis, metabolism, bioavailability, delivery, health benefits, and applications. *Critical Reviews in Food Science and Nutrition*, 1–21.

Zhao, A., Liu, N., Yao, M., Zhang, Y., Yao, Z., Feng, Y., Liu, J., & Zhou, G. (2022). A review of neuroprotective effects and mechanisms of ginsenosides from Panax ginseng in treating ischemic stroke. *Frontiers in Pharmacology*, *13*, 946752.

Zhu, J., Xu, F., Lai, H., Yuan, H., Li, X.-Y., Hu, J., Li, W., Liu, L., & Wang, C. (2023). ACO_2 deficiency increases vulnerability to Parkinson's disease via dysregulating mitochondrial function and histone acetylation-mediated transcription of autophagy genes. *Communications Biology*, *6*(1), 1201.

20 Therapeutic Nutrition Strategies in Alzheimer's Disease

Esra Tansu Sarıyer and Meral Yüksel

20.1 INTRODUCTION

Alzheimer's disease (AD) is a chronic, irreversible, progressive age-related brain disease expressed with neurodegeneration by the irreversible damage of basal forebrain cholinergic neurons and cognitive dysfunction (Kowalski & Mulak, 2019). Worldwide, millions of patients and their families severely impair the quality of life because clinical features such as dementia and especially memory impairment in AD are the most prevalent neurodegenerative problem (Georges et al., 2020; Patterson, 2018). In 2018, global prevalence of dementia was estimated at approximately 50 million people. This figure is expected to triple by 2050 and two-thirds of cases are found in low- and middle-income countries. Other recent data show that the rates of dementia in Europe may double by 2050 (Georges et al., 2020; Patterson, 2018).

Such common clinical symptoms of AD are impaired planning, learning, problem-solving, calculation, memory, speech, and decision-making. While the pathology of AD is not completely elucidated, intracellular neurofibrillary tangles and extracellular amyloid plaques are implicated as key contributors. Other common features include dystrophic neurites, neuropil threads, microglial activation, and astrogliosis, often associated with cerebral amyloid angiopathy. Such activities are purported to underlie all these non-vascular pathological processes, finally leading to neurodegeneration with neuronal loss and synaptic damage and, in the late chronic stages, macroscopic atrophy (Yildirim & Eke, 2024). Risk factors for AD are characterized as follows:

- *Genetics*: Gene mutations in the presenilin 1 (PSEN1 gene), presenilin 2 (PSEN2 gene), and amyloid precursor protein (APP), Down syndrome, individuals having apolipoprotein E4 (ApoE4) allele.
- *Environmental factors*: Trauma to the head, smoking, alcohol, atmospheric pollution, exposure to heavy metals.
- *Lifestyle factors*: Age, poor nutrition, low mental and physical activity, and cardiovascular and metabolic disorders like hypertension, heart disease, and diabetes.
- *Demographic characteristics*: Older age, female sex (Yildirim and Eke, 2024; Scheltens et al., 2021).

Although AD treatment with medications is available to alleviate some of the symptoms, currently, no treatment can halt or modify the course of the disease, and neurodegeneration cannot be completely prevented. As a result, different non-pharmacological measures are important in increasing the quality of life in individuals suffering from AD (Morris, 2009; Pivi et al., 2017).

Management and alleviation of AD symptoms requires interdisciplinary support by nutrition specialists, psychologists, physiotherapists, speech therapists, and others who can provide individual attention. Moreover, what is known about the modifiable characteristics associated with the disease is limited. Conversely, the literature is increasingly documenting the possible protective and therapeutic role of nutrition in AD (Morris, 2009; Pivi et al., 2017). Nutrition and its impact on AD prevention and management is under exploration. The Task Force on Nutrition and Cognitive Decline of the International Academy on Nutrition and Aging concluded that though there are a number of hypotheses supporting the implication of nutrition and nutrients in cognitive decline, the evidence available is still inconclusive to substantiate these claims (Gillette-Guyonnet et al., 2007).

Malnutrition is common in old individuals, especially in nursing homes, and is linked with cognitive decline. Over the past few years, various studies have been done on the impact of diet-related hormones (ghrelin and cholecystokinin), foods and nutrients (vitamins, antioxidants, and various supplements such as *Ginkgo biloba*), docosahexaenoic acid (DHA), blueberries, and olive oil on cognition and memory. Nutrition-related research has also explored the effects of oxidative damage and inflammation on AD pathogenesis. Whole food can be a rich source of antioxidant molecules and anti-inflammatory nutrients such as flavonoids, polyphenols, carotenoids, and vitamins (vitamins E, D, and C), and the role of diet has been highlighted in terms of AD (Shah, 2013). The protective effects of minerals and vitamins on brain health as well as cognitive functions have been widely documented. Therefore, the micronutrients and dietary supplements in the dietary system may help to preserve excellent cognitive activity or perhaps counteract the progression of dementia and AD. It has been noted in new studies that potential diets holding benefits when it comes to AD prevention/treatment include Mediterranean diet (MeDi), Mediterranean-DASH Intervention for Neurodegenerative Delay (MIND) diet, Dietary Approaches to Stop Hypertension (DASH) diet,

DOI: 10.1201/9781003602477-20

ketogenic diet (KD), and intermittent fasting (IF) (Pritam et al., 2022; Shah, 2013; Xu Lou et al., 2023). The features, indications, and contraindications of these diets are summarized in Table 20.1 (Dyńka et al., 2022; Stefaniak et al., 2022; Zhao et al., 2022). This chapter argues the possible protective effects of various dietary strategies for AD treatment.

20.2 MEDITERRANEAN DIET (MEDI)

The classical MeDi was first given a scientific definition by Ancel Benjamin Keys in the 1960s. The Seven Countries Study which studied the diets of people from the Mediterranean area was the first study to relate diet to cardiovascular disease. This population study was the start of

TABLE 20.1
Components and Characteristics of the MeDi, DASH, MIND, Ketogenic Diets (KDs), and Intermittent Fasting (IF)

Diets	Major Aspects	Recommendations	Restrictions/Prohibitions
MeDi	• The plant-based diet consists of minimally processed products based on the traditional diet of individuals living in the Mediterranean region. • Regular physical activity, adequate rest, conviviality.	• Vegetables • Fruits • Nuts • Legumes • Whole grains • Extra virgin olive oil olives • Fish and seafood • Moderate consumption of fermented dairy • Red wine in limited quantities	• Red meat • Sweets and sugars
DASH	• The plant-based diet consists of all minimally processed products used for cardiometabolic conditions, especially reduced hypertension. • It focuses on reducing the intake of saturated fatty acids, trans-fats, and sodium in the diet. • Higher in magnesium, potassium, calcium, and fiber food sources.	• Vegetables • Fruits • Nuts • Legumes • Whole grain products • Low-fat dairy products	• Total fat intake • Red meat • Salt and sodium-rich products • Sweets and sugars
MIND	• Combination of the Mediterranean diet and the DASH diet. • The plant-based diet consists of all minimally processed products used to improve brain cognition and reduce the risk of certain age-related neurodegenerative diseases.	• Especially green leafy vegetables • Other vegetables • Nuts • Berries • Pods • Whole grains • Poultry (moderate) • Fish and seafood (moderate) • Olive oil (basic oil)	• Red meat • Butter and margarine • Cheese • Cakes and sweets • Fried and fast-food products
KD	• It is a high fat, moderate protein, low carbohydrate diet. • The body is intended to enter ketosis. Although it was initially developed for resistant epilepsy, it is also used in neurodegenerative diseases caused by the brain using ketone bodies as an energy source.	• *Seafood*: Tuna, shrimp, sardines, and all seafood • *Meat and Poultry*: Chicken, egg, bacon, beef, pork • *Dairy*: Yogurt, almond milk, coconut milk, paneer, butter, mozzarella, heavy cream, mayonnaise • *Vegetables*: Cabbage, broccoli, cauliflower, onion, spinach, tomato, cucumber, bell pepper, mushroom, eggplant, lady finger, and all low carb veggies • *Nuts and Seeds*: Almond, walnut, peanut, pistachio, flax seed, sunflower seed, pumpkin seed • *Healthy Fats*: Coconut oil, olive oil, avocado oil, MCT oil	• High-carbohydrate foods (grains, starch, some fruit and vegetables, sugar)
IF	• IF diets are a dietary pattern in which fasting cycles alternate with periods of normal food intake for periods ranging from 12 to 24 hours. Different dietary patterns are applied in intermittent fasting.	• No specific food/dietary recommendation • Calorie or feeding hours are restricted according to the type of IF applied	• No specific dietary restrictions, but some types (e.g., 5:2 IF) have significant caloric restriction

Source: As adapted from Stefaniak et al., Nutrients (2022).

the increasingly recognized health benefits of MeDi, which led to its recommendation as a healthy nutritional model (Keys et al., 1986; Willett et al., 1995).

The MeDi is a classic nutritional framework, including olive oil, whole grains, fresh or dried vegetables and fruits, fish, milk, meat, multiple spices, and regular consumption of wine. This classic MeDi has some main characteristics:

- Consumption of monounsaturated fats to avoid consuming saturated fats. Generally olive products such as virgin and extra virgin oil.
- Consuming high amounts of legumes.
- A high intake of unrefined grains, such as bread.
- Eat plentiful amounts of berries and fruits.
- Consumption of high amounts of raw or cooked vegetables.
- Red and white meat should be consumed limitedly.
- Limited consumption of daily produced foods such as yogurt and cheese.
- Drink wine with your meals (and moderate consumption)—it has antioxidants.
- Emphasis on plant-based foods.
- Desserts are mainly fresh fruits and it is recommended to consume honey- and sugar-containing desserts in moderation.
- Eggs are eaten a maximum of 0–4 times per week.

The MeDi is rich in monounsaturated fatty acids (MUFA). It is compared to other dietary models in which saturated fatty acids meet 7%–8% of the total energy need while the ratio of total fat is between 25% and 40% (Guasch-Ferré & Willett, 2021; Willett et al., 1995). In fact, scientific data have shown the positive impact of the MeDi in the management and prevention of obesity-based diseases, for example, high blood pressure, heart diseases, metabolic syndrome, stroke, type 2 diabetes mellitus, fatty liver diseases, cardiovascular diseases, and obesity-based cancer types such as breast, colorectal, esophageal, pancreatic, and liver cancers (Gardener & Caunca, 2018; Guasch-Ferré & Willett, 2021; Sofi et al., 2013). It is very impressive how MeDi can prevent and treat neurodegenerative diseases. A plethora of solid evidence establishes a cardinal role of MeDi in mitigating the risk and delaying the onset of AD (García-Casares et al., 2021; Gu et al., 2010; Sofi et al., 2010). The therapeutic effects of MeDi on brain and mental health are mostly due to food that is high in brain-health-enhancing nutrients. These include polyphenols from wine and nuts; omega-3 fatty acids from fish, salmon, spinach, and walnuts; and antioxidants from berries, fruits, vegetables, grains, and spices (Stefaniak et al., 2022). There are a lot of components that explain the detrimental outcomes of MeDi on cognitive impairments. One of the major mechanisms is the elevated oxidative stress/antioxidant system ratio in favor of release of reactive oxygen species, which is a central event in the pathophysiological process of formation of AD. Other risk factors include inflammation, amyloid peptide accumulation, activation of astrocytes, migration of microglial cells, deterioration of vascular health, and more advanced stages of brain damage. Antioxidant and anti-inflammatory properties of MeDi show that MeDi decreases the risk of AD progression. Through the nutritional contribution it provides, it counters neuronal damage, mitochondrial dysfunction, caspase activation, Aβ accumulation, and neurotransmitter loss, and exerts relatively protective effects against diverse neurodegenerative diseases, especially AD. Moreover, the polyphenols contained in MeDi promote neurological health in several ways, namely by improving blood flow in the cerebrovascular system, increasing synthesis of the important neurotrophic factor brain-derived neurotrophic factor (BDNF), modulating neuronal signaling, and upregulating neurogenesis (Sánchez-Muniz et al., 2019). Mostly, scientific studies emphasize that the MeDi plays a role in reducing AD. In general, studies have found neutral or favorable outcomes on the relationship between MeDi compliance and the risk of AD. In a pilot study in patients with AD, cognitive performance was improved after 21 days of daily consumption of MeDi enriched with coconut oil (de la Rubia Ortí et al., 2018). Another study compared MeDi adherers' proportion between cognitively normal subjects and individuals at increased risk for AD (e.g., subjects with a family history of AD, cognitive decline, or mild cognitive impairment) and studied its association with AD biomarkers. The results suggest that a better compliance to MeDi was positively correlated with higher mediotemporal gray matter volume, better memory performance, and lower amyloid and pTau181 pathology (Ballarini et al., 2021). Cross-sectional neuroimaging results in people at high AD risk showed higher brain atrophy in the frontal and temporal cortices to correlate with a lower adherence to the MeDi. The researchers proposed that MeDi could provide protective effect against the loss of brain tissue and that dietary interventions might be significant in AD prevention (Mosconi et al., 2014). Moreover, a meta-analysis of 11 studies on the associations among MeDi, AD, and cognitive impairment showed that each one-unit increase in MeDi adherence scores reduced the risk of AD by 11% (García-Casares et al., 2021). In a longitudinal study with 1865 older people aged 64 years and over, Anastasiou et al. found that, to be precise, 90 progressed to dementia, 223 to mild cognitive impairment, and 68 to AD during the study period. Better performance in general cognitive functions, language, visual-spatial perception, and memory was achieved with higher adherence to MeDi. The opposite was true of those with dementia, who were found to eat less vegetables, fruit, and fish. Specifically, one serving of fish per week was associated with an approximately 10% reduced risk of dementia (Anastasiou et al., 2017). MeDi is recommended for use in individuals with normal cognitive capacity and with mild cognitive impairment, as a strategy to prevent cognitive decline and dementia, according to a review on this area provided by the WHO. In the case of AD, more studies are needed on therapeutic effects of MeDi (World Health Organization 2021).

20.3 DIETARY APPROACHES TO STOP HYPERTENSION (DASH) DIET

The DASH diet, which stands for Dietary Approaches to Stop Hypertension, was created by the National Institutes of Health (NIH) in 1992. Both the DASH diet and MeDi involve low consumption of red meat and high consumption of plant-based foods. DASH diet promotes potassium (K), calcium (Ca), magnesium (Mg), and fiber-rich foods while minimizing saturated fat and sugar-rich foods, which help to lower blood pressure. This also emphasizes to limit sodium (Na) intake to ≤2300 mg/day. The DASH diet also focuses more on low-fat dairy products with respect to MeDi (Melekoğlu & Rakıcıoğlu, 2020).

The DASH diet was created in the 1990s to help treat and/or prevent hypertension. The literature shows that it has beneficial effects on systolic and diastolic blood pressure, cardiovascular risk factors, and blood total cholesterol levels. Given that hypertension and high levels of low-density lipoprotein (LDL) cholesterol are well-known risk factors for AD, a reduced risk of AD has been involved with the DASH diet (Khandayataray & Murthy, 2024; Van Den Brink et al., 2019). Additionally, the DASH diet exerts an effect on key pathological processes involved in the pathophysiological changes seen in AD, including oxidative stress, inflammation, and insulin resistance. It also insulates from cardiovascular risk factors, for example, hypertension and elevated LDL cholesterol that predispose to dementia and AD (Cremonini et al., 2019). DASH, on the other hand, does nothing to mitigate AD symptoms; its only effect is preventing disease onset. This dietary pattern has mostly positive effects regarding dementia, but only if the MeDi is followed before the onset of symptoms of dementia, indicating that the Mediterranean pattern of diet may be a preventive approach rather than a therapeutic one regarding cognitive dysfunctions (Śliwińska & Jeziorek, 2021). However, the DASH diet reduces disease risk rather than AD symptoms and research studies focus on its preventive effects. In the Nurses' Health Study, which included 16,144 older women (70 years and older), long-term compliance with the DASH diet was associated with better cognitive function in patients. This effect appeared to be unrelated to apolipoprotein E-ε4 allele status (Berendsen et al., 2017) In a similar study of 160 older individuals with cognitive impairment, cardiovascular risk factors and ideal compliance to the DASH diet were linked with better executive functioning, better processing speed, and better verbal memory (Blumenthal et al., 2017). Nevertheless, other studies have suggested mixed results in the association of DASH diet with cognitive behavior. A study with six-month randomized controlled trial among sedentary individuals confirmed that the DASH diet alone was not effective at improving cognitive impairment. Importantly, the benefits were more pronounced for improved executive function when it was combined with jogging and swimming-like aerobic exercise (Blumenthal et al., 2019). Also, the Women's Health Initiative Memory Study, a longitudinal prospective study, found no association between the DASH diet arrangement and cognitive function in older women (Haring et al., 2016).

These consequences suggest that the neuroprotective effect of DASH diet could be enhanced in tandem with other non-dietary interventions, such as physical exercise, possibly in a synergic manner (Ellouze et al., 2023). However, better-designed studies are warranted to provide a full understanding of the efficacy of the DASH diet for AD prevention and treatment.

20.4 MEDITERRANEAN-DASH INTERVENTION FOR NEURODEGENERATIVE DELAY (MIND) DIET

The MIND diet, formed in 2015 by researchers at Rush University Medical Center in Chicago, IL, and the Harvard School of Public Health in Boston, MA, combines elements of the Mediterranean and DASH diets. This dietary model highlights brain-healthy foods (various vegetables, especially glossy green leafy vegetables, berries, nuts, legumes, whole grains, poultry, seafood, red wine, and olive oil) and discourages unhealthy foods (bakery products, sweets, fried foods, and fast foods) (Cremonini et al., 2019; Morris et al., 2015a). In contrast to the Mediterranean and DASH diets which recommend increased amounts of vegetables and fruits in the diet, the MIND diet is defined by increased consumption of green leafy vegetables and berries. Similar to DASH, the MIND diet encourages only one serving of fish a week, which is less than the amount of fish consumed in the Mediterranean diet. This relatively moderate fish intake is consistent with elegy studies on the fish in AD prevention (Cremonini et al., 2019; Larrieu et al., 2004; Morris et al., 2003). In a research study to deliver data on the MIND diet, 923 older individuals aged 58–98 years were followed for 4.5 years within a cohort study. Results demonstrated that moderate dietary changes in MIND diet were associated with a 35% reduced risk of AD, and high adherence to the MIND diet was associated with a 53% reduced risk of AD (Morris et al., 2015b). Evidence for a positive association at the population level came from an investigation of three prospective cohort studies of 18,136 individuals and a meta-analysis of 11 cohorts of 224,049 individuals. Chen et al. reported that high adherence to the MIND diet was associated with lower dementia risk among middle-aged and older people. The high adherence exhibits a 17% reduced risk of dementia compared to those with low adherence to MIND diet (Chen et al., 2023). Likewise, another 12-year follow-up study conducted in Australia on 1220 participants of 60–64 years old showed that higher adherence to the MIND diet was linked with a 19% decreased odds of developing mild cognitive impairment (MCI)/dementia and a 53% reduced odds of developing MCI in the high consumption tertile. Notably, the MeDi was not significantly associated with dementia risk in the current study (Hosking et al., 2019). A meta-analysis investigated the link of MeDi, DASH, and MIND diets with cognitive decline in AD risk. When compared to the other dietary models, the MIND diet showed the most pronounced effect in mitigating disease risk (Van Den Brink et al., 2019). These clinical analyses show the effectiveness of MIND diet for AD prevention and treatment is few. In a study evaluating how the MIND diet affects cognitive performance and brain structure, 50 obese women were randomly assigned to a control group (conventional calorie restriction diet) and separated into a MIND diet group for a three-month period. It was found that working memory, verbal recognition memory, and attention improved at a much higher rate for the MIND diet group than for the calorie-restricted control group. Also, an increase in surface area of the inferior frontal gyrus and greater reductions were seen in the cerebellum-white matter and cerebellum-cortex (Arjmand et al., 2022). In a two-center randomized controlled trial with three-year follow-up in older obese adults with no cognitive impairment but with a family history of dementia, there were no differences in changes in cognitive function or brain MRI findings between the MIND diet and the control group (receiving a mildly calorie-restricted diet) (Barnes et al., 2023). Clinical pathological autopsy data in two studies obtained from the Rush Memory and Aging Project result with no association of MIND diet adherence and AD pathology or β-amyloid depletion (Dhana et al., 2021). In contrast, the second study found that adherence to the MIND diet was associated with less depletion of β-amyloid. When examined separately from the other components of diet, higher intake of green leafy vegetables was significantly associated with lower AD pathology (Agarwal et al., 2023). The second study did not find a link between tau tangles and the MIND diet adherence (Agarwal et al., 2023; Dhana et al., 2021). The MIND diet has a high content of potentially antioxidant, anti-inflammatory, and neuroprotective dietary components to support brain health. A mainstay of the MIND diet is green leafy vegetables like spinach, cabbage, and radish, and fruits, especially berries like blackberry, strawberry, and cranberry. Green leafy vegetables are a source of β-carotene, lutein-zeaxanthin, flavonoids, folate, and vitamin E, all of which contribute to cognitive function. Vitamin E is an antioxidant that is found in high amounts in green leafy vegetables and nuts and is crucial for protecting neurons from oxidative stress-mediated damage caused by free radicals (Agarwal et al., 2019; Dhana et al., 2021). The dietary components also allegedly have antioxidant, anti-inflammatory, and neuroprotective characteristics, which make the MIND diet potentially beneficial in maintaining brain functions and health. A key tenet of the MIND diet is the promotion of green leafy vegetables and fruits, especially berries. Many mechanisms have been proposed for this, including neurogenesis, enhanced IGF-1 signaling, and decreased oxidative stress and inflammation leading to reduced β-amyloid production, delayed neuronal aging, enhanced cognitive function, and a decreased trace of AD (Krikorian et al., 2023; Miller et al., 2021; Navarro-Hortal et al., 2022; Shukitt-Hale et al., 2015). Because neuronal inflammation and oxidative stress are the two key biomarkers of AD, incorporating more berries into the diet, as recommended by the MIND diet, is a great way to target alleviating such symptoms (Agarwal et al., 2019; Shukitt-Hale et al., 2015). Last of all, despite the relatively recent introduction of the MIND diet when compared to the Mediterranean and DASH diets, a number of studies indicate that it is more effective in reducing the risk of AD when checked to either of the other two dietary patterns (Bhuiyan et al., 2023; Doroszkiewicz et al., 2023).

20.5 KETOGENIC DIET (KD)

KD is a low carbohydrate, moderate protein, and high-fat dietary pattern which was arranged in the 1920s for treatment of intractable epilepsy. The KD was later adapted for improvement of obesity, dyslipidemia, cardiovascular diseases, diabetes, and other diseases associated with aging. The KD is one of the starvation/fasting models, where the predominant source of energy for the body changes from carbohydrates to fat. Glycogen stores can be more abundant, however, when glucose is available, and more than sufficient amount of glucose is taken up by cells, depleting glycogen stores when carbohydrate intake is low, triggering β-oxidation of fat to ketones in the liver. These water-soluble ketone bodies (acetoacetate, β-hydroxybutyrate, and acetone) are released into the bloodstream and trigger ketosis, acting as mediators for the resulting physiologic or pathologic processes. In cases of glucose deficiency, the brain uses ketone bodies as an alternative energy substrate. Moreover, to energy production, ketone bodies are also implicated in brain development, cell membrane maintenance, lipid biosynthesis, and cell signaling (Jiang et al., 2022; Kara & Kilinc, 2021). Mounting evidence shows that KD and/or exogenous ketone supplementation may be useful in the treatment of epilepsy, seizure, stroke, Parkinson's diseases, or even AD. In a case presentation study, the KD was followed for ten weeks by a 71-year-old woman patient with AD and ApoE4 allelic variant. HbA1c, HOMA-IR index, and triglyceride levels that are also cognitive assessment scores are lowered as major outcomes of this case control study (Morrill & Gibas, 2019). In another study a randomized controlled trial of 122 patients with MCI, participants received a ketogenic drink or a placebo twice daily for six months. To measure it, the researchers compared cognitive performance and blood ketones between participants who drank the ketogenic drink and those who received a placebo. The group that drank the ketogenic drink present significantly better on cognitive tests, and they had elevated blood ketones compared with the placebo group (Fortier et al., 2021). As a further matter, patients with AD (mild to moderate levels) showed remarkable improvements in immediate and delayed logical memory tests after three months of a medium-chain triglyceride (MCT)-based ketogenic formula compared with baseline values. Patients achieved meaningful improvements from baseline scores in the cognitive and memory tests at 12 weeks of treatment (Ota et al., 2019). A randomized double-blind pilot study on the effects of a ketogenic agent caprylidene supplementation over 45 days in AD patients with mild to moderate symptoms has demonstrated enhanced blood flow in specific brain regions in patients who do not possess the APOEε4 allele (Torosyan et al., 2018). Separately, a 90-day treatment with KD (70% fat, 30% protein) supplemented with MCT was given to 15 patients with very mild, mild, and moderate AD, followed by a 30-day normal diet period to analyze the long-term effects of KD. The results showed that improvements in cognitive assessments during the ketogenic phase reverted to baseline levels after KD cessation with normal diet resumption (Taylor et al., 2017). Neuroinflammation in neurodegenerative diseases is a multifaceted phenomenon that results from the activation of diverse molecular and cellular pathways by the interaction of metabolism, gut microbiota, and other mechanisms. The KD mediates these therapeutic effects in the above conditions through several pathways, including modulation of both central and peripheral metabolism, decreasing oxidative stress and inflammation, increasing mitochondrial function and antioxidant enzyme activities, enhancing autophagy, and modulating gut microbiome function (Doroszkiewicz et al., 2023). The over-prone aggregation of tau protein into neurofibrillary tangles is a key mechanism in the etiology of AD. This causes a loss of neurons in subcortical regions, especially in the cerebral cortex, besides the deposition of amyloid-β plaques. Literature has indicated that KDs may decrease the quantity and toxicity of Aβ and tau protein aggregates. Yet this seems to do little more than prevents the formation of new plaques rather than sink already existing ones. Thus, KD is suggested as one of the adjuvant treatment that could supplement this process, relying on its ability to delay the disease development and counteract the cognitive impairment (Jiang et al., 2022; Pietrzak et al., 2022; Ulamek-Koziol & Pluta, 2020). It has been shown that hyperglycemia correlates with decreased amount of brain gray matter and hippocampal volume. In fact, impaired glucose metabolism within the brain has been related to the gradually progressive dysfunction of cognitive and motor functions in AD patients (Pietrzak et al., 2022). Insulin resistance and accompanying hypoglycemia may contribute to the emergence of neurodegenerative diseases by inducing detrimental changes in neurons of the brain (Kerti et al., 2013; Weinstein et al., 2015). A reduction in cerebral glucose utilization and significant impairment of insulin signaling occurs in both patients with AD and at-risk individuals (people with pre-symptomatic AD). During ketosis, ketone bodies are formed and used as an alternative energy substrate through the β-oxidation pathway, as they produce more adenosine triphosphate (ATP) than glucose. One of the prime mechanisms of action of ketone bodies to produce neuroprotective effects is enhanced mitochondrial respiration. Ketosis leads to a series of biochemical changes that enhance mitochondrial function, lessen glucose reliance, and elevate insulin sensitivity, thus providing neuroprotection (Kara & Kilinc, 2021). In the context of AD, there are various neurological disorders, and energy deprivation is a fast-evolving process. Appropriately administered, ketone bodies may serve as an effective intervention against bioenergetics disruption (causing energy deprivation) as they are capable of normalizing mitochondrial energy metabolism. Thus, KD is a good supporter for the adjuvant therapy, but its specific applicability largely depends on the severity and type of the disease (Doroszkiewicz et al., 2023; Pietrzak et al., 2022). Even though KD has been proved to have promising therapeutic potential in AD based on scientific evidence, it should be noted that there are obstacles and risks related to KD that should be considered when applying this dietary approach long term. KD is high-fat diet, and long-term maintenance of KD may cause complications; moreover, compliance may be difficult owing to the dietary restriction.

20.6 INTERMITTENT FASTING

IF has been an approach to dietary restriction for the past few years. The most important feature of IF is the pattern of fasting and eating normally in alternating cycles, usually every 12–24 hours. IF is popular because the eating strategy can be regulated by various fasting times. There are different types of IF as follows: time-restricted eating (TRE) in which you only eat during a specific time, alternate-day fasting (ADF), the periodic fasting (PF) or 5:2 diet, and the fasting mimicking diet (FMD). These forms of IF are explained in the following:

- *TRE*: The 16:8 diet. Limits your feeding time without decreasing daily nutrient or calorie consumption.
- *ADF*: Fasting every other day and eating what you want every other day, usually through a number of different schedules.
- *PF or 5:2 Diet*: Works by fasting on two consecutive or non-consecutive days of the week where no food is consumed, and unrestricted food intake is allowed for the other five days.
- *FMD*: Similar to ADF but relies on a low caloric intake (15%–20% of daily caloric needs) on fasting days.

Moreover, religious fasting practices, e.g., Ramadan intermittent fasting (RIF), are other forms of IF (Chen et al., 2023; Zhao et al., 2022). Of late, the positive effects of IF on the aging process have drawn a lot of interest. IF was reported to have neuroprotective effects for neurodegeneration, aging, and several diseases related to neurodegenerative diseases (like Parkinson's disease and AD) (Cremonini et al., 2019). IF modulates neurochemistry and neuronal activity widely, especially in brain areas critical for metabolic control, including the hippocampus, striatum, hypothalamus, and brainstem. IF may help to decrease oxidative stress, increase intracellular antioxidant enzyme expressions, enhance mitochondrial physiology, drive anti-inflammatory processes, improve neurogenesis, and increase synaptic plasticity—all of which could help to decrease brain aging and neurodegeneration. The clinical benefits of IF in AD are believed to involve multiple molecular action mechanisms, such as decreased neuroinflammation, rearrangement of glial cells, improved insulin sensitivity, augmentation of cell autophagy, stimulation of nerve regeneration, alteration of the gut microbiome, and reduction of amyloid deposits (Chen et al., 2023; Zeng et al., 2024). It has been reported that mitochondrial dysfunction plays a critical role in the pathogenesis of AD. Maintaining neuronal activity and protecting neurons from oxidative damage is critically dependent on healthy mitochondrial metabolism. Indeed, IF improves mitochondrial biogenesis, regulates mitochondrial functions, and provides protective properties against AD (Chen et al., 2023; Zhao et al., 2022). Moreover, elevated levels of ketone bodies during IF are associated with reduced levels of β-amyloid, indicating a potentially useful role of IF in the control of AD. Glycogen stores in the liver break down into glucose, and it is primarily this glucose that enters neurons during the first 10–14 hours of fasting. As starvation progresses, release of fatty acids from adipose tissues are transformed to ketone bodies, in particular β-hydroxybutyrate and acetoacetate. These ketone bodies, produced by the liver, become the primary fuel source for neurons during prolonged fasting (Cremonini et al., 2019). In some IF models with calorie restriction, elevations in β-amyloid levels and tau protein have led to improved conditions associated with AD. Yet, IF does not appear to reduce amyloid load much in AD patients at a high genetic risk. Vascular health is another possible mechanism by which IF may affect AD. High likelihood AD is spared in the AD brain health and cognition by IF through improving vascular integrity, stroke risk reduction, and cardiovascular and cerebrovascular health (Elias et al., 2023). However, clinical studies on the association between IF and AD in humans remain limited. Most of the human studies looking at the impact of IF on neuronal function and brain metabolism were conducted in obese or normal-weight adults. Continuous IF over a 30-day period, from dawn to dusk, decreased blood APP, the precursor of amyloid-β, in healthy subjects. Moreover, IF had a positive effect on metabolic functions such as glucose metabolism and insulin sensitivity, lipid metabolism and ketone body formation, regulation of circadian rhythm, control of DNA repair, activation and function of the immune system, and key regulatory proteins active in cognitive function. In the study mentioned above, IF generates a protective serum proteome which might protect against cancer, metabolic syndrome, inflammation, AD, and many neuropsychiatric disorders (Mindikoglu et al., 2020). A cross-sectional study on 883 older people aged over 50 investigated the association between feeding periods and cognition status. Participants who restricted their eating periods to 10 hours or less were followed for six months. And during the four-year study, those with limited access to food had a lower chance of developing cognitive impairment (Currenti et al., 2021). Besides, in APP/PS1 double transgenic mice (an AD mouse model), a feeding regimen, including ADF—in which mice were fed ad libitum on one day and fasted on the next—induced the restoration of aquaporin-4 polarity, promoted amyloid-β clearing, and ameliorated cognitive impairment (Zhang et al., 2017). Using another experimental AD animal model, ADF significantly enhanced survival rates, spatial learning, and memory capabilities, while no significant changes in the accumulation of amyloid-β were evident (Liu et al., 2019). Furthermore, investigating periodic dietary restriction in transgenic mice as a model of AD for the above outcomes found that six weeks of dietary restriction improved cognitive deficits, amyloid pathological alterations, and microglial activity, supporting neuroprotective effects against AD (Gregosa et al., 2019). Although scientific data show preventive and therapeutic activities of IF and calorie restriction on AD pathogenesis, these results are mostly based on animal studies. Well-designed clinical studies in humans are needed to determine IF for the treatment of AD. This being said, the preliminary data are encouraging and show that IF has the potential to be translated as a new valid therapeutic approach for AD.

20.7 CONCLUSION AND FUTURE REMARKS

Dietary approaches have the potential to exert a range of beneficial effects in the treatment of AD. Dietary patterns influence AD via diverse mechanisms (Figure 20.1). Though these programs provide individual benefits using different nutritional approaches, none of them can provide full treatment of AD and none can eradicate the symptoms. Nevertheless, scientific research indicates that certain dietary plans could delay the progression of AD and pathologies linked to it, potentially maintaining this effect across a lifetime. So, a healthy and balanced diet is the key component of a lifestyle, which is necessary to postpone your neurodegenerative diseases such as AD.

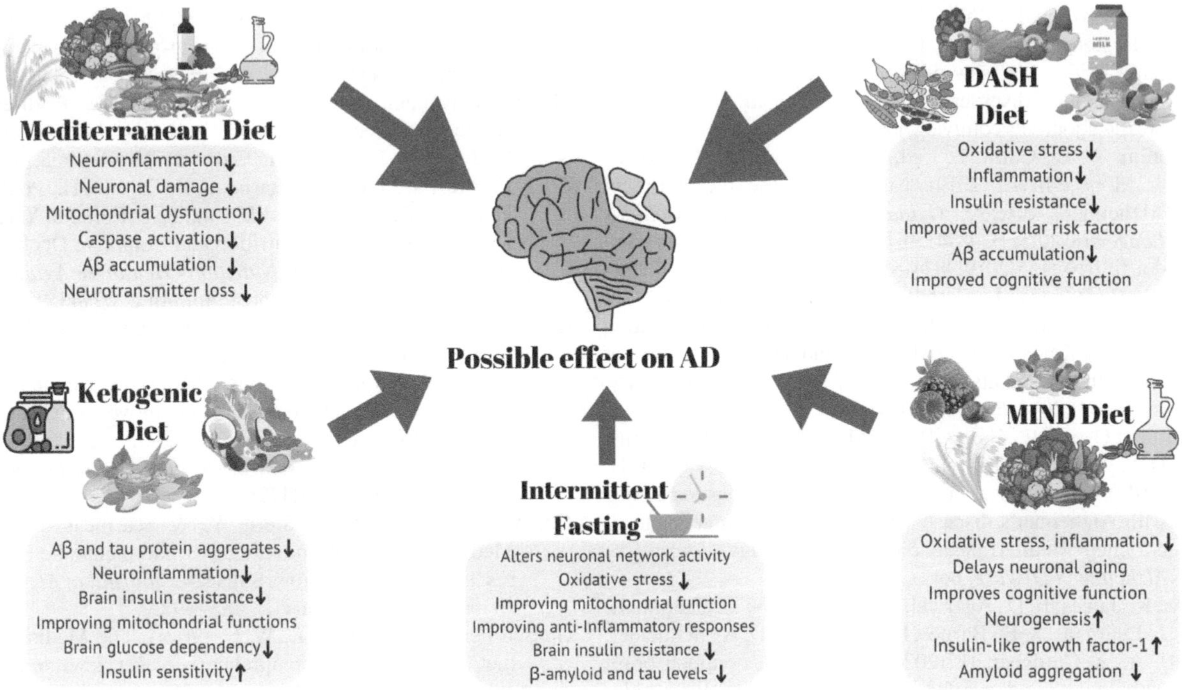

FIGURE 20.1 Summary of the possible impacts of different dietary models on AD.

REFERENCES

Agarwal, P., Leurgans, S. E., Agrawal, S., Aggarwal, N. T., Cherian, L. J., James, B. D., Dhana, K., Barnes, L. L., Bennett, D. A., & Schneider, J. A. (2023). Association of Mediterranean-DASH intervention for neurodegenerative delay and Mediterranean diets with Alzheimer disease pathology. *Neurology*, *100*(22), E2259–E2268.

Agarwal, P., Wang, Y., Holland, T., Bennett, D., & Morris, M. (2019). Strawberry consumption associated with reduced Alzheimer's dementia risk (FS05-06-19). *Current Developments in Nutrition*, *3*(June 1), nzz052.FS05-06-19.

Anastasiou, C. A., Yannakoulia, M., Kosmidis, M. H., Dardiotis, E., Hadjigeorgiou, G. M., Sakka, P., Arampatzi, X., Bougea, A., Labropoulos, I., & Scarmeas, N. (2017). Mediterranean diet and cognitive health: Initial results from the Hellenic longitudinal investigation of ageing and diet. *PloS One 12*(8). https://pubmed.ncbi.nlm.nih.gov/28763509/

Arjmand, G., Abbas-Zadeh, M., & Eftekhari, M. H. (2022). Effect of MIND diet intervention on cognitive performance and brain structure in healthy obese women: A randomized controlled trial. *Scientific Reports*, *2*(1), 1–14. https://www.nature.com/articles/s41598-021-04258-9

Ballarini, T., Lent, D. M. van, Brunner, J., Schröder, A., Wolfsgruber, S., Altenstein, S., Brosseron, F., Buerger, K., Dechent, P., Dobisch, L., Duzel, E., Ertl-Wagner, B., Fliessbach, K., Freiesleben, S. D., Frommann, I., Glanz, W., Hauser, D., Haynes, J. D., Heneka, M. T., & Janowitz, D. (2021). Mediterranean diet, Alzheimer disease biomarkers, and brain atrophy in old age. *Neurology*, *96*(24), E2920–E2932.

Barnes, L. L., Dhana, K., Liu, X., Carey, V. J., Ventrelle, J., Johnson, K., Hollings, C. S., Bishop, L., Laranjo, N., Stubbs, B. J., Reilly, X., Agarwal, P., Zhang, S., Grodstein, F., Tangney, C. C., Holland, T. M., Aggarwal, N. T., Arfanakis, K., Morris, M. C., & Sacks, F. M. (2023). Trial of the MIND diet for prevention of cognitive decline in older persons. *New England Journal of Medicine*, *389*(7), 602–611. https://www.nejm.org/doi/full/10.1056/NEJMoa2302368

Berendsen, A. A. M., van de Rest, O., Feskens, E. J. M., de Groot, L. C. P. G. M., Kang, J. H., Grodstein, F., & Grodstein, F. (2017). The Dietary Approaches to Stop Hypertension diet, cognitive function, and cognitive decline in American older women. *Journal of the American Medical Directors Association*, *18*(5), 427–432. https://pubmed.ncbi.nlm.nih.gov/28108204/

Bhuiyan, N. Z., Hasan, M. K., Mahmud, Z., Hossain, M. S., & Rahman, A. (2023). Prevention of Alzheimer's disease through diet: An exploratory review. *Metabolism Open*, *20*(December 1), 100257.

Blumenthal, J. A., Smith, P. J., Mabe, S., Hinderliter, A., Lin, P.-H., Liao, L., Welsh-Bohmer, K. A., Browndyke, J. N., Kraus, W. E., Doraiswamy, P. M., Burke, J. R., & Sherwood, A. (2017). Lifestyle and neurocognition in older adults with cardiovascular risk factors and cognitive impairment. *Psychosomatic Medicine*, *79*(6), 719–727. https://journals.lww.com/psychosomaticmedicine/fulltext/2017/07000/lifestyle_and_neurocognition_in_older_adults_with.14.aspx

Blumenthal, J. A., Smith, P. J., Mabe, S., Hinderliter, A., Lin, P.-H., Liao, L., Welsh-Bohmer, K. A., Browndyke, J. N., Kraus, W. E., Doraiswamy, P. M., Burke, J. R., & Sherwood, A. (2019). Lifestyle and neurocognition in older adults with cognitive impairments: A randomized trial. *Neurology*, *92*(3), E212–E223.

Huang, L., Tao, Y., Chen, H., Chen, X., Shen, J., Zhao, C., Xu, X., He, M., Zhu, D., Zhang, R., Yang, M., Zheng, Y., & Yuan, C. (2023). Association of the Mediterranean Dietary Approaches to Stop Hypertension intervention for neurodegenerative delay (MIND) diet with the risk of dementia. *JAMA Psychiatry*, *80*(6), 630–638. https://jamanetwork.com/journals/jamapsychiatry/fullarticle/2804268

Chen, W. W., Wang, L., Zhou, A. J., Fan, C., Zhang, Y. D., Wang, Z. P., Rong, S., & Wang, T. (2023). The benefits of intermittent fasting: A review of possible mechanisms on Central neurological disorders. *Acta Alimentaria*, *52*(1), 1–11. https://akjournals.com/view/journals/066/52/1/article-p1.xml

Cremonini, A. L., Caffa, I., Cea, M., Nencioni, A., Odetti, P., & Monacelli, F. (2019). Nutrients in the prevention of Alzheimer's disease. *Oxidative Medicine and Cellular Longevity*, *2019*(1), 9874159. https://onlinelibrary.wiley.com/doi/full/10.1155/2019/9874159

Currenti, W., Godos, J., Castellano, S., Caruso, G., Ferri, R., Caraci, F., Grosso, G., & Galvano, F. (2021). Association between time restricted feeding and cognitive status in older Italian adults. *Nutrients*, *13*(1), 191. https://www.mdpi.com/2072-6643/13/1/191/htm

de la Rubia Ortí, J. E., García-Pardo, M. P., Drehmer, E., Cantus, D. S., Rochina, M. J., Aguilar Calpe, M. A., & Yang, I. H. (2018). Improvement of main cognitive functions in patients with Alzheimer's disease after treatment with coconut oil enriched Mediterranean diet: A pilot study. *Journal of Alzheimer's Disease*, *65*(2), 577–587.

Dhana, K., James, B. D., Agarwal, P., Aggarwal, N. T., Cherian, L. J., Leurgans, S. E., Barnes, L. L., Bennett, D. A., Schneider, J. A., & Gardener, H. (2021). MIND diet, common brain pathologies, and cognition in community-dwelling older adults. *Https://Doi.Org/10.3233/JAD-210107*, *83*(2), 683–692. https://journals.sagepub.com/doi/full/10.3233/JAD-210107

Doroszkiewicz, J., Mroczko, J., Rutkowski, P., & Mroczko, B. (2023). Molecular aspects of a diet as a new pathway in the prevention and treatment of Alzheimer's disease. *International Journal of Molecular Sciences*, *24*(13), 10751. https://www.mdpi.com/1422-0067/24/13/10751/htm

Dyńka, D., Kowalcze, K., & Paziewska, A. (2022). The role of ketogenic diet in the treatment of neurological diseases. *Nutrients*, *14*(23), 5003. https://www.mdpi.com/2072-6643/14/23/5003/htm

Elias, A., Padinjakara, N., & Lautenschlager, N. T. (2023). Effects of intermittent fasting on cognitive health and Alzheimer's disease. *Nutrition Reviews*, *81*(9), 1225. https://pmc.ncbi.nlm.nih.gov/articles/PMC10413426/

Ellouze, I., Sheffler, J., Nagpal, R., & Arjmandi, B. (2023). Dietary patterns and Alzheimer's disease: An updated review linking nutrition to neuroscience. *Nutrients*, *15*(14), 3204. https://www.mdpi.com/2072-6643/15/14/3204/htm

Fortier, M., Castellano, C.-A., St-Pierre, V., Myette-Côté, É., Langlois, F., Roy, M., Morin, M.-C., Bocti, C., Fulop, T., Godin, J.-P., Delannoy, C., Cuenoud, B., & Cunnane, S. C. (2021). A ketogenic drink improves cognition in mild cognitive impairment: Results of a 6-month RCT. *Alzheimer's & Dementia*, *17*(3), 543–552. https://onlinelibrary.wiley.com/doi/full/10.1002/alz.12206

García-Casares, N., Fuentes, P. G., Barbancho, M., López-Gigosos, R., García-Rodríguez, A., & Gutiérrez-Bedmar, M. (2021). Alzheimer's disease, mild cognitive impairment and Mediterranean diet. A systematic review and dose-response meta-analysis. *Journal of Clinical Medicine*, *10*(20), 4642. https://www.mdpi.com/2077-0383/10/20/4642/htm

Gardener, H., & Caunca, M. R. (2018). Mediterranean diet in preventing neurodegenerative diseases. *Current Nutrition Reports*, *7*(1), 10. https://pmc.ncbi.nlm.nih.gov/articles/PMC7212497/

Georges, J., Bintener, C., & Miller, O. 2020. dementia in Europe yearbook 2019: Estimating the prevalence of dementia in Europe. *Alzheimer Europe* (January 28), 1–108. https://www.alzheimer-europe.org/resources/publications/dementia-europe-yearbook-2019-estimating-prevalence-dementia-europe?language_content_entity=en

Gillette Guyonnet, S., Abellan Van Kan, G., Andrieu, S., Barberger Gateau, P., Berr, C., Bonnefoy, M., Dartigues, J. F., de Groot, L., Ferry, M., Galan, P., Hercberg, S., Jeandel, C., Morris, M. C., Nourhashemi, F., Payette, H., Poulain, J. P., Portet, F., Roussel, A. M., Ritz, P., & Rolland, Y. (2007). IANA Task Force on Nutrition and Cognitive Decline with Aging. *The Journal of Nutrition, Health & Aging*, *11*(2), 132–152. https://pubmed.ncbi.nlm.nih.gov/17435956/

Gregosa, A., Vinuesa, Á, Todero, M. F., Pomilio, C., Rossi, S. P., Bentivegna, M., Presa, J., Wenker, S., Saravia, F., & Beauquis, J. (2019). Periodic dietary restriction ameliorates amyloid pathology and cognitive impairment in PDAPP-J20 mice: Potential implication of glial autophagy. *Neurobiology of Disease*, *132*(December 1). https://pubmed.ncbi.nlm.nih.gov/31351172/

Gu, Y., Luchsinger, J. A., Stern, Y., & Scarmeas, N. (2010). Mediterranean diet, inflammatory and metabolic biomarkers, and risk of Alzheimer's disease. *Journal of Alzheimer's Disease*, *22*(2), (January 1), 483–492.

Guasch-Ferré, M., & Willett, W. C. (2021). The Mediterranean diet and health: A comprehensive overview. *Journal of Internal Medicine*, *290*(3), 549–566. https://onlinelibrary.wiley.com/doi/full/10.1111/joim.13333

Haring, B., Wu, C., Mossavar-Rahmani, Y., Snetselaar, L., Brunner, R., Wallace, R. B., Neuhouser, M. L., & Wassertheil-Smoller, S. (2016). No association between dietary patterns and risk for cognitive decline in older women with 9-year follow-up: Data from the women's health initiative memory study. *Journal of the Academy of Nutrition and Dietetics*, *116*(6), 921–930.e1. http://www.jandonline.org/article/S2212267215018158/fulltext

Hosking, D. E., Eramudugolla, R., Cherbuin, N., & Anstey, K. J. (2019). MIND not Mediterranean diet related to 12-year incidence of cognitive impairment in an Australian longitudinal cohort study. *Alzheimer's & Dementia*, *15*(4), 581–589.

Jiang, Z., Yin, X., Wang, M., Chen, T., Wang, Y., Gao, Z., & Wang, Z. (2022). Effects of ketogenic diet on neuroinflammation in neurodegenerative diseases. *Aging and Disease*, *13*(4), 1146. https://pmc.ncbi.nlm.nih.gov/articles/PMC9286903/

Kara, G., & Kilinc, G. E. (2021). Ketogenic Diet Treatment in Alzheimer's Disease (Alzheimer HastalığındaKetojenik Diyet Tedavisi.) *İstanbul Gelişim Üniversitesi Sağlık Bilimleri Dergisi*, *15*, 630–638. https://doi.org/10.38079/igusabder.1007739

Kerti, L., Witte, A. V., Winkler, A., Grittner, U., Rujescu, D., & Flöel, A. (2013). Higher glucose levels associated with lower memory and reduced hippocampal microstructure. *Neurology*, *81*(20), 1746–1752.

Keys, A., Mıenottı, A., Karvonen, M. J., Aravanıs, C., Blackburn, H., Buzına, R., Djordjevıc, B. S., Dontas, A. S., Fıdanza, F., Keys, M. H., Kromhout, D., Nedeljkovıc, S., Punsar, S., Seccareccıa, F., & Toshıma, H. (1986). The diet and 15-year death rate in the seven countries study. *American Journal of Epidemiology*, *124*(6), 903–915. https://pubmed.ncbi.nlm.nih.gov/3776973/

Khandayataray, P., & Murthy, M. K. (2024). Dietary interventions in mitigating the impact of environmental pollutants on Alzheimer's disease – A review. *Neuroscience*, *563*, 148–166.

Kowalski, K., & Mulak, A. (2019). Brain-gut-microbiota axis in Alzheimer's disease. *Journal of Neurogastroenterology and Motility*, *25*(1), 48–60. https://pubmed.ncbi.nlm.nih.gov/30646475/

Krikorian, R., Shidler, M. D., & Summer, S. S. (2023). Early intervention in cognitive aging with strawberry supplementation. *Nutrients*, *15*(20), 4431. https://www.mdpi.com/2072-6643/15/20/4431/htm

Larrieu, S., Letenneur, L., Helmer, C., Dartigues, J. F., & Barberger-Gateau, P. (2004). Nutritional factors and risk of incident dementia in the PAQUID longitudinal cohort. *The Journal of Nutrition, Health & Aging*, *8*(3), 150–154. https://europepmc.org/article/med/15129300

Liu, Y., Cheng, A., Li, Y.-J., Yang, Y., Kishimoto, Y., Zhang, S., Wang, Y., Wan, R., Raefsky, S. M., Lu, D., Saito, T., Saido, T., Zhu, J., Wu, L.-J., & Mattson, M. P. (2019). SIRT3 mediates hippocampal synaptic adaptations to intermittent fasting and ameliorates deficits in APP mutant mice. *Nature Communications*, *10*(1), 1–11. https://www.nature.com/articles/s41467-019-09897-1.

Melekoğlu, E., & Rakıcıoğlu, N. (2020). Dietary patterns associated with the maintenance of cognitive function in elderly (Yaşlılarda bilişsel fonksiyonun korunmasi ile ilişkili diyet modelleri). *Beslenmeve Diyet Dergisi*, *48*(2), 84–92. https://beslenmevediyetdergisi.org/index.php/bdd/article/view/1271

Miller, M. G., Thangthaeng, N., Rutledge, G. A., Scott, T. M., & Shukitt-Hale, B. (2021). Dietary strawberry improves cognition in a randomised, double-blind, placebo-controlled trial in older adults. *British Journal of Nutrition*, *126*(2), 253–263. https://www.cambridge.org/core/journals/british-journal-of-nutrition/article/dietary-strawberry-improves-cognition-in-a-randomised-doubleblind-placebocontrolled-trial-in-older-adults/6DCBA9C5A798C0E6149322BBBCAAD461

Mindikoglu, A. L., Abdulsada, M. M., Jain, A., Choi, J. M., Jalal, P. K., Devaraj, S., Mezzari, M. P., Petrosino, J. F., Opekun, A. R., & Jung, S. Y. (2020). Intermittent fasting from dawn to sunset for 30 consecutive days is associated with anticancer proteomic signature and upregulates key regulatory proteins of glucose and lipid metabolism, circadian clock, DNA repair, cytoskeleton remodeling, immune system and cognitive function in healthy subjects. *Journal of Proteomics*, *217*, 103645.

Morrill, S. J., & Gibas, K. J. (2019). Ketogenic diet rescues cognition in ApoE4+ patient with mild Alzheimer's disease: A case study. *Diabetes & Metabolic Syndrome: Clinical Research & Reviews*, *13*(2), 1187–1191.

Morris, M. C. (2009). The role of nutrition in Alzheimer's disease: Epidemiological evidence. *European Journal of Neurology*, *16*(Suppl 1), 1–7. https://pubmed.ncbi.nlm.nih.gov/19703213/

Morris, M. C., Evans, D. A., Bienias, J. L., Tangney, C. C., Bennett, A., Wilson, R. S., Aggarwal, N., & Schneider, J. (2003). Consumption of fish and n-3 fatty acids and risk of incident Alzheimer disease. *Archives of Neurology*, *60*(7), 940–946. https://pubmed.ncbi.nlm.nih.gov/12873849/

Morris, M. C., Tangney, C. C., Wang, Y., Sacks, F. M., Barnes, L. L., Bennett, D. A., & Aggarwal, N. T. (2015a). MIND diet slows cognitive decline with aging. *Alzheimer's &Dementia : The Journal of the Alzheimer's Association*, *11*(9), 1015. https://pmc.ncbi.nlm.nih.gov/articles/PMC4581900/

Morris, M. C., Tangney, C. C., Wang, Y., Sacks, F. M., Bennett, D. A., & Aggarwal, N. T. (2015b). MIND diet associated with reduced incidence of Alzheimer's disease. *Alzheimer's & Dementia : The Journal of the Alzheimer's Association* *11*(9), 1007–1014. https://pubmed.ncbi.nlm.nih.gov/25681666/

Mosconi, L., Murray, J., Tsui, W. H., Li, Y., Davies, M., Williams, S., Pirraglia, E., Spector, N., Osorio, R. S., Glodzik, L., McHugh, P., & de Leon, M. J., (2014). Mediterranean diet and magnetic resonance imaging-assessed brain atrophy in cognitively normal individuals at risk for Alzheimer's disease. *The Journal of Prevention of Alzheimer's Disease*, *1*(1), 23. https://pmc.ncbi.nlm.nih.gov/articles/PMC4165397/

Navarro-Hortal, M. D., Romero-Márquez, J. M., Esteban-Muñoz, A., Sánchez-González, C., Rivas-García, L., Llopis, J., Cianciosi, D., Giampieri, F., Sumalla-Cano, S., Battino, M., & Quiles, J. L. (2022). Strawberry (*Fragaria × Ananassa* cv. Romina) methanolic extract attenuates Alzheimer's Beta amyloid production and oxidative stress by SKN-1/NRF and DAF-16/FOXO mediated mechanisms in *C. elegans*. *Food Chemistry*, *372*, 131272.

Ota, M., Matsuo, J., Ishida, I., Takano, H., Yokoi, Y., Hori, H., Yoshida, S., Ashida, K., Nakamura, K., Takahashi, T., & Kunugi, H. (2019). Effects of a medium-chain triglyceride-based ketogenic formula on cognitive function in patients with mild-to-moderate Alzheimer's disease. *Neuroscience Letters*, *690*, 232–236.

Patterson, C.. 2018. World Alzheimer Report 2018: The State of the Art of Dementia Research: New Frontiers. https://www.alzint.org/resource/world-alzheimer-report-2018/.

Pietrzak, D., Kasperek, K., Rękawek, P., & Piątkowska-Chmiel, I. (2022). The therapeutic role of ketogenic diet in neurological disorders. *Nutrients*, *14*(9), 1952. https://www.mdpi.com/2072-6643/14/9/1952/htm

Pivi, G. A. K., Vieira, N. M. A., da Ponte, J. B., de Moraes, D. S. C., & Bertolucci, P. H. F. (2017). Nutritional management for Alzheimer's disease in all stages: Mild, moderate, and severe. *Nutrire*, *42*(1), 1–6. https://link.springer.com/article/10.1186/s41110-016-0025-7

Pritam, P., Deka, R., Bhardwaj, A., Srivastava, R., Kumar, D., Jha, A. K., Jha, N. K., Villa, C., & Jha, S. K. (2022). Antioxidants in Alzheimer's disease: Current therapeutic significance and future prospects. *Biology*, *11*(2), 212. https://pmc.ncbi.nlm.nih.gov/articles/PMC8869589/

Sánchez-Muniz, F. J., Macho-González, A., Garcimartin, A., Santos-López, J. A., Benedi, J., Bastida, S., & Gonzsez-Munoz, M. J. (2019). The nutritional components of beer and its relationship with neurodegeneration and Alzheimer's disease. *Nutrients*, *11*(7), 1558. https://doi.org/10.3390/nu11071558

Scheltens, P., De Strooper, B., Kivipelto, M., Holstege, H., Chételat, G., Teunissen, C. E., Cummings, J., & van der Flier, W. M. (2021). Alzheimer's disease. *The Lancet*, *397*(10284), 1577–1590. http://www.thelancet.com/article/S0140673620322054/fulltext

Shah, R. (2013). The role of nutrition and diet in Alzheimer disease: A systematic review. *Journal of the American Medical Directors Association*, *14*(6), 398–402. https://pubmed.ncbi.nlm.nih.gov/23419980/

Shukitt-Hale, B., Bielinski, D. F., Lau, F. C., Willis, L. M., Carey, A. N., & Joseph, J. A. (2015). The beneficial effects of berries on cognition, motor behaviour and neuronal function in ageing. *British Journal of Nutrition*, *114*(10), (October 5): 1542–1549. https://www.cambridge.org/core/journals/british-journal-of-nutrition/article/beneficial-effects-of-berries-on-cognition-motor-behaviour-and-neuronal-function-in-ageing/751902ED1F6B94DFEC9AB7FB8C89C0DA

Śliwińska, S., & Jeziorek, M. (2021). The role of nutrition in Alzheimer's disease. *Roczniki Państwowego Zakładu Higieny*, *72*(1), 29–39. https://doi.org/10.32394/rpzh.2021.0154

Sofi, F., Macchi, C., Abbate, R., Gensini, G. F., & Casini, A. (2010). Effectiveness of the Mediterranean diet: Can it help delay or prevent Alzheimer's disease? *Journal of Alzheimer's Disease*, *20*(3), 795–801.

Sofi, F., Macchi, C., Abbate, R., Gensini, G. F., & Casini, A. (2013). Mediterranean diet and health. *BioFactors*, *39*(4), 335–342. https://onlinelibrary.wiley.com/doi/full/10.1002/biof.1096

Stefaniak, O., Dobrzyńska, M., Drzymała-Czyż, S., & Przysławski, J. (2022). Diet in the prevention of Alzheimer's disease: Current knowledge and future research requirements. *Nutrients*, *14*(21), 4564. https://www.mdpi.com/2072-6643/14/21/4564/htm.

Taylor, M. K., Sullivan, D. K., Mahnken, J. D., Burns, J. M., & Swerdlow, R. H. (2017). Feasibility and efficacy data from a ketogenic diet intervention in Alzheimer's disease, *Alzheimer's & Dementia (New York, N. Y.)*, *4*, 28–36, https://pubmed.ncbi.nlm.nih.gov/29955649/

Torosyan, N., Sethanandha, C., Grill, J. D., Dilley, M. L., Lee, J., Cummings, J. L., Ossinalde, C., & Silverman, D. H. (2018). Changes in regional cerebral blood flow associated with a 45 day course of the ketogenic agent, caprylidene, in patients with mild to moderate Alzheimer's disease: Results of a randomized, double-blinded, pilot study. *Experimental Gerontology*, *111*, 118–121.

Ulamek-Koziol, M., & Pluta, R. (2020). To treat or not to treat Alzheimer's disease by the ketogenic diet? That is the question. *Neural Regeneration Research*, *15*(5), 857–858. https://journals.lww.com/nrronline/fulltext/2020/15050/to_treat_or_not_to_treat_alzheimer_s_disease_by.12.aspx

Van Den Brink, A. C., Brouwer-Brolsma, E. M., Berendsen, A. A. M., & Van De Rest, O. (2019). The Mediterranean, Dietary Approaches to Stop Hypertension (DASH), and Mediterranean-DASH intervention for neurodegenerative delay (MIND) diets are associated with less cognitive decline and a lower risk of Alzheimer's disease—A review. *Advances in Nutrition*, *10*(6), 1040–1065.

Weinstein, G., Maillard, P., Himali, J. J., Beiser, A. S., Au, R., Wolf, P. A., Seshadri, S., & Decarli, C. (2015). Glucose indices are associated with cognitive and structural brain measures in young adults. *Neurology*, *84*(23), 2329–2337.

Willett, W. C., Sacks, F., Trichopoulou, A., Drescher, G., Ferro-Luzzi, A., Helsing, E., & Trichopoulos, D. (1995). Mediterranean diet pyramid: A cultural model for healthy eating. *The American Journal of Clinical Nutrition 61* (6 Suppl). https://pubmed.ncbi.nlm.nih.gov/7754995/

World Health Organization (2021). *Global status report on the public health response to dementia*. World Health Organization. https://iris.who.int/handle/10665/344701.

Xu Lou, I., Ali, K., & Chen, Q. (2023). Effect of nutrition in Alzheimer's disease: A systematic review. *Frontiers in Neuroscience*, *17*, 1147177.

Yildirim, N., & Eke, B. C. (2024). Alzheimer's disease, risk factors and therapy (Alzheimer hastalığı, risk faktörleri ve tedavi). *Ankara Üniversitesi Eczacılık Fakültesi Dergisi*, *48*(2), 766–790. https://dergipark.org.tr/tr/pub/jfpanu/issue/83110/1441827

Zeng, Z., Zhang, H., Wang, X., Shen, J., & Chen, D. (2024). Fasting, a potential intervention in Alzheimer's disease. *Journal of Integrative Neuroscience*, *23*(3), 50. https://www.imrpress.com/journal/JIN/23/3/10.31083/j.jin2303050/htm

Zhang, J., Zhan, Z., Li, X., Xing, A., Jiang, C., Chen, Y., Shi, W., & An, L. (2017). Intermittent fasting protects against Alzheimer's disease possible through restoring aquaporin-4 polarity. *Frontiers in Molecular Neuroscience*, *10*, 293192.

Zhao, Y., Jia, M., Chen, W., & Liu, Z. (2022). The neuroprotective effects of intermittent fasting on brain aging and neurodegenerative diseases via regulating mitochondrial function. *Free Radical Biology and Medicine*, *182*, 206–218.

21 The Role and Importance of Exercise in Alzheimer's Disease

Dilek Özbeyli and Meral Yüksel

21.1 INTRODUCTION

Alzheimer's disease (AD) is a chronic neurodegenerative disease mainly characterized by gradually progressive cognitive impairment and memory loss, as well as gradual deterioration in the function of daily living activities. Although the exact pathogenesis is not fully elucidated, the disease is highly correlated with the major pathologic characteristics of the disease: deposition of amyloid-β (Aβ) plaques and tau protein hyperphosphorylation. Pathological alterations such as oxidative stress, neuroinflammation, and protein aggregation cause an impairment in synaptic integrity, disrupt synaptic communication, and ultimately lead to death of synaptic connections between neurons in the brain (Özbeyli et al., 2017).

The effects of movement and formal exercise protocols are widely known to be fundamental regulatory masters of metabolic homeostasis and whole-body health. Regular physical exercise has complex and synergistic effects in multiple organs and tissues, especially on the crosstalk between muscle-brain-lungs and heart. This interaction regulates gene expression and elicits molecular, biochemical, and physiological adaptations that promote systemic integration. This leads to improved physical fitness and cognitive performance and reduces the risk of chronic diseases. In healthy populations, regular physical activity correlates with enhanced cardiovascular efficiency, optimal glucose metabolism, and neuroprotective pathways that promote cognitive resilience. But in the context of AD, exercise does much more than just provides health benefits. Exercise is a therapeutic measure that counteracts peripheral metabolic dysregulation, prevents cognitive impairment, and may slow the progression of the disease. Physical exercise has also been widely studied as a new potential therapeutic strategy for AD patients (Pinho et al., 2019).

Physical exercise refers to all movement that uses energy and involves the contraction of skeletal muscles. It is essential for maintaining and promoting overall health and providing a number of positive mental and physical effects. Understanding the type of physical activity and what it has to offer is an important piece of the puzzle in designing personalized exercise regimens according to specific needs and goals. Physical activities can mainly be classified into three categories: flexibility exercises, aerobic exercises (AEs), and anaerobic exercises, also called the resistance exercises (REs) (Plowman & Smith, 2014; Siddiqui et al., 2010). Literature shows the countless benefits of each group of exercise to health and fitness, and with wide-ranging and diverse physiological, metabolic (if not psychological) utilities, all of which can deliver a wide and full fitness strategy if you have a plan.

In addition, flexibility work, such as static and certain types of kinetic stretching, improves the mobility of muscles and joints. The advantages of these exercises are great to enhance mobility, prevent injuries, and achieve good posture (Galloza et al., 2017).

AE, also known as cardiovascular or endurance exercises, are continuous, rhythmic movements that use large muscle groups. AE is usually performed at moderate-to-high intensity in steady-state conditions of aerobic energy production, allowing participants to maintain longer training sessions (often 45–60 minutes). As the name implies, the muscles utilized during this form of activity depend on aerobic metabolism to create adenosine triphosphate (ATP) for energy from the metabolism of amino acids, carbohydrates, and fatty acids (Patel et al., 2017). It is suggested that aerobic tempos should engage large muscles during aerobic training for the greatest benefits. Some examples of activities in this group include swimming, cycling, running, dancing, hiking, and brisk walking. Aside from improving the efficiency of the cardiovascular system, those exercises also improve fits metabolic health and overall endurance.

REs such as, weightlifting, jumping, or action with resistance bands aim to increase the strength of muscles by dynamic or static contractions against a resistance (Bird et al., 2005). As per American College of Sports Medicine, anaerobic exercises involve short-duration, high-intensity physical activities that primarily depend on energy phosphates stored in the muscle and less on the utilization of oxygen for ATP production. However, in the absence of oxygen, the body is forced to rely on anaerobic (or lack of oxygen) glycolysis and fermentation to generate ATP, which generates a bicycle of ATP per glucose but produces lactic acid as a by-product. Anaerobic-type exercises mainly require the involvement of fast-twitch muscle fibres like powerlifting, sprinting, and high-intensity interval training (HIIT) (Patel et al., 2017). All of these exercises are the best way to help you gain hypertrophy (muscle growth), increasing muscle strength and bone density. Anaerobic training, which combines both free weights and mechanical resistance, yields targeted advantages for not only the strength of the muscles but also the resilience of the body as a whole and performance capacity (Bird et al., 2005; Mcleod et al., 2024).

DOI: 10.1201/9781003602477-21

Thus, exercise triggers wide-ranging protective mechanisms, opposing predisposing risk factors involved in mechanisms of AD pathophysiology. These are alleviation of oxidative stress, regulation of physiological and psychological stress response, enhancement of cerebral perfusion, reversal of hypomethylation, and suppression of systemic and neuronal inflammation. Moreover, exercise promotes neuroplasticity, acts against tau protein abnormalities, and lowers insoluble Aβ deposition (Abdullahi et al., 2024).

21.2 MECHANISMS UNDERLYING THE BENEFICIAL EFFECTS OFAEROBIC EXERCISE ON AD

AD was proven to have several favourable benefits due to AE by improving cognition and changing the biological mechanisms related to the underlying causes. AE enhances cardiovascular endurance and triggers diverse physiologic adaptations in the body, from increased activity of antioxidant enzymes to higher expression of exerkines such as neurotrophic factors, promoting overall health and optimal brain function (Lu et al., 2023; Valenzuela et al., 2023). Moreover, AE lowers the formation and aggregation of Aβ, aids in the clearance of Aβ and phosphorylated tau proteins (Feng et al., 2023; Wang et al., 2023), and boosts glucose metabolism, which together lead to better neuronal health and resilience.

21.2.1 Improvement of Antioxidant Capacity

The brain is metabolically highly active in an oxidative manner and generates large amounts of reactive oxygen species (ROS) (Niedzielska et al., 2016), while containing comparably low levels of antioxidants. Oxidative stress is one of the well-established etiological contributors to AD (Singh et al., 2019). It leads to damage on lipids (membrane phospholipids), proteins (exerts as protein oxidation), and DNA (oxidative DNA damage such as 8-oxo-2′ deoxyguanosine), progressing to neuronal death and aggravating the AD course. Additionally, oxidative stress induces apoptosis by alteration of ERK1/2 and Nrf2 pathways, with increased expression of GSK-3β and decreased activity of PP2A. Mitochondrial dysfunction results with increased ROS formation by dysregulated electron transport chain. Other ROS generation pathways such as multiple enzymatic reactions augment Aβ aggregation and neurofibrillary tangles, thus driving a vicious cycle of ROS production, capable of self-stimulation (Chen et al., 2023). This emphasizes the importance of treatment that restores oxidants-antioxidants homeostasis, decreases oxidative stress, and enhances the antioxidant defence mechanisms in the brain. AE has been identified as a potent agent with increased antioxidant ability in the brain. In rats and AD humans, it has increased the levels of the intracellular glutathione, which is a potent antioxidant molecule and decreased the malondialdehyde levels, which is an end-product of lipid peroxidation that formed after oxidative injury (Mezghani et al., 2023; Özbeyli et al., 2017). Moreover, in murine models of AD, AE significantly induces the expression of antioxidant enzymes like Cu-Zn-superoxide dismutase, Mn-superoxide dismutase, and catalase (Herring et al., 2010; Um et al., 2008). As a result of all these findings, these data conclude that AE ameliorates oxidative stress and protects from AD by enhancing the brain antioxidant system.

21.2.2 Increased Production of Exerkines Mediated by Brain-Derived Neurotrophic Factor and Neural Growth Factor

Exercise stimulates the production of a wide range of proteins and nucleic acids (termed in general exerkines) from tissues throughout the body that are released locally and into the blood during sustained exercise. One of the most exciting candidates for the neurogenesis-promoting effects of exercise is brain-derived neurotrophic factor (BDNF) due to its far-reaching effects on the health of the brain (Prasad et al., 2024). Because of its indispensable roles in neurogenesis, neuronal viability, synaptic plasticity, and cognitive function, BDNF could take part in neuropathology underlining neurodegenerative diseases such as AD. BDNF is widely studied due to its key roles in brain development and function. During development, it induces neuronal growth, differentiation, and maturation and in the adult brain regulates synaptic plasticity and transmission (Lu et al., 2023; Schmolesky et al., 2013). Neuroplasticity, the ability of the brain to mould itself and rewrite neuronal connections, is a basic property for responding to environmental changes, for learning and recovering from disease. This dynamic process includes changes in synaptic strength, synapse formation, structural and functional changes in neurons, and the generation of new neurons. Search continuity across column dimensions effective synaptic functioning is essential for neuronal communication and serves as the basis of cognition. Synaptic homeostasis stabilizes and regulates synaptic strength in the central nervous system (CNS). Yet the balance between pro-inflammatory and anti-inflammatory cytokines is susceptible to perturbations given that either inflammation, ageing, or infection can trigger an immune dysfunction that alters brain function (Tacke et al., 2024). The BDNF molecule can attached to the tropomyosin receptor kinase B (TrkB) receptors to promote dendritic growth, spine formation, and synaptic function. Subsequently, this binding activates essential signalling cascades, including ERK1/2 and PI3K-mTOR that govern dendritic development and synaptogenesis (Camuso & Canterini, 2023). BDNF is mainly synthesized in the neuronal and glial cell bodies and is transported to presynaptic terminals and postsynaptic dendrites. As might be expected based on its interactions with TrkB at glutamatergic synapses, it plays broad and multifaceted roles in regulating the release of neurotransmitters, the dynamics of ion channels, axonal growth, and neuronal excitability (Schinder & Poo, 2000). BNDF mRNA is prevalent throughout the CNS, with high expression in areas including but not limited to the cortex, hippocampus, substantia nigra, amygdala, and thalamus (Conner et al., 1997). Not only do BDNF mRNA expression levels correlate with learning and memory (as well as synaptic plasticity), but the steady-state expression levels of BDNF mRNA are highest in the hippocampus (Hofer et al., 1990). BDNF is also strongly expressed in the entorhinal cortex, a central communication hub between the cortex and hippocampus and key to learning and memory. Furthermore, it has been demonstrated that the entorhinal cortex synthesizes BDNF that is then transported anterograde towards the hippocampus, enforcing its relevance to sustain the hippocampal activity and synaptic plasticity (Nagahara et al., 2009). Synaptic dysfunction is a characteristic feature of AD that becomes apparent as the disease advances. Increasing evidence supports that disruption of neuronal circuitry and synaptic structure is a central mediatory mechanism linking Aβ tau-pathology to cognitive deficits in AD (Gao et al., 2022). The low level of BDNF has been associated with the Aβ deposition and tau hyperphosphorylation as well as the neuroinflammatory and neurodegenerative hypothesis of AD, and also with decreased neurogenesis (Wang et al., 2019), thus suggesting that BDNF harnesses a role in AD progression. BDNF-activated phosphatidylinositol 3-kinase (PI3K) signalling through TrkB receptor results in dephosphorylation of tau and emphasizes the important role that BDNF plays in tau-related pathology (Elliott et al., 2005). Increasing the expression of BDNF mediated by the ERK/CREB signalling pathway ameliorated Aβ-induced neuronal degeneration and dendritic atrophy, confirming neuroprotection and synaptic resilience (Meng et al., 2013). As a result of the synaptic degeneration seen in AD, BDNF has attracted growing interests for its value as a candidate biomarker and a therapeutic target (Gao et al., 2022).

AE raises BDNF levels through multiple interplaying mechanisms. Neuroplasticity is promoted by exercise-induced neuronal activity-enhanced neurotransmitter signalling, and greater calcium influx leads to greater BDNF expression. Physical activity additionally promotes brain blood flow, allowing for increased oxygen and nutrient delivery to brain tissue that can help maintain neuronal health and the expression of BDNF (Cotman & Berchtold, 2002; Pereira et al., 2007). Exercise stimulates the AMPK/SIRT1 pathway, which upregulates BDNF production. The AMPK/SIRT1 pathway and BDNF synergistically attenuate AD pathology by regulating aberrant intracellular Aβ production, tau phosphorylation, and neurogenesis. This regulation is accomplished through modulation of α-, β and γ-secretases, and glycogen-synthase kinase-3 enzyme (Azimi et al., 2018; Nigam et al., 2017; Wang et al., 2018). AE also induces the release of exerkines—molecules secreted by muscles, brain, liver, kidneys, and other tissues in response to exercise—emphasizing the global nature of exercise-relevant signalling. Skeletal muscle, for example, secretes interleukin-6 (IL-6), the kidney releases nephrokines, and the brain and liver release neurokines and hepatokines, respectively. Such molecules pass through the blood-brain barrier inducing BDNF production (Pedersen, 2011; Prasad et al., 2024) thus facilitating neurogenesis and synaptic plasticity. Two other players are lactate and irisin—both of which are important to the process of BDNF production in response to exercise. Lactate is secreted under anaerobic conditions during physical activity enhances BDNF expression through SIRT1 activation, which facilitates neuroplasticity, learning, and memory in mammals. Lactate crosses the blood-brain barrier and initiates BDNF signalling in the hippocampus, an important region responsible for learning and cognitive functions (El Hayek et al., 2019). Additionally myokine signals like the release of irisin during muscle contractions will multiply the actions already taken. It was associated with elevated BDNF expression in the hippocampus and served as a mediator of exercise-induced brain plasticity and cognitive benefits. Irisin also stimulates neurogenesis and synaptogenesis, leading to improved memory and learning by BDNF pathway modulation (Trettel et al., 2023). Thus, AE increases BDNF level via mainly neurovascular, metabolic, and molecular pathways. Such mechanisms highlight the critical role of AE in enhancing certain aspects of brain function (for example, cognitive reserve) and delivering adaptive potential against neurodegenerations (including AD).

Nerve growth factor (NGF) is a neurotrophin and is required for the survival, maintenance, and differentiation of cholinergic neurons in the basal forebrain, a key region involved in AD pathology. Decreased levels of NGF are a contributing factor in cholinergic neuron degeneration, which is a significant pathophysiological change in AD, and lower levels of NGF correlate with progressive cognitive decline in AD patients (Chen & Mobley, 2019). The lowering of cholinergic neuron degeneration, for instance, is significantly dependent on the presence of these NGFs, which have been shown to be upregulated by AE. Upon high-affinity interactions with its receptor TrkA, NGF activates cytoplasmic networks mediating cell survival and neuronal plasticity (Pahlavani, 2023). Both the aberrant levels of NGF increase the functional activity and survivability of cholinergic neurons, potentially compensating for the loss of these neurons in AD. NGF alters APP metabolism, reducing Aβ generation and plaque deposition, a prominent disease hallmark in AD. NGF regulates neuroinflammation by controlling the function of microglial cells and buffering neuronal injury by chronic inflammation, a key aspect of AD progression. Additionally, NGF promotes hippocampal neurogenesis and is particularly vulnerable to injury caused by AD (Bonanni et al., 2022). As a result, AE augments NGF levels, thus allowing neuronal survival, decreasing Aβ pathology, neuroinflammation, and promoting neurogenesis in hippocampus. These mechanisms underscore why NGF is an attractive target to preserve cognition and slow the trajectory of AD.

21.2.3 Other Exercise-Specific Mechanisms That Inhibit Amyloid-beta Production and Tau Hyperphosphorylation

AE has also been shown in many animal model studies to decrease formation and aggregation of Aβ and to enhance the removal of Aβ and p-tau from neurons, especially from certain brain regions involved in memory and regulating functions (Adlard et al., 2005; Feng et al., 2023; Wang et al., 2023). It has been demonstrated that HIIT can attenuate AD-like pathology through modulation of astrocyte phenotypes and promotion of aquaporin-4 (AQP4) polarization. These adaptations facilitate brain tissue clearance of Aβ and p-tau utilizing the glymphatic system and renal pathways, proposing a mechanistic link between HIIT and neurodegenerative disease management (Feng et al., 2023). Exercise pre-treatment has proven a protective effect against Aβ deposition and tau hyperphosphorylation, as well as preserving spine density, synapses, and presynaptic vesicles (Yang et al., 2022). Other positive effects of exercise include an increase in hippocampal volume and strengthening of synaptic connections, both major determinants of cognitive resilience (Yu et al., 2021). But results in human studies have been mixed. As an example, a randomized controlled trial (RCT) over one year found that AE did not significantly reduce amyloid accumulation in cognitively normal older adults who were already accumulating cerebral amyloid deposits (Vidoni et al., 2021). These results underscore the need to better understand variability in exercise treatment effect between populations and stages of AD in future work.

21.2.4 Neuroinflammation Mitigation

The neuroinflammation hypothesis first formulated in 1992 (McGeer and Rogers) posits that neuroinflammation is a core pathological characteristic in AD. Neuroinflammation is marked by the activation of resident immune cells in brain, specifically microglia and astrocytes, as well as immune cells migrating from the peripheral system (Zhao, 2024). In particular, pro-inflammatory cytokines released by excessive immune activation further activate immune response and include TNF, IL-1β, IL-18, and IL-6; chemokines such as CCL1, CCL5, and CXCL1. Inflammation involves various inflammatory mediators such as prostaglandins, NO, and ROS (DiSabato et al., 2016). Both ageing, the most prominent risk factor for late-onset AD, and obesity, diabetes, and cardiovascular diseases are associated with a systemic chronic low-grade inflammatory state (Newcombe et al., 2018). Promoting the progression of these inflammatory mechanisms can eventually cause the accumulation of amyloid plaques and neurofibrillary tangles with pathologic hyperphosphorylation of tau protein. Such that modulating neuroinflammation is one of the promising strategies that may restore immune equilibrium and minimize neuronal injury in AD. AE is a promising candidate for an intervention that can help reduce neuroinflammation and restore normal brain function. AE has the potential to reverse neuroinflammation, oxidative stress, and synaptic dysregulation in AD models (Zhao, 2024).

In AD, neuroinflammation mediated by chronic activation of microglia and NOD-like receptor protein 3 (NLRP3) inflammasome. In general terms, pathological stimuli from peripheral tissues or within the brain maintain an inflamed microglial state and continually fuel an ongoing inflammatory programme but the mechanisms that sustain such inflammation are unclear. This extended activation leads to oxidative stress and increased signalling activation of microglia, including pyroptosis activation, and other activation pathways of cell death. Microglia-mediated chronic inflammation promotes AD pathology and exacerbates neuronal damage and cognitive decline (Wang et al., 2023). Various components have been proposed to explain the ability of AE to inhibit NLRP3 inflammasome-mediated pyroptosis. AE improves endothelial function by reducing vascular inflammation and oxidative stress in endothelial cells and ameliorates inflammation and insulin sensitivity in adipocytes, which further inhibits pyroptosis in cells. In the CNS, AE suppresses the activation of microglia and TLRs, mitigates neuroinflammation, and prevents Aβ deposition in hippocampal neurons. These results suggest that treatment with AE might represent a possible therapeutic modulator of the NLRP3 inflammasome signalling pathway in the context of pyroptosis-associated neuroinflammatory pathologies (Hu et al., 2022). MicroRNAs (miRNAs) deregulation plays a pivotal role in the pathogenesis of several human pathologies, including AD (Gaudet et al., 2018). Recent studies proposed that miRNAs have a role in the neuroprotection of neurons via exercise by acting as intermediators. For example, four weeks of voluntary AE has been shown to increase the expression of miR-129-5p in the hippocampus of AD mice. The upregulation of miR-129-5p was linked to superior cognition and inflammatory response amelioration and therefore highlights the alleged role of AE in AD treatment (Li et al., 2020).

21.2.5 Improvements in Glucose Metabolism

Zhao and Xu's recent evidence suggests that AD may be a metabolic disorder related to glucose metabolism impairment in the brain (2021). Others have even suggested the potential for AD to be a brain-specific type of "type 3 diabetes" (Janoutová et al., 2022). A healthy human brain depends on glucose to maintain its physiological and metabolic functions (Ishibashi et al., 2016). Nevertheless, glucose metabolism is severely impaired in the brains of all patients diagnosed with AD, especially in those areas involved in memory, such as the hippocampus and posterior cingulate cortex, leading to reduced neuronal excitability and dysfunction of the brain (Ferrari et al., 2019).

Disruption of brain glucose metabolism in AD occurs through various mechanisms such as insulin resistance, downregulation of glucose transport via GLUT transporters, decreased ATP yielding via glycolytic flux, and impaired mitochondrial function. All together, these mechanisms are believed to be vital in the progression of AD (An et al., 2018; Kellar & Craft, 2020). Thus, the neuroprotective effects of physical exercise correlate with changes in glucose metabolism in the brain. AE-formed derivative was able to improve glucose uptake in the brains of patients probable with AD, providing the energy for meeting the demand of different physiological and metabolic pathways. Studies also indicate that parameters such as brain glucose transporter 1 (GLUT-1) protein levels improve after AE practice (Um et al., 2008). Importantly, GLUT-1 molecule is generally diminished in the brains of patients with AD (Kyrtata et al., 2021).

These processes are increased through AE via AMPK/SIRT1 and BDNF. Such factors modulate insulin signalling, glucose transporter activity, and PGC-1α-mediated mitochondrial biogenesis, increasing intracellular glucose uptake in the brain (Hardie et al., 2012; Marosi & Mattson, 2014). Also AMPK is activated by AE to enhance the activity of GLUT-1, GLUT-3, phosphofructokinase (PFK), and LDH, which improves cerebral glucose utilization (Kinni et al., 2011). In conclusion, AE emerges as a promising non-drug intervention for preventing and delaying AD pathology, targeting its multifactorial pathophysiology. By improving the regulation of glucose metabolism, and modulating neuroinflammation, oxidative stress, and synaptic integrity, AE highlights the importance of physical activity as a key component of a multifaceted approach to promoting cognitive health and resilience.

21.3 THE BENEFICIAL MECHANISM OF ANAEROBIC EXERCISE IN AD

AD is often associated with different muscle abnormalities (Wang et al., 2017). Experimental evidence of abnormal APP metabolism and Aβ deposition of skeletal muscle tissues of animal models points towards similar pathological processes occurring between the brain and skeletal muscle of AD patients (Kuo et al., 2000). Dementia patients are more vulnerable to frailty and sarcopenia being characterized by depletion in muscle mass and strength (Waite et al., 2021). While abnormal weight loss and cachexia have been suggested as possible clinical markers for the diagnosis of AD (McKhann et al., 1984), these features are not part of the standard diagnostic criteria. Rather, accelerated lean mass loss in AD is considered risk factor, as in non-demented individuals, it is correlated to skeletal muscle atrophy, brain atrophy, and cognitive decline (Burns et al., 2010). The relationship suggests a need to be concerned about muscle health in the setting of AD. It seems that oxidative and inflammatory stress affect the improved number of cell receptors in the skeletal muscle of patients with AD; however, molecular mechanisms remain unclear in exact (Monteiro-Cardoso et al., 2015). Muscle atrophy and metabolic dysfunction in AD can potentiate brain pathology by compromising the bidirectional relationship between muscle and brain health. Reinforcing skeletal muscle using anaerobic exercise may be a protective strategy that leads to reciprocal benefits in the muscle-brain axis and may slow the disease-related symptoms of AD. Resistance training, an example of anaerobic exercise, has been shown to increase muscle mass, decrease sarcopenia, and promote overall physical resilience. These adaptations may benefit brain health indirectly by lowering systemic inflammation and improving metabolic function and the range of the organism to manage with oxidative stress, factors being implicated in the physiopathology of AD. Despite this the preferential molecular mechanisms used by anaerobic exercise to improve muscle and brain health in patients with AD need to be elaborated through further studies. As such, RE has become a well-established means for achieving favourable changes in whole-body composition, functional capacity, and stability in ageing populations (Smith et al., 2022) that may not only mitigate but reverse sarcopenia. In line with World Health Organization (WHO) (2020), it is recommended that healthy adults are engaged in moderate-to-vigorous RE for at least two days per week (non-consecutively) to improve or maintain over time overall health and physical fitness. The health benefits (Özbeyli et al., 2017), including neurological health benefits of such anaerobic exercises on our health, have been shown to continue on. RE has been effective in improving memory and cognitive performance and possesses neuroprotective and anti-inflammatory properties. The biological benefits of AD will include, but is not limited to, reduced amyloid accrual, decreased amyloid plaques formation, and slower progression of neurodegeneration in AD (Azevedo et al., 2023). RE has been postulated to play several roles in the treatment of progressive forms of neurodegeneration, such as AD, mediating retention of cognitive function and facilitating removal of Aβ (Liu et al., 2020; Özbeyli et al., 2017; Pena et al., 2020). RE has been demonstrated to decrease absolute (Hashiguchi et al., 2020) and relative Aβ plaque load (Liu et al., 2020) and to improve tau pathologies in the brain (Liu et al., 2020). Collectively, these results underscore the efficacy of RE in targeting the underlying pathophysiological AD processes and provide evidence that it is capable of serving as an anti-AD therapeutic modality. Upon deletion of GPER from an experimental model of myogenic differentiation, it has also been proposed that, during exercise in skeletal muscle, RE-activated distinct signalling pathways that were suggested to have conferred or played a role in the neuroprotective benefits of RE. This muscle-brain interaction could help suppress AD pathology, enhance neurogenesis, strengthen synaptic plasticity, and limit neuroinflammation. Exploration of these molecular mechanisms will help to further connect RE with preventing and managing AD.

21.3.1 Reduction of Amyloid-β Formation and Amelioration of Tau Pathology Mechanisms

Such mechanisms include reducing Aβ generation through downregulation of enzymes involved in Aβ generation (Alkadhi & Dao, 2018), microglial activity modulation (Hashiguchi et al., 2020), or astrogliosis reduction (Liu et al., 2020) that activates the Akt/GSK-3β signalling pathway to prevent Aβ deposition and induce tau pathology (Liu et al., 2020). Also the Akt/GSK-3β signalling transduction pathway is pivotal for regulating proliferation, development, and survival of the cells and it represents an important pathway for the pathophysiology of AD (Chu et al., 2017). Importantly, another major mechanism to reduce Aβ plaques and tau hyperphosphorylation is the inhibition of c-Jun N-terminal kinase (JNK) activation. Zhou et al. (2015) presented that using of small-molecule JNK-specific inhibitors attenuated amyloid plaque accumulation, tau hyperphosphorylation, and cognitive impairment in APPswe/PS1dE9 mice. Similarly, RE has been shown to diminish JNK activation in the frontal cortex and hippocampus tissues of 3xTg AD mice (Liu et al., 2020). Apart from molecular signalling, anaerobic exercises liberalize regional cerebral blood perfusion and increase ATP and local metabolites availability to neurons. In turn, such increased perfusion sustains neuronal health and leads to increased IGF-1 and BDNF levels via muscle strength augmentation (Burns et al., 2010; Mavros et al., 2017). While IGF-1 has been primarily associated with the removal of Aβ plaques in the cortex and the hippocampus (Carro et al., 2002; Talbot et al., 2012), BDNF is additionally associated with reduced Aβ production via upregulation of α-secretase-mediated APP processing (Nigam et al., 2017). The negative link between IGF-1 and Aβ deposition observed in these results suggests that high IGF-1 levels are associated with reduced Aβ deposition, while there is a positive correlation between BDNF and inhibited Aβ generation. Combining the effects outlined, these are testament to the potential of RE as a therapeutic strategy to combat the pathological signatures of AD.

21.3.2 Supporting Neuroplasticity: The Role of IGF-1 and BDNF

Anaerobic activities, such as RE, can stimulate synaptic plasticity, likely mediated by elevated levels of neurotropic factors like IGF-1 and BDNF. IGF-1 first attaches to its receptor (IGF1R) during skeletal muscle contractions, thus efficiently activating insulin receptor substrate 1 (IRS-1) in myocytes, which secondarily activates the PI3K signalling pathway. Upon activation, this induces Akt phosphorylation, which affects both the MAPK and ERK signalling pathways. These cascades ultimately activate CREB (pCREB), a transcription factor which regulates BDNF mRNA transcription (Gao et al., 2022). In a RCTs study the total of 13 RCTs were included (participant n = 1352), they show that RE increases plasma BDNF levels in older men (Arazi et al., 2021) and elevates serum BDNF levels in adults over the age of 60 years with a lowered level of depressive symptomatology, improving both PTSD and BDQ scores (Setayesh & Rahimi, 2023). Given that the levels of this neurotrophin are significantly reduced in AD, it has been hypothesized that specific RE interventions may improve BDNF thus providing neuroprotective effects and delaying AD-related neurodegeneration (Cotman et al., 2007). Experimental animal model studies show RE increases the expression of BDNF, NGF, and their respective receptors, TrkA and TrkB, the following eight weeks of training (Jafarzadeh et al., 2021).

The relationship of synaptic vesicle recycling with neuroplasticity and cognitive function is outpaced by synaptic structural integrity. Action potential and calcium influx cause neurotransmitters to release from synaptic terminals through vesicles. The recycling of synaptic vesicles is responsible for the refilling of neurotransmitter stocks maintaining structural homeostasis in the synapse. Disrupted synaptic recycling and decreased synaptic biomarker levels have been identified as one of the earliest alterations in AD brain and a contributor to impaired cognition (Calsolaro & Edison, 2016; Tönnies & Trushina, 2017). Synaptic protein expression, which is crucial for synaptic function, is enhanced by RE. For example, RE upregulates synaptotagmin-1 expression in cortical frontal synaptosome fractions. Synaptotagmin-1 acts as a calcium sensor that specifically promotes synaptic vesicle exocytosis but is also important for synaptic vesicle endocytosis. Synaptic vesicle recycling is disrupted in synaptotagmin-1 loss of function, leading to impaired synaptic vesicle release (Li et al., 2017). Moreover, RE upregulated the levels of synaptobrevin-1 in the synaptosome fractions of hippocampal tissue. Synaptobrevin-1 is essential for membrane fusion and neurotransmitter release. Loss of this protein is correlated with impaired synaptogenesis and progressive neurodegeneration (Haberman et al., 2012). To sum up, these results suggest that RE promotes synaptic integrity to contribute to enhanced neuroplasticity and cognitive performance.

21.3.3 Decreases Neuroinflammation

Ageing and neurodegenerative diseases, notably AD, are marked by chronic neuroinflammation. The neuroinflammation intensity may also be exacerbated by the surgical injury in aged animals and humans, leading to dysfunction of synapses and neuronal apoptosis (Safavynia & Goldstein, 2019). Microglial and astrocytic activation modulates this biological cascade, and chronic glial activation promotes cognitive decline in susceptible populations. Mediators such as cytokines secreted from activated glial cells disrupt mitochondrial function and synaptic transmission that lead to the progression of various neurological problems (Bader & Winklhofer, 2020). In AD, reduced function of mitochondria is largely seen in the hippocampus, a crucial area for the formation of memory and learning due to microglia activation (Wilkins & Swerdlow, 2016).

Research has shown that RE does provide protective activity against neuroinflammation, specifically via a reduction in pro-inflammatory cytokines like TNF-α and an increase of anti-inflammatory cytokines like IL-10 levels (Ayari et al., 2023). With these effects, a more hospitable environment to neuronal functioning, as well as a cell survival environment, is formed in human as well as animal models (de Almeida et al., 2022; Liu et al., 2020; Nicklas et al., 2008). Under the oxidative stress hypothesis, the RE would regulate pro-inflammatory mediators as well as oxidative stress by reinstating redox homeostasis. This protective effect happens through signalling pathways that include irisin, fibroblast growth factor 21 (FGF21), BDNF, and the lipid molecule 12,13-diHOME (Félix-Soriano & Stanford, 2023).

According to the lipid hypothesis, RE is proposed to alter lipid metabolic pathways; lower total cholesterol levels; and modify high-density lipoprotein (HDL), low-density ipoprotein (LDL), and very low-density lipoprotein (VLDL) cholesterol profiles. Such lipid modulations might indirectly reduce neuroinflammation and promote brain health (Teixeira et al., 2020). RE is also known to protect the hippocampus in prolonged stress models. Such protective roles have been mediated through several major mechanisms involving reduced inflammasome activation, increased TrkB expression, and attenuated Akt/mTOR signalling that are essential for neuronal survival and synaptic plasticity (Jung et al., 2023). Collectively, emerging evidence illustrates the protective and restorative properties of anaerobic exercise, like resistance training, against neuroinflammation in AD. RE may serve as a novel non-pharmacological lifestyle intervention to be added as an adjunct to treatment protocols for AD patients as it targets inflammatory pathways, oxidative stress, and lipid metabolism (Figure 21.1).

21.4 THE RECENT FINDINGS OF EXERCISE STUDIES IN ALZHEIMER'S DISEASE IN HUMANS: EXAMPLES FROM THE LITERATURE

Studies on exercise interventions in AD emerged in the last few decades with a focus on metabolic and cognitive outcome (metabolome) and show variable results according to the type, intensity, and duration of exercise. Neuroimaging showed that three months of moderate-intensity AE on a treadmill elevated brain glucose and acetoacetate uptake in grey matter in women with moderate AD. Cognitive improvement was not significant, however (Castellano et al., 2017). In summary, aerobic training improves brain energy metabolism, including ketone uptake and utilization, while preserving brain glucose uptake in mild AD. And these metabolic benefits might explain the small cognitive gains. In a single-blind randomized MRI study in women with AD, 16 weeks of AE (3×60 min/week) had no significant effect on overall brain volume. No such associations were observed, however, for hippocampal volume, frontal cortical thickness, or for the intensity of the exercise regime. Mental speed, attention, and exercise intensity were correlated with cortical thickness. Although there were no significant volumetric differences, reductions in Alzheimer's symptoms were reported, indicating that longer interventions may be necessary to observe structural changes (Frederiksen et al., 2018). In another study, AD patients underwent 16 weeks of moderate- to high-intensity AE, which increased cerebrospinal fluid sTREM2 and plasma IL-6 expression. IL-6 plays a dual role of pro- and anti-inflammatory mediator, which may promote BDNF release without causing nitric oxide. Soluble trigger receptor expressed on myeloid cells

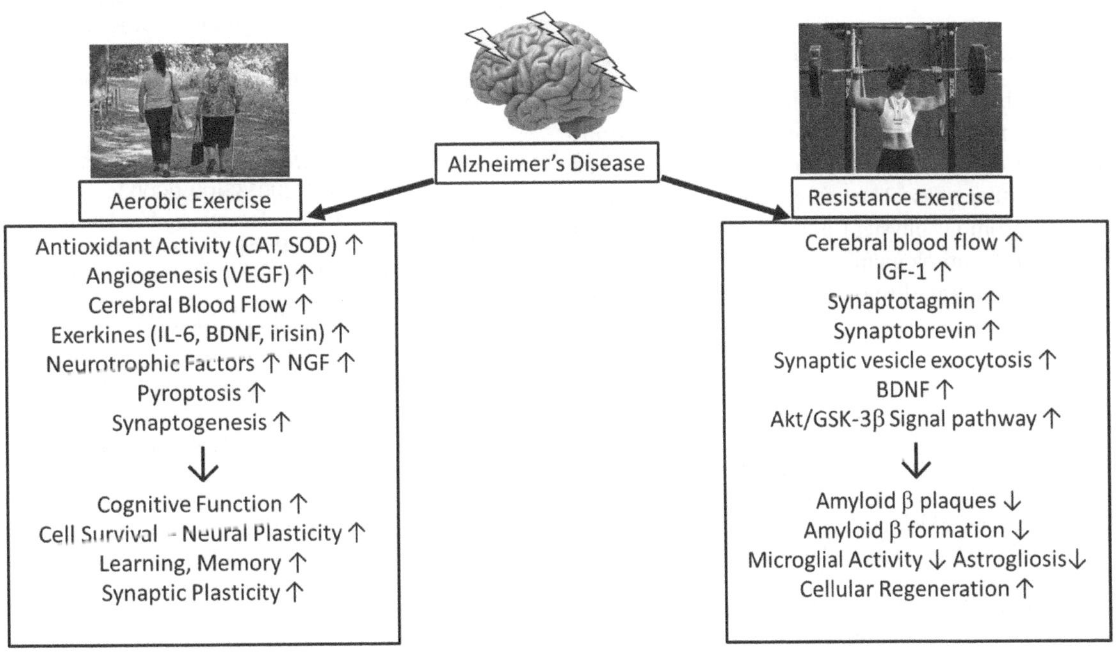

FIGURE 21.1 Positive effects of aerobic exercise and resistance exercise in Alzheimer's disease.

2 (sTREM2), an activation marker for microglial cells, was also elevated post-exercise, suggesting that in early stages of AD this marker reflects a survival response; however, in later stages sTREM2 could play a detrimental role. One way AE may help as an intervention that modulates the immune system from a pro-inflammatory to an anti-inflammatory profile (Jensen et al., 2019). AE for three months increased serum miR-129-5p expression in AD patients that was associated with the amelioration of cognitive function and decreased levels of inflammatory cytokines. It emphasizes the potential molecular pathways why exercise has positive effects on AD patients (Li et al., 2020).

RE has also been associated with direct effects on cognitive functioning in AD patients. A study showed substantial benefits to memory and executive functions among AD patients who regularly engaged in RE. These benefits were ascribed to both the physical and the neurological consequences of exercise (Nagamatsu et al., 2012). RE training for 12 weeks demonstrated an improvement in functional capacity and independence in daily living activities of AD patients. Such low-cost interventions have been advised as part of comprehensive care plans for these patients (Santana-Sosa et al., 2008). Based on a RCT with 155 older women, a 52-week RE programme (1 × wk or 2 × wk) was compared with balance and toning exercises. Executive function and memory were measured at baseline, one-year, and two-year follow-up. At two years, significant improvements in executive function relative to balance and toning exercise were identified for both once- and twice-weekly RE groups. The group, which was administered exercise twice weekly, also exhibited greater memory, lower atrophy of cortical white matter, and improved peak muscle strength, thus demonstrating both cognitive and structural impacts of long-term exercise (Best et al., 2015).

21.5 CONCLUSION AND FUTURE REMARKS

Particularly aerobic and RE can be integrated as a therapeutic intervention for the treatment of AD and other forms of dementia and could serve as a valuable adjunct to the current medical therapies. Therefore, these interventions could not only promote physical and cognitive health but also target central pathological mechanisms behind AD, raising hope for delaying disease progression and improving life quality.

REFERENCES

Abdullahi, A., Wong, T. W., & Ng, S. S. (2024). Understanding the mechanisms of disease modifying effects of aerobic exercise in people with Alzheimer's disease. *Ageing Research Reviews*, *94*, 102202. https://doi.org/10.1016/j.arr.2024.102202

Adlard, P. A., Perreau, V. M., Pop, V., & Cotman, C. W. (2005). Voluntary exercise decreases amyloid load in a transgenic model of Alzheimer's disease. *The Journal of Neuroscience*, *25*(17), 4217–4221. https://doi.org/10.1523/JNEUROSCI.0496-05.2005

Alkadhi, K., & Dao, A. (2018). Exercise decreases BACE and APP levels in the hippocampus of a rat model of Alzheimer's disease. *Molecular and Cellular Neuroscience*, *86*, 25–29. https://doi.org/10.1016/j.mcn.2017.11.008

An, Y., Varma, V. R., Varma, S., Casanova, R., Dammer, E., Pletnikova, O, Chia, C. W., Egan, J. M., Ferrucci, L., Troncoso, J., Levey, A. I., Lah, J., Seyfried, N. T., Legido-Quigley, C., O'Brien, R., & Thambisetty, M. (2018). Evidence for brain glucose dysregulation in Alzheimer's disease. *Alzheimer's & Dementia*, *14*(3), 318–328. https://doi.org/10.1016/j.jalz.2017.09.011

Arazi, H., Babaei, P., & Moghimi, M., & Asadi, A. (2021). Acute effects of strength and endurance exercise on serum BDNF and IGF-1 levels in older men. *BMC Geriatrics*, *21*, 50. https://doi.org/10.1186/s12877-020-01937-6

Ayari, S., Abellard, A., Carayol, M., Guedj, É., & Gavarry, O. A systematic review of exercise modalities that reduce pro-inflammatory cytokines in humans and animals' models with mild cognitive impairment or dementia. *Exp Gerontol.* 2023 May;175:112141. doi: 10.1016/j.exger.2023.112141

Azevedo, C. V., Hashiguchi, D., Campos, H. C., Figueiredo, E. V., Otaviano, S. F. S. D., Penitente, A. R., … & Longo, B. M. (2023). The effects of resistance exercise on cognitive function, amyloidogenesis, and neuroinflammation in Alzheimer's disease. *Frontiers in Neuroscience*, *17*, 1131214. https://doi.org/10.3389/fnins.2023.1131214

Azimi, M., Gharakhanlou, R., Naghdi, N., Khodadadi, D., & Heysieattalab, S. (2018). Moderate treadmill exercise ameliorates amyloid-β-induced learning and memory impairment, possibly via increasing AMPK activity and up-regulation of the PGC-1α/FNDC5/BDNF pathway. *Peptides*, *102*, 78–88.

Bader, V., & Winklhofer, K. F. (2020). Mitochondria at the interface between neurodegeneration and neuroinflammation. *Seminars in Cell & Developmental Biology*, *99*, 163–171. https://doi.org/10.1016/j.semcdb.2019.05.028

Best, J. R., Chiu, B. K., Liang Hsu, C., Nagamatsu, L. S., & Liu-Ambrose, T. (2015). Long-term effects of resistance exercise training on cognition and brain volume in older women: Results from a randomized controlled trial. *Journal of the International Neuropsychological Society*, *21*(10), 745–756. https://doi.org/10.1017/S1355617715000673

Bird, S. P., Tarpenning, K. M., & Marino, F. E. (2005). Designing resistance training programmes to enhance muscular fitness: A review of the acute programme variables. *Sports Medicine*, *35*(10), 841–851. https://doi.org/10.2165/00007256-200535100-00002

Bonanni, R., Cariati, I., Tarantino, U., D'Arcangelo, G., & Tancredi, V. (2022). Physical exercise and health: A focus on its protective role in neurodegenerative diseases. *Journal of Functional Morphology and Kinesiology*, *7*(2), 38. https://doi.org/10.3390/jfmk7020038

Burns, J. M., Johnson, D. K., Watts, A., Swerdlow, R. H., & Brooks, W. M. (2010). Reduced lean mass in early Alzheimer disease and its association with brain atrophy. *Archives of Neurology*, *67*(4), 428–433. https://doi.org/10.1001/archneurol.2010.38

Calsolaro, V., & Edison, P. (2016). Neuroinflammation in Alzheimer's disease: Current evidence and future directions. *Alzheimer's & Dementia: The Journal of the Alzheimer's Association*, *12*(6), 719–732. https://doi.org/10.1016/j.jalz.2016.02.010

Camuso, S., & Canterini, S. (2023). Brain-derived neurotrophic factor in main neurodegenerative diseases. *Neural Regeneration Research*, *18*(3), 554–555. https://doi.org/10.4103/1673-5374.350199

Carro, E., Trejo, J. L., Gomez-Isla, T., LeRoith, D., & Torres-Aleman, I. (2002). Serum insulin-like growth factor I regulates brain amyloid-beta levels. *Nature Medicine*, *8*(12), 1390–1397. https://doi.org/10.1038/nm1202-793

Castellano, C. A., Paquet, N., Dionne, I. J., Imbeault, H., Langlois, F., Croteau, E., … & Cunnane, S. C. (2017). A 3-month aerobic training program improves brain energy metabolism in mild Alzheimer's disease: Preliminary results from a neuroimaging study. *Journal of Alzheimer's Disease*, *56*(4), 1459–1468. https://doi.org/10.3233/jad-161163

Chen, X. Q., & Mobley, W. C. (2019). Exploring the pathogenesis of Alzheimer disease in basal forebrain cholinergic neurons: Converging insights from alternative hypotheses. *Frontiers in Neuroscience*, *13*, 446. https://doi.org/10.3389/fnins.2019.00446

Chen. Z., Yuan, Z., Yang, S., Zhu, Y., Xue, M., Zhang, J., & Leng, L. Brain Energy Metabolism: Astrocytes in Neurodegenerative Diseases. *CNS Neurosci Ther.* 2023 Jan;29(1):24–36. doi: 10.1111/cns.13982

Chu, J., Lauretti, E., & Pratico, D. (2017). Caspase-3-dependent cleavage of Akt modulates tau phosphorylation via GSK3beta kinase: Implications for Alzheimer's disease. *Molecular Psychiatry*, *22*, 1002–1008. https://doi.org/10.1038/mp.2016.214

Conner, J. M., Lauterborn, J. C., Yan, Q., Gall, C. M., & Varon, S. (1997). Distribution of brain-derived neurotrophic factor (BDNF) protein and mRNA in the normal adult rat CNS: Evidence for anterograde axonal transport. *The Journal of Neuroscience*, *17*(7), 2295–2313. https://doi.org/10.1523/JNEUROSCI.17-07-02295.1997

Cotman, C. W., & Berchtold, N. C. (2002). Exercise: A behavioral intervention to enhance brain health and plasticity. *Trends in Neurosciences*, *25*(6), 295–301. https://doi.org/10.1016/s0166-2236(02)02143-4

Cotman, C. W., Berchtold, N. C., & Christie, L. A. (2007). Exercise builds brain health: Key roles of growth factor cascades and inflammation. *Trends in Neurosciences*, *30*(9), 464–472.

de Almeida, E. J. R., Ibrahim, H. J., ChitolinaSchetinger, M. R., de Andrade, C. M., & Cardoso, A. M. (2022). Modulation of inflammatory mediators and microglial activation through physical exercise in Alzheimer's and Parkinson's diseases. *Neurochemical Research*, *47*(11), 3221–3240. https://doi.org/10.1007/s11064-022-03713-x

DiSabato, D. J., Quan, N., & Godbout, J. P. (2016). Neuroinflammation: The devil is in the details. *Journal of Neurochemistry*, *139*(Suppl 2), 136–153. https://doi.org10.1111/jnc.13607.

El Hayek, L., Khalifeh, M., Zibara, V., Abi Assaad, R., Emmanuel, N., Karnib, N., … & Sleiman, S. F. (2019). Lactate mediates the effects of exercise on learning and memory through SIRT1-dependent activation of hippocampal brain-derived neurotrophic factor (BDNF). *The Journal of Neuroscience*, *39*(13), 2369–2382. https://doi.org/10.1523/JNEUROSCI.1661-18.2019

Elliott, E., Atlas, R., Lange, A., & Ginzburg, I. (2005). Brain-derived neurotrophic factor induces a rapid dephosphorylation of tau protein through a PI-3 kinase signalling mechanism. *The European Journal of Neuroscience*, *22*(5), 1081–1088. https://doi.org/10.1111/j.1460-9568.2005.04290.x

Félix-Soriano, E., & Stanford, K. I. (2023). Exerkines and redox homeostasis. *Redox Biology*, *63*, 102748. https://doi.org/10.1016/j.redox.2023.102748

Feng, S., Wu, C., Zou, P., Deng, Q., Chen, Z., Li, M., … & Yang, L. (2023). High-intensity interval training ameliorates Alzheimer's disease-like pathology by regulating astrocyte phenotype-associated AQP4 polarization. *Theranostics*, *13*(10), 3434–3450. https://doi.org/10.7150/thno.81951

Ferrari, B. L., Neto, G. C. C., Nucci, M. P., Mamani, J. B., Lacerda, S. S., Felício, A. C., Amaro, E. Jr., & Gamarra, L. F. (2019). The accuracy of hippocampal volumetry and glucose metabolism for the diagnosis of patients with suspected Alzheimer's disease, using automatic quantitative clinical tools. *Medicine (Baltimore)*, *98*(45), e17824. https://doi.org/10.1097/md.0000000000017824

Frederiksen, K. S., Larsen, C. T., Hasselbalch, S. G., Christensen, A. N., Høgh, P., Wermuth, L., … & Garde, E. (2018). A 16-week aerobic exercise intervention does not affect hippocampal volume and cortical thickness in mild to moderate Alzheimer's disease. *Frontiers in Aging Neuroscience*, *10*, 293. https://doi.org/10.3389/fnagi.2018.00293

Galloza, J., Castillo, B., & Micheo, W. (2017). Benefits of exercise in the older population. *Physical Medicine and Rehabilitation Clinics of North America*, *28*(4), 659–669. https://doi.org/10.1016/j.pmr.2017.06.001

Gao, L., Zhang, Y., Sterling, K., & Song, W. (2022). Brain-derived neurotrophic factor in Alzheimer's disease and its pharmaceutical potential. *Translational Neurodegeneration*, *11*(1), 4. https://doi.org/10.1186/s40035-022-00279-0

Gaudet, A. D., Fonken, L. K., Watkins, L. R., Nelson, R. J., & Popovich, P. G. (2018). MicroRNAs: Roles in regulating neuroinflammation. *The Neuroscientist: A Review Journal Bringing Neurobiology, Neurology and Psychiatry*, *24*(3), 221–245. https://doi.org/10.1177/1073858417721150

Haberman, A., Williamson, W. R., Epstein, D., Wang, D., Rina, S., Meinertzhagen, I. A., & Hiesinger, P. R. (2012). The synaptic vesicle SNARE neuronal synaptobrevin promotes endolysosomal degradation and prevents neurodegeneration. *Journal of Cell Biology*, *96*, 261–276. https://doi.org/10.1083/jcb.201108088

Hardie, D. G., Ross, F. A., & Hawley, S. A. (2012). AMPK: A nutrient and energy sensor that maintains energy homeostasis. *Nature Reviews Molecular Cell Biology*, *13*(4), 251–262.

Hashiguchi, D., Campos, H., Wuo-Silva, R., Faber, J., Gomes da Silva, S., Coppi, A. A., Arida, R. M., & Longo, B. M. (2020). Resistance exercise decreases amyloid load and modulates inflammatory responses in the APP/PS1 mouse model for Alzheimer's disease. *Journal of Alzheimer's Disease*, *73*, 1525–1539.

Herring, A., Blome, M., Ambrée, O., Sachser, N., Paulus, W., & Keyvani, K. (2010). Reduction of cerebral oxidative stress following environmental enrichment in mice with Alzheimer-like pathology. *Brain Pathology*, *20*(1), 166–175. https://doi.org/10.1111/j.1750-3639.2008.00257.x

Hofer, M., Pagliusi, S. R., Hohn, A., Leibrock, J., & Barde, Y. A. Regional distribution of brain-derived neurotrophic factor mRNA in the adult mouse brain. *EMBO J.* 1990 Aug;9(8):2459–64. doi: 10.1002/j.1460-2075.1990.tb07423.x

Hu, S., Wan, X., Li, X., & Wang, X. (2022). Aerobic exercise alleviates pyroptosis-related diseases by regulating NLRP3 inflammasome. *Frontiers in Physiology*, *13*, 965366. https://doi.org/10.3389/fphys.2022.965366

Ishibashi, K., Wagatsuma, K., Ishiwata, K., & Ishii, K. (2016). Alteration of the regional cerebral glucose metabolism in healthy subjects by glucose loading. *Human Brain Mapping*, *37*(8), 2823–2832. https://doi.org/10.1002/hbm.23210

Jafarzadeh, G., Shakerian, S., Farbood, Y., & Ghanbarzadeh, M. Effects of Eight Weeks of Resistance Exercises on Neurotrophins and Trk Receptors in Alzheimer Model Male Wistar Rats. *Basic Clin Neurosci.* 2021 May-Jun;12(3):349–359. doi: 10.32598/bcn.2021.2067.1

Janoutová, J., Machaczka, O., Zatloukalová, A., & Janout, V. (2022). Is Alzheimer's disease a type 3 diabetes? A review. *Central European Journal of Public Health, 30*(3), 139–143. https://doi.org/10.21101/cejph.a7238

Jensen, C. S., Bahl, J. M., Østergaard, L. B., Høgh, P., Wermuth, L., Heslegrave, A., Zetterberg, H., Heegaard, N. H. H., Hasselbalch, S. G., & Simonsen, A. H. (2019). Exercise as a potential modulator of inflammation in patients with Alzheimer's disease measured in cerebrospinal fluid and plasma. *Experimental Gerontology, 121*, 91–98. https://doi.org/10.1016/j.exger.2019.04.003

Jung, J. T. K., Marques, L. S., Zborowski, V. A., Silva, G. L., Nogueira, C. W., & Zeni, G. (2023). Resistance training modulates hippocampal neuroinflammation and protects anxiety-depression-like dyad induced by an emotional single prolonged stress model. *Molecular Neurobiology, 60*(1), 264–276. https://doi.org/10.1007/s12035-022-03069-x

Kellar, D., & Craft, S. (2020). Brain insulin resistance in Alzheimer's disease and related disorders: Mechanisms and therapeutic approaches. *The Lancet Neurology, 19*(9), 758–766.

Kinni, H., Guo, M., Ding, J. Y., Konakondla, S., Dornbos, D. III, Tran, R., … & Ding, Y. (2011). Cerebral metabolism after forced or voluntary physical exercise. *Brain Research, 1388*, 48–55.

Kuo, Y. M., Kokjohn, T. A., Watson, M. D., Woods, A. S., Cotter, R. J., Sue, L. I., Kalback, W. M., Emmerling, M. R., Beach, T. G., & Roher, A. E. Elevated abeta42 in skeletal muscle of Alzheimer disease patients suggests peripheral alterations of AbetaPP metabolism. *Am J Pathol.* 2000 Mar;156(3):797–805. doi: 10.1016/s0002-9440(10)64947-4

Kyrtata, N., Emsley, H. C. A., Sparasci, O., Parkes, L. M., & Dickie, B. R. (2021). A systematic review of glucose transport alterations in Alzheimer's disease. *Frontiers in Neuroscience, 15*, 626636. https://doi.org/10.3389/fnins.2021.626636

Li, Y. C., Chanaday, N. L., Xu, W., & Kavalali, E. T. (2017). Synaptotagmin-1- and synaptotagmin-7-dependent fusion mechanisms target synaptic vesicles to kinetically distinct endocytic pathways, *Neuron, 93*, 616–631. https://doi.org/10.1016/j.neuron.2016.12.010

Li, Z., Chen, Q., Liu, J., & Du, Y. (2020). Physical exercise ameliorates the cognitive function and attenuates the neuroinflammation of Alzheimer's disease via miR-129-5p. *Dementia and Geriatric Cognitive Disorders, 49*(2), 163–169. https://doi.org/10.1159/000507285

Liu, Y., Chu, J., Yan, T., Zhang, Y., Chen, Y., Chang, R. C. C., & Wong, G. T. C. (2020). Short-term resistance exercise inhibits neuroinflammation and attenuates neuropathological changes in 3xTg Alzheimer's disease mice. *Journal of Neuroinflammation 17*, 4. https://doi.org/10.1186/s12974-019-1653-7

Lu, Y., Bu, F. Q., Wang, F., Liu, L., Zhang, S., Wang, G., & Hu, X. Y. (2023). Recent advances on the molecular mechanisms of exercise-induced improvements of cognitive dysfunction. *Translational Neurodegeneration, 12*(1), 9. https://doi.org/10.1186/s40035-023-00341-5

Marosi, K., & Mattson, M. P. (2014). BDNF mediates adaptive brain and body responses to energetic challenges. *Trends in Endocrinology & Metabolism, 25*(2), 89–98.

Mavros, Y., Gates, N., Wilson, G., Jain, N., Meiklejohn, J., Brodaty, H., Ven, W, Singh, N., Baune, B.T., Suo, C., Baker, M. K., Foroughi, N., Wang, Y., & Sachdev, P. S. (2017). Mediation of cognitive function improvements by strength gains after resistance training in older adults with mild cognitive impairment: Outcomes of the study of mental and resistance training. *Journal of the American Geriatrics Society 65*, 550–559. https://doi.org/10.1111/jgs.14542

McGeer, P. L., & Rogers, J. Anti-inflammatory agents as a therapeutic approach to Alzheimer's disease. *Neurology.* 1992 Feb;42(2):447–9. doi: 10.1212/wnl.42.2.447

McKhann, G., Drachman, D., Folstein, M., Katzman, R., Price, D., & Stadlan, E. M. (1984). Clinical diagnosis of Alzheimer's disease: Report of the NINCDS-ADRDA work group under the auspices of department of health and human services task force on Alzheimer's disease. *Neurology, 34*(7), 939–944. https://doi.org/10.1212/wnl.34.7.939

Mcleod, J. C., Currier, B. S., Lowisz, C. V., & Phillips, S. M. (2024). The influence of resistance exercise training prescription variables on skeletal muscle mass, strength, and physical function in healthy adults: An umbrella review. *Journal of Sport and Health Science, 13*(1), 47–60. https://doi.org/10.1016/j.jshs.2023.06.005

Meng, C., He, Z., & Xing, D. (2013). Low-level laser therapy rescues dendrite atrophy via upregulating BDNF expression: Implications for Alzheimer's disease. *The Journal of Neuroscience: The Official Journal of the Society for Neuroscience, 33*(33), 13505–13517. https://doi.org/10.1523/JNEUROSCI.0918-13.2013

Mezghani, I., Turki, M., Ben Yahia, M., Bouzid, A., & Ayadi, F. (2023). Effect of aerobic exercise on oxidative status in patients with Alzheimer's disease [Conference Abstract]. *S815 Clinical Chemistry and Laboratory Medicine, 61.* https://doi.org/10.1515/cclm-2023-7043

Monteiro-Cardoso, V. F., Castro, M, Oliveira, M. M., Moreira, P. I., Peixoto F, & Videira R. A. Age-dependent biochemical dysfunction in skeletal muscle of triple-transgenic mouse model of Alzheimer`s disease. *Curr Alzheimer Res.* 2015;12(2):100–15. doi: 10.2174/1567205012666150204124852.

Nagahara, A. H., Merrill, D. A., Coppola, G., Tsukada, S., Schroeder, B. E., Shaked, G. M., Wang, L., Blesch, A., Kim, A., Conner, J. M., Rockenstein, E., Chao, M. V., Koo, E. H., Geschwind, D., Masliah, E., Chiba, A. A., & Tuszynski, M. H. (2009). Neuroprotective effects of brain-derived neurotrophic factor in rodent and primate models of Alzheimer's disease. *Nature Medicine, 15*(3), 331–337. https://doi.org/10.1038/nm.1912

Nagamatsu, L. S., Handy, T. C., Hsu, C. L., Voss, M. W., & Liu-Ambrose, T. (2012). Resistance training promotes cognitive and functional brain plasticity in seniors with probable mild cognitive impairment. *Archives of Internal Medicine, 172*(8), 666–668.

Newcombe, E. A., Camats-Perna, J., Silva, M. L., Valmas, N., Huat, T. J., & Medeiros, R. (2018). Inflammation: The link between comorbidities, genetics, and Alzheimer's disease. *Journal of Neuroinflammation, 15*(1), 276. https://doi.org/10.1186/s12974-018-1313-3

Nicklas, B. J., Hsu, F. C., Brinkley, T. J., Church, T., Goodpaster, B. H., Kritchevsky, S. B., & Pahor, M. (2008). Exercise training and plasma C-reactive protein and interleukin-6 in elderly people. *Journal of the American Geriatrics Society, 56*(11), 2045–2052. https://doi.org/10.1111/j.1532-5415.2008.01994.x

Niedzielska, E., Smaga, I., Gawlik, M., Moniczewski, A., Stankowicz, P., Pera, J., & Filip, M., 2016. Oxidative stress in neurodegenerative diseases. *Molecular Neurobiology, 53.*

Nigam, S. M., Xu, S., Kritikou, J. S., Marosi, K., Brodin, L., & Mattson, M. P. (2017). Exercise and BDNF reduce Aβ production by enhancing α-secretase processing of APP. *Journal of Neurochemistry, 142*(2), 286–296.

Özbeyli, D., Sari, G., Özkan, N., Karademir, B., Yüksel, M., Kaya, Ö. T. Ç., & Kasımay Çakır, Ö. (2017). Protective effects of different exercise modalities in an Alzheimer's disease-like model. *Behavioural Brain Research*, *328*, 159–177. https://doi.org/10.1016/j.bbr.2017.03.044

Pahlavani, H. A. (2023). Exercise therapy to prevent and treat Alzheimer's disease. *Frontiers in Aging Neuroscience*, *15*, 1243869. https://doi.org/10.3389/fnagi.2023.1243869

Patel, H., Alkhawam, H., Madanieh, R., Shah, N., Kosmas, C. E., & Vittorio, T. J. (2017). Aerobic vs anaerobic exercise training effects on the cardiovascular system. *World Journal of Cardiology*, *9*(2), 134–138. https://doi.org/10.4330/wjc.v9.i2.134

Pedersen, B. K. (2011). Muscles and their myokines. *Journal of Experimental Biology*, *214*(2), 337–346.

Pena, G. S., Paez, H. G., Johnson, T. K., Halle, J. L., Carzoli, J. P., Visavadiya, N. P., Zourdos, M. C., Whitehurst, M. A., & Khamoui, A. V. (2020). Hippocampal growth factor and myokine cathepsin b expression following aerobic and resistance training in 3xTg-AD mice. *International Journal of Chronic Diseases*, *2020*, 5919501. https://doi.org/10.1155/2020/5919501

Pereira, A. C., Huddleston, D. E., & Brickman, A. M. (2007). An in vivo correlate of exercise-induced neurogenesis in the adult dentate gyrus. *Proceedings of the National Academy of Sciences*, *104*(13), 5638–5643. https://doi.org/10.1073/pnas.0611721104

Pinho, R. A., Aguiar, A. S. Jr., & Radák, Z. (2019). Effects of resistance exercise on cerebral redox regulation and cognition: An interplay between muscle and brain. *Antioxidants*, *8*(11), 529. https://doi.org/10.3390/antiox8110529

Plowman, S. A., & Smith, D. L. (2014). *Exercise physiology: For health, fitness, and performance* (4th ed.). Wolters Kluwer/Lippincott Williams & Wilkins.

Prasad, V., Samson, J., & Parvathi, V. (2024). Exercise and exerkine upregulation: Brain-derived neurotrophic factor as a potential non-pharmacological therapeutic strategy for Parkinson's disease. *Biocell*, *48*(5), 693–706. https://doi.org/10.32604/biocell.2024.048776

Safavynia, S. A., & Goldstein, P. A. (2019). The role of neuroinflammation in postoperative cognitive dysfunction: Moving from hypothesis to treatment, *Frontiers in Psychiatry*, *9*, 752, https://doi.org/10.3389/fpsyt.2018.00752

Santana-Sosa, E., Barriopedro, M. I., López-Mojares, L. M., Pérez, M., & Lucia, A. (2008). Exercise training is beneficial for Alzheimer's patients. *International Journal of Sports Medicine*, *29*(10), 845–850. https://doi.org/10.1055/s-2008-1038432

Schinder, A. F., & Poo, M. (2000). The neurotrophin hypothesis for synaptic plasticity. *Trends in Neurosciences*, 23(12), 639–645. https://doi.org/10.1016/s0166-2236(00)01672-6

Schmolesky, M. T., Webb, D. L., & Hansen, R. A. (2013). The effects of aerobic exercise intensity and duration on levels of brain-derived neurotrophic factor in healthy men. *Journal of Sports Science & Medicine*, *12*(3), 502–511.

Setayesh, S., & Mohammad Rahimi, G. R. (2023). The impact of resistance training on brain-derived neurotrophic factor and depression among older adults aged 60 years or older: A systematic review and meta-analysis of randomized controlled trials. *Geriatric Nursing (New York, N.Y.)*, 54, 23–31. https://doi.org/10.1016/j.gerinurse.2023.08.022

Siddiqui, N. I., Nessa, A., & Hossain, M. A. (2010). Regular physical exercise: Way to healthy life. *Mymensingh Medical Journal: MMJ*, *19*(1), 154–158. PMID: 20046192.

Singh, A., Kukreti, R., Saso, L., & Kukreti, S. (2019). Oxidative stress: A key modulator in neurodegenerative diseases. *Molecules (Basel, Switzerland)*, *24*(8), 1583. https://doi.org/10.3390/molecules24081583

Smith, C., Woessner, M. N., Sim, M., & Levinger, I. (2022). Sarcopenia definition: Does it really matter? Implications for resistance training. *Ageing Research Reviews*, *78*, 101617. https://doi.org/10.1016/j.arr.2022.101617

Tacke, C., Landgraf, P., Dieterich, D. C., & Kröger, A. (2024). The fate of neuronal synapse homeostasis in aging, infection, and inflammation. *American Journal of Physiology. Cell Physiology*, *327*(6), C1546–C1563. https://doi.org/10.1152/ajpcell.00466.2024

Talbot, K., Wang, H. Y., Kazi, H., Han, L. Y., Bakshi, K. P., Stucky, A., Fuino, R. L., Kawaguchi, K. R., Samoyedny, A. J., Wilson, R. S., Arvanitakis, Z., Schneider, J. A., Wolf, B. A., Bennett, D. A., Trojanowski, J. Q., & Arnold, S. E. (2012). Demonstrated brain insulin resistance in Alzheimer's disease patients is associated with IGF-1 resistance, IRS-1 dysregulation, and cognitive decline. *The Journal of Clinical Investigation*, *122*(4), 1316–1338. https://doi.org/10.1172/JCI59903

Teixeira, G. R., Mendes, L. O., Veras, A. S. C., Thorpe, H. H. A., Fávaro, W. J., de Almeida Chuffa, L. G., Pinheiro, P. F. F., & Martinez, F. E. (2020). Physical resistance training-induced changes in lipids metabolism pathways and apoptosis in prostate. *Lipids in Health and Disease*, *19*(1), 14. https://doi.org/10.1186/s12944-020-1195-0

Tönnies, E., & Trushina, E. (2017). Oxidative stress, synaptic dysfunction, and Alzheimer's disease. *Journal of Alzheimer's Disease*, *57*(4), 1105–1121.

Trettel, C. D. S., Pelozin, B. R. A., Barros, M. P., Bachi, A. L. L., Braga, P. G. S., Momesso, C. M., Furtado, G. E., Valente, P. A., Oliveira, E. M., Hogervorst, E., & Fernandes, T. (2023). Irisin: An anti-inflammatory exerkine in aging and redox-mediated comorbidities, *Frontiers in Endocrinology*, *14*, 1106529. https://doi.org/10.3389/fendo.2023.1106529.

Um, H. S., Kang, E. B., Leem, Y. H., Cho, I. H., Yang, C. H., Chae, K. R., Hwang, D. Y., & Cho, J. Y. (2008). Exercise training acts as a therapeutic strategy for reduction of the pathogenic phenotypes for Alzheimer's disease in an NSE/APPsw-transgenic model. *International Journal of Molecular Medicine*, *22*(4), 529–539.

Valenzuela, P. L., Ruilope, L. M., Santos-Lozano, A., Wilhelm, M., Kränkel, N., Fiuza-Luces, C., & Lucia, A. (2023). Exercise benefits in cardiovascular diseases: From mechanisms to clinical implementation. *European Heart Journal*, *44*(21), 1874–1889. https://doi.org/10.1093/eurheartj/ehad170

Vidoni, E. D., Morris, J. K., Watts, A., Perry, M., Clutton, J., Van Sciver, A., Kamat, A. S., Mahnken, J., Hunt, S. L., Townley, R., Honea, R., Shaw, A. R., Johnson, D. K., Vacek, J., & Burns, J. M. (2021). Effect of aerobic exercise on amyloid accumulation in preclinical Alzheimer's: A 1-year randomized controlled trial. *PloS One*, *16*(1), e0244893. https://doi.org/10.1371/journal.pone.0244893

Waite, S. J., Maitland, S., Thomas, A., & Yarnall, A. J. (2021). Sarcopenia and frailty in individuals with dementia: A systematic review. *Archives of Gerontology and Geriatrics*, *92*, 104268. https://doi.org/10.1016/j.archger.2020.104268

Wang, J., Gu, B. J., Masters, C. L., & Wang, Y. J. (2017). A systemic view of Alzheimer disease – Insights from amyloid-β metabolism beyond the brain. *Nature Reviews. Neurology*, *13*(10), 612–623. https://doi.org/10.1038/nrneurol.2017.111

For Product Safety Concerns and Information please contact our
EU representative GPSR@taylorandfrancis.com Taylor & Francis
Verlag GmbH, Kaufingerstraße 24, 80331 München, Germany